SECOND EDITION

INTRODUCTORY
BOTANY

PLANTS, PEOPLE, AND THE ENVIRONMENT

LINDA R. BERG, Ph.D.

Former Affiliations:
University of Maryland, College Park
St. Petersburg College

THOMSON

BROOKS/COLE

Australia • Brazil • Canada • Mexico • Singapore • Spain • United Kingdom • United States

Introductory Botany: Plants, People, and the Environment, Second Edition
Linda R. Berg

Acquisitions Editor: Peter Adams
Development Editor: Kari Hopperstead
Assistant Editor: Lauren Oliveira
Editorial Assistant: Rose Barlow
Technology Project Manager: Keli Amann
Marketing Manager: Kara Kindstrom
Marketing Communications Manager: Bryan Vann
Project Manager, Editorial Production: Belinda Krohmer
Creative Director: Rob Hugel
Art Director: John Walker
Print Buyer: Doreen Suruki

Permissions Editor: Bob Kauser
Production Service: Newgen–Austin
Text Designer: Lisa Buckley
Photo Researcher: Abigail Baxter
Copy Editor: Cynthia Lindlof
Cover Designer: Bill Stanton
Cover Image: Tim Gainey/Alamy
Back Cover Image: DK. Khattiya/Alamy
Cover Printer: Courier Kendallville
Compositor: Newgen–Austin
Printer: Courier Kendallville

Thomson Higher Education
10 Davis Drive
Belmont, CA 94002-3098
USA

For more information about our products, contact us at:
Thomson Learning Academic Resource Center
1-800-423-0563
For permission to use material from this text or product,
submit a request online at http://www.thomsonrights.com.
Any additional questions about permissions can
be submitted by e-mail to thomsonrights@thomson.com.

ExamView® and ExamView Pro® are registered trademarks of FSCreations, Inc. Windows is a registered trademark of the Microsoft Corporation used herein under license. Macintosh and Power Macintosh are registered trademarks of Apple Computer, Inc. Used herein under license.

Library of Congress Control Number: 2006939406

ISBN-13: 978-0-534-46669-5
ISBN-10: 0-534-46669-9

This book is dedicated to botany professors and students everywhere.

Crescat scientia vita excolatur.

Brief Contents

Contents

Jeffrey Lepore/Photo Researchers, Inc.

Andrew Ward/Life File/Getty Images

© David Cavagnaro/Visuals Unlimited

Alan & Sandy Carey/Photodisc/Getty Images

Tony Camacho/Photo Researchers, Inc.

C. Borland/PhotoLink/Getty Images

Dan Suzio/Photo Researchers, Inc.

© Jerome Wexler/Visuals Unlimited

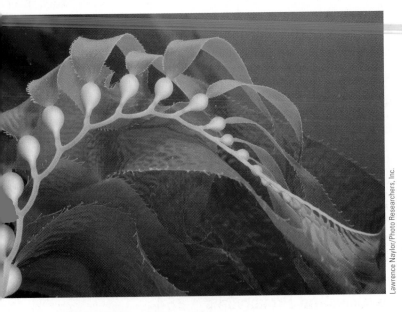

Lawrence Naylor/Photo Researchers, Inc.

Sinclair Stammers/Photo Researchers, Inc.

© Wendell Metzen/Index Stock Imagery, Inc.

Box Features

Preface

The past 10 years have witnessed an explosion of knowledge in all disciplines of plant biology. The second edition of *Introductory Botany: Plants, People, and the Environment*, published 10 years after the first edition, updates and expands the first-edition presentation of basic biological concepts as they relate to plants. Like the first edition, *Introductory Botany: Plants, People, and the Environment*, Second Edition, is intended primarily as an introductory text for undergraduate students, both nonscience and science majors. It is written so that it will be easily understood by students with little or no background in plant biology.

One of my principal goals in developing this edition has been to share with beginning botany students an appreciation of the diverse organisms we call plants, including their remarkable adaptations to the environment and their evolutionary and ecological relationships. Special emphasis has been placed on ecology, particularly the role of plants in the biosphere, and relevant environmental issues have been integrated into many chapters.

Botanical knowledge is important in its own right, and learning about plants can make your life richer and more rewarding. But we live in a complex society, with science and technology affecting every aspect of our lives. For that reason, botany as a scientific process has also been stressed so that students will understand how scientists think, how they approach problems, and how they obtain scientific knowledge about our world.

Pedagogical Features

Learning any science is a challenging endeavor, and botany is no exception. A variety of learning aids are included in the text to help the student achieve mastery of the concepts presented.

1. **Chapter openers** focus on specific plants relevant to the subjects in the chapters. These short essays promote an awareness and appreciation of plants and motivate the student to read further.
2. **Learning objectives** indicate, in behavioral terms, what students must be able to do to demonstrate mastery of material covered in the chapter.
3. **Concept statement subheads** introduce sections, previewing and summarizing the key idea to be discussed in that section.
4. **Three categories of boxes** highlight the environment, the human importance of plants, and topics of general interest, particularly as they relate to the process of science. These boxes are written to spark student interest, familiarize students with current environmental issues, and show the relevance of plants to people.
5. **New terms are boldfaced,** permitting easy identification and providing emphasis. **NEW** A running glossary in the text calls out and defines each chapter's most important terms.
6. **NEW Process of Science icons,** embedded at relevant points in the text, highlight discussions about the work that scientists do and convey that the process of science is ongoing.
7. **NEW Evolution Link icons,** embedded at relevant points in the text, emphasize the links between evolution and the particular subject being discussed.
8. **NEW Figure in Focus statements** provide an overview for diagrams of complex topics. Many of these figures have numbered parts that show sequences of events in botanical processes or life cycles.
9. **NEW Process of Science** figures emphasize the scientific process in both classic and modern research.
10. **NEW Study outlines** at the end of each chapter are organized around the chapter learning objectives. Because the running glossary terms are boldfaced and defined in the outlines, students are given the opportunity to study vocabulary words within the context of related concepts.
11. **Review questions** focus on the chapter learning objectives.
12. **Thought questions** can be used for essay assignments or class discussion. These questions challenge students to apply knowledge learned in the chapter to new situations.
13. The **Glossary** contains the new and improved definitions of all running glossary terms, appropriately cross-listed.

14. **Appendices** include Metric Equivalents and Temperature Conversion; Some Important Biological Events in Geologic Time; and eight new appendices: USDA Plant Hardiness Zones; Annual Flowers; Flowering Perennials; Narrowleaf Evergreens; Broadleaf Evergreens; Deciduous Trees; Plant Selection for Various Conditions; and Biological Control of Insects.

What's New: An Overview of *Introductory Botany: Plants, People, and the Environment,* Second Edition

All of the content has been completely updated and revised for this new edition. A general outline of the chapters, with some specific additions and changes highlighted, follows.

Introduction

Chapter 1, An Introduction to the Science of Botany, introduces important biological concepts and lays the foundation for understanding how science works. **NEW** Plants and People box on spices in history. **NEW** Information on the three domains recognized in classification of organisms.

Part 1 The Plant Cell

Chapter 2 provides the basic chemistry needed to understand botany and introduces the biological molecules that interact in the construction and maintenance of plants. **NEW** Plants and People box on phytochemicals and human health. Chapter 3 focuses on the structure and functions of cells and their membranes. **NEW** EM and paired drawing of plasmodesmata (Figure 3-13). Chapter 4 discusses the energy transactions involved in life processes; both photosynthesis and cellular respiration are introduced with simplified overviews, followed by a more detailed explanation. **NEW** Updated organization, explanations, and definitions in detailed discussions of both photosynthesis and cellular respiration. **NEW** Focus On box on electron transport and heat in plants. **NEW** Improved art throughout. **NEW** Experiment figures: 4-7 (Engelmann's experiment) and 4-9 (Calvin's experiment).

Part 2 Plant Structures and Life Processes

Chapters 5 through 8 discuss the structural adaptations of roots, stems, and leaves. **NEW** Plants and People box on fibers and textiles in Chapter 5. **NEW** Updated section on mycorrhizae in Chapter 6. **NEW** Chapter 7 has two new Figures in Focus: 7-5 (on the development of secondary growth in a dicot stem) and 7-6 (a radial view of a dividing vascular cambium cell). **NEW** Updated section on the physiology of stomatal opening and closing in Chapter 8. Chapter 9 describes reproduction in flowering plants, including discussions of flowers, fruits, and seeds. **NEW** Section on self-incompatibility. Chapter 10 focuses on mineral nutrition and the mechanisms of transport in xylem and phloem. **NEW** Updated section on translocation in phloem. **NEW** Experiment figures: 10-9 (hydroponics experiment) and 10-15 (aphid experiment). Chapter 11 presents plant hormones and responses. **NEW** Chapter completely reorganized and updated. **NEW** Section on plant hormone action by signal transduction.

Part 3 The Continuity of Plant Life

The processes of mitosis and meiosis are explained and contrasted in Chapter 12. **NEW** Chapter introduction on date palms. Chapter 13 presents patterns of inheritance. **NEW** Process of Science section on how Mendel's work was recognized during the early 20th century. The molecular basis of inheritance is discussed in Chapter 14; both transcription and translation are introduced with simplified overviews, followed by a more detailed explanation. **NEW** Section on mutations that involve transposons. Chapter 15 presents concepts on the cutting edge of genetic research. **NEW** Section on genomics. **NEW** Section on applications of genetic engineering includes medicine and pharmacology, DNA fingerprinting, transgenic plants, and safety concerns of DNA technology. **NEW** Focus On box on DNA microarrays. Chapter 16 includes natural selection and the scientific evidence that supports evolution. **NEW** Plants and People box on evolution and bacterial resistance to antibiotics. **NEW** Section on experimental testing of evolutionary hypotheses. The evolution within populations and of new species is discussed in Chapter 17. **NEW** Sections on the founder effect and on stabilizing, directional, and disruptive selection. **NEW** Expanded sections on the Hardy–Weinberg principle and microevolution.

Part 4 Diversity

Chapter 18 describes how biologists classify plants and other organisms. **NEW** Information on author citations in scientific names and on cultivars versus subspecies. **NEW** Updated discussions of recent changes in classification systems. The various groups of organisms traditionally studied in botany are presented in an evolutionary framework in Chapters 19 through 25: archaea and

Preface • XXV

eubacteria in Chapter 19; protists in Chapter 20; fungi in Chapter 21; bryophytes in Chapter 22; seedless vascular plants in Chapter 23; gymnosperms in Chapter 24; and flowering plants in Chapter 25. **NEW** Expanded coverage in Chapter 19 on viruses, including history of discovery of viruses, how viruses are classified, evolutionary origin of viruses, bacteriophages, and viruses that infect plants. **NEW** Chapter 20 was reorganized around evolutionary relationships among the protists and includes new section on evolution and classification of the eukaryotes. **NEW** Sections on chytrids, microsporidia, and glomeromycetes in Chapter 21. **NEW** Chapter 22 has new material that updates discussion of bryophyte evolution. **NEW** Information on the evolution of vascular plants and new table on major groups of seedless vascular plants in Chapter 23. **NEW** Chapter 24 introduction on Wollemi pine. **NEW** Sections on pumpkin family (Cucurbitaceae) and sunflower family (Compositae) in Chapter 25. **NEW** Completely updated and revised section on evolution of flowering plants in Chapter 25.

Part 5 Plant Ecology

Chapter 26 presents the foundations of ecology, such as how organisms interact with one another, how matter cycles through ecosystems, and how energy flows through ecosystems. **NEW** Chapter completely reorganized and expanded. **NEW** Section on population ecology and expanded sections on community ecology and ecosystem ecology. Chapter 27 describes nine biomes and examines two ways that humans are altering the ecosphere. **NEW** Sections on declining biological diversity and conservation biology. **NEW** Major revision of section on global warming.

Supplements Package

To further facilitate learning and teaching, a supplements package has been carefully designed for the student and instructor. It includes:

1. **An Instructor's Manual with Test Bank.** These ancillaries are available on the Multimedia Manager CD. The Test Bank items are also available through ExamView, and the Instructor's Manual can also be accessed on the password-protected instructor's website. These resources contain materials to help the instructor, an expanded test bank that addresses the chapter learning objectives, new labeling questions, and answers to all end-of-chapter questions from the text.

2. **A Laboratory Manual for Introductory Botany.** This manual is designed to complement the main

text. You can also customize this manual through Thomson Custom Publishing to create a resource that matches your course needs. An instructor's guide for the Lab Manual is posted on the password-protected instructor's website.

3. **A Multimedia Manager.** The Multimedia Manager is a comprehensive instructor's resource CD-ROM that includes all of the tables, illustrations, and photographs from the text. It also includes the Instructor's Manual, Test Bank, additional photos, and precreated lecture slides. The Multimedia Manager offers the ability to edit the lecture slides, to present art in segments, to select and enlarge portions of a figure, and to edit or remove labels or to present one label at a time. Selected animations on plant physiology, evolution, and ecology can be linked from the lectures. All of these lecture materials combined on one easy-to-use Multimedia Manager CD will save you time when preparing lectures.

4. **ExamView.** ExamView is an electronic test generator that allows you to create, deliver, and customize tests and study guides (both print and online) in minutes. This easy-to-use assessment and tutorial system offers both a Quick Test Wizard and an Online Test Wizard that guide you step by step through the process of creating tests. Tests of up to 250 questions can be built using up to 12 question types.

5. **Overhead transparency acetates.** This comprehensive set of colorful transparencies includes all of the art and tables that are found in the text.

6. **Website.** The content-rich companion website that accompanies *Introductory Botany: Plants, People, and the Environment,* Second Edition, contains features to help students study, learn vocabulary, and investigate items of interest. This site features self-scoring quizzes, flashcards, weblinks, and InfoTrac₎ College Edition references. In addition, media editions of the text come with access to selected animations on plant physiology, evolution, and ecology, to help students review key concepts from the text.

7. **Virtual Biology Labs.** The labs included in this online resource engage students in gathering data and performing experiments virtually through state-of-the-art simulations. Each experiment is organized into an easy-to-use three-part format, including a general introduction, a series of interactive laboratory exercises, and a series of review questions. The 3.0 version offers four new labs, an improved design, and great flexibility in assessment and submission. A "Best Bets" package of labs has been put together to accompany and supplement the introductory plant biology course.

Acknowledgments

The development and production of *Introductory Botany: Plants, People, and the Environment,* Second Edition, required extensive interaction and cooperation among the author, editors, and reviewers. I appreciate the valuable input and support from my editors, colleagues, and students. I thank my family and friends for their understanding, support, and encouragement as I worked through many revisions and deadlines. I particularly appreciate the many hours Alan Berg spent formatting the manuscript.

The Editorial Environment

Preparing a book of this complexity is challenging and requires a great deal of time and effort. I could not have produced it without the help and support of the outstanding editorial and production staff at Brooks/Cole Thomson Learning. I thank Michelle Julet, Vice President and Editor-in-Chief, and Peter Adams, Publisher and Executive Editor, for their support and help. I am grateful to my Development Editor Kari Hopperstead who provided valuable input in every aspect of the project, from content to photograph selection. Associate Development Editor Suzannah Alexander used her wonderful artistic talents to render new art for the second edition. I thank Photo Editor Abigail Baxter for working hard to find exceptional photographs to enhance the text.

I greatly appreciate the help of Project Manager Belinda Krohmer and Production Coordinator Jamie Armstrong who expertly guided the project through the complexities of production. I also thank Lisa Buckley for creating the book design. I appreciate the efforts of Marketing Managers Stacy Best and Kara Kindstrom, whose expertise ensured that you would know about our new edition. All of these dedicated professionals and many others at Brooks/Cole Thomson Learning made important contributions to the production of *Introductory Botany: Plants, People, and the Environment,* Second Edition. I thank them for their help and support throughout this project.

Many professors and students have provided valuable suggestions. I thank them and ask for their comments as they use this second edition. I can be reached through my editors at Brooks/Cole Thomson Learning.

The Professional Environment—Reviewers

I very much appreciate the input of the many professors and scientists who have reviewed the manuscript during various stages of its preparation and provided me with valuable suggestions for improving it. Their work contributed greatly to the final product. Reviewers include:

Bonnie Amos, Angelo State University
Paul W. Barnes, Texas State University
Robert Bauman, Amarillo College
Charles E. Beard, Clemson University
Alan W. Bjorkman, North Park University
Steven B. Carroll, Truman State University
Cheryld L. Emmons, Alfred University
Robert C. Evans, Rutgers University
Sandi B. Gardner, Triton College
Russell H. Goddard, Valdosta State University
Charles R. Hart, University of Wisconsin–Manitowoc
L. Michael Hill, Bridgewater College
A. Scott Holaday, Texas Tech University
David E. Lemke, Texas State University
Bernard A. Marcus, Genesee Community College
Thomas Lawrence Mellichamp, University of North Carolina–Charlotte
Timothy Metz, Campbell University
Christopher J. Miller, St. Leo University
Carlos R. Ramirez, Southern Connecticut State University
David Reed, Defiance College
Lyndell P. Robinson, Lincoln Land Community College
Manfred Ruddat, University of Chicago
Michael W. Ruhl, Vernon College
Neil Schanker, College of the Siskiyous
Teresa Snyder-Leiby, State University of New York–New Paltz
Enid Spielman, Webster University–Thailand
Leslie R. Towill, Arizona State University
Michael R. Twiss, Clarkson University
Steven J. Wolf, California State University–Stanislaus
Carol Wymer, Morehead State University

Reviewers of the first edition include: Rolf W. Benseler, California State University–Hayward; Lawrence C. Bliss, University of Washington; Frank Bowers, University of Wisconsin–Stevens Point; Maynard Bowers, Northern Michigan State University; Richard C. Bowmer, Idaho State University; Richard Churchill, Southern Maine Technical College; Curtis Clark, California State Polytechnic University–Pomona; W. Dennis Clark, Arizona State University; James T. Dawson,

Pittsburg State University; Roger Del Moral, University of Washington; Darleen DeMason, University of California–Riverside; Michael Farris, Hamline University; Daniel Flisser, Cazenovia College; Doug Friend, University of Hawaii–Manoa; Stephen W. Fuller, Mary Washington College; Michael Gardiner, University of Puget Sound; John H. Green, Nicholls State University; William M. Harris, University of Arkansas; Hon H. Ho, State University of New York–New Paltz; William A. Jensen, Ohio State University; J. Morris Johnson, Western Oregon State College; Robert J. Lebowitz, Mankato State University; Marion Blois Lobstein, Northern Virginia Community College–Manassas; George H. Manaker, Temple University; Michael Howard Marcovitz, Midland Lutheran College; James E. Marler, Louisiana State University–Alexandria; Bill Mathena, Kaskaskia College; W. D. McBryde, Central Texas College; John McMillon, Lee College; Robert S. Mellor, University of Arizona; H. Gordon Morris, University of Tennessee–Martin; Jeanette C. Oliver, Flathead Valley Community College; John Olsen, Rhodes College; Lee R. Parker, California Polytechnic State University–San Luis Obispo; Pat Pendse, California Polytechnic State University–San Luis Obispo; William J. Pietraface, State University of New York–Oneonta; James A. Raines, North Harris College; James A. Rasmussen, Southern Arkansas University; Maralyn A. Renner, College of the Redwoods; Robert Rupp, Ohio State University, Agricultural Technical Institute; Barbara Schumacher, San Jacinto College Central; Richard Sims, Jones County Junior College; Ann Spencer, Charles County Community College; Michael P. Timko, University of Virginia; B. Dwain Vance, University of North Texas; Judith Verbeke, University of Arizona; David Williams, Anne Arundel Community College; and James Winsor, Pennsylvania State University–Altoona.

Linda R. Berg
November 2006

An Introduction to the Science of Botany

LEARNING OBJECTIVES

❶ Briefly describe the field of botany, and give short definitions of at least five subdisciplines of plant biology.

❷ Summarize and discuss the features of plants and other organisms that distinguish them from nonliving things.

❸ Distinguish among the six kingdoms and three domains, and give representative organisms for each.

❹ Summarize the main steps in the scientific method, and explain how science differs from many other human endeavors.

Cartoon artists often depict the sole survivor of a shipwreck stranded on a tiny tropical island that sports a single coconut palm (*Cocos nucifera*). If such a situation were to occur in real life, the shipwreck survivor would be truly fortunate, because this versatile plant provides almost everything a person needs to survive and be comfortable.

A single coconut palm tree usually produces 50 to 100 fruits (called "nuts") each year, and each nut contains a single large seed—the largest seed in the world. The coconuts that we buy in stores have had most of the outer layers of fruit removed (much like a peach pit without the rest of the peach). The outer husks of the coconut fruit can be separated from the nut, softened by soaking for several months in salty water, and shredded into soft fibers that are spun into yarn and twisted to make ropes or woven to make floor mats.

An immature green coconut seed has a highly nutritious liquid center called endosperm. As the seed matures, its endosperm solidifies to form coconut meat. Copra, or dried coconut meat, can be eaten just as it is, or it can be pressed to extract its oil. The oil is then used to make soap and other products. Coconut oil is found in dozens of everyday products, such as margarine, nondairy whipped topping, nondairy creamer, cosmetics, suntan lotions, and hand lotions. Although it is still found in some commercial baked goods, coconut oil is no longer considered desirable in foods because it is highly saturated. (Chapter 2 discusses the chemical explanation for saturated fats and their effects on health.)

The coconut shell that surrounds the seed can be used as an eating utensil, and the trunk of the coconut palm can provide wood for a shelter. Its leaves can be used to thatch a roof or to weave baskets and hats. An adventurous shipwreck survivor could also hollow out the coconut palm's trunk, make a canoe, and leave the island for more populated shores!

The coconut palm may be the world's most useful tree. Small wonder, then, that natives of the tropics call it the "tree of life." The coconut palm is an appropriate introduction to the study of plant biology because it demonstrates how important plants are to us. Indeed, humans depend on plants for food, shelter, fibers, fuel, and countless other products. Life would be impossible without plants.

Coconut palms. The coconut palm (*Cocos nucifera*) provides food, shelter, and a variety of useful products.

The Human Population and Plants

In 2006 the population of the world exceeded 6.5 billion individuals—approximately 82 million more humans than Earth supported in 2005. The 2005 world population, in turn, was approximately 81 million greater than the 2004 population. Although our numbers continue to grow, the global *rate* of population increase is slowing. The United Nations periodically publishes population projections for the 21st century. Their latest figures, published in 2003, forecast that the human population will increase to 8.9 billion (their "most likely" projection) by the year 2050.

As our numbers continue to increase during this century, environmental deterioration, hunger, persistent poverty, and health issues will become more critical. The need for food for the increasing numbers of people living in arid regions around the world has already led to overuse of the land for grazing and crop production. Partly as a result of such overuse, many of these formerly productive lands have been degraded into unproductive deserts.

Although the human population is increasing, not all countries have the same rate of population increase. **Highly developed countries,** which include industrialized countries such as the United States, Canada, Japan, and European countries, have low rates of population growth. **Less developed countries,** such as the relatively unindustrialized countries of Laos, Bangladesh, Nigeria, and Ethiopia, have high rates of population growth. **Moderately developed countries,** such as Mexico, Turkey, Thailand, and most South American countries, fall in the middle, with population growth rates that are higher than in highly developed countries but lower than in less developed countries.

The relationships among population growth, natural resource consumption, and environmental degradation are complex, but two generalizations are useful:

1. In many less developed and moderately developed countries, individual resource use is small. However, a rapidly increasing *number* of people tends to overwhelm and deplete these countries' soils, forests, and other natural resources.

2. In highly developed countries, individual resource demands are large, far above the minimum requirements for survival. To satisfy their desires rather than their basic needs, people in more affluent countries exhaust natural resources and degrade the global environment through extravagant *consumption* and "throwaway" lifestyles. A single child born in a highly developed country such as the United States has a greater impact on the environment and on resource use than a dozen or more children born in a country such as Nigeria.

Thus, the disproportionately great consumption of resources by people in highly developed countries affects natural resources and the environment as much as, or more than, the population explosion in the developing world. In this book, references to population issues refer to both too many people *and* too much consumption per person.

Why is a *botany* textbook starting with a discussion of human population? One reason is that the increase in human numbers affects Earth's organisms, including its plants. A dramatic reduction in biological diversity—that is, the number and kinds of organisms—is occurring worldwide, largely as a result of human endeavors. Many human activities that harm plants and other wildlife—clearing forests to grow crops or graze livestock, for example—are the direct outcomes of population growth (●**Figure 1-1**).

Even some well-educated people do not recognize the seriousness of the current decline in biological diversity. Earth's plants and other organisms contribute to a **sustainable**

FIGURE 1-1 **Overgrazing in Africa.**
Goats, sheep, and cattle have eaten almost all the ground cover. Overgrazing in the Sahel, a semiarid region south of the Sahara Desert, is increasing the amount of unproductive desert. Photographed in Burkina Faso.

Mark Edwards/Peter Arnold, Inc.

PLANTS AND PEOPLE | Taxol

Plants produce many organic chemicals that are vitally important to the pharmaceutical industry. Plants produce the active ingredients in a wide variety of medicines—laxatives, diuretics, heart medicines, ulcer treatments, analgesics, hormones, cancer treatments, and anesthetics, to name a few. In fact, one of every four prescription drugs contains ingredients originally derived from plants.

One of the most effective drugs currently used against cancer is taxol, an organic compound first obtained from the bark of the Pacific yew (*Taxus brevifolia*) (see figure). This tree is found primarily in the old-growth forests of the Pacific Northwest region of North America.

Taxol, which was first tested as a potential anticancer drug in the mid-1960s, inhibits the growth of cancer cells. After extensive clinical trials, it was approved in 1992 for treating advanced stages of ovarian cancer. In clinical trials, taxol caused tumors to shrink in up to 50 percent of patients with advanced ovarian cancer. Taxol was also approved in 1994 for treating breast cancer that has spread to the lymph nodes; about two-thirds of breast cancers in women have spread to the lymph nodes by the time they are first diagnosed. In addition, taxol shows promise in treating other cancers, such as cancers of the cervix, colon, lung, prostate, and skin.

To meet the demand for taxol, drug manufacturers planted yew trees, and several biotechnology firms began growing the taxol-producing yew cells in culture. However, the Pacific yew is one of the slowest-growing trees in the world. Furthermore, harvesting the bark kills the yew tree: Six 100-year-old trees provide enough taxol to treat just one breast-cancer patient. Fortunately, a close relative of the Pacific yew, the English yew (*T. baccata*), produces a chemical similar to taxol that can be obtained in sufficient quantities to treat cancer patients.

Since 1994, taxol has been synthesized in three different ways. Because taxol is such a complicated molecule, its synthesis requires almost 30 steps. Biochemists continue to try to shorten the number of steps in the synthesis of taxol so that it will be less expensive for drug companies to manufacture synthetically. The ability to synthesize taxol also allows chemists to use the steps they have developed to synthesize similar compounds and determine whether they are also effective in treating cancer. Sometimes the discovery of a single plant compound leads to the development of an entire family of drugs.

Taxol is not the only plant chemical under current study. In the 1990s, the National Cancer Institute screened thousands of plants and discovered three—from plants in Malaysia, Cameroon, and Samoa—that produce chemicals that inhibit HIV, the virus that causes AIDS.

The plant kingdom has the potential to provide us with other medicines. Of the approximately 300,000 species of flowering plants on Earth, only about 5000 have been evaluated in laboratories for their pharmaceutical potential. The ability of the world's plants to provide us with useful medicines is a strong case for their preservation. After all, if we destroy forests and natural areas before studying and evaluating the organisms that live there, we have permanently lost a valuable potential resource.

© Peter Ziminski/Visuals Unlimited

■ Bark from the Pacific yew (*Taxus brevifolia*), a tree found primarily in old-growth forests of the Pacific Northwest, is the original source of taxol, an anticancer drug.

environment, one that allows humans (as well as other organisms) to survive without compromising the ability of future generations to meet their needs. For example, humans and other animals depend on plants to produce food by the process of *photosynthesis.* Bacteria and fungi perform the essential role of decomposition, which releases essential minerals (inorganic nutrients) for plants to reuse.

In addition to their essential roles in the environment, plants and other organisms provide us with a variety of products. For example, plants supply us with such diverse products as oils and lubricants, perfumes and fragrances, dyes, paper, lumber, waxes, rubber and other elastic latexes, resins, poisons, cork, fibers, and medicines (see *Plants and People: Taxol*). Although there are more than 330,000 different species of plants, the vast majority have never been evaluated with respect to their potential usefulness! Biological diversity, as long as it remains high, provides an important natural resource for future uses and benefits. A decline in biological diversity results in lost opportunities and lost solutions to current and future problems.

PLANTS AND PEOPLE | Spices: A Brief History

If our food is bland, we do not hesitate to add spices such as pepper to provide pungency or piquancy. **Spices,** aromatic plants used as seasonings and preservatives, are common ingredients in our kitchens, and it is hard to imagine a time when they were exceedingly rare and very expensive. At one time, however, spices were steeped in glamour and mystery and were an important part of human history. Spices lured adventurers with the promised riches of the spice trade, stimulated global exploration, and caused countries to go to war over the possession of spice-producing areas.

Spices and herbs have undoubtedly been used to flavor foods and beverages for thousands of years. (There are no precise botanical definitions of herbs and spices, both of which are used as flavorings. Generally speaking, *herbs* are the leaves of plants from temperate regions, and *spices* are the barks, roots, seeds, or fruits of plants from tropical regions.) No one knows when primitive people began using spices to flavor their food. Perhaps early humans wrapped leaves around meats before cooking them and in the process

learned that certain leaves gave the meat a better flavor.

Before the days of food preservation and refrigeration, people used spices and herbs to help preserve food. Spices also helped people eat food that was spoiled or otherwise unpalatable. In addition to improving the flavor of these unpreserved foods, many spices contain antimicrobial compounds that reduce the incidence of food-borne illness. The flavors of different spices depend on the volatile oils they contain. Most spices consist of dried, ground plant parts that contain the oils. In spices such as vanilla and oil of lemon, however, the volatile oils are extracted and used as a liquid flavoring.

Over time, people discovered that certain spices had medicinal properties. For example, cinnamon oil was used to treat toothaches and ginger to treat diabetes mellitus. Spices also became an important part of religious rituals—consider the incense burned in many churches—and were added to perfumes and cosmetics.

Different spices and herbs are native to different regions of the world. Europe was originally home to dill, marjoram,

parsley, and thyme. India, Ceylon (Sri Lanka), and the Spice Islands (islands in the Malay Archipelago) provided other, more exotic spices, such as cinnamon, cloves, ginger, nutmeg, and peppercorns, which are the dried berries of a vine native to Sumatra.

In the city-states of ancient Greece, trade routes from East Asia brought spices to Alexandria, Egypt, which was then the trade center for the entire Mediterranean region. During the time of the Roman Empire, spice trading expanded throughout Europe, and more people began to use spices and herbs to flavor foods and wine. After the fall of the Greek and Roman empires, Arabia dominated the spice trade, controlling access between Europe and the Far East.

Europeans, beginning with Marco Polo, a Venetian traveler in East Asia, eventually discovered the locations where Asia-originating spices were grown. By the end of the Crusades, the trading activities of Italian cities such as Venice and Genoa linked Europe to the spices of the East. European explorers began searching for new trade routes to spice sources and, in

---

It is now time to embark on a journey of discovery. During the coming weeks, you will learn much about the biology of plants, a fascinating group of organisms very much like you and yet very different. In doing so, you will not only broaden your own intellectual horizons but develop an awareness of many environmental issues that link plants and people.

What Is Botany?

The study of **botany,** or **plant biology,** encompasses the origin, diversity, structure, and internal processes of plants as well as their relationships with other organisms and with the nonliving physical environment. The time frame encompassed by botany extends from the present back to almost 3.5 billion years ago, the age of the earliest fossilized cells. The scope of botany is extensive: Some plant biologists study how global climate affects

plants, whereas others examine the molecules that make up plant cells.

Today is arguably the most exciting time to be studying plant biology. Scientists are beginning to unravel some of the most fundamental mysteries of life. Moreover, some of our most pressing contemporary human problems have a botanical component. Some plant biologists confront the challenge of producing enough food to support the ever-expanding world population, whereas others identify plants that might be important sources of future drugs to treat diseases such as cancer or AIDS. Certain botanists choose to study plants that are in danger of extinction, with the ultimate goal of keeping these organisms from disappearing forever from our planet. The effect of human-produced pollution on plants is another focus for many plant biologists. Some of these pollution problems are regional or global in scope—such as acid precipitation, global warming, and the destruction of the ozone layer in the upper atmosphere.

the process, discovered North and South America. South America and the islands of the Caribbean provided many new spices—for example, allspice, cayenne, paprika, and vanilla—to the rest of the world.

Because of their dominance in early sea explorations, the Portuguese controlled the spice trade for many years. The Dutch took over the Spice Islands between 1605 and 1621 and ruled the spice trade for almost 200 years. Each of these countries obtained great wealth and experienced a flowering of culture as a result of their involvement with the spice trade. The British Empire slowly wrested control of the spice trade from Holland in about 1800. However, the importance of the spice trade had begun to slowly wane because many spice plants had been smuggled from their native regions and planted around the world, making spices a more common commodity.

Today, spices and herbs are grown, packaged, and priced so that nearly everyone can afford a well-stocked spice shelf.

Barry Wong/The Image Bank/Getty Images

■ Representative spices and herbs. (Top, from left to right) Fennel seeds, whole chili de arbol peppers, and ground cumin. (Middle, from left to right) Ground cayenne pepper, ground cinnamon, and fenugreek seeds. (Bottom, from left to right) Whole coriander, ground turmeric, and Mexican oregano leaves.

Botany comprises many disciplines

Because of its immense scope, plant biology comprises a number of specialties. Botanists who specialize in **plant molecular biology** study the structures and functions of important biological molecules such as proteins and nucleic acids. **Plant biochemistry** is the study of the chemical interactions within plants, including the variety of chemicals that plants produce. **Plant cell biology** encompasses the structures, functions, and life processes of plant cells; **plant anatomy** is microscopic plant structure (cells and tissues); and **plant morphology** refers to the structures of plant parts such as leaves, roots, and stems, including their evolution and development. Plant biologists who specialize in **plant physiology** study such processes as photosynthesis and mineral nutrition to understand how plants function. Plant heredity and variation make up the specialty of **plant genetics. Plant ecology** is the study of the interrelationships among plants and between plants and their environment. **Plant systemat-** ics encompasses the evolutionary relationships among different plant groups. **Plant taxonomy,** a subdiscipline of systematics, deals with the description, naming, and classification of plants. **Paleobotany** is the study of the biology and the evolution of plants in the geologic past.

In addition to the specialties of plant biology just described, many botanists specialize in particular types of plants. For example, **bryology** is the study of mosses and similar plants. Also, many areas of applied plant biology exist, including **agronomy** (field crops and soils), **horticulture** (ornamental plants and fruit and vegetable crops), **forestry** (forest conservation and forest products such as lumber), and **economic botany** (plants with commercial importance; see *Plants and People: Spices: A Brief History*).

BOTANY The scientific study of plants; also called *plant biology.*

Characteristics of Plants

Although plants are a dominant part of our landscape, they are easy to overlook or take for granted because they appear so passive. Plants do not appear to "live" in the sense that animals live. Plants do not run or swim or slither or fly; they do not eat other plants or animal prey; nor do they reproduce by an obvious coupling of two partners. Plants have adapted to life on land in ways that seem completely different from the *adaptations* of humans and other animals. Despite the perceived differences between plants and animals, however, plants share many important characteristics with other organisms. In this section we examine in detail these general characteristics of life.

Plants are highly organized

Like all living things, plants have a complex organization. Although a palm tree looks different from a tulip, both are composed of basic building blocks called cells. The **cell,** which is microscopic, is the smallest unit that can perform all the activities associated with life. Some organisms are unicellular (single celled), whereas many plants are composed of trillions of cells (●**Figure 1-2**). Those trillions of cells interact in a coordinated way to carry out the life processes required by the multicellular organism.

The biological world is organized on more than just a cellular level. The elements (basic chemical substances) of which cells are composed constitute the simplest level of organization in the biological world. An **atom** is the smallest particle of an element that possesses the properties of that element (●**Figure 1-3**). An atom of carbon, for example, is the smallest quantity of carbon possible. Atoms combine chemically by forming bonds to produce **molecules.** A molecule can be composed of two or more atoms of a single element (for example, O_2, molecular oxygen) or different elements (for example, CO_2, carbon dioxide). Molecules, in turn, may be organized into **macromolecules,** large biological molecules such as proteins and nucleic acids. Macromolecules associate with one another to form compartments called **organelles** within cells. A cell's nucleus, which contains the hereditary material of the cell, is an example of an organelle. Organelles associate to form the cell.

In most multicellular organisms, cells are organized into **tissues,** associations of cells that perform specific functions. Examples of plant tissues include epidermis, the protective tissue that covers the plant's surface, and xylem, the water-conducting tissue. Tissues, in turn, are organized into **organs,** functional units that perform specific roles. Leaves and roots are examples of plant organs. Several types of organs, functioning with great precision, make up a multicellular plant. Plants are **organisms**—distinct, living entities.

Several higher levels of biological organization occur at the ecological level, above the level of the organism. Organisms are arranged into **populations,** groups of members of the same species that live together in the same area at the same time, such as a population of daisies in a field. Populations, in turn, are organized into communities. A **community** consists of all the populations of different organisms that live and interact within an area. The organisms living in a pond and a forest are two examples of communities; each might contain hundreds of types of organisms. *Ecosystem* is a more inclusive term than *community,* because an **ecosystem** is a community together with its nonliving environment. Thus, an ecosystem encompasses not only all interactions among organisms but also the interactions between organisms and their physical environment. All of

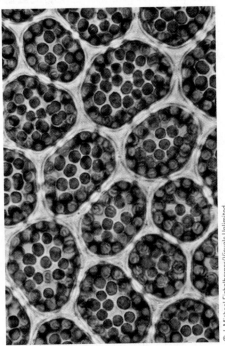

(a) Moss (*Rhizomnium*) "leaf" cells contain many small, green chloroplasts.

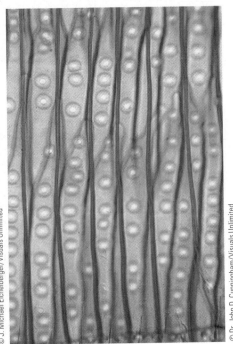

(b) A cluster of water-conducting cells (tracheids) in a pine (*Pinus*) stem. Plants and other complex multicellular organisms are composed of billions or even trillions of cells.

FIGURE 1-2 **Plants are composed of cells.**

The study of biology ranges from atoms, which are the basic chemical building blocks of the natural world, to the global ecosystem known as the biosphere.

FIGURE 1-3 Levels of biological organization.

These levels, from atoms to the biosphere, are explained in the text.

Annie Reynolds/PhotoLink/Getty Images

FIGURE 1-4 Sunlight filters through a tree.
Plants absorb radiant energy for photosynthesis.

Earth's ecosystems are collectively called the **biosphere.** The organisms of the biosphere depend on one another and on Earth's physical environment—its water (liquid and frozen, fresh and salty), its crust (soil and rock), and its gaseous atmosphere.

Plants take in and use energy

A continuous input of energy from the sun enables life to exist on Earth. All organisms require energy for their activities, which include growth, repair, reproduction, and maintenance. The two most important energy-related activities in the living world are photosynthesis and cellular respiration.

Plants, algae, and a few bacteria perform **photosynthesis,** which is essential for almost all life. In photosynthesis, radiant (light) energy from the environment is converted to chemical energy that is stored in molecules such as the sugar glucose (•**Figure 1-4**). Photosynthesis provides plants, algae, and photosynthetic bacteria with

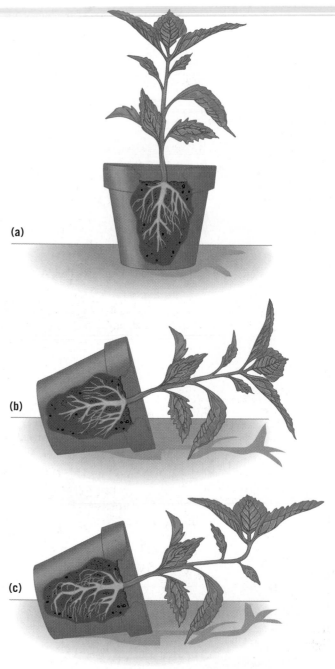

FIGURE 1-5 Root growth in the direction of gravity.
When a potted plant **(a)** is placed on its side **(b),** making the roots horizontal, the root tips change the direction of their growth so that they are again growing downward **(c).** The roots other than the tips do not change their direction, because they are no longer growing; the root tips are the only parts that increase in length.

a ready supply of energy (in molecules of glucose) that they can use as the need arises. The energy can also be transferred from one organism to another—for example, from plants to animals that eat plants, and to animals that eat other animals that eat plants. Thus, animals receive their energy from the sun indirectly by eating

(a) A Venus flytrap with open traps.

(b) A closed trap and insect prey. After enzymes produced by the plant digest as much of the insect as is digestible, the trap will reopen.

FIGURE 1-6 The modified leaves of a Venus flytrap snap shut on its prey.
Venus flytraps (*Dionaea muscipula*) grow in bogs in North and South Carolina.

plants. Photosynthesis also produces oxygen, which most organisms require when they break down food.

The chemical energy that plants store in food molecules is released within the cells of plants, animals, and other organisms through **cellular respiration.** In cellular respiration, food molecules such as glucose are broken down, usually in the presence of oxygen. Cellular respiration makes the chemical energy stored in food molecules available to the cell for biological work. *All* organisms, plants included, obtain energy via cellular respiration.

Plants respond to stimuli

Plants and other organisms respond to **stimuli,** changes in their environment. Stimuli to which plants respond include changes in the direction, color, or luminosity of light; in temperature or in the orientation toward gravity; and in the chemical composition of the surrounding soil, air, or water.

The responses of plants to environmental stimuli are often not as obvious as those of animals. For example, when a coyote charges a rabbit, the rabbit responds by darting away. In contrast, plants often respond to stimuli by the differential growth of parts of their bodies. Roots, for example, normally grow in the direction of gravity. When a potted plant is placed on its side, its root tips respond by changing their direction of growth so that the root is again growing downward (•**Figure 1-5**). Stems normally grow upward, so when the plant is placed on its side, its stem tip (the site of growth in a stem) bends upward.

Some plants respond to stimuli in a dramatic fashion. The Venus flytrap, with its hinged leaves that re-

semble tiny bear traps, is exceptionally sensitive to touch and can catch insects (•**Figure 1-6**). Tiny "trigger" hairs on the leaf detect when an insect alights, and the leaf rapidly folds shut, preventing the escape of the prey. The leaf then secretes enzymes to digest the insect. Venus flytraps and similar plants usually live in soil that is deficient in nitrogen; they capture and "eat" insects to obtain part of the nitrogen they require.

Plants grow and develop

Growth is an increase in the size and mass of an organism. In plants, growth results from both an increase in the *number* of cells and an increase in the *size* of cells. Many plants—most trees, for example—continue to grow larger throughout their life. In contrast, growth ceases when humans and many other animals reach adulthood.

Growth is a part of **development,** which includes all the changes in a plant or other organism from the start of its life through its immature (juvenile) stage, through its mature (adult) stage, to its death. An oak tree begins life as a fertilized egg, or *zygote,* which grows and develops into a multicellular embryo—with an embryonic

PHOTOSYNTHESIS The biological process that includes the capture of light energy and its transformation into chemical energy of organic molecules that are manufactured from carbon dioxide and water.

CELLULAR RESPIRATION The cellular process in which energy of organic molecules is released for biological work.

root, stem, and leaves—within a seed, in this case, an acorn. Development continues when the seed germinates and the young plant (called a seedling) emerges from the acorn (•**Figure 1-7**). The seedling eventually develops into an adult oak tree.

Plants reproduce

One of the basic concepts of biology is that living organisms come from previously existing organisms. **Reproduction,** the formation of a new individual by sexual or asexual means, is the most distinctive characteristic of life. Although some organisms exist for a long time—individual bristlecone pines have lived for thousands of years—all organisms eventually die. Reproduction enables an organism to perpetuate its traits beyond an individual's own death.

In many organisms, reproduction is **asexual** and does not involve the union of gametes (reproductive cells). In asexual reproduction, one parent gives rise to offspring that are virtually identical to it (•**Figure 1-8**). **Sexual reproduction** in plants involves the union of gametes that may or may not come from two separate individuals. The gametes—an egg cell and a sperm cell—unite to form a zygote, which develops into the new offspring. In sexual reproduction, the offspring are not exact copies of a single parent but are products of the unique combination of traits inherited from both parents.

Plant DNA transmits information from one generation to the next

The characteristics of an organism are encoded in its **genes,** which are the units of hereditary information. The genes of a peach tree provide information about all aspects of the plant, such as how tall the tree grows, what color its flowers are, whether it can tolerate full sun or shade, and how many seeds each fruit produces. Of course, an organism's environment also plays a crucial role in determining its characteristics, but always within that organism's genetic constraints.

Genes are composed of **deoxyribonucleic acid (DNA),** the organic molecule that stores and carries important genetic information in cells. DNA, which stands for *deoxyribonucleic acid*, is the molecule of inheritance for all organisms, from beans to giraffes to bacteria.

Information encoded in genes—that is, in the DNA that composes genes—is transmitted from one genera-

Michael P. Gadomski/Photo Researchers, Inc.

FIGURE 1-7 Germinating acorn.
A northern red oak (*Quercus rubra*) acorn has germinated, and the emerging root is growing into the forest soil. The young shoot (stem and leaves) will emerge next.

tion to the next during reproduction. Genes ensure that a bean plant produces seeds that grow into bean plants, not into roses or cucumbers.

EVOLUTION LINK   Plant populations undergo genetic changes over time

The ability of a plant population to adapt to its environment helps it exist in a changing world. **Adaptations** are characteristics that enable an organism to better survive in a certain environment. Adaptations may involve changes in structure, form, or function. The thick, succulent (fleshy, water-storing) stem of a cactus is an adaptation that enables it to live in dry regions where precipitation is low (•**Figure 1-9**). Likewise, the prickly spines covering the surface of the cactus stem are an adaptation that discourages desert animals from eating the succulent stem tissue.

Every organism contains many interdependent adaptations that help it survive in the particular environment to which it is adapted. **Evolution,** the process by which organisms adapt to their environment over time, is the genetic change in a population of organisms from generation to generation. Evolutionary processes typically require long periods of time and occur over many generations.

Charles Darwin and Alfred Wallace first suggested a plausible mechanism, **natural selection,** to explain evolution (see Chapter 16). In his book, *On the Origin of Species by Means of Natural Selection,* published in 1859, Darwin brought together many new findings in geology and biology. He presented a wealth of evidence

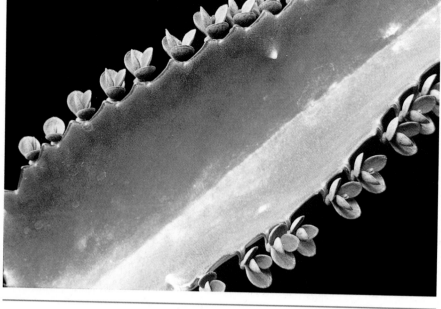

Dennis Drenner

FIGURE 1-8 The mother-in-law plant.
This plant (*Kalanchoe pinnata*) produces young plants asexually along the margins of the leaves. When the young plants attain a certain size, they drop off and root in the ground. Each young plant is genetically identical to its parent.

Biologists consider evolution to be the unifying theory of biology because it encompasses all the changes that have transformed life from the first primitive cells into the millions of different organisms that exist today. Evolution helps us understand why organisms are the way they are and how they are related, both to one another and to earlier forms of life. ■

Biological Diversity

We do not know exactly how many kinds of organisms exist, but most biologists estimate that there are at least 5 million to 10 million species. So far, more than 330,000 plant species and more than 1 million animal species have been identified. The variation among organisms staggers the imagination and has required the development of a system to organize this diversity.

that organisms existing today descended, with modifications, from previously existing life-forms.

Darwin's theory of evolution by natural selection has influenced the biological sciences to the present. His work has generated many scientific observations and research experiments that have provided additional evidence that evolution is responsible for the great diversity of life-forms on Earth. Even today, botanists and other biologists continue to focus much research and discussion on details of the evolutionary process.

Many biologists assign organisms into six kingdoms

For hundreds of years, biologists viewed organisms as belonging to two broad groups, the plant and animal **kingdoms.** As organisms underwent closer study, it became increasingly obvious that many did not fit well into either the plant or animal kingdom. For example, *bacteria,* single-celled organisms that lack cell nuclei, were originally considered plants. The fact that all bacteria lack nuclei is now considered a fundamental difference between them and other organisms—a difference more significant than the differences between plants and animals. With our present knowledge, it is difficult to consider bacteria as plants.

James Mauseth, University of Texas

FIGURE 1-9 The cactus possesses adaptations that help it survive in dry climates.
The stem of a cactus (a species of *Melocactus* is shown) functions for both photosynthesis and water storage. The leaves of cacti are modified into spines for protection.

DEOXYRIBONUCLEIC ACID (DNA) A nucleic acid present in a cell's chromosomes that contains genetic information.

EVOLUTION Cumulative genetic changes in a population of organisms from generation to generation.

NATURAL SELECTION The mechanism of evolution proposed by Charles Darwin; the tendency of organisms that have favorable adaptations to their environment to survive and become the parents of the next generation.

KINGDOM A broad taxonomic category made up of related phyla; many biologists currently recognize six kingdoms of living organisms.

In addition, certain organisms—*Euglena*, for example—seem to possess characteristics of plants *and* animals (•Figure 1-10). In fact, because many unicellular organisms have more in common with one another than with either plants or animals, biologists assigned them to a separate kingdom.

These and other considerations have led to the six-kingdom system of classification that many biologists use today. The kingdoms currently recognized are Archaea, Bacteria, Protista, Fungi, Plantae, and Animalia (•Figure 1-11).

Prokaryotes, organisms that lack nuclei and other membrane-bounded organelles, are now divided into two groups on the basis of significant biochemical differences. Members of the **kingdom Archaea** are often adapted to harsh conditions and frequently live in oxygen-deficient environments, including hot springs, salt ponds, and hydrothermal vents in the ocean depths. The thousands of kinds of other prokaryotes are classified in the **kingdom Bacteria.**

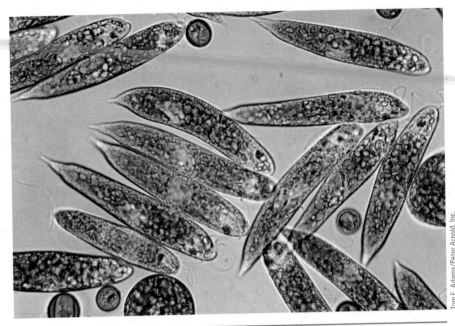

Tom E. Adams/Peter Arnold, Inc.

FIGURE 1-10 Numerous euglenoids crowded into a drop of water.
Euglena has both plantlike and animal-like traits. This unicellular organism is plantlike because it can photosynthesize and is animal-like because it can move about from place to place. In the past, *Euglena* was classified in the plant kingdom (when euglenoids were considered plants) *and* in the animal kingdom (when protozoa, including euglenoids, were considered animals). Today, most biologists place *Euglena* in the kingdom Protista.

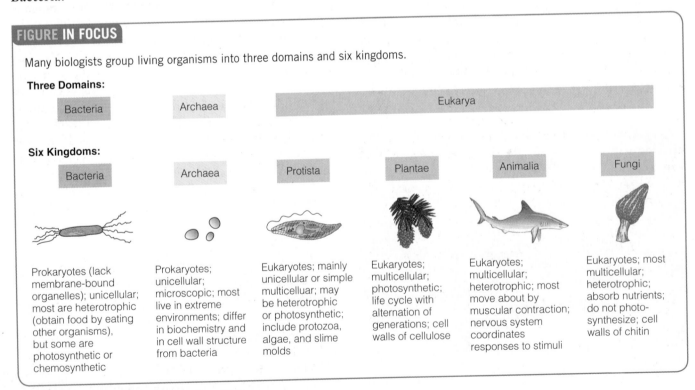

FIGURE IN FOCUS

Many biologists group living organisms into three domains and six kingdoms.

Three Domains:

| Bacteria | Archaea | Eukarya |

Six Kingdoms:

| Bacteria | Archaea | Protista | Plantae | Animalia | Fungi |

Prokaryotes (lack membrane-bound organelles); unicellular; most are heterotrophic (obtain food by eating other organisms), but some are photosynthetic or chemosynthetic

Prokaryotes; unicellular; microscopic; most live in extreme environments; differ in biochemistry and in cell wall structure from bacteria

Eukaryotes; mainly unicellular or simple multicelluar; may be heterotrophic or photosynthetic; include protozoa, algae, and slime molds

Eukaryotes; multicellular; photosynthetic; life cycle with alternation of generations; cell walls of cellulose

Eukaryotes; multicellular; heterotrophic; most move about by muscular contraction; nervous system coordinates responses to stimuli

Eukaryotes; most multicellular; heterotrophic; absorb nutrients; do not photosynthesize; cell walls of chitin

FIGURE 1-11 The kingdoms and domains of life.
The six-kingdom system of classification includes the Bacteria, Archaea, Protista, Plantae, Animalia, and Fungi. The three domains, which are color-coded to match their respective kingdoms, consist of Bacteria, Archaea, and Eukarya.

TABLE 1-1 The Classification of Corn

Domain Eukarya: several million species
Organisms with a eukaryotic cell structure

Kingdom Plantae: about 330,000 species
Terrestrial, multicellular, photosynthetic eukaryotes

Phylum Anthophyta: about 300,000 species
Vascular plants with flowers, fruits, and seeds

Class Monocotyledones: more than 90,000 species
Monocots: flowering plants with one seed leaf
(cotyledon)

Order Commelinales: about 17,300 species
Monocots with reduced flower parts, elongated
leaves, and dry, one-seeded fruits

Family Poaceae: about 9000 species
Grasses with hollow stems, fruit that is a grain,
and abundant endosperm in seed

Genus *Zea*: 2 species
Tall annual grass with separate male and female flowers

Species *Zea mays*: 1 species
Corn: one of two species in genus

Carlyn Iverson

a level of classification above the kingdom and is based on fundamental differences among organisms. The kingdoms Archaea and Bacteria correspond to two of the domains: kingdom Archaea corresponds to **domain Archaea,** and kingdom Bacteria corresponds to **domain Bacteria.** The remaining four kingdoms are assigned to **domain Eukarya.**

PROCESS OF SCIENCE Biologists who classify organisms in domains may or may not recognize the kingdom level. Differing views on how to classify organisms reflect the creative and dynamic process of science. Biologists are responsive to new data and consequently the classification of organisms at all levels is continually being reevaluated. Biologists try to base classification on evolutionary relationships so they can use classification to reconstruct the history of life. ■

The **kingdom Protista** contains protozoa, algae, water molds, and slime molds. These organisms are single-celled or simple multicellular organisms. The **kingdom Fungi** consists of molds, yeasts, and mushrooms, which obtain their nutrients by secreting digestive enzymes into food and then absorbing the predigested nutrients. Animals belong to the **kingdom Animalia.** Animals are multicellular organisms that must eat other organisms to obtain nourishment. They possess muscular systems, which enable them to move, and nervous systems, such as the brain, spinal cord, and sense organs such as eyes.

Plants, multicellular organisms that typically photosynthesize, are placed in the **kingdom Plantae.** Plants generally possess a **cuticle** (a waxy covering over their outer parts that reduces water loss), **stomata** (tiny openings in leaves and stems for gas exchange),[1] and **multicellular gametangia** (reproductive organs that protect gametes). Plants are usually rooted in one place and, unlike animals, do not possess a nervous system. The kingdom Plantae includes vascular plants—such as ferns, conifers, and flowering plants—which possess tissues that conduct food and water throughout the plant body, and nonvascular plants—such as mosses—which lack these tissues.

Many biologists also assign organisms into three domains

As scientists report new data, the classification of organisms changes. Many biologists now assign all organisms to three domains (see Figure 1-11). The **domain** is

Classification is hierarchical

The broadest classification category is the domain, and the narrowest is the species. Between them is a range of categories that forms a hierarchy (•Table 1-1). A **species** is a group of similar organisms that interbreed in their natural environment but do not interbreed with other species. Similar species are assigned to the same **genus** (pl., *genera*), and similar genera are grouped in the same **family.** Families are grouped into **orders,** orders into **classes,** classes into **phyla** (sing., *phylum*), phyla into kingdoms, and kingdoms into domains. Plants and fungi were traditionally classified into *divisions* rather than phyla. The International Botanical Congress, however, approved the phylum designation for plants and fungi in 1993.

Each species is named using a binomial system

In the 18th century, Carolus Linnaeus, a Swedish botanist, simplified the naming of organisms (see Figure 18-2).

DOMAIN A taxonomic category that includes one or more kingdoms.

SPECIES A group of organisms with similar structural and functional characteristics that in nature breed only with one another and have a close common ancestry.

---

[1]*Liverworts, the simplest of plants, lack stomata.*

In Linnaeus's system, called the **binomial system of nomenclature,** each species receives a two-part name. The first word designates the genus to which the organism is assigned, and the second word is a *specific epithet,* that is, a descriptive word that characterizes the organism. The specific epithet is always used with the full or abbreviated generic name preceding it. For example, the white oak (*Quercus alba,* sometimes abbreviated *Q. alba*) and the red oak (*Quercus rubra*) belong to the same genus, *Quercus.* The American beech (*Fagus grandifolia*) belongs to a different genus. The genus is capitalized, and the specific epithet is usually not capitalized; both names are italicized.

PROCESS OF SCIENCE The Scientific Method

Most people think of science as a body of knowledge—a collection of facts about the natural world. However, science is also a dynamic *process.* It is a particular way to investigate nature in an attempt to understand the universe. Science seeks to reduce the apparent complexity of our world to general principles, which help solve problems or provide new insights. In other words, science is devoted to discovering the general principles that govern the operation of the natural world.

The information collected by scientists is called **data** (sing., *datum*). Data are collected by observation and experimentation and then analyzed or interpreted. Scientific conclusions are inferred from the available data and are not based on faith, emotion, or intuition.

Science is an ongoing enterprise, and scientific concepts must be reevaluated in light of newly discovered data. Thus, scientists can never claim to know the "final answer" about anything, because scientific knowledge changes.

The processes that scientists use to answer questions or solve problems are collectively called the **scientific method** (see *Focus On: Using the Scientific Method*). Although there are many versions of the scientific method, it basically involves five steps:

1. Recognize a question or an unexplained occurrence in the natural world. Science is based on knowledge accumulated previously. Therefore, after one recognizes a problem, one investigates relevant scientific literature to determine what is already known about it.
2. Develop a **hypothesis,** or educated guess, to explain the problem. A good hypothesis makes a prediction that can be tested and possibly disproved. The same factual evidence can be used to formulate several alternative hypotheses, each of which must be tested.

3. Design and perform an experiment to test the hypothesis. An experiment involves the collection of data by making careful observations and measurements. Much of the creativity in science lies in designing experiments that sort out confusion caused by competing hypotheses.
4. Analyze and interpret the data to reach a conclusion. Does the evidence match the prediction stated in the hypothesis? In other words, do the data support or refute the hypothesis? Should the hypothesis be modified or rejected based on the data?
5. Share new knowledge with the scientific community. The scientist does this by publishing articles in scientific journals or books and by presenting the information at scientific meetings. Sharing new knowledge with the scientific community permits other scientists to repeat the experiment or design new experiments that either verify or refute the work.

Although the scientific method is usually described as a linear sequence of steps, it is rarely that straightforward or tidy. Good science requires creativity in recognizing questions, developing hypotheses, and designing experiments. A hypothesis that cannot be rejected is tentatively accepted, which explains why scientific theories are subject to modification. Science progresses by trial and error. Many creative ideas result in dead ends, and there are often temporary setbacks as knowledge progresses. Often the "big picture" emerges slowly from confusing and seemingly contradictory information.

An experiment must have a control

Most often, the question we wish to answer is influenced by many factors, which are called **variables.** To test a particular hypothesis about *one* variable, it is necessary to hold all other variables constant so that we are not misguided by them. This is done by performing two forms of the experiment. In the main experiment, the conditions are altered in one way (one variable is adjusted). The second experiment, the **control,** is identical to the main experiment in all respects except that conditions are *not* altered. By designing an experiment with a control, the researcher ensures that any difference in the outcomes between the main experiment and its control must be the result of the variable.

Scientists use inductive and deductive reasoning

Discovery of general concepts by the careful examination of specific cases is called **inductive reasoning.** The scientist begins by organizing data into manageable cat-

FOCUS ON | Using the Scientific Method

PROCESS OF SCIENCE As stated earlier in this chapter, roots grow downward, in the direction of gravity. Such a reaction of plants to Earth's gravitational pull is known as **gravitropism,** that is, growing in response to gravity. Roots are considered to be positively gravitropic because they grow downward, toward a gravitational pull, and some stems are negatively gravitropic because they grow upward, opposite the direction of gravity.

Suppose you wished to conduct an experiment using the scientific method to determine whether the young roots of germinating (sprouting) corn kernels exhibit the downward growth of positive gravitropism. How would you begin?

1. Recognize a question: Do the roots of germinating corn kernels exhibit positive gravitropism? After stating the question, you would do some library research related to the problem. You might determine, for example, that each corn kernel has a top and a bottom; the young root emerges from the bottom of the germinating kernel; and the young shoot (stem and leaves), from the top. You would have to learn the environmental requirements for seed germination. For example, seeds must absorb water before they begin growing. You would also have to investigate ways to grow seeds so that their roots could be observed. Normally, of course, the roots are hidden from view in the soil.

2. Develop a hypothesis (in the form of a simple declarative sentence): Regardless of the direction in which a corn kernel is planted, the emerging root always grows downward, toward the pull of gravity.

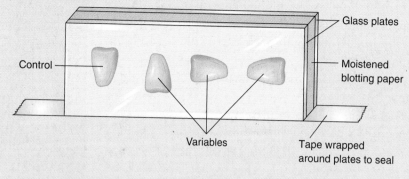

Glass plates

Control

Moistened blotting paper

Variables

Tape wrapped around plates to seal

KERNEL NO.	INITIAL ORIENTATION OF KERNEL	DIRECTION OF ROOT, DAY 1*	DIRECTION OF ROOT, DAY 2*	DIRECTION OF ROOT, DAY 3*	DIRECTION OF ROOT, DAY 4*
1 (control)					
2					
3					
4					

*After germination.

3. Design and perform an experiment to test the hypothesis. There are several ways of designing an experiment to test this hypothesis. One way is to place the corn kernels on moistened blotting paper between two plates of glass (see figure).

Every experiment has a control. In this experiment, the control would be a corn kernel oriented in the normal direction, that is, with its "bottom" side downward. The variables in this experiment would be other kernel positions, which are oriented in various directions.

After setting up the experiment, you would collect data, probably by observing the experiment over a period of several days. You would confirm that the emerging root always grows out of the bottom of the kernel regardless of the kernel's orientation. A few days after the kernel germinates, you would notice that the roots of all the kernels were growing downward, even those coming from an upside-down kernel. It is helpful to present the data in a table.

4. Analyze and interpret the resulting data to reach a conclusion. Examine the data to determine whether the evidence matches the prediction stated in the hypothesis. What is your conclusion?

5. Share new knowledge with the scientific community. Because you are a student in a college class, you might write a laboratory report. Most laboratory reports have a format similar to those of papers that scientists submit for publication in scientific journals. ■

egories and asking the question "What does this information have in common?" He or she then seeks a unifying explanation for the data. Inductive reasoning is the basis of modern experimental science.

HYPOTHESIS An educated guess (based on previous observations) that may be true and is testable by observation and experimentation.

As an example of inductive reasoning, consider the following:

Datum: Poppies obtain their energy by photosynthesis.
Datum: Daisies obtain their energy by photosynthesis.
Datum: Roses obtain their energy by photosynthesis.
Conclusion: All flowering plants obtain their energy by photosynthesis.

Even if inductive reasoning makes use of correct data, the conclusion may be either true or false. As new data come to light, they may show that the generalization arrived at through inductive reasoning is false. Science has shown, for example, that the flowering plant Indian pipe (*Monotropa uniflora*) is nonphotosynthetic and relies on fungi living in the soil for nutrient energy (•**Figure 1-12**); the fungi, in turn, are usually associated with the roots of nearby trees, from which they obtain nutrients. When one adds this information to the preceding list, one must formulate a different conclusion. Inductive reasoning, then, produces new knowledge but is error-prone.

Science also makes use of **deductive reasoning,** which proceeds from generalities to specifics. Deductive reasoning adds nothing new to knowledge, but it can make relationships among data more apparent. For example:

General rule: A plant's aerial parts (those exposed to air) are covered by a waxy cuticle that helps reduce water loss.
Specific example: Corn is a plant.
Conclusion: Corn plants possess a cuticle.

This is a valid argument. The conclusion that corn plants possess a cuticle follows inevitably from the information given. Scientists use deductive reasoning to determine the type of experiment or observations that are necessary to test a hypothesis.

Many observations and experiments support theories

Theories explain scientific laws. A **theory** is an integrated explanation of numerous hypotheses, each supported by a large body of observations and experiments. A good theory condenses and simplifies many data that previously appeared unrelated. A theory grows as additional information becomes known. It predicts new data and suggests new relationships among a range of natural phenomena.

By demonstrating relationships among classes of data, a theory simplifies and clarifies our understanding

Carlyn Iverson

FIGURE 1-12 The Indian pipe.
Indian pipes (*Monotropa uniflora*) do not contain chlorophyll or photosynthesize, and they live underground except for their flowers (*shown*).

of the natural world. Theories are the solid ground of science; they are the explanations of which we are most sure. This definition contrasts sharply with the general public's usage of the word *theory,* implying *lack* of knowledge, or a guess (as in "I have a theory about the presence of UFOs"). In this book, we *always* use the word *theory* in its scientific sense, to refer to a broadly conceived, logically coherent, and well-supported explanation.

Some theories—for example, the cell theory that every organism is composed of one or more cells and that all cells come from preexisting cells—are so strongly supported that the likelihood of their being rejected in the future is small. Because these theories have withstood repeated testing and are the strongest statements we can make about the natural world, the scientific community promotes them to a higher level of confidence and calls them scientific **principles.**

Yet there is no absolute truth in science—only varying degrees of uncertainty. Science is continually evolving as new evidence comes to light, and its conclusions are always provisional. The possibility always remains that future evidence will contradict a prevailing theory, requiring its revision to better explain the scientific rules governing the natural world.

Practicing science requires honesty

Scientific investigation depends on a commitment to truthfulness. Consider the great (although temporary) damage done when an unprincipled researcher, whose career may depend on the publication of a research study, knowingly disseminates false data. Until the deception is uncovered, other scientists may devote research funds and labor to futile lines of research inspired by erroneous reports. Fortunately, science tends to correct itself through consistent use of the scientific method. Sooner or later, someone's experimental results will cast doubt on false data.

In addition to being honest in reporting their work, scientists sometimes face ethical issues surrounding their research—in areas such as stem cell research, cloning, and human and animal experimentation. For example, some stem cells that show potential for treating human diseases, such as stroke, Parkinson's disease, and Alzheimer's disease, come from human embryos at a very early stage of development. Scientists and society will need to determine if the potential benefits of these kinds of research outweigh their ethical concerns. ■

THEORY A widely accepted explanation that is supported by a large body of observations and experiments.

STUDY OUTLINE

❶ **Briefly describe the field of botany, and give short definitions of at least five subdisciplines of plant biology.**
Botany, or **plant biology,** is the scientific study of plants. Botanists who specialize in **plant molecular biology** study the structures and functions of important biological molecules such as proteins and nucleic acids. **Plant cell biology** encompasses the structures, functions, and life processes of plant cells. Plant biologists who specialize in **plant physiology** study such processes as photosynthesis and mineral nutrition to understand how plants function. Plant heredity and variation make up the specialty of **plant genetics. Plant ecology** is the study of the interrelationships among plants and between plants and their environment.

❷ **Summarize and discuss the features of plants and other organisms that distinguish them from nonliving things.**
Plants and other organisms are highly organized, with **cells** as their basic building blocks. Like other organisms, plants take in and use energy (by the processes of **photosynthesis** and **cellular respiration**), respond to **stimuli** in their environment, and undergo **growth** and **development.** Plants form new individuals by **asexual** or **sexual reproduction. Deoxyribonucleic acid (DNA)** is the molecule that transmits genetic information from one generation to the next in plants and other organisms. Plants and other organisms evolve: In the process of **evolution,** populations change, or **adapt,** to survive in changing environments. **Natural selection,** the mechanism of evolution proposed by Charles Darwin, is the tendency of organisms that have favorable adaptations to their environment to survive and become the parents of the next generation.

❸ **Distinguish among the six kingdoms and three domains, and give representative organisms for each.**
Many biologists classify organisms into six **kingdoms: Archaea, Bacteria, Protista** (protozoa, algae, water molds, and slime molds), **Fungi** (molds and yeasts), **Animalia,** and **Plantae.** Many biologists also assign all organisms to three domains. The **domain** is a level of classification above the kingdom and is based on fundamental differences among organisms. The kingdoms Archaea and Bacteria correspond to two of the domains: kingdom Archaea corresponds to **domain Archaea,** and kingdom Bacteria corresponds to **domain Bacteria.** The remaining four kingdoms are assigned to **domain Eukarya.**

❹ **Summarize the main steps in the scientific method, and explain how science differs from many other human endeavors.**
There are many versions of the **scientific method,** but it basically encompasses five steps.

1. Recognize a problem or an unanswered question.
2. Develop a **hypothesis,** an educated guess that is testable, to explain the problem.
3. Design and perform an experiment to test the hypothesis.
4. Analyze and interpret the **data** to reach a conclusion.
5. Share new knowledge with the scientific community.

Scientific data are collected by observation and experimentation and then analyzed or interpreted. Unlike what occurs in many other human endeavors, scientific conclusions are inferred from available data and are not based on faith, emotion, or intuition.

REVIEW QUESTIONS

1. How is the conspicuous consumption by people in highly developed countries related to environmental problems?
2. What is botany? List at least five subdisciplines of botany, and briefly describe each.
3. What features distinguish plants from nonliving things?
4. Outline the hierarchical levels of biological organization, from atoms to the biosphere.
5. A cactus has a thick, succulent stem and spines that are modified leaves. Explain why these adaptations may have evolved over time.
6. List the six kingdoms of life and the three domains. Briefly describe the distinguishing features of the kingdom Plantae.
7. Outline the hierarchical levels of classification, from species to domain.
8. Two plants are classified in the same genus, and two other plants are classified in the same class. Which pair of plants is more similar? Why?
9. Define the binomial system used to name species, and give an example.
10. How is science different from other human endeavors?
11. Contrast inductive and deductive reasoning. Contrast a hypothesis with a theory.
12. What is a control? Why is it an important part of an experiment?
13. Fill in the levels of biological organization in the figure on page 21. Check your answers with Figure 1-3.

THOUGHT QUESTIONS

1. Make a list of the natural resources you use in a single day. How would this list compare to a similar list made by a poor person in a less developed country?
2. Which parts of the scientific method involve inductive reasoning? Deductive reasoning?
3. The cell theory, which states that organisms are composed of cells, was arrived at by what type of reasoning? List at least five observations that the scientists who developed the cell theory might have used.
4. Explain the following statement: In science a hypothesis is hypothetical, but a theory is *not* theoretical.
5. Design a suitably controlled experiment to test the effect of low temperature on a bean seedling's rate of growth. What is the hypothesis? What is the control? The variable?
6. Why could photosynthesis be considered the single most important biological process in the living world?
7. Many meat recipes in Norway use only 2 spices and would be characterized as bland. Many meat recipes in India use at least 10 spices and are characterized as spicy. Based on the geographic locations of these two countries, suggest a reason why Indian food is so spicy.

Visit us on the web at http://www.thomsonedu.com/biology/berg/ for additional resources, such as flashcards, tutorial quizzes, further readings, and web links.

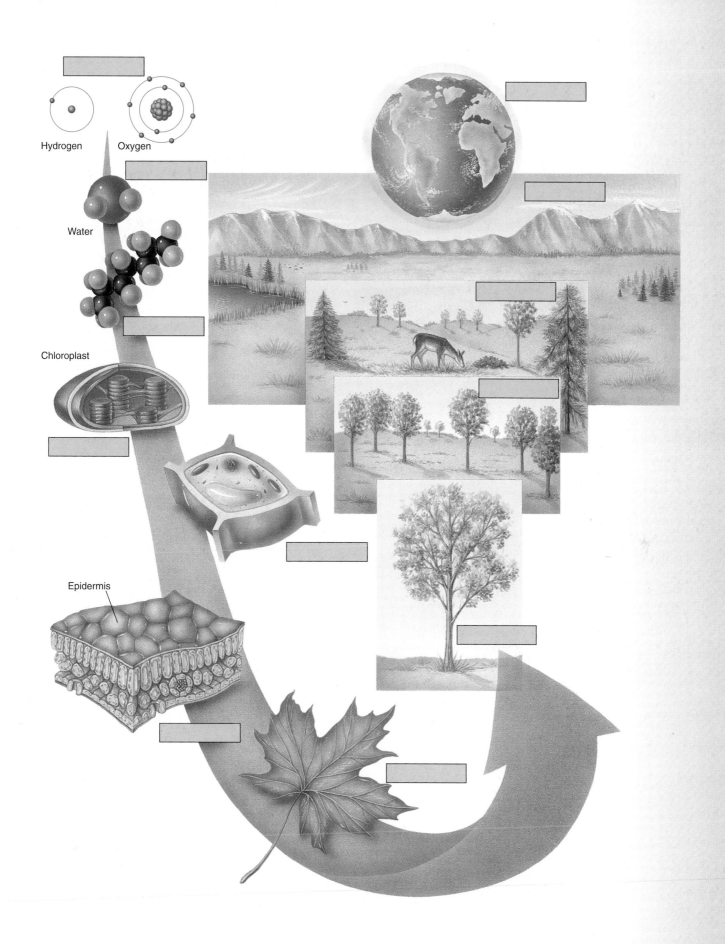

Hydrogen   Oxygen

Water

Chloroplast

Epidermis

The Chemical Composition of Cells

A mong the novelties early Spanish explorers brought to Europe from South America were bouncing balls made from the heated, solidified sap of certain trees. Although this material, later named rubber, was considered an interesting curiosity, its potential uses were not appreciated for about 300 years. Rubber was used in the early 19th century to coat and waterproof clothing, but it had certain drawbacks: it got sticky at warmer temperatures and brittle at cooler temperatures. In the mid-19th century Charles Goodyear developed *vulcanization,* a process that combines rubber with sulfur and eliminates these undesirable properties.

Rubber came into its own in the early 20th century with the invention and mass production of the automobile, which required rubber for its tires, hoses, gaskets, and so on. Demand outstripped supply, and large rubber plantations were established in Asia. When World War II disrupted the supply of rubber, chemists in the United States developed a synthetic rubber from petroleum. Today, rubber tires contain predominantly synthetic rubber, mainly because it costs less to produce synthetic rubber than to grow and harvest natural rubber.

However, synthetic rubber does not have all the desirable properties of natural rubber. For example, natural rubber is more resilient than synthetic rubber; that is, it returns to its original shape after being stretched—a characteristic required in high-performance radial tires. As a result, natural rubber is a major ingredient in radial tires.

Although many plants produce rubber, most of our natural supply comes from the rubber tree (*Hevea brasiliensis*), which is native to South America. The bark of its stem contains a network of canals that hold latex, a milky substance that is a mixture of many chemical compounds, some of which have elastic properties. Latex is tapped from the tree and used to make rubber, but its role in the rubber tree is not known. Some evidence indicates that latex discourages animals from eating the tissues—latex compounds taste bad, and many of them are poisonous. Other biologists think that latex compounds are waste products that are simply stored in the canals.

Because plants produce a substantial number of different chemical compounds, a basic knowledge of chemistry is required to understand plants and how they function. This chapter provides the foundation for understanding how the structure of atoms determines the way they form chemical bonds to produce complex compounds such as those in latex.

Latex is tapped from a rubber tree (*Hevea brasiliensis*) in Cameroon, West Africa.

An Introduction to Basic Chemistry

Earth contains an astonishing variety of materials in its water, rocks, minerals, soil, atmosphere, and organisms, but there is an underlying order to this profusion. All materials are made of matter, and all matter, living and nonliving, is composed of chemical **elements,** substances that cannot be broken down into simpler substances by chemical changes. There are 92 naturally occurring elements, ranging from hydrogen (the lightest) to uranium (the heaviest).

Instead of writing out the name of each element, chemists use a system of abbreviations called **chemical symbols.** The symbol for an element is usually the first one or two letters of the element's English or Latin name. For example, O is the symbol for oxygen, C for carbon, Cl for chlorine, and Fe for iron (the Latin name for iron is *ferrum*). •**Table 2-1** gives the chemical symbols and functions of some common elements found in plants.

Atoms are the smallest particles of elements

If you take a piece of an element—for example, a bit of iron—and divide it into smaller and smaller pieces, eventually you will get to a particle of iron that cannot be divided further. The smallest possible particle of an element that still retains the properties of that element is called an **atom** (from the Greek *atomos,* "indivisible").

During the 19th century, numerous experiments led scientists to conclude that atoms themselves are composed of units called subatomic particles (•**Figure 2-1**). A **proton** is a subatomic particle that has a positive electric charge and a small amount of mass (*mass* is a measure of the quantity of matter in an object). A **neutron** is an uncharged particle with about the same mass as a proton. An **electron** is a particle with a negative electric

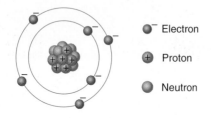

FIGURE 2-1 Model of a carbon atom.
Positively charged protons and electrically neutral neutrons are situated in the center (the nucleus) of the atom. Negatively charged electrons move about the nucleus. (This simple representation of an atom is easy to visualize and is accurate enough for our discussion of plant biology, but it does not accurately reflect the locations of electrons.)

TABLE 2-1 Some Common Elements Found in Plants		
NAME	SYMBOL	FUNCTIONS
Carbon	C	Backbone of organic molecules
Oxygen	O	Present in most organic molecules; required for aerobic respiration
Hydrogen	H	Present in most organic molecules
Nitrogen	N	Present in all proteins and nucleic acids; present in chlorophyll
Phosphorus	P	Present in nucleic acids and energy transfer molecules such as ATP
Potassium	K	Helps provide ionic balance in cells
Magnesium	Mg	Present in chlorophyll
Iron	Fe	Component of certain enzymes
Calcium	Ca	Constituent of cell walls; required for some energy transfer reactions

charge and an extremely small mass (each electron has about 1/1800 the mass of a proton). Protons and neutrons, which are located in the center of each atom (the atomic nucleus), make up almost all the mass of an atom. An atom's electrons spin about in the space surrounding the nucleus.

Each atom has an atomic number and an atomic mass

Any given atom of each kind of element has a fixed number of protons in its atomic nucleus. The number of protons in an atom is its **atomic number.** Because a single atom is electrically neutral (has the same number of electrons as protons), the atomic number also designates the number of electrons in an atom. The atomic number is usually written as a subscript to the left of the chemical symbol. Thus, hydrogen, with an atomic number of 1 (indicating one proton in its atomic nucleus), is designated $_1$H. The atomic number determines the chemical identity of an element; that is, the only element with the atomic number 1 is hydrogen, the only element with the atomic number 6 is carbon, and so on.

Because electrons have so little mass, almost the entire mass of an atom is in the protons and neutrons that make up its nucleus. The total number of protons plus neutrons in the nucleus is known as the **atomic mass,** and it is usually written as a superscript to the left of the chemical symbol. The common form of carbon, for example, is designated $_6^{12}$C, indicating that it has 6 protons and 6 neutrons (6 + 6 = 12).

Although atoms of the same element always possess the same number of protons, they can vary in the number of neutrons. Atoms of the same element that contain different numbers of neutrons and thus have different atomic masses are called **isotopes** (•**Figure 2-2**).

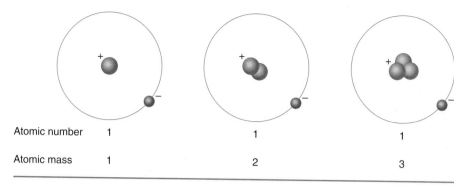

FIGURE 2-2 Isotopes of hydrogen.
Isotopes differ in the number of neutrons per atom; the numbers of protons and electrons are the same. These isotopes are all atoms of the element hydrogen, and their chemical properties are identical.

A sample of an element taken from nature is often a mixture of its isotopes, although one isotope usually predominates.

Electrons move around the nucleus, but their exact positions are uncertain

Although the precise location of an electron at any given time is uncertain, we can describe its approximate location—that is, where it will probably be found 90 percent of the time. An atom's **electron configuration** is the arrangement of electrons around its nucleus. Because electrons are negatively charged, they are attracted to the positively charged nucleus (opposite electric charges attract). At the same time, electrons repel one another. These considerations help determine an atom's electron configuration.

An atom may have several energy levels in which electrons are located. The first energy level is the one closest to the nucleus; only two electrons can occupy this energy level. The second energy level can hold a maximum of eight electrons. The third and fourth energy levels can each hold more than eight electrons, but they are most stable when only eight are present. Thus, the first energy level is considered complete when it contains two electrons, and every other energy level is complete when it contains eight electrons. Although the simple diagrams of electron configuration shown in •**Figure 2-3** help us understand atomic structure, they are highly oversimplified. Electrons move around the nucleus in an unpredictable way, first closer and then farther away.

An electron's distance from the nucleus depends on the electron's energy level. When an electron is provided with more energy—in the form of light, for example—it can move farther from the nucleus (such a move requires energy). When an electron gives up energy, it drops back to a lower energy level closer to the nucleus. We say more about electrons absorbing light energy to move

from lower to higher energy levels in the discussion of photosynthesis in Chapter 4.

Chemical bonds hold atoms together

In addition to existing as single atoms, elements unite in fixed ratios to form chemical **compounds.** Examples of compounds include water (H_2O), which is composed of the elements hydrogen and oxygen in a ratio of 2:1, and glucose ($C_6H_{12}O_6$), composed of 6 carbons, 12 hydrogens, and 6 oxygens.

A **chemical bond** is the attractive force that holds two or more atoms together in a compound. The atoms of each element form only a particular number of chem-

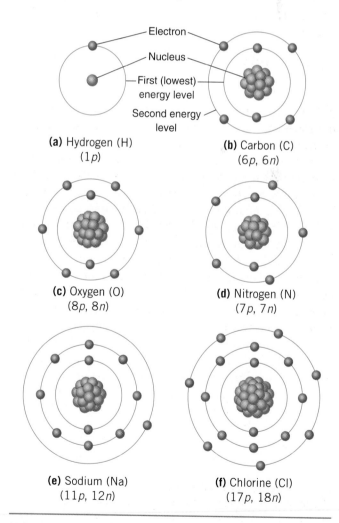

(a) Hydrogen (H)
(1*p*)

(b) Carbon (C)
(6*p*, 6*n*)

(c) Oxygen (O)
(8*p*, 8*n*)

(d) Nitrogen (N)
(7*p*, 7*n*)

(e) Sodium (Na)
(11*p*, 12*n*)

(f) Chlorine (Cl)
(17*p*, 18*n*)

FIGURE 2-3 Electron configurations of some biologically important atoms.
Each circle represents an energy level in which electrons occur; *p* = proton and *n* = neutron.

ical bonds, which is determined by the number and arrangement of the atom's electrons in the outermost energy level. When an atom's outer energy level contains fewer than eight electrons, the atom tends to gain, lose, or share electrons to achieve an outer energy level of 8. (The exceptions are hydrogen and helium, the only two elements with a single energy level, which holds a maximum of two electrons.) To complete its outer energy level, an atom forms one of two types of chemical bonds: ionic and covalent.

Atoms gain or lose electrons to form ionic bonds

In attempting to fill their outer energy levels, some atoms completely pull one or more electrons away from another atom. Because the number of protons in the nuclei of the two atoms remains the same, the gain or loss of electrons results in an **ion,** an atom with a negative or positive electric charge. An atom that has one, two, or three electrons in its outer energy level may donate those electrons to other atoms and become a positively charged ion in the process (because it has more positively charged protons than negatively charged electrons). An atom with five, six, or seven electrons in its outer energy level may gain electrons from other atoms, which makes it a negatively charged ion (because it has more electrons than protons). The force of attraction between two oppositely charged ions is called an **ionic bond.**

Sodium chloride (table salt) is an example of a compound formed by ionic bonding (•**Figure 2-4**). The sodium atom donates its outermost electron to chlorine. As a result, the outer energy levels of both sodium and chlorine are complete. The positively charged sodium ion is attracted to the negatively charged chloride ion to form an ionic bond. Ionic compounds are not composed of individual molecules. A crystal of sodium chloride, for example, is composed of millions of sodium and chloride ions in a 1:1 ratio (see bottom of Figure 2-4). Each sodium ion is surrounded by a number of chloride ions and is equally attracted to each of them. Likewise, each chloride ion is surrounded by and attracted to several sodium ions. Thus, a single sodium ion is not bonded to a single chloride ion to form a molecule.

Atoms share electrons to form covalent bonds

A **covalent bond** forms when two atoms share a pair of electrons to complete their outermost energy levels. When two or more atoms join to one another by covalent bonding, a molecule forms. A **molecule** is the smallest unit of a covalent compound.

The simplest example of a covalent bond is the one that joins two hydrogen atoms to form a molecule of hydrogen gas (•**Figure 2-5**). Each atom of hydrogen

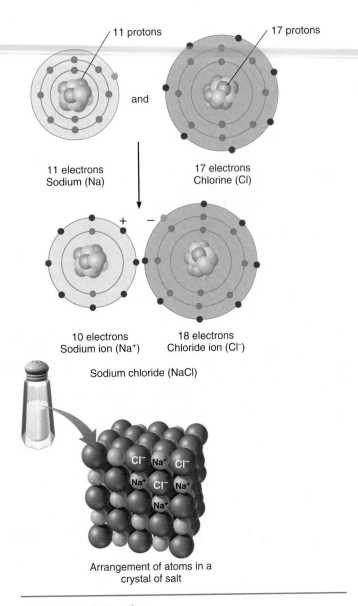

FIGURE 2-4 Ionic bonding.
Sodium becomes a positively charged ion when it donates an electron to chlorine. With this additional electron, chlorine becomes a negatively charged chloride ion. Sodium and chloride ions, when attracted to one another by their unlike electric charges, form the ionic compound sodium chloride.

contains a single electron in its first energy level, which needs two electrons to be complete. The two hydrogen atoms share their single electrons so that each of the two electrons is attracted simultaneously to the two hydrogen nuclei. Because the two electrons are under the influence of both atomic nuclei, the two hydrogen atoms are joined, or bonded.

The carbon atom has four electrons in its outer energy level, but it would be most stable if it possessed eight (recall that atoms tend to gain, lose, or share electrons to achieve an outer energy level of 8). Each of the

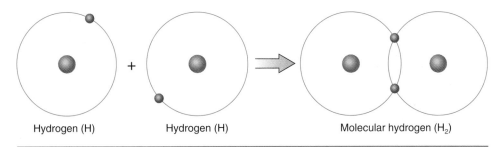

Hydrogen (H)   Hydrogen (H)   Molecular hydrogen (H₂)

FIGURE 2-5 Covalent bonding between two hydrogen atoms.
When two hydrogens share a pair of electrons between them, they form a molecule of hydrogen.

four electrons in carbon's outer energy level is available to form a covalent bond. If each of carbon's outer electrons forms a covalent bond with a hydrogen atom, a molecule of methane (CH_4) is formed (•**Figure 2-6**). Each atom in a covalently bonded molecule shares only the electrons in its outer energy level. In methane, this sharing results in a full first energy level (two electrons) for each hydrogen and a full second energy level (eight electrons) for carbon. Each line in the following structural formula designates a covalent bond formed by a single pair of shared electrons.

$$\begin{array}{c} \text{H} \\ | \\ \text{H---C---H} \\ | \\ \text{H} \end{array}$$

We refer to the sharing of a single pair of electrons as a **single bond.** Some atoms share *two* pairs of electrons, thereby forming **double bonds.** The ability of carbon to form double bonds with other carbon atoms is evident in many biologically important molecules, such as unsaturated fatty acids (discussed shortly). In structural formulas, two lines between carbon atoms (C=C) designate a double bond.

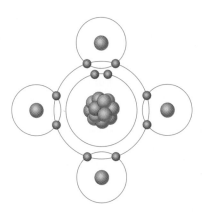

FIGURE 2-6 Covalent bonding between a carbon atom and four hydrogen atoms.
This bonding results in the formation of a molecule of methane.

Covalent bonds in which the electrons are equally shared, such as in the hydrogen molecule (H_2), are called **nonpolar covalent bonds.** However, some atoms, such as oxygen and nitrogen, have a stronger attraction for the shared electrons in a covalent bond, so the electrons are unequally shared between the bonded atoms. Such a bond is known as a **polar covalent bond.** In such a bond, electrons are much closer to one atom than to the other, and as a result of this unequal sharing, different regions of the molecule are slightly charged.

Hydrogen bonds are attractions between adjacent molecules

A **hydrogen bond** is an attraction between a positively charged hydrogen atom in one polar molecule and a negatively charged oxygen or nitrogen atom in another polar molecule. Hydrogen bonds, which are weak compared to covalent bonds, form and break readily. Although a single hydrogen bond is only about one-tenth or one-fifteenth as strong as a single covalent bond, the collective force of many hydrogen bonds is significant. In organisms, hydrogen bonding affects the shape and function of protein and nucleic acid molecules (discussed shortly) and is important in determining the properties of water (discussed in the next section).

Inorganic Compounds

Chemical compounds are divided into two broad groups, inorganic and organic. **Inorganic compounds** are composed of elements other than carbon. Among the biologically important groups of inorganic compounds are water, acids, and bases. (For a discussion of how life may have originated from inorganic compounds, see *Focus On: How Did Life Begin?*). **Organic compounds** contain carbon and usually hydrogen. In an organic

IONIC BOND An electrostatic attraction between oppositely charged ions.

COVALENT BOND A chemical bond involving one or more shared pairs of electrons.

HYDROGEN BOND An attraction between a slightly positive hydrogen atom in one molecule and a slightly negative atom (usually oxygen) in another molecule.

FOCUS ON | How Did Life Begin?

EVOLUTION LINK Scientists generally accept the hypothesis that the first living cells developed from nonliving matter. This development process, called **chemical evolution,** probably involved several stages. First, small organic molecules formed spontaneously and accumulated over time. Then organic macromolecules, such as proteins and nucleic acids, assembled from smaller molecules and gradually accumulated. As the macromolecules interacted with one another, they combined into more complicated structures that could eventually carry on chemical reactions and duplicate themselves.

These macromolecular assemblages developed into cell-like structures that ultimately became the first true cells. It is thought that life originated only once and that this process occurred under environmental conditions quite different from those we experience today.

Although we will never be certain of the exact conditions on Earth when life arose, evidence from many sources provides us with valuable clues (see figure). The atmosphere of early Earth, which contained little or no molecular (or free) oxygen, included carbon dioxide (CO_2), water vapor (H_2O), carbon monoxide (CO), hydrogen (H_2), and nitrogen (N_2). The early atmosphere may have also contained some ammonia (NH_3), hydrogen sulfide (H_2S), and methane (CH_4), although ultraviolet radiation from the sun may have broken down these molecules rapidly. As Earth's temperature cooled, water vapor may have condensed, causing torrential rains to fall that formed the ocean. The falling rain would have eroded Earth's surface, adding minerals to the ocean and making it salty.

There are four requirements for chemical evolution to have occurred: no free oxygen, a source of energy, the availability of chemical building blocks, and time. First, life could have begun only in the absence of free oxygen because oxygen is reactive and would have broken down the organic molecules that were the basis of life's origin. Evidence indicates that Earth's early atmosphere was strongly reducing; that is, any free oxygen would have reacted with other elements so that the oxygen would be tied up in chemical compounds.

A second requirement for life to have evolved from nonliving chemicals was energy. Early Earth was a place of high energy, with violent thunderstorms, volcanoes, and intense radiation from the sun. The young sun probably produced more ultraviolet radiation than it does today, and Earth had no protective ozone layer to block much of this radiation. The radiation provided the energy to form organic molecules from simple inorganic compounds.

Third, the chemicals that would be the building blocks for chemical evolution must have been present. These included water, dissolved inorganic minerals (present as ions), and the gases present in the early atmosphere.

The fourth requirement was time—time for molecules to accumulate and react. The age of Earth provides adequate time for chemical evolution. Earth is approximately 4.6 billion years old, and there is geologic evidence that simple organisms appeared about 3.5 billion years ago. Once the first cells originated, they evolved over millions of years into the rich biological diversity on Earth today. ∎

---

compound, many carbon atoms are usually bonded to one another to form a kind of molecular backbone that may consist of a straight chain, a branched chain, or rings. Examples of biologically important organic compounds are carbohydrates, lipids, and proteins.

Water, an inorganic compound, is essential to plants

The view of planet Earth from outer space reveals that it is different from other planets in the solar system. Earth is a predominantly blue planet because of the water that covers three-fourths of its surface. Water, which exists as a solid (ice), liquid, or vapor (steam), has a great effect on our planet: it helps shape the continents, moderates our climate, and allows organisms to exist and survive. All life-forms, from simple bacteria to multicellular plants and animals, contain water.

Water is vital to plants and other organisms because it carries dissolved nutrients and other important materials to cells. Solutions of water and dissolved materials surround every living cell and are present inside every cell. Water is also essential to life because almost all chemical reactions that sustain life occur in aqueous solutions; many of these chemical reactions involve water directly. Water also affects the organization of biological molecules that compose cells. For example, the interaction of lipids (fats and oils) with water shapes cell membranes. Because water influences plants and other organisms so profoundly, we now examine some of its properties.

Water's polarity causes many of its properties

Water is a polar molecule; that is, the electric charge is unevenly distributed. The negative area (the oxygen part) of one water molecule is attracted to the positive area (the hydrogen part) of another water molecule, so

PROCESS OF SCIENCE

QUESTION: Could organic molecules have formed in the conditions of early Earth?

HYPOTHESIS: Organic molecules can form in a reducing atmosphere similar to that found on early Earth.

EXPERIMENT: The apparatus that Miller and Urey used to simulate the reducing atmosphere of early Earth contained nitrogen, hydrogen, methane, ammonia, and water. An electric spark was produced in the upper right flask to simulate lightning.

RESULTS AND CONCLUSION: The gases present in the flask reacted together, forming a variety of simple organic compounds, such as amino acids, that accumulated in the trap at the bottom. Thus, the formation of organic molecules—the first step in the origin of life—could have occurred in the conditions present on early Earth. ■

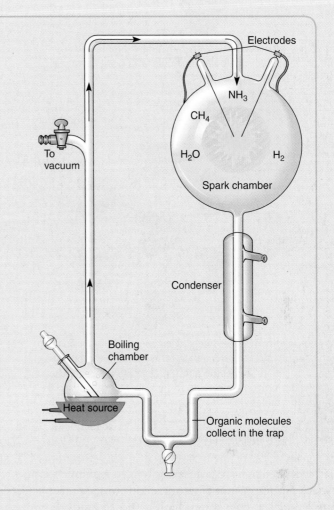

■ **Miller and Urey's experiment in chemical evolution.**

a hydrogen bond forms between the two molecules (•**Figure 2-7**).

The hydrogen bonds in water are the basis for many of its physical properties, including its high melting/freezing point (0°C, 32°F) and high boiling point (100°C, 212°F). Most of Earth's surface has a temperature between 0°C and 100°C, and as a result, most water exists as the liquid on which organisms depend.

Because of hydrogen bonding, water must absorb a lot of heat (to break hydrogen bonds) before it vaporizes, or changes from liquid to vapor. When it does evaporate, water vapor carries the heat with it. Thus, evaporating water has a cooling effect. For this reason, evaporation of perspiration from your skin cools your body. Leaves are cooled in a similar way as water evaporates from their surfaces.

Water is sometimes called the universal solvent, and although this is an exaggeration, many materials do dissolve in water as a result of its polarity. In nature, water is never completely pure, because it contains dissolved gases from the atmosphere and dissolved mineral salts from Earth. Seawater, for example, is a solution of water and a variety of dissolved salts, including sodium chloride, magnesium chloride, magnesium sulfate, calcium sulfate, and potassium chloride. Water's dissolving ability has one major drawback, however: many pollutants (such as fertilizers, road salt, and acids) also dissolve in water.

Two properties of water, cohesion and adhesion, are particularly important in the transport of water and dissolved minerals (inorganic nutrients) in plants. The tendency of *like* molecules to adhere, or stick together, is known as **cohesion**. Water is cohesive, and this strong attraction of water molecules to one another results from the hydrogen bonds among them. **Adhesion** is the tendency of *unlike* molecules to adhere to one another. Water is strongly adhesive with many other materials, particularly those that are polar. The ability of water to make things wet is the result of adhesion.

(a) Water, which consists of two hydrogen atoms and one oxygen atom, is a polar molecule with positively and negatively charged areas.

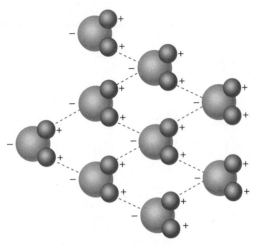

(b) The polarity of water molecules causes hydrogen bonds (*represented by dashed lines*) to form between the positive area of one water molecule and the negative areas of others. Each water molecule forms up to four hydrogen bonds with other water molecules.

FIGURE 2-7 Chemical structure of water.

Hydrogen bonding causes ice to have a unique property that has important environmental consequences: ice floats on denser, cold water. Unlike most liquids, liquid water expands as it freezes because the hydrogen bonds joining the water molecules in the crystalline lattice of ice keep the water molecules far enough apart to give ice a lower density than the density of liquid water. When ice is warmed enough to melt, the hydrogen bonds are broken, and the water molecules slip closer together. Therefore, the density of water is greatest at 4°C (39°F). At temperatures warmer than 4°C, water begins to expand as the speed of its molecules increases.

When a deep body of water cools, it becomes covered with floating ice. The ice insulates the liquid water below it, thus retarding freezing and permitting organisms to survive below the icy surface. If ice had a greater density than water, ice would sink. Eventually all ponds, lakes, and even the ocean would freeze solid from the bottom to the surface, making life impossible.

Acids and Bases

The chemical makeup of a plant's internal environment and the external environment in which it lives affect its overall health and well-being. One of the most impor-

tant aspects of either environment is how acidic or basic (alkaline) it is. The degree of acidity or alkalinity inside a plant cell must remain fairly stable, for example, or important chemical reactions will not take place. Likewise, certain plants decline and may even die if the acidity or alkalinity of the soil is too high.

Acids and bases dissociate when dissolved in water

An **acid** is a compound that dissociates, or breaks up, in a solution of water to form hydrogen ions (H^+, or protons) and negatively charged ions. Hydrochloric acid (HCl) is an example of an acid.

$$HCl \xrightarrow{\text{(in water)}} H^+ + Cl^-$$

Hydrochloric acid   Hydrogen ion   Chloride ion

Some acids are known as strong acids because they dissociate almost completely in water. Hydrochloric acid is a strong acid because most of its molecules dissociate, producing hydrogen ions and chloride ions. Other acids, called weak acids, dissociate only slightly. Vinegar, which is a dilute solution of acetic acid, is an example of a weak acid.

A compound that dissociates in water to produce negatively charged hydroxide ions (OH^-) and positively charged ions is known as a **base**. Sodium hydroxide (NaOH) is an example of a base.

$$NaOH \xrightarrow{\text{(in water)}} Na^+ + OH^-$$

Sodium hydroxide   Sodium ion   Hydroxide ion

Acids and bases react with one another, in the process neutralizing the chemical effect that each group had originally. Each hydrogen ion of an acid combines with a hydroxide ion of a base to form a water molecule:

$$H^+ + OH^- \longrightarrow H_2O$$

A solution's acidity or alkalinity is expressed in terms of pH

The **pH scale** measures the relative concentrations of hydrogen ions and hydroxide ions in a solution. The pH scale extends from 0, the pH of a strong acid such as HCl, to 14, the pH of a strong base such as NaOH. Pure water, which is neutral (neither acidic nor basic), has a pH of 7. A solution that contains a greater concentration of H^+ ions than OH^- ions has a pH less than 7 and is acidic. A solution that contains a greater concentration of OH^- ions than H^+ ions has a pH greater than 7 and is basic.

For purposes of comparison, pure water has a pH of 7, tomato juice has a pH of 4, vinegar has a pH of 3,

PLANTS ᴬᴺᴰᴛʜᴱ ENVIRONMENT | Acid Rain

During the past few decades, acid rain has received considerable media coverage. Although the extent and acidity of acid rain vary widely from one region to another, precipitation in certain parts of eastern North America, Europe, Japan, China, and Korea is sometimes as acidic as vinegar! Acid rain corrodes materials such as metal and stone and poses a serious threat to the environment. It kills aquatic organisms such as fish and is linked to widespread damage to forests, perhaps by changing soil chemistry (see figure).

Acid rain is the result of accumulation of sulfur oxides and nitrogen oxides in the atmosphere. This accumulation can arise from natural events such as volcanic activity, which adds large amounts of sulfuric acid to the atmosphere. However, in recent years, industrial activity and automobile exhaust have become important contributors to acid rain. Motor vehicles are a major source of nitrogen oxides. Electrical power plants and heavy industrial equipment, both of which burn coal, are the main sources of sulfur dioxide emissions and produce substantial amounts of nitrogen oxides as well.

During their stay in the atmosphere, sulfur oxides and nitrogen oxides combine with water to produce dilute solutions of sulfuric acid and nitric acid. Precipitation such as rain or snow returns these acids to the ground, bathing plants in acid and causing the soil to become acidic.

Scientists are studying how acid rain changes the chemistry of soils

■ Damage from acid rain and acid fog. These trees, on Mount Mitchell, North Carolina, are damaged from the increase in soil acidity that results when pollutants from fossil-fuel power plants and motor vehicle exhaust are converted into acids in the atmosphere and fall as precipitation.

and interferes with the root uptake of essential minerals. Some essential minerals, such as calcium and magnesium, wash out of acidic soil. Other minerals, including heavy metals such as aluminum, accumulate in acidic soil and may then be absorbed by plants in toxic amounts.

A practice that has increased and complicated the problem of acid rain is the construction of tall smokestacks to release the sulfur-rich smoke high up in the atmosphere. Unfortunately, the wind blows the pollutants long distances. For example, tall smokestacks allow the midwestern United States to "export" its acid emissions hundreds of

kilometers away, to New England and Canada.

The pollutants in smokestack emissions can be captured and removed, but this requires the installation of pollution-control devices and thus makes prevention of acid rain expensive. The Clean Air Act amendments of 1990 require coal-burning power plants in the United States to reduce their sulfur and nitrogen oxide emissions. However, most scientists maintain that air-pollution controls that are stricter than those mandated by the Clean Air Act amendments of 1990 will have to be initiated if forests and lakes are to recover in the near future.

and lemon juice has a pH of 2. Normally, rainfall, with a pH of 5 to 6, is slightly acidic because carbon dioxide and other materials in the air dissolve in rainwater to form dilute acids. However, the pH of precipitation in the northeastern United States and Canada averages 4 and is often 3 or even lower. Such acidic precipitation is commonly called **acid rain** (see *Plants and the Environment: Acid Rain*).

Organic Compounds

Organic compounds are the foundations on which the structures of plants and other organisms are built. Within cells, organic compounds participate in thousands of chemical reactions, including those that provide energy to sustain life. Four groups of organic compounds are essential for all organisms: carbohydrates, lipids, proteins, and nucleic acids. In addition, plants

PLANTS AND PEOPLE | Phytochemicals and Human Health

Have you eaten your blueberries, spinach, asparagus, and sweet potatoes today? Many studies have shown that diets rich in fruits and vegetables lower the incidence of heart disease and of certain types of cancer. For example, a 2004 study published in the *Journal of the National Cancer Institute* analyzed the diets of more than 100,000 health-care professionals and concluded that those who ate diets rich in fruits and vegetables had significantly less chronic illness, particularly heart disease. In Asian countries where diets are low in fat and high in fruits and vegetables, including soy and green tea, the incidence of breast, prostate, and colorectal cancer is low. However, nutritionists estimate that in the United States only about 1 in 11 people eats the daily recommended three to five servings of vegetables and two to three servings of fruits.

A diet that relies on dietary supplements to provide the body's essential nutrients does not provide the same health benefits as one that is rich in fruits and vegetables. The missing ingredients appear to be **phytochemicals** (from the Greek *phyto,* "plant"), plant compounds that play important roles in preventing certain diseases. Many phytochemicals are **antioxidants,** chemicals that destroy naturally forming *free radicals,* reactive molecules that damage the DNA, proteins, and unsaturated fatty acids in cells. Nutritionists are

C Squared Studios/Getty Images

■ **Blueberries are rich in phytochemicals called flavonoids.**

currently investigating the benefits of phytochemicals, and new research is reported almost every week.

Fruits and vegetables with bright colors generally contain the highest levels of phytochemicals. Among the important classes of phytochemicals are the carotenoids and flavonoids. Fruits and vegetables rich in **carotenoids** are generally yellow, orange, red, or green. Evidence suggests that diets rich in carotenoids decrease the risk of heart disease, cancer, and stroke; carotenoids may also slow the onset of diabetes and help reduce some of the damaging effects of this disease.

Fruits and vegetables rich in **flavonoids** are often colored red, blue, purple, or black. Like carotenoids, flavonoids are antioxidants that appear to reduce the risk of heart disease,

cancer, and other health problems associated with aging. In a study with lab animals, for example, the flavonoids in blueberries were linked to learning and memory improvements (see figure).

Let us go back to my opening question: Have you eaten your blueberries, spinach, asparagus, and sweet potatoes today? If you have eaten blueberries, you may have gotten protection against certain kinds of cancer and against high cholesterol in the blood. Spinach in your diet may have slowed the aging process and reduced the risk of certain types of cancer; a diet rich in spinach may also reduce the risk of cataracts and macular degeneration, a leading cause of visual impairment in people older than 50 years. If you ate asparagus, you provided your cells with high levels of antioxidants; asparagus is also rich in folate, an essential nutrient that is critical for pregnant women because it protects against neural tube defects in developing fetuses. Sweet potatoes are rich in beta-carotene, a carotenoid associated with slowing the aging process, reducing the risk of cancer, and reducing the damage of diabetes.

Blueberries, spinach, asparagus, and sweet potatoes are just a few examples of fruits and vegetables that contain beneficial phytochemicals. So make sure you eat a variety of colorful produce every day!

produce a variety of organic compounds that promote human health (see *Plants and People: Phytochemicals and Human Health*).

Sugars, starches, and cellulose are examples of carbohydrates

Carbohydrates are organic compounds, such as sugars, starches, and cellulose, that plants use as fuel molecules, as constituents of other important compounds such as nucleic acids, and as structural components of cells. Carbohydrates are composed of carbon (C), hydrogen (H), and oxygen (O) atoms in an approximate ratio of

1C:2H:1O. The general equation for carbohydrates is $(CH_2O)_n$, where n refers to any number from 3 to several thousand. Three kinds of carbohydrates occur: monosaccharides, disaccharides, and polysaccharides.

Monosaccharides, or simple sugars, usually contain three to six carbon atoms (●**Figure 2-8**). **Glucose** ($C_6H_{12}O_6$), a six-carbon sugar, is commonly called *blood sugar* because it is the form of sugar that is transported in the bloodstreams of humans and many other animals. Plants produce glucose by photosynthesis, but most of the glucose is immediately converted to other compounds. Another six-carbon sugar, **fructose** ($C_6H_{12}O_6$), is commonly called *fruit sugar* because it is often found

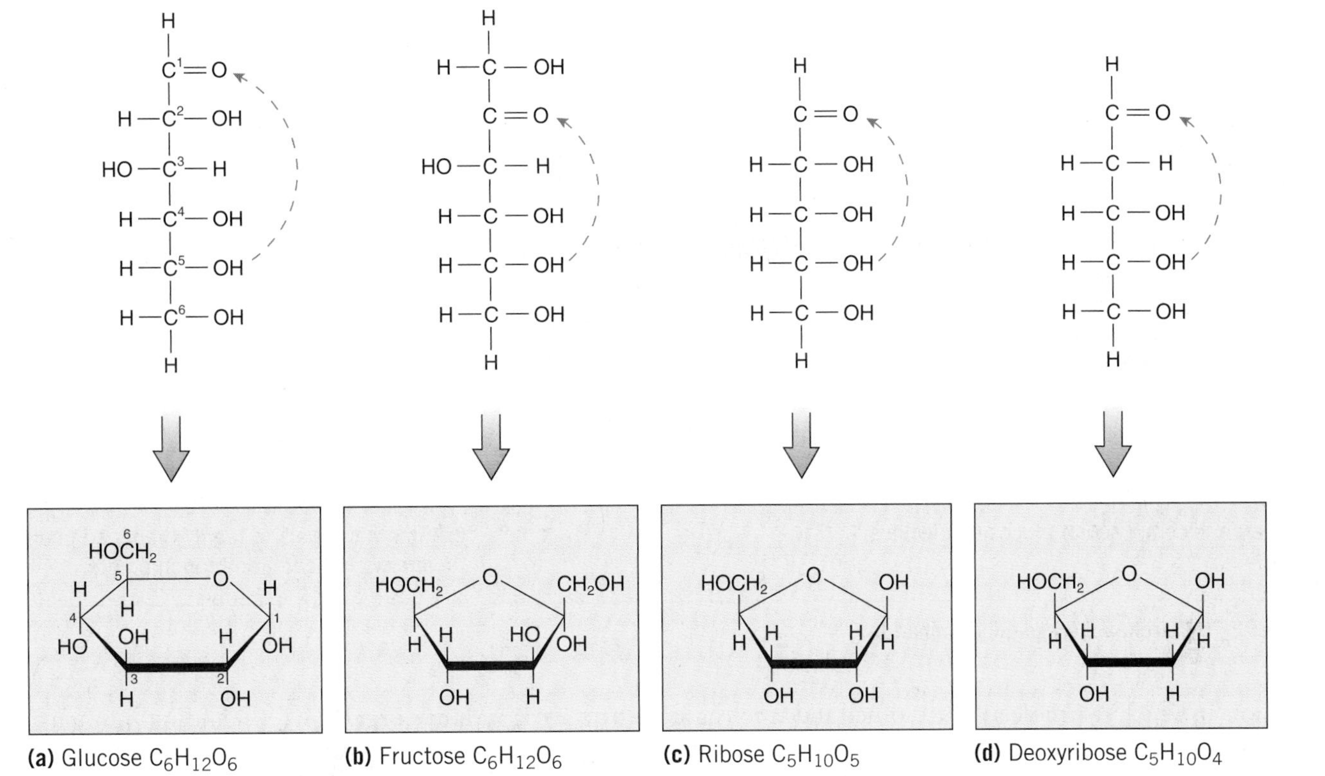

FIGURE 2-8 Structures of some common monosaccharides.
The linear form (*top row*) of each is shown because it is easier to visualize bonds and count the number of carbon, hydrogen, and oxygen atoms. In solution, these molecules have a ring form (*bottom row*). The six-carbon monosaccharides, glucose (a) and fructose (b), are important fuel molecules, whereas the five-carbon monosaccharides, ribose (c) and deoxyribose (d), are components of nucleic acids.

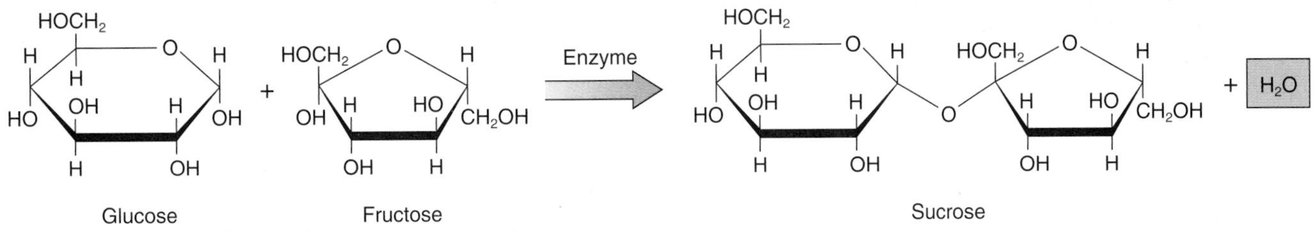

FIGURE 2-9 Synthesis of sucrose.
The monosaccharides glucose and fructose combine covalently by a condensation reaction to form the disaccharide sucrose and a molecule of water.

in fruits, along with other sugars. Glucose, fructose, and other monosaccharides are fuel molecules that cells break down to obtain energy for cellular activities. Although glucose and fructose have the same chemical formula, $C_6H_{12}O_6$, their atoms are arranged differently, and as a result, they have different properties. Fructose is sweeter than glucose, for example.

A **disaccharide** consists of two bonded monosaccharide units. The disaccharide sucrose ($C_{12}H_{22}O_{11}$), known as *common table sugar*, is the carbohydrate stored in sugarcane (*Saccharum officinarum*) and sugar beets (*Beta vulgaris*). Sucrose is also the form of sugar transported in a plant's vascular system (conducting tissue). Sucrose consists of a molecule of glucose combined with a molecule of fructose (•**Figure 2-9**).

$$C_6H_{12}O_6 + C_6H_{12}O_6 \longrightarrow C_{12}H_{22}O_{11} + H_2O$$

Glucose   Fructose   Sucrose   Water

Note that the formation of sucrose from glucose and fructose involves the removal of a molecule of water. Such a

CARBOHYDRATE An organic compound containing carbon, hydrogen, and oxygen in the approximate ratio of 1C:2H:1O.

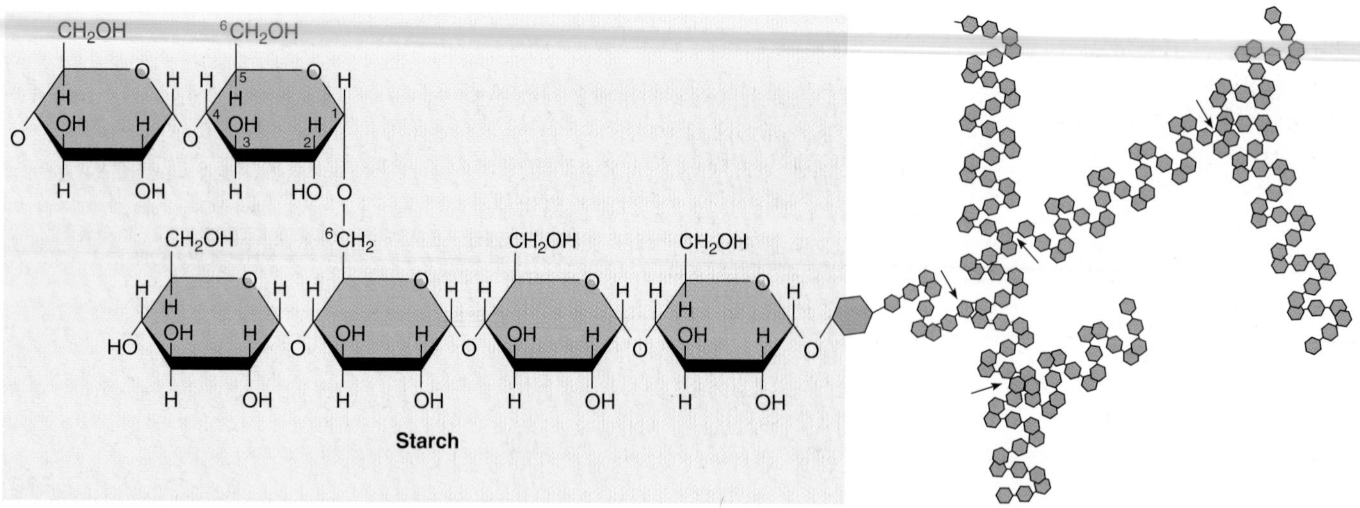

(a) Starch is composed of glucose molecules.

(b) Starch consists of highly branched chains; the arrows indicate the branch points.

FIGURE 2-10 Starch, a storage polysaccharide.

reaction, in which two molecules are joined and a molecule of water is removed, is called a **condensation reaction.**

Starches and cellulose are the most important **polysaccharides,** carbohydrates composed of many sugar units. Each polysaccharide molecule consists of a long chain, either branched or unbranched, of thousands of monosaccharide units. **Starches,** enormous polysaccharide molecules composed of thousands of glucose units, are the main storage carbohydrates in plants (•Figure 2-10). Plants build up their energy reserves by storing starch. Many of the foods humans consume, such as potatoes (*Solanum tuberosum*), corn (*Zea mays*), and rice (*Oryza sativa*), contain abundant starch.

Cellulose, a major component of plant cell walls, is an important structural polysaccharide. Cellulose, the most abundant carbohydrate on Earth, accounts for about 50 percent by mass of the organic compounds in plants. Although cellulose, like starch, is composed of glucose units, the bonds that join the glucose units in cellulose form differently from those in starch (•**Fig-**

ure 2-11). Many organisms (including humans) can digest starch, but few can digest cellulose. Plant fiber, which consists mainly of cellulose, is an important part of the human diet. Although it cannot be digested to provide nutrients, cellulose adds bulk to the materials moving through the intestines and aids in bowel function.

Fats and oils are examples of lipids

Lipids are organic compounds that have a greasy or oily consistency and do not readily dissolve in water. Lipids are composed primarily of carbon and hydrogen, although they also contain some oxygen. They function in all cells as fuel molecules and as essential components of cell membranes. In plants, lipids also function as a waterproof covering (that is, the *cuticle*) over the plant body and as light-gathering molecules (that is, *chlorophylls* and *carotenoids*) for photosynthesis. Lipids include neutral fats and oils, phospholipids, steroids, cer-

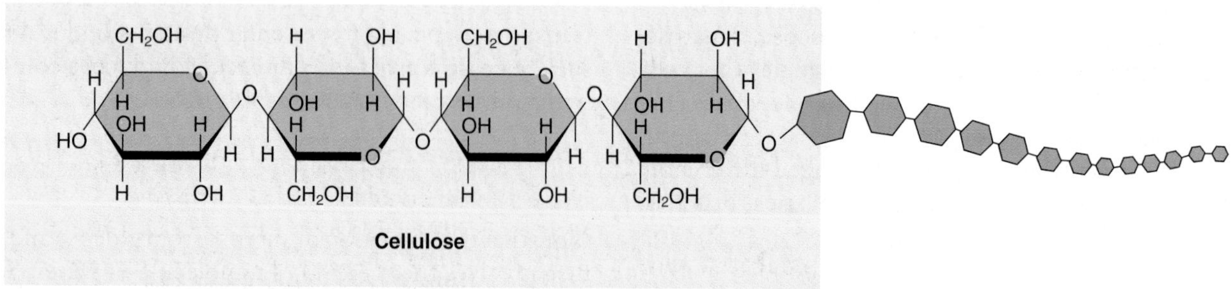

FIGURE 2-11 Cellulose, a structural polysaccharide.
The cellulose molecule is an unbranched polysaccharide consisting of about 10,000 glucose units.

tain pigments, and waxes. We confine our discussion here to neutral fats and oils, phospholipids, and waxes; other sections of the text discuss other types of lipids.

Neutral fats and **oils** are easily distinguished: a fat is solid at room temperature, and an oil is liquid at room temperature. These lipids, which are used as fuel molecules, provide a lot of energy when broken down: a gram of fat contains more than twice as much energy as a gram of carbohydrate. Neutral fats and oils each consist of a molecule of glycerol joined to one, two, or three fatty acids (•**Figure 2-12**). **Glycerol** is a three-carbon compound that contains three hydroxyl (—OH) groups. A **fatty acid** is a long, un-branched hydrocarbon (composed of hydrogen and carbon) chain with a carboxyl (—COOH) group at one end. Fatty acids are typically composed of even numbers of carbon atoms and range in length from 4 to 20 carbons or even longer. The most common fatty acid, oleic acid, contains 18 carbon atoms.

A fatty acid is either saturated or unsaturated. **Saturated fatty acids** contain the maximum number of hydrogen atoms possible because they contain no carbon–carbon double bonds. For example, the 12-carbon fatty acid shown here is saturated.

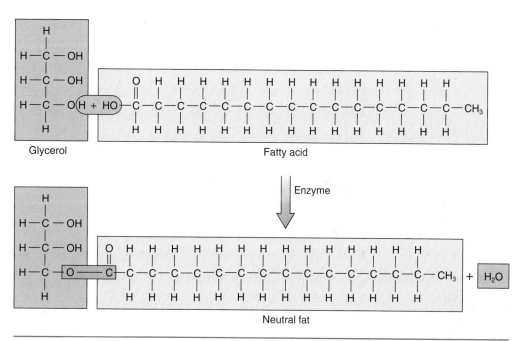

FIGURE 2-12 Formation of a neutral fat or oil.
Glycerol combines with one (*shown*), two, or three fatty acids in a condensation reaction. The green boxed area of the neutral fat indicates where the fatty acid and glycerol combined.

If a fatty acid contains one or more carbon–carbon double bonds, the molecule is **unsaturated** (that is, not fully saturated with hydrogens). The following 12-carbon fatty acid is unsaturated.

Fats, which contain high percentages of saturated fatty acids, are generally produced by animals; butter and lard are examples. Oils contain high percentages of unsaturated fatty acids and are produced by plants; examples include peanut oil, soybean oil, safflower oil, olive oil, and corn oil. (A notable exception is coconut oil, which contains few unsaturated fatty acids.)

Diets high in saturated fatty acids and cholesterol (another lipid) tend to raise the blood cholesterol level. These lipids are associated with heart disease, particularly *atherosclerosis*, a progressive disease in which the arteries become blocked with fatty material. On the other hand, ingestion of unsaturated fats, particularly polyunsaturated fats,[1] tends to decrease the blood cholesterol level. For this reason, many people now cook with vegetable oils rather than with butter and lard, drink skim milk rather than whole milk, and eat frozen yogurt or ice milk instead of ice cream.

Phospholipids are a group of lipids important as components of cell membranes. Each phospholipid molecule consists of a glycerol molecule attached at one end to two fatty acids and at the other end to a phosphate group linked to an organic compound. The fatty acid portion of the molecule is not soluble in water, but the portion composed of glycerol, phosphate, and the organic molecule is soluble in water. As a result of these differences in solubility, phospholipids form double layers, or bilayers, in water environments such as cells (see Figure 3-13). Thus, phospholipids are uniquely suited as the basic component of cell membranes.

---

[1]"Polyunsaturated" refers to the presence of three or more carbon–carbon double bonds in a fatty acid.

LIPID Any of a group of organic compounds that are insoluble in water but soluble in fat solvents.

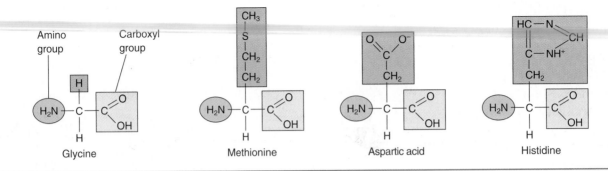

FIGURE 2-13 Four representative amino acids.
These amino acids differ in their R groups (*green boxed areas*). Note that each amino acid
contains an amino group (—NH_2) and a carboxyl group (—COOH).

Other important lipids are the *waxes*. **Cutin,** a waxy substance found in the outer walls of epidermal cells, forms the *cuticle* that covers the aerial portions of plant leaves and herbaceous (nonwoody) stems. **Suberin** is a waxy substance found in the walls of cork cells (the outer bark of woody plants). Both cutin and suberin protect the plant's aerial surfaces from excess water loss (see Chapter 5).

Proteins are large organic molecules composed of amino acids

Proteins are macromolecules composed of carbon, oxygen, hydrogen, nitrogen, and usually sulfur that serve as structural components of cells and tissues. Proteins also regulate biochemical processes in plants and other organisms. Virtually every chemical reaction that occurs in living cells is controlled by its own specific *enzyme*, a protein that affects the rate at which a chemical reaction occurs (discussed shortly). The importance of enzymes cannot be overstated: *Enzymes are crucial for an organism's survival.*

EVOLUTION LINK The types of proteins found in a cell determine to a large extent what type of cell it is—how it looks and how it functions. The slight variation in proteins from one species to another partly explains differences among species. A rose is a rose because it produces proteins characteristic of roses. Protein composi-

tion also reflects evolutionary relationships: the proteins of closely related species have more similarities than those of species that are only distantly related. Information on the kinds of proteins that organisms possess is encoded in their hereditary material (discussed shortly) and passed from one generation to the next. ■

Proteins are composed of hundreds of units called **amino acids.** Each amino acid contains a carbon atom bonded to an amino group (—NH_2), a carboxyl group (—COOH), and a side chain, designated R. The general formula for amino acids is

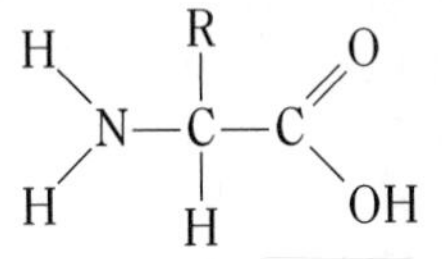

About 20 different amino acids are found in proteins; they differ in their R groups (•**Figure 2-13**). Most plants synthesize the various amino acids that they need from simpler substances. Animals, including humans, can manufacture some amino acids, but they must obtain other amino acids, referred to as **essential amino acids,** in their diets.

Amino acids are not randomly linked to form proteins. Just as the order of letters determines the meanings of words, the order of amino acids in a protein determines its structure and function. The amino acid chains of many proteins fold up on themselves to

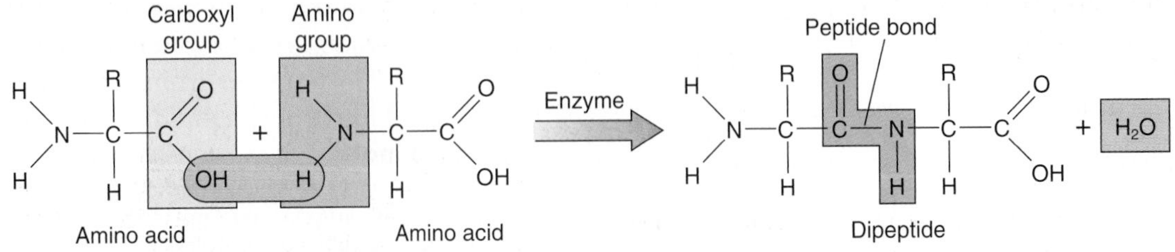

FIGURE 2-14 Formation of a peptide bond.
Two amino acids combine by forming a peptide bond between the carboxyl group of one amino acid and the amino group of another. The reaction is a condensation reaction.

Proteins are composed of simple building blocks called amino acids, but they are not simply straight chains of amino acids. Proteins coil and fold into specific conformations, or three-dimensional shapes.

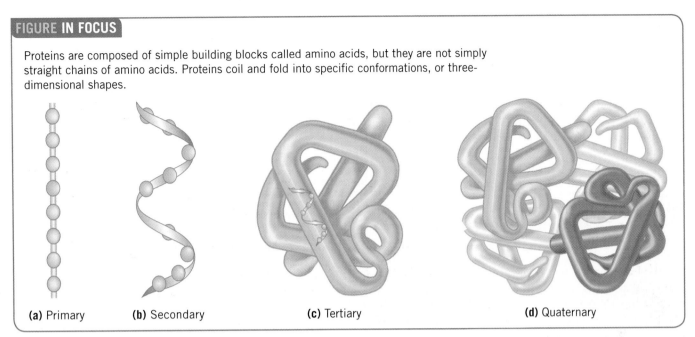

(a) Primary   **(b)** Secondary   **(c)** Tertiary   **(d)** Quaternary

FIGURE 2-15 **The four levels of organization of protein molecules.**
Proteins have a primary **(a)**, secondary **(b)**, tertiary **(c)**, and sometimes quaternary **(d)** level of organization. In this example the quaternary structure consists of four separate polypeptide chains.

form the three-dimensional shapes that determine their functions.

The bond linking one amino acid to another forms between the carboxyl carbon of one amino acid and the amino nitrogen of another (•**Figure 2-14**). This covalent bond, called a **peptide bond,** forms as a result of a condensation reaction. Each additional amino acid added to the growing chain likewise forms a peptide bond between itself and the *polypeptide chain.* Some proteins are composed of a single polypeptide chain, whereas other proteins contain two or more polypeptide chains.

Protein structure has three—and sometimes four—levels of organization (•**Figure 2-15**). *Primary structure* is the linear sequence of amino acids. *Secondary structure* is a regular shape, such as a spiral helix, that is caused by rotation of the polypeptide chain; hydrogen bonding among different parts of the polypeptide chain causes secondary structure. *Tertiary structure* is the overall shape of the polypeptide chain, as determined by interactions of the side chains of amino acids. Some proteins also have a *quaternary structure,* in which two or more polypeptide chains associate to form the final protein molecule.

Enzymes are proteins that accelerate chemical reactions

Enzymes are protein molecules that function as catalysts; that is, they increase the rate at which chemical reactions occur but are not used up in the reactions. Most enzymes are highly specific and catalyze only a single chemical reaction or a few closely related chemical reac-

tions. The enzyme sucrase, for example, catalyzes only the breakdown of sucrose into glucose and fructose. The material on which the enzyme works, in this case sucrose, is known as the **substrate.**

Enzymes greatly increase the rates of chemical reactions in cells. In the absence of enzymes, most chemical reactions in cells would occur at a rate too slow to support life. The disaccharide sucrose, for example, is a stable molecule that in the absence of sucrase would take *years* to break down into glucose and fructose. However, if a tiny amount of sucrase is present, the reaction occurs very quickly. Each molecule of sucrase catalyzes the breakdown of thousands of sucrose molecules *per second!*

How do sucrase and other enzymes work? Before new chemical bonds form during a chemical reaction, old ones must be broken. The process of breaking old bonds requires an input of **activation energy.** An enzyme works by lowering the activation energy needed to initiate a chemical reaction (•**Figure 2-16**).

Each enzyme has one or more **active sites,** areas that fit particular substrates and enable the enzyme to form

PROTEIN A large, complex organic compound composed of amino acid subunits.

ENZYME An organic catalyst, produced within an organism, that accelerates specific chemical reactions.

ACTIVATION ENERGY The energy required to initiate a chemical reaction.

An enzyme speeds up a reaction by lowering the activation energy.

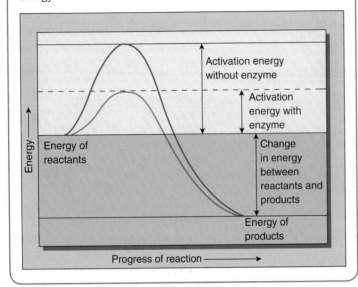

FIGURE 2-16 Enzymes and activation energy.
A reaction catalyzed by an enzyme (*red curve*) proceeds faster than an uncatalyzed reaction (*blue curve*) because a catalyzed reaction requires less activation energy to initiate.

a temporary *complex* with them. During the course of a chemical reaction, the substrate molecules occupying these sites are brought close together and react with one another. The products then separate from the enzyme, and the enzyme is free to bind to other substrate molecules (•**Figure 2-17**).

Enzyme + substrate(s) ⟶ enzyme–substrate complex
⟶ enzyme + product(s)

DNA and RNA are nucleic acids

Nucleic acids are macromolecules composed of carbon, oxygen, hydrogen, nitrogen, and phosphorus. There are only two types of nucleic acids: **deoxyribonucleic acid (DNA)** and **ribonucleic acid (RNA)**. Together, these two nucleic acids control all the life processes of an organism and are involved in the transmission of hereditary information from one generation to the next.

DNA contains the instructions for making all the proteins that an organism needs. These instructions are encoded in **genes,** units of hereditary information that consist of DNA and are part of the chromosomes. RNA functions in the process of protein synthesis.

Nucleic acids are composed of repeating units called **nucleotides.** Each nucleotide molecule is composed of three parts: (1) a nitrogenous base, (2) a five-carbon sugar, and (3) a phosphoric acid (phosphate) molecule (•**Figure 2-18**). In a nucleic acid molecule, the phosphate portion of one nucleotide is attached to the sugar of the next nucleotide. The sequence of nucleotides in the nucleotide chain determines the specific information encoded in a nucleic acid molecule. RNA and DNA are considered in detail in Chapter 14.

Energy is temporarily stored in ATP

In addition to being the building blocks of nucleic acids, some nucleotides serve other important functions in cells. **Adenosine triphosphate (ATP)**, for example, is a modified nucleotide compound composed of the base adenine, the sugar ribose, and three phosphate molecules. ATP is present in *all* living cells as their "energy

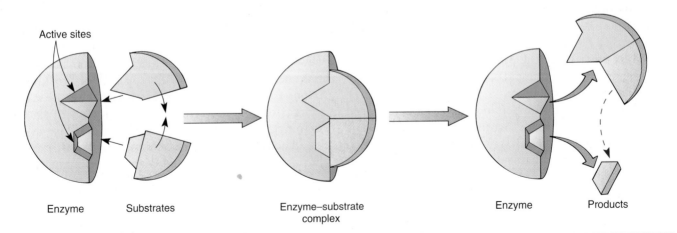

Active sites

Enzyme   Substrates   Enzyme–substrate complex   Enzyme   Products

FIGURE 2-17 Enzyme–substrate complex.
A substrate forms a temporary complex with the specific enzyme that catalyzes its reaction. When the products separate from the enzyme, the enzyme can bind to additional substrate molecules.

currency." Unstable bonds join the two terminal phosphate groups in the ATP molecule to each other. The biologically usable energy in these bonds is transferred to other molecules when the bonds are broken. Much of the chemical energy of a cell is temporarily stored in the bonds of ATP, ready to be released to do biological work, such as catalyzing chemical reactions, when a phosphate group is transferred to another molecule.

NUCLEIC ACID A large, complex organic molecule composed of nucleotides; the two nucleic acids are deoxyribonucleic (DNA) and ribonucleic acid (RNA).

ADENOSINE TRIPHOSPHATE (ATP) An organic compound of prime importance for energy transfers in biological systems.

Energy and Biological Work

Energy is the capacity or ability to do work. In organisms, the biological work that requires energy includes processes such as growing, reproducing, and repairing damaged tissues.

Energy exists in several forms: heat, radiant energy from the sun, chemical energy in the chemical bonds of molecules, mechanical energy, and electrical energy. Energy exists as stored energy—called **potential energy**—or as **kinetic energy**, the energy of motion. You can think of potential energy as an arrow on a drawn bow (●**Figure 2-19**). When the string is released and the arrow shoots through the air, the potential energy is

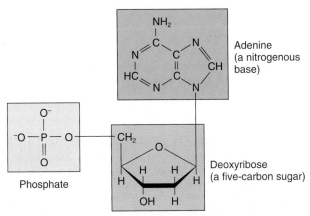

(a) Nucleic acids are composed of nucleotides. Like all nucleotides, this DNA nucleotide has three parts: a nitrogenous base, a five-carbon sugar, and a phosphate group.

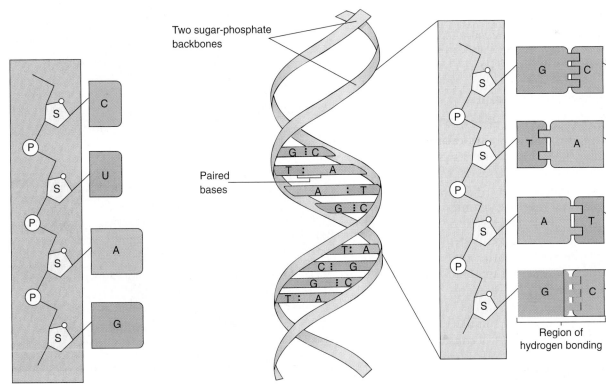

(b) This small part of an RNA molecule has four nucleotides.

(c) The DNA molecule is a double helix consisting of two nucleotide chains joined by their paired bases.

(d) A small part of a DNA molecule is unwound to show how the bases pair (the region of hydrogen bonding).

FIGURE 2-18 Nucleic acids.

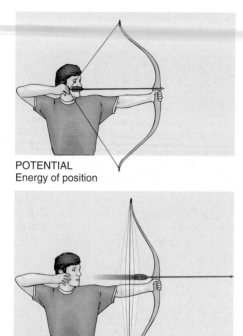

FIGURE 2-19 Potential and kinetic energy.
Potential energy is stored in the drawn bow and is converted to kinetic energy as the arrow speeds toward its target.

converted to kinetic energy. Thus, energy can change from one form to another.

The laws of thermodynamics govern energy transformations

The study of energy and its transformations is called **thermodynamics.** There are two laws about energy that apply to all things in the universe: the first and second laws of thermodynamics.

The **first law of thermodynamics** can be stated as follows: Energy cannot be created or destroyed, although it can be transferred or converted from one form to another. As far as we know, the total energy present in the universe at its formation, approximately 15 billion years ago, equals the amount of energy present in the universe today and represents all the energy that will ever be present in the universe. Similarly, the energy of any object and its surroundings is constant. An object may absorb energy from its surroundings, or it may give up some energy to its surroundings, but the total energy content of that object plus its surroundings is always the same.

As stipulated by the first law of thermodynamics, plants and other organisms cannot create the energy that

they require to live. Instead, they must capture energy from the environment and transform it to a form that can be used to do biological work. Through the process of photosynthesis, for example, plants absorb the radiant energy of the sun and convert it to the chemical energy contained in the bonds of fuel molecules such as carbohydrates.

As each energy transformation occurs, some of the energy changes to heat energy that is then given off into the surroundings. No organism can ever use this energy again for biological work, but because of the first law of thermodynamics, the energy is not "gone"; it still exists in the surroundings. The **second law of thermodynamics** can be stated as follows: When energy is converted from one form to another, some usable energy—that is,

FIGURE 2-20 Maintaining order requires energy.
A grazing cow consumes grasses. Plants photosynthesize and animals eat to replace the energy that they continuously give up to the surroundings as they carry on life processes.

energy available to do work—degrades into a less usable form, usually heat, that disperses into the surroundings. As a result, the amount of usable energy available to do work in the universe decreases over time.

The second law of thermodynamics is consistent with the first law. That is, the total amount of energy in the universe is *not* decreasing with time. However, the total amount of energy available to do work is decreasing over time.

Less-usable energy is more dilute, or disorganized. **Entropy** is a measure of this disorder, or randomness; organized, usable energy such as radiant, electrical, and chemical energy has a low entropy, whereas disorganized energy such as heat has a high entropy. Entropy is continuously increasing in the universe in all natural processes. It may be that at some time billions of years from now, all energy will exist as heat uniformly distributed throughout the universe. If that happens, the universe will cease to operate because no work will be possible. Everything will be at the same temperature, so there will be no way to convert the thermal energy of the universe into usable mechanical energy. Another

way to explain the second law of thermodynamics, then, is that entropy, or disorder, in a system tends to increase over time.

Living things are highly organized and at first glance appear to refute the second law of thermodynamics. That is, as plants and other organisms grow and develop, they maintain a high level of order and do not appear to become more disorganized. However, organisms maintain their degree of order over time only with the constant input of energy from their surroundings. For this reason, plants must photosynthesize (to obtain radiant energy) and animals must eat (to obtain chemical energy) (●**Figure 2-20**).

FIRST LAW OF THERMODYNAMICS Energy cannot be created or destroyed, although it can be transformed from one form to another.

SECOND LAW OF THERMODYNAMICS When energy is converted from one form to another, some of it is degraded into a lower-quality, less useful form.

STUDY OUTLINE

❶ Describe the basic structure of an atom, and explain ionic, covalent, and hydrogen bonds.
Atoms are composed of protons, neutrons, and electrons. A **proton** has a positive electric charge and a small amount of mass, a **neutron** is uncharged and has about the same mass as a proton, and an **electron** is negatively charged and has an extremely small mass. Electrons move around the nucleus in different energy levels. Elements combine chemically in fixed ratios to form chemical **compounds**. Some atoms tend to gain or lose electrons, which results in the formation of **ions; ionic bonds** are attractions between two oppositely charged ions. Some atoms share electrons with other atoms; a **covalent bond** involves the sharing of a pair of electrons. Electrons are equally shared in **nonpolar covalent bonds**, whereas electrons are unequally shared in **polar covalent bonds.** The attraction between the slightly positively charged hydrogen of one polar molecule and the slightly negatively charged oxygen or nitrogen of another is called a **hydrogen bond.**

❷ Discuss the properties of water, and explain the importance of water to life.
Water has a strong dissolving ability. Water molecules exhibit the property of **cohesion** because they form hy-

drogen bonds with one another. Water molecules also exhibit **adhesion** through hydrogen bonding to substances with ionic or polar regions. All living things require water to survive, in part because almost all chemical reactions that sustain life occur in aqueous solution.

❸ Distinguish between acids and bases, and describe the pH scale.
Acids dissociate in water to form hydrogen ions (H^+, or protons); **bases** dissociate in water to yield negatively charged hydroxide ions (OH^-). A solution's acidity or alkalinity is expressed in terms of the **pH scale,** a measure of the relative concentrations of H^+ and OH^- in a solution.

❹ Describe the chemical compositions and functions of carbohydrates, lipids, proteins, and nucleic acids.
Carbohydrates, which include sugars, starches, and cellulose, are important as fuel molecules, as constituents of compounds such as nucleic acids, and as a component of cell walls. Simple sugars are **monosaccharides; disaccharides** are composed of two monosaccharide units; **polysaccharides** are composed of many monosaccharide units. **Lipids,** which have a greasy consistency and do not readily dissolve in water, are important as

fuel molecules, as components of cell membranes, as waterproof coverings over plant surfaces, and as light-gathering molecules for photosynthesis. A **neutral fat** or **oil** molecule is composed of a molecule of **glycerol** plus one, two, or three **fatty acids**. A **protein** is a macro-molecule composed of **amino acids** joined by **peptide bonds**; the order of amino acids determines the structure and function of a protein molecule. **Enzymes** are proteins that increase the rate of chemical reactions; other proteins are important as structural molecules. **Deoxyribonucleic acid (DNA)** and **ribonucleic acid (RNA)** are **nucleic acids**. Nucleic acids control the cell's life processes: DNA transmits hereditary information from one generation to the next, and RNA is involved in protein synthesis. Nucleic acids are composed of repeating units called **nucleotides**; the order of nucleotides in a nucleic acid chain determines the specific information it encodes. **Adenosine triphosphate (ATP)** is a modified nucleotide compound important in energy transfers in biological systems.

⑤ Discuss the role of enzymes in cells.
An **enzyme** speeds up a chemical reaction by lowering its **activation energy**, the energy needed to initiate the

reaction. Most enzymes are highly specific and catalyze only a single chemical reaction. Without enzymes, most chemical reactions in cells would occur at a rate too slow to support life.

⑥ State the first and second laws of thermodynamics, and describe how each applies to plants and other organisms.
Energy is the ability to do work. The **first law of thermodynamics** states that energy can be converted from one form to another, but it can be neither created nor destroyed. As stipulated by the first law of thermodynamics, plants and other organisms cannot create the energy that they require to live but must capture energy from the environment and use it to do biological work. The **second law of thermodynamics** states that **entropy** continuously increases in the universe as usable energy is converted to a lower-quality, less usable form—usually heat. As each energy transformation occurs in organisms, some of the energy changes to heat energy that is then given off into the surroundings. No organism can ever use this energy again for biological work.

REVIEW QUESTIONS

1. What is the difference between an element and a compound? An atom and a molecule? An atom and an ion?
2. Diagram the structure of a carbon atom ($^{12}_{6}C$), showing the protons, neutrons, and electrons.
3. Distinguish between an element's atomic number and its atomic mass.
4. For the element phosphorus, $^{31}_{15}P$: (a) How many protons, neutrons, and electrons are present in one atom? (b) What is the atomic number? (c) What is the atomic mass?
5. Distinguish between a covalent bond and an ionic bond.
6. Describe some of the unique properties of water.
7. What is the pH scale?
8. Contrast a monosaccharide such as glucose with a disaccharide such as sucrose. Contrast a monosaccharide with a polysaccharide such as starch.
9. What are the molecular components of neutral fats and oils?
10. Draw the structure of an amino acid, and indicate which portions correspond to the amino and carboxyl groups.

11. What are enzymes? Why are they essential to plants and other organisms?
12. What are the two nucleic acids?
13. State the first and second laws of thermodynamics.
14. Label the following figure. Check your answers against Figure 2-16.

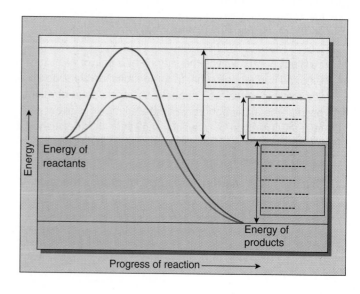

THOUGHT QUESTIONS

1. "When one form of energy is changed into another form, some of the energy is turned into heat." Which law of thermodynamics does this statement describe? Explain your answer.

2. From experience you know that fats and oils are not very soluble in water. Explain that fact based on the chemical structure of fats and oils. (*Hint:* Are fatty acids polar or nonpolar?)

3. What connection is there between a cell's DNA and the enzymes that catalyze its biochemical reactions?

4. How can life exhibit so much order in an increasingly disordered universe?

Visit us on the web at http://www.thomsonedu.com/biology/berg/ for additional resources, such as flashcards, tutorial quizzes, further readings, and web links.

Plant Cells

❶ Contrast prokaryotic and eukaryotic cells.

❷ Describe the functions of the following 10 parts of a plant cell: plasma membrane, nucleus, chloroplasts, mitochondria, ribosomes, endoplasmic reticulum, Golgi apparatus, vacuole, cytoskeleton, and cell wall.

❸ Summarize the similarities and differences between plant cells and animal cells.

❹ Explain the basic structure of the fluid mosaic model of a membrane.

❺ Define the following processes that are important to the cell: diffusion, osmosis, facilitated diffusion, and active transport.

Submerged along the rocky shores of warm seas lives a most remarkable organism, the mermaid's wineglass (*Acetabularia*). This little organism has a body composed of three parts: a delicate base that anchors it to a rock or piece of coral, a long slender stalk, and a cap. Different species of mermaid's wineglass have caps with different shapes. One (*A. mediterranea*) is smooth and cup shaped, and another (*A. crenulata*) is rounded with a series of fingerlike projections. Green in color, the mermaid's wineglass is an alga, an aquatic organism that obtains energy by photosynthesis; its green color is due to the presence of chlorophyll, the light-trapping pigment found in most photosynthetic organisms.

Why is the mermaid's wineglass remarkable? One reason is that this dainty alga, from 2 to 10 centimeters (0.8 to 4 inches) tall, is composed of a single gigantic cell. The mermaid's wineglass has also had a significant role in the development of *cell biology,* because biologists have used it since the 1930s to answer basic questions about how cells function.

When the cap of *Acetabularia* is cut off at a certain time in the cell's life cycle, the cell base grows a new cap identical in appearance to the original. (An organism's replacement of a lost part is called *regeneration.*) Biologists wondered what kind of cap would form if they grafted the base of one species onto the stalk of another. When the base of *A. mediterranea* was grafted onto the stalk of *A. crenulata,* the cap that first regenerated on the *A. crenulata* stalk was an *A. crenulata* cap. If the regenerated cap was removed, however, the second regenerated cap always matched the base (*A. mediterranea*). No matter how many times the regenerated caps were removed, new ones continued to regenerate to match the base.

Because the base contains the cell's nucleus, biologists concluded that the nucleus controls the type of cap the organism makes. They suggested that the nucleus produces some type of temporary messenger substance that is sent into the stalk to tell the stalk what type of cap to grow. Because the newly grafted stalk initially contained some of the messenger substance produced by its old base, the first cap to regenerate resembled the old shape. As the nucleus in the new base began sending messenger substance into the grafted stalk, the stalk regenerated caps with the new shape.

These experiments have provided biologists with some of the first evidence that the cell's nucleus contains hereditary information—a master plan of everything about the cell. The experiments also have indicated that the nucleus sends out a messenger substance, which we know today as messenger RNA (discussed in Chapter 12).

The marine alga *Acetabularia* is a unicellular organism consisting of a base, a stalk, and a cap.

PROCESS OF SCIENCE **Methods to Study Cells**

The invention of the microscope proved to be one of the most important contributions to expanding biological knowledge. The first microscopes, invented in the late 1500s, were not useful instruments. Although they could magnify specimens, the magnified images were blurry and details were indiscernible. In the 1600s, Anton van Leeuwenhoek (1632–1723) perfected the art of grinding lenses (pieces of glass with curved surfaces) and used them in microscopes of his own design to produce clear, magnified images. Leeuwenhoek made many microscopes and peered through them to discover a world of organisms that were previously unknown.

Robert Hooke, an English physicist and microscopist of the 1600s, made significant discoveries with the compound microscope (a microscope with two lenses, the eyepiece and the objective). In 1665 Hooke used a microscope to examine a sliver of cork from the bark of a certain tree (•**Figure 3-1**). Hooke saw that cork was composed of tiny boxes or compartments, which he named cells. Hooke recognized that he was looking at *dead* cells, of which all that remained were the cell walls. As Hooke and later biologists observed the contents of living cells, they began to realize that the cell's *interior* was also an important part of the cell.

Microscopes continued to improve during the 19th century so that biologists were able to observe tiny structures within cells, which they called **organelles** (little organs). The biologists envisioned these structures carrying out special functions for the cell much as organs, such as the heart and stomach or leaves and roots, carry out specific jobs for a multicellular organism. In 1830 Robert Brown, a Scottish botanist, first identified and named the cell's nucleus (an organelle that serves as its control center).

By the late 1830s, biologists had examined enough different kinds of tissues to conclude that all organisms are composed of cells. Two German biologists, Matthias Schleiden, a botanist, and Theodor Schwann, a zoologist, published separate papers in 1838 and 1839, respectively, that clearly stated that cells are the structural units of life. This statement has come to be known as the **cell theory**.

Another German scientist, Rudolf Virchow, extended the cell theory in 1855 by stating that all cells come from preexisting cells. That is, cells divide to give rise to new cells. In 1880 August Weismann pointed out that since cells come from preexisting cells, all cells in existence today trace their origins back to ancient cells.

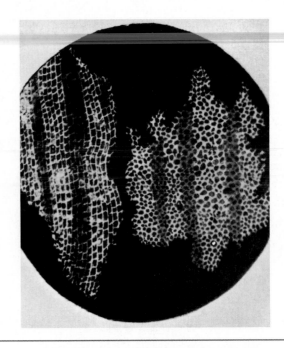

FIGURE 3-1 Robert Hooke's drawing of a thin slice of cork. This drawing appeared in the book *Micrographia*, published in 1665, in which Hooke described many of the objects he had viewed with his compound microscope.

Plant biologists today use a variety of methods to study cells

As the preceding material indicates, the knowledge of cells—their existence, structure, and composition—advanced as microscope technology improved. Today, scientists use a variety of microscopes in their studies of various organisms. •**Figure 3-2** shows how three different microscopes work. Many of the photographs in this book, especially in this chapter, were shot with the assistance of microscopes.

The **light microscope,** which is much improved since Leeuwenhoek's time, focuses a beam of visible light through a transparent sample (see Figure 3-2a). Light microscopes provide **magnification,** an increase in the apparent size of an object, of up to about 1000 times; they also provide **resolving power,** the ability to reveal fine detail, up to 500 times that of the human eye.

The **transmission electron microscope (TEM),** which has much greater resolving power than a light microscope, passes a beam of electrons rather than light through the sample being studied (see Figure 3-2b and d). The electron microscope can magnify an object 250,000 times or more and has resolving power up to 500,000 times that of the human eye.

In another type of electron microscope, the **scanning electron microscope (SEM),** the electron beam

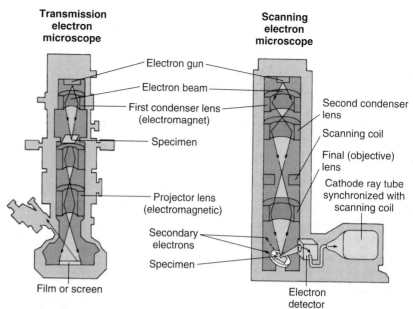

Light microscope

- Light beam
- Ocular lens
- Objective lens
- Specimen
- Condenser lens
- Light source

Transmission electron microscope

- Electron gun
- Electron beam
- First condenser lens (electromagnet)
- Specimen
- Projector lens (electromagnetic)
- Secondary electrons
- Specimen
- Film or screen

Scanning electron microscope

- Second condenser lens
- Scanning coil
- Final (objective) lens
- Cathode ray tube synchronized with scanning coil
- Electron detector

(a) A light microscope focuses a beam of light through the sample.

(b) A transmission electron microscope (TEM) directs a beam of electrons through the sample. Lenses in the electron microscope are actually magnets that bend the beam of electrons.

(c) The scanning electron microscope (SEM) is used to provide a clear view of surface features.

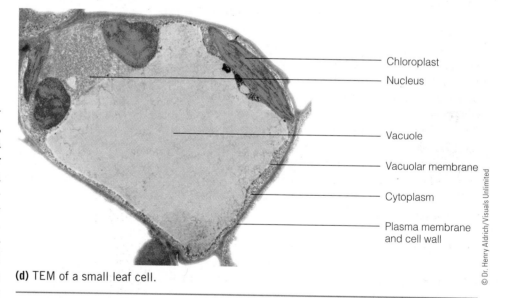

- Chloroplast
- Nucleus
- Vacuole
- Vacuolar membrane
- Cytoplasm
- Plasma membrane and cell wall

© Dr. Henry Aldrich/Visuals Unlimited

(d) TEM of a small leaf cell.

FIGURE 3-2 Light and electron microscopes.

does not pass through the specimen (see Figure 3-2c). Instead, the specimen is coated with a thin film of gold or some other metal. When the electron beam strikes various points on the surface of the specimen, secondary electrons are emitted whose intensity varies with the contour of the surface. Thus, an SEM provides information about the shape and external features of the specimen that cannot be obtained with the TEM.

Scientists also study cells by using a variety of chemical and physical methods. For example, cells can be broken apart, placed in tubes, and centrifuged (spun) at high speeds to separate the cellular organelles. In centrifuges, centrifugal force (the force that makes rotating bodies move away from the center of rotation) causes larger, heavier organelles to settle out at the bottom of a tube, while smaller, lighter components remain suspended in solution. Once various cellular organelles are separated from one another by centrifugation, researchers study them to determine their chemical compositions and their activities. ■

CELL THEORY The theory that the cell is the basic unit of life, of which all living things are composed, and that all cells are derived from preexisting cells.

Eukaryotic and Prokaryotic Cells

Cells consist of a small mass of jellylike living material called **cytoplasm** surrounded by a *plasma membrane,* the outer boundary of the cell. (A cell wall encloses the plasma membrane of a plant cell, but for reasons to be discussed later in the chapter, the cell wall is *not* considered the outer boundary of the cell.)

All cells contain genetic material, the DNA that encodes instructions for the cell's activities. In some cells, called **eukaryotic cells**, the genetic material is located in a special structure, the *nucleus,* which is bounded by a membrane. In cells that lack nuclei, called **prokaryotic cells**, the genetic material is in the cytoplasm.

Prokaryotic cells are generally smaller and simpler than eukaryotic cells. A prokaryotic cell has a simple internal organization: it lacks the nucleus and other membrane-bounded organelles commonly found in eukaryotic cells (see Figure 19-6). The word *prokaryotic* is derived from the Greek *pro,* "before," and *kary,* "nucleus"; evidence indicates that prokaryotic cells evolved before eukaryotic cells. Archaea and bacteria, introduced in Chapter 1, are the only **prokaryotes.** Prokaryotic cells are covered in more detail in Chapter 19.

Eukaryotic cells are generally larger and have more complex structures than prokaryotic cells. The term *eukaryotic* is derived from the Greek *eu,* "true," and *kary,* "nucleus," reflecting the fact that the genetic material of eukaryotic cells is located in a membrane-bounded nucleus. The cytoplasm of eukaryotic cells also contains membrane-bounded organelles not found in prokaryotic cells. All organisms other than prokaryotes—including algae, fungi, plants, and animals—are composed of eukaryotic cells. Organisms with a eukaryotic cell structure are sometimes called **eukaryotes.**

The Structure of Plant Cells

When early biologists examined living plant cells under a microscope, the contents of the cells' interiors appeared to be uniform. As light microscopes were improved, however, it became clear that the cytoplasm was filled with numerous organelles. With the invention of the electron microscope, the interiors of these organelles, as well as organelles that were too small to be seen with the light microscope, became visible.

Each plant cell is like a tiny state; it possesses a control center, power plants, factories that make products, packaging and transport systems, a communication system, and a waste removal system (•**Figure 3-3**). The plant cell's organelles perform these various functions.

The plasma membrane is the outer boundary of the cell

For a cell to exist, it must keep its contents together and separated from the environment. The **plasma membrane** is a physical boundary that confines the contents of the cell to an internal compartment. Although an outer cell wall surrounds a plant cell, the plasma membrane defines the boundary of a cell, because the plasma membrane regulates the flow of materials into and out of the cell. This regulation is *selective:* some substances pass through the plasma membrane unimpeded, other materials are allowed passage in a controlled fashion, and some substances are denied entrance or exit. Thus, the plasma membrane is a selective barrier that allows the interior of the cell to have a chemical composition quite different from that of the outside environment. The importance of the plasma membrane and other cellular membranes is discussed in more detail later in the chapter.

The nucleus is the cell's control center

Housed within the **nucleus** is a complete set of genetic plans—in the form of DNA—for everything about the cell and its activities. These plans are not simply stored in the nucleus for safekeeping but are used continually to direct the activities of the cell. When a cell divides, its DNA is carefully replicated (copied) so that the two new daughter cells contain identical plans.

The nucleus has a remarkably complex structure (•**Figure 3-4**). It is separated from the rest of the cell by a double membrane, the **nuclear envelope,** which contains pores lined with protein molecules. Substances enter and leave the nucleus through these pores, but the pores are selective, and only certain materials can pass through the nuclear envelope. The interior of the nucleus, called **nucleoplasm,** contains the DNA, which is associated with certain protein molecules to form chromatin, a threadlike material. Although chromatin is normally not visible under the light microscope, during cell division it coils and thickens and becomes visible as distinct structures called **chromosomes** (see Figure 12-3).

The **nucleoli** (sing., *nucleolus*), visible within the nucleus, are involved in making and assembling the subunits of *ribosomes,* important organelles in the cytoplasm (ribosomes are discussed shortly).

FIGURE IN FOCUS

Plant cells are eukaryotic and have highly organized, membrane-enclosed organelles.

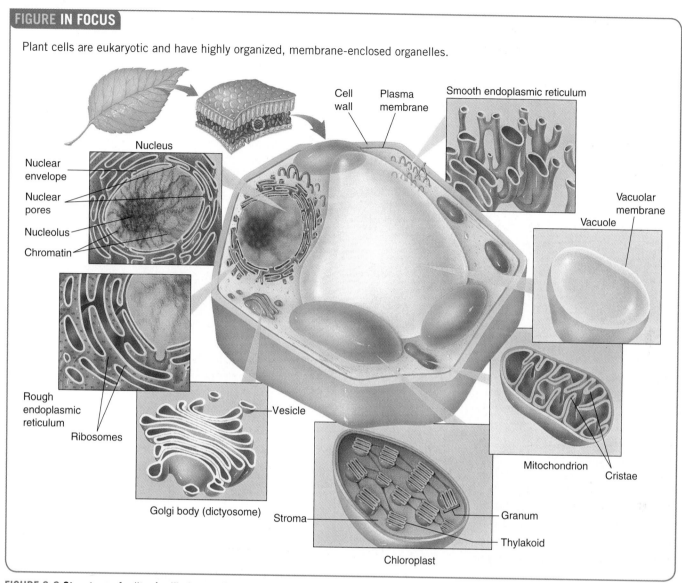

FIGURE 3-3 Structure of a "typical" plant cell.
Some plant cells do not possess all the organelles depicted here. Root cells, for example,
do not contain chloroplasts.

Chloroplasts convert light energy to chemical energy

Algal and plant cells contain large organelles called **plastids,** each of which is surrounded by a double membrane. Although there are several kinds of plastids, the most common are **chloroplasts,** plastids that have a photosynthetic function and occur in certain leaf and stem cells. Recall from Chapter 1 that during photosynthesis plants use the energy of light to convert carbon dioxide and water to carbohydrates such as glucose. Chloroplasts contain the enzymes necessary for photosynthesis plus the green pigment **chlorophyll,** a molecule with the vital role of absorbing light energy.

EUKARYOTIC CELL A cell that possesses a nucleus and other organelles surrounded by membranes.

PROKARYOTIC CELL A cell that lacks nuclei and other membrane-bounded organelles; the archaea and bacteria.

PLASMA MEMBRANE The living surface membrane of a cell that acts as a selective barrier to the passage of materials into and out of the cell.

NUCLEUS A cellular organelle that contains DNA and serves as the control center of the cell.

PLASTID A group of membrane-bounded organelles occurring in photosynthetic eukaryotic cells; includes chloroplasts, leucoplasts, and chromoplasts.

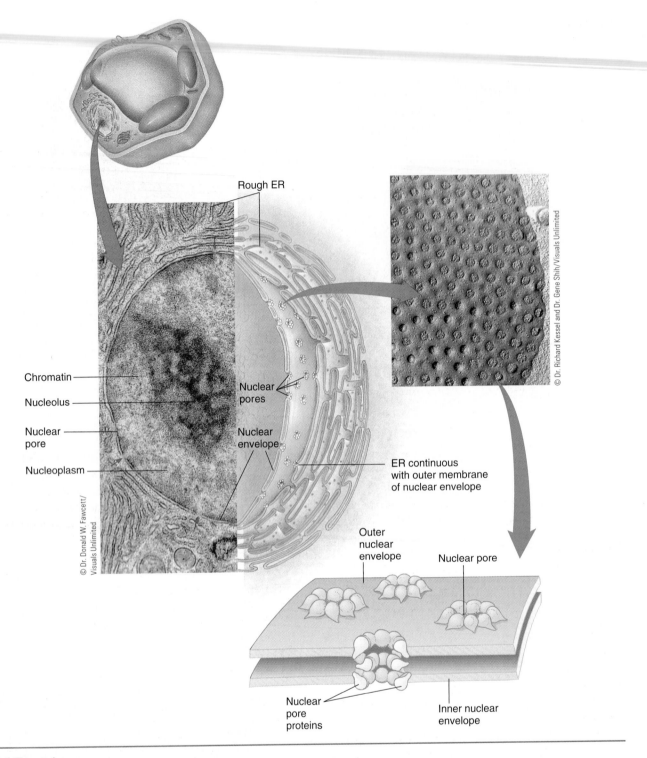

Rough ER

Chromatin

Nucleolus

Nuclear pore

Nucleoplasm

© Dr. Donald W. Fawcett/ Visuals Unlimited

© Dr. Richard Kessel and Dr. Gene Shih/Visuals Unlimited

Nuclear pores

Nuclear envelope

ER continuous with outer membrane of nuclear envelope

Outer nuclear envelope

Nuclear pore

Nuclear pore proteins

Inner nuclear envelope

FIGURE 3-4 The nucleus.
(a) The electron micrograph and interpretive drawing show that a nuclear envelope, composed of two membranes, is perforated by nuclear pores (*see red arrows*). The outer membrane of the nuclear envelope is continuous with the membrane of the endo- plasmic reticulum. **(b)** Electron micrograph of nuclear pores. **(c)** The nuclear pores, which are made up of proteins, form channels between the interior of the nucleus and the cytoplasm.

Chloroplasts are usually disc shaped in plant cells but occur in a variety of shapes in algae. •**Figure 3-5** shows the structure of a chloroplast. In addition to in- ner and outer membranes, the interior of a chloroplast contains thylakoids, membranous stacks of thin, flat, circular plates; a stack of **thylakoids** is called a **granum** (pl., *grana*). The grana are embedded in a jellylike fluid, the **stroma,** which contains enzymes that catalyze the

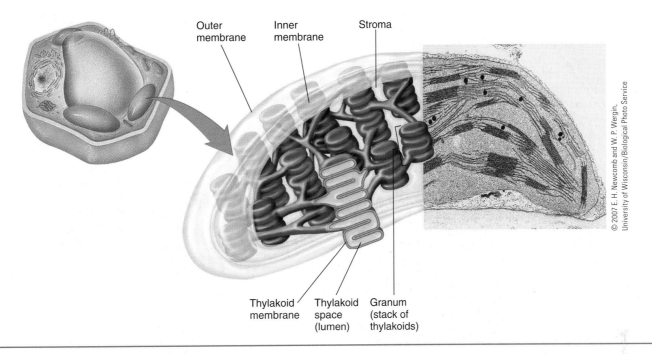

Outer membrane
Inner membrane
Stroma
Thylakoid membrane
Thylakoid space (lumen)
Granum (stack of thylakoids)

© 2007 E. H. Newcomb and W. P. Wergin, University of Wisconsin/Biological Photo Service

FIGURE 3-5 The chloroplast.
Electron micrograph (*right*) and interpretive drawing (*left*) of a chloroplast from a corn leaf cell. The thylakoids—flat, disclike sacs—are arranged in stacks called grana.

chemical reactions of photosynthesis that convert carbon dioxide to carbohydrate.

Each chloroplast also contains a small amount of DNA and a few ribosomes. The presence of DNA and ribosomes in chloroplasts is significant because it indicates that they had free-living ancestors (see *Focus On: The Evolution of Eukaryotic Cells*). Chapter 4 considers chloroplasts and photosynthesis in detail.

Plant cells also contain **leucoplasts,** colorless plastids that form and store starch, oils, or proteins. Leucoplasts are common in seeds and in roots and stems modified for food storage. When leucoplasts are exposed to light, they can synthesize chlorophyll and function as chloroplasts; this happens, for example, when potato tubers are exposed to light.

A third type of plastid, known as a **chromoplast,** contains pigments that provide yellow, orange, and red colors to certain flowers, such as marigolds, and to ripe fruit, such as tomatoes and red peppers. Chromoplasts often form from chloroplasts when chlorophyll breaks down; this happens, for example, when green tomatoes ripen and turn red.

Mitochondria convert the chemical energy in food molecules to ATP

The powerhouses of eukaryotic cells are the **mitochondria** (sing., *mitochondrion*), tiny organelles bounded by double membranes. Cellular respiration, a series of

chemical reactions in which fuel molecules are broken down into carbon dioxide and water with the release of energy, occurs in mitochondria. The released energy is temporarily packaged in the chemical bonds of adenosine triphosphate (ATP) molecules, which are distributed to whatever part of the cell needs chemical energy.

Mitochondria vary in shape but often appear as tiny rods. They are too small to be seen under a student-grade light microscope. Their internal structure is complex. Each mitochondrion is bounded by a double membrane. The inner membrane, because it has a larger surface area than the outer membrane, folds inward, with the folds, called **cristae,** projecting into the interior of the mitochondrion (●**Figure 3-6**). Some of the enzymes for cellular respiration are arranged along the cristae; other respiratory enzymes are found in the **matrix,** the fluid inside the inner mitochondrial membrane.

Each mitochondrion also contains a small amount of DNA and a few ribosomes. As in chloroplasts, the presence of DNA and ribosomes in mitochondria is significant (see *Focus On: The Evolution of Eukaryotic Cells*). Chapter 4 considers mitochondria and cellular respiration in detail.

MITOCHONDRION An intracellular organelle associated with respiration; provides the cell with ATP.

FOCUS ON | The Evolution of Eukaryotic Cells

EVOLUTION LINK According to fossil records, eukaryotic cells appeared approximately 1.9 billion to 2.1 billion years ago, long after the evolution of prokaryotic cells some 3.5 billion years ago. Fossil evidence suggests that eukaryotic cells evolved from prokaryotic cells.

The hypothesis of **serial endosymbiosis** offers an explanation of how eukaryotic cells obtained organelles such as chloroplasts and mitochondria, but it does not explain how the genetic material in the nucleus became surrounded by a nuclear envelope. According to the hypothesis of serial endosymbiosis, organelles such as chloroplasts and mitochondria originated from mutually advantageous symbiotic relationships between two prokaryotic organisms (see figure). (A *symbiotic relationship* is an intimate relationship between two different kinds of organisms.) Chloroplasts are thought to have evolved from *cyanobacteria* (a type of photosynthetic bacteria) that lived inside other, larger prokaryotes. Mitochondria are thought to have evolved from *aerobic* (oxygen-requiring) *prokaryotes* that lived inside larger prokaryotic cells.

How did these prokaryotes come to be **endosymbionts**, which are organisms that live symbiotically inside a host cell? Probably they were originally ingested by the host cell but not digested by it. Once incorporated, they survived and reproduced along with the host cell so that subsequent generations of the host also contained endosymbionts. The two organisms became dependent on each other, and eventually the endosymbiont lost the ability to exist outside its host, and the host cell lost the ability to survive without its endosymbionts.

The principal evidence in favor of serial endosymbiosis is that chloroplasts and mitochondria possess some (although not all) of their own DNA. They also possess ribosomes, a heritage received from their free-living ancestors. Both chloroplasts and mitochondria conduct protein synthesis on a limited scale, independently of the nucleus. Additional evidence for serial endosymbiosis comes from the many endosymbiotic relationships that exist today—for example, certain algae (zooxanthellae) that reside in the bodies of corals and other marine animals (discussed in Chapter 20). ■

ORIGINAL PROKARYOTIC HOST CELL — DNA

Aerobic bacteria

Multiple invaginations of the plasma membrane

Aerobic bacteria become mitochondria

Endoplasmic reticulum and nuclear envelope form from the plasma membrane invaginations (This idea is not part of the hypothesis of serial endosymbiosis)

Photosynthetic bacteria...

...become chloroplasts

EUKARYOTIC CELLS: PLANTS, SOME PROTISTS

EUKARYOTIC CELLS: ANIMALS, FUNGI, SOME PROTISTS

■ The hypothesis of serial endosymbiosis explains the origin of eukaryotes.

Ribosomes are sites of protein synthesis

Ribosomes are small organelles that are protein-manufacturing centers of the cell. Specifically, ribosomes use instructions from DNA in the nucleus to assemble proteins by joining amino acids in precise sequences. Some of these proteins are used inside the cell; others are exported for use outside the cell.

Each ribosome is composed of two subunits; each subunit, in turn, consists of RNA and protein molecules. Although ribosomes occur in the nucleus, plastids, and mitochondria, they are most numerous in the cytoplasm, where they are found free—not associated with a particular organelle—or bound to the endoplasmic reticulum.

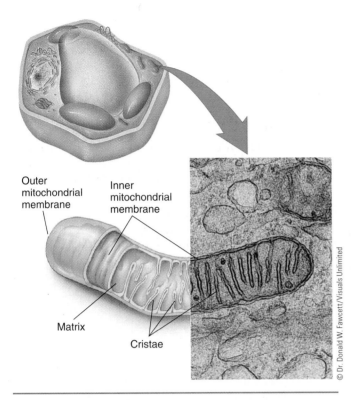

Outer mitochondrial membrane

Inner mitochondrial membrane

Matrix

Cristae

© Dr. Donald W. Fawcett/Visuals Unlimited

FIGURE 3-6 The mitochondrion.
Electron micrograph (*right*) and interpretive drawing (*left*) of a mitochondrion, showing details of the interior, including the matrix and cristae.

The endoplasmic reticulum has many functions

The **endoplasmic reticulum (ER)** is an extensive network of parallel membranes that extends throughout the cell's interior (•**Figure 3-7**). The ER is probably continuous with both the plasma membrane and the nuclear envelope surrounding the nucleus.

The ER is one of the major manufacturing centers of the cell. Many enzymes are associated with the ER membranes; these enzymes catalyze chemical reactions that synthesize biologically important molecules. ER that has ribosomes attached to it is called **rough ER** and is a site of protein synthesis; ER without ribosomes is known as **smooth ER** and is associated with lipid synthesis.

ER also synthesizes the membranes for various organelles throughout the cell, including the nuclear envelope (recall that the ER is continuous with the nuclear envelope) and other cellular organelles such as the Golgi apparatus.

The Golgi apparatus is the cell's collecting and packaging center

A **Golgi body,** or **dictyosome,** is a factory for processing and packaging proteins and polysaccharides. It consists of several flattened sacs, each of which is surrounded

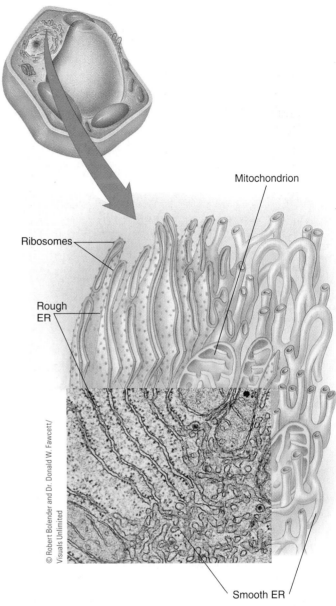

Mitochondrion

Ribosomes

Rough ER

Smooth ER

© Robert Bolender and Dr. Donald W. Fawcett/ Visuals Unlimited

FIGURE 3-7 Endoplasmic reticulum (ER).
The electron micrograph and its paired interpretive drawing show both smooth and rough ER. The smooth ER is more tubular, whereas the rough ER is more flattened. The rough ER is studded with ribosomes.

RIBOSOME A cellular organelle that is the site of protein synthesis.

ENDOPLASMIC RETICULUM (ER) An organelle composed of an interconnected network of internal membranes within eukaryotic cells; rough ER is associated with ribosomes, whereas smooth ER lacks ribosomes.

GOLGI BODY An organelle composed of a stack of flattened membranous sacs that modifies, packages, and sorts proteins that will be secreted or sent to the plasma membrane or other organelles.

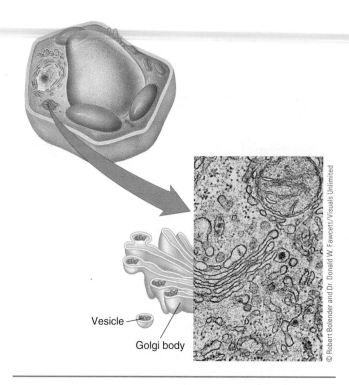

Vesicle

Golgi body

© Robert Bolender and Dr. Donald W. Fawcett/Visuals Unlimited

FIGURE 3-8 Golgi body (dictyosome).
(*Right*) Electron micrograph of Golgi body in a mucilage-secreting plant cell. (*Left*) Three-dimensional view of a Golgi body. Note the small vesicles pinching off the flattened sacs.

by a membrane (•**Figure 3-8**). The edges of the sacs often bulge out and detach as **vesicles,** sacs that contain cellular products. The vesicles transport materials to the plasma membrane, to the outside of the cell, or to other organelles within the cell. The collective term for all the Golgi bodies (dictyosomes) in a cell is the **Golgi apparatus.**

The Golgi apparatus collects and processes materials that are to be exported from the cell. In plant cells, for example, the Golgi apparatus produces and transports some of the polysaccharides that make up the cell wall. The Golgi apparatus also collects materials that are stored inside large, membrane-bounded sacs called *vacuoles.*

The manufacture and export of materials from the cell involve several cellular organelles. For example, consider the secretion of proteins. Proteins that are manufactured by the ribosomes along the rough ER are sealed in vesicles and transported to a Golgi body. The vesicle from the ER then fuses with the membrane of the Golgi body and deposits its contents inside. Once inside, the proteins are modified in various ways. The finished products are then sealed inside a vesicle that pinches off the edge of the Golgi body. The vesicle from the Golgi body then migrates to the cell's plasma membrane, fuses with it, and deposits its contents outside the cell.

Vacuoles are large, membrane-bounded sacs

A **vacuole** is a membrane-bounded sac filled with a liquid that contains a variety of materials in addition to water—dissolved salts, ions, pigments, and waste products. Vacuoles are present in many types of cells but are most common in plant cells and the cells of certain protists. In certain mature plant cells, the vacuole may occupy as much as 90 percent of the volume of the cell (•**Figure 3-9**). The vacuolar membrane (also called the *tonoplast*) surrounds each vacuole and is similar in many respects to the plasma membrane.

The vacuole performs several important functions for a plant cell. One of the most significant is that the vacuole helps the cell maintain its shape by making it turgid (from the Latin *turg,* "swollen"). A **turgid** cell is one that is swollen or firm due to water uptake. The large concentration of ions and other materials dissolved in the vacuole causes water to accumulate. The vacuole swells and presses against the cytoplasm, which in turn presses against the cell's plasma membrane and cell wall. Collectively, vacuoles provide strength for nonwoody plants. These plants are erect because the vacuoles pressing against the cell walls give them rigidity. Without the collective strength of their turgid vacuoles, plants would wilt.

The vacuole also serves as a temporary storage area; excess materials such as calcium ions are stored in the vacuole until the cell needs additional calcium. Water-soluble pigments such as *anthocyanins,* which are blue, purple, or red, are often stored in the vacuole. For example, anthocyanin is stored in the vacuoles of red onion cells, giving them their characteristic color. Waste products, malformed proteins, and the like also enter the vacuole, where they may be disassembled so that their component parts can be used again. Some wastes accumulate and form small crystals that are visible under the light microscope.

The cytoskeleton is composed of protein fibers

The **cytoskeleton** is a network of fibers that extends throughout the cytoplasm and provides structure to a eukaryotic cell. The cytoskeleton, which is also important in cell movement, includes two types of fibers, microtubules and microfilaments.

Microtubules are involved in the addition of cellulose to the cell wall (discussed in the next section). Microtubules also make up the spindle, a special structure that moves chromosomes during cell division. Other microtubules are a part of flagella and cilia, hairlike extensions of certain cells that aid in locomotion; flagella are longer than cilia and occur in smaller numbers. Flagella

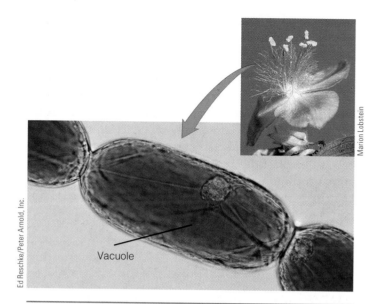

FIGURE 3-9 The vacuole.
A pigmented vacuole of a stamen hair of spiderwort (*Tradescantia virginiana*) dramatically conveys the size of the vacuole. The nucleus and cytoplasm are restricted to the edges of the cell. (The nucleus looks as though it is in the middle of the vacuole because it is lying in a thin layer of cytoplasm on *top* of the vacuole.)

and cilia are never associated with the cells of flowering plants, but they are important structures in algae and in male reproductive cells of other plants.

Microfilaments, which are much thinner than microtubules, can contract and are responsible for **cytoplasmic streaming,** the movement of cytoplasm within the cell. Cytoplasmic streaming has a variety of purposes; for example, the movement of cytoplasm in leaf cells helps orient the chloroplasts for optimal exposure to light, which strikes the leaf cells at different angles during the day as the sun crosses the sky.

Cell walls surround plant cells

Although *every* cell is contained by a plasma membrane, a plant cell also has a firm **cell wall** *outside* the plasma membrane. The cell wall, which is a coating secreted by the cell, supports and protects each plant cell while providing routes for water and dissolved materials to pass to and from the cell. Collectively, cell walls provide strength to the entire plant; a massive tree stands tall and does not collapse on itself because of the combined strength of its cell walls.

Although animal cells do not have cell walls, the cells of many other organisms—prokaryotes, algae and some other protists, and fungi—do. The chemical compositions of cell walls in these organisms often differ from those in plants, however. Plant cell walls are com-

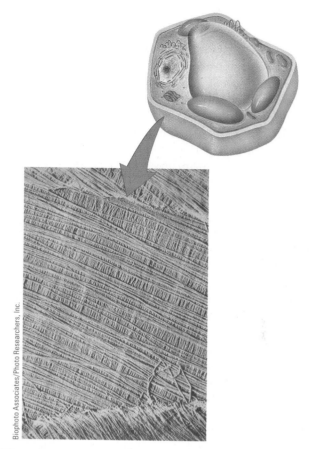

FIGURE 3-10 Electron micrograph of cellulose fibers in a cell wall. Each of these fibers consists of many long, thin cellulose molecules.

posed largely of **cellulose,** a long-stranded polysaccharide that consists of as many as several thousand linked glucose molecules. Cellulose and other cell-wall components are produced within the cell and transported out of the cytoplasm within vesicles from the Golgi apparatus. Microtubules appear to control the movement of these vesicles to the plasma membrane.

Cellulose forms bundles of fibers that are held together by other polysaccharides, including **pectin** (the material that thickens jellies). Each layer of cellulose fibers in a plant cell wall runs in a different direction from the adjacent layer, thereby providing strength to the wall (•**Figure 3-10**).

> **VACUOLE** A fluid-filled, membrane-bounded sac within the cytoplasm that contains a solution of salts, ions, pigments, and other materials.
>
> **CELL WALL** A comparatively rigid supporting wall exterior to the plasma membrane in plants, fungi, prokaryotes, and certain protists.

A growing plant cell secretes a thin **primary cell wall,** which stretches and expands as the cell increases in size. After the cell stops growing, additional wall material may be secreted that thickens and solidifies the primary cell wall. The **middle lamella,** a layer of pectic compounds, cements the primary cell walls of adjacent cells together.

After growth ceases, multiple layers of a **secondary cell wall** with a different chemical composition may form between the primary cell wall and the plasma membrane (•**Figure 3-11**). In addition to cellulose, secondary cell walls usually contain **lignin,** a hard substance in which the cellulose fibers become embedded. (Lignin may also be found in primary cell walls.) Lignin gives wood, which consists largely of secondary cell walls, many of its distinctive properties.

Cells in a multicellular plant need to communicate among themselves, a process that is known as **cell signaling.** Most commonly, cells communicate with chemical signals, either molecules or ions. For this reason, plant cells have connections called **plasmodesmata** (sing., *plasmodesma*), which are tiny channels through adjacent cell walls that connect the cytoplasm of neighboring cells (•**Figure 3-12**). The plasma membranes of adjacent cells are continuous with each other through the plasmodesmata, which generally allow molecules and ions but not organelles to pass from cell to cell.

Plant cells and animal cells are more alike than different

Because plants and animals are eukaryotes, their cell structures are fundamentally the same. Plasma membranes enclose both plant and animal cells, and both have nuclei, mitochondria, ribosomes, ER, the Golgi apparatus, and a cytoskeleton. Plant cells differ from animal cells in several respects, however. Plant cells have cell walls, plastids, and conspicuous vacuoles, whereas animal cells do not. In addition, animal cells contain centrioles that function in cell division and lysosomes that are involved in digestion; plant cells lack both of these organelles.

Biological Membranes

You have seen that every cell must have an outer boundary that separates it from its surroundings and defines its limits. The plasma membrane serves this function in all cells. Most of a eukaryotic cell's organelles also possess specialized membranes that are intimately connected with their structures and functions.

The fluid mosaic model describes membrane structure

The **fluid mosaic model** characterizes the plasma membrane and other cell membranes as consisting of a double layer, or bilayer, of lipid molecules (•**Figure 3-13**). A number of proteins are embedded in the lipid bilayer in a way that resembles a *mosaic* pattern. The membrane structure is *fluid* rather than motionless, and the lipids (and protein molecules, to a lesser extent) move laterally (sideways) within the membrane.

One of the important lipid components of membranes is **phospholipid,** composed of a glycerol molecule to which are attached two fatty acids and a molecule containing a phosphate group. The phosphate end of the phospholipid mol-

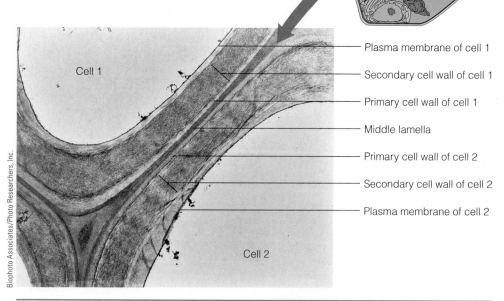

Cell 1

Cell 2

Plasma membrane of cell 1
Secondary cell wall of cell 1
Primary cell wall of cell 1
Middle lamella
Primary cell wall of cell 2
Secondary cell wall of cell 2
Plasma membrane of cell 2

Biophoto Associates/Photo Researchers, Inc.

FIGURE 3-11 Electron micrograph showing layers of the plant cell wall.
The primary cell wall is the first wall formed. When present, the secondary cell wall is laid down between the plasma membrane and the primary cell wall. The thickness of plant cell walls varies greatly.

Communication between plant cells—despite the presence of rigid cell walls between the plasma membranes of adjacent cells—is accomplished by specialized channels called plasmodesmata.

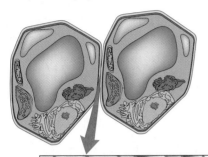

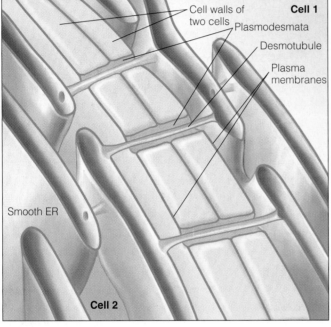

Cell walls of two cells — **Cell 1**

Plasmodesmata

Desmotubule

Plasma membranes

Smooth ER

Cell 2

FIGURE 3-12 Plasmodesmata.
Plasmodesmata are channels through the cell walls of adjacent plant cells that allow passage of water, ions, and small molecules. The channels are lined with desmotubules, which connect the ER of the two cells.

ecule is polar—that is, slightly charged—whereas the fatty acid chains are nonpolar (see Figure 3-13, inset). The polar "head" is **hydrophilic** (from the Greek *hydro,* "water," and *phil,* "love"); that is, it has an affinity for water. The nonpolar "tail" is **hydrophobic** (from the Greek *hydro,* "water," and *phobos,* "fear"); that is, it has an aversion to water.

Because the cell and its surroundings are composed largely of water, phospholipids and other lipid components of membranes spontaneously arrange themselves in a double layer. The hydrophilic heads are positioned on the outer edges of each side of the layer—toward the watery surroundings inside *and* outside the cell or or-

ganelle. The hydrophobic tails form the inside of the double layer (away from the water). No chemical reactions occur to hold the membrane molecules together, nor do covalent bonds connect adjacent molecules. The forces of hydrophilic portions that are attracted to water and of hydrophobic regions that are repelled by water are strong enough to form and maintain the membrane's structure.

Membranes perform many functions

Membranes are not just passive boundaries around cells and their organelles; they play many important roles in cellular function. First, membranes regulate the passage of materials because they are **selectively permeable;** that is, they prevent the entrance or exit of certain materials while permitting—and even helping—the entrance or exit of other materials. For example, the lipid bilayer is impermeable to ions and polar molecules. Thus, charged ions such as Na^+ and Cl^- are not allowed to pass into or out of a cell on their own. The membrane's regulation of the passage of materials enables the cell to maintain **homeostasis,** a relatively constant set of internal conditions. All materials that the cell's cytoplasm requires for survival must pass through the plasma membrane. Nutrient molecules, carbon dioxide, and oxygen must enter the cell through the plasma membrane.

Second, membranes—particularly the plasma membrane—receive information from their surroundings, including other cells (recall the earlier discussion of cell signaling). Chemical messengers such as hormones often bind to special molecules in a membrane and set off some type of response in the cell. Thus, membranes help the cell respond to its environment.

Some of the proteins in membranes are involved in important energy and enzymatic functions. Mitochondria, for example, cannot store energy in ATP without intact, functioning membranes, nor can chloroplasts use the sun's energy without thylakoid membranes. Membranes of other organelles, such as the ER, are the sites of enzymatic activity.

FLUID MOSAIC MODEL The current model for the structure of the plasma membrane and other cell membranes in which protein molecules "float" in a fluid phospholipid bilayer.

A cell membrane is composed of a fluid bilayer of phospholipids in which membrane proteins move about like icebergs in a sea.

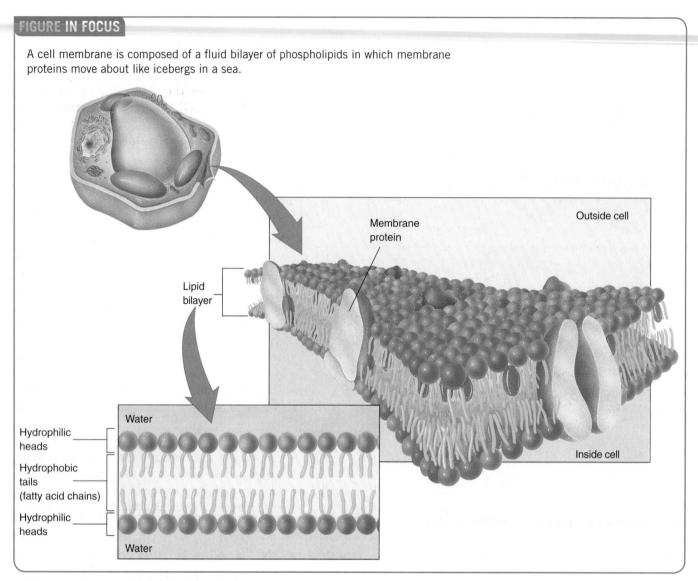

FIGURE 3-13 Structure of a membrane.
The fluid mosaic model describes membranes as consisting of a lipid bilayer in which are embedded a number of proteins. (*Inset*) A phospholipid bilayer forms in watery surroundings. The hydrophilic heads point outward toward the watery surroundings, and the hydrophobic tails point inward so that they are not exposed to the water. In diagrams of phospholipid bilayers, the hydrophilic heads are usually represented by circles and the hydrophobic tails by two lines.

Passage of Materials across Biological Membranes

Some materials move passively through membranes by physical processes such as diffusion and osmosis. Other materials are moved in and out of the cell by processes such as active transport, which requires the cell to expend energy.

Diffusion is the movement of a substance from a region of higher concentration to a region of lower concentration

Some materials pass into and out of cells by simple diffusion (•**Figure 3-14**). During diffusion, atoms and molecules move along a **concentration gradient**—that is, from where they are more concentrated to where they are less concentrated.

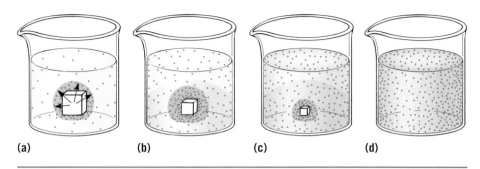

FIGURE 3-14 Diffusion.
(a) When a lump of sugar is placed in a beaker of water, the sugar begins to dissolve and diffuse into the water. **(b, c)** The sugar molecules continue to diffuse. **(d)** Eventually, the sugar is evenly distributed throughout the water.

Diffusion occurs because atoms and molecules are in constant random motion that makes them collide with other particles. As a particle diffuses, it moves in a straight line until it collides with some other particle. This collision causes the original particle to rebound in another direction. Eventually, the particles are uniformly distributed throughout a particular space. Even though the particles continue to move and collide, they remain uniformly distributed, because as fast as some particles move in one direction, other particles travel in the opposite direction.

Diffusion is important to cellular function because it is responsible for the movement of many materials throughout the cytoplasm and into and out of cells. Oxygen, carbon dioxide, and water, for example, diffuse readily into and out of cells. Other materials, however, cannot pass through a cell's plasma membrane by diffusion because they are either too large or too polar.

Osmosis is the diffusion of water across a selectively permeable membrane

Osmosis, a special kind of diffusion, is the movement of water through a selectively permeable membrane from a solution with a higher concentration of water to a solution with a lower concentration of water. In biological systems a **solution** is a mixture in which salts, sugars, and other materials are dissolved in water. The substances that are dissolved in water are referred to as **solutes,** and the water is referred to as the **solvent.**

A cell's plasma membrane is relatively impermeable to sugars and salts, but water moves across the membrane freely in either direction. When a cell is placed in a solution with a solute concentration equal to that inside the cell, water molecules diffuse through the plasma membrane equally in both directions. Such solutions are said to be **isotonic** (from the Greek *iso,* "equal")—that is, they have a solute concentration equal to that in the cell.

When a cell is placed in a solution with a solute concentration higher than that within the cell, the solution is said to be **hypertonic** (from the Greek *hyper,* "over") to the cell. In such a situation, water flows out of the cell and into the surrounding solution. Suppose, for example, that cells are placed in a concentrated sugar solution. Because the sugar solution has more solute molecules than the cell does, the solution has proportionately *fewer* water molecules than the cell's cytoplasm. Water passes freely through the plasma membrane from the cell to the surrounding sugar solution, but the sugar does not move from the solution into the cell because it cannot get through the membrane. Because water always moves from an area of higher concentration (of water) to an area of lower concentration (of water), water moves out of the cell, and the cytoplasm shrinks.

When a cell is placed in a solution with a solute concentration lower than that within the cell, the solution is said to be **hypotonic** (from the Greek *hypo,* "under") to the cell. In such a situation, water flows into the cell from the surrounding solution. For example, suppose that cells are placed in distilled water—that is, 100 percent water with no solute molecules. In this situation, the cell contains more solute molecules and has proportionately *fewer* water molecules than the water in which it was placed. Because water passes freely through the plasma membrane and always moves from an area of higher concentration (of water) to an area of lower concentration (of water), water moves into the cell, and the cell swells. •**Table 3-1** summarizes the movement of water into and out of a solution (or cell) depending on relative solute concentrations.

Normally, the roots of plants are exposed to soil water that is actually a dilute solution of inorganic mineral salts (•**Figure 3-15**). Because soil water has a lower concentration of solutes than the cytoplasm of a root cell does, the soil–water solution that bathes plant roots is hypotonic to the root cells. As water moves

DIFFUSION The net movement of particles (atoms, molecules, or ions) from a region of higher concentration to a region of lower concentration as a result of random motion.

OSMOSIS The net movement of water (the principal solvent in biological systems) by diffusion through a selectively permeable membrane.

TABLE 3-1 **Osmotic Terminology**

SOLUTE CONCENTRATION IN SOLUTION A	SOLUTE CONCENTRATION IN SOLUTION B	TONICITY	DIRECTION OF NET MOVEMENT OF WATER
Greater 	Less	A hypertonic to B; B hypotonic to A	B to A
Less	Greater	B hypertonic to A; A hypotonic to B	A to B
Equal	Equal	A and B are isotonic to each other	No net movement

into the root cells, their cell walls enable them to withstand the building pressure caused by the incoming water. This internal pressure of water against the cell wall is known as **turgor pressure.** As turgor pressure increases, an equilibrium is reached in which the turgor pressure forces water molecules out of the cell in numbers equal to that of the molecules coming in by osmosis.

Some substances cross membranes by facilitated diffusion or active transport

In **facilitated diffusion,** materials diffuse from a region of higher concentration to a region of lower concentration through special passageways in the membrane (•**Figure 3-16**). These passageways are actually membrane proteins called **carrier proteins,** which are channels in the membrane that function as conveyor belts.

Facilitated diffusion enhances the diffusion of certain materials, such as ions, through cell membranes, but *always* in the direction of the concentration gradient—that is, from high to low concentration of that material. Without facilitated diffusion, these materials would not be able to move through the plasma membrane.

Active transport is the assisted movement of a substance from a *lower* concentration to a *higher* concentration of that substance (see Figure 3-16). In other words, during active transport substances move *against* the concentration gradient. Working against the concentration gradient requires a direct expenditure of energy, usually supplied by ATP, the cell's energy-carrying molecule. Active transport occurs with the assistance of carrier proteins in the membrane that move the substance from one side of the membrane to the other.

Why does the cell expend energy for active transport? Cells require some materials—potassium ions (K^+), for example—in greater concentrations than are found in their surroundings. Active transport

Dennis Drenner

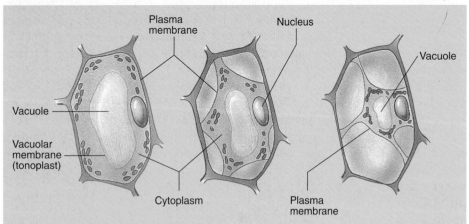

(a) In hypotonic surroundings, the vacuole fills with water, but the rigid cell wall prevents the cell from expanding. The cells of this begonia are turgid.

(b) After a salt solution (that is, a hypertonic solution) is added to the pot, the cells lose water, and their cytoplasm shrinks. The plant begins to wilt.

(c) After a brief period, the begonia is completely wilted (the cells have lost their turgor pressure). Eventually, the plant dies.

FIGURE 3-15 Turgor pressure and various solutions.

FIGURE **IN FOCUS**

Diffusion and facilitated diffusion are passive processes that allow specific substances to move down their concentration gradients. Active transport requires the expenditure of the cell's energy to move specific substances against their concentration gradients.

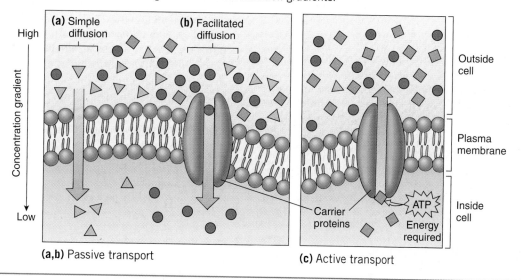

(a,b) Passive transport

(c) Active transport

FIGURE 3-16 Diffusion, facilitated diffusion, and active transport. **(a)** In simple diffusion, materials move directly through the lipid bilayer from high to low concentration. **(b, c)** Carrier proteins are involved in transporting materials across a membrane in both facilitated diffusion and active transport. **(b)** The movement of materials in facilitated diffusion is along the concentration gradient, from high to low concentration. **(c)** The movement of materials in active transport is *against* the concentration gradient, from low to high concentration. The cell expends energy for active transport.

allows these materials to build up inside the cell. Other materials, such as Na^+ and H^+, are found in the surroundings in greater concentrations than cells can tolerate. They move into the cell by diffusion or facilitated diffusion and are then pumped out of the cell by active transport.

STUDY OUTLINE

❶ Contrast prokaryotic and eukaryotic cells.
Two fundamentally different types of cells exist: prokaryotic and eukaryotic. A **prokaryotic cell** has a simple internal organization; it lacks a nucleus and other membrane-bounded **organelles**. A **eukaryotic cell** possesses many cellular organelles, including a nucleus.

❷ Describe the functions of the following 10 parts of a plant cell: plasma membrane, nucleus, chloroplasts, mitochondria, ribosomes, endoplasmic reticulum, Golgi apparatus, vacuole, cytoskeleton, and cell wall.
The **plasma membrane** separates the cell from its surroundings and controls what materials enter and leave the cell. The **nucleus** is the cell's control center. **Chloroplasts** are the sites of photosynthesis. **Mitochondria** perform cellular respiration, in which the chemical energy in fuel molecules is transferred to ATP. **Ribosomes** are sites of protein synthesis. The **endoplasmic reticulum**

(ER) is the site of enzymatic activity and synthesizes membranes such as the nuclear envelope; **smooth ER** is the site of lipid synthesis, and **rough ER** is studded with ribosomes and is associated with protein synthesis. The **Golgi apparatus** collects, modifies, and packages material for export from the cell. **Vacuoles** are large, membrane-bounded sacs that contain dissolved salts, ions, pigments, and waste products. **Microtubules** and **microfilaments** form the cell's **cytoskeleton**; they maintain the shape of the cell and have a role in cellular movement. The **cell wall** is exterior to the plasma membrane and supports and protects the cell.

❸ Summarize the similarities and differences between plant cells and animal cells.
Plant and animal cells have the following structures in common: plasma membrane, nucleus, mitochondria, ribosomes, ER, Golgi apparatus, and cytoskeleton. Plant

cells possess plastids, cell walls, and large vacuoles, all of which animal cells lack. Animal cells have centrioles and lysosomes, both of which plant cells lack.

❹ **Explain the basic structure of the fluid mosaic model of a membrane.**
The **fluid mosaic model** explains membrane structure. According to this model, each membrane is composed of a **phospholipid bilayer** in which varying proteins are embedded. The nonpolar, **hydrophobic** fatty acid chains of the phospholipids project into the interior of the double-layered membrane. The polar, **hydrophilic** heads are located on the two surfaces of the double-layered membrane.

❺ **Define the following processes that are important to the cell: diffusion, osmosis, facilitated diffusion, and active transport.**
Diffusion is the net movement of a substance along the **concentration gradient** from an area of higher concentration to an area of lower concentration. **Osmosis** is the diffusion of water across a **selectively permeable** membrane. In **facilitated diffusion,** a **carrier protein** helps move a material across a membrane in the direction of the concentration gradient, from high to low concentration. In **active transport,** energy is expended to move a material against the concentration gradient, from low to high concentration.

REVIEW QUESTIONS

1. Distinguish between a microscope's magnification and its resolving power.
2. How do prokaryotic and eukaryotic cells differ? Give an example of each kind of cell.
3. Which part of a plant cell is its control center? What does the cell's control center do?
4. How are mitochondria and chloroplasts alike? How are they different?
5. A plant cell secretes a protein. Trace the protein's production, packaging, and release from the cell.
6. The animal body obtains support and strength from its skeletal system; plants lack skeletons and yet attain great size. Explain this phenomenon for both woody and nonwoody plants.
7. How are plant and animal cells alike? How are they different?
8. Describe the fluid mosaic model of membrane structure. Make sure you use the following terms: *hydrophilic, hydrophobic, phospholipid, protein, bilayer, fluid,* and *mosaic.*
9. Distinguish between diffusion and osmosis and between facilitated diffusion and active transport.
10. Describe what happens to a plant cell placed in a hypertonic solution and why.
11. Carrier proteins are used in both facilitated diffusion and active transport, yet these processes are different. Explain their differences.
12. Label the following diagram of a plant cell. Check your answers against Figure 3-3.

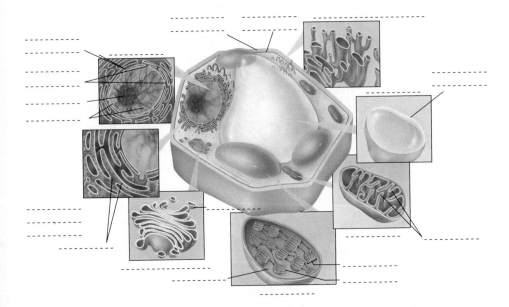

13. Label the following diagram of a membrane. Check your answers against Figure 3-13.

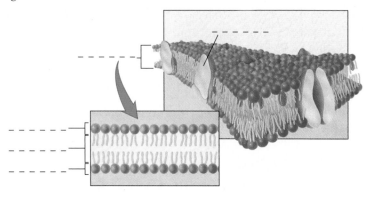

THOUGHT QUESTIONS

1. In a related experiment involving *Acetabularia* (discussed in the chapter introduction), the caps are removed from two species, and the bases are joined; they subsequently regenerate a common cap.

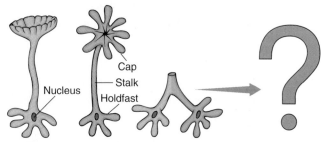

A. mediterranea   A. crenulata

Predict the appearance of the cap that regenerates. State your answer in the form of a hypothesis.

2. What would be the fate of a mutant cell that lacked mitochondria? Why?
3. What would be the fate of a plant cell that possessed a plasma membrane but not a cell wall? Why?
4. What would be the fate of a plant cell that possessed a cell wall but not a plasma membrane? Why?
5. Why do carrot and celery sticks become limp over time? How could they be made crisp once more? Explain your answers in terms of turgor pressure.

Visit us on the web at http://www.thomsonedu.com/biology/berg/ for additional resources, such as flashcards, tutorial quizzes, further readings, and web links.

Metabolism in Cells

LEARNING OBJECTIVES

❶ Relate the transfer of electrons (or hydrogen atoms) to the transfer of energy.

❷ Define *photosynthesis,* and describe how photosynthesis is important not only to plants but to the entire web of life on planet Earth.

❸ Write a summary reaction for photosynthesis, explaining the origin and fate of each compound involved.

❹ Summarize the events of the light-dependent reactions of photosynthesis, including the role of light in the activation of chlorophyll. Describe how a proton gradient allows the formation of ATP according to the chemiosmotic model.

❺ Summarize the events of the carbon fixation reactions of photosynthesis.

❻ Write a summary reaction for aerobic respiration, giving the origin and fate of each substance involved.

❼ List and give a brief overview of the four stages of aerobic respiration.

❽ Distinguish between alcohol fermentation and lactate fermentation.

Rice (*Oryza sativa*) is an important human food that feeds more people than any other crop. It is a staple crop for half the world's population, particularly people in the humid tropics and subtropics of Asia, Africa, and Latin America. Like wheat and corn, two other important human foods, rice is a cereal crop. Cereals are grasses cultivated for their hard, dry "seeds." (Although rice kernels and other cereal grains are called seeds, botanically they are fruits.) Rice is unusual among important food crops in that it is generally grown submerged in 10 to 15 centimeters (4 to 6 inches) of water. Most crops would die under those conditions, because their roots would not get enough oxygen in the waterlogged soil. In contrast, rice roots thrive underwater.

Although rice is nutritious, it contains a relatively small amount of protein (about 10 percent). As a result, a balanced diet with rice as the principal food must be supplemented with meat, fish, or a protein-rich plant such as beans. Modern milling procedures, which strip off the protein- and vitamin-rich bran and germ (embryo), thereby converting brown rice to white rice, further reduce the nutritional value. Although many people prefer white rice, white rice is far less nutritious than brown rice.

In this chapter we examine photosynthesis, the chemical process in which rice and other plants capture energy from their environment and use it to manufacture carbohydrates that support not only humans but almost the entire biosphere. Chapter 4 also considers cellular respiration, the essential process in which plants and all other organisms release energy from carbohydrates and other fuel molecules. You will encounter rice again in the discussion of cellular respiration, because rice root cells carry out an unusual (for plants) type of cellular respiration.

Rice heads ready to be harvested. Rice is one of the world's most important food crops.

Metabolism

The chemical processes that occur in a cell are collectively referred to as its **metabolism.** Two kinds of chemical reactions occur in a cell's metabolism: anabolic and catabolic. **Anabolic reactions** (also called **anabolism**) are chemical reactions in which energy is stored in molecules. In anabolism, a building-up process, large, complex molecules are synthesized from simpler molecules. During photosynthesis, an anabolic reaction, carbon dioxide (CO_2) and water are used to synthesize carbohydrate molecules such as glucose. **Catabolic reactions** (also called **catabolism**) are chemical reactions in which energy is released from molecules. In catabolism, a breaking-down process, large molecules are split apart, or degraded, into simpler ones to release energy. The cellular respiration of glucose, in which glucose is broken down into CO_2 and water with the release of biologically useful energy, is an example of catabolism.

Oxidation–reduction reactions occur in metabolism

The processing of energy by cells involves a transfer of energy through the flow of electrons. **Oxidation** is a chemical process in which a substance *loses* electrons, and **reduction** is a chemical process in which a substance *gains* electrons. An oxidation reaction is always accompanied by a reduction reaction because there are no free electrons in living cells—all electrons are in atoms. If a substance gains an electron in a reduction reaction, that electron must come from another substance that has lost an electron.

As quickly as electrons are released, another atom or molecule accepts them. At the same time as the oxidized molecule gives up electrons, it also gives up energy associated with the electrons; the reduced molecule receives energy when it gains electrons. Oxidation–reduction reactions are characteristic of many cellular processes, including photosynthesis and cellular respiration. Generally, a sequence of oxidation–reduction reactions takes place as electrons associated with hydrogen atoms are transferred from one compound to another.

Electrons associated with hydrogen are transferred to electron acceptor molecules

You may recall from Chapter 2 that when atoms share a pair of electrons, a covalent bond forms. Because such bonds are stable, removing electrons from covalent compounds is difficult unless an entire atom is removed. In a cell, oxidation usually involves the removal of a hydrogen atom and its single electron from a compound. Reduction usually involves a gain of a hydrogen atom (and thus a gain in an electron).

When hydrogen atoms are removed from an organic compound, they take with them some of the energy that was stored in their electrons. The electrons are transferred to an **electron acceptor molecule,** which temporarily accepts them until they move along to the next acceptor molecule. The shuttling of these electrons from one acceptor to another is known as an **electron transport chain.** As we will see shortly, oxidation–reduction reactions and electron transport chains are of crucial importance to both photosynthesis and cellular respiration.

Photosynthesis

For billions of years, almost all life in the biosphere has run on solar energy. Of all organisms, however, only plants, algae, and certain prokaryotes are capable of absorbing and converting light energy from the sun to chemical energy through the process of **photosynthesis** (•**Figure 4-1**). The end products of photosynthesis are carbohydrates (formed from the simple raw materials water and carbon dioxide) and oxygen (O_2).

Plants are of central importance to life. Without plants and other photosynthetic organisms, there would be no energy for animals or other nonphotosynthetic organisms. Animals, humans included, depend on plants for food, which provides energy, and for oxygen.

Humans also burn **biomass,** which is any organic material used as fuel (see *Plants and the Environment: The Environmental Effects of Using Biomass for Fuel*). Biomass, which includes wood, agricultural wastes, and fast-growing plants, contains chemical energy that can be traced to solar energy captured during photosynthesis.

Light exhibits properties of both waves and particles

To understand how plants convert light energy to chemical energy in photosynthesis, it is necessary to know a little about the nature of light. Light makes up a small portion of the **electromagnetic spectrum,** which is a vast, continuous range of electromagnetic radiations propagated through space and matter (•**Figure 4-2a**). All radiations in the electromagnetic spectrum behave as though they travel in waves. A **wavelength** is the distance from one wave peak to the next. At one end of the spectrum are gamma rays with extremely short wavelengths, measured in nanometers (nm), billionths of a meter. At the other end of the spectrum are radio waves with wavelengths so long that they are measured in full meters.

The portion of the electromagnetic spectrum from 380 and 760 nanometers is called the *visible spectrum* because humans can see it. The visible spectrum includes all the colors of the rainbow; violet light has the short-

(a)

(b)

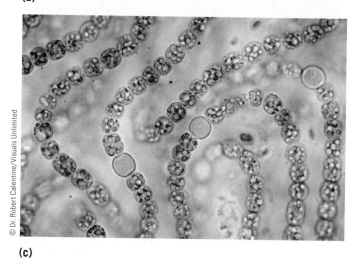

(c)

FIGURE 4-1 Photosynthetic organisms.
(a) Plants, such as hosta (*Hosta*); **(b)** algae, such as this kelp (*Macrocystis pyrifera*); and **(c)** cyanobacteria (*Nostoc*) are photosynthetic organisms. Plants are primarily terrestrial, whereas algae are primarily aquatic. Algae range in size and complexity from microscopic unicellular organisms to large seaweeds that make up underwater "forests." Cyanobacteria are prokaryotic organisms that photosynthesize as plants and algae do.

est wavelength, and red light has the longest. Ultraviolet (UV) radiation, which is invisible to the human eye, has a shorter range of wavelengths than visible light, and infrared (IR) radiation, also invisible, has a longer range.

Light is composed of small particles, or packets, of energy called **photons.** The amount of energy in a photon depends on the wavelength of light. The shorter the wavelength is, the more energy there is per photon; the longer the wavelength, the less energy per photon. Thus, the energy of a photon is inversely proportional to its wavelength.

Photosynthesis depends on visible light rather than some other wavelength of radiation. The reason may be that most of the radiation reaching our planet from the sun is within this portion of the electromagnetic spectrum (•**Figure 4-2b**). Only radiation in the visible-light portion of the spectrum excites certain types of biological molecules, causing their electrons to jump into higher energy levels. Wavelengths of radiation that are longer than visible light (for example, microwaves, television waves, and radio waves) do not possess enough energy to excite biological molecules. Wavelengths shorter than visible light (for example, ultraviolet radiation, X-rays, and gamma rays) possess so much energy that they disrupt biological molecules by breaking chemical bonds.

The interaction between photons and atoms depends on the arrangement of electrons in the atoms. Recall from Chapter 2 that an atom consists of an atomic nucleus (protons and neutrons) surrounded by electrons positioned in one or more energy levels. The lowest energy state an electron possesses is called the **ground state,** but energy can be added to an electron to boost it to a higher energy level. When an electron is raised to a higher energy level than its ground state, the electron is said to be **energized.**

When a molecule absorbs a photon of light energy, one of its electrons becomes energized; that is, it is raised to a higher energy level. One of two things then happens, depending on the atom and its surroundings (•**Figure 4-3**). The atom may return to its ground state, which is the condition in which all its electrons are in their normal, lowest-energy levels. When an electron re-

OXIDATION The loss of electrons from a compound.

REDUCTION The gain of electrons by a compound.

ELECTRON TRANSPORT CHAIN A series of chemical reactions during which hydrogens or their electrons are passed from one acceptor molecule to another, with the release of energy.

PHOTOSYNTHESIS The biological process that includes the capture of light energy and its transformation into chemical energy of organic molecules (such as glucose), which are manufactured from carbon dioxide and water.

PLANTS AND THE ENVIRONMENT | The Environmental Effects of Using Biomass for Fuel

Humans burn biomass fuel to release its energy, which green plants originally converted from solar energy during photosynthesis. Biomass fuel can be a solid, liquid, or gas. Solid biomass, such as wood, is burned directly for energy. Biomass is also converted to liquid fuels such as ethyl alcohol, which is burned in internal-combustion engines. However, a major disadvantage of alcohol fuels is that 30 to 40 percent of the energy in the starting material is lost during the conversion to alcohol. Some experts have calculated that alcohol production from a crop such as corn consumes *more* energy than it yields, when all energy costs associated with growing and harvesting a crop for alcohol production are assessed.

Biomass—particularly firewood, charcoal (wood that is turned into coal by partial burning), animal dung, and peat (partly decayed plant matter found in bogs and swamps)—meets a substantial portion of worldwide energy requirements (see figure). At least half of the world's people rely on biomass as their main source of energy. In developing countries, for example, wood is the primary fuel for cooking.

In many areas of the world, wood is burned faster than it can regrow. Intensive use of wood has resulted in severe damage to the environment, including soil erosion, deforestation (loss of forests), and desertification (expansion of deserts). (See Chapters 6 and 7 for discussions of deforestation and desertification.)

■ Congolese women gather firewood, a form of biomass. Firewood is the major energy source for most of the developing world.

Air pollution, particularly CO_2 emissions, is a problem when biomass is burned as a fuel. (See Chapters 8 and 27 for discussions of air pollution and CO_2 emissions.) However, burning biomass produces less sulfur and ash than burning certain other fuels, such as coal. Tree planting can offset the CO_2 released into the atmosphere by biomass combustion. As trees photosynthesize, they absorb atmospheric CO_2 and lock it up in organic molecules that make up the body of the tree. Thus, if biomass is regenerated to replace the biomass burned as a fuel, there is no net contribution of CO_2 to the atmosphere.

Biomass production requires land and water. Because the use of agricultural land for energy crops competes with the cultivation of food crops, shifting the balance toward energy production might decrease food production, leading to higher food prices. For this reason, some scientists are interested in the commercial development of certain desert shrubs that produce oils that could be used for fuel. Such shrubs do not require prime agricultural land, although the growers would have to take care to ensure that the desert soils were not degraded or eroded by overuse.

turns to its ground state, the "extra" energy is released as heat or as an emission of light of a longer wavelength than the absorbed light (that is, light with less energy). Alternatively, the energized electron may leave the atom and be accepted by an electron acceptor molecule, which becomes reduced in the process; this process occurs in photosynthesis.

Now that you understand some of the properties of light, let us consider the organelles that use light for photosynthesis.

In plants and algae, photosynthesis takes place in chloroplasts

If you examine a section of leaf tissue in a microscope, you see that the green pigment, **chlorophyll,** is not uniformly distributed in the cell but is confined to organelles called **chloroplasts.** In plants, chlorophyll lies mainly inside the leaf in the cells of the **mesophyll,** a layer with many air spaces and a high concentration of water vapor (●**Figure 4-4a**). The interior of the leaf exchanges gases with the outside through microscopic

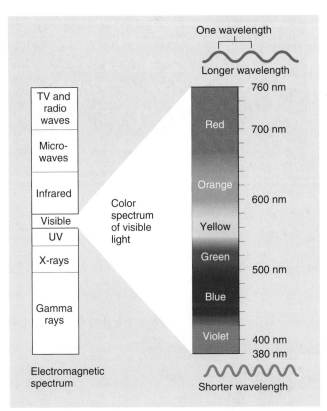

(a) Visible light is that portion of the electromagnetic spectrum that consists of wavelengths from 380 to 760 nm.

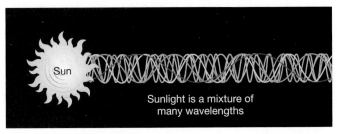

(b) Electromagnetic radiation from the sun includes ultraviolet radiation and visible light of varying colors and wavelengths.

FIGURE 4-2 The electromagnetic spectrum.
The energy of visible light is used in photosynthesis.

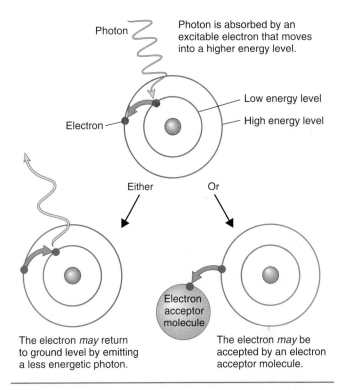

FIGURE 4-3 Interactions between light and atoms or molecules.
(*Top*) When a photon of light energy strikes an atom or a molecule of which the atom is a part, the energy of the photon may push an electron to a higher energy level. (*Lower left*) If the electron returns to the more stable energy level, the energy may be released as a less energetic, longer-wavelength photon (*shown*) or as heat. (*Lower right*) If the appropriate electron acceptors are available, the electron may leave the atom.

pores called **stomata** (sing., *stoma*). Each mesophyll cell has 20 to 100 chloroplasts (•**Figure 4-4b**).

The chloroplast is enclosed by outer and inner membranes (•**Figure 4-4c**). The inner membrane encloses a fluid-filled region, the **stroma,** which contains most of the enzymes required to produce carbohydrate molecules. Suspended in the stroma is a third system of membranes that forms an interconnected set of flat, disc-like sacs called **thylakoids.** The thylakoid membrane encloses a fluid-filled interior space, the *thylakoid lumen.* In some regions of the chloroplast, thylakoid sacs are arranged in stacks called **grana** (sing., *granum*). Each granum looks something like a stack of coins, with each "coin" being a thylakoid.

Thylakoid membranes contain several kinds of pigments, which are substances that absorb visible light. Different pigments absorb different wavelengths. Chlorophyll, the main pigment of photosynthesis, absorbs light primarily in the blue and red regions of the visible spectrum. Green light is not appreciably absorbed by chlorophyll. Plants usually appear green because some of the green light that strikes them is scattered or reflected to your eyes.

There are several kinds of chlorophyll. The most important is chlorophyll *a,* the pigment that initiates photosynthesis (•**Figure 4-5**). Chlorophyll *b* is an accessory pigment that also participates in photosynthesis. It differs slightly from the structure of the chlorophyll *a* molecule. This difference shifts the wavelengths of light absorbed and reflected by chlorophyll *b,* making it appear yellow-green, whereas chlorophyll *a* appears bright green.

Plants and algae also have accessory photosynthetic pigments, such as **carotenoids,** which are yellow and

CHLOROPHYLL One of a group of light-trapping green pigments found in most photosynthetic organisms.

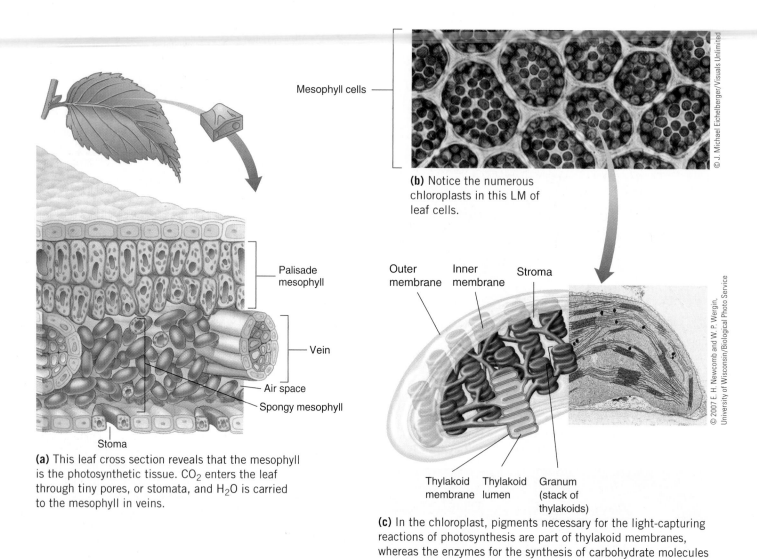

Mesophyll cells

(b) Notice the numerous chloroplasts in this LM of leaf cells.

Palisade mesophyll

Vein

Air space

Spongy mesophyll

Stoma

Outer membrane   Inner membrane   Stroma

Thylakoid membrane   Thylakoid lumen   Granum (stack of thylakoids)

(a) This leaf cross section reveals that the mesophyll is the photosynthetic tissue. CO_2 enters the leaf through tiny pores, or stomata, and H_2O is carried to the mesophyll in veins.

(c) In the chloroplast, pigments necessary for the light-capturing reactions of photosynthesis are part of thylakoid membranes, whereas the enzymes for the synthesis of carbohydrate molecules are in the stroma.

FIGURE 4-4 The site of photosynthesis.

orange. Carotenoids absorb wavelengths of light different from those absorbed by chlorophyll, thereby expanding the spectrum of light that provides energy for photosynthesis. The large quantity of chlorophyll in most leaves usually masks the presence of carotenoids in spring and summer; in autumn, when the chlorophyll breaks down, other pigments, including carotenoids, become visible.

Chlorophyll may be energized by light directly (by energy from the light source) or indirectly (by energy passed by the accessory pigments that have become energized by light). When a carotenoid molecule is energized, its energy can be transferred to chlorophyll *a*. Carotenoids also protect chlorophyll and other parts of the thylakoid membrane from excess light energy that could easily damage the photosynthetic components. (High light intensities often occur in nature.)

Chlorophyll is the main photosynthetic pigment

As you have seen, the thylakoid membrane contains more than one kind of pigment. An instrument called a *spectrophotometer* measures the relative abilities of different pigments to absorb different wavelengths of light. The absorption spectrum of a pigment is a plot of its absorption of different wavelengths of light. •**Figure 4-6a** shows the absorption spectra for chlorophylls *a* and *b*.

An action spectrum of photosynthesis is a graph of the relative effectiveness of different wavelengths of light. To obtain an action spectrum, scientists measure the rate of photosynthesis at each wavelength for leaf cells or tissues exposed to monochromatic light (light of one wavelength) (•**Figure 4-6b**).

In a classic biology experiment, the German biologist T. W. Engelmann obtained the first action spec-

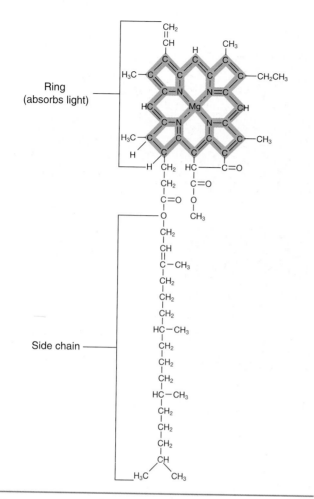

FIGURE 4-5 Structure of the chlorophyll *a* molecule.
Chlorophyll consists of a ring and a side chain. The ring, with a magnesium atom in its center, contains alternating double and single bonds; these are commonly found in molecules that strongly absorb visible light.

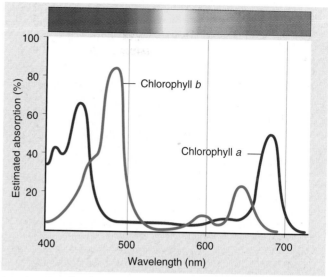

(a) Chlorophylls *a* and *b* absorb light mainly in the blue (422 to 492 nm) and red (647 to 760 nm) regions.

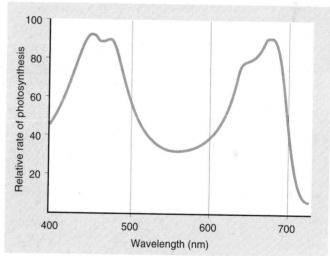

(b) The action spectrum of photosynthesis indicates the effectiveness of various wavelengths of light in powering photosynthesis. Many plant species have action spectra for photosynthesis that resemble the generalized action spectrum shown here.

FIGURE 4-6 A comparison of the absorption spectra for chlorophylls *a* and *b* with the action spectrum for photosynthesis.

trum in 1883. Engelmann's experiment, described in •**Figure 4-7,** took advantage of the shape of the chloroplast in the green alga *Spirogyra.* Its long, filamentous strands are found in freshwater habitats, especially slow-moving or still waters. *Spirogyra* cells each contain a long, spiral-shaped, emerald green chloroplast embedded in the cytoplasm. Engelmann exposed these cells to a color spectrum produced by passing light through a prism. He hypothesized that if chlorophyll were indeed responsible for photosynthesis, the process would take place most rapidly in the areas where the chloroplast was illuminated by the colors most strongly absorbed by chlorophyll.

How could photosynthesis be measured in those technologically unsophisticated days? Engelmann knew that photosynthesis produces oxygen and that certain motile bacteria are attracted to areas of high oxygen concentration. He determined the action spectrum of

photosynthesis by observing that the bacteria swam toward the parts of *Spirogyra* filaments in the blue and red regions of the spectrum. How did Engelmann know that the bacteria were not simply attracted to blue or red light? Engelmann performed a control experiment by exposing bacteria to the spectrum of visible light in the absence of *Spirogyra.* The bacteria showed no preference for any particular wavelength of light. Because the action spectrum of photosynthesis closely matched the absorption spectrum of chlorophyll, Engelmann con-

QUESTION: Is a pigment in the chloroplast responsible for photosynthesis?

HYPOTHESIS: Engelmann hypothesized that chlorophyll was the main photosynthetic pigment. Accordingly, he predicted that he would observe differences in the amount of photosynthesis, as measured by the amount of oxygen produced, depending on the wavelengths of light used and that these wavelengths would be consistent with the known absorption spectrum of chlorophyll.

EXPERIMENT: The photograph **(a)** shows cells of the filamentous alga *Spirogyra*, which has a long spiral-shaped chloroplast. The drawing **(b)** shows how Engelmann used a prism to expose the cells to light that had been separated into various wavelengths. He estimated the formation of oxygen (which he knew was a product of photosynthesis) by exploiting the fact that certain aerobic bacteria would be attracted to the oxygen. As a control (*not shown*), he also exposed the bacteria to the spectrum of light in the absence of *Spirogyra* cells.

RESULTS AND CONCLUSION: Although the bacteria alone (*control*) showed no preference for any particular wavelength, large numbers were attracted to the photosynthesizing cells in red or blue light, wavelengths that are strongly absorbed by chlorophyll (see Figure 4-6). Thus, Engelmann concluded that chlorophyll is responsible for photosynthesis. ■

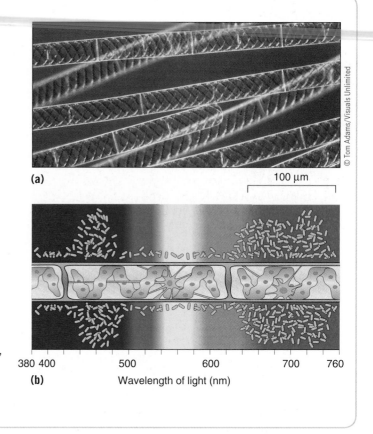

FIGURE 4-7 **The first action spectrum of photosynthesis.**

cluded that chlorophyll in the chloroplasts is responsible for photosynthesis. Numerous studies using sophisticated instruments have since confirmed Engelmann's conclusion.

If you examine Figure 4-6 closely, you will observe that the action spectrum of photosynthesis does not match the absorption spectrum of chlorophyll exactly. This difference occurs because accessory pigments, such as carotenoids, transfer some of the energy of excitation produced by green light to chlorophyll molecules. The presence of these accessory photosynthetic pigments is obvious in temperate climates when leaves change color in the fall. Toward the end of the growing season, chlorophyll breaks down (and its magnesium is stored in the permanent tissues of the tree), leaving orange and yellow accessory pigments in the leaves.

In photosynthesis, plants convert light energy to chemical energy stored in carbohydrate molecules

The principal raw materials for photosynthesis are water and carbon dioxide. The energy that chlorophyll molecules absorb from sunlight is expended to split water, which releases oxygen and hydrogen (that is, the electrons and protons of hydrogens). Ultimately, these hydrogens join with carbon dioxide (although they are not transferred directly) to form carbohydrate molecules such as the six-carbon sugar, glucose. Photosynthesis, which consists of many steps, is summarized as follows:

$$6CO_2 + 12H_2O \xrightarrow[chlorophyll]{Light\ energy,} C_6H_{12}O_6 + 6O_2 + 6H_2O$$

Carbon dioxide   Water   Glucose   Oxygen   Water

This equation describes *what* happens during photosynthesis but not *how* it happens. The reactions of photosynthesis occur in two stages: the light-dependent reactions (the *photo* part of photosynthesis) and the carbon fixation reactions (the *synthesis* part of photosynthesis). Each set of the reactions occurs in a different part of the chloroplast: the light-dependent reactions in association with the thylakoids, and the carbon fixation reactions in the stroma (•**Figure 4-8**).

The light-dependent reactions capture energy

Light energy is converted to chemical energy in the **light-dependent reactions,** which are associated with the thylakoids. The light-dependent reactions begin as chlorophyll absorbs light energy, which causes one of its electrons to move to a higher energy state. The energized electron is transferred to an acceptor molecule and is replaced by an electron from water. During this process, water is split and molecular oxygen (O_2) is released. Some of the energy of the energized electrons is used to make **adenosine triphosphate (ATP),** a temporary energy storage molecule. In addition, **nicotinamide adenine dinucleotide phosphate (NADP⁺)** becomes reduced, forming **NADPH.** The products of the light-dependent reactions, ATP and NADPH, are both needed in the energy-requiring carbon fixation reactions.

The carbon fixation reactions produce carbohydrates

The ATP and NADPH molecules produced during the light-dependent phase are suited for transferring chemical energy but not for long-term energy storage. For this reason, some of their energy is transferred to chemical bonds in carbohydrates, which can be produced in large quantities and stored for future use. Known as **carbon fixation,** these reactions "fix" carbon atoms from CO_2 to existing skeletons of organic molecules. Because the carbon fixation reactions have no direct requirement for light, they were previously referred to as the "dark" reactions. However, they do not require darkness; in fact, many of the enzymes involved in carbon fixation are more active in the light than in the dark. Furthermore, carbon fixation reactions depend on the products of the light-dependent reactions. Carbon fixation reactions take place in the stroma of the chloroplast.

The most common carbon fixation pathway is known as the **Calvin cycle,** named after Melvin Calvin (1911–1997) (•**Figure 4-9**). Calvin, who worked at the University of California with Andrew Benson and others to determine the details of this cycle, was awarded a Nobel Prize in 1961 for this significant scientific contribution.

Now that we have presented an overview of photosynthesis, let us examine the entire process more closely.

Photosystems I and II are light-harvesting units of the light-dependent reactions

The light-dependent reactions of photosynthesis begin when chlorophyll *a* or accessory pigments absorb light. According to the currently accepted model, chlorophyll molecules and accessory pigment molecules are organized with pigment-binding proteins in thylakoid

The light-dependent reactions in the thylakoids capture energy as ATP and NADPH, which power the carbon fixation reactions in the stroma.

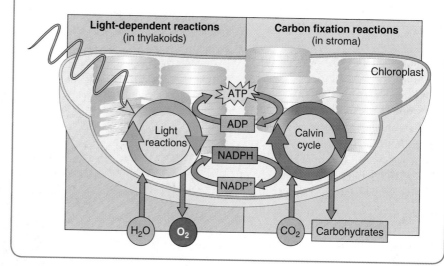

FIGURE 4-8 An overview of photosynthesis.

membranes into units called **antenna complexes.** The pigments and associated proteins are arranged as highly ordered groups of about 250 chlorophyll molecules associated with specific enzymes and other proteins. Each antenna complex absorbs light energy and transfers it to the **reaction center,** which consists of chlorophyll molecules and proteins that participate directly in photosynthesis. Light energy is converted to chemical energy in the reaction centers by a series of electron transfer reactions.

Two types of photosynthetic units, designated photosystem I and photosystem II, are involved in photosynthesis (•**Figure 4-10**). Their reaction centers are distinguishable because they are associated with proteins in a way that causes a slight shift in their absorption spectra. The reaction center of *photosystem I* contains a pair of chlorophyll *a* molecules with an absorption peak of 700 nm; this reaction center is also known as P700. In contrast, the reaction center of *photosystem II* consists of a pair of chlorophyll *a* molecules with an absorption peak of 680 nm and is referred to as P680. An electron transport chain—that is, a chain of alternately oxidized and reduced compounds—links photosystems I and II.

As you read the following details of the light-dependent reactions, refer to the diagram in •**Figure 4-11,** beginning with photosystem II. A pigment molecule in an antenna complex absorbs one photon of light. The absorbed energy is transferred from one pigment mol-

CARBON FIXATION A cyclic series of reactions that fixes carbon dioxide and produces carbohydrate.

ecule to another until the energy reaches the reaction center, where it energizes an electron in a molecule of P680. This energized electron is accepted by a primary electron acceptor and then passes along an electron transport chain until it is donated to P700 in photosystem I.

The electrons that leave photosystem II are replaced by electrons from the splitting of water. When P680 in the reaction center of photosystem II absorbs light energy, P680 becomes positively charged and exerts a strong pull on the electrons in water molecules. Water is split into its components: protons (H^+), electrons, and oxygen. Most of the oxygen split from the water is released into the atmosphere as molecular O_2.

Like photosystem II, photosystem I becomes activated when a pigment molecule in an antenna complex absorbs a photon of light energy. The energy is transferred to the reaction center, where it causes an electron in a molecule of P700 to move to a higher energy level. The energized electron is passed along the electron transport chain until it is transferred to $NADP^+$. When $NADP^+$ accepts two electrons, they unite with a proton (H^+) in the chloroplast to form NADPH, the reduced form of $NADP^+$.

The entire process just described, which produces O_2, NADPH, and ATP, is known as **noncyclic electron transport**. Noncyclic electron transport is *noncyclic* because there is a continuous, one-way flow of electrons from water (the ultimate electron source) to $NADP^+$ (the terminal electron acceptor). To summarize, the

PROCESS OF SCIENCE

QUESTION: What are the steps of carbon fixation (light-independent reactions) in photosynthesis?

HYPOTHESIS: Individual steps in carbon fixation can be elucidated using radioactively labeled carbon dioxide.

EXPERIMENT: Melvin Calvin and his colleague Andrew Benson grew algae in the green "lollipop." Radioactively labeled $^{14}CO_2$ was bubbled through the algae, which were periodically killed by pouring the "lollipop" contents into a beaker of boiling alcohol. The metabolic compounds were separated by means of paper chromatography and then overlaid with X-ray film to locate the radioactive label.

RESULTS AND CONCLUSION: The chromatogram made after 3 seconds of exposure to $^{14}CO_2$ showed radioactive ^{14}C in 3-phosphoglycerate, indicating that this was the first intermediate product of CO_2 fixation. By identifying which compounds contained the radioactive ^{14}C at different times, Calvin determined all of the steps of CO_2 fixation in photosynthesis. ■

FIGURE 4-9 Calvin's classic experiment.

FIGURE 4-10 How a photosystem traps light energy.
Chlorophyll molecules (*green circles*) and accessory pigments (*not shown*) are arranged in light-harvesting arrays, or antenna complexes. When a molecule in an antenna complex absorbs a photon, energy derived from that photon is readily passed from one pigment molecule to another. When this energy reaches one of the two chlorophyll molecules in the reaction center (*green diamonds*), an electron becomes energized and is accepted by a primary electron acceptor.

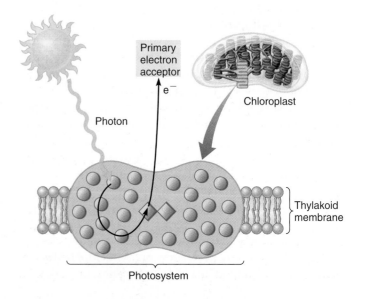

Noncyclic electron transport converts light energy to chemical energy as ATP and NADPH.

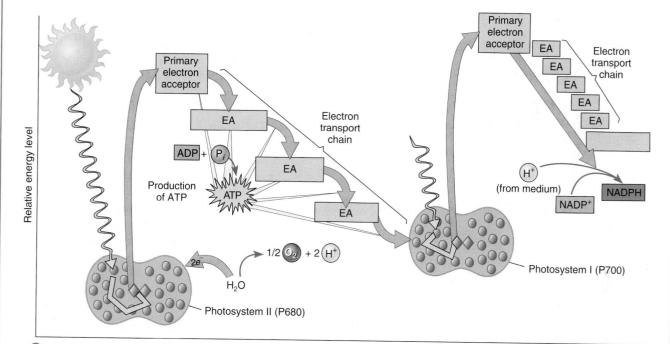

1 Electrons are supplied to system from the splitting of H_2O by photosystem II, with release of O_2 as a by-product. When photosystem II is activated by absorbing photons, electrons are passed along electron transport chain and are eventually donated to photosystem I.

2 Electrons in photosystem I are "reenergized" by absorption of additional light energy and are passed to $NADP^+$, forming NADPH.

FIGURE 4-11 Noncyclic electron transport.
Follow the orange arrows starting at H_2O (*lower left*) to see the flow of energized electrons in one direction, from the splitting of water to the formation of NADPH (*middle right*). ATP formation is also connected to the electron flow. (In the diagram, EA = electron acceptor, one of the molecules in an electron transport chain.)

electrons flow from $H_2O \longrightarrow$ photosystem II $\longrightarrow$ photosystem I $\longrightarrow NADP^+$.

As electrons are transferred along the electron transport chain that connects photosystem II to photosystem I, they lose energy. Some of the energy released in this way is used to synthesize ATP from ADP (see next section).

The light-dependent reactions of photosynthesis also include **cyclic electron transport,** a cyclic flow of electrons through photosystem I that produces ATP. Photosystem II is not involved in cyclic electron transport. In the presence of light, the energized electrons that originate in photosystem I do not go to $NADP^+$ but pass from one acceptor to another as they cycle back to photosystem I. As they pass from one acceptor to another, the electrons lose energy, some of which is used to synthesize ATP. Water is not split, O_2 does not form, and NADPH is not produced. The biological significance of cyclic electron transport in chloroplasts is unclear; recent evidence suggests that cyclic electron

transport provides additional ATP needed to drive the carbon fixation reactions (discussed shortly).

ATP synthesis occurs by chemiosmosis
The electron transport chain that links photosystem II to photosystem I and is involved in both noncyclic and cyclic electron transport is embedded in the thylakoid membrane. Some of the energy released from electrons as they travel through the electron transport chain is used to pump protons (that is, H^+) from the stroma of the chloroplast across the thylakoid membrane and into the thylakoid lumen (•**Figure 4-12**). As protons accumulate within the thylakoids, the pH of the thylakoid interior becomes more acidic. The difference in concen-

NONCYCLIC ELECTRON TRANSPORT In photosynthesis, the linear flow of electrons, produced by the splitting of water molecules, through photosystems I and II; results in the formation of ATP, NADPH, and O_2.

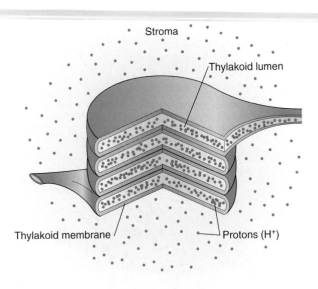

Stroma

Thylakoid lumen

Thylakoid membrane

Protons (H⁺)

FIGURE 4-12 The accumulation of protons in the thylakoid lumen. As electrons move down the electron transport chain, protons (H⁺) move from the stroma to the thylakoid lumen, creating a proton gradient. The greater concentration of H⁺ in the thylakoid interior lowers the pH.

tration of protons between the thylakoid lumen and the stroma, referred to as a **proton gradient,** represents potential energy that can be used to form ATP from ADP.

ATP synthase, an enzyme complex embedded in the thylakoid membrane, forms a channel through which the accumulated protons leave the thylakoid. As the protons pass through ATP synthase, some of their energy is used to form ATP from ADP and inorganic phosphate. The synthesis of ATP by first producing a proton gradient, using the energy released during electron transport, is called **chemiosmosis** (●**Figure 4-13**).[1]

Most plants use the Calvin cycle to fix carbon

The reactions of the Calvin cycle (●**Figure 4-14**), which occur in the stroma, are divided into three phases: ❶ CO_2 uptake, ❷ carbon reduction, and ❸ RuBP regeneration. Let us begin with *ribulose bisphosphate (RuBP),* a five-carbon sugar that was activated by the addition of a phosphate group.

1. *CO_2 uptake.* First, a key enzyme, rubisco, combines a molecule of CO_2 with RuBP. (*Rubisco* is an acronym for the enzyme ribulose bisphosphate carboxylate/oxygenase.) Instantly, this six-carbon mol-

ecule splits into two three-carbon molecules called *phosphoglycerate (PGA).* Because the first detectable molecules to be formed contain three carbon atoms, the Calvin cycle is also referred to as the C_3 pathway.

2. *Carbon reduction.* The PGA molecules are then converted to *glyceraldehyde-3-phosphate (G3P)* using NADPH and ATP. (Keep in mind that the ATP and NADPH that are used throughout the Calvin cycle were formed during the light-dependent reactions.) For every six turns of the Calvin cycle, 2 of the 12 G3P molecules "leave" the cycle to be used in carbohydrate synthesis. Each three-carbon molecule of G3P is essentially half of a six-carbon sugar molecule; G3P molecules are joined in pairs to produce glucose or fructose.

    In some plants, glucose and fructose then combine to form the disaccharide sucrose, or common table sugar. This we harvest from sugarcane, sugar beets, or maple sap. Plants also use glucose to produce the polysaccharides, cellulose (a constituent of plant cell walls), and starch (the principal storage compound in plants).

3. *RuBP regeneration.* Note that although 2 of the 12 G3P molecules formed during six turns of the cycle were removed from the cycle, 10 G3Ps remain, containing 30 carbon atoms in all. Through a series of reactions, these 30 carbons and their associated atoms are rearranged into six molecules of the five-carbon compound *ribulose phosphate (RP).* ATP from the light-dependent reactions is expended to add a second phosphate to ribulose phosphate. This reaction converts ribulose phosphate to ribulose bisphosphate (RuBP), which is where the cycle began.

    In summary, the inputs required for the carbon fixation reactions are CO_2, NADPH, and ATP. The light-dependent reactions of photosynthesis provide the ATP and NADPH. In the end, carbohydrate molecules are manufactured.

Photorespiration reduces the efficiency of the C_3 pathway

Many plants, including certain agriculturally important crops such as soybeans, wheat, and potatoes, do not yield as much carbohydrate from photosynthesis as might be expected. This reduced yield is especially significant during hot periods in summer, which cause water stress in plants. Water-stressed plants lose more water by evaporation from their surfaces than the roots replace by drawing from the soil. To conserve water, plants close the tiny pores (*stomata*) in the leaf surfaces through which CO_2 enters and O_2 exits during photosynthesis. Once the stomata close, photosynthesis in the

---

[1]*The term* chemiosmosis *is somewhat misleading in that* chemiosmosis *has nothing to do with osmosis.*

ATP production in both chloroplasts and mitochondria is linked to the diffusion of protons (H$^+$) across a membrane that exhibits a proton gradient.

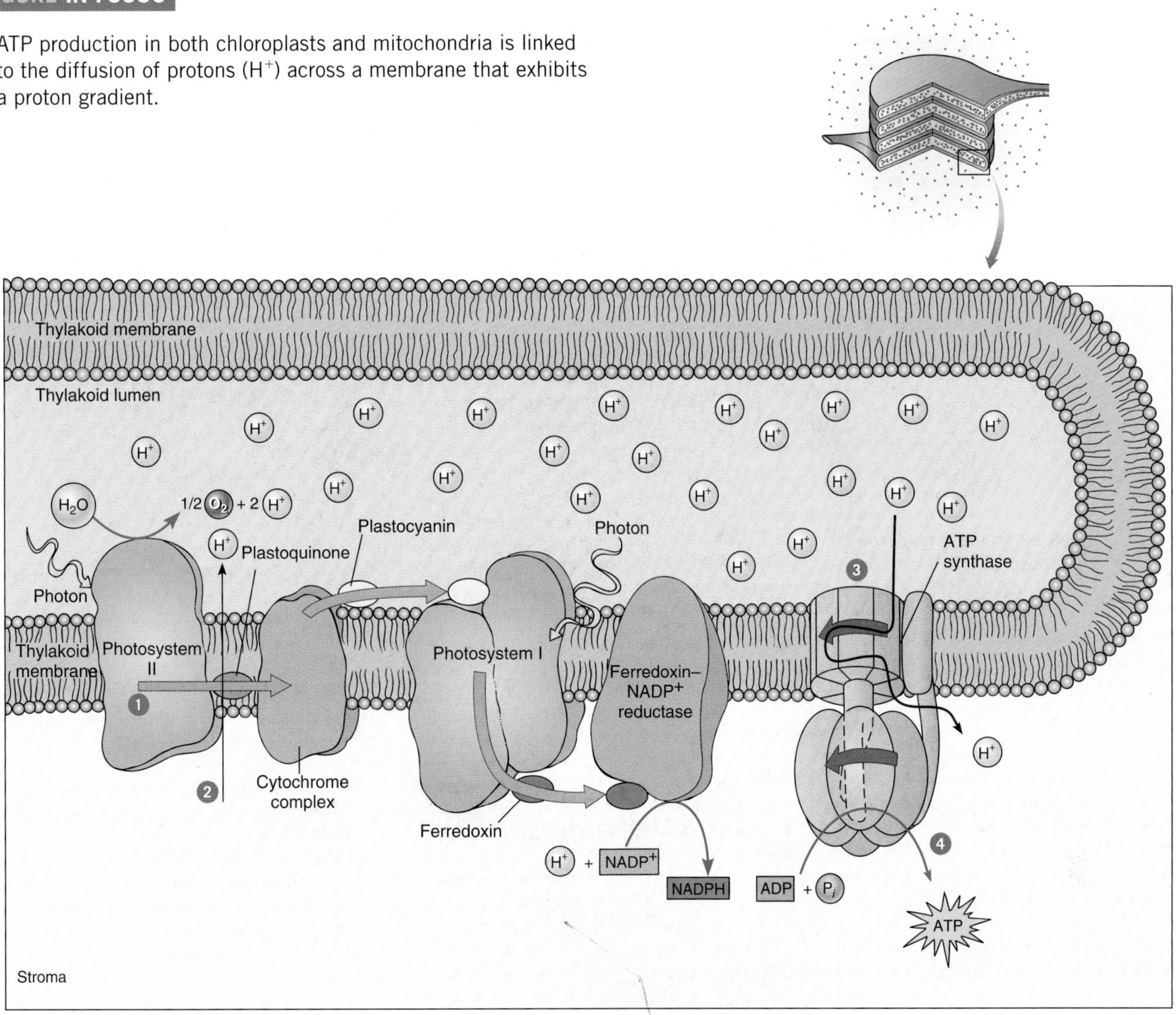

FIGURE 4-13 **A detailed look at electron transport and chemiosmosis.**

① Orange arrows indicate pathway of electrons along electron transport chain in thylakoid membrane. Electron carriers within membrane become alternately reduced and oxidized as they accept and donate electrons.

② Energy released during electron transport is used to transport H$^+$ from the stroma to thylakoid lumen, where high concentration of H$^+$ accumulates.

③ H$^+$ are prevented from diffusing back into stroma except through special channels in ATP synthase in thylakoid membrane.

④ H$^+$ flows through ATP synthase, generating ATP.

chloroplasts rapidly uses up the CO$_2$ remaining in the leaf, and the O$_2$ produced in photosynthesis accumulates in the chloroplasts.

When these conditions exist, a series of reactions involving RuBP occurs in which rubisco (the enzyme responsible for attaching CO$_2$ to RuBP in the Calvin cycle) binds RuBP to O$_2$ instead of to CO$_2$. During these reactions, some of the RuBP involved in the Calvin cycle is degraded, with the release of CO$_2$. This deg-

ATP SYNTHASE An enzyme complex that synthesizes ATP from ADP, using the energy of a proton gradient; located in thylakoid membranes of chloroplasts and in the inner mitochondrial membrane.

CHEMIOSMOSIS The synthesis of ATP using the energy of a proton gradient established across a membrane; occurs during electron transport in both photosynthesis and aerobic respiration.

ATP and NADPH provide the energy that drives carbon fixation in the Calvin cycle.

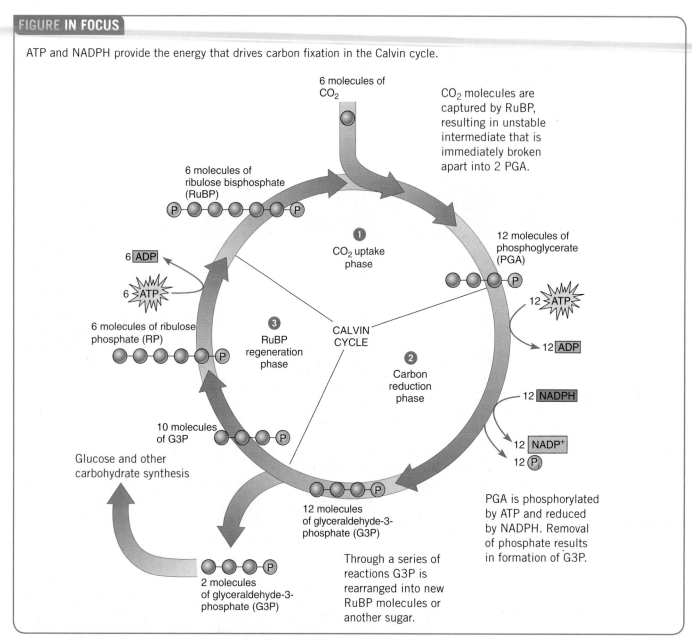

FIGURE 4-14 The carbon fixation reactions of photosynthesis, also known as the Calvin cycle. This diagram, in which carbon atoms are black balls, shows that 6 molecules of CO_2 must be "fixed" (incorporated into preexisting carbon skeletons) in the carbon uptake phase to produce 1 molecule of a six-carbon sugar such as glucose. (The steps indicated by the circled numbers are explained in the text.)

radation process, called **photorespiration**, occurs during intense daylight (hence the prefix *photo-*) and, like cellular respiration, requires oxygen and produces CO_2 and H_2O. Unlike respiration, however, photorespiration does *not* produce biologically useful energy.

Photorespiration reduces photosynthetic efficiency because CO_2 must compete with O_2 for the reaction with RuBP and because the CO_2 released by photorespiration must be reassimilated by the Calvin cycle. The reason that plants carry out photorespiration is un-

known. From a human viewpoint, the process is wasteful, particularly because it reduces the yields of important crop plants.

Many plants with tropical origins fix carbon using the C_4 pathway

Not all plants have reduced photosynthetic efficiency due to photorespiration. Many plants with tropical origins, including crabgrass, corn, and sugarcane, have

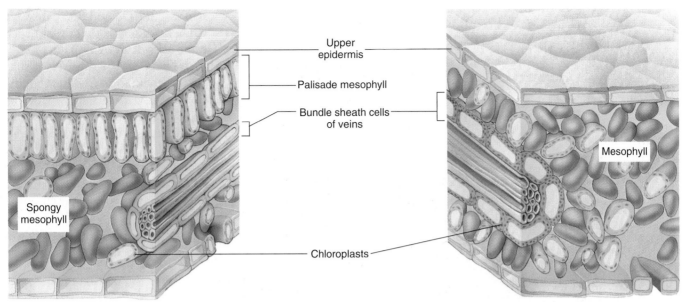

(a) In C_3 plants, the Calvin cycle takes place in the mesophyll cells, and the bundle sheath cells are nonphotosynthetic.

(b) In C_4 plants, reactions that fix CO_2 into four-carbon compounds take place in the mesophyll cells. The four-carbon compounds are transferred from the mesophyll cells to the photosynthetic bundle sheath cells, where the Calvin cycle takes place.

FIGURE 4-15 C_3 and C_4 plant structure compared.

evolved ways to bypass photorespiration. One example is the **C_4 pathway.** The C_4 plants get their name from the fact that the first detectable molecule formed during carbon fixation is a four-carbon compound (produced by joining CO_2 to a three-carbon compound), rather than the three-carbon compound formed in the Calvin (C_3) cycle. This four-carbon compound is then rapidly transported to special cells, called **bundle sheath cells,** that surround veins (conducting tissue) of the leaf (•**Figure 4-15**). In the bundle sheath cells, the CO_2 is removed from the four-carbon molecule and fixed into sugar by the regular C_3 pathway (that is, by the Calvin cycle) (•**Figure 4-16**). The C_4 pathway, in effect, concentrates the amount of CO_2 in the bundle sheath cells, making it greater than it could possibly be as a result of diffusion of CO_2 from the atmosphere.

Photorespiration rarely occurs in C_4 plants, because the concentration of CO_2 in bundle sheath cells (where the enzyme rubisco works) is always high. Some botanists are attempting to transfer the C_4 pathway to crops such as soybeans and wheat. If they succeed, these plants will produce a lot more carbohydrates during hot weather, thus increasing crop yields.

Many desert plants fix carbon using the CAM pathway

Plants living in almost continuously dry conditions have many special adaptations that enable them to sur-

vive. For example, their stomata may open during the cool night hours and close during the hot daylight hours to reduce water loss. (In contrast, most plants have stomata that open during the day and close at night.) Desert plants that close their stomata during the day, however, cannot exchange gases (CO_2 and O_2) during photosynthesis. (Most plants fix CO_2 during the day when sunlight is available to produce ATP and NADPH.)

Many desert plants have a special photosynthetic pathway called **crassulacean acid metabolism (CAM)** that, in effect, solves this dilemma. The name comes from the stonecrop plant family, the Crassulaceae, although it is only one of the more than 25 plant families that possess the CAM pathway (•**Figure 4-17**).

CAM plants fix CO_2 during the night, when their stomata are open for gas exchange, by combining it with a three-carbon compound to form a four-carbon compound. This compound is temporarily stored in the vacuoles of leaf cells. During the day, when stomata are closed and gas exchange cannot occur between the plant and the atmosphere, CO_2 is removed from the four-carbon compound. Thus, the CO_2 is available from the *leaf itself* to be fixed into sugar by the usual C_3 pathway (that is, by the Calvin cycle).

CAM photosynthesis may seem familiar because it is similar to the C_4 pathway, but there are important differences. The C_4 plants initially fix CO_2 into four-carbon compounds in leaf *mesophyll* cells. The four-carbon compounds are then transported to the bundle

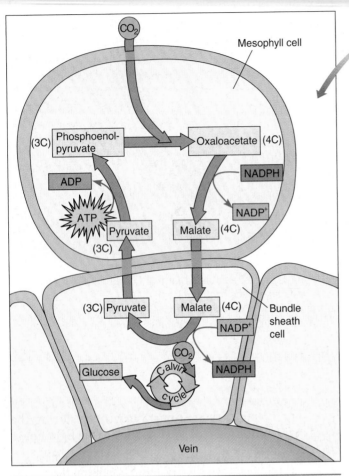

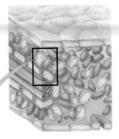

FIGURE 4-16 Summary of the C$_4$ pathway.
CO_2 combines with phosphoenolpyruvate (PEP) in the chloroplasts of mesophyll cells, forming a four-carbon compound that is converted to malate. Malate goes to the chloroplasts of bundle sheath cells, where it is decarboxylated. The CO_2 released in the bundle sheath cell is used to make carbohydrate by way of the Calvin cycle.

sheath cells, where the CO_2 is removed and fixed by the C$_3$ pathway. Thus, the C$_4$ and C$_3$ pathways occur in *different locations* within the leaf of a C$_4$ plant.

In CAM plants, the initial fixation of CO_2 occurs at night. Removal of CO_2 from the four-carbon compound and subsequent production of sugar from the CO_2 by the C$_3$ pathway occur during the day. In other

words, the C$_4$ and C$_3$ pathways occur at *different times* within the same cells of a CAM plant.

The CAM pathway is a successful adaptation to desert conditions. CAM plants carry out gas exchange for photosynthesis *and* significantly reduce water loss during hot daylight hours. Plants with CAM photosynthesis survive in deserts where neither C$_3$ nor C$_4$ plants can. However, CAM plants grow slowly and are therefore limited to arid environments where they do not have to compete with C$_3$ and C$_4$ plants.

Aerobic and Anaerobic Pathways

Every organism must extract energy from the organic fuel molecules that it either manufactures (for example, when plants photosynthesize) or captures from the environment (for example, when an animal eats another organism). These fuel molecules are transported to all the cells of a multicellular organism, where they are broken down to provide the energy for cellular work.

Thus, every plant cell must extract energy from its long-term storage molecules—the organic fuel molecules it manufactures by photosynthesis. Within each plant cell, glucose and other fuel molecules are broken down during **cellular respiration**, a series of chemical reactions that break apart fuel molecules and transfer the energy stored in their bonds to **adenosine triphosphate (ATP)** for use in cellular work.

Cells use three different catabolic pathways to extract energy from fuel molecules: aerobic respiration, anaerobic respiration, and fermentation (another type of anaerobic pathway). The type of environment a cell inhabits may determine which catabolic pathway it uses to break down fuel molecules. Cells that live in environments where oxygen is plentiful use an **aerobic** pathway, one that requires O_2. Cells that inhabit waterlogged soil or polluted water where oxygen is absent must use **anaerobic** pathways (either anaerobic respiration or fermentation) that do not require O_2.

Aerobic Respiration

Plant cells and most other cells extract energy from fuel molecules, such as glucose, fatty acids, and other organic compounds, by **aerobic respiration**. This process involves a sequence of 30 or more chemical reactions, each regulated by a specific enzyme. During aerobic respira-

AEROBIC RESPIRATION The process by which cells use oxygen to break down organic molecules, with the release of energy that can be used for biological work.

GLYCOLYSIS The first stage of cellular respiration, in which glucose is split into two molecules of pyruvate with the production of a small amount of ATP.

tion, energy is released as fuel molecules are catabolized to CO_2 and water. One of the most common pathways of aerobic respiration involves the breakdown of the nutrient glucose. The overall reaction for the aerobic respiration of glucose is summarized as follows:

$$C_6H_{12}O_6 + 6O_2 + 6H_2O \longrightarrow 6CO_2 + 12H_2O + Energy$$

Glucose   Oxygen   Water   Carbon   Water
                                              dioxide

(See •**Table 4-1** for a comparison of photosynthesis and aerobic respiration.)

Aerobic respiration occurs in four stages: An overview

The chemical reactions of aerobic respiration are grouped into four stages: (1) glycolysis, (2) formation of acetyl coenzyme A, (3) the citric acid cycle, and (4) electron transport and chemiosmosis (•**Figure 4-18**).

1. *Glycolysis.* The term **glycolysis** is derived from two Greek words that, taken together, mean "splitting sugar." During glycolysis, the six-carbon glucose molecule is split into two three-carbon molecules of *pyruvate.* During this sequence of reactions, the electrons and protons of hydrogens removed from glucose combine with nicotinamide adenine dinucleotide (NAD^+) to form NADH (reduced NAD^+); a small amount of ATP is also produced. Glycolysis, which takes place in the cytoplasm,

does not require O_2 and therefore proceeds under aerobic or anaerobic conditions.

2. *Formation of acetyl coenzyme A.* The second stage of aerobic respiration, the formation of acetyl coenzyme A, links glycolysis to the citric acid cycle. Each molecule of pyruvate produced during glycolysis passes from the cytoplasm into the mitochondrion (the "power plant" organelle), where it is degraded to a two-carbon fuel molecule (an acetyl group) that combines with coenzyme A (a carrier molecule), forming *acetyl coenzyme A (acetyl CoA).* The carbon removed from pyruvate is released as CO_2, and the

(a) Prickly pear cactus (*Opuntia*) is a CAM plant. More than 200 species of *Opuntia* are found in arid habitats in North and South America.

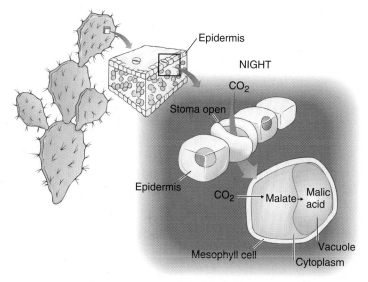

(b) CAM plants open their stomata at night, and CO_2 enters. In mesophyll cells, the CO_2 is converted into malate, which is stored in cell vacuoles as malic acid.

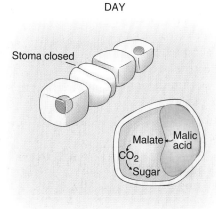

(c) During the day, when stomata are closed, CO_2 is removed from malate and becomes available to be fixed into sugar by the Calvin cycle in the chloroplasts (located in the cytoplasm).

FIGURE 4-17 The CAM pathway.

TABLE 4-1 A Comparison of Photosynthesis and Aerobic Respiration

	PHOTOSYNTHESIS	AEROBIC RESPIRATION
Raw materials	CO_2, H_2O	$C_6H_{12}O_6$, O_2
End products	$C_6H_{12}O_6$, O_2	CO_2, H_2O
Plant cells that have these processes	Plant cells that contain chlorophyll	All plant cells
Organelles involved	Chloroplast	Cytoplasm, mitochondrion
Pathway of energy	Light energy ⟶ NADPH/ATP ⟶ energy stored in carbohydrate molecules	Energy stored in fuel molecules ⟶ NADH/ATP ⟶ energy for work in cell

electrons and protons of the hydrogens released combine with NAD^+ to produce more NADH. The formation of acetyl CoA is an important intermediate step in the aerobic respiration of energy-rich fuel molecules because it prepares the fuel molecule to enter the citric acid cycle (as acetyl CoA).

3. *Citric acid cycle.* In the **citric acid cycle,** which occurs in the mitochondrion, each acetyl CoA combines with a four-carbon molecule to form a six-carbon molecule, citric acid (or, more properly, *citrate*). Citrate eventually degrades into the original four-carbon molecule plus two molecules of CO_2. A small amount of ATP is produced, but most important, the remaining hydrogens are removed from the fuel molecule and picked up by NAD^+ and flavin adenine dinucleotide (FAD), forming NADH and $FADH_2$ (reduced FAD).

4. *Electron transport and chemiosmosis.* The NADH and $FADH_2$ produced in the preceding three stages transfer the electrons that they have accepted to a chain of electron acceptor molecules within the inner membrane of the mitochondrion. As electrons pass along the electron transport chain, energy is released. This energy is used to pump protons (H^+) across the inner mitochondrial membrane, from the interior of the mitochondrion to the space between the inner and outer mitochondrial membranes (•**Figure 4-19**). The difference in concentration of protons across

the inner mitochondrial membrane, referred to as a proton gradient, represents potential energy much as water held behind a dam does. During chemiosmosis the protons diffuse across the membrane, and their energy is used to produce ATP, much as the energy of water spilling over a dam is used to generate electricity. Chemiosmosis in mitochondria is essentially the same process as occurs in chloroplasts (discussed earlier in the chapter).

Aerobic respiration is a complex process

The brief overview of aerobic respiration in the preceding section is sufficient for some introductory botany courses. The following material and •**Figure 4-20** provide greater detail.

FIGURE IN FOCUS

The stages of aerobic respiration occur in specific locations in the cell.

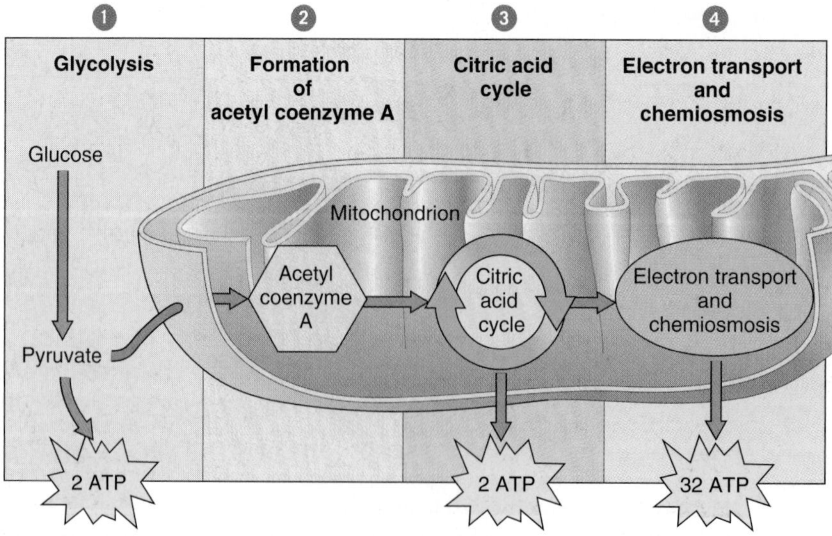

FIGURE 4-18 Overview of aerobic respiration, which has four stages.
❶ Glycolysis, the first stage of aerobic respiration, occurs in the cytoplasm. ❷ Pyruvate, the product of glycolysis, enters a mitochondrion, where cellular respiration continues with the formation of acetyl CoA, ❸ the citric acid cycle, and ❹ electron transport and chemiosmosis. Most ATP is synthesized by chemiosmosis during electron transport.

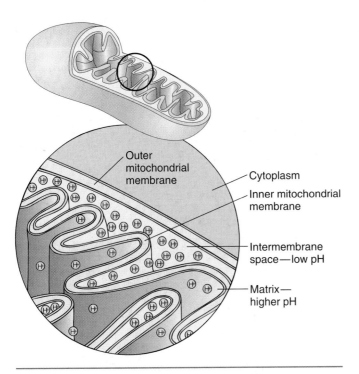

Outer
mitochondrial
membrane

Cytoplasm

Inner mitochondrial
membrane

Intermembrane
space—low pH

Matrix—
higher pH

FIGURE 4-19 The accumulation of protons (H$^+$) within the mitochondrion's intermembrane space.
As electrons move down the electron transport chain, protons (H$^+$) are pumped from the matrix to the intermembrane space, creating a proton gradient. The high concentration of H$^+$ in the intermembrane space lowers the pH.

In glycolysis, glucose is converted to pyruvate

The reactions of glycolysis can be divided into two phases. The first phase requires the input of energy and phosphates from two molecules of ATP (see Figure 4-20a). The two phosphates are added to glucose. The phosphorylated sugar molecule is then split, forming two molecules of the three-carbon compound *glyceraldehyde-3-phosphate (G3P)*.

In the second phase of glycolysis (see Figure 4-20b), each G3P molecule is oxidized with the removal of two hydrogens, and certain other atoms are rearranged so that each molecule of G3P is transformed into a molecule of pyruvate. During these reactions, enough chemical energy is released from the original sugar molecule to produce four ATP molecules. Because the first phase of glycolysis required two molecules of ATP, but the second phase produced four molecules of ATP, glycolysis yields a *net* energy profit of two ATPs per molecule of glucose.

The two hydrogens removed from each G3P immediately combine with NAD$^+$, forming NADH. The fate of these hydrogens is discussed in the section on the electron transport system.

Pyruvate is converted to acetyl CoA

Pyruvate, the end product of glycolysis, contains most of the energy that was present in the original glucose molecule. Pyruvate molecules move into the mitochondrion, where all subsequent reactions of aerobic respiration take place.

Once pyruvate enters the mitochondrion, the pyruvate is converted in a multistep, enzyme-controlled reaction to the two-carbon compound acetyl CoA (see Figure 4-20c). First, a carbon is removed and released as CO_2. Then, the two-carbon fragment remaining is oxidized; the hydrogens removed during this oxidation are accepted by NAD$^+$, forming NADH. Finally the oxidized fragment, an acetyl group, is attached to coenzyme A.

Note that the original six-carbon glucose molecule is now oxidized to two acetyl groups and two CO_2 molecules. Hydrogens have been removed and accepted by NAD$^+$, reducing it to NADH. Two NADH molecules have been formed during glycolysis, and two more during the oxidation of pyruvate.

The citric acid cycle oxidizes acetyl CoA

PROCESS OF SCIENCE The citric acid cycle (see Figure 4-20d) is also known as the **Krebs cycle** after Sir Hans Krebs, a British biochemist who assembled the accumulated contributions of many scientists and worked out the details of the pathway in the 1930s. Krebs received a Nobel Prize in 1953 for this contribution. ■

The citric acid cycle is the common pathway for the final oxidation reactions of the cell's fuel molecules—glucose, fatty acids, and the carbon chains of amino acids—with the carbons being released as CO_2. The citric acid cycle also takes place in the mitochondrion and consists of eight reactions, each catalyzed by a specific enzyme.

The first reaction of the citric acid cycle occurs when acetyl CoA transfers its two-carbon acetyl group to the four-carbon compound *oxaloacetate*, forming citrate, a six-carbon compound. Citrate then goes through a series of chemical transformations, losing first one, then another carboxyl (—COOH) group as CO_2. The two CO_2 molecules produced thus account for the two carbon atoms that entered the citric acid cycle as acetyl CoA. Eventually, the original starting molecule, oxaloacetate, is re-formed; hence, the reactions are part of a *cycle*.

Most of the energy made available by the oxidation steps of the citric acid cycle is transferred as energy-rich electrons to NAD$^+$. For each acetyl group that enters the citric acid cycle, three molecules of NAD$^+$ are reduced to NADH. In addition, electrons are transferred

CITRIC ACID CYCLE A series of aerobic chemical reactions in which fuel molecules are degraded to carbon dioxide and water, with the release of metabolic energy used to produce ATP; also known as the Krebs cycle.

FIGURE IN FOCUS

Most plant cells use aerobic respiration to harvest energy from carbohydrates and other fuel molecules. This energy is then available for the cell's many energy-requiring activities.

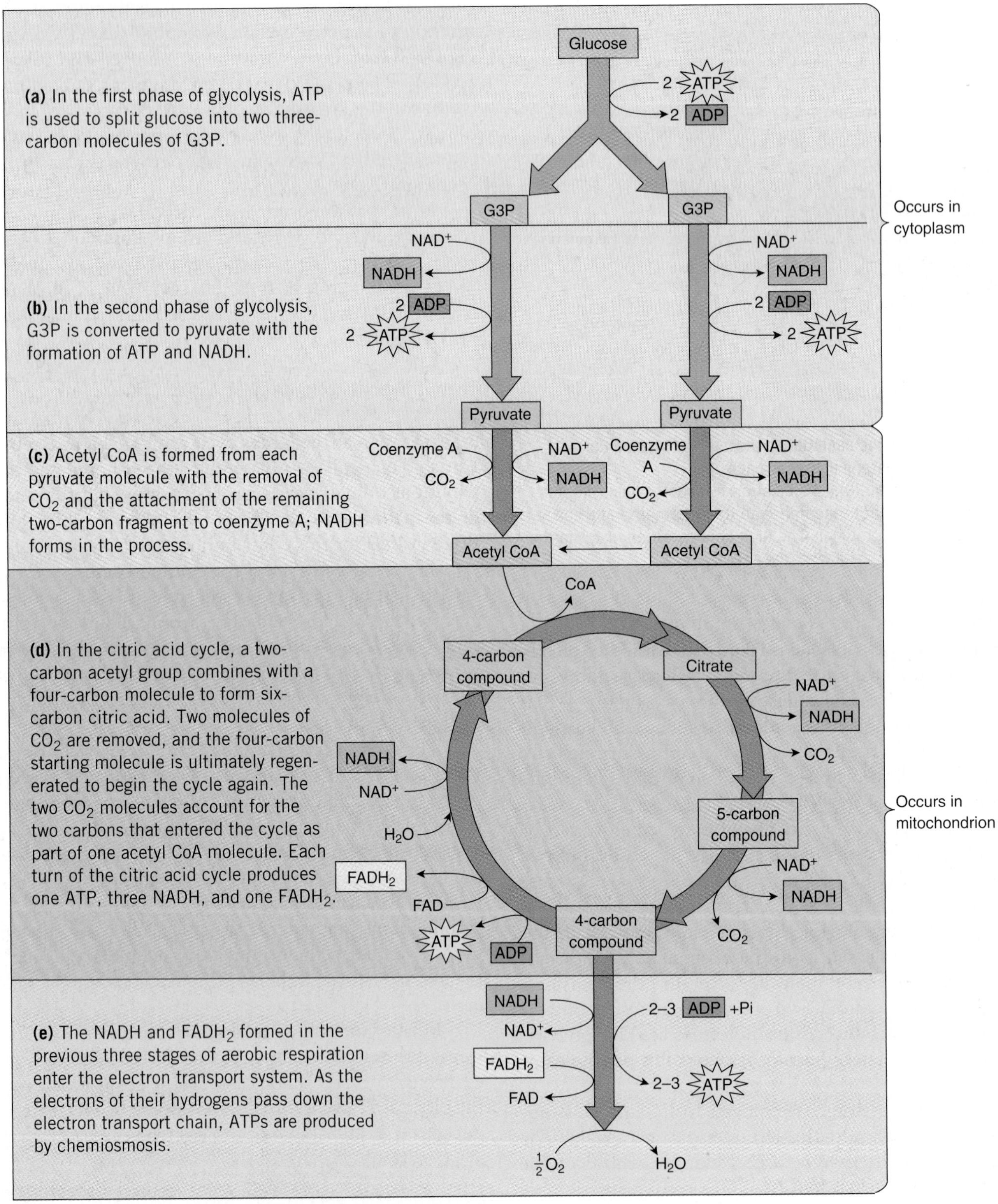
(a) In the first phase of glycolysis, ATP is used to split glucose into two three-carbon molecules of G3P.
Glucose
2 ATP
2 ADP
G3P
G3P
Occurs in cytoplasm
(b) In the second phase of glycolysis, G3P is converted to pyruvate with the formation of ATP and NADH.
NAD+
NADH
2 ADP
2 ATP
Pyruvate
Pyruvate
(c) Acetyl CoA is formed from each pyruvate molecule with the removal of CO_2 and the attachment of the remaining two-carbon fragment to coenzyme A; NADH forms in the process.
Coenzyme A
CO_2
NAD+
NADH
Acetyl CoA
Acetyl CoA
CoA
(d) In the citric acid cycle, a two-carbon acetyl group combines with a four-carbon molecule to form six-carbon citric acid. Two molecules of CO_2 are removed, and the four-carbon starting molecule is ultimately regenerated to begin the cycle again. The two CO_2 molecules account for the two carbons that entered the cycle as part of one acetyl CoA molecule. Each turn of the citric acid cycle produces one ATP, three NADH, and one $FADH_2$.
4-carbon compound
Citrate
5-carbon compound
4-carbon compound
NADH
NAD+
H_2O
$FADH_2$
FAD
ATP
ADP
CO_2
Occurs in mitochondrion
(e) The NADH and $FADH_2$ formed in the previous three stages of aerobic respiration enter the electron transport system. As the electrons of their hydrogens pass down the electron transport chain, ATPs are produced by chemiosmosis.
NADH
NAD+
$FADH_2$
FAD
2–3 ADP +Pi
2–3 ATP
$\frac{1}{2}O_2$
H_2O

FIGURE 4-20 A detailed summary of aerobic respiration.
The reactions of glycolysis can be divided into two phases.

to FAD, forming one molecule of $FADH_2$ for each acetyl group entering the cycle.

Because two acetyl CoA molecules are produced from each glucose molecule, the citric acid cycle must turn twice to process each glucose. At the end of each turn of the cycle, a four-carbon oxaloacetate is all that is left, and the cycle is ready for another turn.

Only one molecule of ATP is produced directly with each turn of the citric acid cycle. Thus, at this point in aerobic respiration, the energy of one glucose molecule has resulted in the formation of only four ATPs (two ATPs from glycolysis and two ATPs from two turns of the citric acid cycle). To maintain their highly ordered state, most cells require much more energy than these four ATPs provide. How, then, is most of the ATP produced?

The electron transport system produces most of the ATP

Now we consider the fate of all the hydrogens removed from the fuel molecule during glycolysis, acetyl CoA formation, and the citric acid cycle. Recall that these hydrogens (and the energy of their electrons) were transferred to NAD^+ and FAD, forming NADH and $FADH_2$. What becomes of the hydrogens?

The electron transport chain accepts hydrogens or their electrons from the previous three stages (see Figure 4-20e). The electron transport system is a chain of electron acceptor molecules embedded in the inner membrane of the mitochondrion. The electrons associated with the hydrogens released from NADH (and $FADH_2$) are passed along the chain of acceptors in a series of oxidation–reduction reactions. Electrons pass from NADH to the first electron acceptor molecule in the chain. That acceptor molecule then transfers electrons to another acceptor molecule, which passes them on to yet another.

The electrons entering the electron transport system have a relatively high energy content. As they pass along the chain of electron acceptor molecules, they lose much of their energy. Some of this energy is used to make ATP (discussed in the next section.) Finally, the last electron acceptor molecule in the electron transport chain passes the two (now low-energy) electrons to oxygen. Simultaneously, the electrons reunite with protons (H^+) in the surrounding environment to form hydrogen. The hydrogen and oxygen combine chemically to produce water. That oxygen is the final electron acceptor in the electron transport system explains why O_2 is required for aerobic respiration.

The passage of each pair of electrons down the electron transport chain from NADH to oxygen is thought to yield enough energy to produce a maximum of three ATPs. Although the flow of electrons in electron transport is usually tightly coupled to the production of ATP, some organisms, such as hibernating animals and certain plants, uncouple the two processes to produce heat (see *Plants and the Environment: Electron Transport and Heat in Plants*).

Chemiosmosis

PROCESS OF SCIENCE Biologists knew for many years that the transfer of electrons from NADH to oxygen resulted in the production of ATP. Just *how* ATPs were synthesized remained a mystery until 1961, when Peter Mitchell proposed the **chemiosmotic model,** which states that the link between the electron transport chain and ATP synthesis is a proton gradient established across the inner mitochondrial membrane. Mitchell was awarded a Nobel Prize in 1978 for this scientific contribution. ■

As electrons from NADH and $FADH_2$ are transferred from one acceptor molecule to another in the electron transport chain, some of the energy released by the electron transport chain is used to pump the protons from the interior of the mitochondrion across its inner membrane to the intermembrane space (the space between the inner and outer mitochondrial membranes). The proton pumps result in the establishment of a proton gradient between the intermembrane space and the interior of the mitochondrion (see Figure 4-19). The difference in concentration of protons across the inner mitochondrial membrane represents potential energy that is used to synthesize ATP.

Because the inner mitochondrial membrane is impermeable to protons, the protons flow back to the interior of the mitochondrion only through special protein channels in the membrane. As in photosynthesis, the protein channels occur within the enzyme complex **ATP synthase.** The protons move down the energy gradient—that is, through ATP synthase from the area where they are highly concentrated (the space between the inner and outer mitochondrial membranes) to the matrix (the interior of the mitochondrion), where they are present in a lower concentration. ATP synthase uses the energy released by the flow of protons to produce ATP. The complete aerobic respiration of one molecule of glucose may produce a *maximum* total of 36 to 38 ATPs.

Anaerobic Pathways

Prokaryotes and some other organisms that inhabit waterlogged soil or stagnant ponds where oxygen is absent must engage in anaerobic respiration. In **anaerobic respiration,** energy is released from glucose and other fuel molecules without O_2; that is, oxygen is not the final electron acceptor in the electron transport chain. In-

PLANTS AND THE ENVIRONMENT | Electron Transport and Heat in Plants

Certain plants have the ability to produce large amounts of heat. Skunk cabbage, for example, lives in North American swamps and wet woodlands and generally flowers during February and March when the ground is still covered with snow (see figure). The flower temperature of skunk cabbage is 15°C to 22°C (59°F to 72°F) when the air surrounding it is −15°C to 10°C (5°F to 50°F). Skunk cabbage flowers maintain this temperature for 2 weeks or more. Other plants, such as splitleaf philodendron (*Philodendron selloum*) and sacred lotus (*Nelumbo nucifera*), also generate heat when they bloom and maintain their temperatures within precise limits.

What is the source of this heat? Essentially, it is a by-product of various chemical reactions, especially those involving the electron transport chains in the mitochondria. Skunk cabbage and other plants produce unusually large amounts of heat by uncoupling electron transport from ATP production. The inner mitochondrial membranes of these mitochondria contain an *uncoupling protein* that produces a passive proton channel through which protons flow into the mitochondrial matrix. As a consequence, most of the energy of glucose is converted to heat rather than to

■ Skunk cabbage (*Symplocarpus foetidus*) flowers surrounded by snow. Skunk cabbage not only produces a significant amount of heat when it flowers but also regulates its temperature within a specific range.

© David Cavagnaro/Visuals Unlimited

chemical energy in ATP. Its uncoupled mitochondria generate large amounts of heat, enabling the plant to melt the snow and attract insect pollinators by vaporizing certain odiferous molecules into the surrounding air.

Some plants generate as much or more heat per gram of tissue than animals in flight, which have long been

considered the greatest heat producers in the living world. The European plant lords-and-ladies (*Arum maculatum*), for example, produces 0.4 joule (0.1 calorie) of heat per second per gram of tissue, whereas a hummingbird in flight produces 0.24 joule (0.06 calorie) per second per gram of tissue.

---

stead, an inorganic compound, such as nitrate (NO_3^-) or sulfate (SO_4^{2-}), serves as the final acceptor of electrons.

Fermentation, another anaerobic pathway, also degrades glucose and other organic molecules without oxygen. Like aerobic respiration, fermentation depends on the reactions of glycolysis. The final acceptor of hydrogen in fermentation is an *organic molecule.* Two common types of fermentation are alcohol fermentation and lactate fermentation.

Yeasts (unicellular fungi) and certain plant cells carry out a type of fermentation known as **alcohol fermentation** (•**Figure 4-21a and b**). First, they degrade glucose to pyruvate through the process of glycolysis. When deprived of O_2, these cells split CO_2 off from pyruvate, eventually forming *ethyl alcohol.* Alcohol fermentation is the basis for the production of beer, wine,

and other alcoholic beverages using yeast, which is also used in baking to produce the CO_2 that causes dough to rise (the alcohol evaporates during baking). The root cells of rice plants grown in flooded conditions (see chapter introduction) also carry out extensive alcohol fermentation.

Certain fungi and prokaryotes carry on **lactate fermentation** (•**Figure 4-21c**). In this pathway, pyruvate produced during glycolysis is converted to lactate. Lactate fermentation occurs when bacteria cause milk to sour or ferment cabbage to form sauerkraut. Under conditions of insufficient oxygen, human muscle cells also use lactate fermentation to obtain limited amounts of energy. Lactate buildup may cause the fatigue and muscle cramps that people sometimes experience during heavy exercise.

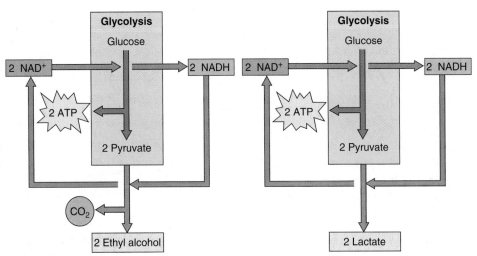

(a) Live yeast cells as seen under a light microscope. Yeast cells possess mitochondria and carry on aerobic respiration when O_2 is present. In the absence of O_2, yeast cells carry on alcohol fermentation.

(b) In alcohol fermentation, the two-carbon ethyl alcohol is the end product.

(c) In lactate fermentation, the final product is the three-carbon compound lactate.

FIGURE 4-21 Glycolysis is part of fermentation reactions.
In both alcohol fermentation and lactate fermentation, there is a net gain of only two ATPs per molecule of glucose.

Fermentation is inefficient

Fermentation is a more inefficient method than aerobic respiration for extracting energy from fuel molecules, because the fuel is only partially oxidized. Alcohol, the end product of fermentation, contains a lot of energy and can even be burned as automobile fuel. Lactate, a three-carbon compound, contains even more energy than the two-carbon alcohol. During aerobic respiration, *all* available energy is removed because fuel molecules are completely oxidized to CO_2. Fermentation produces a net profit of only 2 ATPs from one molecule of glucose, compared with an estimated maximum of 36 to 38 ATPs when oxygen is available.

The inefficiency of fermentation necessitates a large supply of fuel. By rapidly degrading many fuel molecules, a cell compensates somewhat for the small amount of energy that is gained from each molecule. To perform the same amount of work, an anaerobic cell must consume up to 20 times as much glucose or other carbohydrate as a cell metabolizing aerobically.

STUDY OUTLINE

❶ **Relate the transfer of electrons (or hydrogen atoms) to the transfer of energy.**
Oxidation is the loss of electrons from a compound. **Reduction** is the gain of electrons by a compound. An oxidation reaction is always accompanied by a reduction reaction because there are no free electrons in living cells—all electrons are in atoms. If a substance gains an electron in a reduction reaction, that electron must come from another substance that has lost an electron. An **electron transport chain** is a series of chemical reactions during which hydrogens or their electrons are passed from one acceptor molecule to another, with the release of energy.

❷ **Define *photosynthesis*, and describe how photosynthesis is important not only to plants but to the entire web of life on planet Earth.**
Photosynthesis is the biological process that includes the capture of light energy and its transformation into chemical energy of organic molecules (such as glucose), which are manufactured from carbon dioxide and water. Without photosynthesis, there would be no energy for plants, animals, or other organisms. Humans and other animals depend on plants for food, which provides energy, and for oxygen.

❸ **Write a summary reaction for photosynthesis, explaining the origin and fate of each compound involved.**

$$6CO_2 + 12H_2O \xrightarrow[\text{chlorophyll}]{\text{Light energy,}} C_6H_{12}O_6 + 6O_2 + 6H_2O$$

Carbon dioxide Water   Glucose Oxygen Water

Chlorophyll is one of a group of light-trapping green pigments found in most photosynthetic organisms. During photosynthesis, light energy from the sun is captured by chlorophyll and used to chemically combine the hydrogen from water with CO_2 from the atmosphere to produce carbohydrates. Oxygen is released as a by-product of photosynthesis.

❹ **Summarize the events of the light-dependent reactions of photosynthesis, including the role of light in the activation of chlorophyll. Describe how a proton gradient allows the formation of ATP according to the chemiosmotic model.**

During the light-dependent reactions of photosynthesis, chlorophyll absorbs light and becomes energized. Some of this energy is used to make **adenosine triphosphate (ATP)**, and some is used to split water. Hydrogen from the water is transferred to **nicotinamide adenine dinucleotide phosphate (NADP⁺)**, forming **NADPH. Noncyclic electron transport** is the linear flow of electrons, produced by the splitting of water molecules, through photosystems I and II and results in the formation of ATP, NADPH, and O_2. In **cyclic electron transport,** ATP is formed; water is not split, and NADPH is not formed. ATP is synthesized by **chemiosmosis** in both noncyclic and cyclic electron transport using the energy of a **proton gradient** established across a membrane. Protons flow through a membrane channel within the enzyme **ATP synthase,** an enzyme complex that synthesizes ATP from ADP, using the energy of a proton gradient.

❺ **Summarize the events of the carbon fixation reactions of photosynthesis.**
Carbon fixation is a cyclic series of reactions that fixes carbon dioxide and produces carbohydrate. The most common pathway of carbon fixation is the **Calvin cycle,** also called the **C₃ pathway.** During carbon fixation, energy stored within ATP and NADPH during the light-dependent reactions is used to chemically fix CO_2. In the Calvin cycle, CO_2 is combined with ribulose bisphosphate, a five-carbon sugar. With each turn of the cycle, one carbon atom enters the cycle. Six turns of the cycle result in the synthesis of two molecules of a three-carbon compound (G3P), which combine to produce a molecule of glucose.

❻ **Write a summary reaction for aerobic respiration, giving the origin and fate of each substance involved.**

$$C_6H_{12}O_6 + 6O_2 + 6H_2O \longrightarrow 6CO_2 + 12H_2O + \text{Energy}$$

Glucose Oxygen Water   Carbon dioxide Water

Aerobic respiration is the process by which cells use oxygen to break down organic molecules, with the release of energy that can be used for biological work. During aerobic respiration, a fuel molecule such as glucose is oxidized, forming CO_2 and water, with the release of energy stored in the bonds of ATP molecules.

❼ **List and give a brief overview of the four stages of aerobic respiration.**
Aerobic respiration occurs in four stages: glycolysis, formation of acetyl CoA, the citric acid cycle, and the electron transport system with its associated chemiosmosis. In **glycolysis,** glucose is split into two molecules of pyruvate with the production of a small amount of ATP; hydrogen atoms removed from the fuel molecule are transferred to **nicotinamide adenine dinucleotide (NAD⁺),** forming **NADH.** During the formation of acetyl CoA, the two pyruvate molecules each lose a molecule of CO_2, and the remaining acetyl groups combine with coenzyme A, producing acetyl coenzyme A; one NADH is formed as each pyruvate is converted to acetyl CoA. The **citric acid cycle,** also known as the **Krebs cycle,** is a series of aerobic chemical reactions in which fuel molecules are degraded to carbon dioxide and water, with the release of metabolic energy used to produce ATP. Each acetyl CoA enters the citric acid cycle by combining with a four-carbon compound to form citrate, a six-carbon compound; with two turns of the citric acid cycle, the two molecules of acetyl CoA are completely degraded, and four CO_2 molecules are released. Hydrogens are transferred to NAD⁺ and FAD, forming NADH and FADH$_2$; only one ATP is produced directly with each turn of the citric acid cycle. During **electron transport** and **chemiosmosis,** hydrogen atoms (or their electrons) removed from fuel molecules are transferred from one electron acceptor molecule to another; the final acceptor in the chain is O_2, which combines with the hydrogen to form water. According to the **chemiosmotic model,** energy liberated in the electron transport chain is used to establish a **proton gradient** across the inner mitochondrial membrane. Protons flow through a membrane channel within the enzyme **ATP synthase;** energy released is used to synthesize ATP.

❽ **Distinguish between alcohol fermentation and lactate fermentation.**
Fermentation is an anaerobic pathway that degrades glucose and other organic molecules without oxygen.

Fermentation makes use of glycolysis, but the final acceptor of hydrogen is an organic molecule. In **alcohol fermentation** yeasts and certain other organisms degrade glucose to pyruvate, then split CO_2 off, forming ethyl alcohol. Certain fungi and prokaryotes carry on **lactate fermentation,** in which pyruvate produced during glycolysis is converted to lactate.

REVIEW QUESTIONS

1. Explain the roles of light and chlorophyll in photosynthesis.
2. How are almost all organisms on planet Earth dependent on the process of photosynthesis?
3. Write the overall equation for photosynthesis.
4. Summarize the light-dependent reactions of photosynthesis.
5. Summarize the events of the Calvin cycle.
6. Explain how ATP is produced according to the chemiosmotic model. How does a proton gradient contribute to ATP synthesis?
7. Justify referring to mitochondria as the "power plants" of the cell. Be specific, using information you have acquired from this chapter.
8. Write the overall equation for aerobic respiration.
9. Trace the fate of hydrogens removed from the fuel molecule during glycolysis when O_2 is present.
10. What is the specific role of O_2 in cells? What happens when such cells are deprived of O_2?
11. True or false: Cellular respiration occurs simultaneously in cells that also photosynthesize. Explain your answer.

12. Label the following overview of photosynthesis. Check your answers against Figure 4-8.

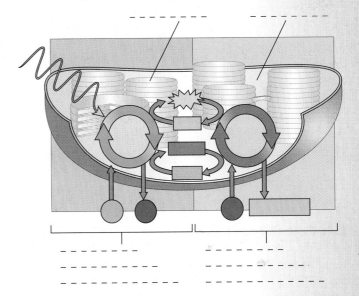

THOUGHT QUESTIONS

1. Would placing a plant under green light increase or decrease its rate of photosynthesis? (The plant was previously growing in sunlight.) Why?
2. You could compare the chemiosmotic model (generation of ATP using the energy of a proton gradient) to hydroelectric power (generation of electricity using the energy of water flowing over a dam). Explain the analogy.
3. Kentucky bluegrass thrives in eastern lawns during the spring months but is usually overrun by crabgrass during the hot summer months. Based on what you know about the C_3 and C_4 pathways, which plant would you suppose possesses the C_3 pathway? The C_4 pathway? Explain your reasoning.
4. The poison cyanide blocks the electron transport system. On the basis of what you have learned in this chapter, why is cyanide poisoning fatal?

5. Louis Pasteur, an important biologist in the 19th century, noted that yeast cells added to a closed container of grape juice broke down the sugar slowly as long as O_2 was present in the container. Once the O_2 was used up, however, the yeast cells consumed the grape sugar quickly, producing ethyl alcohol in the process. Explain this observation on the basis of what you know about the relative efficiencies of aerobic respiration and alcohol fermentation.

Visit us on the web at http://www.thomsonedu.com/biology/berg/ for additional resources, such as flashcards, tutorial quizzes, further readings, and web links.

Plant Tissues and the Multicellular Plant Body

Squirrels often temporarily store acorns in the ground to provide winter food. Many of these are never retrieved, so a new oak often begins its life after a squirrel has planted an acorn. The seed within the acorn first absorbs water from the surrounding soil. Then *germination* occurs as a root emerges and works its way into the soil to anchor the young plant. A miniature shoot grows upward and breaks through the soil. At this point the young oak is a seedling—it has just emerged from the seed and still depends on the food supply stored within the seed.

The stem continues to elongate, and small leaves develop, expand, and begin to photosynthesize. The young plant is now independently established—it is anchored in the ground, it absorbs water and dissolved minerals (inorganic nutrients) from the soil, and it uses energy from organic molecules that it has produced by photosynthesis.

Smaller roots branch off the original root, and the young tree grows taller, always by growth at the tips of its branches. As it ages, the tree also increases in girth (circumference) by forming additional wood and bark. Thus, the oak progressively accumulates more wood and bark, more stem and root tissues, and more leaves.

Growth and expansion of both the root and shoot systems continue throughout the oak's life. As in all plants, growth is flexible, enabling the oak to respond to its environment. For example, the young oak may grow slowly for several years if taller trees shade it. When conditions change—perhaps when an older tree nearby dies and falls to the ground, permitting direct sunlight to reach the oak—the small tree grows rapidly.

After several years of growth, the oak tree becomes reproductively mature. Oaks produce separate male and female flowers in the spring. The minute male flowers occur in slender, drooping, caterpillar-shaped clusters. In contrast, the tiny female flowers are often solitary. After the wind transports pollen from the male flower to the female flower, a sperm cell fertilizes the egg and an acorn develops. Within the acorn is a seed containing the embryo of a tiny miniature oak along with a supply of stored food. Thus, the cycle of life repeats itself, as squirrels collect these acorns and cache them for later retrieval.

PROCESS OF SCIENCE In this chapter we examine the external structure of the flowering plant body; the organization of its cells, tissues, and tissue systems; and its basic growth patterns. Information of this type, which is based on observation and description, rarely receives as much attention as experimental science. However, observation and description are crucial aspects of the scientific process. ▯

Massive oak (*Quercus*) trees grow from tiny acorns. Shown are two young oak seedlings.

Plant Structure and Life Span

Remarkable variety exists among the approximately 300,000 species of flowering plants that live in and are adapted to Earth's many environments.[1] Yet all of these—from desert cacti with enormously swollen stems to cattails partly submerged in marshes to orchids growing in the uppermost branches of lush tropical forest trees—are recognizable as plants. From the tiny floating water-meal, the smallest flowering plant known, to the giant sequoias, conifers that are Earth's largest plants, all plants have the same basic body plan.

The plant body is organized into a root system and a shoot system (•**Figure 5-1**). The **root system** is generally underground. The aerial portion, the **shoot system,** consists of a vertical stem bearing leaves and, in flowering plants, flowers and fruits that contain seeds.

Each plant grows in two different environments: the dark, moist soil and the illuminated, relatively dry air. A plant has both roots and shoots because it needs resources from both environments. Thus, roots branch extensively through the soil, forming a network that anchors the plant firmly in place and absorbs water and dissolved minerals from the soil. Leaves, the flattened organs for photosynthesis, are attached regularly on the stem, where they absorb the sunlight and atmospheric CO_2 used in photosynthesis to produce carbohydrates.

Plants are either herbaceous or woody. **Herbaceous plants** do not develop persistent woody parts aboveground, whereas **woody plants** (trees and shrubs) do. In temperate climates, the aerial parts (stems and leaves) of herbaceous plants die back to the ground at the end of the growing season. In contrast, the aerial stems of woody plants persist.

Annuals are herbaceous plants (such as corn, geranium, and marigold) that grow, reproduce, and die in 1 year or season. Other herbaceous plants (such as carrot, cabbage, and Queen Anne's lace) are **biennials** and take 2 years to complete their life cycles (see Figure 11-19). During their first season, biennials produce extra carbohydrates, which they store and use during their second year, when they typically form flowers and reproduce.

Perennials are woody or herbaceous plants that live for more than 2 years. In temperate climates, the

FIGURE IN FOCUS

During the evolution of plants, the root and shoot systems specialized to obtain resources from the soil and air, respectively.

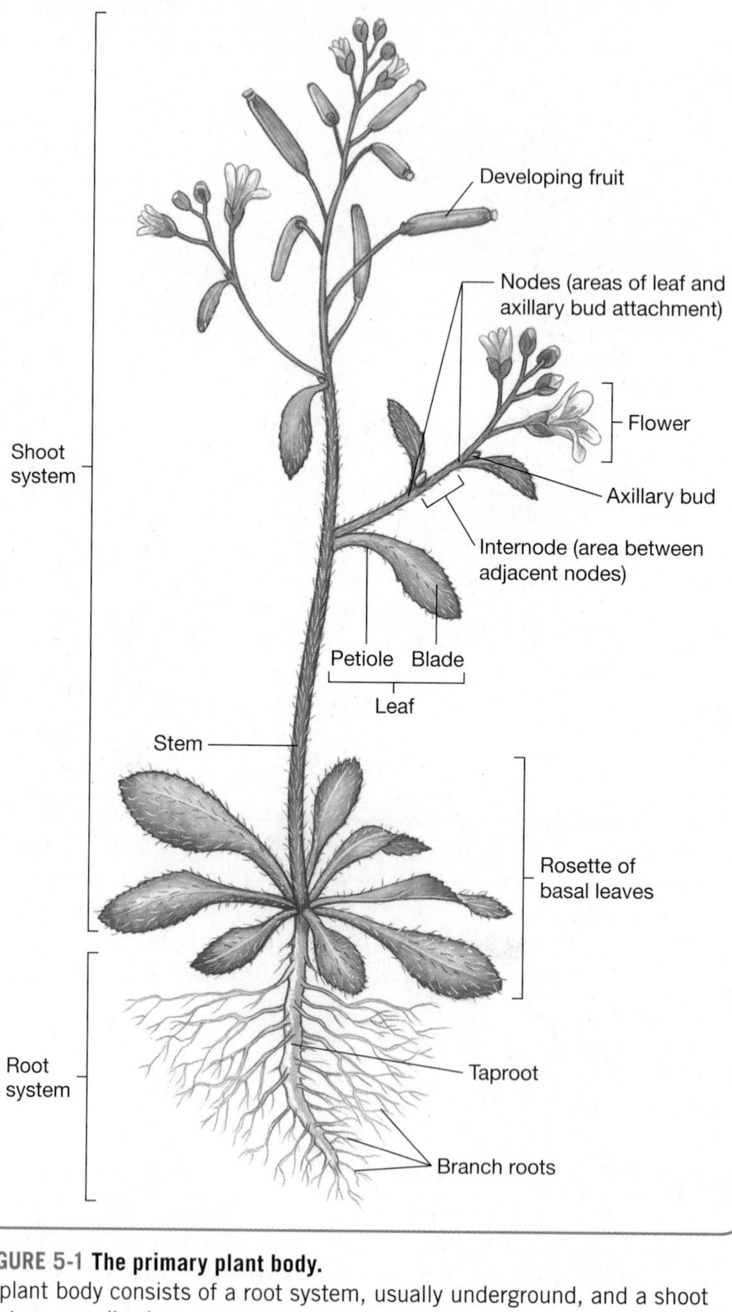

FIGURE 5-1 The primary plant body.
A plant body consists of a root system, usually underground, and a shoot system, usually aboveground. Shown is mouse-ear cress (*Arabidopsis thaliana*), a small plant in the mustard family that is a model plant for biological research. *Arabidopsis* is native to North Africa and Eurasia and has been naturalized (was introduced and is now growing in the wild) in California and the eastern half of the United States.

[1]*About 90 percent of plants are flowering plants, which are characterized by the presence of flowers and of seeds enclosed within fruits. Because flowering plants are the largest group of plants, they are the focus of much of this and subsequent chapters.*

PLANTS AND THE ENVIRONMENT | Plant Life History Strategies

Plants exhibit a variety of patterns of growth, reproduction, and longevity. Woody perennials often live for hundreds of years, whereas some herbaceous annuals may live for only a few weeks or months. Under what conditions is it more favorable for a species to be long-lived or short-lived? Botanists have considered the relative advantages of each **life history strategy,** including any trade-offs involved in allocating various resources. It appears that in some environments a longer life span is advantageous, whereas in others a shorter life span increases a species' chances for reproductive success—that is, for long-term survival of the species.

When an environment is relatively favorable, it is filled with plants that compete for the available space and resources. Because such an environment is so crowded, it has few open

spots where new plants can become established. When a plant dies, the empty area it leaves behind is quickly filled by another plant, but not necessarily by the same species as before. Thus, an adult perennial survives well, but young plants, whether perennials or annuals, do not. A plant with a long life span thrives in this type of environment because it can occupy an area of soil and continue to produce seeds for many years. In a tropical rain forest, for example, such competition prevents most young plants from becoming established; hence, woody perennials predominate.

In a relatively unfavorable environment or one that is frequently disturbed, many growing sites are usually available. This type of environment is not crowded, and young plants usually do not have to compete against large, fully established plants. Here,

smaller, short-lived plants have the reproductive advantage. These plants are opportunists—they maximize their rate of photosynthesis to grow and mature quickly during the brief periods when environmental conditions are most favorable. As a result, their resources are primarily directed into producing as many seeds as possible before dying. Following a rainy spell in a desert, for example, annuals are more prevalent than woody perennials.

Thus, each species has its own characteristic life history strategy, with some plants adapted to variable environments (in which they grow fast and die early) and others adapted to stable environments (in which they grow slowly and survive a long time). The longer life span conferred on woody perennials is just one of several successful life history plans.

---

aerial shoots of herbaceous perennials (for example, iris, rhubarb, onion, and asparagus) die back each winter. Their underground parts (roots and underground stems) become dormant during the winter and send out new growth each spring. (In *dormancy,* an organism reduces its metabolic state to a minimum level to survive unfavorable conditions.) Likewise, in certain tropical climates with pronounced wet and dry seasons, the aerial parts of herbaceous perennials (for example, many grasses) die back and the underground parts become dormant during the dry season. Other tropical plants, such as orchids, are herbaceous perennials that grow year-round.

All woody plants are perennials, and some of them live for hundreds or even thousands of years (see *Plants and the Environment: Plant Life History Strategies*). In temperate climates, the aerial stems of woody plants become dormant during the winter. Many temperate woody perennials shed their leaves before winter and produce new stem tissue with new leaves the following spring. Other woody perennials are evergreen and shed their leaves over a long period so that some leaves are always present. Because they have permanent woody stems that are the starting points for new growth the following season, many trees attain massive sizes.

Cells and Tissues of the Plant Body

As in other organisms, the cell is the basic structural and functional unit of plants. During the course of evolution, plants evolved a diversity of cell types, each specialized for particular functions.

Like animal cells, plant cells are organized into tissues. A **tissue** is a group of cells that forms a structural and functional unit. Some plant tissues (*simple tissues*) are composed of only one kind of cell, whereas other plant tissues (*complex tissues*) have two or more kinds of cells. In vascular plants, tissues are organized into three tissue systems, each of which extends throughout the plant body (•**Figure 5-2**). Each tissue system contains two or more kinds of tissues (•**Table 5-1**).

Most of the plant body is composed of the **ground tissue system,** which has a variety of functions, including photosynthesis, storage, and support. The **vascular tissue system,** an intricate plumbing system that ex-

GROUND TISSUE SYSTEM All of the tissues of the plant body other than the vascular tissues and the dermal tissues.

VASCULAR TISSUE SYSTEM The tissue system that conducts materials throughout the plant body.

The tissue systems are continuous throughout the plant. For example, the vascular tissue system in a leaf is continuous with the vascular tissue system in the stem to which it is attached.

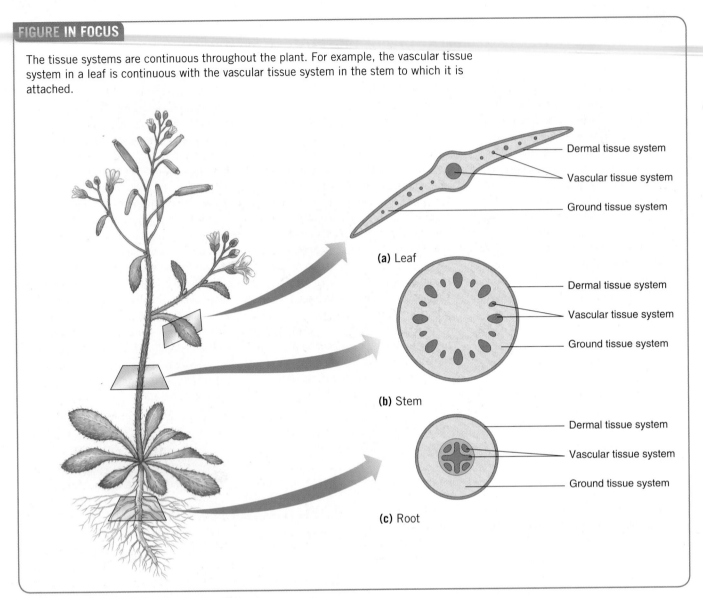

(a) Leaf

- Dermal tissue system
- Vascular tissue system
- Ground tissue system

(b) Stem

- Dermal tissue system
- Vascular tissue system
- Ground tissue system

(c) Root

- Dermal tissue system
- Vascular tissue system
- Ground tissue system

FIGURE 5-2 The three tissue systems in the plant body.
This illustration shows the distribution of the ground tissue system, vascular tissue system, and dermal tissue system in the **(a)** leaves, **(b)** stems, and **(c)** roots of an herbaceous plant such as *Arabidopsis*.

tends throughout the plant body, conducts various substances, including water, dissolved minerals, and food (dissolved sugar). The vascular tissue system also functions in strengthening and supporting the plant. The **dermal tissue system** covers the plant body.

Roots, stems, leaves, flower parts, and fruits are referred to as **organs** because each is composed of all three tissue systems. The tissue systems of different plant organs form an interconnected network throughout the plant. For example, the vascular tissue of a leaf is continuous with the vascular tissue of the stem to which the leaf is attached, and the vascular tissue of the stem is continuous with the vascular tissue of the root.

The ground tissue system is composed of three simple tissues

The bulk of an herbaceous plant is its ground tissue system, which is composed of three tissues: *parenchyma*, *collenchyma*, and *sclerenchyma*. Their cell wall structures distinguish these tissues. Recall that a cell wall that provides structural support surrounds each plant cell (see Chapter 3). A growing cell secretes a thin **primary cell wall**, which stretches and expands as the cell increases in size. After the cell stops growing, it sometimes secretes a thick, strong **secondary cell wall**, which is deposited *inside* the primary cell wall—that is, between the primary cell wall and the plasma membrane.

TABLE 5-1 Tissue Systems, Tissues, and Cell Types of Flowering Plants

TISSUE SYSTEM	TISSUES	CELL TYPES
Ground tissue system	Parenchyma tissue	Parenchyma cells
	Collenchyma tissue	Collenchyma cells
	Sclerenchyma tissue	Sclerenchyma cells (sclereids, fibers)
Vascular tissue system	Xylem	Tracheids, vessel elements, parenchyma cells, fibers
	Phloem	Sieve-tube elements, companion cells, parenchyma cells, fibers
Dermal tissue system	Epidermis	Parenchyma cells, guard cells, trichomes
	Periderm	Cork cells, cork cambium cells, cork parenchyma

Parenchyma cells have thin primary cell walls

Parenchyma tissue, a simple tissue composed of **parenchyma cells,** is found throughout the plant body and is the most common type of cell and tissue (•**Figure 5-3**). The soft parts of a plant, such as the edible part of an apple or a potato, consist largely of parenchyma.

Parenchyma cells perform a number of important functions for plants, such as photosynthesis, storage, and secretion. Parenchyma cells that function in photosynthesis contain chloroplasts, whereas nonphotosynthetic parenchyma cells lack chloroplasts. Materials stored in parenchyma cells include starch grains, oil droplets, water, and salts (sometimes visible as crystals). Resins, tannins, hormones, enzymes, and sugary nectar are examples of substances that parenchyma cells may secrete. The various functions of parenchyma require that they be living and metabolizing cells.

Parenchyma cells have the ability to **differentiate** into other kinds of cells, particularly when a plant is injured. If xylem (water-conducting cells) is severed, for example, adjacent parenchyma calls may divide and differentiate into new xylem cells within a few days.

Collenchyma cells have unevenly thickened primary cell walls

Collenchyma tissue, a simple tissue composed of **collenchyma cells,** is a flexible tissue that provides much of the support in soft, nonwoody plant organs (•**Figure 5-4**). Support is a crucial function in plants, in part because it allows plants to grow upward, thus enabling them to compete with other plants for available sunlight in a plant-crowded area. Instead of the bony skeletal system that is typical of many animals, individual cells, including collenchyma cells, provide support for the plant body.

Collenchyma cells are usually elongated. Their primary cell walls are unevenly thickened and are especially thick in the corners. Collenchyma is not found uniformly throughout the plant and often occurs as long strands near stem surfaces and along leaf veins. The "strings" in a celery stalk (petiole), for example, consist of collenchyma.

Sclerenchyma cells have both primary cell walls and thick secondary cell walls

A second simple tissue specialized for structural support is **sclerenchyma tissue,** whose cells have both primary and secondary cell walls. The root of the word *sclerenchyma* is the Greek word root *sclero,* "hard." The secondary cell walls of **sclerenchyma cells** become strong and hard due to extreme thickening. At functional maturity, when sclerenchyma tissue is providing support for the plant body, sclerenchyma cells are often dead.

Sclerenchyma tissue may occur in several areas of the plant body. Two types of sclerenchyma cells are sclereids and fibers. **Sclereids,** short cells that are variable in shape, are common in the shells of nuts and the stones of fruits, such as cherries and peaches (•**Figure 5-5a**). Pears owe their slightly gritty texture to the presence of clusters of sclereids. **Fibers,** which are long, tapered cells that often occur in groups or clumps, are particularly abundant in the wood, inner bark, and leaf ribs (veins) of flowering plants (•**Figure 5-5b**). (Also see *Plants and People: Fibers and Textiles* for a discussion of flax fibers.)

The vascular tissue system consists of two complex tissues

The vascular tissue system, which is embedded in the ground tissue, transports needed materials throughout the plant via two complex tissues: xylem (pronounced "ZYE-lem") and phloem (pronounced "FLOW-em"). (Chapter 10 discusses the mechanisms of transport in xylem and phloem.)

DERMAL TISSUE SYSTEM The tissue system that provides an outer covering for the plant body.

PARENCHYMA CELL A plant cell that is relatively unspecialized, is thin walled, may contain chlorophyll, and is typically rather loosely packed.

COLLENCHYMA CELL A living plant cell with moderately but unevenly thickened primary walls.

SCLERENCHYMA CELL A plant cell with extremely thick walls that provides strength and support to the plant body.

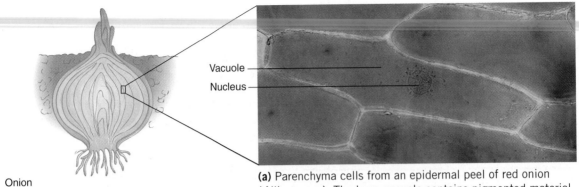

Onion

Vacuole
Nucleus

Ed Reschke

(a) Parenchyma cells from an epidermal peel of red onion (*Allium cepa*). The large vacuole contains pigmented material and occupies most of the cell. The nucleus and cytoplasmic strands are positioned under and on top of the vacuole, between it and the plasma membrane.

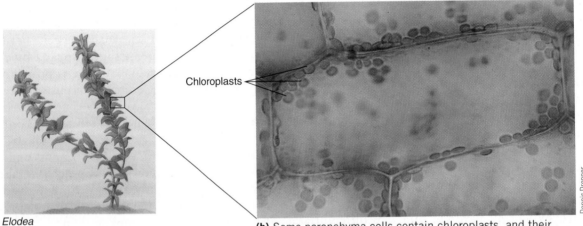

Elodea

Chloroplasts

Dennis Drenner

(b) Some parenchyma cells contain chloroplasts, and their primary function is photosynthesis. These parenchyma cells are from a waterweed (*Elodea*) leaf.

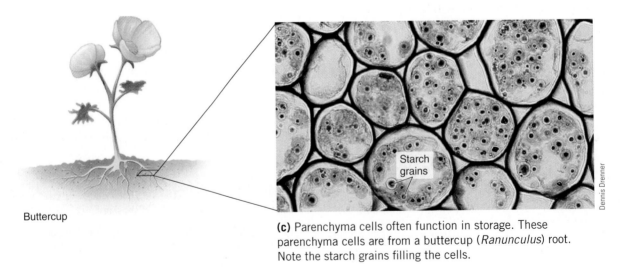

Buttercup

Starch grains

Dennis Drenner

(c) Parenchyma cells often function in storage. These parenchyma cells are from a buttercup (*Ranunculus*) root. Note the starch grains filling the cells.

FIGURE 5-3 Parenchyma cells.

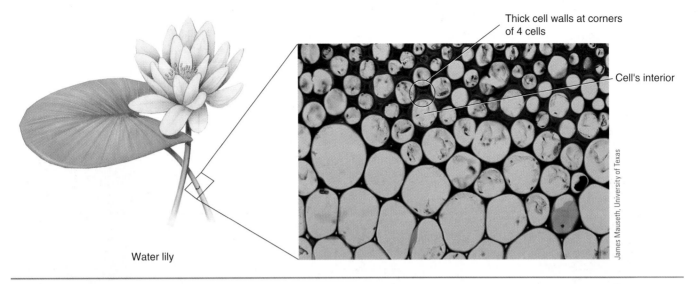

Thick cell walls at corners of 4 cells

Cell's interior

Water lily

James Mauseth, University of Texas

FIGURE 5-4 Collenchyma cells.
Note the unevenly thickened cell walls that are especially thick in the corners, making the cell contents appear spherical in cross section. These collenchyma cells are from a water lily (*Nymphaea*) petiole.

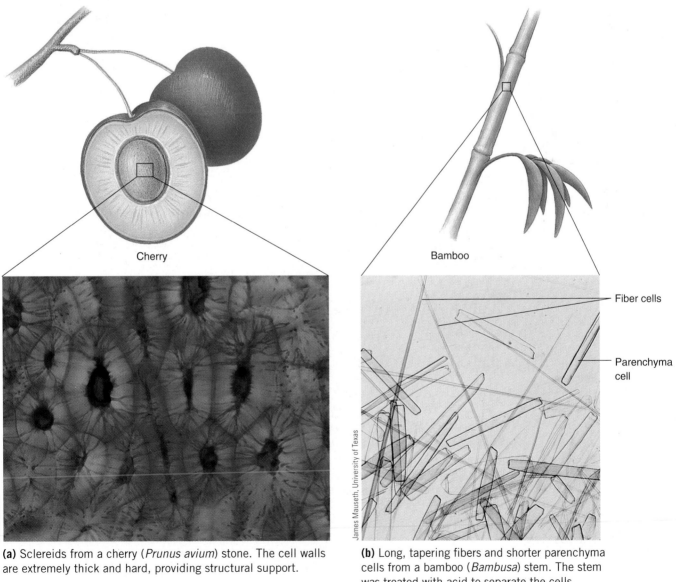

Cherry

Bamboo

Fiber cells

Parenchyma cell

Dennis Drenner

James Mauseth, University of Texas

(a) Sclereids from a cherry (*Prunus avium*) stone. The cell walls are extremely thick and hard, providing structural support.

(b) Long, tapering fibers and shorter parenchyma cells from a bamboo (*Bambusa*) stem. The stem was treated with acid to separate the cells.

FIGURE 5-5 Sclerenchyma cells.

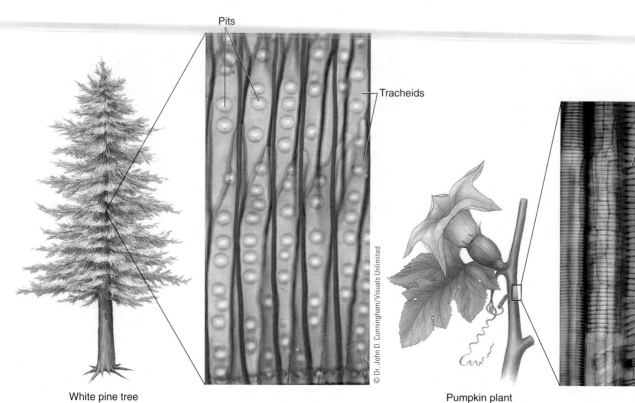

Pits

Tracheids

© Dr. John D. Cunningham/Visuals Unlimited

White pine tree

(a) Tracheids from a white pine (*Pinus strobus*) stem in longitudinal section (that is, cut lengthwise). These cells, which occur in clumps, transport water and dissolved minerals. Water passes readily from tracheid to tracheid through pits, thin places in the cell wall.

Vessel elements

Perennou Nuridsany/Photo Researchers, Inc.

Pumpkin plant

(b) Vessel elements from a pumpkin (*Cucurbita mixta*) stem in longitudinal section. The blue-stained regions are various patterns of the secondary walls in the vessel elements. Perforation plates are not visible in this micrograph.

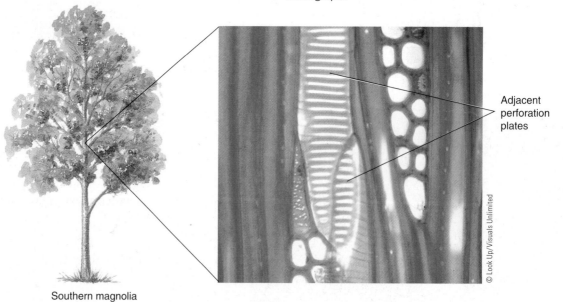

Adjacent perforation plates

© Look Up/Visuals Unlimited

Southern magnolia

(c) The end walls of vessel elements, called perforation plates, have large holes. Water passes through the perforation plate from one vessel element to the next. Shown are adjacent perforation plates from a southern magnolia (*Magnolia grandiflora*) stem; in this species, the perforation plates are at an angle in longitudinal section.

FIGURE 5-6 Conducting cells in xylem tissue.

The conducting cells in xylem are tracheids and vessel elements

Xylem conducts water and dissolved minerals from the roots to the stems and leaves and provides structural support. In flowering plants, xylem is a complex tissue composed of four different cell types: tracheids, vessel elements, parenchyma cells, and fibers. Two of the four cell types found in xylem—the tracheids and vessel elements— actually conduct water and dissolved minerals (•**Figure 5-6**). In addition to these cells, xylem contains parenchyma cells, known as *xylem parenchyma*, that perform storage functions, and xylem fibers that provide support.

Tracheids and vessel elements are highly specialized for conduction of water and minerals. At maturity, both cell types are dead and therefore hollow; only their cell walls remain. **Tracheids,** the chief water-conducting cells in gymnosperms and seedless vascular plants such as ferns, are long, tapering cells located in patches or clumps.[2] Water is conducted upward from roots to shoots, passing from one tracheid into another through **pits,** which are thin areas in the tracheids' cell walls where a secondary cell wall did not form. Pits always occur in pairs, one on each side of the primary cell walls of adjacent cells (•**Figure 5-7**).

Flowering plants possess efficient water-conducting cells called **vessel elements,** in addition to relatively few tracheids. The cell diameters of vessel elements are usually greater than those of tracheids. Vessel elements are hollow, but unlike tracheids, they have holes in their end walls known as **perforations,** or the end walls are entirely dissolved away. Vessel elements are stacked one on top of the other, and water is conducted readily from one vessel element into the next. A stack of vessel elements, called a **vessel,** resembles a miniature water pipe. Vessel elements also have pits in their side walls that permit the lateral transport (sideways movement) of water from one vessel to another.

Sieve-tube elements are the conducting cells of phloem

Phloem conducts food materials—that is, carbohydrates formed in photosynthesis—throughout the plant and provides structural support. In flowering plants, phloem is a complex tissue composed of four different cell types: sieve-tube elements, companion cells, phloem fibers, and phloem parenchyma cells (•**Figure 5-8**). Fibers are frequently extensive in the phloem of flowering plants, providing additional structural support for the plant body.

Food materials are conducted *in solution*—that is, dissolved in water—through the **sieve-tube elements,** which are among the most specialized living cells in nature. Sieve-tube elements are long, thin cells that are stacked end on end to form long **sieve tubes.** The cell's end walls, called **sieve plates,** have a series of holes through which cytoplasm extends from one sieve-tube element into the next. Sieve-tube elements are alive at maturity, but many of their organelles, including the nucleus, vacuole, mitochondria, and ribosomes, disintegrate as they mature.

Sieve-tube elements are among the few eukaryotic cells that function without nuclei, although they typically live for less than a year. There are, however, notable exceptions: certain palms have sieve-tube elements that remain alive approximately 100 years!

Adjacent to each sieve-tube element is a **companion cell** that assists in the functioning of the sieve-tube element. The companion cell is a living cell, complete with a nucleus. This nucleus is thought to direct the activities of both the companion cell and the sieve-tube element. Numerous **plasmodesmata** (sing., *plasmodesma*) occur between a companion cell and its sieve-tube element. Recall that plasmodesmata are cytoplasmic channels through the cell walls of adjacent plant cells (see Figure 3-12). Although the companion cell does not conduct nutrients itself, it plays an essential role in loading food materials into the sieve-tube elements for transport to other parts of the plant.

The dermal tissue system consists of two complex tissues

The dermal tissue system—epidermis and periderm—provides a protective covering over plant parts. In herbaceous plants, the dermal tissue system is a layer of cells called the *epidermis.* Woody plants initially produce an epidermis, but it splits apart as the plant increases in girth as a result of the production of additional woody tissues underneath the epidermis. *Periderm,* a tissue several to many cell layers thick, provides a new protective covering as the epidermis is destroyed. Periderm, which replaces the epidermis in the stems and roots of older woody plants, composes the outer bark.

Epidermis is the outermost layer of an herbaceous plant

The **epidermis** is a complex tissue composed primarily of relatively unspecialized living cells. Dispersed

---

[2]*Gymnosperms are seed-bearing plants that do not produce flowers. They include cone-bearing trees such as pines, hemlocks, and firs as well as ginkgoes and cycads.*

XYLEM A complex vascular tissue that conducts water and dissolved minerals throughout the plant body.

PHLOEM A complex vascular tissue that conducts food (carbohydrate) throughout the plant body.

EPIDERMIS The outermost tissue layer, usually one cell thick, that covers the primary plant body—that is, leaves and young stems and roots.

PLANTS AND PEOPLE | Fibers and Textiles

Many textile fabrics are made from plant fibers found in the stems, leaves, seeds, or fruits of flowering plants. Plant fibers are often divided into three groups. Many textile fibers are **bast fibers,** also called **soft fibers.** Bast fibers, which include flax (linen), hemp, jute, and ramie, are generally taken from the stems of eudicot plants (see footnote 3 on page 105 for examples of eudicots and monocots). These fibers typically come from stem tissues located outside the xylem—including phloem fibers, cortex fibers (discussed in Chapter 7), and periderm fibers. Botanically, bast fiber cells are true fibers. In many cases, they occur as clusters, and the entire cluster is often spun into thread for fabric.

Leaf fibers, also called **hard fibers,** are a second type of plant fiber. Botanically, leaf fibers consist of xylem cells, phloem cells, and fibers that are clustered together in a leaf (or sometimes a stem) of monocot plants. Leaf fibers, which are less durable than bast fibers, are not as important as bast fibers in making fabrics. However, leaf fibers such as Manila hemp and sisal are important cordage fibers (for rope making).

A third type of textile fiber is **surface fibers,** which occur on the surface of seeds and fruits. Cotton, the most important surface fiber, is associated with seeds. Botanically, cotton fibers are not true fibers but are *trichomes,* or hairs, that are attached to the seeds. One pound of cotton contains about 90 million seed hairs.

The two most important textile fibers are cotton (*Gossypium*) and flax (*Linum usitatissimum*). Humans have cultivated cotton in warmer regions of the world for thousands of years (see figure a; also see Figure 15-1). Cotton did not become important in Europe until the Moors invaded Spain about 1000 years ago and introduced cotton cloth to Europe. By the 14th century, cotton grown in India was being traded in Europe, although weaving the fabric remained

(a) Several species of cotton (*Gossypium*) are grown around the world. Some cotton species are annuals, whereas others are perennials. The cotton plant is a shrub reaching a height of 0.6 to 3 meters (2 to 10 feet). It produces white, yellow, or red flowers, followed by seed pods, or bolls, which split open to reveal the cotton fibers and seeds.

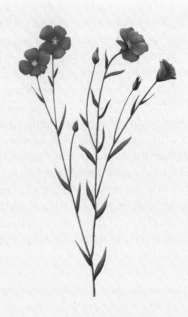

(b) Flax (*Linum usitatissimum*) is a slender annual plant that grows to 1.2 meters (4 feet). Its leaves are small and narrow, and its flowers are a striking shade of blue.

■ **Economically important fiber plants.**

a small-scale cottage industry for the next two centuries. The invention of spinning and looming machines in the early- to mid-19th century increased the European demand for cotton cloth. The southern United States, which had an economic system based on slave labor, supplied much of Europe's cotton. The American Civil War cut off Europe's supply of inexpensive cotton, and Great Britain began growing cotton in its various colonies around the world. Today, the United States still grows cotton, but mostly on irrigated land west of the Mississippi River.

Fibers from the slender stems of flax are woven to produce linen cloth, which is the highest quality (in terms of strength, durability, and beauty) of any cloth from bast fibers (see figure b). Like cotton, flax has been grown since prehistoric times; ancient Hebrews and Egyptians wore linen clothes at least 4000 years ago. Until the Industrial Revolution, linen was more common than cotton in both Europe and America, although the American colonists preferred a sturdy cloth called "linsey-woolsey," which was made by weaving linen and wool together. Cotton did not overtake linen as the fabric of choice until the 19th century. Today, most flax is grown in cool climates, such as Russia and other European countries. (The flax plant originated in Europe or western Asia.) Ireland is particularly known for its linen products, from bed linens to hand towels and handkerchiefs.

among these cells are more specialized guard cells and outgrowths called trichomes (discussed shortly). In most plants, the epidermis consists of a single layer of cells. Epidermal cell walls are somewhat thicker toward the outside of the plant to provide protection. Epidermal cells generally contain no chloroplasts and are therefore transparent, so light can penetrate into the interior tissues of stems and leaves. In both stems and leaves, photosynthetic tissues lie *beneath* the epidermis.

An important requirement of aerial shoots (stems, leaves, flowers, and fruits) is the ability to prevent water loss. Epidermal cells of stems and leaves secrete a waxy layer called a **cuticle** over the surfaces of their walls; this wax greatly restricts the loss of water from plant surfaces.

Although the cuticle prevents most water loss through epidermal cells, it also prevents the carbon dioxide required for photosynthesis from diffusing from the atmosphere into the leaf or stem. **Stomata** (sing., *stoma*) facilitate the diffusion of carbon dioxide. Stomata are tiny pores in the epidermis between two cells called **guard cells** (•**Figure 5-9**). Many gases, including carbon dioxide, oxygen, and water vapor, pass through stomata by *diffusion* (see Chapter 3). Stomata are generally open during the day, when photosynthesis is occurring. The loss of water that also takes place when stomata are open provides some evaporative cooling to the leaf. During the night, stomata usually close. During drought conditions, the requirement to conserve water overrides the need to cool the leaves and exchange gases. Thus, during a drought, the stomata close in the daytime. Stomata are discussed in greater detail in Chapter 8.

The epidermis may also contain special outgrowths, or hairs, called **trichomes,** which occur in many sizes and shapes and have a variety of functions (see *Plants and People: Fibers and Textiles* for a discussion of cotton trichomes). Plants that tolerate salty environments such as the seashore often have specialized trichomes on their leaves to remove excess salt that has accumulated in the plant. The presence of trichomes on the aerial parts of desert plants may increase the reflection of light off the plants, thereby keeping the internal tissues cooler and decreasing water loss. Other trichomes have a protective function. For example, the trichomes on stinging nettle leaves and stems contain irritating substances that discourage herbivorous animals from eating the plant. **Root hairs** are simple, unbranched trichomes that

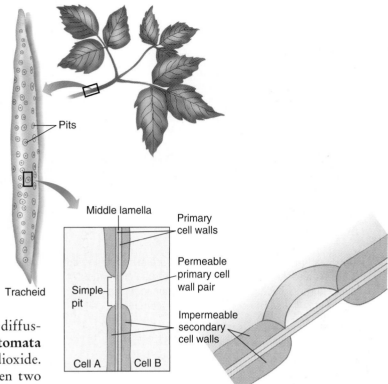

Pits

Tracheid

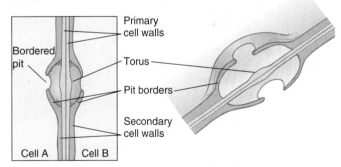

Middle lamella

Primary cell walls

Permeable primary cell wall pair

Impermeable secondary cell walls

Simple pit

Cell A   Cell B

(a) A simple pit pair has an interruption in the secondary cell wall. The primary cell wall in a simple pit pair is permeable to water.

Primary cell walls

Bordered pit

Torus

Pit borders

Secondary cell walls

Cell A   Cell B

(b) A bordered pit pair has a smaller opening into a section of the secondary cell wall that bulges out on both sides. The bordered pit pair functions like a valve.

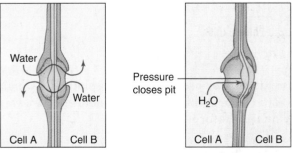

Water

Water

Pressure closes pit

H_2O

Cell A   Cell B   Cell A   Cell B

(c) (*Left*) When water pressure is equal between the two cells (A and B), the bordered pit is open, and water flow is unrestricted. (*Right*) When the pressure is greater in cell A than in cell B, the torus, a thickening in the primary cell walls, blocks the opening, restricting water movement through the pit pair.

FIGURE 5-7 Pit pairs.

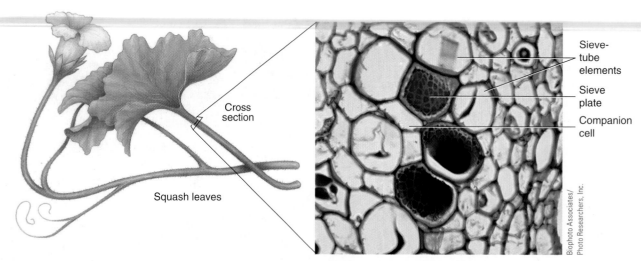

(a) Phloem tissue from a squash (*Cucurbita*) petiole in cross section. Note the sieve plates, the end walls of the sieve-tube elements. Most sieve-tube elements appear empty because they were sectioned in the middle of the cells rather than at the end walls. The smaller cells are companion cells.

(b) Phloem tissue from a squash (*Cucurbita*) petiole in longitudinal section.

FIGURE 5-8 Phloem tissue.

increase the surface area of the root epidermis (which comes into contact with the soil) for more effective water and mineral absorption.

PERIDERM The outermost layer of cells covering a woody stem or root—that is, the outer bark that replaces epidermis when it is destroyed during secondary growth.

Periderm replaces epidermis in woody plants

As a woody plant begins to increase in girth, its epidermis sloughs off and is replaced by **periderm.** Periderm forms the protective outer bark of older stems and roots (•**Figure 5-10**). It is a complex tissue composed mainly of cork cells and cork parenchyma cells. **Cork cells** are dead at maturity, and their walls are heavily coated with a waterproof substance called *suberin*, which helps reduce water loss. **Cork parenchyma cells** (also called

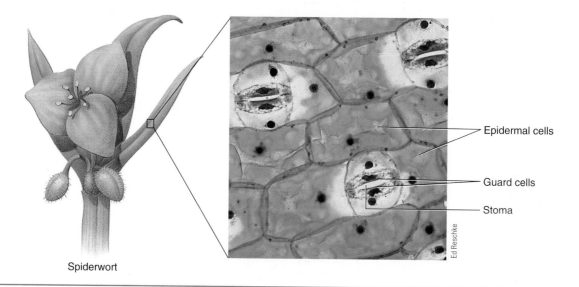

Spiderwort

Epidermal cells

Guard cells

Stoma

Ed Reschke

FIGURE 5-9 Epidermis.
This light micrograph shows the leaf epidermis of spiderwort (*Tradescantia virginiana*). Note the stomata, minute pores formed by guard cells (*stained pink*). Nuclei of all epidermal cells are stained a dark magenta.

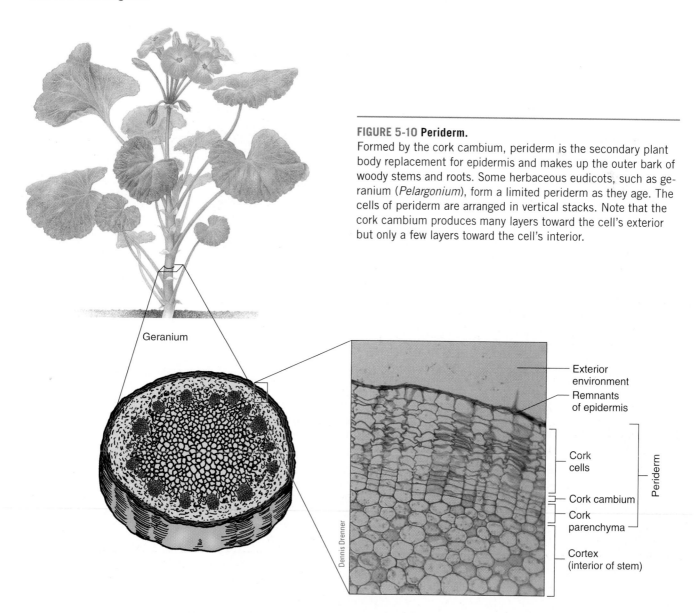

Geranium

FIGURE 5-10 Periderm.
Formed by the cork cambium, periderm is the secondary plant body replacement for epidermis and makes up the outer bark of woody stems and roots. Some herbaceous eudicots, such as geranium (*Pelargonium*), form a limited periderm as they age. The cells of periderm are arranged in vertical stacks. Note that the cork cambium produces many layers toward the cell's exterior but only a few layers toward the cell's interior.

Exterior environment

Remnants of epidermis

Cork cells

Cork cambium

Cork parenchyma

Periderm

Cortex (interior of stem)

Dennis Drenner

TABLE 5-2 A Summary of Selected Cell Types

CELL TYPE	DESCRIPTION	FUNCTION	LOCATION
Parenchyma cell	Living; thin primary walls	Secretion, storage, photosynthesis	Throughout the plant body
Collenchyma cell	Living; unevenly thickened primary walls	Support	Just under stem epidermis; along leaf veins
Sclerenchyma cell (fiber)	Often dead at maturity; thick secondary walls	Support	Throughout the plant body; common in stems and certain leaves
 Tracheid	Dead at maturity; lacks secondary wall at pits	Conduction of water and minerals; support	Xylem
 Vessel element	Dead at maturity; end walls have perforations; lacks secondary wall at pits	Conduction of water and minerals; support	Xylem
Sieve-tube element	Living; end walls are sieve plates; lacks nuclei and other organelles at maturity	Conduction of dissolved food materials (carbohydrates)	Phloem
Companion cell (not shown)	Living; cytoplasmic connections with sieve-tube element	Aids sieve-tube element	Phloem
Epidermal cells	Living; cells fit tightly together	Protective covering over surface	Outermost layer of cells

phelloderm) function primarily in storage. •**Table 5-2** summarizes selected cell types found in the ground, vascular, and dermal tissue systems.

Plant Meristems

Plant growth involves three different processes: cell division, cell elongation (the lengthening of a cell), and cell differentiation. Cell division is an essential part of growth that results in an increase in the number of cells. These new cells elongate as the cytoplasm grows and the vacuole fills with water, which exerts pressure on the cell wall and causes it to expand. For example, in an onion root cell, the vacuole increases in size 30 to 150 times during elongation. Plant cells also **differentiate,** or specialize, into the various cell types just discussed. These cell types compose the mature plant body and perform the various functions required in a multicellular organism. Although differentiation does not contribute to an increase in size, it is considered an important aspect of growth because it is essential for tissue formation.

One difference between plants and animals is the *location* of growth. When a young animal is growing, all parts of its body grow, although not necessarily at the same rate. In contrast, when plants grow, their cells divide only in specific areas, called **meristems,** which are composed of cells whose primary function is the formation of new cells. Meristematic cells do not differentiate. Instead, they retain the ability to divide by mitosis (see Chapter 12), a trait that many differentiated cells lose. The persistence of meristems means that plants, unlike most animals, retain the capability for growth throughout their entire life span.

Two kinds of meristematic growth may occur in plants. **Primary growth** is an increase in the length of a plant. All plants have primary growth, which produces the entire plant body in herbaceous plants and the young, soft shoot tips and root tips of woody trees and shrubs. **Secondary growth** is an increase in the girth of a plant. For the most part, only gymnosperms and woody eudicots have extensive secondary growth.[3] Tissues produced by secondary growth compose the wood and bark, which make up most of the bulk of trees and shrubs. A few annuals (for example, sunflower and geranium) have limited secondary growth even though they lack obvious wood and bark tissues.

---

[3]*Flowering plants are divided into two main groups, informally called eudicots and monocots. Oak, sycamore, ash, cherry, apple, and maple are examples of woody eudicots, whereas bean, daisy, and snapdragon are examples of herbaceous eudicots. Palm, corn, bluegrass, lily, and tulip are examples of monocots.*

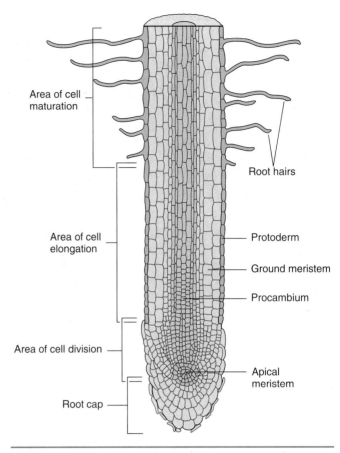

FIGURE 5-11 The root tip.
A root cap protects the root apical meristem, where cells divide and thus increase in number. Farther from the tip is an area of cell elongation, where cells enlarge and begin to differentiate. The area of cell maturation has fully mature, differentiated cells. Note the root hairs in this area.

Primary growth takes place at apical meristems

Primary growth occurs as a result of the activity of **apical meristems,** areas at the tips of roots and shoots. Primary growth is evident when a root tip is examined (•**Figure 5-11**). A protective layer of cells called the **root cap** covers the root tip. Directly behind the root cap, in the *area of cell division,* is the *root apical meristem,*

PRIMARY GROWTH An increase in a plant's length, which occurs at the tips of stems and roots due to the activity of apical meristems.

SECONDARY GROWTH An increase in a plant's girth due to the activity of lateral meristems (the vascular cambium and cork cambium).

APICAL MERISTEM An area of cell division at the tip of a stem or root in a plant; produces primary tissues.

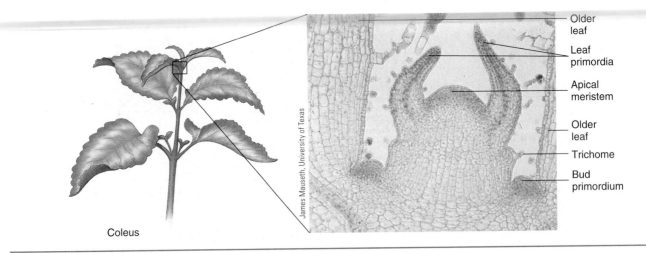

Coleus

FIGURE 5-12 The stem tip.
This longitudinal section through a terminal bud of coleus (*Coleus*) shows the stem apical meristem, leaf primordia, and bud primordia.

which consists of meristematic cells. Meristematic cells, which are small and "boxy" in shape, remain small because they are continually dividing.

Farther back from the tip of the root, just behind the area of cell division, is an *area of cell elongation* where the cells have been displaced from the meristem. Here the cells are no longer dividing but instead growing longer, pushing the root tip ahead of them, deeper into the soil. Some differentiation begins in the area of cell elongation, and immature tissues, such as differentiating xylem and phloem, become evident.

Three *primary meristems* (immature tissues) are found in the area of cell elongation: protoderm, procambium, and ground meristem. The **protoderm** is young, undifferentiated tissue of a root or stem that eventually develops into epidermis. **Procambium** is meristematic tissue that eventually develops into xylem and phloem. **Ground meristem** is meristematic tissue that gives rise to cortex, pith, and ground tissue.

The primary meristems continue to develop and differentiate into primary tissues (epidermis, xylem, phloem, and ground tissues) of the adult plant. Farther back from the tip, in the *area of cell maturation*, the cells have completely differentiated and are fully mature. Root hairs are evident in this area.

A shoot apex—a terminal **bud**, for example—is quite different in appearance from a root tip (•**Figure 5-12**). Within every bud is a dome of tiny, regularly arranged meristematic cells, the *shoot apical meristem*. **Leaf primordia** (developing leaves) and **bud primordia** (develop-

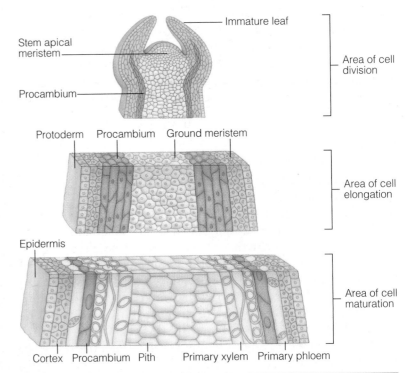

FIGURE 5-13 Development in the stem tip.
This diagram shows the kinds of cells in the area of cell division, the area of cell elongation, and the area of cell maturation. The three primary meristems (protoderm, procambium, and ground meristem) in the area of cell elongation give rise to the mature cell types in the area of cell maturation. (Procambium cells shown in the area of cell maturation—between the primary xylem and primary phloem—may give rise to vascular cambium later in development.)

ing buds) emerge from the shoot apical meristem. The leaf primordia cover and protect the shoot apical meristem. As the cells formed by the shoot apical meristem elongate, the shoot apical meristem is pushed upward. Subsequent cell divisions produce additional stem tissue and new leaf and bud primordia. Farther back from the tip of the stem, the immature cells enlarge and differentiate into the three tissue systems of the mature plant body (•**Figure 5-13**).

Secondary growth takes place at lateral meristems

In addition to primary growth (elongation), trees and shrubs have secondary growth. These plants increase in length by primary growth and increase in girth by secondary growth. Secondary growth is due to cell divisions that occur in **lateral meristems,** areas that extend along the entire lengths of stems and roots, except at the tips. Two lateral meristems are responsible for secondary growth: the vascular cambium and the cork cambium. Secondary growth forms secondary tissues—that is, secondary xylem, secondary phloem, and periderm (•**Figure 5-14**).

   The **vascular cambium** is a layer of meristematic cells that forms a thin, continuous cylinder within the stem and root. It is located between the wood and bark of a woody plant. Cells of the vascular cambium divide, adding more cells to the wood (secondary xylem) and to the inner bark (secondary phloem).

   The **cork cambium** is a thin cylinder or irregular arrangement of meristematic cells in the outer bark region. Cells of the cork cambium divide to form the **cork cells** toward the outside and one or more underlying layers of **cork parenchyma** cells that function in storage.

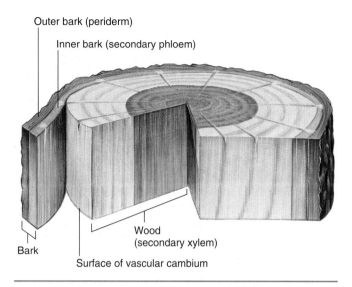

Outer bark (periderm)

Inner bark (secondary phloem)

Wood (secondary xylem)

Bark

Surface of vascular cambium

FIGURE 5-14 Lateral meristems and secondary growth. The vascular cambium, a thin layer of cells sandwiched between the wood and bark, produces secondary vascular tissues: the wood, which is secondary xylem, and inner bark, which is secondary phloem. The cork cambium produces the periderm, the outer bark tissues that replace the epidermis in the secondary plant body.

Collectively, cork cells, cork cambium, and cork parenchyma make up the periderm. Chapters 6 and 7 contain a more comprehensive discussion of secondary growth in roots and stems, respectively.

BUD A dormant embryonic shoot that eventually develops into an apical meristem.

LATERAL MERISTEM An area of cell division on the side of a vascular plant; the two lateral meristems (vascular cambium and cork cambium) give rise to secondary tissues.

STUDY OUTLINE

❶ **Discuss the plant body, including the root system and shoot system.**
   A plant body typically consists of a root system and a shoot system. The **root system** is generally underground and obtains water and dissolved minerals for the plant. Roots also anchor the plant firmly in place. The **shoot system** is generally aerial and obtains sunlight and carbon dioxide for the plant. The shoot system consists of a vertical stem that bears leaves (the main organs of photosynthesis) and flowers and fruits (reproductive structures). Buds (undeveloped embryonic shoots) develop on stems. Although separate **organs** (roots, stems, and leaves) exist in the plant, many **tissues** are integrated throughout the plant body, providing continuity from organ to organ.

❷ **Describe the ground tissue system (parenchyma tissue, collenchyma tissue, and sclerenchyma tissue) of plants.**
   The **ground tissue system** consists of three tissues. **Parenchyma tissue** is composed of living **parenchyma cells** that possess thin **primary cell walls.** Functions include photosynthesis, storage, and secretion. **Collenchyma tissue** is composed of **collenchyma cells** with unevenly thickened primary cell walls. This tissue provides flexible structural support. **Sclerenchyma tissue** is composed of

sclerenchyma cells that have both primary and secondary cell walls. Sclerenchyma cells are often dead at maturity, but they provide structural support.

❸ Outline the structure and function of the vascular tissue system (xylem and phloem) of plants.
The **vascular tissue system** conducts materials throughout the plant body and provides strength and support. **Xylem** is a complex tissue that conducts water and dissolved minerals. The actual conducting cells of xylem are **tracheids** and **vessel elements. Phloem** is a complex tissue that conducts food materials (carbohydrates) in solution. **Sieve-tube elements** are the conducting cells of phloem; they are assisted by **companion cells.**

❹ Describe the dermal tissue system (epidermis and periderm) of plants.
The **dermal tissue system** is the outer protective covering of the plant body. The **epidermis** is a complex tissue that covers the herbaceous plant body. The epidermis that covers aerial parts secretes a layer of wax, called the **cuticle,** that reduces water loss. Gas exchange between the interior of the shoot system and the surrounding atmosphere occurs through **stomata.** The **periderm** is a complex tissue that forms the outer bark in woody plants.

❺ Discuss what is meant by growth in plants and how it differs from growth in animals.
Growth in plants involves cell division, cell elongation, and cell differentiation. One difference between plants and animals is the location of growth. When a young animal is growing, all parts of its body grow, although not necessarily at the same rate. In contrast, plants grow only in specific areas called **meristems,** which are composed of cells that do not differentiate.

❻ Distinguish between primary and secondary growth.
Primary growth, an increase in stem and root length, occurs in all plants due to the activity of **apical meristems** at the tips of roots and at the buds of stems. **Secondary growth** is an increase in stem and root girth. In addition to primary growth, woody plants have secondary growth. Secondary growth is localized, typically as long cylinders of active growth throughout the lengths of older stems and roots. The two **lateral meristems** responsible for secondary growth are the **vascular cambium** and the **cork cambium.**

REVIEW QUESTIONS

1. What are some functions of roots? Of shoots?
2. What are the three tissue systems in plants? Describe the functions of each.
3. Compare the cellular structures and functions of parenchyma, collenchyma, and sclerenchyma cells.
4. What are the functions of xylem and phloem?
5. Compare and contrast epidermis and periderm.
6. What is the role of meristems in plants?
7. Distinguish between primary and secondary growth.
8. Distinguish between apical and lateral meristems.
9. How does growth in plants differ from growth in animals?
10. Label the following figure of the plant body. Check your answers against Figure 5-1.

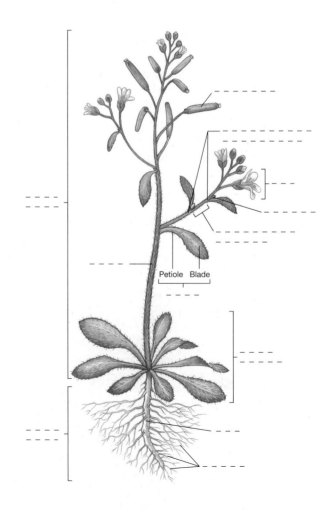

Petiole   Blade

THOUGHT QUESTIONS

1. Grasses have special meristems at the bases of the leaves. Relate this information to what you know about mowing the lawn.

2. Why is meristematic tissue so important to a plant? What would happen if all the meristematic regions of a plant were removed?

3. A couple carved a heart with their initials in a tree trunk, 4 feet above ground level; the tree was 25 feet tall at the time. Twenty years later, the tree was 50 feet tall. How far above the ground were the initials? Explain your answer.

4. Sclerenchyma in plants is the functional equivalent of bone in animals (that is, both sclerenchyma and bone provide support). However, sclerenchyma is dead, and bone is living tissue. What are some of the advantages to a plant of having dead support cells? Can you think of any disadvantages?

Visit us on the web at http://www.thomsonedu.com/biology/berg/ for additional resources, such as flashcards, tutorial quizzes, further readings, and web links.

Plant Organs: Roots

LEARNING OBJECTIVES

❶ Describe the functions of roots, and describe two features of roots that shoots lack.

❷ Contrast the structure of a primary eudicot root and a monocot root, and describe the functions of each tissue.

❸ Trace the pathway of water from the soil through the various root tissues.

❹ Describe several roots that are modified to perform unusual functions.

❺ Discuss the significance of roots to humans.

I f you chanced upon an American ginseng (*Panax quinquefolius*) plant growing in the eastern woodlands of North America, you would probably not pay much attention to it. This small, herbaceous plant consists of a short, erect stem bearing a whorl of three compound leaves, each divided into five leaflets, and a rounded cluster of tiny, greenish white flowers that form bright red fruits. Ginseng is attractive and is sometimes grown as a ground cover in cool-temperate woodland gardens. From only a superficial look at the plant, however, it would be difficult to fathom why it has been collected—at one time almost to the point of extinction—since European colonists discovered it in the early 1700s. Furthermore, it would be hard to understand why the American ginseng is cultivated commercially today, with much of it being used in the United States, Russia, Europe, and China.

Although the aboveground portion of American ginseng is unremarkable, the underground part is quite extraordinary. The root of the American ginseng is thick and often forked. To an imaginative mind, it sometimes resembles a tiny human. Depending on how it branches, the root may possess two armlike branch roots, two leglike branch roots, and sometimes a branch root between the "legs" that resembles male genitals. The root of a related species of ginseng (*P. ginseng*) that is native to China and Korea also possesses this remarkable appearance.

Because of its resemblance to the human body, ginseng root is thought by some to have curative powers for many human conditions and illnesses. This idea is based on the *Doctrine of Signatures,* a belief that was popular in the 1500s and 1600s but was later discredited. According to the Doctrine of Signatures, God created plants for human use and gave each a visible sign, or "signature," to indicate its purpose. Ginseng, considered a powerful Doctrine of Signatures plant, has been used as a tonic for a wide variety of conditions, including cancer, rheumatism, diabetes, impotence, sterility, and aging. Indeed, its scientific name, *Panax,* is Latin for "cure-all." Some people dry ginseng root, grind it into a powder, and use it to brew a tea that is taken as a general tonic or as an aphrodisiac. Although the Chinese have used ginseng root medicinally for centuries and many Americans extol its virtues, the therapeutic value of ginseng has not been scientifically verified.

Ginseng provides an interesting example of the human relevance of roots. For plants, however, roots are indispensable to their survival. This chapter examines the structure and major functions of roots.

American ginseng.
American ginseng (*Panax quinquefolius*) is a
woodland plant that grows in eastern North America.
(*Inset*) Ginseng root.

Root Structure and Function

Roots are underground and out of sight, and therefore people do not always appreciate the important functions they perform. A plant's underground root system, which branches out into the soil from the plant's central axis, is often more extensive than its aerial parts. The roots of a corn plant, for example, may grow to a depth of 2.5 meters (about 8 feet) and spread outward 1.2 meters (4 feet) from the stem. Desert-dwelling tamarisk trees reportedly have roots that grow to a depth of 50 meters (163 feet) to tap underground water. The total root length, not counting root hairs, of a 4-month-old rye plant was found to exceed 500 kilometers (310 miles)! The extent of a plant's root depth and spread varies considerably among species and even among different individuals in the same species. Soil conditions greatly affect the extent of root growth (see discussion of soils in Chapter 10). (See also *Plants and the Environment: Soil Erosion*, on page 113, and *Plants and the Environment: Desertification*, on page 115, for discussions of two soil problems that affect root growth.)

Roots generally grow downward, in the direction of gravity. Two types of root systems—a taproot system and a fibrous root system—may develop from the embryonic root (the *radicle*) in the seed (•**Figure 6-1**). A **taproot system** consists of one main root (formed from the enlarging radicle) with many smaller lateral (branch) roots coming out of it. Lateral roots often occur initially in regular rows along the length of the main root. Taproots are characteristic of many eudicots and gymnosperms.[1] A dandelion is a good example of a common herbaceous eudicot with a taproot system. A few trees, such as hickory, retain their taproots, which become quite massive as the plants age. Most trees, however, have taproots when young and later develop large, shallow lateral roots from which other roots branch off and grow downward.

---

[1]*Gymnosperms are seed-bearing plants that do not produce flowers; cone-bearing trees such as pines, hemlocks, and firs are examples of gymnosperms. Flowering plants are divided into two main groups, informally called eudicots and monocots. Oak, cherry, bean, and daisy are examples of eudicots. Palm, corn, bluegrass, and lily are examples of monocots.*

TAPROOT SYSTEM A root system consisting of one prominent main root with smaller lateral roots branching from it.

FIBROUS ROOT SYSTEM A root system consisting of several adventitious roots of approximately equal size that arise from the base of the stem.

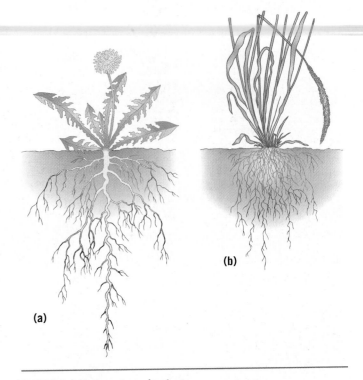

FIGURE 6-1 Root systems in plants.
(a) A taproot system develops from the embryonic root in the seed. **(b)** The roots of a fibrous root system are adventitious and develop from stem tissue.

A **fibrous root system** has several to many roots of the same size that develop from the end of the stem, with smaller lateral roots branching off these roots. Fibrous root systems form in plants in which the embryonic root is short-lived. The roots originate initially from the base of the embryonic root and later from stem tissue. Because fibrous roots do not arise from preexisting roots but from the stem, they are called *adventitious*. **Adventitious** organs occur in unusual locations, such as roots that develop on a stem or buds that develop on roots. Onions, crabgrass, and other monocots have fibrous root systems.

The two types of root systems, the taproot system and the fibrous root system, are adapted to obtain water in a variety of ways. For example, taproot systems often extend down into the soil to obtain water deep underground, whereas fibrous root systems, which are relatively close to the ground surface, are adapted to obtain rainwater from a larger area as it drains into the soil.

Roots provide anchorage, absorption, conduction, and storage

Roots perform several important functions for plants. First, as anyone who has ever pulled weeds can attest, roots anchor a plant securely in the soil. A plant remains in one location throughout its life and needs a solid foundation from which to grow. Firm anchorage is

Soil is a valuable natural resource on which humans depend for food. Water, wind, ice, and other agents cause **soil erosion,** the wearing away or removal of soil from the land. Water and wind are particularly effective in removing soil: rainfall loosens soil particles that are then transported away by moving water (see figure), whereas wind loosens soil and blows it away, particularly if the soil is exposed and dry.

The U.S. Department of Agriculture estimates that approximately one-fifth of U.S. cropland is vulnerable to soil erosion damage. Because erosion reduces the amount of soil in an area, it limits the growth of plants. Erosion also causes a loss in soil fertility, because essential minerals and organic matter that are part of the soil are removed. As a result of these losses, the productivity of eroded agricultural lands declines. Soil erosion has a detrimental impact on other natural resources. Sediments that get into streams, rivers, and lakes, for example, affect water quality and fish habitats. If the sediments contain pesticide and fertilizer residues, as they often do, they further pollute the water.

Sufficient plant cover limits soil erosion: leaves and stems cushion the impact of rainfall, and roots help hold the soil in place. Although soil erosion is a natural process, plant cover makes it negligible in many natural ecosystems.

Wind Erosion in Grasslands The Great Plains of North America extend from Texas north to Alberta and from the base of the Rocky Mountains east for about 650 kilometers (400 miles). Semiarid lands such as the Great Plains have low annual precipitation rates and are subject to periodic droughts that may last for extended periods. Prairie grasses, the plants that grow best in semiarid lands, are adapted to survive droughts. Although the aboveground portions of the plant may die, the dormant root systems can survive several years of drought. When the rains return, the root systems send up new shoots. Soil erosion is minimal because the living, dormant root systems hold the soil in place against the assault by wind.

USDA NRCS (Linda Betts)

■ **Soil erosion caused by water. Gullies, which form as a result of erosion damage from water, expand rapidly because the erosion accelerates as they serve as channels for the runoff of rainwater.**

Problems arise, however, when large areas of land are cleared for crops or when the land is overgrazed by too many animals. The removal of the natural plant cover and the corresponding loss of root systems open the way for climate conditions to "attack" the soil, which gradually deteriorates from the onslaught of hot summer sun, occasional violent rainstorms, and wind. If a prolonged drought occurs under such conditions, disaster can strike.

The American Dust Bowl The inhabitants of a wide region of the central United States vividly experienced the effects of wind on soil erosion during the 1930s. Throughout the late 19th and early 20th centuries, farmers removed much of the native grasses to plant wheat. Then, between 1930 and 1937, the semiarid lands stretching from Oklahoma and Texas into Canada received 65 percent less annual precipitation than was normal. The rugged prairie grasses could have survived these conditions, but there was not enough water to support wheat. The prolonged drought caused crop failures, which left fields barren and vulnerable to wind erosion.

Winds from the west swept across the barren, exposed soil, causing dust storms of incredible magnitude. Colorado and Oklahoma topsoil was blown eastward for hundreds of kilometers. Women hanging out clean laundry in Georgia went outside later to find it covered with dust. Bakers in New York City had to keep freshly baked bread away from open windows and the flying dust, which even discolored the ocean several hundred kilometers off the Atlantic coast.

The Dust Bowl occurred during the Great Depression, and many farmers went bankrupt. Farmers abandoned their dust-choked land and dead livestock and migrated west to the promise of California. The novel *The Grapes of Wrath* by John Steinbeck movingly portrays their plight.

Although the United States no longer has a Dust Bowl, the Great Plains is still subject to droughts and soil erosion. For example, a severe 5-year drought in the early 2000s ruined crops in parts of Montana, Wyoming, Colorado, Kansas, Texas, and New Mexico. Dust storms (brownouts) reoccurred on the High Plains, reminding people of the "Dirty Thirties."

Dennis Drenner

FIGURE 6-2 Storage roots.
Beets (*Beta vulgaris*) and other storage roots are important sources of human food because of the accumulated sugars and starches in the roots.

essential to a plant's survival so that the stem remains upright, enabling leaves to absorb sunlight efficiently.

Second, roots absorb water and dissolved minerals (inorganic nutrients), such as nitrates, phosphates, and sulfates, from the soil. These materials, many of which are necessary for synthesizing important organic molecules, are then transported throughout the plant in the xylem.

Third, roots function for storage. Surplus carbohydrates produced in the leaves by photosynthesis are transported in the phloem, as sugar, to the roots for storage, usually as sugar or starch, until needed. Carrot roots, for example, have an extensive phloem for this purpose. Although roots use some photosynthetic products for their own respiratory needs, most are stored and later transported out of the roots when the plant needs it. Either type of root system, taproot (beets, carrots, radishes, and turnips) or fibrous root (sweet potatoes), may be modified for storing food (•**Figure 6-2;** also see *Plants and People: Important Root Crops*). Storage taproots are usually **biennials** that, as part of the strategy to survive winter, store their food reserves in the root during the first year's growth and use these reserves to reproduce during the second year's growth. Other plants, particularly those living in dry regions, possess storage roots adapted to store water.

In certain species, roots are modified for functions other than anchorage, absorption, conduction, and storage. Later in this chapter we will discuss roots that are specialized to perform unusual functions.

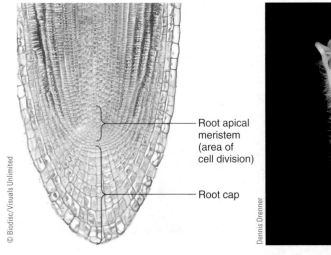

© Biodisc/Visuals Unlimited

Root apical meristem (area of cell division)

Root cap

(a) The root cap of an onion (*Allium cepa*) root. The root cap protects the root's apical meristem.

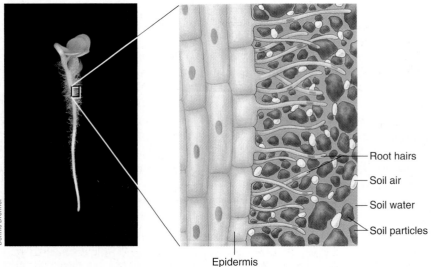

Dennis Drenner

Root hairs

Soil air

Soil water

Soil particles

Epidermis

(b) Root hairs on a radish (*Raphanus sativus*) seedling. Each delicate hair is a unicellular extension of the root epidermis. Root hairs increase the surface area in contact with the soil.

FIGURE 6-3 Structures unique to roots.

PLANTS AND THE ENVIRONMENT | Desertification

Rangelands are grasslands, in both temperate and tropical climates, that serve as important areas of food production for humans by providing fodder for domestic animals such as cattle, sheep, and goats. Grasses, the predominant vegetation of rangelands, have a fibrous root system, in which many diffuse roots anchor the plant in the soil. Such plants hold the soil in place quite well, thereby reducing soil erosion. If animals eat only the upper portion of the grass, the bottom part continues to develop, enabling the plant to recover and grow to its original size again.

The **carrying capacity** of a rangeland is the maximum number of animals that the rangeland plants can support in a sustainable fashion. When the carrying capacity of a rangeland is exceeded, grasses and other plants are **overgrazed;** that is, the grazing animals consume so much of the plant that it cannot recover and therefore dies. Overgrazing results in barren, exposed soil that is susceptible to erosion.

Most of the world's rangelands occur in semiarid areas that have extended natural periods of drought. Under normal conditions, native grasses can survive a drought; if the drought becomes severe, the top portion of the plant dies back while the underground part of the plant remains alive. Although the grasses of a drought-stricken rangeland appear dead, their extensive, living root systems hold the soil in place. When the rains return, the underground parts send forth new stems and leaves.

When overgrazing occurs in combination with a period of extended drought, once-fertile rangeland can be converted to desert. The lack of plant cover due to overgrazing allows winds to erode the soil. Even when the rains return, the land is so degraded that it cannot recover. Water erosion removes the little remaining topsoil, and the land turns into sand dunes. This conversion of rangeland to desert is called **desertification.** Desertification ruins economically productive land and destroys wildlife habitat.

Desertification is related to overpopulation. From 1980 to 1986, a disastrous drought struck the arid lands of East Africa, particularly in Ethiopia, Sudan, and Mozambique. Approximately a decade earlier, a devastating drought had occurred in the African Sahel, which extends south of the Sahara from Senegal to Somalia. In both cases, the people living in these lands suffered greatly. Many children and adults starved to death after their crops failed. These arid lands have always had periodic droughts, but what made the droughts during this period so devastating was that too many people (and livestock) were attempting to live on such ecologically fragile land. The people overwhelmed the land and degraded it; they chopped down most of the trees for firewood, and their livestock severely overgrazed the rangeland. Had many fewer people been farming and ranching in such a marginally productive area, they might have been spared the horrors of starvation.

Roots possess a root cap and root hairs

Because roots are adapted to the soil environment instead of the atmospheric environment, they have several structures, such as root caps and root hairs, that shoots lack. Although stems and leaves have various types of hairs, they are distinct from root hairs in structure and function.

Each root tip is covered by a **root cap,** a protective thimblelike layer many cells thick that covers the delicate root **apical meristem** (•**Figure 6-3a;** also see Figure 5-11). As the root grows, pushing its way through the soil, parenchyma cells of the root cap slough off by the frictional resistance of the soil particles and are replaced by new cells formed by the root apical meristem toward its outer side. The root cap cells secrete lubricating polysaccharides that reduce friction as the root passes through the soil.

The root cap may also be involved in orienting the root so that it grows downward (see discussion of gravitropism in Chapter 11). When a root cap is removed experimentally, the root apical meristem grows a new cap.

However, until the root cap has regenerated, the root fails to sense gravity as it grows.

Root hairs are short-lived, unicellular extensions of epidermal cells near the growing root tip. Root hairs form continually in the area of cell maturation closest to the root tip to replace those that are dying off at the more mature end of the root hair zone. Each root hair is short (typically less than 1 centimeter, or 0.4 inch, in length), but root hairs are quite numerous. They greatly raise the absorptive capacity of the root by increasing the surface area in contact with the moist soil (•**Figure 6-3b**). Soil particles are coated with a microscopically thin layer of water in which inorganic nutrient minerals are dissolved. The root hairs establish intimate contact with soil particles, which allows absorption of much of the water and dissolved materials.

ROOT CAP A covering of cells over the root tip that protects the delicate meristematic tissue directly behind it.

ROOT HAIR An extension of an epidermal cell of a root that increases the absorptive capacity of the root.

Unlike stems, roots lack nodes and internodes and do not usually produce leaves or buds. Although herbaceous roots have certain primary tissues found in herbaceous stems, these tissues are arranged quite differently.

Roots of Herbaceous Eudicots and Monocots

Considerable variation exists in roots, but they all possess an outer protective covering (the epidermis), a cortex for storage of starch and other organic molecules, and vascular tissues for conduction. Let us first consider the structure of herbaceous eudicot roots.

The central core in most herbaceous eudicot roots is vascular tissue

The buttercup root is a representative eudicot root with primary growth (•**Figure 6-4**). Like other parts of this herbaceous eudicot, a single layer of protective tissue, the **epidermis,** covers its roots. The root hairs are a modification of the root epidermis that enables it to absorb more water from the soil. The root epidermis does not secrete a thick, waxy cuticle—particularly in the region of root hairs, where a cuticle would impede the absorption of water from the soil. Both the lack of a cuticle and the presence of root hairs increase absorption.

Most of the water that enters the root moves along the path of least resistance—that is, along the cell walls rather than entering the cells. One of the major components of cell walls is cellulose, which absorbs water as a sponge does. Cotton balls, which are almost pure cellulose, illustrate the absorptive property of this material. (See Chapter 10 for an explanation of *why* water moves from the soil into the root.)

The **cortex** of an herbaceous eudicot root, which is composed primarily of loosely arranged **parenchyma cells** with large intercellular (*between*-cells) spaces, makes up the bulk of the root. The root cortex usually lacks supporting **collenchyma cells,** although it may develop some supporting **sclerenchyma cells** as it ages (see Chapter 5). The primary function of the root cortex is storage. A microscopic examination of the parenchyma cells that form the cortex often reveals numerous starch grains. Starch, an insoluble carbohydrate composed of glucose units, is the most common form of stored food in plants. When used at a later time, these reserves provide energy for such activities as growth following winter or cell replacement following an injury.

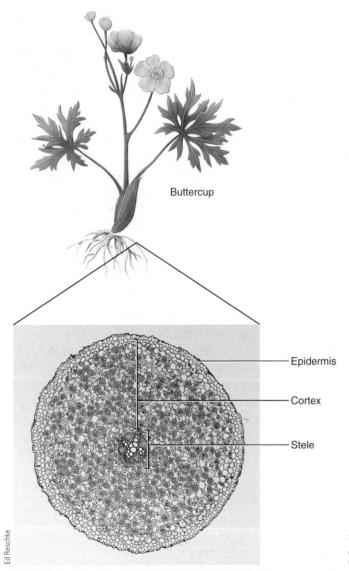

Buttercup

Epidermis

Cortex

Stele

Ed Reschke

(a) Cross section of a buttercup (*Ranunculus*) root. Note that the bulk of the root is the cortex.

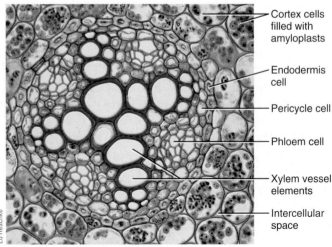

Cortex cells filled with amyloplasts

Endodermis cell

Pericycle cell

Phloem cell

Xylem vessel elements

Intercellular space

Ed Reschke

(b) A close-up of the stele of the buttercup root. Note the solid core of vascular tissues.

FIGURE 6-4 Structure of an herbaceous eudicot root.

PLANTS AND PEOPLE | Important Root Crops

Many roots are storage organs, which store the products of photosynthesis and are important sources of food for human consumption. Root crops are predominantly taproots, including carrots, beets, sugar beets, parsnips, turnips, rutabagas, and radishes.* These taproot crops are *biennials,* plants that accumulate starch and sugar in their roots during the first year's growth and then use it to reproduce during the second year. However, the roots are harvested as annuals—that is, after the first year's growth. Only a few root crops—sweet potatoes and cassava—are fibrous roots.

Worldwide, more cassava and sweet potatoes are grown than all other roots combined. Cassava (*Manihot esculenta*), also called manioc, is a tropical American plant that is grown in many tropical countries worldwide for its edible starchy roots, which resemble large sweet potato roots. It is a mainstay in the diets of millions of people. Tapioca is a granular starch squeezed out of cassava roots and used to make puddings and to thicken soups, primarily in North America and Europe. Sweet potatoes (*Ipomoea batatas*), also American in origin, are widely planted not only in the Americas but also in West Africa, China, India, and the Pacific Islands. Sweet potatoes are often confused with

*In general usage, "root crops" refer to any food crop that grows underground, including tubers (stems) such as white potatoes and bulbs such as onions. We restrict the discussion here to root crops that are true roots.

■ Cassava (*Manihot esculenta*) roots. Cassava originated in Central or South America, but little is known about when humans first brought it into cultivation.

© John Van Hasselt/Corbis

yams, a food crop belonging to a different genus (*Dioscorea*).

Because roots are generally rich in carbohydrates and certain vitamins but poor in proteins, people must consume them with other foods to achieve a balanced diet. The most nutritious roots are sweet potatoes, which contain about 5 percent protein and are rich in vitamins A and D in addition to iron, calcium, and other minerals.

The sugar beet (*Beta vulgaris*) is an important agricultural product that provides 35 percent of the world's sugar. Sugar beets were developed during the 18th century from a white variety of the common beet. Using **selective breeding,** the process by which humans

deliberately enhance desirable features over time by selecting which individuals to cross, scientists increased the sugar content of these beet roots from about 2 to 20 percent. Interestingly, Napoleon supported the development of sugar beets because he wanted to end England's monopoly on cane sugar in France. The sugar produced by sugar beets is sucrose, the same as that in sugarcane.

In addition to their use as important food crops, some roots are used as flavorings. For example, root beer gets its characteristic flavor from dried greenbrier (*Smilax*) roots, as does sarsaparilla, which was once widely used as a medicinal tonic.

The large intercellular air spaces, a common feature of root cortex, provide a pathway for water uptake and allow for aeration of the root. The oxygen that root cells need for aerobic respiration diffuses from air spaces in the soil to the intercellular spaces of the cortex, and from there to the cells of the root.

The inner layer of the cortex, the **endodermis,** controls the amounts and kinds of water and dissolved materials that enter the xylem in the root's center. Structurally, the endodermis differs from the rest of the cortex. Endodermal cells fit snugly against one another, and

each cell has a special bandlike region, called a **Casparian strip,** on its radial (side) and transverse (upper and

ENDODERMIS The innermost layer of the cortex of the root that prevents water and dissolved materials from entering the xylem by passing between cells.

CASPARIAN STRIP A band of waterproof material around the radial and transverse cells of the endodermis; ensures that water and minerals enter the xylem only by passing through the endodermal cells.

The endodermis regulates the kinds of minerals that are absorbed from the soil and conducted to the rest of the plant body.

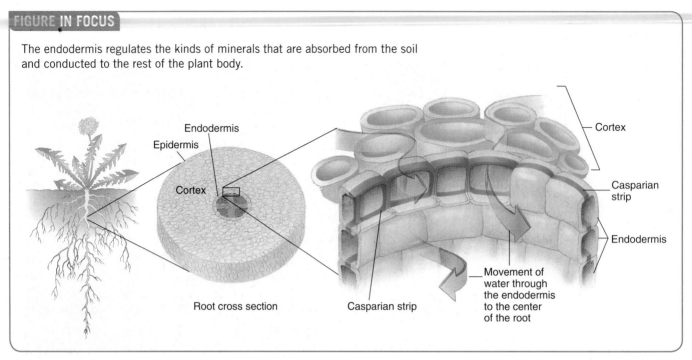

Root cross section

FIGURE 6-5 The endodermis and mineral uptake.
Note the Casparian strip around the radial and transverse walls that prevents water and dissolved minerals from passing into the stele along endodermal cell walls. To reach the vascular tissues, water and dissolved minerals must pass through the plasma membranes of endodermal cells.

lower) walls (•**Figure 6-5**). Casparian strips contain *suberin,* a fatty material that is waterproof.

The water and dissolved minerals that enter the root cortex from the epidermis move in solution along two pathways, the symplast and apoplast, until they reach the endodermis (•**Figure 6-6**). The **symplast** is a continuum of living cytoplasm, which is connected from one cell to the next by cytoplasmic connections called **plasmodesmata.** Some dissolved mineral ions move from the epidermis through the cortex via the symplast. The **apoplast** consists of the interconnected porous cell walls of a plant, along which water and inorganic mineral ions move freely. The water and mineral ions diffuse across the apoplast of the cortex without ever entering a living cell.

Until the endodermis is reached, most of the water and dissolved minerals have traveled along the apoplast and therefore have not passed through a plasma membrane or entered the cytoplasm of a root cell. However, the waterproof Casparian strip on the radial and transverse walls of endodermal cells prevents water and minerals from continuing to move passively along the cell walls. For substances to pass farther into the root interior, they must move from the cell walls into the cytoplasm of the endodermal cells. Water enters endodermal cells by **osmosis,** whereas inorganic minerals en-

ter by passing through **carrier proteins** in their plasma membranes. Thus, the endodermis controls what kinds of dissolved minerals and how much of each kind move from the soil into the vascular tissue of the root (and from there to the rest of the plant body).

At the center of a primary eudicot root is a cylinder of vascular tissues known as a **stele** (see Figure 6-4a). The outermost layer of the stele is a single layer of cells called the **pericycle,** which is just inside the endodermis. The pericycle, which is composed of parenchyma cells that remain meristematic, gives rise to lateral roots (•**Figure 6-7**). Lateral roots originate when a portion of the pericycle starts dividing. As it grows, the lateral root pushes through several layers of root tissue (endodermis, cortex, and epidermis) before entering the soil. Each lateral root has all the structures and features of the larger root from which it emerges: root cap, root hairs, epidermis, cortex, endodermis, pericycle, xylem, and phloem. In addition to producing lateral roots, the pericycle is involved in forming the lateral meristems that produce secondary growth in woody roots (discussed shortly).

Xylem, the centermost tissue of the stele, often has two, three, four, or more extensions, or "xylem arms." **Phloem** is located in patches between the xylem arms. The xylem and phloem of the root have the same functions and kinds of cells as in the rest of the plant:

The symplast and apoplast make up the entire plant body. The symplast consists of the cytoplasm of all the living cells in the plant body, which are interconnected by plasmodesmata, and the apoplast consists of all the cell walls and intercellular spaces within the plant body.

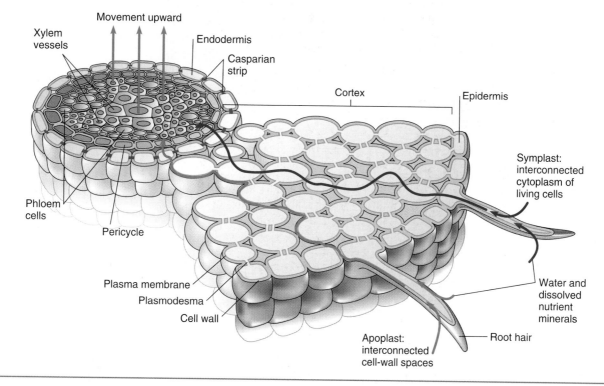

FIGURE 6-6 The symplast and apoplast.
Water and dissolved minerals that enter the root travel from one cell's cytoplasm to another through cytoplasmic connections (the symplast) or from cell to cell along the interconnected porous cell walls (the apoplast). On reaching the endodermis, water and minerals can continue to move into the root's center if they pass through a plasma membrane and enter an endodermal cell. The Casparian strip blocks the passage of water and minerals along the cell walls between adjoining endodermal cells.

tracheids and **vessel elements** of xylem conduct water and dissolved minerals, and **sieve-tube elements** of phloem conduct carbohydrates (sucrose).

After passing through the endodermal cells, water enters the root xylem, often at one of the xylem arms. Up to this point the pathway of water has been horizontal, from the soil into the center of the root:

Root hair ⟶ epidermis ⟶ cortex (symplast or apoplast pathway) ⟶ endodermis ⟶ pericycle ⟶ xylem of root

Once water enters the xylem, it is transported upward through root xylem into stem xylem and from there to the rest of the plant.

One direction of phloem conduction is from the leaves, where sugar is made by photosynthesis, to the root, where sugar is used for growth and maintenance of root tissues or stored, usually as starch. Another direction of phloem conduction is from the root, where sugar is stored as starch, to other parts of the plant, where sugar is used for growth and maintenance of tissues.

The **vascular cambium,** which gives rise to secondary tissues in woody plants, is sandwiched between the xylem and phloem. The primary eudicot root lacks a **pith,** a ground tissue found in the centers of many stems and roots.

SYMPLAST A continuum consisting of the cytoplasm of many plant cells, connected from one cell to the next by plasmodesmata.

APOPLAST A continuum consisting of the interconnected, porous plant cell walls, along which water moves freely.

PERICYCLE A layer of cells just inside the endodermis of the root; gives rise to lateral roots.

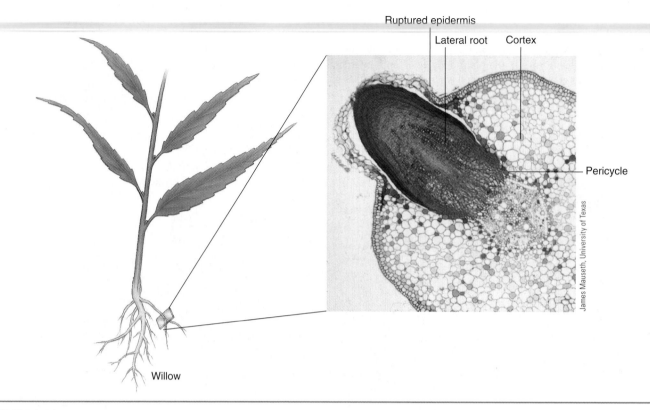

FIGURE 6-7 Lateral root.
A multicellular lateral root emerges from a larger root. Lateral roots originate at the pericycle.

The central tissue in some monocot roots is pith

Monocot roots are considerably more varied in internal structure than eudicot roots are. Starting at the outside of a greenbrier root (a representative monocot root) and moving inward are epidermis, then cortex, endodermis, and pericycle (•**Figure 6-8**). Unlike the xylem in herbaceous eudicot roots, the xylem in a monocot root does not form a solid cylinder of vascular tissues in the center. Instead, the phloem and xylem are in separate alternating strands that in cross section are arranged in a circle around the centrally located pith, which consists of parenchyma cells.

Because monocots do not have true secondary growth, no vascular cambium exists in monocot roots. Despite their lack of secondary growth, long-lived monocots, such as palms, often have thickened roots produced by a modified form of primary growth in which parenchyma cells in the cortex divide and enlarge. Thus, the cortex expands in such plants.

Roots of Woody Plants

Plants that produce stems with secondary growth—that is, wood and bark—also produce roots with secondary growth. Recall from Chapter 5 that gymnosperms and woody eudicots have primary growth at apical meristems, whereas secondary growth occurs at lateral meristems. The production of secondary tissues in the roots of woody plants occurs some distance back from the root tips and is the result of the activity of two lateral meristems, the *vascular cambium* and the *cork cambium.* Major roots of trees are often massive and have both wood and bark. In temperate climates, the wood of both roots and stems exhibits annual rings in cross section. (Chapter 7 discusses secondary growth in greater detail.)

Roots with Unusual Functions

Adventitious roots often arise from stem nodes (regions of the stem where leaves are attached). Many aerial adventitious roots are adapted for functions other than anchorage, absorption, conduction, or storage. **Prop roots** are adventitious roots that develop from branches or from a vertical stem and grow downward into the soil to help support the plant in an upright position (•**Figure 6-9**). Prop roots are more common in monocots than in eudicots. Corn and sorghum, both monocots, are herbaceous plants that produce prop roots. Many tropical and subtropical eudicot trees, such as red mangrove and banyan, also produce prop roots.

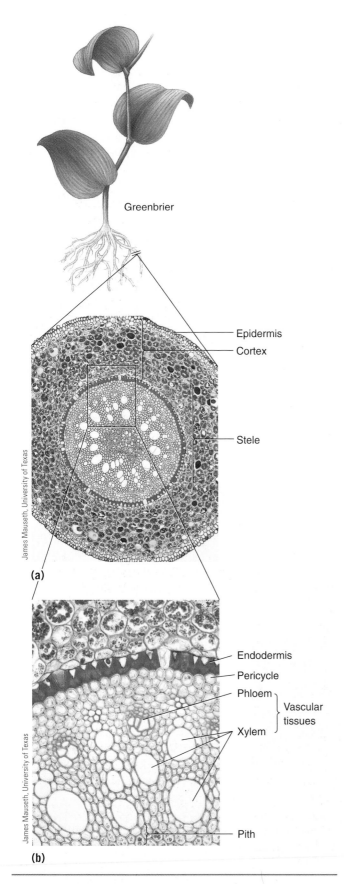

Greenbrier

Epidermis

Cortex

Stele

James Mauseth, University of Texas

(a)

Endodermis

Pericycle

Phloem ⎫
⎬ Vascular
Xylem ⎭ tissues

Pith

James Mauseth, University of Texas

(b)

FIGURE 6-8 Structure of a monocot root.
(a) Cross section of a greenbrier (*Smilax*) root. Note the extensive cortex. **(b)** Close-up of a portion of the center of the root, showing the vascular tissues and pith.

Prop roots

Linda R. Berg

FIGURE 6-9 Prop roots.
Pandanus has an elaborate set of aerial prop roots, adventitious roots that arise near the base of the stem and provide additional support. Photographed in Kauai, Hawaii.

The roots of many tropical rainforest trees are shallow and concentrated near the surface in a mat. The root mat catches and absorbs almost all inorganic minerals released from leaves by decomposers. Swollen bases or braces called **buttress roots** hold the trees upright and aid in the extensive distribution of the shallow roots (•**Figure 6-10**).

In swampy or tidal environments where the soil is flooded or waterlogged, some roots grow upward until they are above the high-tide level. Even though roots live in the soil, they still require oxygen for aerobic respiration. A flooded soil is depleted of oxygen, so these aerial "breathing" roots, known as **pneumatophores,** may assist in getting oxygen to the submerged roots. Pneumatophores, which also anchor the plant, have a well-developed system of internal air spaces that is continuous with the submerged parts of the root, presumably allowing gas exchange. Black mangrove and white

PROP ROOT An adventitious root that arises from the stem and provides additional support for the plant.

PNEUMATOPHORE A specialized aerial root produced by certain trees living in swampy habitats; may facilitate gas exchange between the atmosphere and submerged roots.

FIGURE 6-10 Buttress roots.
Tropical rainforest trees typically possess elaborate buttresses that support them in the shallow, often wet soil. Shown are buttress roots on Australian banyan (*Ficus macrophylla*).

FIGURE 6-11 Pneumatophores.
White mangrove (*Laguncularia racemosa*) produces pencil-like wooden pneumatophores that may provide oxygen for roots buried in anaerobic (oxygen-deficient) mud.

mangrove are examples of plants with pneumatophores (•Figure 6-11).

Climbing plants and *epiphytes,* which are plants that grow attached to other plants, have aerial roots that anchor the plant to the bark, branch, or other surface on which it grows. Some epiphytes have aerial roots specialized for functions other than anchorage. Certain epiphytic orchids, for example, have **photosynthetic roots** (see Figure 25-19a). Epiphytic roots may absorb moisture as well.

The strangler fig starts its life as an epiphyte that produces long, hanging aerial roots that eventually reach the ground and anchor the plant in the soil. The tree on which the strangler fig originally grew is often killed as the strangler fig grows around it, competing with it for light and other resources and crushing its secondary phloem. Eventually, the strangler fig becomes a self-supporting tree.

Some parasites, such as mistletoe, have roots that penetrate the host-plant tissues and absorb water and dissolved minerals from the host's xylem (•Figure 6-12). It may surprise you to know that mistletoe does not take sugars from its host plant. Other parasitic plants, however, take both water and sugars from their hosts.

Plants that produce *corms* or *bulbs* (underground stems or buds specialized for asexual reproduction) often have wiry **contractile roots.** The contractile roots grow into the soil and then contract (the cortical cells shorten or totally collapse on themselves), thus pulling the corm or bulb deeper into the soil (•Figure 6-13). The contraction is irreversible. Contractile roots are necessary for bulbs and corms, because each succeeding year's growth is *on top of* the preceding year's growth. As a result, bulbs and corms tend to move upward in the soil over time. Without contractile roots, they would eventually be exposed at the soil's surface.

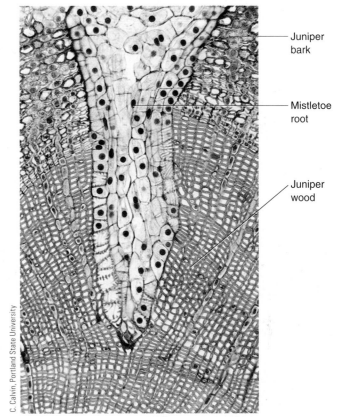

Juniper bark

Mistletoe root

Juniper wood

(b) Micrograph of a parasitized juniper (*Juniperus*) branch, showing a mistletoe root penetrating the wood (secondary xylem) of the juniper.

(a) Clumps of mistletoe (*Phoradendron flavescens*) have invaded a host sycamore (*Platanus wrightii*) tree in Arizona. Mistletoe is a parasitic shrub that obtains water and dissolved minerals from the host on which it lives.

FIGURE 6-12 **Specialized roots of parasitic epiphytes.**

Contractile roots are more common in monocots, but certain eudicots and ferns also possess them.

Some roots reproduce asexually by producing **suckers,** which are aboveground stems that develop from adventitious buds on the roots. Each sucker grows additional roots and becomes an independent plant when the parent plant dies. Examples of plants that form suckers include black locust, pear, apple, cherry, red raspberry, and blackberry. Some weeds—field bindweed, for example—produce many suckers. These plants are difficult to control, because pulling the plant out of the soil seldom removes all the roots. The roots of field bindweed grow as deep as 3 meters (10 feet) in the soil. In fact, in response to a wound the roots produce additional suckers, which is a considerable nuisance.

Root Associations with Other Species

As roots of certain trees grow through the soil, they sometimes encounter roots of other trees of the same or different species. When this occurs, the two types of roots may grow together by secondary growth to form a natural **graft**—as, for example, between a birch and a maple. Because their vascular tissues are connected in the graft, dissolved sugars and other materials such as hormones pass between the two trees. Root grafts have been observed in more than 160 tree species.

The roots of most plant species form a mutually beneficial relationship with certain soil fungi. These subterranean associations, known as **mycorrhizae,** permit the transfer of materials (such as sugars) from the roots

CONTRACTILE ROOT A specialized root, often found on bulbs or corms, that contracts and pulls the plant to a desirable depth in the soil.

MYCORRHIZA A mutually beneficial association between a fungus and a root that helps the plant absorb essential minerals from the soil.

(a) Plants that produce corms or bulbs often have contractile roots that lose much of their length as root cells shorten and broaden.

(b) During succeeding seasons, contractile roots pull the corm (*shown*) or bulb deeper and deeper until it reaches some optimum depth.

Corm

Contractile roots

Courtesy of Judith Jernstedt, University of California, Davis

FIGURE 6-13 Contractile roots.

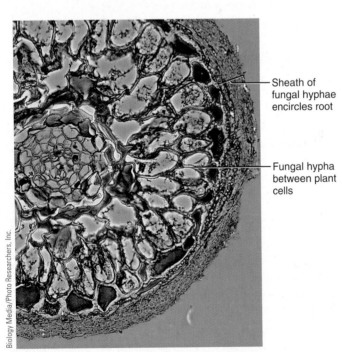

Biology Media/Photo Researchers, Inc.

Sheath of fungal hyphae encircles root

Fungal hypha between plant cells

(a) Cross section of root showing ectomycorrhizae, fungal associations that form a sheath around the root. The fungal hyphae penetrate the root between cortical cells but do not enter the cells.

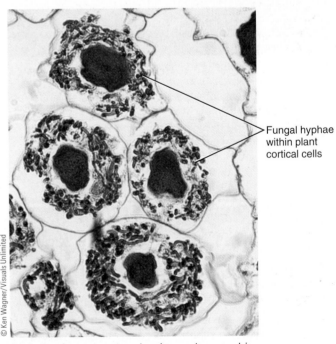

© Ken Wagner/Visuals Unlimited

Fungal hyphae within plant cortical cells

(b) Cells of a root cortex showing endomycorrhizae, fungal associations in which the fungal hyphae penetrate root cells of the cortex to aid in delivering and receiving nutrients. Endomycorrhizae colonize roots of most vascular plant species.

FIGURE 6-14 Mycorrhizae.
Mycorrhizae enhance plant growth by providing essential minerals to the roots.

to the fungus. At the same time, essential minerals such as phosphorus move from the fungus to the roots of the host plant. The threadlike body of the fungal partner extends into the soil, extracting minerals well beyond the reach of the plant's roots. In some mycorrhizae, called *ectomycorrhizae*, the fungal mycelium (nonreproductive body) encircles the root like a sheath (•**Figure 6-14a**). In others, known as *endomycorrhizae*, the fungus penetrates root cells (•**Figure 6-14b**). The relationship is mutually beneficial because when mycorrhizae are not present, neither the fungus nor the plant grows as well. Some evidence suggests that the fungal network of mycorrhizae simultaneously interconnects different plant species in the plant community and that carbon compounds may flow from one plant to another through their mutual fungal partner.

Certain nitrogen-fixing bacteria, collectively called **rhizobia,** form associations with the roots of legumi-nous plants—clover, peas, and soybeans, for example. Swellings called **nodules** develop on the roots and house millions of the rhizobia (see Figure 26-20a). Like mycorrhizae, the association between nitrogen-fixing bacteria and the roots of plants is mutually beneficial. The bacteria receive the products of photosynthesis from the plants while helping the plant meet its nitrogen requirements by producing ammonia (NH_3) from atmospheric nitrogen. Because nitrogen is an essential part of biologically important molecules such as proteins and nucleic acids, organisms—plants included—must have nitrogen to survive.

NODULE A small swelling on the root of a leguminous plant in which beneficial nitrogen-fixing bacteria (*Rhizobium*) live.

STUDY OUTLINE

❶ Describe the functions of roots, and describe two features of roots that shoots lack.

The main functions of roots are anchorage, absorption, conduction, and storage. Roots have several unique structures. Each root tip has a **root cap,** a protective thimblelike layer many cells thick that covers the delicate root apical meristem. The root cap also may orient the root so that it grows downward. **Root hairs** are short-lived, unicellular extensions of the epidermal cells near the growing root tip. Root hairs increase the surface area of the root that is in contact with the moist soil, increasing the root's absorptive capacity.

❷ Contrast the structure of a primary eudicot root and a monocot root, and describe the functions of each tissue.

Primary roots possess an outer protective covering (epidermis), ground tissues (cortex and, in certain roots, pith), and vascular tissues (xylem and phloem). The **epidermis** protects the root, and its root hairs aid in absorption of water and dissolved minerals. The **cortex** consists of parenchyma cells that often store starch. The **endodermis,** the innermost layer of the cortex, controls the uptake of water and dissolved minerals into the xylem of the root. The cells of the endodermis possess a **Casparian strip** around their radial and transverse walls that is impermeable to water and dissolved minerals. The **pericycle** is the origin of lateral roots. The **xylem** conducts water and dissolved minerals, and the **phloem** conducts dissolved sugar. There are some structural differences between monocot and herbaceous eudicot roots. The xylem and phloem of herbaceous eudicot roots form a solid mass in the center of the root. In contrast, monocot roots often have a **pith** in the center of the root. Monocot roots lack a **vascular cambium** and therefore do not have secondary growth.

❸ Trace the pathway of water from the soil through the various root tissues.

In a primary eudicot root, for example, water moves from the soil into the center of the root: Root hair ⟶ epidermis ⟶ cortex (symplast or apoplast pathway) ⟶ endodermis ⟶ pericycle ⟶ xylem of root. Once water enters the xylem, it is transported upward through root xylem into stem xylem and throughout the rest of the plant. The **symplast** is a continuum of interconnected cell interiors—that is, a pathway from the cytoplasm of one cell to that of an adjacent cell by cytoplasmic connections called **plasmodesmata.** The **apoplast** is a pathway that consists of the interconnected porous cell walls along which water flows freely.

❹ Describe several roots that are modified to perform unusual functions.

Prop roots are adventitious roots that develop from branches or from a vertical stem and grow downward into the soil to help support the plant in an upright position. Some tropical rainforest trees have swollen bases or braces called **buttress roots** that hold the trees upright and aid in the extensive distribution of the shallow roots. **Pneumatophores** are aerial "breathing" roots that may assist in getting oxygen to the submerged roots. Certain epiphytes have roots that are modified to photosynthesize. Corms and bulbs often have **contractile**

roots that grow into the soil and then contract, thereby pulling the corm or bulb deeper into the soil. Some roots reproduce asexually by producing **suckers,** aboveground stems that develop from adventitious buds on the roots.

❺ **Discuss the significance of roots to humans.**
Many roots are storage organs, which store the products of photosynthesis and are important sources of food for human consumption. Root crops are predominantly **taproots,** including carrots, beets, sugar beets, parsnips, turnips, rutabagas, and radishes. Only a few root crops—sweet potatoes and cassava—are **fibrous roots.** Some roots are used as flavorings; for example, root beer flavoring comes from dried greenbrier roots.

REVIEW QUESTIONS

1. What are the two types of root systems? Give examples of each.
2. List several functions of roots, and describe the tissue(s) responsible for each function.
3. How would you distinguish between a root hair and a small lateral root?
4. If you were examining a cross section of a primary root of a flowering plant, how would you determine whether the plant is a eudicot or a monocot?
5. Trace the pathway of water from the soil through the various root tissues in an herbaceous eudicot root.
6. Contrast the symplast and apoplast of the plant body.
7. Describe the function of the endodermis and of the pericycle.
8. Distinguish among the following specialized roots: storage root, prop root, buttress root, pneumatophore, photosynthetic root, and contractile root.
9. What are the world's two most important root crops? Where did these two plants originate?

10. Label the tissues of this herbaceous eudicot root. Give at least one function for each tissue. (Refer to Figure 6-4 to check your answers.)

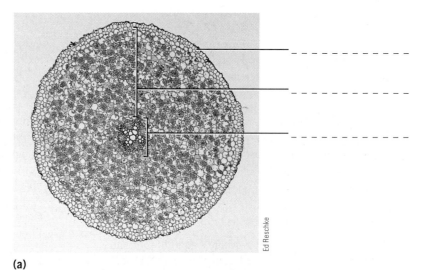

(a)

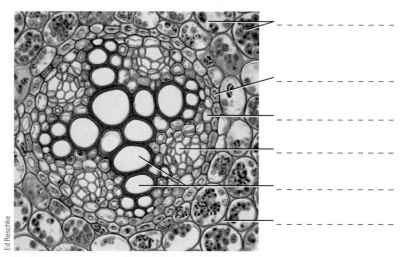

(b)

11. Label the tissues of this monocot root. Give at least one function for each tissue. (Refer to Figure 6-8 to check your answers.)

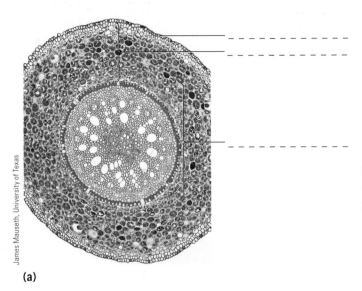

(a)

James Mauseth, University of Texas

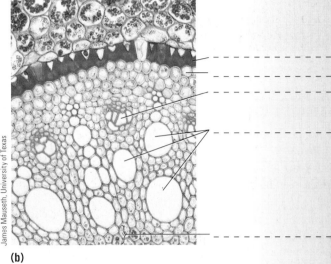

(b)

James Mauseth, University of Texas

THOUGHT QUESTIONS

1. A mesquite root is found penetrating a mine shaft about 150 feet below the surface of the soil. How might you determine *when* the root first grew into the shaft? (*Hint:* Mesquite is a woody plant.)

2. A barrel cactus that is 2 feet tall and 1 foot in diameter has roots more than 10 feet long. However, all the plant's roots are found in the soil at a depth of 2 to 6 inches. What possible adaptive value does such a shallow root system confer on a desert plant?

3. Someone gives you a plant structure that was found growing in the soil and asks you to determine whether it is root or an underground stem. How would you identify the plant part without a microscope? With a microscope?

4. The root cap is thought to orient the root so that it grows downward, in the direction of gravity. Why might it be advantageous for roots to grow downward?

5. Some biologists have suggested that the roots of all the plants living in a forest are a single functional unit. Explain.

6. The American Dust Bowl is sometimes portrayed as a natural disaster brought on by drought and high winds. Present a case for the point of view that this disaster was caused less by nature than by humans.

Visit us on the web at http://www.thomsonedu.com/biology/berg/ for additional resources, such as flashcards, tutorial quizzes, further readings, and web links.

Plant Organs: Stems

During colonial times, one of the most highly prized trees in the forests of eastern North America was the eastern white pine (*Pinus strobus*). Because its main stem grows so straight and tall, the eastern white pine made an ideal mast for sailing ships, and representatives of the English Crown reserved the best trees for the Royal Navy. The wood of eastern white pine stems is lightweight and straight grained. It contains less resin (a clear to yellowish viscous liquid that exudes from some trees, possibly to deter plant-eating insects) than the wood of other pine species. These characteristics make the wood suitable for house construction—frames, interior woodwork, flooring, and paneling—as well as for fashioning crates, boxes, inexpensive furniture, railroad ties, and many other items.

Because of the species' suitability for masts and home construction, Europeans who settled in North America took their first lumber from the vast forests of eastern white pine. Although eastern white pines in virgin (uncut) forests often grew 61 meters (200 feet) or more in height, they were logged to such an extent that few of these giant trees remain today. By the end of the 19th century, overlogging with no attention to conservation had taken its toll, and almost all virgin tracts of eastern white pine were gone. Since then, certain areas have been reforested extensively with eastern white pine, but none of these trees approaches the majestic size of those in the original forests. Today, most eastern white pines are less than 200 years old.

Pines are evergreen conifers (cone-bearing trees and shrubs). About 64 species of pines are native to North America, and the stems of many of them, like the eastern white pine, are valuable sources of lumber for everything from matchsticks to telephone poles. Today, other pines (for example, southern pines such as loblolly pine and western pines such as Ponderosa pine) are much more extensively logged than eastern white pine. Pine stems are also an important source of the pulp used in the manufacture of paper. The sticky resin in pine wood provides turpentine (a solvent for oil-based paints) and rosin (a material used to make varnishes, inks, sealants, and soaps). Many pines are also grown for ornamental purposes, as well as to provide shade or windbreaks. In the natural environment, pines provide food and shelter for wildlife.

This chapter considers the basic structure and functions of stems, including those of pines and other woody plants.

A pine plantation. Most of these eastern white pines (*Pinus strobus*) are 40 to 50 years old.

Stem Structure and Function

A vegetative (not sexually reproductive) vascular plant has three parts: roots, stems, and leaves. Roots anchor the plant and absorb materials from the soil, whereas leaves are primarily for photosynthesis, that is, they convert radiant energy into the chemical energy of carbohydrate molecules. **Stems,** the focus of this chapter, link a plant's roots to its leaves and are usually aerial, although many plants have underground stems. Stems exhibit varied forms, ranging from ropelike vines to massive tree trunks (•**Figure 7-1**). Stems are either herbaceous (consisting of soft, nonwoody tissues) or woody (with extensive hard tissues of wood and bark). Most stems are circular in cross section, although a few, such as mint stems, are square.

Stems perform three main functions in plants. First, stems of most species support leaves and reproductive structures. The upright position of most stems and the arrangement of leaves on them enable each leaf to absorb maximum light for use in photosynthesis. Reproductive structures (flowers and fruits) are located on stems in areas accessible for insects, birds, and air currents, which transfer pollen from flower to flower and help disperse seeds and fruits.

Second, stems provide internal transport. They conduct water and dissolved minerals (inorganic nutrients) from the roots, where they were absorbed from the soil, to the leaves and other plant parts. Stems also conduct the carbohydrates produced in the leaves by photosynthesis to the roots and other parts of the plant. Remember, however, that stems are not the only plant organs that conduct materials. The vascular system is

(a) Morning glory (*Ipomoea purpurea*), with its trailing, twining stem, is often grown to cover a fence or trellis.

(b) The massive trunks of baobab (*Adansonia*) trees are adapted to store water and starch. The tree produces leaves only during the rainy season. Photographed in Madagascar.

Carlyn Iverson

Tony Camacho/Photo Researchers, Inc.

FIGURE 7-1 Variation in stems.

continuous throughout all parts of the plant, and conduction occurs in roots, stems, leaves, and reproductive structures.

Third, stems produce new living tissue, as roots do. They continue to grow throughout a plant's life, forming *buds* that develop into stems with new leaves and/or reproductive structures. You may recall from Chapter 5 that plants undergo two types of growth. Primary growth, an increase in the length of a plant, occurs at **apical meristems** at the tips of stems and roots. Secondary growth, an increase in the girth (circumference) of a plant, is due to the activity of **lateral meristems** along the sides of stems and roots. The new tissues formed by the lateral meristems are called *secondary tissues* to distinguish them from the *primary tissues* produced by apical meristems.

All plants have primary growth; some plants also have secondary growth. Stems with only primary growth are herbaceous, whereas those with both primary and secondary growth are woody. (Some herbaceous stems—for example, geranium—also have limited amounts of secondary growth.) A woody plant increases in length by primary growth at the tips of its stems (and roots), while the older stems (and roots) farther from the tips increase in girth by secondary growth. In other words, at the same time that secondary growth is adding wood and bark (thereby causing the stem to thicken), primary growth (which increases the stem's length) continues.

In addition to performing the main functions of support, conduction, and production of new tissue, some stems are modified for asexual reproduction (discussed later in the chapter). Also, certain stems are specialized to store food or, if green, to manufacture carbohydrates by photosynthesis.

A woody twig exemplifies the external structure of all stems

Although stems exhibit great variation in structure and growth, they all have **buds.** At the tip of a stem is a **terminal bud.** When a terminal bud is dormant (that is, unopened and not actively growing), it is covered and protected by an outer protective layer of **bud scales,** which are modified leaves. **Axillary buds,** also called *lateral buds,* are found in the *axils*—the upper angles between leaves and the stem to which they are attached. When terminal and axillary buds grow, they form stems that bear leaves and/or flowers. The area on a stem where each leaf is attached is called a **node,** and the region of a stem between two successive nodes is an **internode.**

A woody twig of a deciduous tree (a tree that sheds its leaves annually) can provide a model of stem structures (•**Figure 7-2**). Bud scales cover the terminal bud

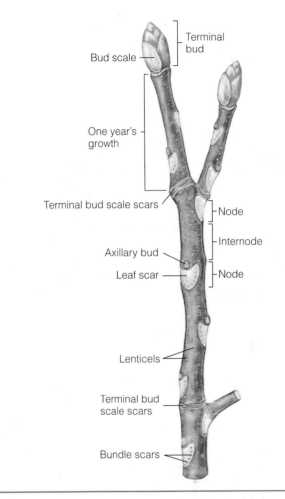

FIGURE 7-2 The external structure of a woody twig.
The age of a woody twig is determined by counting the number of groups of terminal bud scale scars (don't count those on side branches). How old is this twig?

and protect its delicate tip during dormancy. When the bud resumes growth, the bud scales fall off, leaving **bud scale scars** on the stem where they were attached. Woody twigs of temperate plants form terminal buds once a year, at the end of the growing season. Therefore, the number of groups of terminal bud scale scars on a twig reveals its age.

A **leaf scar** shows where a leaf was attached on the stem, and the vascular (conducting) tissue that extends from the stem out into the leaf forms **bundle scars** within the leaf scar. Axillary buds develop above the

---

BUD An undeveloped shoot that contains an embryonic meristem; may be terminal (at the tip of the stem) or axillary (on the side of the stem).

NODE The area on a stem where one or more leaves is attached; stems have nodes, but roots do not.

INTERNODE The area on a stem between two successive nodes.

leaf scars. Also, the bark of a woody twig has **lenticels,** sites of loosely arranged cells that allow gas exchange to occur. Lenticels look like tiny marks, or specks, on the bark of a twig and are often used as an aid in identifying the plant.

Stems of Herbaceous Eudicots and Monocots

Although stems vary considerably, they all possess an outer protective covering, one or more types of ground tissue, and vascular tissues (xylem and phloem). Let us first consider the structures of herbaceous eudicot stems and then of monocot stems.

The vascular bundles of herbaceous eudicot stems are arranged in a circle around a central pith

A young sunflower stem is a representative herbaceous eudicot stem that exhibits primary growth (•**Figure 7-3**). The **epidermis,** an outer covering, provides protection in herbaceous stems as it does in leaves and herbaceous roots. A **cuticle,** a waxy layer that reduces water loss from the stem surface, usually covers the epidermis.

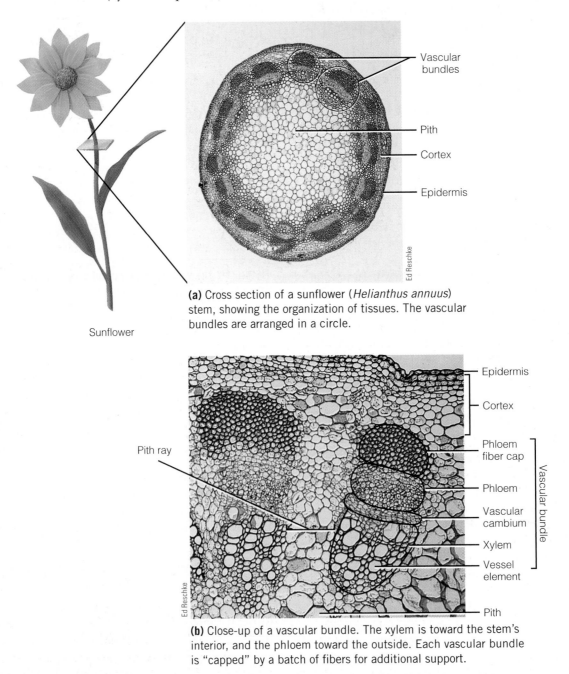

Sunflower

(a) Cross section of a sunflower (*Helianthus annuus*) stem, showing the organization of tissues. The vascular bundles are arranged in a circle.

(b) Close-up of a vascular bundle. The xylem is toward the stem's interior, and the phloem toward the outside. Each vascular bundle is "capped" by a batch of fibers for additional support.

FIGURE 7-3 Primary growth in a eudicot stem.

TABLE 7-1 General Differences between Herbaceous Eudicot Roots and Stems

ROOTS	STEMS
No nodes or internodes	Nodes and internodes
No leaves or buds	Leaves and buds
Nonphotosynthetic	Photosynthetic
No pith	Pith
No cuticle	Cuticle
Root cap	No cap
Root hairs	Trichomes
Pericycle	No pericycle
Endodermis	Endodermis rare
Branches form internally from the pericycle	Branches form externally from lateral buds

Note: Some exceptions to these general differences exist.

Inside the epidermis is the **cortex,** a cylinder several cells thick that is part of a plant's ground tissue system. The cortex is a complex tissue that may contain *parenchyma, collenchyma,* and *sclerenchyma cells* (see Chapter 5). As might be expected from the various types of cells that it contains, the cortex in an herbaceous eudicot stem can have several functions, such as photosynthesis, storage, and support. If the stem is green, photosynthesis occurs within chloroplast-containing parenchyma cells in the cortex. Parenchyma cells in the cortex also store starch grains and crystals. Collenchyma and sclerenchyma in the cortex confer strength and structural support for the stem.

The vascular tissues provide conduction and support. When an herbaceous eudicot stem is viewed in cross section, the vascular tissues appear as **vascular bundles** arranged in a circle. These vascular bundles extend as long strands throughout the length of the stem and are continuous with the vascular tissues of both roots and leaves. The central cylinder of a stem, which includes xylem, phloem, and often pith, is called a **stele.**

Each vascular bundle contains both **xylem,** which transports water and dissolved minerals from the roots to the leaves, and **phloem,** which transports dissolved carbohydrates (sucrose), often from the leaves to the roots. The xylem is usually on the inner side of the vascular bundle, and the phloem is found toward the outside. Sandwiched between the xylem and the phloem is a single layer of cells, the *vascular cambium,* a lateral meristem that is responsible for secondary growth.

Because most stems support the aerial plant body, they are much stronger than roots. The thick walls of *tracheids* and *vessel elements* in xylem help support the plant. *Fibers* also occur in both xylem and phloem, although they are usually more extensive in phloem. These fibers add considerable strength to the herbaceous stem. In sunflowers and certain other herbaceous eudi-

cot stems, the phloem contains a cluster of fibers called a **phloem fiber cap,** which helps strengthen the stem. The phloem fiber cap is not found in all herbaceous eudicot stems.

At the center of the herbaceous eudicot stem is **pith,** a ground tissue composed of large, thin-walled parenchyma cells that function primarily for storage. Due to the arrangement of the vascular tissues in bundles, there is no distinct separation of cortex and pith between the vascular bundles. The areas of parenchyma between the vascular bundles are often referred to as **pith rays.** •Table 7-1 summarizes the major differences between roots and stems in herbaceous eudicots.

Vascular bundles are scattered throughout monocot stems

A monocot stem, such as the herbaceous stem of corn, is covered by an epidermis with its waxy cuticle. As in herbaceous eudicot stems, the vascular tissues run in strands throughout the length of the stem (•Figure 7-4). In cross section, the vascular bundles, which contain xylem toward the inside and phloem toward the outside, are not arranged in a circle, as they are in herbaceous eudicot stems, but are scattered throughout the stem. Each vascular bundle is usually enclosed in a **bundle sheath** of sclerenchyma cells for support. The ground tissue in which the vascular tissues are embedded performs the same functions as cortex and pith in herbaceous eudicot stems. The monocot stem does not have distinct areas of cortex and pith.

Monocot stems do not possess lateral meristems (vascular cambium and cork cambium) that give rise to secondary growth. Monocots have primary growth only and do not produce the secondary tissues wood and bark. Although some treelike monocots such as palms attain considerable size, they do so by a modified form of primary growth rather than by secondary growth. The stems of monocots such as bamboo and palms contain a great deal of sclerenchyma tissue, which makes them hard and woodlike in appearance.

Stems of Woody Plants

Woody plants undergo secondary growth, an increase in the girth of stems and roots. Secondary growth results from the activity of two lateral meristems, the *vascular cambium* and the *cork cambium.* Among flowering plants, only woody eudicots (such as apple, hickory, and maple) have secondary growth. Gymnosperms (such as the conifers pine, juniper, and spruce) also have secondary growth. •Table 7-2 summarizes the relationships of meristems and tissues in a woody eudicot stem.

TABLE 7-2 **Development in a Woody Eudicot Stem**

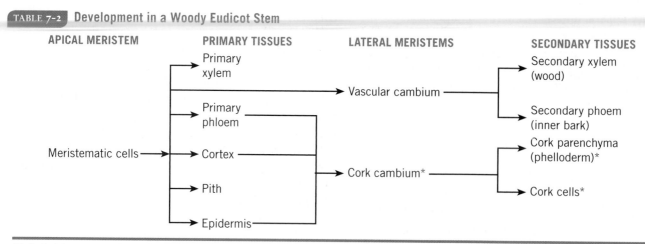

APICAL MERISTEM	PRIMARY TISSUES	LATERAL MERISTEMS	SECONDARY TISSUES

Meristematic cells →
- Primary xylem
- Primary phloem
- Cortex
- Pith
- Epidermis

→ Vascular cambium →
- Secondary xylem (wood)
- Secondary phoem (inner bark)

→ Cork cambium* →
- Cork parenchyma (phelloderm)*
- Cork cells*

Collectively known as periderm (outer bark).

Cells in the **vascular cambium** divide and produce two complex tissues: secondary xylem (wood), to replace primary xylem; and secondary phloem (inner bark), to replace primary phloem. Primary xylem and primary phloem cannot transport materials indefinitely and are replaced in plants that have extended life spans.

Cells of the outer lateral meristem, the **cork cambium,** divide to produce cork cells and cork parenchyma (*phelloderm*). The cork cambium and the tissues it produces are collectively called **periderm** (outer bark). Periderm functions as a replacement for the epidermis, which splits apart as the stem increases in girth.

Vascular cambium gives rise to secondary xylem and secondary phloem

Primary tissues in young stems of woody plants are organized similarly to those in herbaceous eudicot stems, and the vascular cambium is a thin layer of cells sandwiched between the xylem and the phloem in the vascular bundles. Once secondary growth commences, however, the internal structure of the stem changes considerably (•**Figure 7-5**). Although the vascular cambium is not initially a solid cylinder of cells, it becomes continuous when production of secondary tissues begins. This continuity develops because certain parenchyma

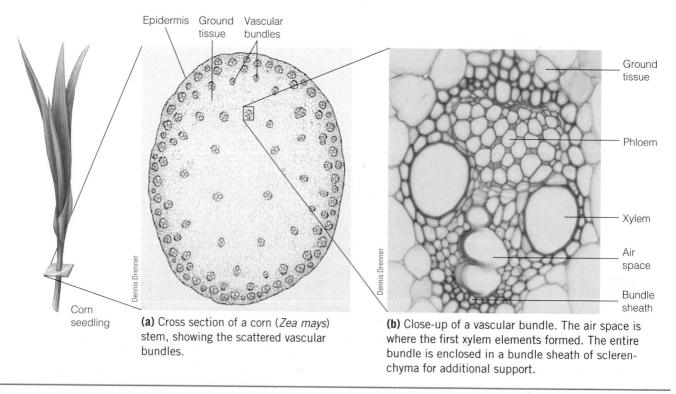

(a) Cross section of a corn (*Zea mays*) stem, showing the scattered vascular bundles.

(b) Close-up of a vascular bundle. The air space is where the first xylem elements formed. The entire bundle is enclosed in a bundle sheath of sclerenchyma for additional support.

Labels in (a): Epidermis, Ground tissue, Vascular bundles, Corn seedling, Dennis Drenner

Labels in (b): Ground tissue, Phloem, Xylem, Air space, Bundle sheath, Dennis Drenner

FIGURE 7-4 Arrangement of tissues in a monocot stem.

FIGURE IN FOCUS

Although a woody plant may live for hundreds of years, plant cells may live for only 2 or 3 years. Secondary growth produces replacement cells in the plant body so that newly formed cells in root and shoot tips remain connected.

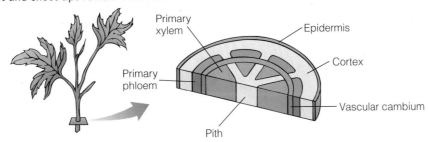

(a) At the onset of secondary growth, vascular cambium arises in the parenchyma between the vascular bundles (that is, in the pith rays), forming a cylinder of meristematic tissue (*blue circle in cross section*).

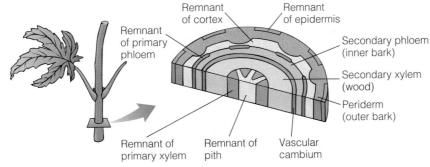

(b) Vascular cambium begins to divide, forming secondary xylem on the inside and secondary phloem on the outside.

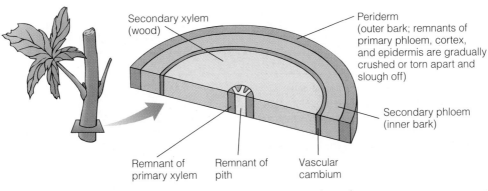

(c) A young woody stem. Vascular cambium produces more secondary xylem than secondary phloem.

FIGURE 7-5 Development of secondary growth in a eudicot stem.
Vascular cambium and the tissues it produces are shown; cork cambium is not depicted. (The figures change in scale from part **a** to part **c** because of space limitations; pith and primary xylem are actually the same size in all three diagrams, but the change in scale makes those tissues appear to shrink.)

cells between the vascular bundles retain the ability to divide. These cells connect to the vascular cambium cells in each vascular bundle to form a complete ring of vascular cambium.

Cells in the vascular cambium divide to produce daughter cells in two directions. The cells formed from

VASCULAR CAMBIUM A lateral meristem that produces secondary xylem (wood) and secondary phloem (inner bark).

CORK CAMBIUM A lateral meristem that produces cork cells and cork parenchyma; cork cambium and the tissues it produces make up the outer bark of a woody plant.

FIGURE IN FOCUS

As secondary xylem accumulates, the vascular cambium "moves" outward, and the woody stem increases in diameter.

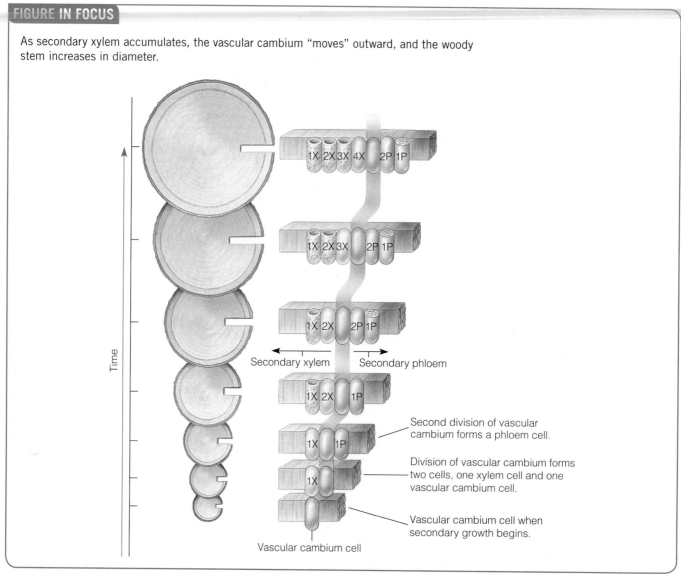

FIGURE 7-6 Radial view of a dividing vascular cambium cell.
To study this figure, start at the bottom and move up. Note that the vascular cambium (*blue cell*) divides in two directions, forming secondary xylem (X) to the inside and secondary phloem (P) to the outside. These cells, which are numbered in the order in which they are produced, differentiate to form the mature cell types associated with xylem and phloem.

the dividing vascular cambium are located either *inside* the ring of vascular cambium (to become secondary xylem, or wood) or *outside* it (to become secondary phloem, or inner bark) (●**Figure 7-6**). The vascular cambium is thus a thin layer of cells sandwiched between the wood and inner bark, the two tissues it produces (●**Figure 7-7**).

When a cell in the vascular cambium divides, one of the two cells produced by the division remains meristematic—that is, it remains as a part of the vascular cambium. The other cell may divide again several times, but it eventually stops dividing and develops into mature secondary tissue.

As the stem increases in circumference, the number of cells in the vascular cambium also increases. This increase in cells occurs by an occasional radial division of a vascular cambium cell, at right angles to its normal direction of division. In this case, both daughter cells remain meristematic.

What happens to the original primary tissues of the stem once secondary growth occurs? As the stem increases in thickness, the relative orientations of the original primary tissues change. For example, secondary xylem and secondary phloem are laid down between the primary xylem and primary phloem within each vascular bundle. Therefore, as the vascular cambium continues

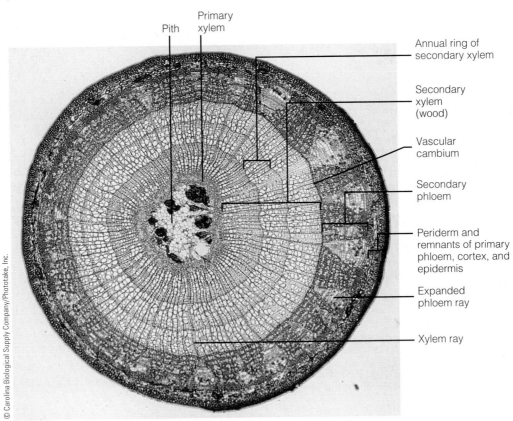

Pith Primary xylem

Annual ring of secondary xylem

Secondary xylem (wood)

Vascular cambium

Secondary phloem

Periderm and remnants of primary phloem, cortex, and epidermis

Expanded phloem ray

Xylem ray

(a) LM of cross section of basswood (*Tilia americana*) stem. Note the location of the vascular cambium between the secondary xylem (wood) and secondary phloem (inner bark).

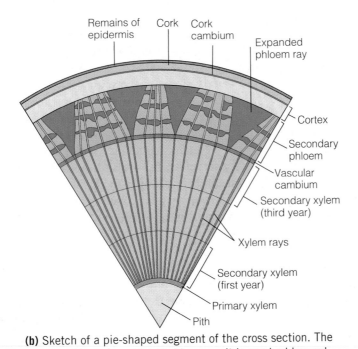

Remains of epidermis Cork Cork cambium

Expanded phloem ray

Cortex

Secondary phloem

Vascular cambium

Secondary xylem (third year)

Xylem rays

Secondary xylem (first year)

Primary xylem

Pith

(b) Sketch of a pie-shaped segment of the cross section. The primary phloem is not labeled because it is crushed beyond recognition.

FIGURE 7-7 Three-year-old stem in cross section.
It may be easier to study the parts labeled on the sketch and then locate them on the micrograph.

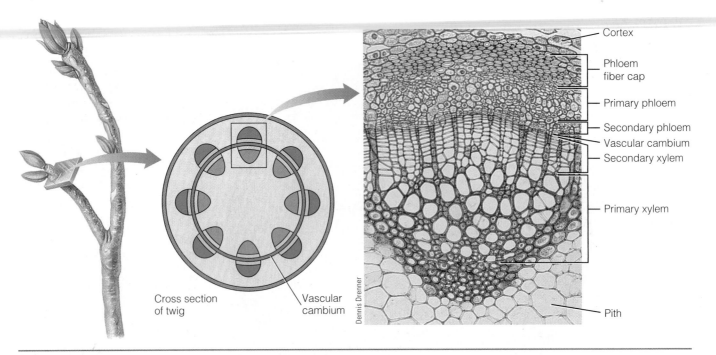

FIGURE 7-8 Onset of secondary growth.
This cross section through part of a magnolia (*Magnolia*) stem shows a vascular bundle that was split apart by secondary growth. Compare this vascular bundle with the one in Figure 7-3b, which has primary growth only.

to form secondary tissues, the primary xylem and primary phloem in each vascular bundle are separated from each other (•**Figure 7-8**). The primary tissues outside the cylinder of secondary growth (that is, the primary phloem, cortex, and epidermis) are subjected to the pressures produced by secondary growth and are gradually crushed or torn apart and sloughed off.

Secondary tissues replace the primary tissues in function. In the woody plant, secondary xylem conducts water and dissolved minerals from roots to leaves and contains the same types of cells found in primary xylem: water-conducting tracheids, water-conducting vessel elements, parenchyma cells, and fibers. The arrangement of the different cell types in secondary xylem results in distinctive wood characteristics for each species (see *Plants and People: The Importance of Wood* for some of the uses of wood derived from these distinctive characteristics).

Secondary phloem conducts dissolved carbohydrates (sucrose)—for example, from the place of manufacture in the leaves to a place of storage in the stem or roots. The same types of cells found in primary phloem are also found in secondary phloem: sieve-tube elements, companion cells, parenchyma cells, and fibers. There are usually more fibers in secondary phloem than in primary phloem, however.

Whereas the secondary xylem and secondary phloem transport water, minerals, and carbohydrates vertically throughout the plant, materials must also move horizontally (that is, sideways, or laterally). These materials move laterally through **rays,** which are chains of parenchyma cells that radiate out from the center of the woody stem. The vascular cambium forms the rays, which are often continuous from the secondary xylem to the secondary phloem. Water and dissolved minerals travel laterally through rays from the secondary xylem to the secondary phloem. Likewise, rays form pathways for the lateral transport of dissolved carbohydrates from the secondary phloem to the secondary xylem, and of waste products to the center, or heart, of the tree (discussed later in the chapter).

Cork cambium produces periderm

The cork cambium, which usually arises from parenchyma cells in the outer cortex, produces *periderm,* the functional replacement for the epidermis. The cork cambium is either a continuous cylinder of dividing cells (similar to the vascular cambium) or a series of overlapping arcs of meristematic cells that form from parenchyma cells in successively deeper layers of the cortex and, eventually, secondary phloem. Varia-

Carlyn Iverson

(a) Bur oak (*Quercus macrocarpa*) bark is deeply fissured.

R. C. Bonner

(b) Shagbark hickory (*Carya ovata*) has a rough, "shaggy" bark.

Carlyn Iverson

(c) Bark from Norway pine (*Pinus resinosa*) is scaly.

Carlyn Iverson

(d) Paper birch (*Betula papyrifera*) has a peeling bark.

FIGURE 7-9 Variation in bark.

tion in cork cambia and their rates of division explain why the outer bark of different tree species is fissured, rough and shaggy, scaly, or smooth and peeling (•**Figure 7-9**).

Like the vascular cambium, the cork cambium divides to form new tissues in two directions, to its inside and to its outside. Cork cells, which form to the outside of the cork cambium, are dead at maturity and have heavily suberized, or waterproofed, walls. These cork cells protect the woody stem against mechanical injury,

mild fires, attacks by insects and fungi, temperature extremes, and water loss. To its inside, the cork cambium produces cork parenchyma, a tissue that stores water and food (as starch granules). Cork parenchyma is only one to several cells thick, much thinner than the cork cell layer.

Cork cells are impermeable to water and gases, yet the internal tissues of the woody stem must exchange gases with the surrounding atmosphere. As a stem thickens from secondary growth, the epidermis, including the stomata that allowed gas exchange for the herbaceous stem, dies. Stomata are replaced by lenticels, areas of the cork in which the cork cells are loosely arranged, permitting gas exchange through the periderm (•**Figure 7-10**).

The bark of cork oak (*Quercus suber*), native to the western Mediterranean region, is commercially important. Harvesters carefully strip away the outer bark so as not to damage the cork cambium, which regenerates new cork cells in sufficient amounts to be harvested every 10 years. The main use of cork today is to stopper bottles; cork is also found in gaskets, life preservers, floats for fishing lines, flooring, and insulation.

Common terms associated with wood are based on plant structure

If you have ever examined different types of lumber, you may have noticed that some trees have wood with two different colors (•**Figure 7-11**). The functional secondary xylem—that is, the part that conducts water and dissolved minerals—is the **sapwood**, the younger, lighter-colored wood closest to the bark. **Heartwood**, the older wood in the center of the trunk, is typically a brownish red. A microscopic examination of heartwood reveals that its vessels and tracheids are plugged with pigments, tannins, gums, resins, and other materials. Therefore, heartwood no longer functions in conduction. Heartwood is denser than sapwood and provides structural support for trees. There is some evidence that heartwood is also more resistant to decay.

PLANTS AND PEOPLE | The Importance of Wood

Not counting firewood, which is still important on a global basis, the most common wood product is lumber. Lumber is formed when logs are sawed into square-edged boards that are cut to standard widths and lengths (see figure a). The pieces are graded and sorted and then seasoned. In seasoning, lumber is air-dried for several months to minimize shrinking and warping.

Lumber from a particular tree species has unique qualities that make it suitable for certain types of products. For example, oak, which is heavy and strong, is fashioned into furniture, doors, coffins, and flooring. Ash is used in baseball bats, oars, skis, tool handles, and bentwood furniture. Maple is valuable for bowling pins and bowling alleys, dance floors, and toys. Birch is used to make clothespins, toothpicks, and spools. Wood paneling, hockey sticks, church pews, and gymnasium equipment are often made from elm. Violins and pianos contain red spruce, which has exceptional resonance. Fences, shingles, clothes closets, and pencils are made from cedar; major uses of teak include furniture, chests, and shipbuilding. Shingles, siding, and garden furniture such as picnic tables continue to be made from redwood, even though these important trees are endangered from overharvesting.

The second most common use of wood after lumber is for paper pulp. Almost 90 percent of the world's pulpwood (wood used to make paper) comes from conifers. Sawmills convert wood to pulp by mechanically grinding it up and adding strong chemicals to free the "fibers" (which are actually tracheids). The slushy pulp is spread on a moving wire screen to dry and is then pressed, trimmed, and wound into large rolls (see figure b). Some of the many types of paper are newsprint; book paper; absorbent paper for blotters and filters; tissue paper for toilet tissue, paper towels, and paper napkins; and paperboard for boxes and containers.

Plywood, an important wood product often used in place of lumber, consists of two or more veneers (thin layers of wood sliced off a log in sheet form) that are glued together. The grains of adjacent layers are oriented at right angles to one another. Such plywood panels are strong, lightweight, and less likely to split than solid wood. Plywood is also available at greater widths than ordinary lumber. Most furniture made today contains some plywood covered by a veneer of a more valuable wood.

Sawmills used to discard crooked trees, branches, wood chips, and shavings. Today, such materials are pressed together to make sturdy boards such as fiberboard (made from pulp) and particle board (made from wood chips). Fiberboard is used for containers and insulation, whereas particle board has applications similar to those of plywood. Milling operations even market bark, which is sold primarily for mulches and soil conditioners.

(a)

(b)

■ (a) Cutting lumber in a sawmill. Computers direct sawmill operations, including providing precise specifications for sawing logs. (b) Rolls of newly made paper. In paper mills the paper is wound into large rolls that will be cut to size and rerolled before being shipped to customers.

The cellulose in wood is used to make products such as industrial alcohols, synthetic fibers (for example, rayon), and cellophane. Other chemical products produced from wood are acetic acid, wood tar, and adhesives.

Almost everyone has heard of hardwood and softwood. Botanically speaking, **hardwood** is the wood of flowering plants—that is, woody eudicots—and **softwood** is the wood of conifers (cone-bearing gymnosperms). The wood of pine and other conifers typically lacks fibers (with their thick secondary cell walls) and vessel elements. The only conducting cells in conifer wood are tracheids. These cell differences generally make conifer woods softer than the woods of flowering plants, although substantial variation occurs from one species to another. The balsa (*Ochroma pyramidale*) tree, for example, is a flowering plant (and therefore a hardwood) whose extremely soft, lightweight wood is used for products ranging from

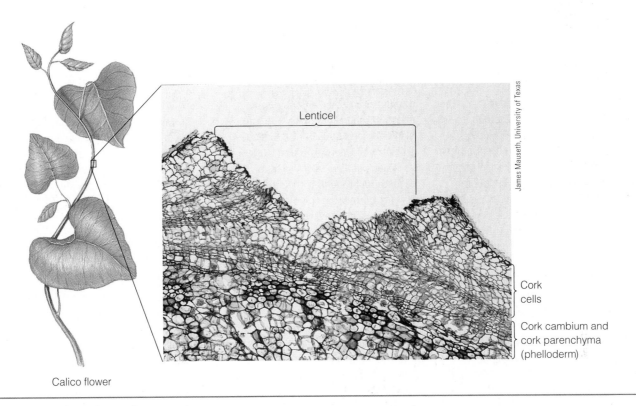

Calico flower

FIGURE 7-10 Structure of a lenticel.
This cross section through the periderm of a calico flower (*Aristolochia elegans*) stem shows a lenticel.

model airplanes to insulation material for ships.

Woody plants that grow in temperate climates where there is a growing period (during spring and summer) and a period of dormancy (during winter), exhibit **annual rings,** or **growth increments,** concentric circles found in cross sections of wood (•**Figure 7-12** on page 144). To determine the age of a woody stem in a temperate zone, simply count the annual rings. In the tropics, environmental conditions, particularly seasonal or year-round precipitation patterns, determine the presence or absence of rings: trees growing in the moist, humid tropics do not produce rings, whereas those growing in areas with pronounced wet and dry seasons do. Thus, rings are not reliable for determining age in most tropical trees.

Examination of annual rings with a magnifying lens reveals that there is no "ring," or line, separating one year's growth from the next. The appearance of a ring in cross sec-

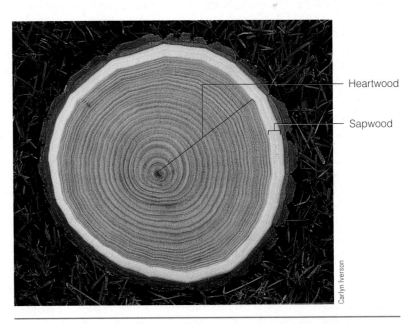

FIGURE 7-11 Heartwood and sapwood.
This cross section through a butternut (*Juglans cinerea*) trunk reveals the darker heartwood in the center of the tree and the lighter sapwood. The sapwood is the functioning xylem that conducts water and dissolved minerals.

FOCUS ON | Tree-Ring Analysis

PROCESS OF SCIENCE In temperate climates, counting a tree's number of annual rings establishes its age. The size of each ring varies according to local weather conditions, including precipitation and temperature. Sometimes the variation in tree rings can be attributed to a single environmental factor, and similar patterns occur in the rings of different tree species over a large geographic area. For example, trees in the Southwest of the United States have similar ring patterns due to variations in the amount of annual precipitation. Years in which adequate precipitation occurs produce wider rings of growth, whereas years of drought produce much narrower rings.

It is possible to study ring sequences that go back several thousand years. First, a **master chronology,** a complete sample of rings dating back as far as possible, is developed (see figure). To obtain a sample of rings, scientists bore out a small core of wood from the trunk of an old, living tree. The oldest rings (those toward the center of the tree) are matched with the youngest rings (those toward the outside) of an older tree or even an old piece of lumber from a house. Scientists develop a master chronology of the area by using successively older and older sections of wood, even those found in prehistoric dwellings, and by overlapping their matching ring sequences. The longest master chronology is of bristlecone pines in the western United States; it goes back almost 9000 years.

Dendrochronology, the study of both visible and microscopic details of tree rings, is used extensively in several fields. Tree-ring analysis has been extremely useful in investigations of prehistoric sites of Native Americans in the Southwest. For example, tree-ring analysis at the Cliff Palace in Mesa Verde National Park, which dates back to the year 1073 CE, indicates that an extended drought forced the original inhabitants to abandon their homes. Tree-ring data are also useful in other disciplines, including ecology (to study changes in a forest community over time), environmental science (to study the effects of air pollution on tree growth), and geology (to date earthquakes and volcanic eruptions).

Climatologists are increasingly using tree-ring data to study past climate patterns. Annual ring widths of certain tree species that grow at high elevations are sensitive to yearly temperature variations. The rings of these trees are wider in warm years and narrower in cool years. Studying tree rings across several thousand years helps researchers determine the natural pattern of global temperature fluctuations. This information is particularly important because of concerns about the human influence on global climate. Scientists generally agree that Earth has warmed in recent decades, and there is little doubt that human production of "greenhouse gases" such as carbon dioxide has contributed to this warming (see Chapter 27).

Researchers are not sure how much of the recent warming is the result of human influence rather than natural climate variability. Tree-ring analysis may help answer this vital question. Scientists in nine European countries are cooperating in a large tree-ring analysis to construct an annual history of temperatures across Northern Europe and Asia since the end of the last Ice Age, about 10,000 years ago. When the 10,000-year-old master chronology is completed (sometime before the year 2010), it should be possible to determine conclusively if today's climate patterns are distinctly different from the natural climate patterns of the past. ■

---

tion is due to differences in cell size and cell-wall thickness between secondary xylem formed at the end of one year's growth and that formed at the beginning of the next year's growth. In the spring, when water is plentiful, wood formed by the vascular cambium has thin-walled, large-diameter conducting cells (tracheids and vessel elements) and few fibers and is appropriately called *springwood.* As summer progresses and water becomes less plentiful, the wood that forms, known as *late summerwood,* has thicker-walled, narrower conducting cells and many fibers. The difference between the late summerwood of one year and the springwood of the following year gives the impression of a ring. We can obtain a lot of information about climates in past times from the study of the annual rings of ancient trees (see *Focus On: Tree-Ring Analysis*).

Wood looks different according to how it is cut (●**Figure 7-13**). In *cross section,* annual rings appear as concentric rings, and rays appear as straight lines radiating from the center of the stem. In *tangential section,* annual rings are vertical lines that often come together in a V shape, and rays are specks or short vertical lines. In *radial section,* annual rings appear as lines running the length of the wood; and rays, as horizontal strips.

As a woody stem increases in girth over the years, the branches that it bears grow along with it as long as they are alive. If a branch dies, it no longer continues to grow with the stem. In time, as the stem increases in

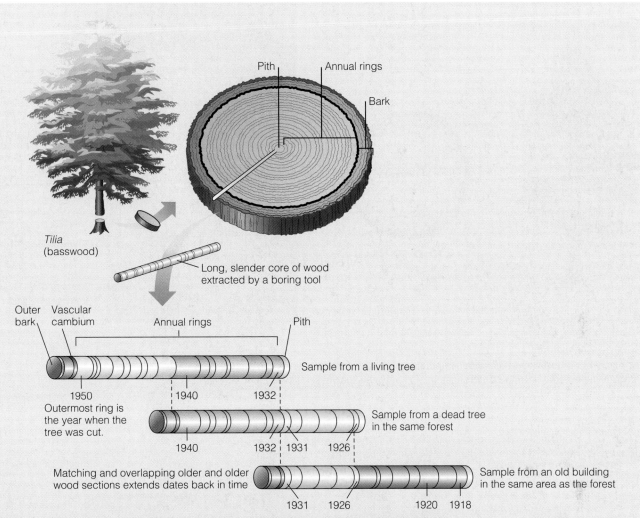

■ Tree-ring dating. Scientists develop a master chronology by using progressively older pieces of wood from the same geographic area. They can then accurately determine the age of a wood sample by using a computer to match its rings to the master chronology.

girth, it surrounds the base of the dead branch. The basal portion of an embedded dead branch is called a **knot.** It is possible for a knot to contain bark as well as wood. The presence of knots in wood lessens its commercial value, except when the knots are desired for ornamental purposes. Some plants, such as knotty pine, are valued for their high production of knots.

Forests are ecologically important

Forests, which cover approximately one-third of Earth's total land area, play an essential role in the global carbon cycle. For example, photosynthesis by trees removes large quantities of carbon dioxide from the atmosphere

and fixes it into carbon compounds. At the same time, the trees release oxygen into the atmosphere. Tree roots hold vast tracts of soil in place and thus reduce erosion. Forests absorb, hold, and slowly release water, thereby regulating the flow of water, even during dry periods, and helping control floods. In addition, forests provide essential wildlife habitat.

The greatest problem facing world forests today is **deforestation,** which is the removal of forest with-

DEFORESTATION The temporary or permanent clearance of large expanses of forests for agriculture or other uses.

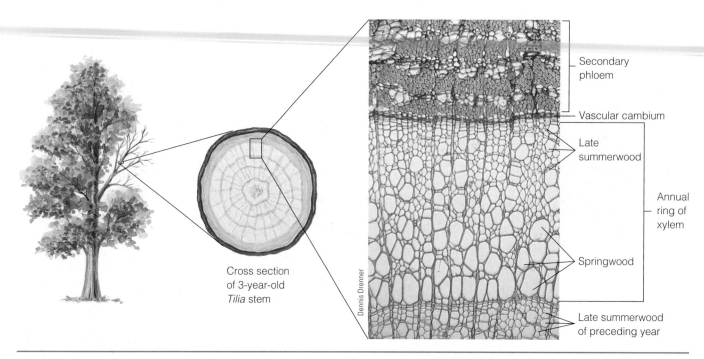

FIGURE 7-12 Annual ring.
This portion of a basswood (*Tilia americana*) stem, in cross section, shows one annual ring, or growth increment. Note the differences in size between the vessel elements of springwood and late summerwood.

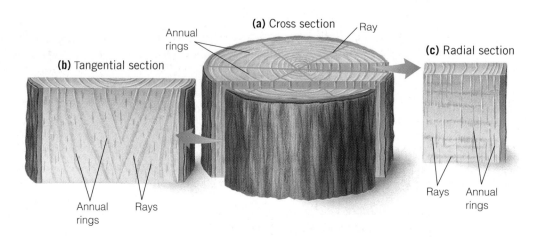

FIGURE 7-13 A block of wood showing (a) cross, (b) tangential, and (c) radial sections.
Both tangential and radial sections are longitudinal sections. A tangential section is cut in a plane that does not pass through the center of the stem but instead passes at a right angle to a radius. A radial section is cut in a plane that passes through the center along a radius of the stem. Rays and annual rings are distinctive for each section.

out sufficient replanting. Deforestation results in the replacement of forest with agricultural and other nonforest land. Only about-one third of harvested trees are replaced, usually with young seedlings of fast-growing species. Although deforestation is occurring around the world, it is particularly acute in the humid tropics (see *Plants and the Environment: The Deforestation of Tropical Rain Forests*).

PLANTS AND THE ENVIRONMENT | The Deforestation of Tropical Rain Forests

Tropical rain forest prevails in tropical places where the climate is moist throughout the year—about 200 or more centimeters (at least 79 inches) of precipitation annually. Tropical rain forests are found in Central and South America, Africa, and Southeast Asia, but almost half of them are in just three countries: Brazil, Democratic Republic of the Congo, and Indonesia (see figure). All over the world, tropical rain forests are being cleared—cut for timber or firewood or burned to make pasture or agricultural land.

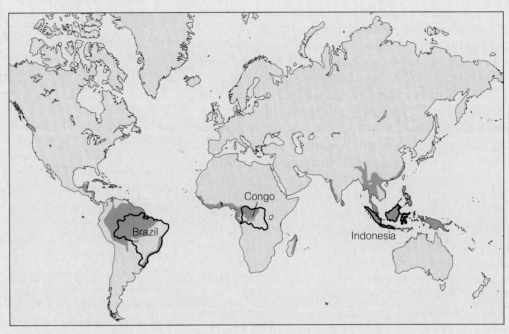

■ The distribution of tropical rain forests. Rain forests (*green areas*) are located in Central and South America, Africa, and Southeast Asia.

When tropical rain forests are harvested or destroyed, they no longer make valuable contributions to the environment or to the people who depend on them. Destruction of tropical rain forest particularly threatens native people, whose cultures and ways of life depend on the forests. Deforestation results in decreased soil fertility and increased soil erosion. When forest is removed, the total amount of surface water that flows into rivers and streams increases. However, because the forest no longer regulates this water flow, the affected region experiences alternating periods of flood and drought.

Deforestation also causes the extinction of plant and animal species. Tropical species often live within a limited area in a forest, so they are particularly vulnerable to habitat destruction or modification. Wildlife in temperate areas, including migratory birds and butterflies, also suffers from tropical deforestation, because the animals spend part of each year in these tropical areas.

The three main causes of tropical deforestation are subsistence agriculture, commercial logging, and cattle ranching. We restrict our discussion here to subsistence agriculture because it is the most important cause. **Subsistence agriculture,** in which each family produces enough food to feed itself, accounts for more than half of tropical rainforest loss. In many developing countries where tropical rain forests occur, most people do not own the land on which they live and work. Many subsistence farmers have no place to go except into the forest, which they clear to grow food. Subsistence farmers often practice **slash-and-burn agriculture,** in which they first cut down the forest and allow it to dry, then burn the area and immediately plant crops. The yield from the first crop is often quite high, because the minerals that were in the burned trees are now available in the soil. However, soil productivity declines rapidly so that subsequent crops are poor. In a short time, the people farming the land must move to new forest and repeat the process.

Slash-and-burn agriculture on a small scale, with plenty of forest to shift around in, is *sustainable*—that is, it can be carried out indefinitely without overusing the soil and other natural resources. But when millions of people try to obtain a living in this way, the land is not allowed to lie uncultivated for adequate recovery periods, and the forest does not recover. Globally, at least 200 million subsistence farmers obtain a living from slash-and-burn agriculture, and the number is growing. Moreover, there is only half as much forest available today as there was 50 years ago.

Vines

Vines are weak-stemmed plants that depend on other plants for support (see Figure 7-1a). Because vines do not have to expend resources to produce structurally strong stems, they can grow extremely rapidly, ascending through the shaded understory to the sunlit canopy of the forest, where they produce luxuriant foliage.

Vines have a variety of adaptations that fit their climbing lifestyle. As newly germinated plants, many vine seedlings grow *away* from sunlight rather than toward it. In growing toward the darkest part of its environment, a vine seedling usually encounters a large tree, which it then ascends. Woody vines (known as *lianas*) often produce special roots with adhesive pads that stick to the bark of the host tree. Herbaceous vines frequently have *tendrils*, modified leaves or stems that wrap around supports, whereas other vines are *twiners*, with stems that grow spirally around their host as they ascend it.

Vines are most numerous and diverse in tropical forests, particularly tropical rain forests, where both herbaceous vines and lianas flourish (see Figure 22-11b). Lianas grow from the upper branches of one forest tree to those of another, connecting the tops of the trees and providing walkways for many of the canopy's animal residents. They and other vines provide nectar and fruit to many tree-dwelling animals.

Temperate forests, boreal (far-northern) forests, and island forests have far fewer vines than do tropical rain forests. Some biologists have suggested that vines are less common in temperate and boreal forests because vines are less resistant to droughts and fires than are the trees that grow there. Vines in tropical rain forests generally do not have to adapt to long droughts or forest fires. Vines may be relatively uncommon on islands because the seeds of most vines, which are dispersed by wind, cannot reach islands to colonize them.

VINE A plant with a long, thin, often climbing stem.

RHIZOME A horizontal underground stem that often serves as a storage organ and a means of sexual reproduction.

TUBER The thickened end of a rhizome that is fleshy and enlarged for food storage.

BULB A rounded, fleshy underground bud that consists of a short stem with fleshy leaves.

CORM A short, thickened underground stem specialized for food storage and asexual reproduction.

STOLON An aerial horizontal stem with long internodes; often forms buds that develop into separate plants.

Asexual Reproduction of Stems

In **asexual reproduction,** a single individual may split, bud, or fragment, giving rise to offspring that are genetically similar to the parent. Asexual reproduction in flowering plants does not usually involve the formation of flowers, fruits, and seeds. In plants that reproduce asexually, the stems, leaves, and roots can form offspring, usually when part of an existing plant becomes separated from the rest of the plant. This part subsequently grows to form a complete, independent plant. Asexual reproduction always involves only one parent, and there is no fusion of reproductive cells (gametes); the offspring are virtually identical to one another and to the parent plant.

Flowering plants have evolved many methods of asexual reproduction, several of which involve modified stems: rhizomes, tubers, bulbs, corms, and stolons. A **rhizome** is a horizontal, underground stem that may or may not be fleshy. Fleshiness indicates that the stem is used for storing food materials such as starch (•**Figure 7-14a**). Although rhizomes resemble roots, they are really stems, as indicated by the presence of scale-like leaves, buds, nodes, and internodes. Rhizomes frequently branch in different directions. Over time, the old portion of the rhizome dies, and the two branches eventually separate to become two distinct plants. Irises, bamboos, ginger, and many grasses are examples of plants that reproduce asexually by forming rhizomes.

Some rhizomes produce greatly thickened ends called **tubers,** which are fleshy underground stems enlarged for food storage. When the attachment between a tuber and its parent plant breaks, often as a result of the death of the parent plant, the tuber grows into a separate plant. White potatoes and elephant's ear (*Caladium*) are examples of plants that produce tubers (•**Figure 7-14b**). The "eyes" of a potato are actually axillary buds, evidence that the tuber is an underground stem rather than a storage root like sweet potatoes or carrots.

A **bulb** is a modified underground bud in which fleshy storage leaves are attached to a short stem (•**Figure 7-14c;** also see Figure 8-15e). A bulb is rounded and is covered by paperlike bulb scales, which are modified leaves. It frequently forms small daughter bulbs (bulblets). These new bulbs are initially attached to the parent bulb, but when the parent bulb dies and rots away, each daughter bulb can become established as a separate plant. Lilies, tulips, onions, and daffodils are some of the plants that form bulbs.

A **corm** is a short, erect underground stem that superficially resembles a bulb (•**Figure 7-14d**). Unlike the bulb whose food is stored in underground leaves, the corm's storage organ is a thickened underground stem covered by papery scales (modified leaves). Axillary buds that will give rise to new corms frequently develop on a corm; the death of the parent corm separates these daughter corms, which then become established as separate plants. Familiar garden plants that produce corms include crocus, gladiolus, and cyclamen.

Stolons, or **runners,** are horizontal aerial stems that grow along the ground's surface and are characterized by long internodes (•**Figure 7-14e**). Buds develop along the stolon, and each bud gives rise to a new plant that roots in the ground. When the stolon dies, the daughter plants live separately. The strawberry plant produces stolons.

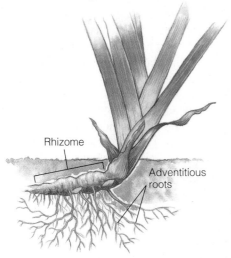

(a) Irises have horizontal underground stems called rhizomes. New aerial shoots arise from buds that develop along the rhizome.

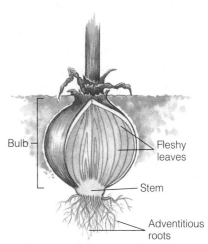

(b) Potato plants form rhizomes, which enlarge into tubers (the potatoes) at the ends.

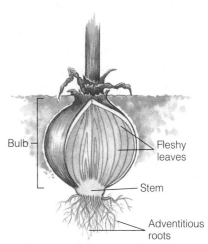

(c) A bulb is a short underground stem to which overlapping, fleshy leaves are attached; most of the bulb consists of leaves.

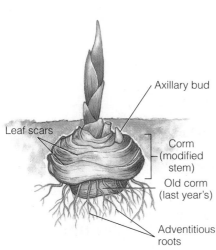

(d) A corm is an underground stem that is almost entirely stem tissue surrounded by a few papery scales.

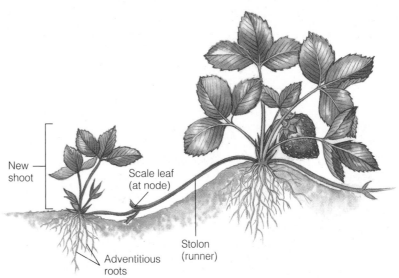

(e) Strawberries reproduce asexually by forming stolons, or runners. New plants (shoots and roots) are produced at every other node.

FIGURE 7-14 Modified stems.

STUDY OUTLINE

❶ Describe three functions of stems.

Stems support leaves and reproductive structures. Like other plant organs, stems conduct water, dissolved minerals, and carbohydrates. Stems produce new living tissues at apical meristems and, in plants with secondary growth, at lateral meristems.

❷ Relate the functions of each tissue in an herbaceous stem.

An herbaceous stem consists of an epidermis, vascular tissue, and cortex and pith, or ground tissue. The **epidermis,** a protective outer layer, is covered by a water-conserving **cuticle. Xylem** conducts water and dissolved minerals, and **phloem** conducts dissolved carbohydrates (sucrose). The **cortex, pith,** and ground tissue function primarily for storage.

❸ Contrast the structures of an herbaceous eudicot stem and a monocot stem.

Although herbaceous stems all have the same basic tissues, the arrangement of the tissues varies considerably. Herbaceous eudicot stems have the **vascular bundles** arranged in a circle in cross section and have a distinct cortex and pith. Monocot stems have scattered vascular bundles and ground tissue instead of a distinct cortex and pith.

❹ Distinguish between the structures of stems and roots.

Unlike roots, stems have **nodes** and **internodes,** and leaves and **buds.** Unlike stems, roots have root caps and root hairs. Internally, herbaceous roots possess an endodermis and pericycle; stems lack a pericycle and rarely have an endodermis.

❺ Outline the transition from primary growth to secondary growth in a woody stem. List the two lateral meristems, and describe the tissues that arise from each.

Secondary growth occurs in woody eudicots and conifers. The **vascular cambium** occurs between the primary xylem and primary phloem. Although the vascular cambium is not initially a solid cylinder of cells, it becomes continuous when production of secondary tissues begins. (Certain parenchyma cells between the vascular bundles, which retain the ability to divide, connect to the vascular cambium cells in each vascular bundle, forming a complete ring of vascular cambium.) Vascular cambium produces secondary xylem (wood) to the inside and secondary phloem (inner bark) to the outside. The **cork cambium,** which arises near the stem's surface, is either a continuous cylinder of dividing cells or a series of overlapping arcs of meristematic cells that form from parenchyma cells in successively deeper layers of the cortex and, eventually, secondary phloem. The cork cambium produces cork parenchyma to the inside and cork cells to the outside.

❻ Contrast the various stems that are specialized for asexual reproduction.

A **rhizome** (example: iris) is a horizontal underground stem. A **tuber** (example: white potato) is an underground stem greatly enlarged for storing food. A **bulb** (example: onion) is an underground bud with a shortened stem and fleshy storage leaves. A **corm** (example: crocus) is an underground stem that superficially resembles a bulb. **Stolons** (example: strawberry) are aerial horizontal stems with long internodes.

REVIEW QUESTIONS

1. List several functions of stems, and describe the tissue(s) responsible for each function.
2. What is the difference between terminal and axillary buds?
3. What is the function of bud scales? Of lenticels?
4. If you were examining a cross section of an herbaceous stem of a flowering plant, how would you determine whether it is a eudicot or a monocot?
5. List two root structures that stems lack. List two stem structures that roots lack.
6. What happens to the primary tissues of a stem when secondary growth occurs?
7. When a strip of bark is peeled off a tree branch, what tissues are removed?
8. Distinguish between the vascular cambium and the cork cambium, and describe the tissues that arise from each.
9. What is an annual ring, and how is it formed?
10. How are rhizomes and tubers involved in asexual reproduction?
11. How are corms and bulbs similar? How are they different?

12. Label the tissues of this herbaceous eudicot stem. Give at least one function for each tissue. (Refer to Figure 7-3a to check your answers.)

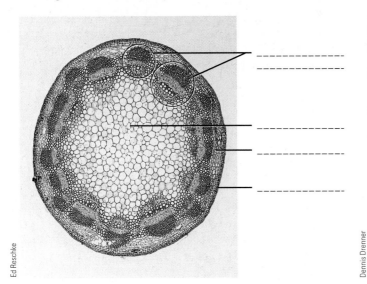

Ed Reschke

13. Label the tissues of this monocot stem. Give at least one function for each tissue. (Refer to Figure 7-4a to check your answers.)

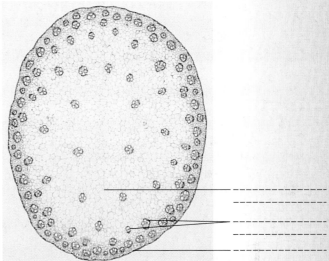

Dennis Drenner

THOUGHT QUESTIONS

1. When secondary growth commences, certain cells become meristematic and begin to divide. Could a tracheid ever do this? A sieve-tube element? Why or why not?
2. Why does the wood of many tropical trees lack annual rings?
3. Why is hardwood more desirable than softwood for making furniture? Explain on the basis of the structural differences between hardwood and softwood.
4. Someone gives you a plant structure that was found growing in the soil and asks you to determine whether it is a root or an underground stem. How would you do this without a microscope? With a microscope?
5. The term *dendroclimatology* was not used in this chapter. Based on what you have learned, propose a definition for this term.

Visit us on the web at http://www.thomsonedu.com/biology/berg/ for additional resources, such as flashcards, tutorial quizzes, further readings, and web links.

Plant Organs: Leaves

What leaf has had the greatest impact on human history? On human health? Many scholars would agree that the answer to both questions is tobacco (*Nicotiana tabacum*), a leaf crop that humans have cultivated for centuries.

Christopher Columbus and other early European explorers discovered the native peoples of North and South America growing tobacco, drying and curing the leaves, and then smoking, chewing, or snuffing them. The tobacco was used for medicinal purposes as well as for religious rituals. Shamans (medicine men) in the Amazon, for example, fasted and smoked large quantities of tobacco to produce hallucinations, which they interpreted as visits from spirits.

When Columbus introduced tobacco to Europe, it did not gain immediate acceptance. Within a century, however, many Europeans had adopted the habit of smoking, its popularity enhanced by tobacco's supposed medicinal values. It was touted as a cure for everything from headaches to snakebites and was also used as an aphrodisiac. Although it has no medicinal uses today, tobacco's popularity has continued to the present. Currently, millions of people in virtually every country in the world smoke tobacco.

In the 1600s farmers in Virginia and other English colonies began raising tobacco leaf to meet the demand in Europe, and many of them became wealthy. As a "cash crop," tobacco provided the grower with more profit per acre than any food crop. Today, tobacco leaf still commands a high price per pound, and in the United States alone tens of thousands of farmers raise tobacco on more than 800,000 hectares (2 million acres) of land.

The physiologically active ingredients of tobacco are nicotine and related *alkaloids,* which are colorless, bitter chemicals that affect the central nervous system. When inhaled, nicotine, which is habit forming, is absorbed through the lung membranes and produces a stimulating sensation. Pure nicotine, however, is poisonous; absorbed in large amounts, it produces nausea, dizziness, and hallucinations. Medical evidence is overwhelming that tobacco smoking contributes to serious health problems, including heart disease, stroke, emphysema, chronic bronchitis, male impotence, and cancer. If everyone in the United States stopped smoking, more than 300,000 lives would be saved each year.

Tobacco (*Nicotiana tabacum*) flowering in a field.

Leaf Form and Structure

Leaves are the most variable plant organ—so variable that plant biologists had to develop an entire set of terminology to describe their shapes, margins (edges), vein patterns, and the way they attach to stems. Because each leaf is characteristic of the species on which it grows, many plants can be identified by their leaves alone. Leaves may be round, needlelike, scalelike, cylindrical, heart shaped, fan shaped, or thin and narrow. They vary in size from those of the raffia palm (*Raphia ruffia*), whose leaves often grow to more than 20 meters (65 feet) long, to those of water-meal (*Wolffia*), whose leaves are so small that 16 of them laid end to end measure 2.5 centimeters (1 inch).

Most leaves are composed of two parts, a blade and a petiole (•**Figure 8-1;** also see *Plants and People: Important Edible Leaf Crops*). The broad, flat portion of a leaf is the **blade,** and the stalk that attaches the blade to the stem is the **petiole.** Many leaves also have **stipules,** which are leaflike outgrowths usually present in pairs at the base of the petiole. Some leaves do not have petioles or stipules. A *sessile* leaf lacks a petiole and has a blade directly attached to the stem, often by a *sheath* that encircles the stem.

Leaves may be **simple** (having a single blade) or **compound** (having a blade divided into two or more leaflets) (•**Figure 8-2a**). Compound leaves are **pinnately compound** (the leaflets are borne on an axis that is a continuation of the petiole) or **palmately compound** (the leaflets arise from a common point at the end of the petiole).

Sometimes it is difficult to tell whether a plant has formed one compound leaf or a small stem bearing several simple leaves. One easy way to determine whether a plant has simple or compound leaves is to look for axillary buds, so called because each develops in a leaf *axil* (see Figure 8-1). Axillary buds form at the base of a leaf, whether it is simple or compound. However, axillary buds never develop at the base of leaflets. Also, the leaflets of a compound leaf lie in a single plane (you can lay a compound leaf flat on a table), whereas simple leaves are usually not arranged in one plane on a stem.

Leaves are arranged on a stem in one of three possible ways (•**Figure 8-2b**). Plants such as beeches and walnuts have an **alternate leaf arrangement,** with one leaf at each node. *Nodes* are the areas on a stem where leaves are attached. In an **opposite leaf arrangement,** as occurs in lilacs, maples, and ashes, two leaves grow at each node. In a **whorled leaf arrangement,** as occurs in catalpa trees, three or more leaves grow at each node.

Leaf blades may possess parallel or netted venation (•**Figure 8-2c**). In **parallel venation** the many primary veins—strands of vascular tissue—run approximately

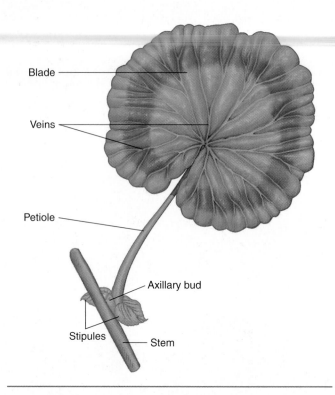

FIGURE 8-1 A "typical" leaf.
The geranium leaf consists of a blade and petiole. Two stipules emerge at the base of the leaf. Note the axillary bud in the leaf axil.

parallel to one another along the leaf's long axis, with smaller veins forming connections between the primary veins. In **netted venation** smaller and smaller veins branch off the larger veins in such a way that they resemble a net. Parallel veins are generally characteristic of monocots, whereas netted veins are generally characteristic of eudicots. Netted veins can be **palmately netted,** in which several major veins radiate out from one point, or **pinnately netted,** in which major veins branch off along the entire length of the midvein (the main or central vein of a leaf).

Epidermis, mesophyll, xylem, and phloem are the major tissues of the leaf

The leaf is a complex organ composed of several tissues organized to optimize photosynthesis (•**Figure 8-3**). The leaf blade has upper and lower surfaces consisting of an epidermal layer. The **upper epidermis** covers the upper surface, and the **lower epidermis** covers the lower surface. Most cells in these layers are living parenchyma cells that lack chloroplasts and are relatively transparent. One interesting feature of leaf epidermal cells is that the cell wall facing toward the outside environment is somewhat thicker than the cell wall facing inward. This

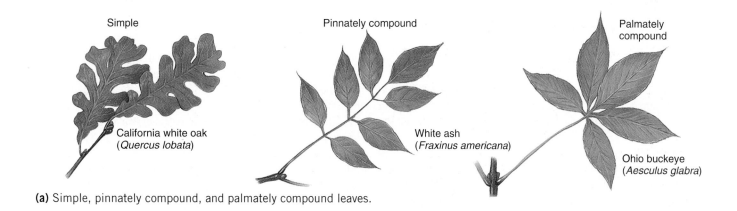

(a) Simple, pinnately compound, and palmately compound leaves.

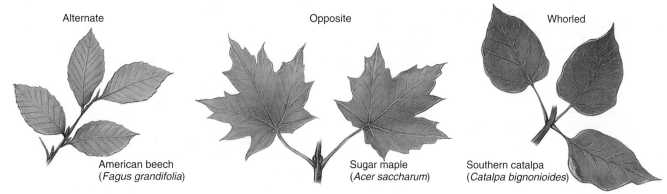

(b) Leaf arrangement may be alternate, opposite, or whorled. (*Note*: Catalpa trees may have opposite or whorled leaf arrangement; only whorled is shown.)

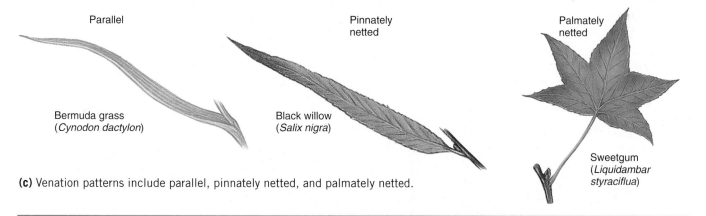

(c) Venation patterns include parallel, pinnately netted, and palmately netted.

FIGURE 8-2 Leaf morphology.
All leaves shown are woody eudicot trees from North America, except Bermuda grass, which is an herbaceous monocot native to Europe and Asia.

extra thickness may provide the plant with additional protection against injury or water loss.

    A leaf has a large surface area exposed to the atmosphere; as a result, water loss by evaporation from the leaf's surface is unavoidable. However, epidermal cells secrete a waxy layer, the **cuticle,** that reduces water loss

BLADE The broad, flat part of a leaf.

PETIOLE The part of a leaf that attaches the blade to the stem.

CUTICLE A waxy covering over the epidermis of the aerial parts (leaves and stems) of a plant.

PLANTS AND PEOPLE | Important Edible Leaf Crops

Numerous vegetable crops are grown for their edible leaves. Leaves, especially dark green leaves, are particularly good sources of vitamins A and C and the minerals iron and calcium. Cabbage, lettuce, spinach, celery, and rhubarb are representative leaf crops. Also, the leaves of many herbs contain pungent chemicals that make these leaves useful for flavoring food.

Cabbage (*Brassica oleracea*) is a biennial that is harvested as an annual. Probably native to the eastern Mediterranean region, cabbage has been grown as a vegetable crop for more than 4000 years. A head of cabbage consists of a shortened stem with numerous short-petioled leaves overlapping a small terminal bud. Some cabbage varieties are green, and others are purple. Cabbage is eaten raw, cooked, or fermented (as sauerkraut).

Lettuce (*Latuca sativa*) is an annual, native to the Mediterranean region, that has been cultivated for more than 2500 years. There are three types of lettuce: head lettuce (for example, iceberg lettuce), which forms a tight, rounded head; cos or romaine type (for example, romaine lettuce), which forms a loose cylindrical head; and leaf type (for example, red lettuce and Bibb lettuce), which does not form a head. Lettuce is eaten fresh as a salad vegetable.

Spinach (*Spinacia oleracea*) is an annual that is native to the eastern Mediterranean region. It has been cultivated for at least 2000 years for its dark green leaves. The two main types of spinach are smooth leaf and wrinkled leaf. Spinach is eaten fresh as a salad vegetable or cooked.

Celery (*Apium graveolens*) is a biennial that is harvested as an annual. Native to Europe, celery was first used for medicinal purposes; folklore indicates that it may be useful in treating diabetes. Only later was it accepted as a food. Celery has been a common leaf crop since the early 1600s. The edible part is the thick, fleshy petiole, which is eaten raw or cooked.

Rhubarb (*Rheum rhaponticum*), native to Siberia, is an herbaceous perennial. Rhubarb was formerly taken as a folk medicine to expel parasitic worms from the intestinal tract. Today, it is cultivated for its thick, fleshy petioles, which are stewed and eaten in sauces and pies. The leaf blades, however, contain toxic levels of oxalic acid and should not be consumed.

Many herbs that people use to flavor food come from leaves—for example, basil, bay leaves, marjoram, and tarragon. Basil (*Ocimum basilicum*) is native to India. When dried, basil leaves have a distinctive aroma that complements many foods, particularly those containing tomatoes. Bay leaves (*Laurus nobilis*) come from an evergreen tree native to the Mediterranean. Bay leaves add a pungent flavor to soups, chowders, and fish. Marjoram (*Majorana hortensis*), which is native to the Mediterranean region, has been used medicinally for centuries. Today, the dried leaves are added to almost any food except sweets. Tarragon (*Artemisia dracunculus*) comes from a perennial shrub that originated in Asia. Tarragon is responsible for the distinctive taste of béarnaise sauce.

Spinach   Celery   Iceberg lettuce   Rhubarb   Cabbage

from their exterior walls. The cuticle, which consists primarily of a waxy substance called **cutin,** varies in thickness in different plants, partly as a result of environmental conditions. As one might expect, the leaves of plants adapted to hot, dry climates have thick cuticles. Furthermore, a leaf's exposed (and warmer) upper epidermis generally has a thicker cuticle than the shaded (and cooler) lower epidermis.

The epidermis contains minute openings, or **stomata** (sing., *stoma*), for gas exchange. Each stoma is flanked by two specialized epidermal **guard cells** (see Figure 5-9). Changes in the shape of each pair of guard cells open and close the stoma. Guard cells are usually the only epidermal cells with chloroplasts.

Guard cells are associated with special epidermal cells called **subsidiary cells** that are often structurally

FIGURE IN FOCUS

Leaves contain all three tissue systems found in plants: The dermal tissue system, represented by the upper and lower epidermis; the ground tissue system, represented by the mesophyll; and the vascular tissue system, represented by the xylem and phloem in the veins.

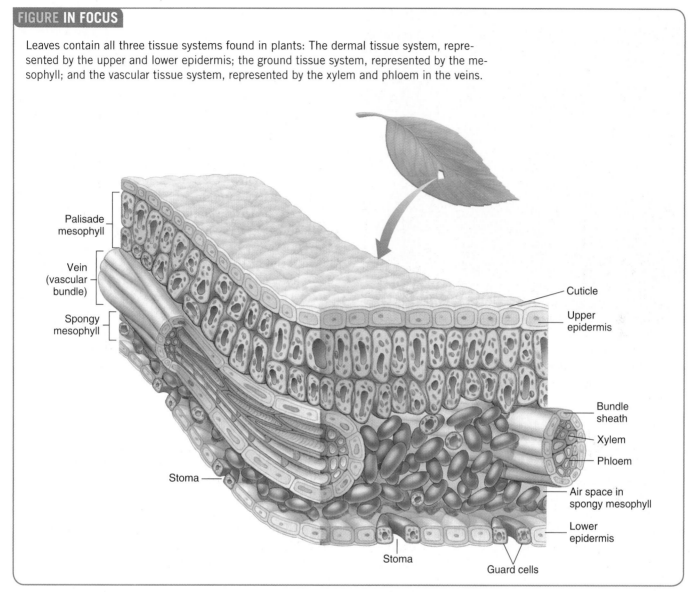

FIGURE 8-3 Tissues in a typical leaf blade.
The blade is covered by an upper and lower epidermis. The photosynthetic tissue, called mesophyll, is often arranged into palisade and spongy layers. Veins branch throughout the mesophyll.

different from other epidermal cells. Subsidiary cells provide a reservoir of water and ions that move into and out of the guard cells as they change shape during stomatal opening and closing.

Stomata are especially numerous on the lower epidermis of horizontally oriented leaves—an average of about 100 stomata per square millimeter, although the actual number varies widely—and in many species are located *only* on the lower surface. The lower epidermis of apple (*Malus sylvestris*) leaves, for example, has almost 400 stomata per square millimeter, whereas the upper epidermis has none. This adaptation reduces wa-

ter loss, in part because stomata on the lower epidermis are shielded from direct sunlight and are therefore cooler than those on the upper epidermis.

The epidermis of many leaves is covered with various hairlike structures called **trichomes** (•**Figure 8-4**). Some leaves, such as those of the popular cultivated plant

STOMA A small pore flanked by guard cells in the epidermis.

GUARD CELL A cell in the epidermis of a stem or leaf; two guard cells form a pore, called a stoma, for gas exchange.

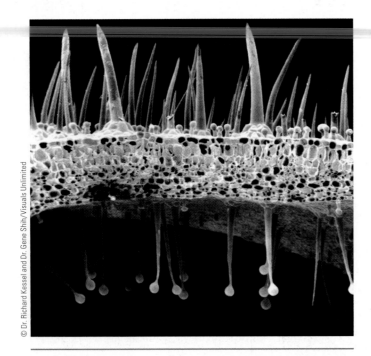

© Dr. Richard Kessel and Dr. Gene Shih/Visuals Unlimited

FIGURE 8-4 Trichomes.
The epidermis is often covered with trichomes that may limit the transpiration of water, discourage herbivores, sting, or perform other functions. This scanning electron micrograph shows a geranium (*Pelargonium*) leaf with trichomes on both the upper and lower epidermis.

lamb's ears (*Stachys byzantina*), have so many trichomes that they feel fuzzy. Trichomes of some plants help reduce water loss from the leaf surface by retaining a layer of moist air next to the leaf and by reflecting excessive sunlight, thereby protecting the plant from overheating. Some trichomes secrete stinging irritants that deter animals that feed on plants. In addition, a leaf covered with trichomes is difficult for an insect to walk over or eat. Other trichomes excrete excess salts absorbed from a salty soil.

The photosynthetic ground tissue of the leaf, the **mesophyll,** is sandwiched between the upper epidermis and the lower epidermis. The word *mesophyll* comes from Greek and means "the middle of the leaf." Mesophyll cells, which are parenchyma cells packed with chloroplasts, are loosely arranged with many air spaces between them that facilitate gas exchange. These intercellular air spaces account for as much as 70 percent of the leaf's volume.

In many plants, the mesophyll is divided into two sublayers. Toward the upper epidermis, the columnar cells are stacked closely together in a layer called **palisade mesophyll.** In the lower portion, the cells are more loosely and more irregularly arranged, in a layer called **spongy mesophyll.** The two layers have different functions. Palisade mesophyll is the main site of photosynthesis in the leaf. Photosynthesis also occurs in the spongy mesophyll, but the primary function of the spongy mesophyll is to allow diffusion of gases, particularly CO_2, throughout the leaf's interior.

Palisade mesophyll may be further organized into one, two, three, or even more layers of cells. The presence of additional layers of palisade mesophyll is at least partly an adaptation to environmental conditions. Leaves exposed to direct sunlight contain more layers of palisade mesophyll than do shaded leaves on the same plant. In direct sunlight, the light is strong enough to effectively penetrate multiple layers of palisade mesophyll, allowing all layers to photosynthesize efficiently.

The **veins,** or **vascular bundles,** of a leaf extend through the mesophyll. Branching is extensive, and no mesophyll cell is more than two or three cells away from a vein. Therefore, the slow process of diffusion does not limit the movement of needed resources between mesophyll cells and veins. Each vein contains two types of vascular tissue: xylem and phloem. **Xylem,** which conducts water and dissolved minerals (inorganic nutrients), is usually located on the upper side of a vein, toward the upper epidermis, whereas **phloem,** which conducts dissolved sugars, is usually confined to the lower side of a vein.

One or more layers of nonvascular cells surround the larger veins and make up the **bundle sheath.** Bundle sheaths are composed of parenchyma or sclerenchyma cells (see Chapter 5). Frequently, the bundle sheath has support columns, called **bundle sheath extensions,** that extend through the mesophyll from the upper epidermis to the lower epidermis (•**Figure 8-5**). Bundle sheath extensions may be composed of parenchyma, collenchyma, or sclerenchyma cells.

Leaf structure differs in eudicots and monocots

A eudicot leaf is usually composed of a broad, flattened blade and a petiole. As mentioned previously, eudicot leaves typically have netted venation. In contrast, monocot leaves often lack a petiole; they are narrow, and the base of the leaf often wraps around the stem, forming a sheath. Parallel venation is characteristic of monocot leaves.

The internal leaf anatomies of eudicots and certain monocots also differ (•**Figure 8-6**). Although most eudicots and monocots have both palisade and spongy layers, some monocots (corn and other grasses) do not have mesophyll differentiated into distinct palisade and spongy layers. Because eudicots have netted veins, a cross section of a eudicot blade often shows veins in both cross-sectional and lengthwise views. In a cross section of a monocot leaf, the parallel venation pattern

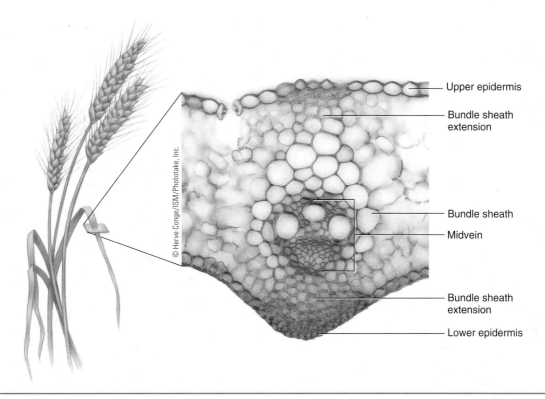

FIGURE 8-5 Bundle sheath extensions.
Cross section of a wheat (*Triticum aestivum*) midvein, showing a bundle sheath extension; wheat is a monocot. Xylem makes up the top part of the midvein, and phloem makes up the bottom part.

produces evenly spaced veins, all of which appear in cross section.

The upper epidermises of the leaves of certain grasses have large, thin-walled cells called **bulliform cells** on both sides of the midvein (•**Figure 8-7**). These cells appear to be involved in the rolling or folding inward of the leaf during drought. When water is plentiful, bulliform cells are turgid (swollen with water) and the leaf is open. When bulliform cells lose water (as they may during a drought), the leaf folds inward, decreasing its surface area exposed to the air, an action that reduces water loss by evaporation.

Differences between the guard cells in eudicot and certain monocot leaves also occur (•**Figure 8-8** on page 160). The guard cells of eudicots and many monocots are shaped like kidney beans. Other monocot leaves (those of grasses, reeds, and sedges) have guard cells shaped like dumbbells. These structural differences affect how the cells swell or shrink to open or close the stoma.

Leaf structure is related to function

The primary function of leaves is to collect radiant energy and convert it to the chemical energy stored in the bonds of organic molecules such as glucose. During this process, called **photosynthesis,** plants take relatively simple inorganic molecules (carbon dioxide and water) and convert them to sugar. Oxygen is given off as a by-product.

How is leaf structure related to the leaf's primary function, photosynthesis? Most leaves are thin and flat, a shape that allows maximum absorption of light energy and efficient internal diffusion of gases. As a result of their ordered arrangement on the stem, leaves efficiently catch the sun's rays. The leaves of plants form an intricate green mosaic, bathed in sunlight and atmospheric gases.

MESOPHYLL The photosynthetic tissue in the interior of a leaf.

BUNDLE SHEATH A ring of parenchyma or sclerenchyma cells surrounding the vascular bundle in a leaf.

PHOTOSYNTHESIS The biological process that includes the capture of light energy and its transformation into chemical energy of organic molecules (such as glucose), which are manufactured from carbon dioxide and water.

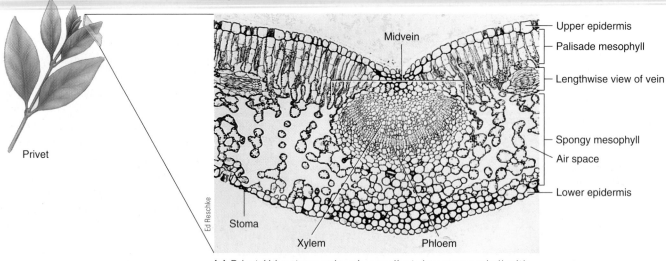

Midvein

Upper epidermis

Palisade mesophyll

Lengthwise view of vein

Spongy mesophyll

Air space

Lower epidermis

Privet

Ed Reschke

Stoma

Xylem

Phloem

(a) Privet (*Ligustrum vulgare*), a eudicot, has a mesophyll with distinct palisade and spongy sections.

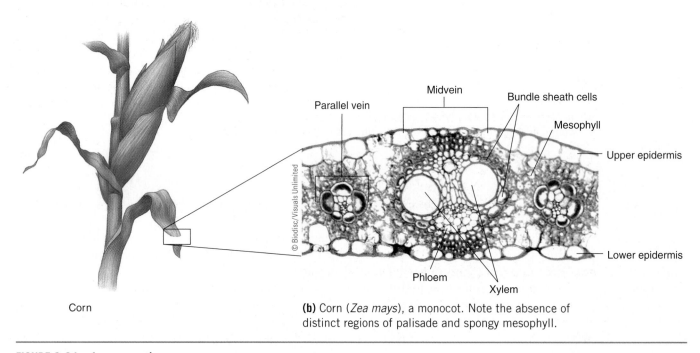

Midvein

Bundle sheath cells

Parallel vein

Mesophyll

Upper epidermis

© Biodisc/Visuals Unlimited

Lower epidermis

Phloem

Xylem

Corn

(b) Corn (*Zea mays*), a monocot. Note the absence of distinct regions of palisade and spongy mesophyll.

FIGURE 8-6 Leaf cross sections.

The epidermis of a leaf is relatively transparent and allows light to penetrate to the interior of the leaf, where the photosynthetic ground tissue, the mesophyll, is located. Water required for photosynthesis is obtained from the soil and transported in the xylem to the leaf, where it diffuses into the mesophyll and moistens the surfaces of mesophyll cells. The loose arrangement of the mesophyll cells, with air spaces between cells, allows for rapid diffusion of carbon dioxide to the mesophyll cell surfaces; there it dissolves in a film of water before diffusing into the cells.

The veins not only supply the photosynthetic ground tissue with water and minerals (from the roots, by way of the xylem) but carry (in the phloem) the dissolved sugar produced during photosynthesis to all parts of the plant. Bundle sheaths and bundle sheath extensions associated with the veins provide additional support to prevent the leaf, which is structurally weak because of the large amount of air space in the mesophyll, from collapsing.

Stomata, which dot the leaf surfaces, permit the exchange of gases between the atmosphere and the inside of the leaf. Carbon dioxide, a raw material of photosynthesis, diffuses through stomata from the surrounding atmosphere into the leaf, and the oxygen produced during photosynthesis diffuses out of the leaf through stomata. Stomata also permit other gases, including air pollutants, to enter the leaf (see *Plants and the Environment: The Effects of Air Pollution on Leaves*).

Stomatal opening and closing are controlled by the amount of water in the guard cells

Stomata are adjustable pores that are usually open during the day when carbon dioxide is required for photosynthesis (the CO_2 diffuses into the leaf through the pore from the atmosphere) and closed at night when photosynthesis is shut down. The opening and closing of stomata are controlled by changes in the shape of the two guard cells that surround each pore. When water moves into guard cells from surrounding cells, they become turgid (swollen) and bend, producing a pore. When water leaves the guard cells, they become flaccid (limp) and collapse against one another, closing the pore.

Although daylight and darkness trigger the opening and closing of guard cells, other environmental factors are also involved, including carbon dioxide concentration. A low concentration of CO_2 in the leaf induces stomata to open, even in the dark. The effects of light and CO_2 concentration on stomatal opening are interrelated. Photosynthesis, which occurs in the presence of light, reduces the internal concentration

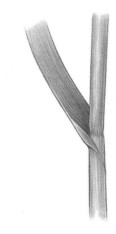
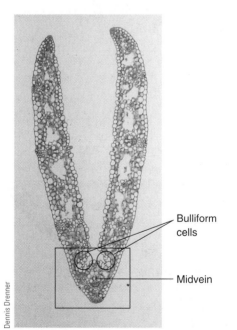
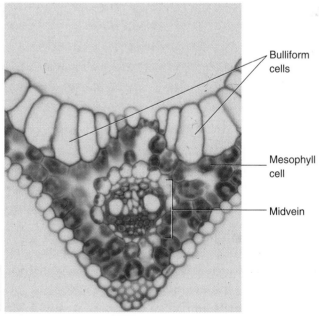

(a) A folded leaf blade. The inconspicuous bulliform cells occur in the upper epidermis on both sides of the midvein.

(b) An expanded leaf blade. A higher magnification of the midvein region shows the enlarged, turgid bulliform cells.

FIGURE 8-7 Bulliform cells.
Shown are leaves of Kentucky bluegrass (*Poa pratense*), a monocot.

Open   Closed   Open   Closed

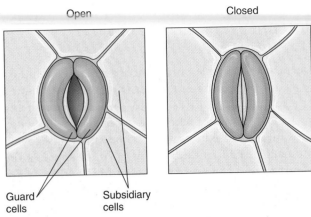

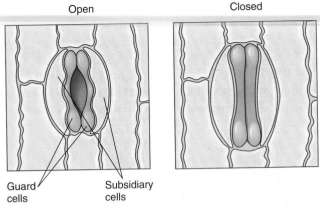

(a) Guard cells of eudicots and many monocots are bean shaped.

(b) Some monocot guard cells (those of grasses, reeds, and sedges) are narrow in the center and thicker at each end.

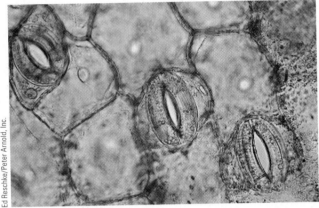

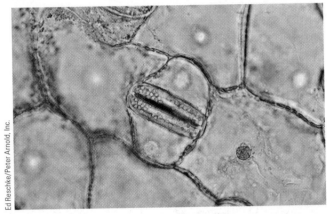

(c) LM of open *Tradescantia* guard cells.

(d) LM of closed *Tradescantia* guard cell.

FIGURE 8-8 Variation in guard cells.
Guard cells are associated with special epidermal cells called subsidiary cells.

of CO_2 in the leaf, thus triggering stomatal opening. Another environmental factor that affects stomatal opening and closing is dehydration (water stress). During a prolonged drought, stomata remain closed, even during the day. Stomatal opening and closing are also under hormonal control.

An internal biological clock that measures time may regulate opening and closing of stomata. After plants are placed in continuous darkness, their stomata continue to open and close at more or less the same time each day. Such rhythms that follow an approximate 24-hour cycle are known as **circadian rhythms.**

Blue light triggers stomatal opening

Both the gas exchange needed for photosynthesis and the loss of water by transpiration (discussed shortly) occur through open stomata. These processes occur during the day, when stomata are open. At night, stomata typically close.

Data from numerous experiments and observations are beginning to explain the details of stomatal movements. Let us begin with stomatal opening, which occurs when the plant detects light from the rising sun. You already know that light is a form of energy; plants absorb light and convert it to chemical energy in the process of photosynthesis. However, light is also an important *environmental signal* for plants—that is, light provides plants with information about their environment that they use to modify various activities at the molecular and cellular levels.

In stomatal opening and several other plant responses, **blue light** is an environmental signal. Any plant response to light must involve a **pigment,** a molecule that absorbs the light before the induction of a particular biological response. Evidence suggests the pigment involved in stomatal opening and closing is yellow (yellow pigments strongly absorb blue light) and located in the plasma membranes of the guard cells.

PLANTS AND THE ENVIRONMENT | The Effects of Air Pollution on Leaves

The air we breathe is often dirty and contaminated with many pollutants, particularly in urban areas. **Air pollution** consists of gases, liquids, or solids present in the atmosphere at high enough levels to harm humans, other animals, plants, or materials. Although air pollutants can come from natural sources—as, for example, a lightning-caused fire or a volcanic eruption—human activities are a major cause of global air pollution. Motor vehicles and industry are the two main human-caused sources of air pollution.

Air pollution can damage all parts of a plant, but leaves are particularly susceptible because of their structure and function (see figure). The thin blade provides a large surface area that comes into contact with the surrounding air. The thousands of stomatal pores that dot the epidermis and allow gas exchange with the atmosphere also permit pollutants to diffuse into the leaf. Just as lungs, the organs of gas exchange in humans, often show the effects of air pollution, so, too, the leaves of a plant are most affected.

Many studies have shown that high levels of most forms of air pollution reduce the overall productivity of crop plants. The worst pollutant in terms of yield loss is ozone, a toxic gas produced when sunlight catalyzes a reaction between pollutants emitted by motor vehicles and industries. Ozone has been observed to reduce soybean yields by as much as 35 percent, and the average yield reduction from exposure to ozone is 15 percent. Ozone inhibits photosynthesis because it damages the mesophyll cells, probably by altering the permeability of their cell membranes. Exposure to low levels of air pollution often causes a decline in photosynthesis without other symptoms of injury. Lesions on leaves and other

© Science VU/Visuals Unlimited

■ **Effects of air pollution on plants.** Plants exposed to air pollution exhibit a variety of symptoms, including reduced growth. Compare the size and overall vigor of the radish (*Raphanus sativus*) grown in clear air (*left*) and the radish damaged by air pollution (*right*). Note the extensive leaf damage.

obvious symptoms appear at much higher levels of air pollution.

When air pollution is combined with other environmental stressors (such as low winter temperatures; prolonged droughts; insects; and bacterial, fungal, and viral diseases), it can cause plants to decline and die (see *Plants and the Environment: Acid Rain,* in Chapter 2). More than half of the red spruce trees in the mountains of the northeastern United States have died since the mid-1970s, and sugar maples in eastern Canada and the United States are also dying. Many still-living trees are exhibiting symptoms of **forest decline,** characterized by gradual deterioration and eventual death of trees. The general symptoms of forest decline are reduced vigor and growth, but some plants exhibit specific symptoms, such as yellowing of needles in conifers.

Forest decline is more pronounced at higher elevations, possibly because most trees growing at high elevations are at the limit of their normal range and are therefore more susceptible to wind and low temperatures.

Many factors can interact to decrease the health of trees, and no single factor accounts for the recent instances of forest decline. Several air pollutants have been implicated, including acid rain, ozone, and toxic heavy metals such as lead, cadmium, and copper. To complicate matters further, the actual causes of forest decline may vary from one tree species to another and from one location to another. Thus, forest decline appears to result from the combination of multiple stressors. When one or more stressors weaken a tree, an additional stressor may be enough to cause its death.

Water follows potassium and chloride ions into guard cells, which causes the guard cells to bow apart and form a stoma.

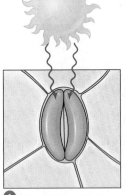 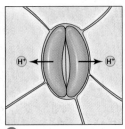 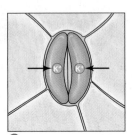 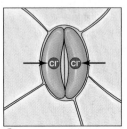 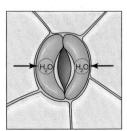

① Blue light activates proton pumps.

② Protons are pumped out of guard cells, forming proton gradient.

③ Potassium ions enter guard cells through voltage-activated ion channels.

④ Chloride ions also enter guard cells through ion channels.

⑤ Water enters guard cells by osmosis, and stoma opens.

FIGURE 8-9 Mechanism of stomatal opening.
A proton (H^+) gradient drives the accumulation of osmotically active ions (K^+ and Cl^-) in the guard cells.

In step **①** of •**Figure 8-9,** blue light, which is a component of sunlight, triggers the activation of proton pumps located in the guard-cell plasma membrane.

In step **②**, the proton pumps use ATP energy to actively transport protons (H^+) out of the guard cells. The H^+ that are pumped are formed when malic acid produced in the guard cells ionizes to form H^+ and negatively charged malate ions. As the proton pumps in the plasma membranes of guard cells transport protons out of the guard cells, a **proton gradient** with a charge and concentration difference forms on the two sides of the guard-cell plasma membrane.

In step **③**, the resulting gradient of H^+ drives the **facilitated diffusion** of large numbers of potassium ions *into* guard cells. This movement occurs through **voltage-activated ion channels,** which open when a certain difference in charge between the two sides of the guard-cell plasma membrane is attained. As shown in step **④**, chloride ions are also taken into the guard cells through ion channels in the guard-cell plasma membrane. The negatively charged chloride ions help electrically balance the positively charged potassium ions.

The potassium and chloride ions accumulate in the vacuoles of the guard cells, increasing the solute concentration in the vacuoles. You may recall from the discussion of osmosis in Chapter 3 that when a cell has a solute concentration greater than that of surrounding cells, water flows *into* the cell. Thus, in step **⑤**, water enters the guard cells from surrounding epidermal cells by **osmosis.** The increased turgidity of the guard cells changes their shape, because the thickened inner cell walls do not expand as much as the outer walls, so the stoma open. To summarize:

Blue light activates proton pumps ⟶ proton pump moves H^+ out of guard cells ⟶ K^+ and Cl^- move into guard cells through voltage-activated ion channels ⟶ water diffuses by osmosis into guard cells ⟶ guard cells change shape, and stoma opens

In the late afternoon or early evening, stomata close, but not by an exact reversal of the opening process. Recent studies have demonstrated that the concentration of potassium ions in guard cells slowly decreases during the day. However, the guard-cell concentration of sucrose (common table sugar), another osmotically active substance, increases during the day—maintaining the open pore—and then slowly decreases as evening approaches. This sucrose comes from the splitting of starch, which is stored in the guard-cell chloroplasts. As evening approaches, the sucrose concentration in the guard cells declines as sucrose is converted back to starch (which is osmotically inactive), water leaves by osmosis, the guard cells lose their turgidity, and the pore closes.

In summary, different mechanisms may regulate the opening and closing of stomata. The uptake of potassium and chloride ions is mainly associated with stomatal opening, and the declining concentration of sucrose is mainly associated with stomatal closing. (In Chapter 11, we discuss other blue-light responses in plants.)

Each type of plant has leaves adapted to its environment

Leaf structure reflects the environment to which a particular plant is adapted. Although both aquatic plants and plants adapted to dry conditions perform photosynthesis and have the same basic leaf anatomy, their leaves are modified to enable them to survive different environmental conditions. The leaves of water lilies, for example, have stomata on the upper epidermis, which is exposed to the air, rather than the lower epidermis, which touches the water. The petioles of water lilies are long enough to allow the blade to float on the water's surface (•**Figure 8-10**), and if the water level rises, their petioles elongate. Large air spaces in the mesophyll make the floating blade buoyant. These petioles and other submerged parts have an internal system of air ducts; oxygen-containing air moves through these ducts from the floating leaves to the underwater roots and stems, which live in a poorly aerated environment.

The leaves of **conifers**—an important group of trees and shrubs that includes pine, spruce, fir, redwood, and cedar—are waxy needles. Most conifers are evergreen, which means that they lose leaves throughout the year rather than during certain seasons. Conifers dominate a large portion of Earth's land area, particularly in northern forests and mountains. Their needles have structural adaptations that help them survive winter, the driest part of the year. Winter is arid even in areas of heavy snows, because roots cannot absorb water from soil when the soil temperature is low. Indeed, many of the structural features of needles are also found in many desert plants.

Blade   Petiole

© Tony Bomford/Jupiterimages

FIGURE 8-10 Water lily leaves.
This view of water lily (*Nymphaea*) leaves shows their petioles as well as their blades, which float on the water's surface. Water lilies have special adaptations, such as stomata on the upper epidermis of the blade rather than on the lower epidermis, that enable them to thrive in a watery environment.

•**Figure 8-11** shows a cross section of a pine needle. Note that the needle is somewhat thickened rather than flat and bladelike. The needle's relative thickness, which results in less surface area exposed to the air, reduces water loss. Other features that help conserve water include the thick, waxy cuticle and sunken stomata; these permit gas exchange while minimizing water loss. Thus, needles help conifers tolerate the dry (low relative humidity) winds that occur during winter. With the warming of spring, water becomes available again, and the needles quickly resume photosynthesis. In contrast, plants that are **deciduous**—that is, shed all their leaves during a particular season—have a lag period before resuming growth, because they first have to grow new leaves.

PROTON GRADIENT The difference in concentration of protons on the two sides of a cell membrane; contains potential energy that can be used to form ATP or do work in the cell.

FACILITATED DIFFUSION The diffusion of materials from a region of higher concentration to a region of lower concentration through special passageways in the membrane.

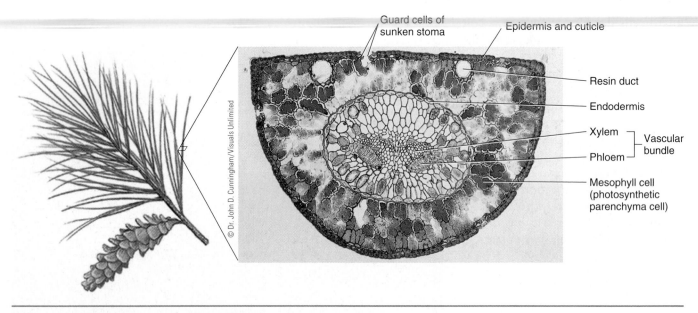

FIGURE 8-11 Cross section of a pine (*Pinus*) needle.
Note the sunken stomata, which trap a layer of moist air over the pore.

Transpiration and Guttation

Despite leaf adaptations such as the cuticle, approximately 99 percent of the water that a plant absorbs from the soil is lost by evaporation from the leaves and, to a lesser extent, the stems. Loss of water vapor from aerial plant parts is called **transpiration.**

The cuticle is extremely effective in reducing water loss from transpiration. It is estimated that only 1 to 3 percent of the water lost from a plant passes directly through the cuticle. Most transpiration occurs through open stomata. The numerous stomatal pores that are so effective in gas exchange for photosynthesis also provide openings through which water vapor escapes. In addition, the loose arrangement of the cells of the spongy mesophyll provides a large surface area within the leaf from which water can evaporate.

Several environmental factors influence the rate of transpiration. More water is lost from plant surfaces at higher air temperatures. Light increases the transpiration rate, in part because it triggers the opening of stomata and in part because it increases the leaf's temperature. Wind and dry air increase transpiration, but humid air *decreases* transpiration because the air is already saturated, or nearly so, with water vapor.

Although transpiration may seem wasteful, particularly to farmers in arid lands, it is an essential process that has adaptive value. Transpiration is responsible for water movement in plants, and without it, water from the soil would not reach the leaves (see Chapter 10). The large amount of water that plants lose by transpiration may provide some additional benefits for plants. Transpiration, like sweating in humans, cools the leaves and stems. When water changes from a liquid state to a vapor, it absorbs a great deal of heat. When the water molecules leave the plant as water vapor, they carry this heat with them. Thus, the cooling effect of transpiration may prevent the plant from overheating, particularly in direct sunlight. On a hot summer day, for example, the internal temperature of leaves is measurably less than that of the surrounding air. (See *Plants and the Environment: Trees, Transpiration, and Climate.*)

A second benefit of transpiration is that it moves essential minerals from the roots to stems and leaves. The water a plant transpires is initially absorbed from the soil, where it is present not as pure water but as a dilute solution of dissolved mineral salts. The water and dissolved minerals are then transported in the xylem throughout the plant body, including its leaves. Water moves from the plant to the atmosphere during transpiration, but minerals remain in plant tissues. Many of these minerals are required for the plant's growth. It has been suggested that transpiration enables a plant to take in sufficient water to provide enough essential minerals and that plants cannot satisfy their mineral requirements if the transpiration rate is not high enough.

There is no doubt, however, that under certain circumstances transpiration can be harmful to a plant. On hot summer days, plants frequently lose more water by

PLANTS ᴀɴᴅ ᴛʜᴇ ENVIRONMENT | Trees, Transpiration, and Climate

Climate is the average weather conditions that occur in an area over a period of years. Factors that determine an area's climate include temperature and precipitation—both of which are influenced by transpiration, the loss of water vapor from the aerial portions of plants.

Forests affect local and regional climate conditions. Transpiration by trees, for example, influences the local temperature of a forest. If you walk into a forest on a hot summer day, you will notice that the air there is cooler and moister than it is outside the forest. This difference occurs because transpiration, like human sweating, is a biological cooling process. Water must absorb a lot of heat before it changes from a liquid to a vapor. When water evaporates through the stomata, the water vapor carries this heat with it. Thus, water transpired from the surfaces of leaves and stems has a cooling effect, not only for the plant but for the local area around the plant.

Transpiration is an important part of the **hydrologic cycle,** in which water cycles from the ocean and land to the atmosphere and back to the ocean and land. As a result of transpiration, water evaporates from leaves and stems to form clouds in the atmosphere. Thus, transpiration eventually results in precipitation because forest trees release substantial amounts of moisture into the air by transpiration (see figure).

Studies suggest that the local climate has become drier in parts of Brazil where tracts of rain forest have been burned. Where deforestation has occurred in Central America, the nearby cloud forests have lost much of their moisture-providing clouds. Temperatures may also rise slightly in a deforested area because there is less evaporative cooling from the trees.

75%
Water recycled
by transpiration
and
evaporation

25% Water
seeps into
ground or
runs off
to rivers, streams,
and lakes

■ **Trees and the hydrologic cycle. Forests play an important role in the hydrologic cycle by returning much precipitation water to the atmosphere by transpiration.**

transpiration than they take in from the soil. Their cells experience a loss of turgor, and the plant wilts (•**Figure 8-12**). If a plant recovers overnight, because of the combination of negligible transpiration (recall that stomata are closed) and absorption of water from the soil, the plant is said to have experienced *temporary wilting.* Most plants recover from temporary wilting with no ill effects. In cases of prolonged drought, however, the soil may not contain sufficient moisture to permit recovery from wilting. A plant that cannot recover is said to be *permanently wilted* and will die.

TRANSPIRATION Loss of water vapor from a plant's aerial parts.

Some plants exude water as a liquid

Many leaves have **hydathodes,** openings at the tips of leaf veins through which liquid water is literally forced out. This loss of liquid water, known as **guttation,** occurs when transpiration is negligible and available soil moisture is high. Guttation typically occurs at night, because the stomata are closed, but water continues to move into the roots by osmosis. People sometimes think erroneously that the early-morning water droplets on leaf margins are dew rather than guttation (•**Figure 8-13**). Unlike dew, which condenses from cool night air, guttation droplets come from within the plant.

Leaf Abscission

All trees shed leaves. Many conifers shed their needles in small numbers year-round. The leaves of deciduous plants turn color and **abscise,** or fall off, once a year—as winter approaches in temperate climates, or at the beginning of the dry period in tropical climates with pronounced wet and dry seasons. In temperate forests, most woody plants with broad leaves shed their leaves to survive the low temperatures of winter. During winter, the plant's metabolism, including its rate of photosynthesis, slows down or halts temporarily. As a result, plants have little need for leaves at that time.

Another reason for abscission is related to a plant's water requirements, which become critical during the physiological drought of winter. As mentioned previously, as the ground chills, absorption of water by the roots is inhibited. When the ground freezes, *no* absorption occurs. If the broad leaves were to stay on the plant during the winter, the plant would continue to lose water by transpiration but could not replace it with water absorbed from the soil.

Leaf abscission involves many physiological changes that are initiated and orchestrated by changes in the levels of plant hormones. As autumn approaches, sugars, amino acids, and many essential minerals—such as nitrogen, phosphorus, and possibly potassium—are mobilized and transported from the leaves to other plant parts. Chlorophyll breaks down, allowing the **carotenoids** (the orange or yellow carotenes and xanthophylls), some of the accessory pigments in the chloroplasts of leaf cells, to become evident. These pigments are

Carlyn Iverson

(a) In the late afternoon of a hot day, the leaves wilted because of water loss. Wilting helps reduce the surface area from which transpiration occurs.

(b) The following morning, water in the leaves had been replenished from the soil. This occurred during the night when transpiration was negligible.

FIGURE 8-12 Temporary wilting in squash (*Cucurbita pepo*) leaves. Shown is the same squash plant within a 24-hour period.

Ed Reschke/Peter Arnold, Inc.

FIGURE 8-13 Guttation. Many people mistake guttation for early-morning dew. Shown is a compound leaf of strawberry (*Fragaria*).

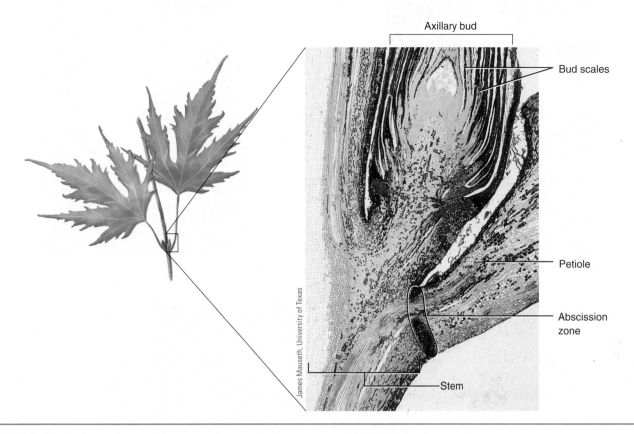

FIGURE 8-14 **Abscission zone.**
This longitudinal section through a silver maple (*Acer saccharinum*) branch shows the base
of the petiole. Note the abscission zone, where the leaf will abscise from the stem. An axillary bud with its protective bud scales is evident above the petiole.

always present in the leaf but are masked by the green of
the chlorophyll. In addition, red water-soluble pigments
called **anthocyanins** are synthesized and stored in the
vacuoles of epidermal leaf cells in some species; anthocyanins may protect leaves against damage by ultraviolet radiation. The various combinations of carotenoids
and anthocyanins are responsible for the brilliant colors found in autumn landscapes in temperate climates.

In many leaves, abscission occurs at an abscission zone near the base of the petiole

The area where a leaf petiole detaches from the stem,
called the **abscission zone,** is structurally different
from surrounding tissues. Composed primarily of thin-
walled parenchyma cells, the abscission zone is anatom-
ically weak because it contains few fibers (•**Figure 8-14**).
As autumn approaches, a protective layer of cork cells
develops on the stem side of the abscission zone. These
cells have **suberin,** a waxy, waterproof material, impregnated in their walls. Enzymes then dissolve the **middle
lamella** (the "cement" that holds the primary cell walls
of adjacent cells together) in the abscission zone.

Once this process is completed, nothing holds the
leaf to the stem but a few xylem cells. A sudden breeze
is enough to make the final break, and the leaf detaches.
The protective layer of cork remains, sealing off the area
and forming a leaf scar.

Modified Leaves

Although photosynthesis is the main function of leaves,
certain leaves have special modifications for other functions. The winter buds of a dormant woody plant are
covered by protective **bud scales,** modified leaves that
protect the delicate meristematic tissue of the bud from
injury and drying out (•**Figure 8-15a;** also see Figure 7-2).

ABSCISSION The normal (usually seasonal) falling off of
leaves or other plant parts, such as fruits or flowers.

BUD SCALE A modified leaf that covers and protects winter
buds.

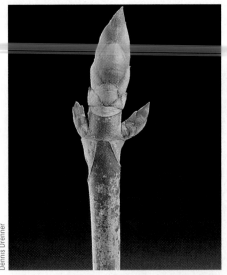

(a) Overlapping bud scales protect buds. Shown here are a terminal bud and two axillary buds of a maple (*Acer*) twig.

(b) Close-up of the leaves of a barrel cactus (*Ferrocactus*) that are modified as spines for protection.

(c) Showy red bracts surround each poinsettia (*Euphorbia pulcherrima*) inflorescence; the inconspicuous flowers are in the center.

(d) Tendrils of a sweet pea (*Lathyrus odoratus*) are modified leaves that aid in climbing.

(e) The leaves of bulbs such as the onion (*Allium cepa*) are fleshy for storage of food materials and water.

(f) Stone plants (*Lithops karasmontana*) have thick, succulent leaves modified for water storage as well as photosynthesis.

FIGURE 8-15 Leaf modifications.

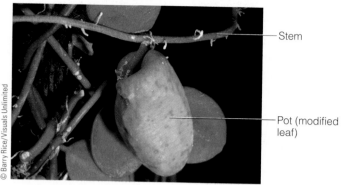

Stem

Pot (modified leaf)

© Barry Rice/Visuals Unlimited

(a) The leaves of the flowerpot plant (*Dischidia rafflesiana*) are modified to hold water and organic material carried in by ants.

FIGURE 8-16 **The flowerpot plant.**

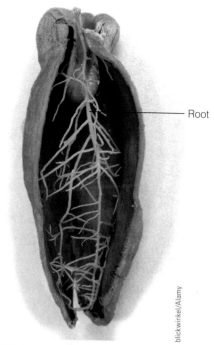

Root

blickwinkel/Alamy

(b) A cutaway view of a pot removed from a plant reveals the special root that absorbs water and dissolved minerals inside the pot.

Some plants have leaves specialized for deterring animals that eat plants. **Spines,** modified leaves that are hard and pointed, are found on many desert plants, such as cacti (•**Figure 8-15b;** also see Figures 1-9 and 25-11a). In the cactus, the main organ of photosynthesis is the stem rather than the leaf. Spines discourage animals from eating the succulent (water-filled) stem tissue.

Some leaves associated with flower clusters (inflorescences) are modified as **bracts.** In flowering dogwood, the inconspicuous flowers are clustered in the center of each inflorescence, and what appear to be four white-to-pink petals are actually bracts. Similarly, the red "petals" of poinsettia are not petals at all but bracts (•**Figure 8-15c**).

Vines are climbing plants whose stems cannot support their own weight, so they often have **tendrils** that help keep the vine attached to the structure on which it is growing (•**Figure 8-15d**). The tendrils of some plants, such as peas and sweet peas, are specialized leaves. Some tendrils, such as those of ivy, Virginia creeper, and grape, are specialized stems.

Some leaves are modified for storage of water or food. For example, a **bulb** is a short underground stem to which large, fleshy leaves are attached (•**Figure 8-15e**). Onions and tulips form bulbs. Many plants adapted to arid conditions—for example, jade plant, medicinal aloe, and stone plants—have fleshy, succulent leaves for water storage. Stone plants, which are native to Africa, grow in clumps with only the tips of their leaves, which occur in pairs, exposed (•**Figure 8-15f**). These leaves function in photosynthesis.

Plants living in unusual environments often have specialized foliar adaptations. The flowerpot plant (*Di-*

schidia rafflesiana) is a small tropical plant that is an *epiphyte*—a plant that grows attached to another, larger plant, which it uses for support. The flowerpot plant lives high in the canopy of the tropical rain forest, attached to the bark of a tree trunk. It has two kinds of leaves; some are normal, and others are rolled up to form hollow containers, or "flowerpots," about 10 centimeters (4 inches) deep (•**Figure 8-16**). Water collects in the pot from condensation of water vapor that enters the leaf through stomata. Ant colonies that live in the pots collect organic debris that releases minerals as it decomposes. The plant grows adventitious roots into its own pots from nearby stems or petioles and in this way absorbs water and dissolved minerals high above the ground.

SPINE A leaf modified for protection, such as a cactus spine.

BRACT A modified leaf associated with a flower or inflorescence but not part of the flower itself.

TENDRIL A leaf or stem that is modified for holding on or attaching to objects.

BULB A rounded, fleshy, underground bud that consists of a short stem with fleshy leaves.

Modified leaves of carnivorous plants capture insects

Carnivorous plants are plants that capture insects. Most carnivorous plants grow in acidic bogs with poor soil that is deficient in essential minerals, particularly nitrogen and phosphorus. These plants digest insects and other small animals to meet some of their mineral requirements. The leaves of carnivorous plants are adapted to attract, capture, and digest their animal prey.

Some carnivorous plants have passive traps. The leaves of a pitcher plant, for example, are shaped so that rainwater collects and forms a reservoir that also contains acid secreted by the plant (•**Figure 8-17a**). Some pitchers are quite large; in the tropics, for example, a pitcher plant may be large enough to hold 1 liter (approximately 1 quart) or more of liquid.

An insect attracted by the odor or nectar of the pitcher may lean over the edge and fall in. Although it may make repeated attempts to escape, the insect is prevented from crawling out by the slippery sides and the rows of stiff spines that point downward around the lip of the pitcher. The insect eventually drowns, and part of its body is digested and absorbed.

Although most insects are killed in pitcher plants, the larvae of several insects (certain fly, midge, and mosquito species), as well as a large community of microorganisms, actually live inside the pitchers. These insect species obtain their food from the insect carcasses, and the pitcher plant digests what remains. It is not known how these insects survive the acidic environment inside the pitcher.

The Venus flytrap is a carnivorous plant with active traps. Its leaves, which produce a sweet-smelling nectar, resemble tiny bear traps (•**Figure 8-17b**). Each side of the leaf blade contains three to six small, stiff "trigger" hairs. If an insect alights and brushes against two of the trigger hairs (or touches one hair two times) within a 20-second interval, the trap springs shut with amazing rapidity. The interlocking spines along the margins of the blades fit closely together to prevent the insect from

(a) The pitcher plant (*Sarracenia purpurea*) has leaves modified to form a pitcher that collects water, in which the plant's prey drown. Notice the dead insect in the "pitcher."

(b) Hairs on the leaf surface of the Venus flytrap (*Dionaea muscipula*) detect the touch of an insect, and the leaf responds by folding. This fly is about to be trapped.

FIGURE 8-17 Carnivorous plants.

escaping. Digestive glands on the interior surface of the trap secrete enzymes in response to the insect pressing against them. After the insect has died and the soft parts have been digested and absorbed into the leaf, the trap reopens (usually 5 to 12 days after the initial capture), and the indigestible remains fall out. Each trap captures prey only three to five times; then it photosynthesizes for a few more months before falling off the plant.

Sundews are carnivorous plants with leaves modified as active traps. Sundew leaves are covered with many large, glandular hairs that can slowly change direction when needed. The glands produce sticky materials and digestive enzymes that accumulate at the tips of the hairs. When an insect alights on a sundew leaf, the insect becomes stuck. As it struggles, additional hairs bend over and adhere to the insect, enclosing it to speed digestion.

STUDY OUTLINE

❶ **Describe the major tissues of the leaf (epidermis, mesophyll, xylem, and phloem), and relate the structure of the leaf to its function of photosynthesis.**
Leaf structure is adapted for its primary function, **photosynthesis.** The transparent **epidermis** allows light to penetrate into the mesophyll, where photosynthesis occurs. **Stomata** are small pores in the epidermis that permit gas exchange for both photosynthesis and transpiration. Stomata typically open during the day, when photosynthesis takes place, and close at night. The waxy **cuticle** coats the epidermis, enabling the plant to survive in the dry conditions of a terrestrial environment. The photosynthetic ground tissue, or **mesophyll,** contains air spaces that permit rapid diffusion of carbon dioxide and water into, and oxygen out of, mesophyll cells. Leaf **veins** have **xylem** to conduct water and essential minerals to the leaf, and **phloem** to conduct sugar produced by photosynthesis to the rest of the plant. One or more layers of nonvascular cells surround the larger veins and make up the **bundle sheath.**

❷ **Contrast leaf structure in eudicots and monocots.**
Monocot and eudicot leaves are distinguished on the basis of external structure and internal anatomy. Monocot leaves are often narrow, wrap around the stem in a sheath, and have **parallel venation.** Eudicot leaves often have a broad, flattened **blade** and **netted venation.** The upper epidermises of the leaves of certain monocots (grasses) have large, thin-walled cells called **bulliform cells** on both sides of the midvein that may be involved in the rolling or folding inward of the leaf during drought.

❸ **Outline the physiological changes that accompany stomatal opening and closing.**
In stomatal opening, **blue light** triggers the activation of proton pumps located in the guard-cell plasma membrane. Protons (H⁺) are pumped out of the **guard cells,** forming a **proton gradient** with a charge and concentration difference on the two sides of the guard-cell plasma membrane. The resulting gradient drives the **facilitated diffusion** of potassium ions into guard cells. Chloride ions are also taken into the guard cells. These ions accumulate in the vacuoles of the guard cells, increasing the solute concentration so it is greater than that of surrounding cells, and water enters the guard cells from surrounding epidermal cells by **osmosis.** The increased turgidity of the guard cells changes their shape, causing the stoma to open. In the late afternoon or early evening, stomata close, but not by an exact reversal of the opening process. As evening approaches, the sucrose concentration in the guard cells declines as sucrose is converted to starch (which is osmotically inactive), water leaves by osmosis, the guard cells lose their turgidity, and the pore closes.

❹ **Discuss transpiration and its effects on the plant.**
Transpiration is the loss of water vapor from plants. Transpiration occurs primarily through the stomata. Environmental factors such as temperature, wind, and relative humidity affect the rate of transpiration. Transpiration may be both beneficial and harmful to the plant.

❺ **Define** *leaf abscission,* **and explain why it occurs and what physiological and anatomical changes precede it.**
In temperate climates, most woody plants with broad leaves shed their leaves in the fall; this helps them survive the low temperatures of winter. Leaf **abscission** (falling off of leaves from a plant) involves physiological and anatomical changes. As autumn approaches, the plant reabsorbs sugar, and many essential minerals are transported out of the leaves. Chlorophyll is broken down, and red water-soluble pigments are synthesized and stored in the vacuoles of leaf cells in some species. The area where a leaf petiole detaches from the stem, called the **abscission zone,** is composed primarily of thin-walled parenchyma cells. As autumn approaches, a protective layer of cork cells develops on the stem side of the abscission zone. Enzymes then dissolve the **middle lamella** (the "cement" that holds the primary cell walls of adjacent cells together) in the abscission zone. After the leaf detaches, the protective layer of cork seals off the area, forming a leaf scar.

⑥ **List at least five modified leaves, and give the function of each.**

Some leaves are modified for functions other than photosynthesis. **Bud scales** are leaves modified to protect the delicate meristematic tissue of buds. **Spines** are leaves adapted to provide protection. **Bracts** are modified leaves associated with a flower or inflorescence but not part of the flower itself. **Tendrils** are leaves or stems modified for grasping and holding on to other structures (to support weak stems). **Bulbs** are short underground stems with fleshy leaves specialized for storage. Leaves of carnivorous plants are modified to trap insects.

REVIEW QUESTIONS

1. Draw diagrams to demonstrate simple versus compound leaves; alternate, opposite, and whorled leaf arrangement; and parallel, palmately netted, and pinnately netted venation.
2. Give the general equation for photosynthesis (see Chapter 4), and discuss how the leaf is organized to deliver the raw materials and products of photosynthesis.
3. What are two ways to distinguish between eudicot and monocot leaves under the microscope?
4. What leaf structure is related to both photosynthesis and transpiration? How is it tied to each physiological process?
5. Relate how availability of water is involved in stomatal opening and closing.
6. Briefly explain the sequence of events that occurs when guard cells are exposed to blue light.
7. Discuss at least two ways in which certain leaves evolved to conserve water.
8. What is transpiration? How do environmental factors (sunlight, temperature, humidity, and wind) influence the rate of transpiration?
9. How does the environment influence stomatal opening and closing?
10. What is leaf abscission? Why do many woody plants lose their leaves in autumn?
11. Distinguish among spines, tendrils, bud scales, and bulbs.
12. Discuss the specialized features of the leaves of carnivorous plants.
13. Discuss some of the possible causes of tree decline.
14. Label the following diagram. (Refer to Figure 8-3 to check your answers.)

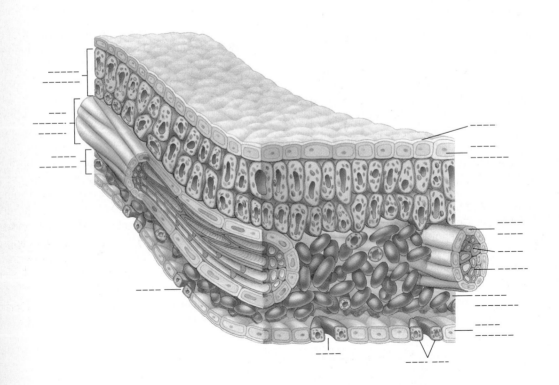

THOUGHT QUESTIONS

1. Stomata open in response to light. Could the pigment involved in stomatal opening be chlorophyll? Why or why not?

2. Given that xylem is situated toward the upper epidermis in leaf veins and phloem toward the lower epidermis, and the vascular tissue of a leaf is continuous with that of the stem, suggest one possible arrangement of vascular tissues in the stem that accounts for the arrangement of vascular tissue in the leaf.

3. Suppose you observe a micrograph of a leaf cross section and are asked to determine which side is covered by the upper epidermis and which side by the lower epidermis. What characteristics would you look for to make this decision?

4. What might be some advantages of a plant that has a few large leaves? Disadvantages? What might be some advantages of a plant that has many small leaves? Disadvantages? In what kinds of habitats might you expect to find such plants?

Visit us on the web at http://www.thomsonedu.com/biology/berg/ for additional resources, such as flashcards, tutorial quizzes, further readings, and web links.

Flowers, Fruits, and Seeds

LEARNING OBJECTIVES

❶ Name the parts of a flower, and describe the functions of each part.

❷ Distinguish between pollination and fertilization.

❸ Compare the general characteristics of flowers pollinated in different ways (by insects, birds, bats, and wind).

❹ Define *coevolution,* and give examples of ways in which plants and their animal pollinators have affected one another's evolution.

❺ List and define the main parts of a seed.

❻ Distinguish among simple, aggregate, multiple, and accessory fruits; give examples of each type; and cite several different methods of seed and fruit dispersal.

❼ Summarize the influence of environmental factors on the germination of seeds.

Ragweed (*Ambrosia*) is one of the most frequently encountered weeds in the northeastern and central parts of North America. Found along roadsides, in fields, in woods, and in backyards, this plant grows to about 1.5 meters (5 feet) in height and has deeply dissected leaves and clusters of minute greenish flowers borne on a slender, hairy stem (see photograph). During late summer, ragweed makes many people miserable with allergies, because it releases large quantities of pollen grains into the air as it reproduces. (Pollen comes from flowers, the organs of reproduction in most plants.)

If you suffer from allergic rhinitis, commonly called hay fever, you are not alone. About 40 million people in the United States endure the sneezing and the itchy, watery eyes associated with this condition. Because one of the primary causes of hay fever is pollen grains suspended in the air, many sufferers blame their discomforts on any plant that happens to be in bloom when they exhibit symptoms. Consequently, roses, sunflowers, and goldenrod are often accused of causing allergies. Contrary to popular opinion, however, the pollen grains of these showy flowering plants generally do not cause allergic reactions. Insects or other animals pollinate plants with large, colorful petals; their pollen grains do not get into the air in appreciable quantities, so they do not cause hay fever.

Hay fever is caused by inhalation of the pollen grains of certain wind-pollinated plants, that is, plants that rely on wind to transfer their pollen grains from one flower to another. When inhaled, the pollen grains stimulate the body to release histamine and other substances, which cause inflammation and other symptoms of allergy. Wind-pollinated plants must produce vast amounts of tiny pollen grains to ensure that at least some of them will land on flowers of the same species and result in successful reproduction. Not all pollen grains of wind-pollinated plants cause allergic reactions, however. Pines, for example, are wind pollinated, yet allergies to pine pollen are rare.

People with allergies suffer at different times during the growing season, depending on which plants are releasing pollen grains and on whether they are sensitive to the pollen grains of those plants. In early spring, many trees release pollen grains before their leaves are fully developed. Trees that cause allergic reactions include oaks, ashes, maples, alders, and elms. If you suffer from allergies in late spring and early summer, you are probably allergic to pollen grains from grasses such as bluegrass, timothy, and redtop. Interestingly, most of our major grass crops (corn, rice, and wheat) do not cause allergies in humans. In late summer and early fall, ragweed is the primary culprit in all areas of the United States except the West.

Ragweed. The spikes at the tips of ragweed (*Ambrosia*) contain numerous clusters of inconspicuous, pollen-producing flowers.

Most people are born with some resistance to pollen allergies, but many become sensitized by repeated contact. For that reason, a move to a different geographic location often temporarily halts the suffering from one pollen allergy, although it frequently gives rise to another!

This chapter examines sexual reproduction in flowering plants, including floral adaptations that are important in pollination; seed and fruit structure and dispersal; and germination and early growth.

Reproductive Flexibility of Flowering Plants

Flowering plants, which include about 300,000 species, are the largest, most successful group of plants. One reason for the success of flowering plants, or **angiosperms,** is that they reproduce both sexually and asexually. You may have admired flowers for their fragrances as well as their appealing colors and varied shapes. The biological function of flowers, however, is sexual reproduction. Their varied colors, shapes, and fragrances are **adaptations** (evolutionary modifications) that increase the likelihood that *pollen grains* will be carried from plant to plant.

Sexual reproduction entails the fusion of reproductive cells: eggs and sperm cells, collectively called **gametes.** The union of gametes, which is called **fertilization,** occurs within the flower's ovary.

The offspring of plants that reproduce sexually show considerable genetic variation. They may resemble one of the parent plants, both of the parents, or neither of the parents. Sexual reproduction offers the advantage of new combinations of genes (the units of heredity) not found in either parent. These new gene combinations may make an individual plant better suited to its environment. (Chapters 12 to 14 give the details of how this variation occurs.)

Many flowering plants also reproduce asexually. Asexual reproduction does not usually involve the formation of flowers, seeds, and fruits. Instead, offspring generally form asexually when a vegetative structure—stems, leaves, or roots—of an existing plant expands, grows, and then becomes separated from the rest of the plant, often by the death of tissues. This part then forms a complete, independent plant. Because asexual reproduction involves only one parent and no fusion of gametes occurs, the offspring of asexual reproduction are virtually genetically identical to each other and to the parent plant from which they came.

Flowers

A flower is a reproductive shoot usually consisting of four kinds of organs—sepals, petals, stamens, and carpels—arranged in whorls (circles) on the end of a flower stalk, or **peduncle** (•**Figure 9-1**). The peduncle may terminate in a single flower or a cluster of flowers known as an **inflorescence** (•**Figure 9-2**). In flowers with all four organs, the normal order of whorls from the flower's periphery to the center (or from the flower's base upward) is the following:

Sepals $\longrightarrow$ petals $\longrightarrow$ stamens $\longrightarrow$ carpels

The tip of the peduncle enlarges to form a **receptacle** that bears some or all of the flower parts. All four floral parts are important in the reproductive process, but only the stamens and carpels participate directly in sexual reproduction. Sepals and petals are sterile.

A flower that has all four parts—sepals, petals, stamens, and carpels—is said to be a **complete flower;** an **incomplete flower** lacks one or more of these four parts. A flower that has both stamens *and* carpels is described as a **perfect flower;** an **imperfect flower** has stamens *or* carpels, but not both. Thus, an imperfect flower is also an incomplete flower. However, a perfect flower may be complete (if it has both sepals and petals) or incomplete (if it lacks sepals or petals).

Sepals, which constitute the outermost and lowest whorl on a floral shoot, are leaflike in shape and form and are often green. Some sepals, such as those in lily flowers, resemble petals. Sepals cover and protect the flower parts when the flower is a bud (•**Figure 9-3a**). As the blossom opens from a bud, the sepals fold back to reveal the more conspicuous petals. The collective term for all the sepals of a flower is **calyx.**

The whorl just inside and above the sepals consists of **petals,** which are broad, flat, and thin (like sepals and leaves) but tremendously varied in shape and frequently brightly colored, which attracts pollinators. As discussed later, petals play an important role in ensuring that sexual reproduction will occur. Sometimes petals fuse to form a tube (for example, trumpet honeysuckle flowers, •**Figure 9-3b**) or other floral shape (for example, snapdragons). The collective term for all the petals of a flower is **corolla.**

Just inside and above the petals are the **stamens** (•**Figure 9-3c**). Each stamen has a thin stalk called a **filament,** at the top of which is an **anther,** a saclike structure in which **pollen grains** form. For sexual reproduction to occur, pollen grains must be transferred from the anther to the carpel, usually of another flower of the same species. At first, each pollen grain consists of two cells surrounded by a tough outer wall. One cell generates two male gametes, known as **sperm cells,** and the other produces a **pollen tube** through which the sperm cells travel to reach the ovule.

In the center or top of most flowers is one or more **carpels.** Carpels bear **ovules,** which are structures that have the potential to develop into seeds. The carpels of a flower may be separate or fused into a single structure. A single carpel or a group of fused carpels is sometimes called a *pistil*. A pistil may consist of a single carpel (making it a *simple* pistil) or a group of fused carpels (making it a *compound* pistil) (•**Figure 9-4**). In most flowers, each carpel or group of fused carpels has three sections: a **stigma,** on which the pollen grains land; a **style,** a necklike structure through which the pollen

Darwin Dale/Photo Researchers, Inc.

(a) An *Arabidopsis thaliana* flower.

FERTILIZATION Fusion of male and female gametes. After fertilization, flowering plants produce *seeds* inside fruits.

SEPAL One of the outermost parts of a flower, usually leaf-like in appearance, that protect the flower as a bud.

PETAL One of the often conspicuously colored parts of a flower attached inside the whorl of sepals.

STAMEN The pollen-producing part of a flower.

CARPEL The ovule-bearing reproductive unit of a flower.

OVULE The structure in the ovary that develops into a seed after fertilization.

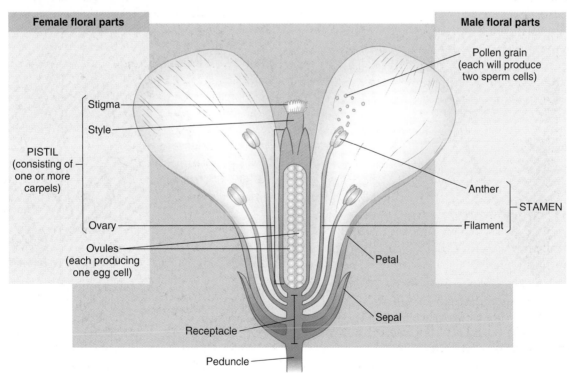

(b) Cutaway view of an *Arabidopsis* flower. Each flower has four sepals (*two are shown*), four petals (*two are shown*), six stamens, and one pistil. Four of the stamens are long, and two are short (*two long and two short are shown*). Pollen grains develop within sacs in the anthers. In *Arabidopsis*, the pistil consists of two fused carpels that each contain numerous ovules.

FIGURE 9-1 Flower structure.

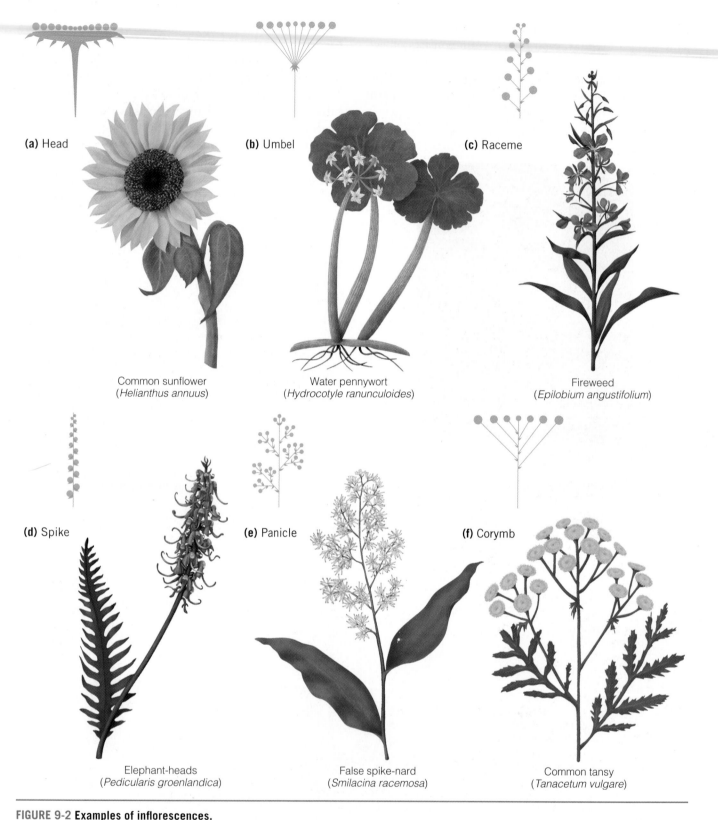

(a) Head

Common sunflower
(*Helianthus annuus*)

(b) Umbel

Water pennywort
(*Hydrocotyle ranunculoides*)

(c) Raceme

Fireweed
(*Epilobium angustifolium*)

(d) Spike

Elephant-heads
(*Pedicularis groenlandica*)

(e) Panicle

False spike-nard
(*Smilacina racemosa*)

(f) Corymb

Common tansy
(*Tanacetum vulgare*)

FIGURE 9-2 Examples of inflorescences.
Many arrangements of inflorescences are recognized.

tube grows; and an **ovary,** a juglike structure that contains one or more ovules and can develop into a fruit (see Figure 9-3c). Each ovule contains a **female gameto-phyte,** also known as an **embryo sac,** in which develop one female gamete (an **egg**) and two **polar nuclei.** As

discussed in Chapter 25, the egg and both polar nuclei participate directly in fertilization.

An ovary is designated as superior or inferior depending on its location relative to other flower parts; this character is used a great deal in the classification of

Petals   Sepals

James Mauseth, University of Texas

(a) The leaflike sepals of a rose flower cover and protect the inner flower parts.

Marion Lobstein

Stamen

Pistil

(c) A twinleaf (*Jeffersonia diphylla*) flower has eight stamens. Note the rounded green ovary in the center of the flower.

© Jeffrey Howe/Visuals Unlimited

(b) The five petals of each trumpet honeysuckle (*Lonicera sempervirens*) flower are fused to form a tubular corolla.

FIGURE 9-3 The four parts of a flower.

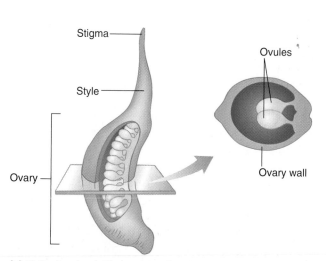

Stigma

Style

Ovary

Ovules

Ovary wall

(a) This simple pistil consists of a single carpel.

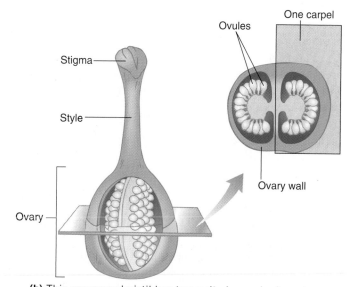

One carpel

Ovules

Stigma

Style

Ovary

Ovary wall

(b) This compound pistil has two united carpels. In most flowers with single pistils, the pistils are compound, consisting of two or more fused carpels.

FIGURE 9-4 Simple and compound pistils.

flowering plants. A **superior ovary** is one that has the other floral organs (sepals, petals, and stamens) free from the ovary and attached at the ovary's base. An **inferior ovary** is one that is located below the point at which the other floral organs are attached (see Figure 25-16a).

Pollination

Before fertilization occurs, pollen grains must travel from the anther (where they form) to the stigma. The transfer of pollen grains from anther to stigma is known as **pollination.** Plants are *self-pollinated* if pollination occurs within the same flower or within a different flower on the same individual plant. *Cross-pollination* occurs when pollen grains are transferred to a flower on another individual of the same species.

Flowering plants accomplish pollination in a variety of ways. Beetles, bees, flies, butterflies, moths, wasps, and other insects pollinate many flowers. Other animals, such as birds, bats, snails, and small nonflying mammals (rodents, primates, and marsupials) also pollinate plants. Wind is an agent of pollination for certain flowers, whereas water transfers pollen grains in a few aquatic flowers.

Many plant species have mechanisms to prevent self-pollination

In plant sexual reproduction, the two gametes that unite to form a zygote (fertilized egg) may be from the same plant or from two different parents. The combination of gametes from two different parents increases the variation in offspring, and this variation may confer a selective advantage. Some offspring, for example, may survive environmental changes better than either of their parents.

Plants have evolved a variety of mechanisms that prevent self-pollination and thus prevent **inbreeding,** which is the mating of genetically similar individuals. Inbreeding is generally undesirable because it can increase the concentration of harmful genes in the offspring. To avoid inbreeding, some species have individuals with staminate flowers that lack carpels and other individuals with carpellate flowers that lack stamens. Other species have flowers with both stamens and carpels, but the pollen grains are shed from a given flower before or after the time that the stigma of that flower is receptive to pollen grains.

Many species have genes for **self-incompatibility,** a genetic condition in which pollen grains are ineffective in fertilizing the same flower or other flowers on the same individual plant. In other words, an individual plant can identify and reject its own pollen grains. Genes for self-incompatibility usually inhibit the growth of the pollen tube in the stigma and style, thereby preventing sperm

cell delivery to the ovule. Self-incompatibility, which is more common in wild species than in cultivated plants, ensures that reproduction occurs only if the pollen grains come from a genetically different individual.

PROCESS OF SCIENCE The molecular basis of self-incompatibility in *Arabidopsis* and related plants is an area of active botanical interest. Research suggests that self-incompatibility in these plants is the ancestral (normal) condition. Self-fertile individuals contain mutations (changes in the deoxyribonucleic acid, or DNA) in one or more genes involved in self-incompatibility. ∎

EVOLUTION LINK ## Flowering plants and their animal pollinators have coevolved

Animal pollinators and the plants they pollinate have had such a close, interdependent relationship over time that they have affected the evolution of certain physical and behavioral features in one another. The term **coevolution** describes such reciprocal adaptation, in which two species interact so closely that they become increasingly adapted to each other as each undergoes evolutionary change by **natural selection** (see Chapter 16). Let us examine some of the features of flowers and their pollinators that may be the products of coevolution.

Flowers pollinated by animals have various features to attract them, including showy petals (a visual attractant) and scent (an olfactory attractant). One reward for the animal pollinator is food. Some flowers produce nectar, a sugary solution, in special floral glands called **nectaries.** Pollinators use nectar as an energy-rich food. Pollen grains are also a protein-rich food for many animals. As they move from flower to flower searching for food, pollinators inadvertently carry pollen grains on their body parts, facilitating sexual reproduction in plants.

Botanists estimate that insects pollinate about 70 percent of all flowering plant species. Bees are particularly important as pollinators of crop plants (●Figure 9-5a). Bee-pollinated crops provide about 30 percent of human food. Plants pollinated by insects often have blue or yellow petals. The insect eye does not see color the same way the human eye does. Most insects see well in the violet, blue, and yellow ranges of visible light but do not perceive red as a distinct color. Consequently, insect-pollinated flowers are not usually red. Insects also see ultraviolet radiation, wavelengths that are invisible to the human eye. Insects see ultraviolet radiation as a color called *bee's purple.* Many flowers have dramatic ultraviolet markings called **nectar guides** that may or may not be visible to humans but that direct insects to the center of the flower where the pollen grains and nectar are (●**Figure 9-6**).

Insects have a well-developed sense of smell, and many insect-pollinated flowers have strong scents that

(a) A honeybee pollinates an Indian blanket (*Gaillardia pulchella*) flower. The bee's body is covered in hairs that pick up pollen grains.

(b) A broad-billed hummingbird obtains nectar from a desert flower, ocotillo (*Fouquieria splendens*), in Arizona. The pollen grains on the bird's feathers are carried to the next plant.

FIGURE 9-5 Animal pollinators.

(c) An Underwood's long-tongued bat obtains nectar from a cluster of macuna (*Macuna rostrata*) flowers. Macuna grows in tropical rain forests. The flowers hang down so the bats can easily get to them while in flight. As the bat moves from flower cluster to flower cluster, it transfers pollen.

may be pleasant or foul to humans. The carrion flower, for example, is pollinated by flies and smells like the rotting flesh in which flies lay their eggs (•**Figure 9-7**). As flies move from one reeking flower to another looking for a place to lay their eggs, they transfer pollen grains. (Should a fly lay her eggs on the carrion flower, the larvae will starve to death when they hatch.)

Birds such as hummingbirds are important pollinators (•**Figure 9-5b**). Bird-pollinated flowers are usually red, orange, or yellow because birds see well in this region of visible light. Because birds do not have a strong sense of smell, bird-pollinated flowers usually lack a scent. Hummingbird-pollinated flowers have a long, tubular corolla with nectar glands at the bottom. The bird hovers beside a flower and inserts its beak and long tongue inside to lap up the nectar. Hummingbird-pollinated flowers produce more nectar than insect-pollinated flowers, because hummingbirds are larger animals than insects and therefore require more food.

Bats, which feed at night and do not see well, are important pollinators, particularly in the tropics, where they are most abundant (•**Figure 9-5c**). Bat-pollinated flowers are night blooming and often have dull white petals and a strong scent, usually of fermented fruit. Nectar-feeding bats are attracted to the flowers by their scent; they lap up the nectar with their long, extensible tongues. As they move from flower to flower, they transfer pollen grains. At least one bat-pollinated flower (the tropical vine *Mucana holtonii*) has evolved an unusual adaptation to encourage pollination by bats. When the pollen grains in a given flower are mature, a concave petal lifts up. The petal bounces the echo from the bat's echo-locating calls back to the bat, helping it find the flower.

POLLINATION In seed plants, the transfer of pollen grains from the anther to the stigma.

(a) An evening primrose (*Oenothera parviflora*) flower as seen by the human eye is solid yellow.

(b) The same flower viewed under ultraviolet radiation provides clues about how the insect eye perceives it. The outer, light-appearing portions of the petals appear purple to a bee's eyes, whereas the dark blue inner parts appear yellow.

FIGURE 9-6 Nectar guides.

Many insect-pollinated flowers have nectar guides that are invisible to humans but conspicuous to insects. These differences in color draw attention to the center of the flower, where the pollen grains and nectar are located.

During the time when plants were coevolving specialized features such as petals, scent, and nectar to attract pollinators, animal pollinators also coevolved. Specialized body parts and behaviors adapted animals to obtain nectar and pollen and, coincidently, to assist the plant in transferring pollen grains.

For example, coevolution may have led to the long, curved beaks of the 'i'iwi, one of the Hawaiian honeycreepers. An 'i'iwi inserts its beak into tubular flowers to obtain nectar (•**Figure 9-8**). The long, tubular corollas of the flowers that 'i'iwis visit probably also came about through coevolution. During the 20th century, some of the tubular flower species (such as lobelias) became rare, largely as a result of grazing by nonnative cows and feral goats, and about 25 percent of lobelia species have become extinct. The 'i'iwi now feeds largely on the flowers of the 'ohi'a tree, which lacks petals, and the 'i'iwi bill may be slowly adapting to this change in feeding preference. Comparison of the bills of 'i'iwi museum specimens collected in 1902 with the bills of live birds captured in the 1990s showed that the 'i'iwi bills were about 3 percent shorter in the 1990s than in 1902.

Animal behavior has also coevolved, sometimes in bizarre and complex ways. The flowers of certain orchids (*Ophrys*) resemble female bees in coloring and shape (see Figure 16-8a). The unpollinated flowers also secrete a scent similar to that produced by female bees, and the males are irresistibly attracted to it. The resemblance between *Ophrys* flowers and female bees is so strong that male bees mount the flowers and try to copulate with them. During this *pseudocopulation*, a pollen sac usually becomes attached to the bee's back. When the bee departs and tries to copulate with another orchid flower, pollen grains are transferred to that flower. Interestingly, once a flower has been pollinated, it emits a scent like that released by female bees that have already mated. Male bees have no interest in visiting flowers that have been pollinated (just as they lose interest in female bees that have already been inseminated). ∎

FIGURE 9-7 Carrion flower.

This desert plant is sometimes called the carrion flower (*Stapelia*) because of its flower color, which resembles dried blood, and its putrid scent. Photographed at the Fairchild Tropical Garden in Miami, Florida.

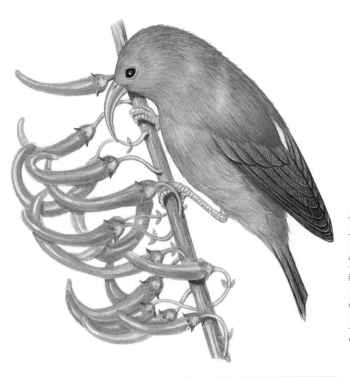

FIGURE 9-8 Coevolution.
The gracefully curved bill of the 'i'iwi, one of the Hawaiian honeycreepers, enables it to sip nectar from flowers of the lobelia (*Lobelia*). The 'i'iwi bill fits perfectly into the long, tubular lobelia flowers.

Dr. Jeremy Burgess/Photo Researchers, Inc.

FIGURE 9-9 Wind pollination.
Clusters of male oak (*Quercus*) flowers, which lack petals, dangle from a tree branch and shed a shower of pollen grains when the wind blows.

Some flowering plants depend on wind to disperse pollen grains

Some flowering plants, such as grasses, ragweed, maples, and oaks, are pollinated by wind. Wind-pollinated plants produce many small, inconspicuous flowers (•**Figure 9-9**). They do not produce large, colorful petals, scent, or nectar. Some have large, feathery stigmas, presumably to trap wind-borne pollen grains. Because wind pollination is a hit-or-miss affair, the likelihood of a pollen grain landing on a stigma of the same species of flower is slim. Wind-pollinated plants produce large quantities of pollen grains, which increases the likelihood that some pollen grains will land on the appropriate stigma.

Fertilization and Seed and Fruit Development

Once pollen grains have been transferred from anther to stigma, the *tube cell,* one of the two cells in the pollen grain, grows a thin *pollen tube* down through the style and into an ovule in the ovary (•**Figure 9-10**). How does the pollen tube "know" where to grow? Botanists have found that molecular signals from the ovule guide the growing pollen tube toward the ovule. Once a pollen

tube penetrates the ovule, the attracting signals cease. As a result, only one pollen tube enters each ovule.

The second cell (the *generative cell*) within the pollen grain divides to form two male gametes (the sperm cells), which move down the pollen tube and enter the ovule. After fertilization has occurred, the ovule develops into a seed, and the ovary surrounding it develops into a fruit. (Chapter 25 considers the fertilization process and the flowering plant life cycle in greater detail.)

Embryonic development in seeds is orderly and predictable

Flowering plants package a young plant embryo complete with stored nutrients in a compact **seed,** which develops from the ovule after fertilization. The nutrients in seeds not only are used by *germinating* plant embryos but are eaten by animals, including humans (see *Plants and People: Seed Banks*). Development of the embryo following fertilization is possible because of the constant flow of nutrients from the parent plant into the developing seed.

The mature seed contains an embryonic plant and storage materials

A mature seed contains an embryonic plant and food stored in either the endosperm or the cotyledons.

For fertilization to take place, a pollen grain must germinate on the stigma, and a pollen tube must grow through the style and ovary to the female gametophyte (embryo sac) in the ovule.

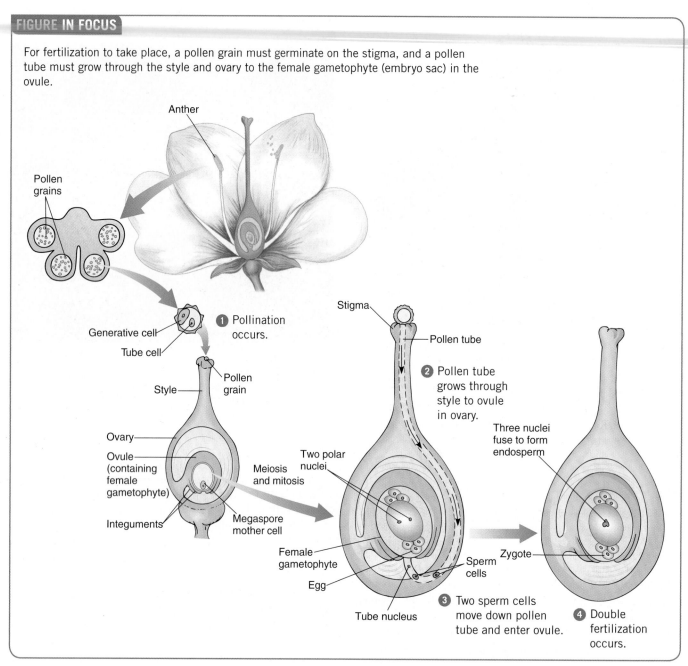

FIGURE 9-10 **Pollination, pollen tube growth, and fertilization.**
See Figure 25-3 for additional details.

Endosperm is the nutritive tissue that surrounds the embryonic plant in a seed. The seed, in turn, is surrounded by a tough, protective **seed coat,** derived from the outermost layers (the *integuments*) of the ovule, and enclosed within a fruit.

The mature embryo within the seed consists of a short embryonic root, or **radicle;** an embryonic shoot; and one or two seed leaves, or **cotyledons.** Monocots have a single cotyledon, and eudicots have two (•**Figure 9-11**). The short portion of the embryonic shoot connecting the radicle to one or two cotyledons is the **hypocotyl.** The shoot apex above the point of attachment of the cotyledon(s) is the **plumule** (also known as

the *epicotyl*). After the radicle, hypocotyl, cotyledon(s), and plumule have formed, the young plant's development is arrested, usually by desiccation (drying out) or dormancy.

When conditions are right for continuing the developmental program, the seed **germinates,** or sprouts, and the embryo resumes growth. Because the embryonic plant is nonphotosynthetic, it must be nourished during germination until it becomes photosynthetic and therefore self-sufficient. The cotyledons of many plants function as storage organs and become large, thick, and fleshy as they absorb the food reserves (starches, oils, and proteins) initially produced as endosperm. Seeds

PLANTS AND PEOPLE | Seed Banks

To preserve diverse varieties of plants, many countries are collecting plant **germplasm,** which is any plant material used in breeding. Germplasm includes seeds, plants, and plant tissues of traditional crop varieties. The International Plant Genetics Resource Institute in Rome, Italy, is the scientific organization that oversees plant germplasm collections worldwide. More than 100 seed collections, called *seed banks,* exist around the world and collectively hold more than 3 million samples of thousands of kinds of plants (see figure). The U.S. National Plant Germplasm System in Fort Collins, Colorado, stores seeds of about 250,000 different species and varieties.

Most seed banks help preserve the genetic variation within different varieties of crops and their wild relatives. Farmers typically discontinue planting local varieties when newer, improved varieties become available. The newer varieties have desirable genetic characteristics, such as a greater yield, but the discontinued local varieties also contain valuable genes. Each local variety's combination of genes gives it distinctive nutritional value, size, color, flavor, resistance to disease, and adaptability to different climates and soil types. Maintaining the genetic diversity present in local crop varieties and their wild relatives helps preserve genes we may need in the future. The gene combinations of local varieties are potentially valuable to agricultural breeders because these genes can be transferred to other varieties, either by traditional breeding methods or by genetic engineering.

Seed banks offer the advantage of storing a large amount of plant genetic material in a small space. Seeds stored in seed banks are safe from habitat destruction, climate changes, diseases, predators, and general neglect. There have even been some instances of seeds from seed banks being used to reintroduce a plant species that has become extinct in the wild. For example, *Bromus,* a grass native to Belgium, became extinct in the mid-1900s but was reintroduced from seeds stored in a seed bank in 2005.

Seed banks have some disadvantages, as well. Seeds do not remain alive indefinitely and must be germinated periodically so that new seeds can be collected. Growing, harvesting, and returning seeds to storage is the most expensive aspect of storing plant material in seed banks. According to the UN Food and Agricultural Organization, many countries establish seed banks but do not provide the funds to pay for periodic replanting.

Many types of plants—avocados and coconuts, for example—cannot be stored as seeds. The seeds of these plants do not tolerate being dried out, which is necessary before they are sealed in moisture-proof containers for storage at −18°C (−0.4°F). If 20 percent or more moisture remains in a seed when it is frozen, it will probably die. Some seeds cannot be stored successfully because they remain viable for only a short period, such as a few months or even just a few days. *Cryopreservation* at −160°C (−256°F) in liquid nitrogen is a new method being developed for certain kinds of seeds. Seeds stored at this temperature survive for longer periods than seeds stored at warmer temperatures.

Perhaps the most important disadvantage of seed banks is that plants stored in this manner remain stagnant

■ **Seeds from a seed bank. These small vials of seeds are from the seed bank in Svalbard, Norway.**

Courtesy of Nordiska Genbanken, Alnarp, Svenge. photo by Stellan Stebe

in an evolutionary sense. Removed from their natural habitats, they are no longer subject to the forces of natural selection. As a result, they may be less fit for survival when they are reintroduced into the wild.

Despite their shortcomings, seed banks are increasingly viewed as an important method of safeguarding seeds for future generations. For example, the Millennium Seed Bank Project at the Royal Botanic Gardens (Kew Gardens) in Great Britain is currently collecting and storing seeds from 10 percent of the world's plant species, including all species native to Great Britain.

---

that store nutrients in cotyledons have little or no endosperm at maturity. Examples of such seeds are peas, beans, squash seeds, sunflower seeds, and peanuts. Other plants—wheat and corn, for example—have thin cotyledons that function primarily to help the young plant digest and absorb food stored in the endosperm.

ENDOSPERM The nutritive tissue that is formed at some point in the development of all flowering plant seeds.

COTYLEDON The seed leaf of a plant embryo that often contains food stored for germination.

PLANTS AND THE ENVIRONMENT | Relating Seed Size to a Plant's Habitat

Assuming that a given plant species invests a fixed amount of its energy in reproduction, is it more advantageous to produce many small seeds or a few big ones? After studying the seed sizes that predominate in different habitats, botanists have suggested that in some environments a smaller seed size may be advantageous, whereas in others a larger seed size may be better.

Seed size for each plant species probably represents a trade-off between the requirements for dispersal and for successful establishment of seedlings. Small-seeded plants can produce more seeds with a given amount of energy than can large-seeded plants. On the other hand, seedlings growing from large seeds may be more likely to survive environmental stresses they encounter because the large seeds contain more stored nutrients. For example, plants that grow in widely scattered open sites (such as old fields) usually produce smaller seeds, perhaps because they can disperse more easily over large areas than can larger seeds. However, wide dispersal is probably less important for plants adapted to densely vegetated areas such as forests. These plants generally produce bigger seeds with an ample food reserve that may confer a greater likelihood of successfully establishing seedlings in a shaded environment. The stored energy may allow the young seedling to grow tall enough to reach adequate sunlight for photosynthesis.

Other ecological factors are associated with seed size. Larger seeds are typical of many plants that live in arid habitats, possibly because the food stored in a large seed lets a young seedling establish an extensive root system quickly, enabling it to survive the dry climate. Island plants also produce larger seeds than similar species on the nearby mainland. Botanists hypothesize that large seeds are less likely to be widely dispersed and therefore less likely to fall into the ocean. Seeds that remain dormant in the soil for a long time tend to be smaller; the exact reason for this tendency is not known.

Despite many associations of seed size with specific environments, no general rule concerning the adaptive advantages of large seeds versus small seeds has emerged. Botanists continue to study the wide divergence in seed sizes that has evolved in present-day plants.

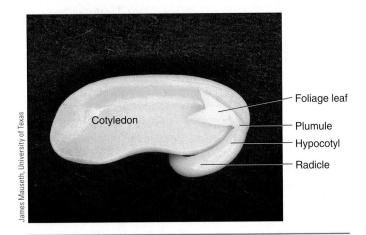

FIGURE 9-11 Parts of a bean seed.
This dissected bean (*Phaseolus vulgaris*) seed shows the radicle, hypocotyl, cotyledon, plumule, and foliage leaf. (The seed coat and one of its two cotyledons were removed). This particular seed has begun germinating, so its radicle is larger than that of a nongerminating seed.

FRUIT In flowering plants, a mature, ripened ovary that often provides protection and dispersal for the enclosed seeds.

SIMPLE FRUIT A fruit that develops from one or several united carpels.

In flowering plants, seed size varies considerably, from the microscopic, dustlike seeds of orchids to the giant seeds of the double coconut (*Lodoicea maldivica*), which weigh as much as 27 kilograms (almost 60 pounds). Despite this variation among species, seed size is a remarkably constant trait within a species (see *Plants and the Environment: Relating Seed Size to a Plant's Habitat*).

Fruits are mature, ripened ovaries

After fertilization takes place within the ovule, the ovule develops into a seed, and the ovary surrounding it develops into a **fruit**. For example, a pea pod is a fruit, and the peas within it are seeds. A fruit may contain one or more seeds; some orchid fruits contain several thousand to a few million seeds! Fruits provide protection for the enclosed seeds and sometimes aid in their dispersal.

There are several types of fruits; their differences result from variations in the structure or arrangement of the flowers from which they were formed. The four basic types of fruits are simple fruits, aggregate fruits, multiple fruits, and accessory fruits (•**Figure 9-12**).

Most fruits are simple fruits. A **simple fruit** develops from a single carpel or several fused carpels. At maturity, simple fruits may be fleshy or dry. Two examples of simple, fleshy fruits are berries and drupes

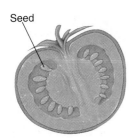

Berry (simple fruit)
A simple, fleshy fruit in which the fruit wall is soft throughout.

Tomato (*Lycopersicon lycopersicum*)

Seed

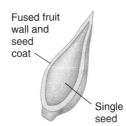

Caryopsis (simple fruit)
A simple, dry fruit in which the fruit wall is fused to the seed coat.

Wheat (*Triticum*)

Fused fruit wall and seed coat

Single seed

Drupe (simple fruit)
A simple, fleshy fruit in which the inner wall of the fruit is a hard stone.

Peach (*Prunus persica*)

Single seed inside stone

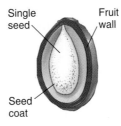

Achene (simple fruit)
A simple, dry fruit in which the fruit wall is separate from the seed coat.

Sunflower (*Helianthus annuus*)

Single seed

Fruit wall

Seed coat

Follicle (simple fruit)
A simple, dry fruit that splits open along one suture to release its seeds; fruit is formed from ovary that consists of a single carpel.

Milkweed (*Asclepias syriaca*)

Seed

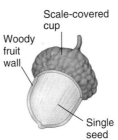

Nut (simple fruit)
A simple, dry fruit that has a stony wall, is usually large, and does not split open at maturity.

Oak (*Quercus*)

Scale-covered cup

Woody fruit wall

Single seed

Aggregate fruit
A fruit that develops from a single flower with several to many pistils (i.e., carpels are not fused into a single pistil).

Blackberry (*Rubus*)

Seed

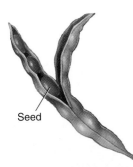

Legume (simple fruit)
A simple, dry fruit that splits open along two sutures to release its seeds; fruit is formed from ovary that consists of a single carpel.

Green bean (*Phaseolus vulgaris*)

Seed

Seed

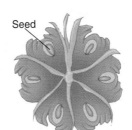

Multiple fruit
A fruit that develops from the ovaries of a group of flowers.

Mulberry (*Morus*)

Seed

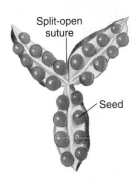

Capsule (simple fruit)
A simple, dry fruit that splits open along two or more sutures or pores to release its seeds; fruit is formed from ovary that consists of two or more carpels.

Iris (*Iris*)

Split-open suture

Seed

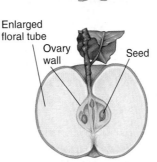

Accessory fruit
A fruit composed primarily of nonovarian tissue (such as the receptacle or floral tube).

Apple (*Malus sylvestris*)

Enlarged floral tube

Ovary wall

Seed

FIGURE 9-12 Representative fruit types.
Fruits are botanically classified into four groups—simple, aggregate, multiple, and accessory fruits—based on their structure and mechanism of seed dispersal.

(•**Figure 9-13**). A **berry** is a fleshy fruit that has soft tissues throughout and contains few to many seeds; a tomato is a berry, as are grapes, blueberries, cranberries, and bananas. Many so-called berries do not fit the botanical definition. Strawberries, raspberries, and mulberries, for example, are not berries; these three fruits are discussed shortly.

A **pepo** is a modified berry in which the fruit wall is a leathery rind. Pumpkin, squash, cucumber, and watermelon fruits are pepos. Another modified berry is a **hesperidium** (pl., *hesperidia*), which has a leathery fruit wall with numerous oil glands surrounding the succulent cavities where the seeds occur. Citrus fruits (lemons, limes, oranges, and grapefruits) are hesperidia.

A **drupe** is a simple, fleshy or fibrous fruit that contains a hard stone surrounding a single seed. Examples of drupes include peaches, cherries, avocados, olives, and almonds. The almond shell is actually the stone, which remains when the rest of the fruit has been removed.

Many simple fruits are dry at maturity. Some of these are *dehiscent* and split open, usually along seams called *sutures,* to release their seeds (•**Figure 9-14**). A milkweed pod is an example of a **follicle,** a simple, dry fruit that splits open along one suture to release its seeds. A **legume** is a simple, dry fruit that splits open along two sutures (the top and bottom). Pea pods are legumes, as are lima bean pods, although both are generally harvested before the fruit has dried out and split open. Pea seeds are usually removed from the fruit and consumed, whereas in green beans the entire fruit and seeds are eaten. A **capsule** is a simple, dry fruit that splits open along multiple sutures or pores. Iris, poppy, buckeye, and cotton fruits are capsules.

Other simple, dry fruits—**caryopses** (sing., *caryopsis*), or **grains**, for example—are *indehiscent* and do not split open at maturity (•**Figure 9-15a**). Each caryopsis contains a single seed. Because the seed coat is fused to the fruit wall, a caryopsis looks like a seed rather than a fruit. Kernels of corn and wheat are fruits of this type.

Nuts are simple, dry fruits that have a stony wall and do not split open at maturity (•**Figure 9-15b**). Nuts are usually large and one-seeded. Examples of nuts include chestnuts, acorns, and hazelnuts. Many so-called nuts do not fit the botanical definition. Peanuts, almonds, and Brazil nuts, for example, are seeds, not nuts.

An **achene** is similar to a caryopsis in that it is simple and dry, does not split open at maturity, and contains a single seed (•**Figure 9-15c**). However, the seed coat of an achene is not fused to the fruit wall. Instead, the single seed is attached to the fruit wall at one point only, permitting an achene to be separated from its seed. The sunflower fruit is an example of an achene. One peels off the fruit wall (the shell) to obtain the sunflower seed within.

(a) The tomato (*Lycopersicon lycopersicum*) is a berry composed of soft tissues throughout.

(b) The blueberry (*Vaccinium*) is another representative berry.

(c) The avocado (*Persea americana*), a drupe, has a hard stone that surrounds the single seed.

FIGURE 9-13 Simple, fleshy fruits.

Aggregate fruits are a second main type of fruit. An **aggregate fruit** is formed from a single flower that contains several separate (free) carpels (•**Figure 9-16**). After fertilization, each ovary from each individual carpel enlarges. As they enlarge, the ovaries may fuse to form

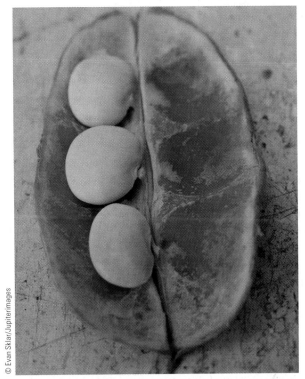

(a) The milkweed (*Asclepias syriaca*) follicle, which is dry at maturity, splits open along one suture to release its seeds.

(b) A lima bean (*Phaseolus limensis*) fruit is a legume that splits open along two sutures at maturity.

a single fruit. Raspberries, blackberries, and magnolia fruits are examples of aggregate fruits.

A third type of fruit is the **multiple fruit,** which forms from the carpels of many flowers that grow close to one another on a common floral stalk. The carpel from each flower fuses with nearby carpels as it develops and enlarges after fertilization. Pineapples, figs, and mulberries are multiple fruits (•**Figure 9-17**).

Accessory fruits are the fourth type. They differ from other fruits in that other plant tissues in addition to ovary tissue make up the fruit. For example, the edible portion of a strawberry is the red, fleshy receptacle. (Each tiny "seed" on a strawberry is actually a fruit—an achene—that contains a single seed.) Apples and pears are accessory fruits called **pomes;** the outer part of each pome is an enlarged **floral tube,** consisting of receptacle tissue, along with portions of the calyx, that surrounds the ovary (•**Figure 9-18**).

Seed dispersal is highly varied

Wind, animals, water, and explosive dehiscence disperse the various seeds and fruits of flowering plants. Effective methods of seed dispersal have made it possible for certain plants to expand their geographic range. In some cases, the seed is the actual agent of dispersal, whereas in others the fruit performs this role. In tumbleweeds, such as Russian thistle, the entire plant is the agent of

(c) These red buckeye (*Aesculus pavia*) capsules split along several sutures.

FIGURE 9-14 Simple, dry fruits that split open to release their seeds.

AGGREGATE FRUIT A fruit that develops from a single flower with several separate carpels that fuse, or grow together.

MULTIPLE FRUIT A fruit that develops from the carpels of closely associated flowers that fuse, or grow together.

ACCESSORY FRUIT A fruit whose fleshy part is composed primarily of tissue other than the ovary.

dispersal because it detaches and blows across the ground, scattering seeds as it bumps along. Tumbleweeds are lightweight and are sometimes blown many kilometers by the wind.

Wind disperses the seeds of many plants (•**Figure 9-19a**). Plants such as maple trees have winged fruits adapted for wind dispersal. Light, feathery plumes enable other seeds or fruits to be transported by wind, often for considerable distances (see Figure 9-14a). Both dandelion fruits and milkweed seeds have this type of adaptation.

Some plants have special structures that aid in the dispersal of their seeds and fruits by animals (•**Figure 9-19b and c**). The spines and barbs of burdock burs and similar fruits often get caught in animal fur and are dispersed as the animal moves about. Fleshy, edible fruits are also adapted for animal dispersal. As an animal eats these fruits, it either discards or swallows the seeds. Many seeds that are swallowed have thick seed coats and are not digested; instead, they pass through the digestive tract and are deposited with the animal's feces some distance from the parent plant. In fact, some seeds will not germinate unless they have passed through an animal's digestive tract. The animal's digestive juices probably aid germination by partially digesting the seed coat. Some edible fruits apparently contain chemicals that function as laxatives to speed seeds through an

Caryopsis — corn

(a) The corn (*Zea mays*) fruit is a caryopsis, or grain. In grains, the fruit wall is fused to the seed coat.

Nut — acorn

(b) An oak (*Quercus*) acorn is a nut. A nut has a hard fruit wall that surrounds a single seed.

Achene — sunflower

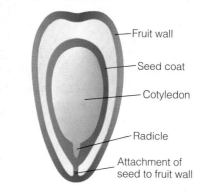

(c) A sunflower (*Helianthus annuus*) fruit is an achene. Its seed coat is attached to the fruit wall at one spot only, and it is possible to peel off the fruit wall, to separate it from the seed.

FIGURE 9-15 Simple, dry fruits that do not split open.

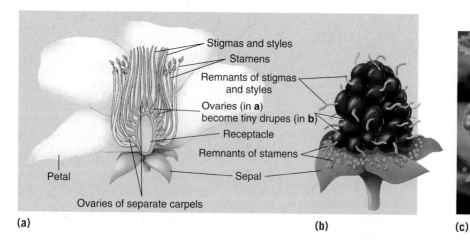

FIGURE 9-16 Aggregate fruit.
(a) Cutaway view of a blackberry (*Rubus*) flower, showing the many separate carpels in the center of the flower. **(b)** A developing blackberry fruit is an aggregate of tiny drupes. The little "hairs" on the blackberry are remnants of stigmas and styles. **(c)** Developing fruits at various stages of maturity.

(a)

Single female flower

Inflorescence (a cluster of flowers on a common floral stalk)

Multiple fruit

(b)

FIGURE 9-17 Multiple fruits.
Both **(a)** pineapple (*Ananas comosus*) and **(b)** mulberry (*Morus*) are formed from the ovaries of many separate flowers that fused to become a multiple fruit. Mulberry flowers are imperfect and contain either stamens or carpels. The inflorescence of carpellate flowers from which the mulberry fruit develops is also shown.

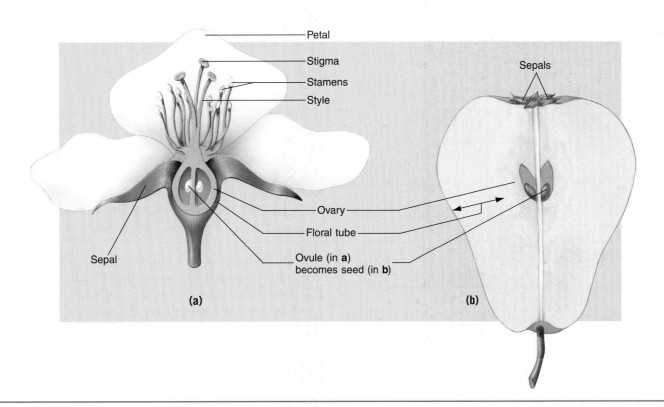

Petal

Stigma

Stamens

Style

Sepals

Sepal

Ovary

Floral tube

Ovule (in **a**) becomes seed (in **b**)

(a)   (b)

FIGURE 9-18 An accessory fruit.
(a) Note the floral tube surrounding the ovary in the pear (*Pyrus*) flower. This tube becomes the major edible portion of the pear. **(b)** Longitudinal section through a pear, showing the fruit tissue derived from both the floral tube and the ovary.

(a) The fruits of maple (*Acer*) have wings for wind dispersal.

Seed   Wing

(b) Great burdock (*Arctium lappa*) flowers develop into hooked fruits that, when mature, may be carried away from the parent plant by becoming matted in animal fur or clothing.

FIGURE 9-19 Methods of seed and fruit dispersal.

(c) Fleshy fruits such as blackberries (*Rubus*) are eaten by animals such as this white-footed mouse. The seeds are frequently swallowed whole and pass unharmed through the animal's digestive tract.

animal's digestive tract. The less time these seeds spend in the gut, the more likely they are to germinate.

Animals such as squirrels and many bird species also help disperse acorns and other fruits and seeds by burying them for winter use. Many buried seeds are never used by the animal and germinate the following spring. (Also see *Plants and the Environment: Seed Dispersal by Ants.*)

The coconut is an example of a fruit adapted for dispersal by water. The coconut has air spaces that make it buoyant and capable of being carried by ocean currents. When it washes ashore, the seed may germinate and grow into a coconut palm tree.

Some seeds are not dispersed by wind, animals, or water. Such seeds are found in fruits that use *explosive dehiscence*, in which the fruit bursts open suddenly, and quite often violently, to forcibly discharge its seeds. Pressures due to differences in **turgor** or to drying out cause these fruits to burst open suddenly. The fruits of plants such as touch-me-not (*Impatiens noli-tangere*)

PLANTS AND THE ENVIRONMENT | Seed Dispersal by Ants

For a plant species to survive, it must disperse its seeds to places where they will successfully germinate and grow. Plants lend themselves to a variety of dispersal methods (such as wind, animals, and water) that increase the chances of their seeds landing in suitable locations. Regardless of how seeds are dispersed from the parent plant, most seeds either land in places that are unsuitable for growth or are eaten by animals, such as mice and squirrels, shortly after being dispersed.

The survivability of the seeds of some plants is enhanced by a dispersal method in which the seeds are buried underground, where they are less likely to be eaten by animals. For many plants, the role of burying seeds is performed by ants, which collect the seeds and take them underground to their nests. Ants disperse and bury seeds for hundreds of plant species in almost every terrestrial environment, from northern coniferous forests to tropical rain forests to deserts.

Both ants and flowering plants benefit from their association. The ants ensure the reproductive success of the plants whose seeds they bury, and the plants supply food to the ants. Many of the seeds that ants collect and take underground have a special protruding structure called an **elaiosome,** or *oil body* (see figure). Ants carry the seeds underground before removing the elaiosomes, which are a nutritious food for them. Once an elaiosome is removed from a seed, the ants discard the undamaged seed in an underground refuse pile, which happens to be rich in organic material (such as ant droppings and dead ants) and thus contains the minerals (inorganic nutrients) required by young seedlings. Hence, ants not only bury seeds away from animals that might eat them but also place them in rich soil that is ideal for seed germination and seedling growth.

Marion Lobstein

■ **Elaiosomes.** Bloodroot (*Sanguinaria canadensis*) seeds have nutrient-rich elaiosomes, or oil bodies. The golden brown part is the seed proper, and the white part is the elaiosome.

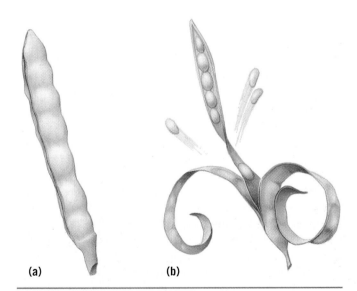

(a) **(b)**

FIGURE 9-20 Explosive dehiscence.
(a) An intact fruit of bitter cress (*Cardamine pratensis*) before it has split open. In this long, dry, dehiscent fruit, called a silique, the two halves split open, leaving a central partition. **(b)** The bitter cress fruit dehisces with explosive force, propelling the seeds some distance from the plant.

and bitter cress (*Cardamine pratensis*) split open so explosively that their seeds are scattered a meter or more (•**Figure 9-20**).

Seed Germination and Early Growth

You have seen that pollination and fertilization are followed by seed and fruit development. Each seed develops from an ovule and contains an embryonic plant and food to provide nourishment for the embryo during **germination,** when the seed sprouts. A mature seed—that is, a seed in which the embryo is fully developed—is often dormant (not actively growing) and may not germinate immediately, even if growing conditions are ideal.

Numerous factors influence whether or not a seed germinates. Many of these are environmental cues, including the presence of water and oxygen, proper temperature, and sometimes the presence of light penetrating the soil surface. No seed germinates, for example,

Marion Lobstein

FIGURE 9-21 **Pinto bean (*Phaseolus vulgaris*) seeds (*left*) before and (*right*) after imbibition.**

unless it has absorbed water. The embryo in a mature seed is dehydrated, and a watery environment in cells is necessary for active metabolism. When a seed germinates, its metabolic machinery is turned on, and numerous materials are synthesized and degraded. Therefore, water is an absolute requirement for germination.

The absorption of water by a dry seed that precedes germination is known as **imbibition.** As a seed imbibes water, it often swells to several times its original, dry size (•**Figure 9-21**). Cells imbibe water by osmosis, which is the movement of water across a membrane from an area of high concentration to an area of low concentration, and by adsorption of water onto and into materials such as cellulose, pectin, and starches within the seed. Water is attracted and bound to these materials by adhesion, the attraction between unlike materials.

Seed germination and subsequent growth also require a great deal of energy. Because plants obtain this energy by converting the energy of fuel molecules stored in the seed's endosperm or cotyledons to ATP by aerobic respiration, oxygen is usually needed during germination. Some plants, such as rice, can respire without oxygen during the early stages of germination and seedling growth. This enables rice plants to grow and become established in flooded soil, an environment that would suffocate most young plants.

Temperature is another environmental factor that affects germination. In a group of seeds of the same species,

some germinate at each temperature in a broad range of temperatures. Each plant species, however, has an optimal, or ideal, temperature at which the largest number of seeds germinates. For most plants, the optimal germination temperature is between 25°C and 30°C (77°F and 86°F). Some seeds, such as those of apples, require prolonged exposure to cold before they germinate at any temperature.

Some of the environmental factors necessary for seed germination help ensure the survival of the young plant. The requirement of a prolonged cold period ensures that seeds germinate in the spring rather than in the autumn. Some plants—especially those with tiny seeds, such as lettuce—require light for germination. A light requirement ensures that a tiny seed germinates only if it is close to the surface of the soil. If such a seed germinates several inches below the soil surface, it may not have enough food reserves to grow to the surface. On the other hand, if this light-dependent seed remains dormant until the soil is disturbed and it is brought to the surface, it has a much greater likelihood of survival.

In certain seeds, internal factors, which are under genetic control, prevent germination even when all external conditions are favorable. Many seeds are dormant either because the embryo is immature and must develop further or because certain chemicals are present. The presence of such chemical inhibitors helps en-

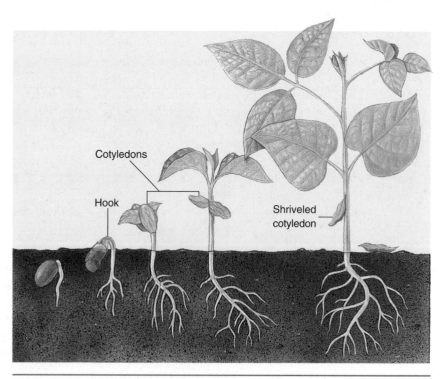

FIGURE 9-22 **Germination and growth of a young soybean (*Glycine max*), a eudicot.**
Note the hook in the stem of the young seedling, which protects the delicate stem tip as it moves up through the soil. Once the stem has emerged from the soil, the hook straightens.

sure the survival of the plant. The seeds of many desert plants, for example, often contain high levels of abscisic acid (discussed in Chapter 11). Abscisic acid is washed out only when rainfall is sufficient to support the plant's growth after the seed germinates. Some seeds, such as those of legumes, have extremely hard, thick seed coats that prevent water and oxygen from entering, thereby inducing dormancy. After these seeds are dispersed into the natural environment, exposure to the elements gradually weakens their seed coats so that germination eventually occurs. *Scarification,* the process of scratching or nicking the seed coat (physically with a knife or chemically with an acid) before sowing it, induces germination in these plants.

Eudicots and monocots exhibit characteristic patterns of early growth

Once conditions are right for seed germination, the first part of the plant to emerge from the seed is the radicle, or embryonic root. As the root grows and forces its way through the soil, it encounters considerable friction from soil particles. The delicate apical meristem of the root tip is protected by a root cap.

The plant shoot is next to emerge from the seed. Stem tips are not protected by anything like a root cap, but plants have ways to protect the delicate stem tip as it grows up through the soil to the surface. The stem of a bean seedling (a eudicot), for instance, curves over to form a hook so that the stem tip and cotyledons are actually pulled up through the soil (•**Figure 9-22**). Corn and other grasses (monocots) have a special sheath of cells called a **coleoptile** that surrounds the young shoot (•**Figure 9-23**). First the coleoptile pushes up through the soil, and then the leaves and stem grow up through the middle of the coleoptile.

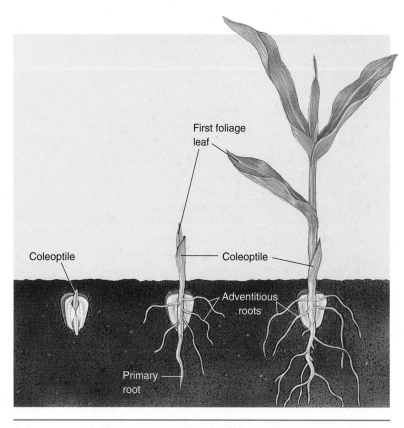

FIGURE 9-23 Germination and growth of a young corn (*Zea mays*) plant. Note the coleoptile, a sheath of cells that emerges first from the soil. The stem and leaves grow up through the middle of the coleoptile.

Certain parts of a plant grow throughout its life. This **indeterminate growth**—the ability to grow indefinitely—is characteristic of stems and roots, both of which arise from apical meristems. Theoretically, stems and roots could continue to grow forever. Other parts of a plant, such as leaves and flowers, have **determinate growth;** that is, they stop growing after reaching a certain size. The size of each of these structures varies from species to species and from individual to individual depending on the plant's genetic programming and on environmental conditions, such as availability of sunlight, water, and essential minerals.

STUDY OUTLINE

❶ Name the parts of a flower, and describe the functions of each part.
The flower is the structure in which sexual reproduction occurs. A flower may contain sepals, petals, stamens, and carpels. **Sepals** cover and protect the parts of the flower when it is a bud. **Petals** play an important role in attracting animal pollinators to the flower. **Stamens** produce **pollen grains.** Each pollen grain contains two cells: one cell generates two **sperm cells,** and the other produces a **pollen tube** through which the sperm cells will reach the ovule. Each **carpel** bears one or more **ovules;** an **egg** and two **polar nuclei** form in the ovule.

❷ **Distinguish between pollination and fertilization.**
Pollination is the transfer of pollen grains from **anther** to **stigma.** After pollination, **fertilization,** or fusion of **gametes,** occurs.

❸ **Compare the general characteristics of flowers pollinated in different ways (by insects, birds, bats, and wind).**
Some plants rely on animals to transfer pollen grains. Flowers pollinated by insects are often yellow or blue and have a scent. Bird-pollinated flowers are often yellow, orange, or red and do not have a strong scent. Bat-pollinated flowers often have dusky white petals and are scented. Plants that are wind pollinated often have smaller petals or lack petals altogether, and they do not produce a scent or nectar. Wind-pollinated flowers make copious amounts of pollen grains.

❹ **Define *coevolution,* and give examples of ways in which plants and their animal pollinators have affected one another's evolution.**
Coevolution occurs when two different organisms (such as flowering plants and their animal pollinators) form such an interdependent relationship that they affect the course of each other's evolution. During the time plants were coevolving specialized features such as petals, scent, and nectar to attract pollinators, animal pollinators coevolved specialized body parts and behaviors that enabled them to aid pollination and obtain nectar and pollen grains as a reward. Coevolution is responsible for the long, curved beaks of honeycreepers, which insert their beaks into tubular flowers to obtain nectar. The long, tubular **corollas** of the flowers that honeycreepers visit also came about through coevolution.

❺ **List and define the main parts of a seed.**
A mature **seed** contains a young plant embryo and nutritive tissue (stored in the **endosperm** or **cotyledons**)

for use during germination. The seed is covered by a protective **seed coat.** The mature flowering plant embryo consists of a **radicle,** a **hypocotyl,** one or two cotyledons, and a **plumule** (epicotyl).

❻ **Distinguish among simple, aggregate, multiple, and accessory fruits; give examples of each type; and cite several different methods of seed and fruit dispersal.**
Seeds of flowering plants are enclosed within **fruits,** which are mature, ripened **ovaries. Simple fruits** develop from one or several united carpels. Some simple fruits are fleshy at maturity, whereas others are dry. **Berries** (grapes) and **drupes** (peaches) are simple, fleshy fruits. **Follicles** (milkweed pods), **legumes** (bean pods), and **capsules** (poppy fruits) are simple, dry fruits that split open at maturity. **Caryopses** (wheat grains), **achenes** (sunflower fruits), and **nuts** (acorns) are simple, dry fruits that do not split open at maturity. An **aggregate fruit** (raspberry) develops from a single flower with many separate ovaries. A **multiple fruit** (pineapple) develops from many ovaries of many flowers growing in proximity on a common axis. In **accessory fruits** (strawberries) the major part of the fruit consists of tissue other than ovary tissue. Seeds and fruits of flowering plants are adapted for various means of dispersal, including wind, animals, water, and explosive dehiscence.

❼ **Summarize the influence of environmental factors on the germination of seeds.**
Both internal and external factors affect seed germination. External environmental factors that may affect seed germination include requirements for oxygen, water, temperature, and light. Internal factors affecting whether a seed germinates include the maturity of the embryo and the presence or absence of chemical inhibitors.

REVIEW QUESTIONS

1. Distinguish between sexual and asexual reproduction.
2. What is a flower? What are the four parts of a flower?
3. Is an imperfect flower also incomplete? Is an incomplete flower also imperfect? Explain your answers.
4. What is the difference between pollination and fertilization? Which process occurs first?
5. On a walk through a field, you find a flower with large yellow petals and a sweet scent. Suggest a possible pollinator for this flower, and explain your reasoning.

6. In the early spring, before the trees have developed large leaves, you notice clusters of tiny flowers on the branches. These flowers do not have any petals, nor do they have a scent, but when you tap them, they shed a large amount of pollen grains. Suggest how these flowers might be pollinated, and explain your reasoning.
7. What is coevolution?
8. Why does the 'i'iwi have a curved bill? Why might its bill be getting shorter?
9. Distinguish between seeds and fruits. From what part of the flower does each develop?

10. What is endosperm? What are cotyledons?

11. Distinguish among simple, aggregate, multiple, and accessory fruits, and give examples of each.

12. Explain some of the features of seeds and fruits that are dispersed by animals.

13. Label the following diagram, and give the function of each part. (Refer to Figure 9-1b to check your answers.)

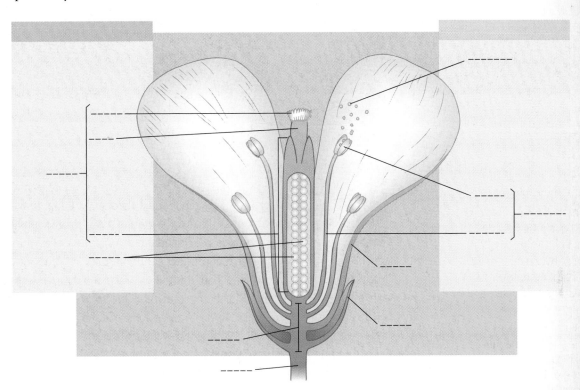

THOUGHT QUESTIONS

1. Might sexual or asexual reproduction be more beneficial in each of the following sets of circumstances, and why?

  a. A perennial (a plant that lives more than 2 years) in a stable environment

  b. An annual (a plant that lives 1 year) in a rapidly changing environment

  c. A plant adapted to an extremely narrow climate range

2. Draw pictures to show the kinds of flowers that might form simple, aggregate, multiple, and accessory fruits.

3. How is sexual reproduction in humans similar to sexual reproduction in flowering plants? How is it different?

4. Could seed dispersal by ants be considered an example of coevolution? Why or why not?

5. If apple seeds are planted in a tropical area where there is no winter, can they germinate successfully? Why or why not?

Visit us on the web at http://www.thomsonedu.com/biology/berg for additional resources, such as flashcards, tutorial quizzes, further readings, and web links.

Mineral Nutrition and Transport in Plants

LEARNING OBJECTIVES

❶ List the four components of soil, and give the ecological significance of each.

❷ Describe the factors involved in soil formation.

❸ Distinguish between macronutrients and micronutrients, and list the essential elements.

❹ Discuss tension–cohesion and root pressure as mechanisms to explain the rise of water in xylem.

❺ Discuss the pressure–flow hypothesis of sugar translocation in phloem.

The coastal redwood (*Sequoia sempervirens*) is an immense evergreen with deeply grooved, spongy, reddish brown bark that may be 30 centimeters (12 inches) thick on a mature tree. It produces flat, bluish gray needles and bears its seeds in small, oval cones that mature in one season; needles and cones are about 2.5 centimeters (1 inch) in length.

Coastal redwoods require the high humidity and cool temperatures provided by ocean fogs and do not grow farther inland than the fogs will penetrate. Consequently, their growth is restricted to a narrow ribbon of land about 800 kilometers (500 miles) long in the foothills along southwestern Oregon and the northern California coast and no more than 30 kilometers (19 miles) from the Pacific Ocean.

Coastal redwoods are among the world's tallest trees: the world's tallest known tree is a coastal redwood that is 113 meters tall (370 feet). Most mature redwood trees are 61 to 83 meters (200 to 275 feet) tall, with trunks measuring 3 meters (10 feet) in diameter. As a coastal redwood ages, the lower branches die and fall off, leaving a tapering, columnar trunk that extends about 30 meters (100 feet) above the ground. The trunk spreads out at its base to provide support.

In addition to being noted for their gigantic size, coastal redwoods are known for their extreme longevity. They mature in 400 to 500 years, and individuals 1500 or more years old are not uncommon. The oldest known coastal redwood is 2200 years old.

Redwoods are named for their reddish heartwood, which is straight grained, attractive, and resistant to decay. Redwood lumber is highly valued for such uses as shingles, furniture, interior paneling, doors, windowsills, and fence posts. These majestic trees have been overexploited for their valuable timber, and most of the original forests have been cut down. As a result, only a few virgin groves of coastal redwoods remain, and conservationists strongly support their protection.

The coastal redwood is an appropriate plant to feature in this chapter because of its immense size. How do water and dissolved minerals (inorganic nutrients) move from the roots to the needles at the top of the coastal redwood? How does sugar produced in photosynthesis move from the needles to the living cells in the roots? We now consider these questions and examine the soil from which plants obtain water and minerals.

The majestic redwoods are some of the world's tallest plants.

Soil

Soil is the ground underfoot, a relatively thin layer of Earth's crust modified by the natural actions of weather, wind, water, and organisms. It is easy to take soil for granted. We walk on and over it throughout our lives, but we rarely stop to think about how important it is to our survival. Vast numbers and kinds of organisms colonize soil and depend on it for shelter and food. Most plants anchor themselves in soil, and from it they draw water and essential minerals. Many elements essential for plant growth are obtained directly from the soil. Most plants cannot survive on their own without soil, and because we depend on plants for our food, neither could humans exist without soil.

Soils are generally formed from rock (called *parent material*) that is gradually broken down, or fragmented, into smaller and smaller particles by biological, chemical, and physical **weathering processes** (•**Figure 10-1**). Two important factors that work together in the weathering of rock are climate and organisms. When plant roots and other organisms living in the soil respire, they produce carbon dioxide (CO_2), which diffuses into the soil and reacts with soil water to form carbonic acid (H_2CO_3). Soil organisms such as *lichens*, which are "dual organisms" composed of a fungus and an alga or cyanobacterium, produce other kinds of acids. These acids etch tiny cracks, or fissures, in rock surfaces; water then seeps into these cracks. If the parent material is located in a temperate climate, the alternate freezing and thawing of the water during winter break off small pieces of rock, causing the cracks to enlarge. Small plants then become established and send their roots into the larger cracks, fracturing the rock further.

Topography, which is a region's surface features, such as the presence or absence of mountains and valleys, is also involved in soil formation. Steep slopes often have little or no soil on them, because gravity continually transports soil and rock down the slopes. Runoff from precipitation tends to amplify erosion on steep slopes. Moderate slopes, on the other hand, may have less downhill transport and therefore form deep soils.

The disintegration of solid rock into finer and finer mineral particles and the accumulation of organic material (discussed in the next section) in the soil take an extremely long time, sometimes thousands of years. Soil forms constantly as the weathering of parent material beneath already formed soil continues to add new soil.

Soil is composed of inorganic minerals, organic matter, air, and water

Four distinct components compose soil: inorganic mineral particles (which make up about 45 percent of a typical soil), organic matter (about 5 percent), water (about 25 percent), and air (about 25 percent) (•**Figure 10-2**). The plants, animals, fungi, and microorganisms that inhabit soil interact with it, and minerals are continually cycled from the soil to organisms, which use them in their biological processes. When the organisms die, bacteria and other soil organisms decompose the remains, so the minerals return to the soil.

The inorganic mineral particles, which come from weathered rock, provide anchorage and essential minerals for plants, as well as pore space for water and air. Because different rocks consist of different minerals, soils vary in mineral composition and chemical properties. Rocks rich in aluminum form acidic soils, for example, whereas rocks that contain silicates of magnesium and iron form soils that may be deficient in calcium, nitrogen, and phosphorus. Also, soils formed from the same kind of parent material may not develop in the same way, because other factors, such as weather, topography, and kinds of organisms, differ.

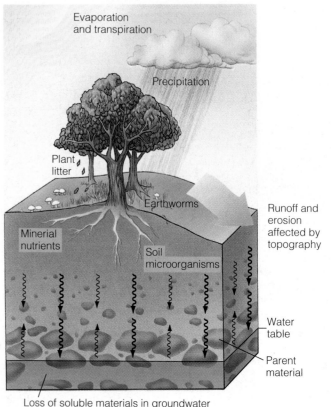

Organic additions (underground) from soil organisms

Fine particles and solubles are washed downward

Capillarity and evaporation cause some materials to rise

FIGURE 10-1 The dynamics of soil formation.
Weather, climate, topography, and organisms interact with Earth's crust to form soil, the material that supports life on land.

to 0.02 millimeter in diameter), and the smallest particles are called **clay** (less than 0.002 millimeter in diameter). Sand particles are large enough to easily see, and silt particles (about the size of flour particles) are barely visible. Most individual clay particles are too small to see with an ordinary light microscope; they can be seen only under an electron microscope.

A soil's texture affects many of that soil's properties, in turn influencing plant growth. The clay component of a soil is particularly important in determining many of its characteristics, because clay particles have the greatest surface area of all soil particles. If the surface areas of about 450 grams (1 pound) of clay particles were laid out side by side, they would occupy 1 hectare (2.5 acres).

Each clay particle has predominantly negative electric charges on its outer surface that attract and reversibly bind **cations,** which are positively charged mineral ions, such as potassium (K^+) and magnesium (Mg^{2+}). Because many cations are essential for plant growth, cation absorption is an important aspect of soil fertility. Roots secrete protons (H^+), which are exchanged for other positively charged mineral ions attracted to the surface of soil particles, in a process known as **cation exchange** (•**Figure 10-3**). The plant's roots absorb these "freed" ions and the water that forms a film around the soil particles.

In contrast, **anions,** which are negatively charged mineral ions, are repelled by the negative surface charges of clay particles and tend to remain in solution. Anions such as nitrate (NO_3^-) are often washed out of the root zone by water moving through the soil.

Soil always contains a mixture of different-sized particles, but the proportions vary from one soil to another. A **loam,** which is an ideal agricultural soil, has an optimal combination of different-sized soil particles: it contains approximately 40 percent each of sand and silt and about 20 percent of clay. Generally, the larger particles provide structural support, aeration, and permeability to the soil, whereas the smaller particles bind together into aggregates, or clumps, and hold minerals and water. Soils with larger proportions of sand are not as desirable for most plants, because they do not hold water and mineral ions well. Plants grown in such soils are more susceptible to drought and mineral deficiencies. Soils

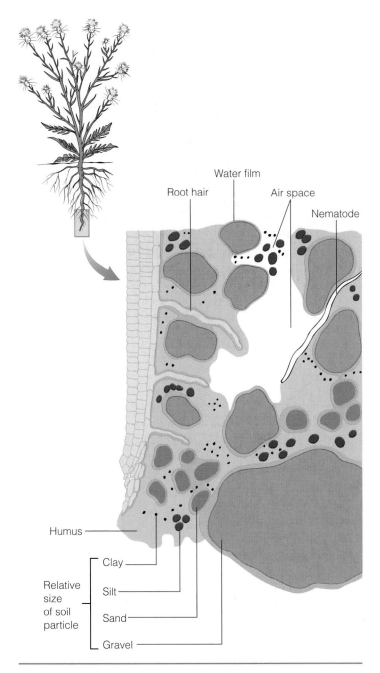

FIGURE 10-2 Soil components.
Soil is composed of inorganic mineral particles (sand, silt, and clay), organic material (humus), soil air, and soil water.

The **texture,** or structural characteristic, of a soil is determined by the percentages (by weight) of the different-sized inorganic mineral particles—sand, silt, and clay—in it. The size assignments for sand, silt, and clay give soil scientists a way to classify soil texture. Any particles larger than 2 millimeters in diameter, called gravel or stones, are not considered soil particles because they do not have any direct value to plants. The largest soil particles are called **sand** (0.02 to 2 millimeters in diameter), the medium-sized particles are called **silt** (0.002

SOIL The surface layer of Earth's crust, consisting primarily of fragmented and weathered grains of rocks; supports terrestrial plants as well as many animals and microorganisms.

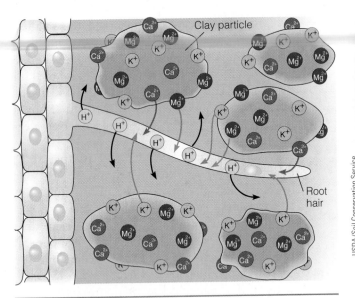

FIGURE 10-3 Cation exchange.
Negatively charged clay particles bind to positively charged mineral cations in the soil, holding the cations in the soil. Roots pump out protons (H^+), which are exchanged for the cations, facilitating cation absorption.

FIGURE 10-4 Humus.
Soil that is rich in humus, partially decomposed organic material (primarily from plant and animal remains), has a loose, somewhat spongy structure with several properties (such as increased water-holding capacity) that are beneficial for plants and other organisms living in it.

with larger proportions of clay are also not desirable for most plants, because they have poor drainage and often do not provide enough oxygen. Clay soils used in agriculture tend to get compacted, which reduces the number of soil spaces filled by water and air.

A soil's organic matter consists of the wastes and remains of soil organisms

The organic matter in soil is composed of litter (dead leaves and branches on the soil's surface); droppings (animal dung); and the remains of dead plants, animals, and microorganisms in various stages of decomposition. Microorganisms, particularly soil bacteria and fungi, decompose organic matter—that is, break it down into simpler materials. During decomposition, essential mineral ions are released into the soil, where they are bound by soil particles or absorbed by plant roots. Organic matter increases the soil's water-holding capacity by acting much like a sponge. For this reason, gardeners often add organic matter to soils, especially sandy soils, which are naturally low in organic matter (see *Plants and People: Managing Your Lawn and Garden at Home*).

The partly decayed organic portion of the soil is referred to as **humus** (•Figure 10-4). Humus, which is not a single chemical compound but a mixture of many organic compounds, binds mineral ions and holds water. On average, humus persists in agricultural soil for about 20 years, although certain components of humus may persist for hundreds of years. Humus is somewhat

resistant to decay, but a succession of microorganisms gradually reduces it to CO_2, water, and mineral ions.

Organic matter dominates some soils. Peat soils, which consist almost entirely of undecayed plant remains, may form where poor drainage limits decomposition by soil-dwelling bacteria.

About 50 percent of soil volume is composed of pore spaces

Soil has numerous pore spaces of different sizes around and among the soil particles. Pore spaces occupy roughly 50 percent of a soil's volume and are filled with varying proportions of soil air and soil water. Both air and water are necessary to produce a moist but aerated soil that sustains plants and other soil-dwelling organisms (•Figure 10-5). Water is usually held in the smaller pores, whereas air is found in the larger pores. After a prolonged rain, almost all the pore spaces are filled with water, but water drains rapidly from the larger pore spaces, drawing air from the atmosphere into those spaces.

Soil air contains the same gases as atmospheric air, although they are usually present in different proportions. As a result of cellular respiration by roots and other soil organisms, there are less oxygen and more carbon dioxide in soil air than in atmospheric air. (Recall from Chapter 4 that aerobic respiration uses oxygen and produces carbon dioxide.) Among the important gases in soil air are oxygen (O_2), required by roots and soil organisms for aerobic respiration; nitrogen (N_2),

PLANTS AND PEOPLE | Managing Your Lawn and Garden at Home

You can put to use some of what you have learned in this chapter when managing your lawn and garden at home. Using compost and mulch can maintain and improve your soil. Gardeners often dispose of grass clippings, leaves, and other plant refuse by bagging it for garbage collection or burning it. Neither action is desirable, however; treating yard wastes as garbage contributes more material to our already overburdened landfills, and the burning of yard wastes contributes to air pollution.

These materials are a valuable resource for making **compost,** a natural soil-and-humus mixture that improves fertility and soil structure. Grass clippings, leaves, weeds, sawdust, coffee grounds, animal manure, ashes from the fireplace or grill, shredded newspapers, fruit peelings and other meat- and cheese-free leftovers, and eggshells are just some of the materials that can be composted, that is, transformed into compost by microbial action.

To make a compost heap, spread a 15- to 30-centimeter (6- to 12-inch) layer of grass clippings, leaves, or other plant material in a shady area, sprinkle it with garden fertilizer or a thin layer of animal manure, and cover it with several centimeters of soil. Add layers as you collect more organic debris. Water the material thoroughly, and turn it over with a pitchfork each month to aerate it. Although it is possible to make compost by just heaping it in layers, it is more efficient to construct a compost enclosure. An enclosed compost heap is also less likely to attract animals. When the compost is uniformly dark in

USDA/Soil Conservation Service

■ Mulches discourage the growth of weeds and help keep the soil damp. Organic mulches, such as shredded bark (*shown*), have the added benefit of gradually decaying, thereby increasing soil fertility.

color, is crumbly, and has a pleasant "woodsy" odor, it is ready to use. The time required for decomposition varies from 1 to 6 months depending on what the temperature is, what materials are used, and how often the heap is turned and watered.

Whereas compost is mixed into soil to improve the soil's fertility, **mulch** is placed on the surface of soil, around the bases of plants (see figure). Mulch helps control weeds and increases the amount of water in the upper levels of the soil by reducing evaporation. It lowers the soil temperature in the summer and extends the growing season slightly by providing protection against cold in the fall. It also decreases soil erosion by lessening precipitation runoff.

Although mulches can consist of inorganic materials such as plastic sheets or gravel, natural mulches of grass clippings, straw, chopped corncobs, or shredded bark have the added benefit of increasing the organic content of the soil. Grass clippings are an effective mulch when placed around the bases of garden plants because they mat together, making it difficult for weeds to become established. You must replace grass mulches often, however, because they decay rapidly. Some gardeners prefer mulches of more expensive materials such as shredded bark, because they take longer to decompose and are more attractive.

used by nitrogen-fixing bacteria; and carbon dioxide, involved in soil weathering.

Soil water originates as precipitation, which drains downward, or as groundwater (water stored in porous underground rock), which rises upward from the water table (the uppermost level of groundwater). Soil water contains low concentrations of dissolved minerals that enter the roots of plants when they absorb water. Wa-

ter not bound to soil particles or absorbed by roots percolates (drains down) through soil, carrying dissolved minerals with it. The removal of dissolved materials from soil by percolating water is called **leaching.** The deposition of leached material in the lower layers of soil is known as **illuviation.** Iron and aluminum compounds, humus, and clay are some of the illuvial materials that gather in the subsurface portion of the soil.

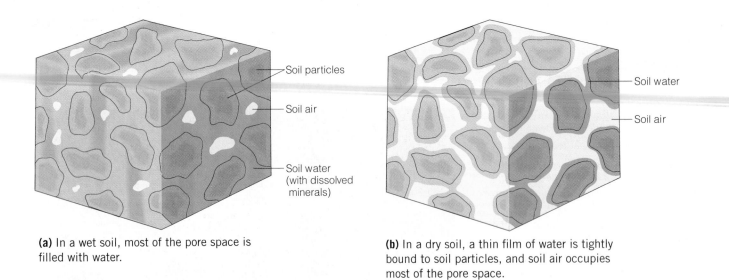

(a) In a wet soil, most of the pore space is filled with water.

(b) In a dry soil, a thin film of water is tightly bound to soil particles, and soil air occupies most of the pore space.

FIGURE 10-5 **Pore space.**

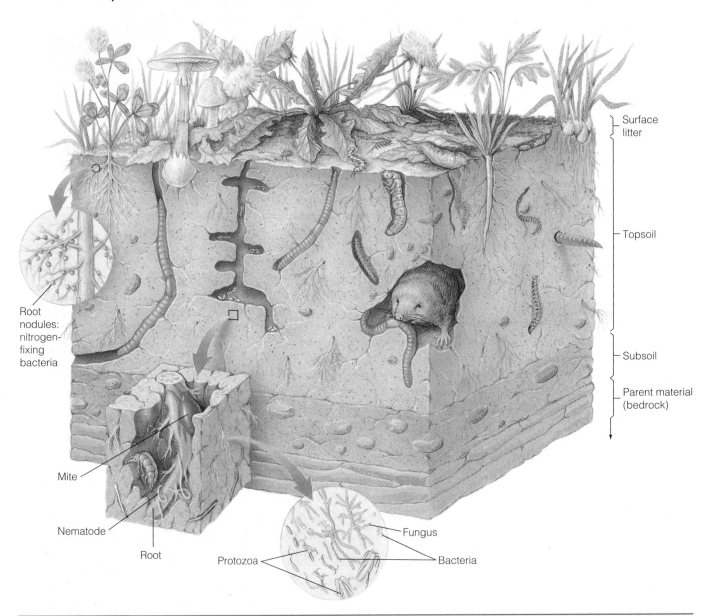

FIGURE 10-6 **Diversity of life in fertile soil.**
Soil-dwelling life-forms include plant roots, algae, fungi, earthworms, flatworms, roundworms, insects, spiders and mites, bacteria, and burrowing animals such as moles and groundhogs.

Some substances are completely leached out of the soil because they are so soluble that they migrate down to the groundwater. It is also possible for water that is moving *upward* through the soil to carry dissolved materials with it.

Soil organisms form a complex community

A single teaspoon of fertile agricultural soil may contain millions of microorganisms such as bacteria, fungi, algae, protozoa, and microscopic worms. Many other organisms also colonize soil, including plant roots, earthworms, insects, moles, gophers, snakes, and groundhogs (•**Figure 10-6**). Most numerous in soil are bacteria, which number in the hundreds of millions per gram of soil. Biologists have identified about 170,000 species of soil organisms, but thousands remain to be identified. Little is known about the roles of soil organisms, in part because it is hard to study their activities under natural conditions.

Bacteria and fungi are important, not only in decomposing dead organic material in the soil but in *nutrient cycling*, the processes by which matter cycles from the living world to the nonliving physical environment and back again. Nutrient cycles break down or build up chemicals that plants need (•**Figure 10-7**). For example, microorganisms are involved in most steps in the nitrogen cycle, which supplies nitrogen, an essential part of proteins and nucleic acids, to organisms (see Figure 26-19).

Worms are some of the most important organisms living in soil. Earthworms, one of the most familiar soil inhabitants, ingest soil and obtain energy and raw materials by digesting humus. *Castings,* bits of soil that have passed through the gut of an earthworm, are deposited on or near the soil surface. In this way, minerals from deeper layers are brought to upper layers. Earthworm tunnels aerate the soil, and the worms' waste products and corpses add organic material to the soil.

Ants live in the soil in enormous numbers, constructing tunnels and chambers that aerate it. Members of soil-dwelling ant colonies forage on the surface for bits of food, which they carry back to their nests. Not all this food is eaten, however, and its eventual decomposition helps increase the organic matter in the soil. Many ants are also indispensable in plant reproduction because they bury seeds in the soil (see *Plants and the Environment: Seed Dispersal by Ants,* in Chapter 9).

Soil pH affects soil characteristics and plant growth

As discussed in Chapter 2, soil acidity is measured with the pH scale, which ranges from 0 (extremely acidic) through 7 (neutral) to 14 (extremely alkaline). The pH of

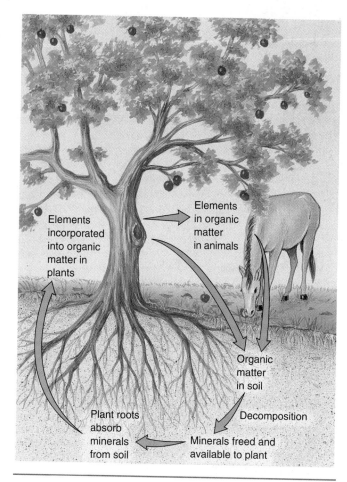

FIGURE 10-7 Nutrient cycles.
Mineral nutrients cycle from the soil to organisms and then back to the soil.

Labels in figure: Elements incorporated into organic matter in plants; Elements in organic matter in animals; Organic matter in soil; Decomposition; Plant roots absorb minerals from soil; Minerals freed and available to plant

most soils ranges from 4 to 8, but some soils are outside this range. The soil of the Pygmy Forest in Mendocino County, California, is extremely acidic (pH 2.8 to 3.9). At the other extreme, certain saline soils in Death Valley, California, have a pH of 8.5.

Soil pH affects plants, partly because the solubility of certain minerals varies with differences in pH. Plants absorb soluble mineral elements but not insoluble forms. At a low pH, for example, the aluminum and manganese in soil water are more soluble and are sometimes absorbed by roots in toxic amounts. (See *Plants and the Environment: Soil Salinization* for an examination of soils with too many minerals.) At a higher pH, certain mineral salts essential for plant growth, such as calcium phosphate, become less soluble and thus less available to plants.

Soil pH greatly affects the leaching of minerals. An acidic soil has less ability than an alkaline soil to bind positively charged ions because the soil particles also bind the abundant protons (•**Figure 10-8**). As a result, certain minerals essential for plant growth, such as po-

PLANTS AND THE ENVIRONMENT | Soil Salinization

Soils found in arid and semiarid areas often contain high concentrations of inorganic mineral salts. In these areas, the amount of water that drains into lower soil layers is minimal because the little precipitation that falls quickly evaporates. In contrast, humid climates have enough precipitation to leach salts out of the soils and into waterways and groundwater. Irrigation of agricultural fields in arid and semiarid areas often results in their becoming increasingly salty. Also, when irrigated soil becomes waterlogged, salts may move upward from groundwater to the soil surface, where they are deposited as a crust.

Most plants cannot obtain all the water they need from salty soil, because such soil produces a water-balance problem. Under normal conditions, the dissolved materials in the watery cytoplasm of plant cells give them a concentration of water lower than that of soil. As a result, water moves by *osmosis* from the soil into plant roots. When soil water contains a large quantity of dissolved salts, however, its concentration of water is lower than that in plant cells; consequently, water moves *out* of plant roots and into the salty soil (see figure). Not surprisingly, most plants cannot survive under these conditions. Plant species that thrive in saline soils have special adaptations that enable them to tolerate the high amount of salt. Most crops, unless they

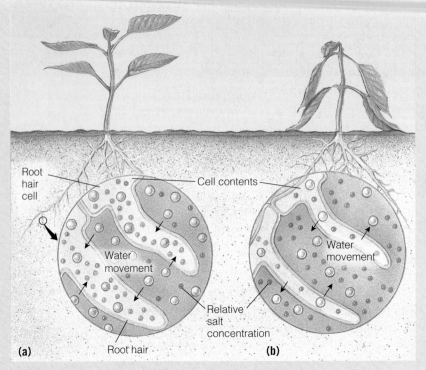

■ The effect of saline soil on water absorption by roots. (a) Normally there is a net movement of water (*black arrows*) into root cells from the soil. (b) When soil contains a high amount of salt, water moves *out* of the roots into the soil, even when the soil is wet.

are genetically selected to tolerate high salt, are not productive in saline soil.

In principle, the way to remove excess salt from saline soils is to add enough water to leach it away. Although this sounds straightforward, it is extremely difficult and in many cases impossible. Recall that saline soils usually occur in areas where water is in

short supply. Also, many soils lack good drainage properties, so adding lots of water simply causes them to become waterlogged. Even if the salt is flushed out of the soil, it has to go somewhere. The excess salt is usually carried to groundwater or to rivers and streams, where it becomes a water pollutant.

---

tassium (K^+), are leached more readily from acidic soil. The optimum soil pH for most plant growth is 6.5 to 7.5, because most essential elements needed by plants are available in that range.

Soil pH affects plants, and plants and other organisms, in turn, influence soil pH. The decomposition of humus and the cellular respiration of soil organisms and roots decrease the pH of the soil.

Acid precipitation, a type of air pollution in which sulfuric and nitric acids produced by human activities fall to the ground as acid rain, sleet, snow, or fog, can

seriously decrease soil pH (see *Plants and the Environment: Acid Rain,* in Chapter 2). Acid precipitation is one of several factors implicated in **forest decline,** the gradual deterioration, and often death, of trees in many European and North American forests (see *Plants and the Environment: The Effects of Air Pollution on Leaves,* in Chapter 8). Forest decline may be partially the result of soil changes caused by acid precipitation. In central European forests that have experienced forest decline, a strong correlation exists between forest damage and soil chemistry altered by acid precipitation.

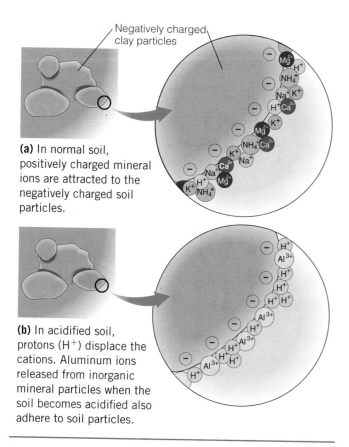

Negatively charged
clay particles

(a) In normal soil, positively charged mineral ions are attracted to the negatively charged soil particles.

(b) In acidified soil, protons (H⁺) displace the cations. Aluminum ions released from inorganic mineral particles when the soil becomes acidified also adhere to soil particles.

FIGURE 10-8 How acid alters soil chemistry.

Soil provides most of the minerals found in plants

More than 90 naturally occurring elements exist on Earth, and more than 60 of these, including elements as common as carbon and as rare as gold, are found in plant tissues. Not all these elements are considered essential for plant growth, however.

Nineteen elements are essential for most, if not all, plants (•**Table 10-1**). Ten of these are required in fairly large quantities (greater than 0.05 percent dry weight) and are therefore known as **macronutrients**. These include carbon, hydrogen, oxygen, nitrogen, potassium, calcium, magnesium, phosphorus, sulfur, and silicon. The remaining nine **micronutrients** are needed in very small, or trace, amounts for normal plant growth and development. These include chlorine, iron, boron, manganese, sodium, zinc, copper, nickel, and molybdenum.

Four of the 19 elements—carbon, oxygen, hydrogen, and nitrogen—come directly or indirectly from soil water or from gases in the atmosphere. Carbon is obtained from carbon dioxide in the atmosphere during photosynthesis. Oxygen is obtained from atmospheric oxygen and water. Water also supplies hydrogen to the plant. Plants absorb their nitrogen from the soil as ions of nitrogen salts—nitrate (NO_3^-) and ammonium (NH_4^+). The nitrogen in the nitrogen salts ultimately comes from atmospheric nitrogen (N_2) and is converted to nitrate and ammonium by various microorganisms in either the soil or the root nodules of certain plants (see Chapter 26). The remaining 15 essential elements are obtained from the soil as dissolved mineral ions. Their ultimate source is the parent material from which the soil was formed.

Let us examine the main functions of essential elements. Carbon, hydrogen, and oxygen are found in the structures of all biologically important molecules, including lipids, carbohydrates, nucleic acids, and proteins. Nitrogen is part of proteins, nucleic acids, and chlorophyll.

Potassium, which plants use in fairly substantial amounts, is not found in a specific organic compound in plant cells. Instead, it remains as free K^+ and plays a key role in maintaining the turgidity of cells because it is osmotically active. The presence of K^+ in cytoplasm causes water to pass through the plasma membrane into the cell by **osmosis**. Potassium is also involved in stomatal opening (see Chapter 8).

Calcium plays a key structural role as a component of the *middle lamella* (the cementing layer between the cell walls of adjacent plant cells). Calcium ions (Ca^{2+}) are also important in a number of physiological roles in plants, such as altering membrane permeability and *cell signaling* (see Chapter 3).

Magnesium is critical for plants because it is part of the chlorophyll molecule. Phosphorus is a component of nucleic acids, phospholipids (an essential part of cell membranes), and energy transfer molecules such as adenosine triphosphate (ATP). Sulfur is essential because it is found in certain amino acids and vitamins. Many plants accumulate silicon in their cell walls and intercellular spaces. Silicon enhances the growth and fertility of some species and may help reinforce cell walls.

Chlorine and sodium are micronutrients that help maintain turgidity of cells. In addition to its osmotic role, chloride (Cl^-) and sodium (Na^+) ions are essential for photosynthesis.

MACRONUTRIENT An essential element that is required in a fairly large amount for normal plant growth.

MICRONUTRIENT An essential element that is required in very small amounts for normal plant growth.

OSMOSIS The net movement of water (the principal solvent in biological systems) by diffusion through a selectively permeable membrane.

TABLE 10-1 Functions of Elements Required by Most Plants

ELEMENT	SOURCE	MAJOR FUNCTIONS
Macronutrients		
Carbon	Air (as CO_2)	In carbohydrate, lipid, protein, and nucleic acid molecules
Hydrogen	Water	In carbohydrate, lipid, protein, and nucleic acid molecules
Oxygen	Water, air (as O_2)	In carbohydrate, lipid, protein, and nucleic acid molecules
Nitrogen	Soil	In proteins, nucleic acids, chlorophyll, certain coenzymes
Potassium	Soil	Osmotic and ionic balance; opening and closing of stomata; enzyme activator (for 40+ enzymes)
Calcium	Soil	In cell walls; involved in membrane permeability; enzyme activator
Magnesium	Soil	In chlorophyll; enzyme activator
Phosphorus	Soil	In nucleic acids, phospholipids, ATP (energy transfer compound)
Sulfur	Soil	In certain amino acids and vitamins
Silicon	Soil	In cell walls
Micronutrients		
Chlorine	Soil	Ionic balance; involved in photosynthesis
Iron	Soil	Involved in photosynthesis and respiration
Boron	Soil	In cell walls; involved in nucleic acid metabolism and in cell growth
Manganese	Soil	In enzymes involved in respiration and nitrogen metabolism; required for photosynthesis
Sodium	Soil	Involved in photosynthesis; substitutes for potassium in osmotic and ionic balance
Zinc	Soil	In enzymes involved in respiration and nitrogen metabolism
Copper	Soil	In enzymes involved in photosynthesis
Nickel	Soil	In enzymes (urease) involved in nitrogen metabolism
Molybdenum	Soil	In enzymes involved in nitrogen metabolism

Six of the micronutrients (iron, manganese, zinc, copper, nickel, and molybdenum) are associated with various plant enzymes, often as enzyme activators, and are involved in certain enzymatic reactions. Boron, present in cell walls, is also involved in nucleic acid metabolism and cell growth.

PROCESS OF SCIENCE How do botanists determine whether an element is essential?

It is impossible to conduct mineral nutrition experiments by growing plants in soil, because soil contains too many elements. Therefore, nutritional studies require different methods. One of the most useful methods of testing whether or not an element is essential is **hydroponics,** which is the cultivation of plants in aerated water to which mineral salts are added (•**Figure 10-9**). Hydroponics has commercial applications in addition to its scientific use (see *Plants and People: Commercial Hydroponics*).

If botanists suspect that a particular element is essential for plant growth, they grow plants in a nutrient solution that contains all known essential elements *except* the one in question. If the plants grown in the absence of that element do not develop normally or complete their life cycle, the element may be essential.

Additional criteria are used to confirm whether an element is essential. For example, botanists must demonstrate that the element has a direct effect on the plant's metabolism and that the element is essential for a wide variety of plants. ■

Fertilizers replace essential elements missing from the soil

Soil is a valuable natural resource on which humans depend for food. Many human activities may generate or aggravate soil problems. For example, mineral depletion may occur in soils that are farmed.

In a natural ecosystem, essential minerals removed from the soil by plants are returned when the plants or the animals that have eaten the plants die and are decomposed. An agricultural system disrupts this pattern of nutrient cycling when the crops are harvested. Plant material containing minerals is removed from the nutrient cycle, and the harvested crops fail to decay and release their nutrients back to the soil. Over time, soil that is farmed inevitably loses its fertility, that is, its ability to produce abundant crops (•**Figure 10-10**). Homeowners often mow their lawns and remove the clippings, similarly preventing decomposition and cycling of minerals that were in the grass blades (see *Plants and People:*

PROCESS OF SCIENCE

QUESTION: How does nitrogen deficiency affect a crop plant such as a tomato?

HYPOTHESIS: Tomatoes grown in the absence of nitrogen exhibit recognizable symptoms.

EXPERIMENT: The plants are grown in a liquid solution of mineral nutrients through which air is bubbled to allow the roots to respire. The liquid solution for the experimental plants contains all known nutrient minerals except nitrogen. The liquid solution for control plants contains all known nutrient minerals, including nitrogen.

RESULTS AND CONCLUSION: The control plant is on the left, and the experimental plant is on the right. Compared to the control, tomatoes grown in the absence of nitrogen exhibit retarded growth and have small leaves that are light green to pale yellow in color. The plants eventually die (*not shown*). ■

FIGURE 10-9 A hydroponics experiment.

Managing Your Lawn and Garden at Home on page 203).

The essential material (water, sunlight, or some essential element) that is in shortest supply usually limits plant growth. This phenomenon is sometimes called the concept of **limiting resources.** The three elements that are most often limiting resources for plants are nitrogen, phosphorus, and potassium. To sustain the productivity of agricultural soils, farmers periodically add fertilizers to depleted soils to replace the minerals that limit plant growth.

The two main types of fertilizers are organic and commercial inorganic. *Organic fertilizers* include such natural materials as cow manure, crop residues, bone meal, and compost. *Green manure,* a special type of organic fertilizer, is a crop such as alfalfa or sweet clover that is planted and then plowed into the soil to decompose instead of being harvested. Frequently, plants grown as green manure have nitrogen-fixing bacteria living in root nodules, thereby increasing the amount of nitrogen in the soil. Organic fertilizers are complex, and their exact compositions vary. The mineral nutrients in organic fertilizers become available to plants only as the organic material decomposes. For that reason, organic fertilizers are slow acting and long lasting. Another advantage of organic fertilizers is that, in ways that are not completely understood, they change the types of organisms living in the soil, sometimes suppressing microorganisms that cause certain plant diseases.

Commercial inorganic fertilizers are manufactured from chemical compounds, and their exact compositions are known. Because they are soluble, they are immediately available to plants. Commercial inorganic fertilizers are available in the soil for only a short period (relative to organic fertilizers), because they quickly leach away, often into groundwater or surface runoff, and pollute the water. Most commercial inorganic fertilizers

PLANTS AND PEOPLE | Commercial Hydroponics

Botanists have long used hydroponics, the practice of growing plants in an aerated solution of mineral salts, to determine which elements are essential for plant growth. Initially, entrepreneurs hailed hydroponics as the scientific way to grow plants in places where soil was poor or unavailable. The expenses involved in commercial cultivation of produce for human consumption, however, prevented hydroponics from becoming more than a curiosity. However, technical improvements have revived interest in commercial hydroponics (see figure).

Hydroponics has great potential in several places. It is being tried experimentally in desert countries in the Middle East, where the soil is too arid to support cultivation and water is unavailable for irrigation. When plants are grown hydroponically in greenhouses, little water is used in comparison with traditional irrigation methods. Hydroponics is also being tried in temperate climates, particularly to produce crops in winter.

Hydroponics has other advantages. Because it is possible to grow crops hydroponically under conditions in which disease-causing microorganisms and insect pests are completely absent, the crops are not exposed to chemical pesticides. Hydroponics is also used to grow crops near their area of use, thus reducing transportation costs.

■ A greenhouse with lettuce growing hydroponically.

Joseph H. Bailey/National Geographic/Getty Images

(Most foods consumed by Americans travel an average of 2100 kilometers, or 1300 miles.)

The main disadvantage of hydroponics is the expense. Plants grown hydroponically must be given a nutrient solution that is continually monitored and adjusted. Heating and lighting costs are high. Aeration of the roots was a major expense until developments such as the nutrient-film technique helped cut costs. In the *nutrient-film technique,* plants are grown in gently sloping trenches through which a thin layer of water and nutrients trickle. Because most of the roots are exposed to the air, they get adequate aeration. The nutrient solution is saved and reused, which cuts down on water and mineral costs.

Although hydroponics will probably never replace traditional agriculture, it is a viable alternative in certain situations. As new techniques are developed, hydroponics may become even more common.

---

contain the three elements—nitrogen, phosphorus, and potassium—that are usually the limiting resources in plant growth. The numbers on fertilizer bags (for example, 10-20-20) tell the relative concentrations of each of the three elements (N, P, and K).

One advantage of inorganic fertilizers over organic fertilizers is that one knows precisely how much of a particular element is being applied to the soil. Varying the relative concentrations of nitrogen, phosphorus, and potassium causes different growth responses in plants. When growing a lettuce crop, for example, it is best to use a fertilizer with a high nitrogen content, because nitrogen stimulates vigorous vegetative growth (growth of leaves, stems, and roots) rather than reproduction (growth of flowers, fruits, and seeds). The application of a fertilizer with a high nitrogen content around tomato plants, however, causes a low production of tomatoes. Although the roots, stems, and leaves of the tomato plants grow vigorously with a high-nitrogen fertilizer, reproduction is *not* stimulated; the lush plants form few flowers and therefore few fruits.

It is environmentally sound to avoid or limit the use of manufactured fertilizers, for several reasons. First, because of their high solubility, inorganic fertiliz-

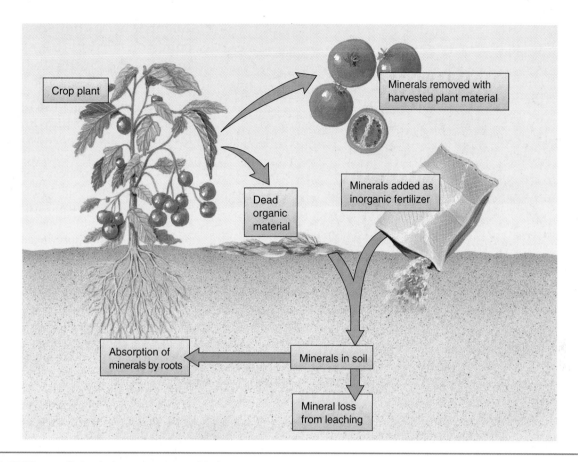

FIGURE 10-10 Why a soil that is farmed needs fertilizer.
As plant and animal remains decompose in natural environments, nutrients are cycled back to the soil for reuse. In agriculture, much of the plant material is harvested. Because the mineral nutrients in the harvested portions are unavailable to the soil, the nutrient cycle is broken. For this reason, fertilizer is added to the soil periodically.

ers cause water pollution. Second, manufactured fertilizers do not improve the water-holding capacity of the soil as organic fertilizers do. Third, the manufacture of inorganic fertilizers requires a great deal of fossil fuels and is therefore not energy conserving.

Transport in the Plant Body

Now that we have discussed soil and its importance to plants, we examine internal transport in the vascular system of the plant (•**Figure 10-11**). Roots obtain water and dissolved minerals from the soil. Once inside roots, these materials are transported upward to stems, leaves, flowers, fruits, and seeds. Furthermore, sugar molecules manufactured in the leaves by photosynthesis are transported in solution (dissolved in water) throughout

the plant, including the subterranean roots. Water and dissolved minerals are transported from roots to other parts of the plant in xylem, whereas dissolved sugar is **translocated** in phloem.

Xylem transport and phloem translocation do not resemble the movement of materials in animals because in plants nothing *circulates* in a system of vessels. Water and minerals transported in xylem travel in one direction only (upward), whereas translocation of dissolved sugar may occur upward or downward in separate phloem cells. In addition, xylem transport and phloem translocation differ from internal transport in animals because movement in both xylem and phloem is driven largely by natural physical processes rather than by a pumping organ, or heart.

How, exactly, do materials travel in the continuous system of the plant's vascular tissues? First we examine water and its movement through the plant, and later we discuss the translocation of dissolved sugar.

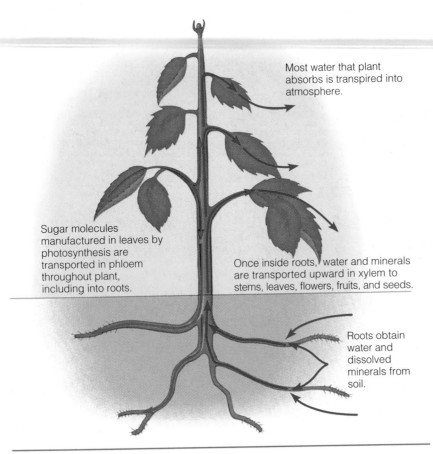

Most water that plant absorbs is transpired into atmosphere.

Sugar molecules manufactured in leaves by photosynthesis are transported in phloem throughout plant, including into roots.

Once inside roots, water and minerals are transported upward in xylem to stems, leaves, flowers, fruits, and seeds.

Roots obtain water and dissolved minerals from soil.

FIGURE 10-11 An overview of transport in vascular plants.
Xylem transport is explained on the right side of the figure (start at the bottom and work upward), whereas phloem transport is on the left side.

Transport in Xylem

Water initially moves horizontally into the roots from the soil, passing through several tissues until it reaches the xylem (see Chapter 6). Once the water is in the tracheids and vessel elements (see Figure 5-6) of the root xylem, it travels upward through a continuous network of xylem cells from root to stem to leaf (•**Figure 10-12**). The dissolved minerals are carried along passively in the water, although their initial uptake from the soil may be active (discussed shortly). The plant does not expend any energy of its own to transport water, which moves as a result of natural physical processes. The movement of water in the xylem is the most rapid of any movement of materials in plants (•**Table 10-2**).

How does water move to the tops of plants? It is either pushed up from the bottom of the plant or pulled up to the top of the plant. Although plants use both mechanisms, current evidence indicates that most water is transported through the xylem by being *pulled* to the top of the plant.

According to the tension-cohesion model, transpiration in the leaves pulls water up the stem

Currently, most botanists consider the tension–cohesion model, first proposed in 1896 by Irish botanist Henry H. Dixon, to be the dominant mechanism of xylem transport in most plants. According to the **tension–cohesion model,** also known as the **transpiration–cohesion model,** water is pulled up the plant as a result of a *tension* produced at the top of the plant (•**Figure 10-13**). The evaporative pull of **transpiration** causes this tension, which resembles sucking a liquid up a straw. Most water loss from transpiration takes place through the numerous stomata present on leaf and stem surfaces. The tension extends from the leaves, where most transpiration occurs, down the stems and into the roots. It draws water up the stem xylem to leaf cells that have lost water as a result of transpiration and pulls water from root xylem into stem xylem. As water is pulled upward, additional water from the soil is drawn into the roots. Thus, the pathway of water movement is as follows:

Soil $\longrightarrow$ root tissues (epidermis, cortex, and so forth) $\longrightarrow$ root xylem $\longrightarrow$ stem xylem $\longrightarrow$ leaf mesophyll $\longrightarrow$ atmosphere

This upward pulling of water is possible only as long as there is an unbroken column of water in the xylem throughout the plant. Water forms an unbroken column in a plant because of the *cohesiveness* of water molecules. Recall from Chapter 2 that water molecules exhibit **cohesion,** that is, they are strongly attracted to one another because of **hydrogen bonding.** In addition, the **adhesion** of water to the walls of the xylem cells, also the result of hydrogen bonding, is an important factor in maintaining an unbroken column of water. Thus, the cohesive and adhesive properties of water enable it to form a continuous column that is pulled up through the xylem.

Is the tension–cohesion model powerful enough to explain the rise of water in the tallest plants? Plant biologists have calculated that the tension produced by transpiration is strong enough to pull water upward 150 meters (500 feet) in tubes the diameter of xylem

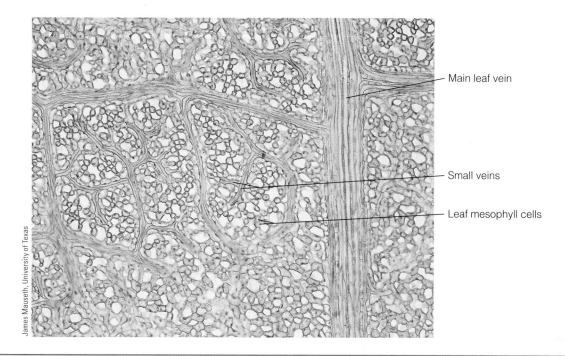

— Main leaf vein

— Small veins

— Leaf mesophyll cells

James Mauseth, University of Texas

FIGURE 10-12 Network of veins in leaf tissue.
Transport of materials throughout the plant body involves xylem and phloem, which are found in leaves as veins that branch extensively. No leaf cell is far from a vein.

vessels. Because the tallest trees are about 117 meters (375 feet) high, the tension–cohesion model easily accounts for the transport of water.

Root pressure pushes water in a root up through the stem

In the less important mechanism for water transport, known as **root pressure,** water that moves into a plant's roots from the soil is *pushed* up through the xylem to-ward the top of the plant. Water moves from soil into root cells by osmosis. The accumulation of water in root tissues produces a pressure that forces the water up through the xylem.

Guttation, the forceful release of liquid water through special openings (*hydathodes*) in leaves (see Chapter 8), results from root pressure. Root pressure exerts an influence in smaller plants, particularly in the spring when the soil is wet, but it clearly does not cause water to rise 100 meters (330 feet) or more in the tall-

TABLE 10-2 Xylem and Phloem Transport Rates in Selected Plants

PLANT	MAXIMUM RATE IN XYLEM (CM/HR)	MAXIMUM RATE IN PHLOEM (CM/HR)
Conifer	120	48
Woody eudicot	4,400	120
Herbaceous eudicot/monocot	6,000	168–660
Herbaceous vine	15,000	72

Source: Adapted from J. D. Mauseth, Botany: An Introduction to Plant Biology, *2nd ed., Philadelphia, Saunders College Publishing, 1995.*

Note: Xylem and phloem rates are from different plants within each general group and should be used for comparative purposes only.

TENSION–COHESION MODEL The mechanism by which water and dissolved minerals may be transported in xylem: water is pulled upward under tension due to transpiration, while maintaining an unbroken column in xylem because of cohesion.

TRANSPIRATION Loss of water vapor from a plant's aerial parts.

COHESION The tendency of like molecules to adhere or stick together.

ADHESION The tendency of unlike molecules to adhere to one another.

ROOT PRESSURE The pressure in xylem sap that occurs as a result of water moving into roots from the soil.

Transpiration is the driving force of the tension–cohesion model.

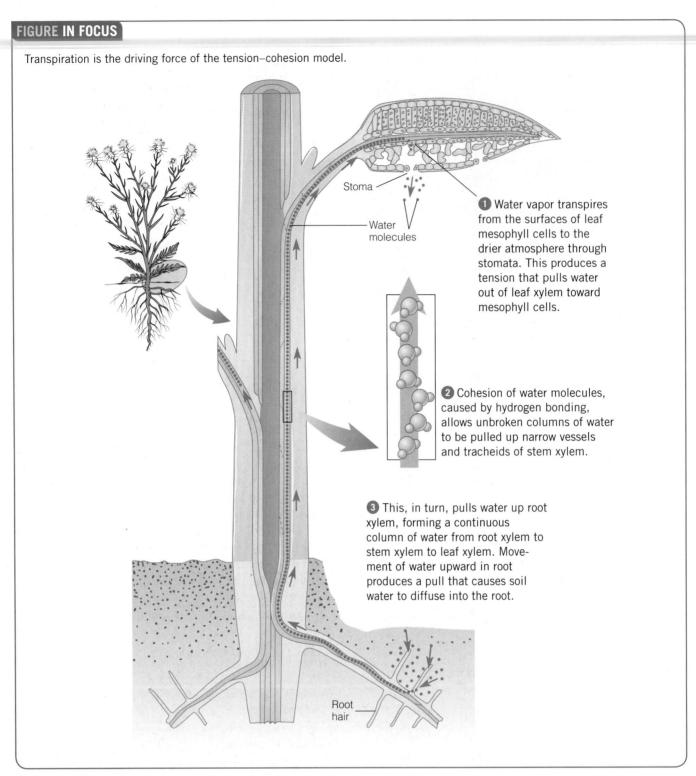

❶ Water vapor transpires from the surfaces of leaf mesophyll cells to the drier atmosphere through stomata. This produces a tension that pulls water out of leaf xylem toward mesophyll cells.

❷ Cohesion of water molecules, caused by hydrogen bonding, allows unbroken columns of water to be pulled up narrow vessels and tracheids of stem xylem.

❸ This, in turn, pulls water up root xylem, forming a continuous column of water from root xylem to stem xylem to leaf xylem. Movement of water upward in root produces a pull that causes soil water to diffuse into the root.

Stoma

Water molecules

Root hair

FIGURE 10-13 The tension–cohesion model.
To follow the figure, start at the top and work downward.

est plants. Furthermore, root pressure does not occur to any appreciable extent in summer (when water is often not plentiful in the soil), yet water transport is greatest during hot summer days.

Roots selectively absorb minerals

Minerals are available to plants as ions dissolved in soil water. Because the concentrations of various mineral ions in xylem sap (fluid in the xylem) are different from those in soil water, plants appear to selectively accumulate the mineral ions they require. Most minerals probably enter the root by passing from one cell's cytoplasm to the next (the *symplast*) or along the interconnected cell walls (the *apoplast*) until they reach the Casparian strip of the endodermis (see Chapter 6). At this point minerals enter the endodermal cells by passing through their plasma membranes.

Dissolved mineral ions pass through plasma membranes by **active transport**. In active transport, the mineral ions move *against* the concentration gradient (that is, from an area of *low* concentration to an area of *high* concentration of that mineral) through special channels in the membrane. One of many reasons root cells require sugar and oxygen for cellular respiration is that active transport of mineral ions across biological membranes requires the expenditure of energy, usually in the form of ATP. From the endodermis, mineral ions enter the xylem of the root and are conducted to the rest of the plant.

Translocation in Phloem

The sugar produced during photosynthesis is converted into *sucrose* before being loaded into the phloem and translocated to the rest of the plant. Sucrose is the predominant photosynthetic product carried in phloem. Phloem sap also contains much smaller amounts of other materials, such as amino acids, organic acids, proteins, hormones, certain minerals, and sometimes disease-causing plant viruses. Translocation in phloem is not as rapid as xylem transport (see Table 10-2).

Fluid within phloem tissue moves both upward and downward. Sucrose is translocated in individual sieve tubes from a *source,* an area of excess sugar supply (usually a leaf), to a *sink,* an area of storage (as starch) or sugar use such as a root, an apical meristem, a fruit, or a seed.

The pressure–flow hypothesis explains translocation in phloem

Current experimental evidence supports the translocation of dissolved sugar in phloem by the **pressure–flow hypothesis,** first proposed in 1930 by the German botanist Ernst Münch. The pressure–flow hypothesis states that dissolved sugar moves in phloem because of a pressure gradient—that is, a difference in pressure. The pressure gradient exists between the source, where the sugar is loaded into the phloem, and the sink, where the sugar is removed from the phloem.

At the source, the dissolved sucrose moves from a leaf's mesophyll cells, where it was manufactured, into the companion cells, which load it into the sieve-tube elements of phloem. This loading occurs by active transport, a process that requires ATP (•**Figure 10-14**). The ATP supplies energy to pump protons out of the sieve-tube elements, producing a proton gradient that drives the uptake of sugar into the sieve-tube elements. The sugar therefore accumulates in the sieve-tube elements at the source. As a result of the increase in dissolved sugars in the sieve-tube elements, water moves by osmosis from the xylem cells into the sieve tubes, increasing the **turgor pressure** (hydrostatic pressure) inside them. Thus, phloem loading at the source is as follows:

> Proton pump moves H$^+$ out of sieve-tube element ⟶ sugar is actively transported into sieve-tube element ⟶ water diffuses from xylem into sieve-tube element ⟶ turgor pressure increases within sieve tube

At its destination (the sink), sugar is unloaded by various methods, both active and passive, from the sieve-tube elements. With the loss in sugar, water moves out of the sieve tubes by osmosis and into surrounding cells. Most of this water diffuses back to the xylem and is transported upward again. This water movement decreases the turgor pressure inside the sieve tubes at the sink. Thus, phloem unloading at the sink proceeds as follows:

> Sugar is transported out of sieve-tube element ⟶ water diffuses out of sieve-tube element and into xylem ⟶ turgor pressure decreases within sieve tube

The pressure–flow hypothesis explains the movement of dissolved sugar in phloem because of a pressure gradient. The difference in sugar concentrations between the source and the sink causes translocation in phloem, as water and dissolved sugar flow along the pressure gradient. This pressure gradient pushes the sugar solution en masse through the phloem much as water is forced

ACTIVE TRANSPORT The energy-requiring movement of a substance across a membrane from a region of lower concentration to a region of higher concentration.

PRESSURE–FLOW HYPOTHESIS The mechanism by which dissolved sugar may be transported in phloem; caused by a pressure gradient between the source (where sugar is loaded into the phloem) and the sink (where sugar is removed from the phloem).

TURGOR PRESSURE The internal pressure of water against the cell wall.

In phloem, sucrose moves from sources to sinks.

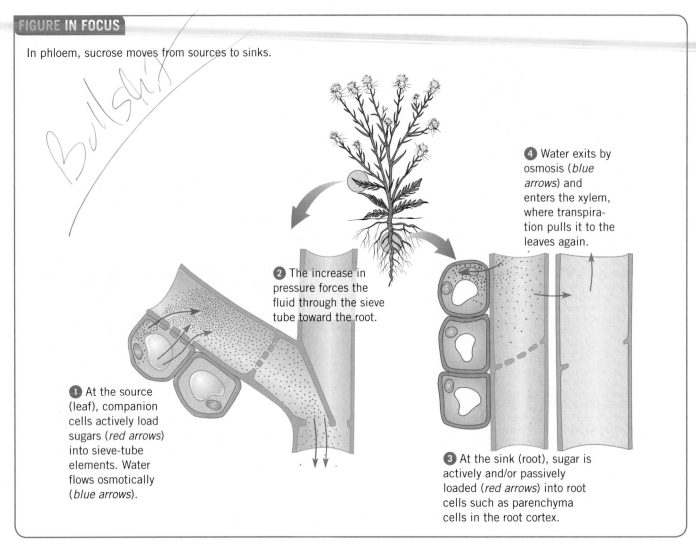

4 Water exits by osmosis (*blue arrows*) and enters the xylem, where transpiration pulls it to the leaves again.

2 The increase in pressure forces the fluid through the sieve tube toward the root.

1 At the source (leaf), companion cells actively load sugars (*red arrows*) into sieve-tube elements. Water flows osmotically (*blue arrows*).

3 At the sink (root), sugar is actively and/or passively loaded (*red arrows*) into root cells such as parenchyma cells in the root cortex.

FIGURE 10-14 The pressure–flow hypothesis.
Sugar is actively loaded into the sieve-tube element at the source. As a result, water moves into the sieve-tube element. At the sink, the sugar is actively or passively unloaded, and water moves from the sieve-tube element to the xylem. The pressure gradient within the sieve tube, from source to sink, causes translocation from the area of higher turgor pressure (the source) to the area of lower turgor pressure (the sink).

through a hose. Such pressure-driven flow of a solution is known as *bulk flow,* or *mass flow.*

The actual translocation of dissolved sugar in the phloem does not require metabolic energy. However, the loading of sugar at the source and the active unloading of sugar at the sink require energy derived from ATP to move the sugar across cell membranes by active transport.

PROCESS OF SCIENCE Although the pressure–flow hypothesis adequately explains current data on phloem translocation, much remains to be learned about this process. Phloem translocation is difficult to study in plants. Be-

cause phloem cells are under pressure, cutting into the phloem to observe it relieves the pressure and causes the contents of the sieve-tube elements (the phloem sap) to exude and mix with the contents of other severed cells that are also unavoidably cut.

In the 1950s botanists began to use a unique research tool to avoid contaminating the phloem sap: aphids, which are small wingless insects that insert their mouthparts into phloem sieve tubes for feeding (•**Figure 10-15**). The pressure in the punctured phloem drives the sugar solution through the aphid's mouthpart and into its digestive system. When the aphid's mouthpart is

PROCESS OF SCIENCE

QUESTION: How can phloem sap be studied without cutting non-phloem cells that would contaminate the sap?

HYPOTHESIS: An aphid mouthpart can be used to penetrate a single sieve-tube element.

EXPERIMENT: After allowing aphids to insert their mouthparts into phloem of a stem, researchers anesthetized the feeding aphids with CO_2 and used a laser to cut their bodies from their mouthparts. The mouthparts remained in the phloem and functioned like miniature pipes.

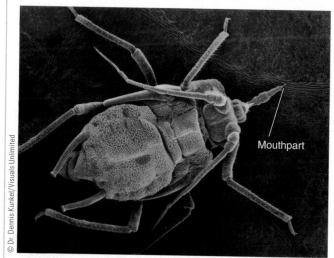

© Dr. Dennis Kunkel/Visuals Unlimited

© 2007 by the American Association for the Advancement of Science

(a) Aphid feeding on a bean leaf; the aphid's mouthpart (*red*) is penetrating a vein.

(b) LM of phloem cells, showing an aphid mouthpart penetrating a sieve-tube element.

RESULTS AND CONCLUSION: About 1 mm³ of phloem sap exuded from each severed mouthpart per hour for several days, so researchers were able to collect and analyze the composition of the sap. ■

FIGURE 10-15 Collecting and analyzing phloem sap.

severed from its body by a laser beam, the sugar solution continues to flow through the mouthpart at a rate proportional to the pressure in the phloem. This rate can be measured, and the effects on phloem transport of different environmental conditions—varying light intensities, darkness, and mineral deficiencies, for example—can be determined.

The identity and proportions of translocated substances are also determined using severed aphid mouthparts. This technique has verified that in most plant species, the sugar sucrose is the primary carbohydrate transported in phloem; however, some species transport other sugars, such as raffinose, or sugar alcohols, such as sorbitol. ■

STUDY OUTLINE

❶ **List the four components of soil, and give the ecological significance of each.**
Four distinct components compose **soil:** inorganic mineral particles, organic matter, water, and air. Inorganic mineral particles (sand, silt, and clay) come from weathered parent material and constitute most of what we call soil. Organic matter consists of litter, droppings, and the remains of dead plants, animals, and microorganisms

that are in various stages of decomposition. Pore spaces of different sizes around and among the soil particles are filled with varying proportions of soil air and soil water; plants and other soil-dwelling organisms need soil minerals, soil air, and soil water.

❷ **Describe the factors involved in soil formation.**
Soils form from rock, called parent material, that is gradually fragmented into smaller and smaller particles by biological, chemical, and physical **weathering processes.** Two factors that work together in the weathering of rock are climate and organisms. Topography, a region's surface features, is also involved in soil formation. Steep slopes often have little or no soil on them because gravity continually transports soil and rock down the slopes; runoff from precipitation tends to amplify erosion on steep slopes. The disintegration of solid rock into small mineral particles and the accumulation of organic material take an extremely long time, sometimes thousands of years.

❸ **Distinguish between macronutrients and micronutrients, and list the essential elements.**
Ten essential elements are required in fairly large quantities and are known as **macronutrients:** carbon, hydrogen, oxygen, nitrogen, potassium; calcium, magnesium, phosphorus, sulfur, and silicon. Nine **micronutrients** are needed in very small amounts for normal plant growth and development: chlorine, iron, boron, manganese, sodium, zinc, copper, nickel, and molybdenum.

❹ **Discuss tension–cohesion and root pressure as mechanisms to explain the rise of water in xylem.**
In the **tension–cohesion model,** water is pulled up the plant as a result of a tension produced at the top of the plant by the evaporative pull of **transpiration,** the loss of water vapor from the aerial parts of plants. The tension draws water up the stem xylem to leaf cells that have lost water as a result of transpiration and pulls water

from root xylem into stem xylem. As water is pulled upward, soil water is drawn into the roots. This upward pulling of water is possible only as long as there is an unbroken column of water in the xylem throughout the plant. Water forms an unbroken column because of the **cohesion** among water molecules, which are strongly attracted to one another by **hydrogen bonding.** The **adhesion** of water to the walls of xylem cells also helps maintain an unbroken column of water. In the less important mechanism for water transport, known as **root pressure,** water that moves into a plant's roots from the soil is pushed up through the xylem toward the top of the plant; the accumulation of water in root tissues produces a pressure that forces the water up through the xylem.

❺ **Discuss the pressure–flow hypothesis of sugar translocation in phloem.**
The **pressure–flow hypothesis** states that dissolved sugar moves in phloem because of a pressure gradient—that is, a difference in pressure. The pressure gradient exists between the source, where the sugar is loaded into the phloem, and the sink, where the sugar is removed from the phloem. At the source, the dissolved sucrose moves from a leaf's mesophyll cells, where it was manufactured, into the companion cells, which load it into the sieve-tube elements of phloem by **active transport,** a process that requires ATP. As a result of the increase in dissolved sugars in the sieve-tube elements, water moves by **osmosis** from the xylem cells into the sieve tubes, increasing the **turgor pressure** (hydrostatic pressure) inside them. At the sink, sugar is unloaded from the sieve-tube elements, and water moves out of the sieve tubes by osmosis and into surrounding cells. Most of this water diffuses back to the xylem. This water movement decreases the turgor pressure inside the sieve tubes at the sink.

REVIEW QUESTIONS

1. List the four components of soil, and tell how each is important to plants.
2. Explain how weathering processes convert rock to soil.
3. What criteria have biologists used to determine which elements are essential for plant growth?
4. Contrast macronutrients and micronutrients.
5. Give a role for each of the following essential elements: nitrogen, magnesium, phosphorus, potassium, and calcium.
6. Compare organic and commercial inorganic fertilizers.
7. Briefly describe how the tension–cohesion model explains the rise of water in the tallest trees.

8. Describe root pressure. What are its limitations?
9. Describe the pressure–flow hypothesis of sugar movement in phloem, including the activities at source and sink.

10. Explain why saline soils are physiologically dry for plants even when the soils are physically wet.

THOUGHT QUESTIONS

1. How would you design an experiment to determine whether gold is essential for plant growth? What would you use for an experimental control?
2. Why does overwatering the soil damage a plant?
3. Why are hydroponic solutions aerated?
4. The San Joaquin Valley in California, with its mild climate and rich soil, is one of the nation's most fertile agricultural areas. It is irrigated, however,

because not much rain falls there. In recent years the productivity of the San Joaquin Valley has declined. Explain the probable cause.

Visit us on the web at http://www.thomsonedu.com/biology/berg for additional resources, such as flashcards, tutorial quizzes, further readings, and web links.

Growth Responses and Regulation of Growth

Grapes are one of the world's most important fruit crops. They are eaten fresh as table grapes, squeezed for juices and wines, dried to make raisins, and processed into jams, jellies, and canned fruit. The European grape (*Vitis vinifera*), which has been cultivated for thousands of years, probably originated in Europe between the Black and Caspian Seas, but humans have introduced it to temperate areas worldwide. More recently, grapes have spread into subtropical areas. Today, more than 5000 different grape varieties exist, all descended from the European grape (see photograph). These different varieties have black, purple, blue, red, pink, golden, pale green, or white fruits.

Although more than 90 percent of cultivated grapes are varieties of the European grape, other important species of the genus *Vitis* originated in North America. The North American species have a more robust odor and taste, their skins separate from the fruit pulp easily, and their roots are resistant to low temperatures and insect and fungal pests. Frequently, disease-susceptible European vines are grafted to disease-resistant American rootstock to produce high-quality European grapes.

The grape plant is a climbing, deciduous, woody vine with coiled tendrils that attach the plant to its supports. Grape flowers are small, green, fragrant, and borne in clusters. After self-pollination and fertilization, clusters of fruits develop that vary in size, form, and color depending on the variety. Botanically, the fruits are berries and usually contain seeds within a soft, juicy pulp. Thompson seedless grapes, first developed in the United States, and other seedless varieties are widely consumed because they are easier to eat.

Growers increase the size of grapes by spraying them with a natural plant hormone called *gibberellin* during their development. Almost all Thompson seedless grapes grown for direct human consumption are treated with gibberellin, which lengthens the branches that bear flowers and thereby gives the fruit clusters additional space in which to grow. Another advantage of gibberellin treatment of grapes is that the longer stems allow air to circulate around the individual berries, keeping them dry and preventing destructive fungi from attacking them.

In this chapter we consider the role of chemicals called hormones in regulating all aspects of plant *growth* (increase in size) and *development* (progressive changes during an organism's life). These processes include not only seed germination and the growth of seedlings into mature plants but also a plant's responses to changes in various aspects of its environment, including temperature, light, gravity, and touch.

A variety of pinot grape (*Vitis vinifera*).

Genetic, Hormonal, and Environmental Controls of Growth and Development

The ultimate control of plant growth and development is genetic. If the genes required for development of a particular trait—for example, the shape of the leaf, the color of the flower, or the type of root system—are not present, that characteristic does not develop. When a particular gene is present, its expression—that is, how it exhibits itself as an observable feature in an organism—is influenced by several factors, including signals from other genes and from the environment. The location of a cell in the young plant body also has a profound effect on gene expression during development. Chemical signals from adjacent cells help a cell "perceive" its location within the plant body. Each cell's spatial environment helps determine what that cell ultimately becomes.

Growth and development, including a plant's responses to various changes in its environment, are regulated by plant *hormones,* organic compounds that act as chemical signals between cells. Environmental cues, such as changing day length and changing temperatures, exert an important influence on gene expression and hormone production, as they do on all aspects of plant growth and development.

The initiation of sexual reproduction is often under environmental control, particularly in temperate latitudes, and plants switch from vegetative growth to reproductive growth after receiving appropriate signals from the environment. Many flowering plants are sensitive to changes in the relative amounts of daylight and darkness that accompany the changing seasons, and these plants flower in response to those changes (•Figure 11-1). Other plants have temperature requirements that induce sexual reproduction. Plants, then, continually perceive information from the environment and use this information to help regulate normal growth and development.

Growth Movements: Tropisms

Plants exhibit movements in response to environmental stimuli such as light, gravity, and touch. A plant may

FIGURE 11-1 Black-eyed Susans.
This plant (*Rudbeckia hirta*) produces flowers in response to the shortening nights of spring and early summer.

PHOTOTROPISM The directional growth of a plant caused by light.

GRAVITROPISM Plant growth in response to the direction of gravity.

THIGMOTROPISM Plant growth in response to contact with a solid object.

respond to such an external stimulus by directional growth—that is, the direction of growth depends on the direction of the stimulus. Such a directional growth response, called a **tropism,** results in a change in the position of a plant part. Tropisms are irreversible and are positive or negative, depending on whether the plant grows toward the stimulus (a positive tropism) or away from it (a negative tropism). Tropisms are under hormonal control, which is discussed in the next section.

Most growing shoot tips exhibit positive **phototropism** by bending (growing) toward light (•Figure 11-2). This growth response increases the likelihood that stems and leaves receive adequate light for photosynthesis. Blue light (wavelengths between 400 and 500 nm) triggers the bending response of phototropism. (You may recall from Chapter 8 that blue light also induces stomata to open.)

For a plant to have a biological response to light, it must contain a light-sensitive substance called a *photoreceptor* to absorb the light. The photoreceptor that absorbs blue light and triggers the phototropic response and other blue-light responses (such as stomatal opening) is a family of yellow pigments called **phototropins.**

Most stem tips exhibit negative **gravitropism** by growing away from the center of Earth, whereas most root tips exhibit positive gravitropism (•**Figure 11-3**). The root cap is the site of gravity perception in roots; when the root cap is removed, the root continues to grow, but it loses any ability to perceive gravity.

Special cells in the root cap have starch-containing **amyloplasts** that collect toward the bottoms of the cells in response to gravity, and several lines of evidence indicate that these amyloplasts initiate the gravitropic response. If the root is reoriented, as when a potted plant is laid on its side, the amyloplasts tumble to a new position, always settling in the direction of gravity. The gravitropic response (bending) occurs shortly thereafter and involves the hormone *auxin* (discussed in the next section).

Despite the movement of amyloplasts in response to gravity, their role in gravitropism remains an open question. Mutant *Arabidopsis* plants with few or no amyloplasts in their root cap cells exhibit some gravitropic response when placed on their sides, suggesting that roots do not necessarily need amyloplasts to perceive gravity. (*A. thaliana* is a model research plant that plant biologists use to study many aspects of plant growth and development.)

Thigmotropism is growth in response to touching an object. An example of thigmotropism is the twining or curling growth of tendrils or stems, which helps attach a climbing plant such as a vine to some type of support. Like phototropism and gravitropism, thigmotropism involves auxin.

Maryann Frazier/Photo Researchers, Inc.

FIGURE 11-2 Phototropism.
A geranium (*Pelargonium*) was cut back, and the plant was placed in a window. The new growth is in the direction of light, and therefore the plant exhibits positive phototropism.

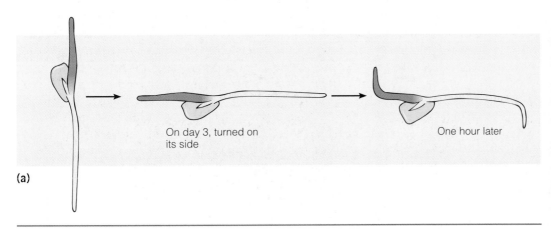

On day 3, turned on its side

One hour later

(a)

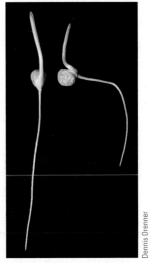

Dennis Drenner

(b)

FIGURE 11-3 Gravitropism.
(a) A corn (*Zea mays*) seedling was turned on its side 3 days after germinating. In 1 hour, the root and shoot tips had curved. **(b)** In 24 hours, the new root growth was downward (positive gravitropism), and the new shoot growth was upward (negative gravitropism). A control seedling germinated at the same time is on the left.

TABLE 11-1 Plant Hormones and Signaling Molecules

HORMONE	SITE OF PRODUCTION	PRINCIPAL ACTIONS
Auxins (e.g., IAA)	Shoot apical meristem, young leaves, seeds	Stem elongation, apical dominance, root initiation, fruit development
Gibberellins (e.g., GA3)	Young leaves and shoot apical meristems, embryo in seed	Seed germination, stem elongation, flowering, fruit development
Cytokinins (e.g., Zeatin)	Roots	Cell division, delay of leaf senescence, inhibition of apical dominance, flower development, embryo development, seed germination
Ethylene	Stem nodes, ripening fruit, damaged or senescing tissue	Fruit ripening, responses to environmental stressors, seed germination, maintenance of apical hook on seedlings, root initiation, senescence and abscission in leaves and flowers, apical dominance
Abscisic acid	Almost all cells that contain plastids (leaves, stems, roots)	Seed dormancy, responses to water stress
Brassinosteroids (e.g., Brassinolide)	Shoots (leaves and flower buds), seeds, fruits	Cell division, cell elongation, light-mediated differentiation, seed germination, vascular development
Oligosaccharins	Unknown	May function in normal cell growth and development, defense responses to disease organisms, inhibition of flowering
Jasmonates (e.g., Jasmonic acid)	Leaves? Probably many tissues	Initiation of defenses against insect predators or disease organisms, pollen development, root growth, fruit ripening, senescence
Salicylic acid	Wound (site of infection)	Resistance to insect predators and disease organisms
Systemin	Wound (site of herbivore or pathogen attack)	Initiation of defenses against predators (herbivores) or disease organisms

Source: From Solomon, Berg, Martin Biology 8/e, Table 37-1, p. 791.

PROCESS OF SCIENCE

QUESTION: Which part of the grass coleoptile absorbs the directional light that triggers phototropic growth?

HYPOTHESIS: The coleoptile tip perceives the directional light.

EXPERIMENT: Some canary grass coleoptiles were uncovered, some were covered only at the tip, some had the tip removed, and some were covered everywhere but at the tip. The covers were impervious to light.

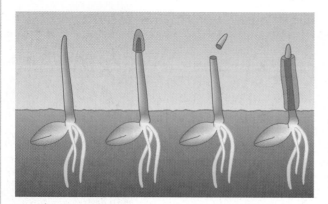 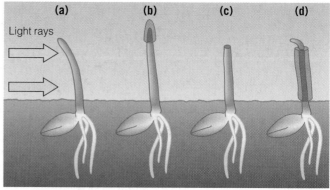

RESULTS AND CONCLUSION: After exposure to light coming from one direction, the uncovered plants **(a)** and the plants with uncovered tips **(d)** grew toward the light. The plants with tips covered **(b)** or removed **(c)** did not bend toward light. The Darwins concluded that some substance was produced in the tip and transmitted to the lower part that caused it to bend. ■

FIGURE 11-4 **The Darwins' phototropism experiments.**

Plant Hormones

A plant **hormone** is an organic chemical compound that acts as a chemical signal, eliciting a variety of responses that regulate growth and development. The study of plant hormones and their effects is challenging because hormones are effective in extremely small amounts. In addition, the effects of different plant hormones overlap, and it is difficult to determine which hormone, if any, is the primary cause of a particular response. Plant hormones may also stimulate a response at one concentration and inhibit that same response at a different concentration.

For many years, biologists studied five major classes of plant hormones: auxin, gibberellin, cytokinin, ethylene, and abscisic acid. More recently, researchers have uncovered compelling evidence for a variety of other signaling molecules, such as brassinosteroids, oligosaccharins, jasmonates, salicylic acid, and systemin. (●**Table 11-1** summarizes plant hormones and hormone-like signaling molecules discussed in this chapter.)

Auxins promote cell elongation

PROCESS OF SCIENCE Charles Darwin, the British naturalist best known for developing the theory of natural selection to explain evolution, also provided the first evidence for the existence of plant hormones. The experiments that Darwin and his son Francis performed in the 1870s involved phototropism in newly germinated canary grass seedlings. As in all grasses, the first part of a canary grass seedling to emerge from the soil is the *coleoptile*, a protective sheath that encircles the stem. When coleoptiles are exposed to light from only one direction, they bend toward the light. The bending occurs close to the tip of the coleoptile.

The Darwins tried to influence this bending in several ways (●**Figure 11-4**). For example, they covered the tip of the coleoptile as soon as it emerged from the soil. When they covered that part of the coleoptile above where the bend would normally occur, the plants did not bend. On other plants, they removed the coleoptile tip and found that bending did not occur. When the bottom of the coleoptile where the curvature would occur was shielded from the light, the coleoptile bent toward light. From these experiments, the Darwins concluded that "some influence is transmitted from the upper to the lower part, causing it to bend."

In the 1920s Frits Went, a young Dutch plant physiologist, isolated this "influence" from oat coleoptiles. To do so, he removed the coleoptile tips and placed them on tiny blocks of agar for a period of time. When he put one of these agar blocks squarely on a decapitated cole-

PROCESS OF SCIENCE

QUESTION: Is there a chemical substance responsible for elongation of coleoptiles, and can it be isolated?

HYPOTHESIS: The factor responsible for coleoptile growth is a diffusible chemical that can be isolated from coleoptile tips.

EXPERIMENT: Coleoptile tips were placed on agar blocks for a period of time **(a)**. The agar block was transferred to a decapitated coleoptile. It was placed off center, and the coleoptile was left in darkness **(b)**.

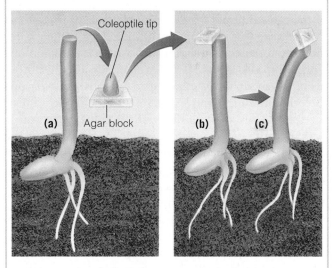

RESULTS AND CONCLUSION: The coleoptile bent (see **c**), indicating that a chemical moved from the original coleoptile tip to the agar block and from there to one side of the decapitated coleoptile. As a result, that side elongated more than the side without the agar block. ■

FIGURE 11-5 Isolating auxin from coleoptiles.

optile, normal growth resumed. When he placed one of these agar blocks to one side of the tip of a decapitated coleoptile in the dark, bending occurred (●**Figure 11-5**). This bending indicated that the substance had diffused from the coleoptile tip into the agar and, later, from the agar into the decapitated coleoptile. Went named this substance **auxin** (from the Greek word for "enlarge" or "increase"). ■

HORMONE An organic chemical messenger that regulates growth and development in plants and other multicellular organisms.

AUXIN A plant hormone involved in growth and development, including stem elongation, apical dominance, and root formation on cuttings.

Several naturally occurring auxins have been identified, and some artificial auxins have been synthesized. The most common naturally occurring auxin is **indoleacetic acid (IAA).** The movement of auxin in the plant is *polar*, or unidirectional. Auxin moves downward along the shoot–root axis from its site of production, usually the shoot apical meristem. Young leaves and seeds are also sites of auxin production.

Auxin's most characteristic action is promotion of cell elongation in stems and coleoptiles. This effect, apparently exerted by acidifying the cell walls, increases their plasticity, enabling them to expand under the force of the cell's internal *turgor pressure*. (Turgor pressure is the pressure that develops against a cell wall when water moves into the cell by osmosis.) Auxin's effect on cell elongation also provides an explanation for phototropism. When a plant is exposed to a light from only one direction, the auxin moves laterally to the shaded side of the stem before moving down the stem by polar transport. Because of the greater auxin concentration on the shaded side of the stem, the cells there elongate more than the cells on the lighted side, and the stem bends toward the light (●**Figure 11-6**). As mentioned earlier, auxin is also involved in gravitropism and thigmotropism.

Auxin exerts other effects on plants. For example, some plants branch out very little when they grow. Growth in these plants occurs almost exclusively from the shoot apical meristem rather than from axillary buds, which do not develop as long as the terminal bud is present. Such plants are said to exhibit **apical dominance**—the inhibition of axillary bud growth by the shoot apical meristem (●**Figure 11-7**). In plants with strong apical dominance, auxin produced in the apical meristem inhibits axillary buds near the apical meristem from developing into actively growing shoots. When the apical meristem is pinched off, the auxin source is removed, and axillary buds grow to form branches. Apical dominance is often quickly reestablished, however, as one branch begins to inhibit the growth of others. Other hormones (ethylene and cytokinin, both discussed later) are also involved in apical dominance. As with many physiological processes, the interactions of these hormones, rather than the action of one hormone alone, may be responsible for apical dominance.

Auxin produced by developing seeds stimulates the development of the fruit. When auxin is applied to certain flowers in which fertilization has not occurred (and, therefore, in which seeds are not developing), the ovary enlarges and develops into a seedless fruit. Seedless tomatoes are produced in this manner.[1] Auxin is not the only hormone involved in fruit development, however.

Some manufactured, or synthetic, auxins have structures similar to that of IAA. A synthetic auxin is

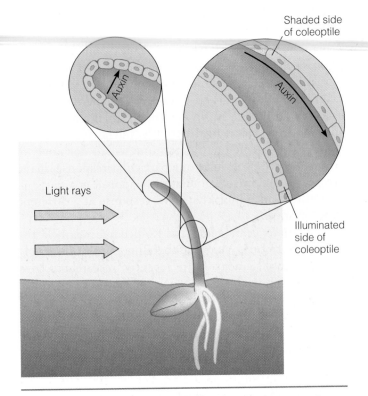

FIGURE 11-6 Unequal distribution of auxin causes phototropism. Auxin travels down the side of the stem or coleoptile away from the light (*black arrow*), causing cells on the shaded side to elongate. Therefore, the stem or coleoptile bends toward light.

used to stimulate root development on stem cuttings for asexual propagation, particularly of woody plants with horticultural importance (●**Figure 11-8**). The synthetic auxins 2,4-D and 2,4,5-T are used as selective herbicides (weed killers). These compounds kill plants with broad leaves but, for reasons not completely understood, do not kill grasses. Although both herbicides are similar in structure to IAA, they disrupt the plants' normal growth processes. Because many of the world's most important crops are grasses (for example, wheat, corn, and rice), both 2,4-D and 2,4,5-T can be used to kill broadleaf weeds that compete with these crops. The use of 2,4,5-T is no longer allowed in the United States because of its association with dioxins, a group of mildly to very toxic compounds formed as by-products during the manufacture of 2,4,5-T.

---

[1]*Not all seedless fruits are produced by treatment with auxin. As discussed in the chapter introduction, Thompson seedless grapes result from treatment of the flowers and young fruits with gibberellins, another group of hormones. In this case, fertilization occurs, but the embryos abort, and therefore the seeds fail to develop.*

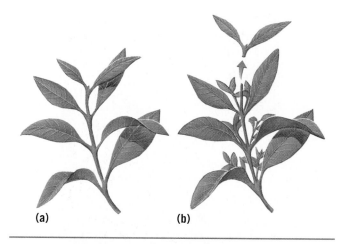

(a) **(b)**

FIGURE 11-7 Auxin inhibits the development of axillary buds.
(a) When the tip of the plant (the source of auxin) is intact, the axillary buds do not develop. **(b)** The tip of the plant was removed. Because auxin is not moving down from the stem tip, axillary buds develop into branches.

© Robert Lyons/Visuals Unlimited

FIGURE 11-8 Honeysuckle cuttings treated with various concentrations of a synthetic auxin.
Many adventitious roots formed on honeysuckle (*Lonicera fragrantissima*) cuttings placed in a higher auxin concentration (*left*), whereas fewer roots formed in a lower auxin concentration (*middle*). Cuttings placed in water (*right*) served as a control and did not form roots in the same period.

One of the many controversial aspects of the Vietnam War was the defoliation program carried out by the United States in South Vietnam. From 1961 to 1971, the United States sprayed Agent Orange, a mixture of the herbicides 2,4-D and 2,4,5-T, over large areas of South Vietnam to expose suspected hiding areas and destroy crops planted by the Vietcong and North Vietnamese troops. The negative impact of these herbicides on the environment is still being felt today. Many ecologically important mangrove forests and commercially valuable hardwood forests were destroyed. In addition, the herbicide sprays caused health problems in the native people and members of the U.S. military who were exposed to them in the Vietnamese jungles. The herbicides contained traces of dioxin, which is now known to cause birth defects in animals. Reportedly, the number of birth defects and stillbirths in Vietnam increased during the herbicide spraying. Also, American veterans who were exposed to high levels of herbicides have more health problems than do other veterans, including certain types of cancer and skin diseases.

Gibberellins promote stem elongation

PROCESS OF SCIENCE In the 1920s, a Japanese biologist was studying a disease of rice in which the young rice seedlings grow extremely tall and spindly, fall over, and die. The cause of the disease, dubbed the "foolish seedling" disease, was a fungus that produces the chemical substance **gibberellin.** Not until after World War II did scientists in Europe and North America learn of this

work, and the hormone was found to be one of a class of compounds collectively called gibberellins. During the 1950s and 1960s, studies in the United States and Great Britain showed that healthy plants also produce gibberellins. ▪

Gibberellins are hormones involved in many normal plant functions. The symptoms of the "foolish seedling" disease are caused by an abnormally high gibberellin concentration in the plant tissue (because both the plant and the fungus are producing gibberellin). Currently, dozens of naturally occurring gibberellins are known, although many are probably inactive forms; there are no synthetic gibberellins.

Gibberellins promote stem elongation in many plants. When gibberellin is applied to a plant, particularly certain dwarf varieties, this elongation may be spectacular. Some corn and pea plants that are dwarfs as a result of one or more mutations (changes in their genetic material) grow to a normal height when treated with gibberellin (•**Figure 11-9a**). Short-stemmed, high-yielding varieties of wheat have short stems because they have a reduced response to gibberellin. Gibberellins are also involved in **bolting,** the rapid elongation of a floral stalk that occurs naturally in many plants when they initiate flowering (•**Figure 11-9b**).

GIBBERELLIN A plant hormone involved in growth and development, including stem elongation, flowering, and seed germination.

Courtesy of B. O. Phinney, University of California, Los Angeles

(a) Effect of the continued application of gibberellin on normal and dwarf corn (*Zea mays*) plants. *From left to right:* dwarf, untreated; dwarf, treated with gibberellin; normal, treated with gibberellin; normal, untreated. This dwarf variety is a mutant with a single recessive gene that impairs gibberellin metabolism.

© Robert Lyons/Visuals Unlimited

(b) Bolting in Indian blanket (*Gaillardia pulchella*). Many biennials grow as a rosette (a circular cluster of leaves close to the ground) during their first year (*left*) and then bolt when they initiate flowering in their second year (*right*). Gibberellin triggers the rapid stem elongation.

FIGURE 11-9 Effects of gibberellin.

Gibberellins cause stem elongation by stimulating cells to divide as well as elongate. The actual mechanism of cell elongation differs from that caused by auxin, however. Recall that IAA-induced cell elongation involves acidifying the cell wall. In gibberellin-induced cell elongation, cell-wall acidification does not occur; instead, gibberellin increases the cell wall's ability to expand.

Gibberellins affect several reproductive processes in plants. They stimulate flowering, particularly in long-day plants (discussed later in the chapter). In addition, gibberellins can substitute for the low temperature that biennials require before they begin flowering. If gibberellins are applied to biennials during their first year of growth, flowering occurs without exposure to a period of low temperature. Gibberellins, like auxin, affect fruit development. As mentioned in the chapter introduction, agriculturists apply gibberellins to several varieties of grapes to produce larger fruits.

Gibberellins are also involved in the seed germination of certain plants (cereals and other grasses). In a classic experiment involving barley seed germina-tion, researchers showed that the release of gibberellin from the embryo triggers the synthesis of an enzyme that digests starch in the endosperm. As a result, glu-cose becomes available for absorption by the embryo as it resumes growth. In addition to mobilizing food re-serves in certain newly germinated seeds, application of gibberellins substitutes for low-temperature or light requirements for germination in seeds such as lettuce, oats, and tobacco.

Cytokinins promote cell division

PROCESS OF SCIENCE During the 1940s and 1950s, research-ers were trying to find substances that might induce plant cells to divide in **tissue culture,** a technique in which cells are isolated from plants and grown in a nu-trient medium. They discovered that cells divided when placed in a culture containing coconut milk. Because coconut milk has a complex chemical composition, in-vestigators did not chemically identify the division-inducing substance for some time. Finally, researchers isolated the active substance from a different source.

They called it **cytokinin** because it induces cell division, or *cytokinesis*. In 1963, researchers identified the first plant cytokinin (in corn), and since then similar molecules have been identified in other plants. Biologists have also synthesized several cytokinins. ▪

Cytokinins promote cell division and differentiation of young, relatively unspecialized cells into mature, more specialized cells in intact plants. They are a required ingredient in any plant tissue culture medium because they must be present in all dividing plant cells. In tissue culture, cytokinins interact with auxin during the formation of plant organs such as roots and stems. For example, in tobacco tissue culture, a high ratio of cytokinin to auxin induces shoots to form, whereas a low ratio of cytokinin to auxin induces roots to form (•**Figure 11-10**).

Cytokinins and auxin also interact in the control of apical dominance, in which the shoot apical meristem suppresses the growth of axillary buds. Here their relationship is antagonistic: auxin inhibits the growth of axillary buds, and cytokinin promotes their growth.

One effect of cytokinins on plant cells is to delay the aging process. Plant cells, like all living cells, go through a natural aging process known as **senescence.**

Senescence is accelerated in cells of plant parts that are cut, such as flower stems. Botanists think plants have to have a continual supply of cytokinins from the roots. Cut stems, of course, lose their source of cytokinins and therefore age rapidly.

Despite their involvement in many aspects of plant growth and development, cytokinins have not been approved for many commercial applications other than plant tissue culture. However, in the mid-1990s, molecular biologists used genetic engineering to develop tobacco plants that produce more cytokinin and therefore live longer (•**Figure 11-11**). This discovery has the potential to increase the longevity and productivity of certain crops.

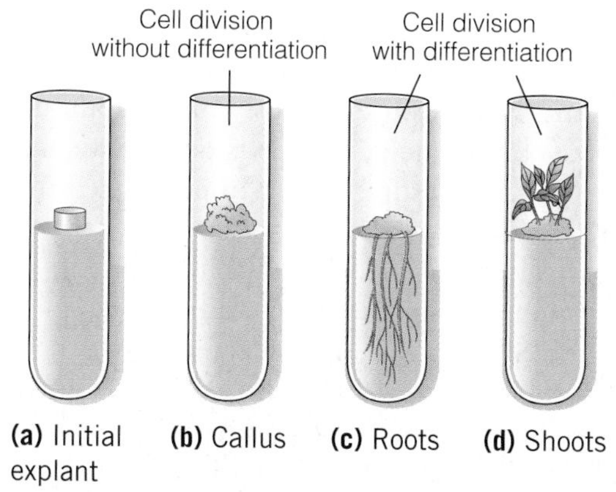

Cell division without differentiation   Cell division with differentiation

(a) Initial explant   **(b)** Callus   **(c)** Roots   **(d)** Shoots

FIGURE 11-10 Hormones and organ formation when propagating tobacco by tissue culture.
(a) A fragment of tissue explant from the center of a tobacco stem is placed in a culture medium. A complete plant can form from the tissue fragment because each cell of the fragment contains all the genetic information for the entire organism. Varying amounts of auxin and cytokinin in the culture media produce different growth responses. **(b)** Nutrient agar containing a moderate amount of both auxin and cytokinin caused cells to divide and form a callus (a clump of disorganized, undifferentiated cells). **(c)** When callus is transplanted to a medium with a higher relative amount of auxin, roots differentiate. **(d)** Shoot growth is stimulated by a medium with a higher relative amount of cytokinin. Plants grown using tissue culture techniques can be transferred to soil and grown normally.

Courtesy of Dr. Richard M. Amasino, University of Wisconsin

FIGURE 11-11 Cytokinin synthesis and delay of senescence.
The tobacco (*Nicotiana tabacum*) on the left was genetically engineered to produce additional cytokinin as it aged, whereas the unaltered tobacco plant of the same age on the right served as a control. Note the extensive senescence and death of older leaves on the control plant.

CYTOKININ A plant hormone involved in growth and development, including cell division and delay of senescence.

Ethylene promotes abscission and fruit ripening

During the early 20th century, scientists observed that the gas **ethylene** has several effects on plant growth, but not until 1934 did they demonstrate that plants produce ethylene. This natural hormone influences many diverse plant processes. Ethylene inhibits cell elongation, promotes seed germination, promotes apical dominance, and is involved in plant responses to wounding or invasion by disease-causing microorganisms.

Ethylene also has a major role in many aspects of development, including fruit ripening. As a fruit ripens, it produces ethylene, which triggers an acceleration of the ripening process. This induces the fruit to produce more ethylene, which further accelerates ripening. The expression "one rotten apple spoils the lot" is true. A rotten apple is one that is overripe and produces large amounts of ethylene, which diffuses and triggers the ripening process in nearby apples. Humans use ethylene commercially to uniformly ripen bananas and tomatoes. These fruits are picked while green and shipped to their destination, where they are exposed to ethylene before they are delivered to grocery stores (●**Figure 11-12**).

Ethylene, along with auxin, is involved in leaf senescence and **abscission.** As a leaf ages (as autumn approaches, for deciduous trees in temperate climates), the level of auxin in the leaf decreases. Concurrently, cells in the abscission layer at the base of the petiole (where the leaf will break away from the stem) begin producing ethylene, which in turn stimulates weakening of the cell walls in this location.

Abscisic acid promotes seed dormancy

Abscisic acid was discovered simultaneously in 1963 by two independent research teams. Despite its name, abscisic acid does not induce abscission in most plants. Instead, abscisic acid is involved in a plant's response to stress and in seed **dormancy,** a temporary state of arrested physiological activity when growth does not occur even when environmental conditions are favorable.

As an environmental stress hormone, abscisic acid particularly promotes changes in plant tissues that are water stressed. The level of abscisic acid increases dramatically in the leaves of plants exposed to severe drought conditions. The high level of abscisic acid in the leaves triggers the closing of stomata, which saves the water that the plant would normally lose by transpiration, thereby increasing the plant's likelihood of survival. As knowledge of cell signals relating to abscisic acid increases, botanists hope to use this information to

FIGURE 11-12 Ethylene and fruit ripening.
Both boxes of tomatoes were picked at the same time, while green. The tomatoes in the box on the right were exposed to an atmosphere containing ethylene for 3 days, whereas the tomatoes on the left were not.

engineer crops and horticultural plants that are more resistant to drought.

The low temperatures of winter are also a type of stress on plants. A winter adaptation that involves abscisic acid is dormancy in seeds. Many seeds have high levels of abscisic acid in their tissues and do not germinate until spring, after winter snows and rain have washed out the abscisic acid. In a corn mutant unable to synthesize abscisic acid, the seeds germinate as soon as the embryos are mature, even while attached to the ear (●**Figure 11-13**).

Abscisic acid is not the only hormone involved in seed dormancy. For example, addition of gibberellin reverses the effects of dormancy. In seeds, the level of abscisic acid decreases during the winter, and the level of gibberellin increases. Cytokinins are also implicated in breaking dormancy. Once again you see that a single activity such as seed dormancy may be controlled by the interaction of several hormones. The plant's actual response may result from changing ratios of hormones rather than the effect of each individual hormone.

Additional signaling molecules affect growth and development

Biologists continue to discover new plant hormones and hormone-like signaling molecules. Many of these signaling molecules are involved in defensive responses of plants to disease organisms and insects. Here we briefly consider five groups—brassinosteroids, oligosaccharins, jasmonates, salicylic acid, and systemin.

Although steroid hormones have crucial roles in animals, we are just beginning to understand their roles in plants. The **brassinosteroids (BRs)** are a group of steroids that function as plant hormones. BRs are involved in several aspects of growth and development. *Arabidopsis* mutants that cannot synthesize BRs are dwarf plants with reduced fertility. Researchers reverse this defect by applying BR. Studies of these mutants suggest that the brassinosteroids are involved in multiple processes such as cell division, cell elongation, light-induced differentiation, seed germination, and vascular development.

Oligosaccharins are carbohydrate fragments that consist of short, branched chains of sugar molecules. They are present in extremely small quantities in cells and active at much lower concentrations (100 to 1000 times as low) than hormones such as auxin. Some oligosaccharins trigger the production of **phytoalexins** (from the Greek *phyto*, "plant," and *alexi*, "to ward off"), antimicrobial compounds that limit the spread of plant disease organisms such as fungi. Other oligosaccharins inhibit flowering and induce vegetative growth (that is, the growth of leaves, stems, and roots).

Jasmonates are lipid-derived hormones that affect several plant processes, such as pollen development, root growth, fruit ripening, and senescence. They are also produced in response to the presence of insect pests and disease-causing organisms. Jasmonates may be of practical value in controlling certain insect pests without the use of chemical pesticides.

For centuries people chewed willow (*Salix*) bark to treat headaches and other types of pain. **Salicylic acid** was first extracted in the 19th century from willow bark and is chemically related to aspirin (acetylsalicylic acid). Recently, biologists have shown that salicylic acid is a signaling molecule that helps plants defend against insect pests and disease-causing agents such as viruses.

Although many animal hormones are polypeptides, the first plant polypeptide with hormonal properties, called **systemin,** was not isolated until 1991. In response to wounding by insects, systemin may trigger the plant to produce molecules that disrupt insect digestion, thereby curbing leaf damage done by caterpillars and other herbivorous (plant-eating) insects. The discovery of systemin in tomato leaves prompted a flurry of research that resulted in the discovery of additional polypeptides in other plants.

PROCESS OF SCIENCE

Progress is being made in identifying the elusive flower-promoting signal

Experiments in which different tobacco species are grafted together indicate that additional flower-promoting and flower-inhibiting substances may exist. *Nicotiana silvestris* is a long-day tobacco plant (it flowers when exposed to short nights; discussed later in the chapter). A variety of *N. tabacum* is a day-neutral tobacco plant (day length does not regulate its flowering; also discussed later in the chapter). When a long-day tobacco is grafted to a day-neutral tobacco and exposed to short nights, both plants flower (•**Figure 11-14**). The day-neutral tobacco plant flowers sooner than it normally would.

For many years biologists have found evidence that a flower-promoting substance, **florigen,** may be induced in the long-day plant and transported to the day-neutral plant through the graft union, causing the day-neutral tobacco plant to flower sooner than expected. An intact plant may produce florigen in the leaves and transport it in the phloem to the shoot apical meristem. There, it induces a transition from vegetative to reproductive development—that is, to a meristem that produces flowers.

When a botanist grafts a long-day tobacco to a day-neutral tobacco and exposes them to long nights, neither plant flowers. As long as these conditions continue,

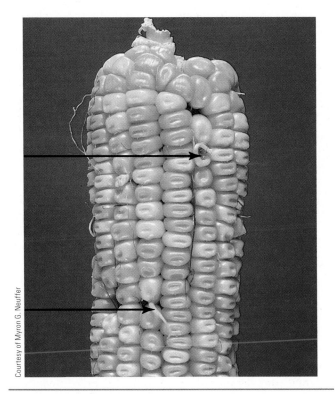

FIGURE 11-13 Abscisic acid and seed germination.
In a corn (*Zea mays*) mutant that does not produce abscisic acid, some of the kernels have germinated while still on the ear, producing roots (*arrows*).

Courtesy of Myron G. Neuffer

ETHYLENE A gaseous plant hormone involved in growth and development, including leaf abscission and fruit ripening.

ABSCISIC ACID A plant hormone involved in growth and development, including dormancy and responses to stress.

the day-neutral plants do not flower even when they would normally do so. In this case, the long-day tobacco may produce a flower-inhibiting substance that is transported to the day-neutral tobacco through the graft union. This substance prevents the day-neutral tobacco from flowering.

Although biologists have not yet isolated and chemically characterized all flower promoters and flower inhibitors, plant biologists working with *Arabidopsis* reported the identification of the elusive florigen signal in 2005. A gene is activated (turned on) in *Arabidopsis* leaves, and a messenger RNA (mRNA) molecule is transcribed from this gene. This mRNA molecule, thought to be florigen, travels in the phloem from the leaf to the shoot apical meristem. There, it is translated into a protein that alters gene expression (turning genes on or off) in cells at the meristem, thereby initiating flower development. ■

Plant hormones act by signal transduction

Researchers have used *Arabidopsis* mutants to better understand the biology of plant hormones. For example, some mutants have defects in signal transduction. In **signal transduction,** a receptor in a cell's plasma membrane converts the hormone, which is outside the cell, into a signal that causes some change inside the cell; that change ultimately leads to a physiological response. Many plant hormones bind to receptors located in the plasma membrane, triggering enzymatic reactions of some sort.

Let us consider a specific example, involving the plant hormone auxin (•**Figure 11-15**). As shown in step ❶, the receptor for auxin is located in the plasma membrane and has a shape that binds to the auxin molecule. The binding of auxin to its receptor catalyzes the attachment of a molecule, called *ubiquitin,* to proteins known to inhibit certain genes (see step ❷). Whenever ubiquitin is attached to protein molecules, the cell destroys those molecules (❸). As a result, the genes that were inhibited by those proteins are now turned on (❹), resulting in changes in cell growth and development.

Photoperiodism

In the early 20th century, Wightman Garner and Henry Allard, two American plant physiologists working at an agricultural research center in Maryland, discovered that day length (or, more correctly, night length) was involved in flowering. **Photoperiodism** is any response of

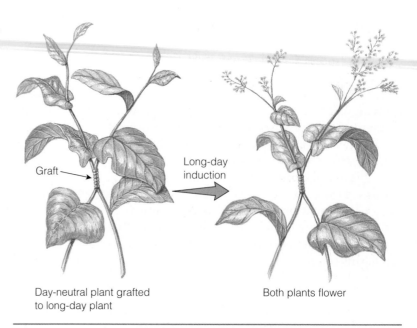

Graft

Long-day induction

Day-neutral plant grafted to long-day plant

Both plants flower

FIGURE 11-14 Evidence of a flower-promoting substance. When a long-day tobacco (*Nicotiana silvestris*) is grafted to a day-neutral tobacco (*N. tabacum*) and both plants are exposed to a long-day, short-night regimen, they both flower. The day-neutral plant flowers sooner than it normally would, presumably because a flower-promoting substance passes from the long-day plant to the day-neutral one through the graft. In 2005, after years of searching, biologists identified florigen as a messenger RNA molecule.

a plant to the relative lengths of daylight and darkness. Initiation of flowering at the shoot apical meristem is one of several processes that are photoperiodic in many plants. Plants are classified into four main groups on the basis of how photoperiodism affects their transition from vegetative growth to flowering: short-day, long-day, intermediate-day, and day-neutral plants.

Short-day plants (also called **long-night plants**) flower when the night length is equal to or greater than some critical period (•**Figure 11-16**). The initiation of flowering in short-day plants is due to the long, uninterrupted period of darkness rather than the short period of daylight (•**Figure 11-17 ❶**). The minimum critical night length varies considerably from one plant species to another but falls between 12 and 14 hours for many. Examples of short-day plants are florist's chrysanthemum, cocklebur, and poinsettia, all of which typically flower in late summer or fall. Poinsettias, for example, initiate flower buds in early October in the Northern Hemisphere and flower about 8 to 10 weeks later, hence their traditional association with Christmas. Short-day plants detect the lengthening nights of late summer or fall, and they flower at that time.

Long-day plants (also called **short-night plants**) flower when the night length is equal to or less than some critical length (•**Figure 11-17 ❷**). Plants such as spinach, black-eyed Susan (see Figure 11-1), and *Arabidopsis* flower in late spring or summer and are long-

Many plant hormones activate genes through signal transduction.

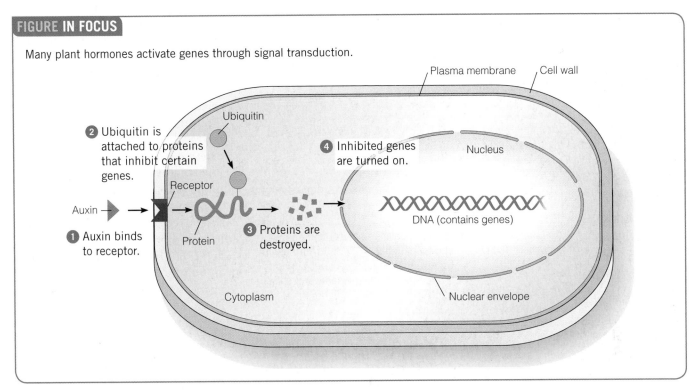

FIGURE 11-15 How auxin works.
The numbered steps are explained in the text.

day plants. These plants detect the shortening nights of spring and early summer, and they flower at that time.

Intermediate-day plants do not flower when day length is either too long or too short. Sugarcane and coleus are intermediate-day plants. These plants flower when they are exposed to days and nights of intermediate length.

Some plants, called **day-neutral plants,** do not initiate flowering in response to seasonal changes in the period of daylight and darkness but instead respond to some other type of stimulus, external or internal. Tomato, dandelion, string bean, and pansy are examples of day-neutral plants. Many of these plants originated in the tropics, where day length does not vary appreciably during the year. (In contrast, short-day, long-day, and intermediate-day plants are temperate species.)

Phytochrome detects day length

The main photoreceptor for photoperiodism and many other light-initiated plant responses is **phytochrome,** a family of about five blue-green pigment proteins. A mixture of phytochrome proteins is present in cells of all vascular plants examined so far.

Much current knowledge of phytochrome is based on various mutant *Arabidopsis* plants that do not pro-

FIGURE 11-16 Short-day plants.
The chrysanthemum (*Chrysanthemum*) is a short-day plant and therefore flowers only when exposed to long nights. Two cuttings from the same plant were planted in separate pots. The plant on the left, which flowered, received short days (8 hours of daylight) and long nights (16 hours of darkness) for several weeks. The plant on the right, which remained vegetative, received long days (16 hours of daylight) and short nights (8 hours of darkness) during the same period.

PHOTOPERIODISM The physiological response (such as flowering) of plants to variations in the length of daylight and darkness.

PHYTOCHROME A blue-green proteinaceous pigment involved in many plant responses to light, independent of photosynthesis.

QUESTION: Is day length or night length the critical factor affecting flowering?

HYPOTHESIS: The plant measures day length, not night length: Short-day plants flower when the day length is short, and long-day plants flower when the day length is long.

EXPERIMENT: Short-day plants (chrysanthemum) and long-day plants (black-eyed Susan) were exposed to different light–dark regimes.

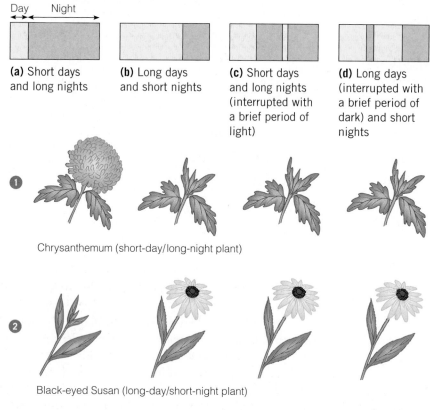

Day   Night

(a) Short days and long nights

(b) Long days and short nights

(c) Short days and long nights (interrupted with a brief period of light)

(d) Long days (interrupted with a brief period of dark) and short nights

Chrysanthemum (short-day/long-night plant)

Black-eyed Susan (long-day/short-night plant)

RESULTS AND CONCLUSION: A short-day plant flowers when it is grown under long-night conditions **(a)**, but it does not flower when exposed to either **(c)** long nights interrupted with a brief period of light or **(d)** long days interrupted with a brief period of darkness. A long-day plant flowers when it is grown under short-night conditions **(b)** but does not flower when grown under long-night conditions **(a)** unless the long night is interrupted with a brief period of light **(c)**. **These data indicate that the plant is actually measuring the dark period, not day length. The original hypothesis is disproved.** ■

FIGURE 11-17 Photoperiodic responses of short-day and long-day plants.

duce a particular type of phytochrome. Biologists studying the responses of plants that do not produce an individual phytochrome have concluded that the individual forms of phytochrome have both unique and overlapping functions.

Each member of the phytochrome family exists in two forms and readily converts from one form to the other after absorption of light of specific wavelengths. One form, designated **Pr** (for *red*-absorbing phytochrome), strongly absorbs red light with a relatively short wave-

length (660 nm). In the process, the shape of the molecule changes to the second form of phytochrome, **Pfr**, so designated because it absorbs *far-red* light, which is red light with a relatively long wavelength (730 nm) (•**Figure 11-18**). When Pfr absorbs far-red light, the Pfr reverts to the original form, Pr. Pfr, the active form of phytochrome, triggers or inhibits responses such as flowering.

What does a pigment that absorbs red light and far-red light have to do with daylight and darkness? Sunlight consists of various amounts of the entire spectrum of visible light, in addition to ultraviolet and infrared radiation. Because sunlight contains more red light than far-red light, however, when a plant is exposed to sunlight, the level of Pfr increases.

Why is phytochrome important to plants? Timing of day length and darkness is the most reliable way for plants to measure the change from one season to the next. This measurement, which synchronizes the stages of plant development, is crucial for survival, particularly in environments where the climate has an annual pattern of favorable and unfavorable seasons.

Phytochrome is involved in many other responses to light

Phytochrome is involved in the light requirement that some seeds have for germination. Seeds with a light requirement must be exposed to light containing red wavelengths. Exposure to red light converts Pr to Pfr, and germination occurs. Many temperate species with small seeds require light for germination. (Larger seeds generally do not have a light requirement.) This adaptation enables the seeds to germinate at the optimal time. During early spring, sunlight, including red light, penetrates the bare branches of overlying deciduous trees and reaches the soil between the trees. As spring temperatures warm, the seeds on the soil absorb red light and germinate. During their early growth, the newly germinated seedlings do not have to compete with the taller trees for sunlight.

PLANTS AND THE ENVIRONMENT | Phytochrome and Shade Avoidance

Can plants sense the presence of nearby plants? The answer is yes: They not only detect the presence of nearby plants, which are potential competitors, but they also change the way they grow and develop in response. Many plants, from small herbs to large trees, compete for light, a response known as **shade avoidance,** in which plants tend to grow taller when closely surrounded by other plants. If successful, the shade-avoiding plant projects its new growth into direct sunlight, increasing its chances of survival.

Since the 1970s, botanists have recognized the environmental factor that triggers shade avoidance: Plants perceive changes in the ratio of red to far-red light that results from the presence of nearby plants. The leaves of neighboring plants absorb much more red light than far-red light. (Recall from Chapter 4 that the green pigment chlorophyll strongly absorbs red light during photosynthesis.) In a densely plant-populated area, the ratio of red light to far-red light (r/fr) decreases, affecting the equilibrium between Pr and

Pfr forms. This signal triggers a series of responses that cause the shade-avoiding plant, which is adapted to full-light environments, to grow taller or flower earlier.

When a plant is using many of its resources for stem elongation, it has fewer resources to allocate for new leaves and branches, storage tissues, or reproductive tissues. However, for a shade-avoiding plant that is shaded by its neighbors, a rapid increase in stem length is advantageous, because once this plant is taller than its neighbors, it obtains a larger share of unfiltered sunlight.

Other processes under the influence of phytochrome include sleep movements in leaves (discussed later); shoot dormancy; leaf abscission; and pigment formation in flowers, fruits, and leaves. (Also see *Plants and the Environment: Phytochrome and Shade Avoidance.*)

FIGURE IN FOCUS

Phytochrome is a unique pigment that undergoes changes in shape on exposure to light of different wavelengths.

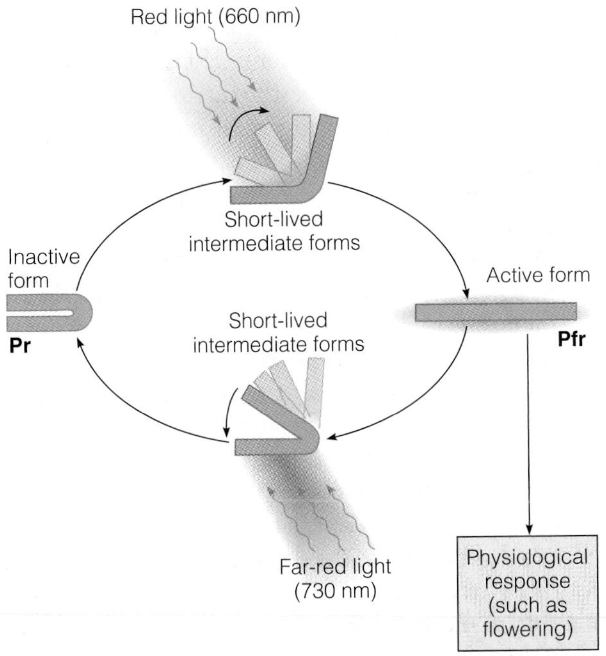

Red light (660 nm)

Short-lived intermediate forms

Inactive form

Pr

Short-lived intermediate forms

Active form

Pfr

Far-red light (730 nm)

Physiological response (such as flowering)

FIGURE 11-18 Phytochrome.
This pigment occurs in two forms, designated Pr and Pfr, and readily converts from one form to the other. Red light converts Pr to Pfr, and far-red light converts Pfr to Pr.

Temperature and Reproduction

You have seen that light is an important mechanism affecting seasonal responses, such as flowering. Temperature also affects seasonal responses: certain plants have a temperature requirement that must be met if they are to flower. The promotion of flowering by exposure to low temperature is known as **vernalization.** The part of the plant that must be exposed to low temperature varies. For some plants, recently germinated seedlings have a "cold" requirement. For other plants, the moist seeds of some plants must be exposed to low temperature for a period of several weeks for flowering to occur after the seeds germinate and grow.

In some plant species, the requirement of a low-temperature period is absolute, meaning that they will not flower unless they have been vernalized. Other species flower sooner if exposed to low temperatures but will still flower at a later date if they are not exposed to low temperatures.

Examples of plants with a low-temperature requirement include annuals such as winter wheat, which grow, reproduce, and die in 1 year, and biennials such as carrots, which take 2 years to complete their life cycles. Winter wheat is planted in the fall and germinates at that time. The young seedlings are exposed to cold during the winter and subsequently flower after resuming growth the following spring.

VERNALIZATION The low-temperature requirement for flowering in some plant species.

Carrots and other biennials grow vegetatively the first year and store surplus food in their roots (•**Figure 11-19**). If the roots are not harvested, the plants flower and reproduce sexually during the second year, after exposure to the low temperatures of winter. Carrots growing in a warm environment and not exposed to low temperatures continue vegetative growth indefinitely and do not initiate sexual reproduction.

An external stimulus to which a plant responds, such as low temperature, may be influenced by internal conditions, such as hormone levels in the plant. It is possible, for example, to eliminate the low-temperature requirement for flowering in biennials by treating the plants with the hormone gibberellin.

(a) The carrot (*Daucus carota*), a biennial plant, grows vegetatively the first year, storing surplus food in its storage root.

(b) After the plant remains dormant through the winter, the energy stored in the root is used during the plant's second year to flower and reproduce.

FIGURE 11-19 Temperature requirements for flowering.

Circadian Rhythms

Almost all organisms, including plants, animals, and microorganisms, appear to have an internal timer, or biological clock, that approximates a 24-hour cycle, the time it takes for Earth to rotate around its own axis. Such internal cycles are known as **circadian rhythms** (from the Latin *circum,* "around," and *diurn,* "daily"). Circadian rhythms help an organism respond to the time of day, whereas photoperiodism enables an organism to detect the time of year.

Why do plants and other organisms have circadian rhythms? Predictable environmental changes, such as the sunrise and sunset, occur during the course of each 24-hour period. These predictable changes may be important to an individual organism by causing it to change its behavior (in the case of animals) or its physiological processes. It is thought that circadian rhythms help an organism synchronize repeatable daily activities so that they occur at the appropriate time each day. If, for example, an insect-pollinated flower does not open at the time of day that pollinating insects are foraging for food, reproduction will be unsuccessful.

Circadian rhythms in plants affect such biological processes as gene expression, the rate of photosynthesis, and the opening and closing of stomata. **Sleep movements** observed in the common bean and other plants are another example of a circadian rhythm (•**Figure 11-20**). During the day, bean leaves are horizontal, possibly for optimal light absorption, but at night the leaves fold down or up, a movement that orients them perpendicular to their daytime position. The biological significance of sleep movements is unknown at this time.

When constant environmental conditions are maintained, circadian rhythms repeat every 20 to 30 hours, at least for several days. In nature, the rising and setting of the sun reset the biological clock so that the cycle repeats every 24 hours. What happens if a plant's circadian clock is mutated so that the day–night cycle is not resynchronized? In 2005, biologists identified mutants of the model plant *Arabidopsis* in which the circadian clock is not synchronized to match the external day–night cycle. The mutant plants contained less chlorophyll, fixed less carbon by photosynthesis, grew more slowly, and had less of a competitive advantage than *Arabidopsis* plants with a normal circadian clock.

CIRCADIAN RHYTHM A biological activity with an internal rhythm that approximates the 24-hour day.

(a) Leaf position at noon. **(b)** Leaf position at midnight.

FIGURE 11-20 Sleep movements in the common bean (*Phaseolus vulgaris*).
It is not known why some plants exhibit sleep movements, but they occur on an approximate 24-hour cycle, regardless of the amount of light or darkness to which the plants are exposed.

For many plants, two photoreceptors—the red light–absorbing phytochrome and the blue/ultraviolet-A light–absorbing **cryptochrome**—are implicated in resetting the biological clock. Certain parts of the phytochrome molecule are like clock proteins in fruit flies, fungi, mammals, and bacteria; this molecular evidence strongly supports the circadian clock role of phytochrome. The evidence for cryptochrome as a clock protein is also convincing. First discovered in plants, cryptochrome counterparts are found in the fruit fly and mouse biological clock proteins. Possibly both photoreceptors are involved in resetting the biological clock in plants: Researchers have evidence that phytochrome and cryptochrome sometimes interact to regulate similar responses.

Turgor Movements

The sensitive plant (*Mimosa pudica*) dramatically folds its leaves and droops in response to touch (or to an electrical, chemical, or thermal stimulus) (•**Figure 11-21**). The response, which typically occurs in a few seconds, spreads throughout the plant even if only one leaflet is initially touched. When a sensitive plant leaf is touched, an electrical impulse moves down the leaf to special cells housed in an organ at the base of the petiole, called

(a) The sensitive plant before being touched.

(b) The sensitive plant several seconds after being touched.

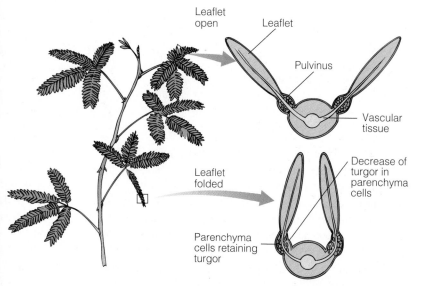

(c) How the folding and drooping occurs. Pulvini occur in three areas: the base of each leaflet, the base of each cluster of leaflets, and the base of each leaf. Only changes in the pulvini at the bases of leaflets are shown. (*Top right*) A section through two leaflets, showing their pulvini when the leaf is undisturbed. (*Bottom right*) A section through the two leaflets, showing how a loss of turgor produces the folding of the leaves.

FIGURE 11-21 Turgor movements in the sensitive plant (*Mimosa pudica*).

the **pulvinus** (pl., *pulvini*). The pulvinus is a somewhat swollen joint that acts as a hinge. When the electrical signal reaches cells in the pulvinus, the electrical signal induces a chemical signal that increases membrane permeability to certain ions. A loss of *turgor* (rigidity or distension caused by absorption of water) occurs in certain pulvinus cells as potassium ions exit through the now-permeable plasma membrane, causing water to leave the cells by osmosis. The sudden change in turgor

> **TURGOR MOVEMENT** A temporary plant movement that results from changes in internal water pressure in a plant part.

causes the leaf movement. Such **turgor movements** are temporary and reversible. The movement of potassium ions and water back into the pulvinus cells causes the plant part to return to its original position, although recovery takes several to many minutes longer than the original movement.

The mechanism by which the Venus flytrap leaf closes is similar to the mechanism of the sensitive plant. An electrical signal, which moves more rapidly than in the sensitive plant, induces a chemical signal that causes a movement of potassium ions out of certain cells, followed by the exit of water. The loss of turgor causes the leaf to snap shut.

STUDY OUTLINE

❶ Discuss genetic and environmental factors that affect plant growth and development.
Plant growth and development are controlled by both internal genetic factors and external environmental factors. The location of a cell in the young plant body affects gene expression during development, causing some genes in that cell to be turned off and others to be turned on. Many factors in the physical environment (such as changing day length, variation in precipitation, and temperature) determine gene expression and affect plant growth and development.

❷ Describe phototropism, gravitropism, and thigmotropism.
Tropisms are directional growth responses and are permanent. **Phototropism** is the directional growth of a plant caused by light. **Gravitropism** is plant growth in response to the direction of gravity. **Thigmotropism** is plant growth in response to contact with a solid object.

❸ List several ways in which each of the following hormones affects plant growth and development: auxin, gibberellin, cytokinin, ethylene, abscisic acid.
Hormones are organic chemical messengers in plants and other multicellular organisms. Biologists have identified five major classes of plant hormones (auxin, gibberellin, cytokinin, ethylene, and abscisic acid) as well as a variety of signaling molecules. **Auxin** is involved in growth and development, including stem elongation, apical dominance, and root formation on cuttings. **Gibberellin** is involved in stem elongation, flowering, and seed germination. **Cytokinin** is involved in cell division and delay of senescence. **Ethylene** is a gaseous hormone involved in leaf abscission and fruit ripening. **Abscisic acid** is involved in dormancy and responses to stress.

❹ Relate which hormone or hormones is/are involved in each of the following biological processes: leaf abscission, seed germination, apical dominance.
Many plant processes may be the result of interactions of several hormones rather than the effect of a single hormone. Ethylene, along with auxin, is involved in leaf **abscission;** as a leaf ages, the level of auxin in the leaf decreases, and the level of ethylene increases. Gibberellins are involved in the seed germination of certain plants (cereals and other grasses); application of gibberellins substitutes for low-temperature or light requirements for germination in seeds such as lettuce, oats, and tobacco. Ethylene also promotes seed germination, whereas abscisic acid inhibits seed germination. **Apical dominance** is the inhibition of axillary bud growth by the apical meristem. In plants with strong apical dominance, auxin produced in the shoot apical meristem inhibits axillary buds near the apical meristem from developing into actively growing shoots. Cytokinins and ethylene also affect apical dominance: cytokinins promote the growth of axillary buds, and ethylene inhibits axillary bud development.

❺ Explain how varying amounts of light and darkness induce flowering, and describe the role of phytochrome.
Photoperiodism is any physiological response (such as flowering) to variations in the length of daylight and darkness. Some plants are **short-day plants,** some are **long-day plants,** and others are **intermediate-day plants.** In these cases, the plant is actually measuring the length of the dark period. In **day-neutral plants,** photoperiod does not affect flowering. **Phytochrome** is a blue-green proteinaceous pigment involved in many plant responses to light, independent of photosynthesis. There are about five different phytochrome proteins; each exists in two forms and readily converts from one form

to the other after absorption of light of specific wavelengths. **Pr** strongly absorbs red light with a relatively short wavelength (660 nm), changing to the second form, **Pfr**, which absorbs red light with a relatively long wavelength (730 nm). Pfr, the active form, triggers or inhibits responses such as flowering.

❻ Explain how temperature affects flower induction in certain plants.
Certain plants have a temperature requirement that must be met if they are to flower. **Vernalization** is the low-temperature requirement for flowering in some plant species.

❼ Define *circadian rhythm,* and give an example.
A **circadian rhythm** is a biological activity with an internal rhythm that approximates the 24-hour day. Circadian rhythms are reset by the rising and setting of the sun. Circadian rhythms in plants affect such biological processes as gene expression, the rate of photosynthesis, and the opening and closing of stomata.

❽ Give an example of a turgor movement, and distinguish between turgor movements and tropisms.
Turgor movements are temporary plant movements that result from changes in internal water pressure in a plant part. In contrast, tropisms are permanent growth responses. Leaves of the sensitive plant and Venus flytrap exhibit dramatic turgor movements.

REVIEW QUESTIONS

1. Why are plant growth and development so sensitive to environmental cues?
2. What is a tropism? Give two examples of tropisms.
3. What is a hormone?
4. How is auxin involved in phototropism?
5. Discuss the various plant hormones that are involved in each of the following processes: (a) germination of seeds, (b) stem elongation, (c) ripening of fruits, (d) abscission of leaves, (e) seed dormancy.

6. Summarize the roles of auxin, gibberellins, cytokinins, ethylene, and abscisic acid.
7. What is phytochrome? Describe its role in flowering.
8. Define *vernalization.*
9. Distinguish between photoperiodism and circadian rhythms.
10. Distinguish between turgor movements and tropisms.

THOUGHT QUESTIONS

1. What evolutionary advantages are conferred on a plant whose stems are positively phototropic and whose roots are positively gravitropic?
2. As a result of apical dominance, axillary buds closest to the shoot apex are inhibited the most and therefore grow the least, whereas those farthest from the shoot apex grow more. Would a "Christmas tree," such as spruce, have this type of apical dominance? How about a "bushy" tree, such as maple? Explain your answers.
3. The nursery industry uses a plant hormone to induce dormancy in late summer so that plant material can be dug at that time rather than in autumn. (It is less traumatic for a plant to be moved when dormant than when it is actively growing.) Based on what you have learned in this chapter, which hormone do you think is used? Explain your answer.

4. Predict whether flowering will occur in each of the following situations. Explain each answer.
   a. A short-day plant is exposed to 15 hours of daylight and 9 hours of darkness.
   b. A short-day plant is exposed to 9 hours of daylight and 15 hours of darkness.
   c. A short-day plant is exposed to 9 hours of daylight and 15 hours of darkness, with a 10-minute flash of red light in the middle of the night.
5. When an insect lands on a *Mimosa* leaf, a turgor movement occurs. In what way is this response beneficial to the plant? Of what importance is it that this movement is only temporary and not permanent?

Visit us on the web at http://www.thomsonedu.com/biology/berg for additional resources such as flashcards, tutorial quizzes, further readings, and web links.

12

Mitosis, Meiosis, and Life Cycles

LEARNING OBJECTIVES

❶ Distinguish between a haploid cell and a diploid cell, and define *homologous chromosomes*.

❷ Identify the phases in the cell cycle, and describe the main events of each phase.

❸ Describe the events that occur in each stage of mitosis.

❹ Explain why meiosis is needed at some point in the life cycle of every sexually reproducing organism, and describe the events that occur during meiosis I and meiosis II.

❺ Compare and contrast mitosis and meiosis.

❻ Define *alternation of generations*.

Date palms (*Phoenix*)—the palms associated with a desert oasis—are probably second only to coconut palms as the world's best-known palm. Valued for their sugar-rich fruits, date palms probably originated in North Africa or the Middle East. They have been cultivated in arid regions for at least 5000 years. In the United States, date palm plantations are found primarily in Arizona and California, although date palms are also grown as ornamentals in Florida.

Individual date palm plants are *dioecious*—that is, each plant is either male or female. Commercial plantings of date palms are mostly of female plants, because the females produce the fruit. A few male palms are needed in each orchard, however, to provide pollen. The female flowers can be pollinated by wind, although growers sometimes dust the female flowers with pollen. Flowers develop in the spring, and the fruit ripens by the end of summer.

Date palms are easily propagated from either seeds or offshoots. Fruits are produced in clusters during sexual reproduction. Each date fruit is botanically a berry and contains a single, woody, oblong seed that germinates readily when planted. However, commercial growers rarely grow date palms from seed because roughly half the plants are male (and you cannot tell the sex of the plant until it is about 5 years old, when it begins to produce flowers). Also, the female palms that grow from seed produce fruits of variable, often inferior, quality.

Instead of growing date palms from seeds, commercial growers use a sharp knife to remove the offshoots from a high-yielding female tree that produces superior dates. *Offshoots* are short branches that develop from axillary buds near the base of the main stem. Many of these offshoots have already developed roots, so they can be planted directly into the soil. If the offshoots lack roots, they can be treated with a rooting hormone and then planted. All offshoots are genetically identical to the original palm, so all offshoots from a female palm will bear dates.

The date palm is an excellent plant to introduce this chapter because it reproduces both asexually by offshoots (which requires mitosis) and sexually by seeds (which requires meiosis as well as mitosis). In this chapter we examine how the genetic material—deoxyribonucleic acid (DNA)—in chromosomes is distributed into daughter cells during cell division. *Mitosis* is a process that ensures a parent cell transmits one copy of every chromosome to each of its two daughter cells. In this way, the chromosome number is preserved through successive mitotic divisions.

The date palm. Dates grow in clusters on female date palms (*Phoenix dactylifera*). Photographed by the Sea of Galilee in Israel.

Most body cells of plants and other eukaryotes divide by mitosis. *Meiosis* is a process that reduces the chromosome number by half. Meiosis is required before sexual repro-

duction occurs and is largely responsible for the variations observed among the offspring of sexual reproduction.

Chromosome Number

In plants and all other eukaryotes, genes are found on **chromosomes,** threadlike structures in a cell's nucleus that are visible under the microscope only during cell division. Each chromosome consists of proteins and a single large molecule of DNA that contains hundreds or thousands of different genes.

The number of chromosomes within a nucleus varies from one species to another, but every somatic cell (nonreproductive body cell) in an organism of a given species contains a characteristic number of chromosomes. Human body cells have 46 chromosomes, for example, and cabbage cells have 20. A desert-adapted daisy (*Haplopappus gracilis*) has only 4 chromosomes per cell, the olive tree (*Olea europaea*) has 46 chromosomes (the same number as in humans), and the adder's-tongue fern (*Ophioglossum reticulatum*) has 1262 chromosomes per cell. No simple relationship exists between the number of chromosomes and the size or complexity of an organism, although most plant and animal species have between 8 and 50 chromosomes per body cell.

In somatic cells of a plant or animal, each chromosome normally has a partner chromosome. The 20 chromosomes of a cabbage cell, for example, occur in 10 pairs. The chromosome pairs in a cell vary enough in size and shape that biologists can distinguish the different chromosomes and match up the pairs. The two members of a pair of chromosomes are referred to as **homologous chromosomes.**

Homologous chromosomes carry information governing the same genetic traits, although this information is not necessarily identical. For example, each member of a homologous pair may carry a gene that specifies flower color, but one chromosome might have the information for red petals and the other specifies white petals. Homologous chromosomes carry different information because one of each pair originally came from the parental sperm cell and the other from the parental egg during sexual reproduction.

A cell in which each chromosome occurs in pairs is called **diploid, or *2n*.** For example, somatic cells of an olive tree, with 46 chromosomes, are diploid and have 23 pairs of chromosomes per cell. The somatic cells of many, but not all, organisms are diploid. A cell that has a single set of unpaired chromosomes is called **haploid, or *n*.** The haploid chromosome number is half of the diploid number. In olives, for example, male and female reproductive cells are haploid (*n* = 23).

The Cell Cycle

In somatic cells that are capable of dividing, the **cell cycle** is the period from the beginning of one division to the beginning of the next division. The cell cycle is often represented as a circle (like the face of a clock) and consists of two main phases, interphase and M phase (•**Figure 12-1**). The period between two successive divisions, represented by a complete revolution of the circle, is the **generation time.** Timing of the cell cycle varies widely from one species to another and from one cell type to another, but in actively growing plant and animal cells, it is usually about 8 to 20 hours.

M phase involves two main processes, mitosis and cytokinesis. **Mitosis** is the division of the nucleus, and **cytokinesis** is the division of the cytoplasm to form two cells. Before a eukaryotic cell divides by cytokinesis, its nucleus must undergo mitosis, a process that precisely distributes complete sets of chromosomes to

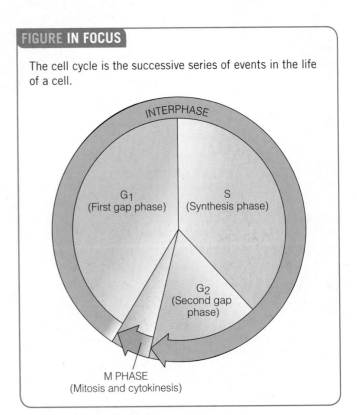

FIGURE IN FOCUS

The cell cycle is the successive series of events in the life of a cell.

INTERPHASE

G₁
(First gap phase)

S
(Synthesis phase)

G₂
(Second gap phase)

M PHASE
(Mitosis and cytokinesis)

FIGURE 12-1 The cell cycle.
The cell cycle includes interphase (G₁, S, and G₂) and M phase (mitosis and cytokinesis). The time required for each stage varies with cell type and species. Most cells spend about 90 percent of their cell cycle in interphase.

each daughter nucleus. As a result of mitosis, each new cell contains the identical number and types of chromosomes present in the original parent cell.

For the most part, mitosis and cytokinesis take place in localized areas of the plant body called **meristems** (from the Greek *meristo,* meaning "divided"). Meristems occur in the shoot and root tips (the **apical meristems**) and, in some plants, in thin cylindrical regions that run the entire lengths of stems and roots except at the tips (the **lateral meristems**) (see Chapter 5). The production and subsequent elongation of new cells by apical meristems causes growing stems and roots to increase in length. Lateral meristems produce additional wood and bark tissues that add girth to—that is, thicken—stems and roots of trees and shrubs. In addition, plants produce temporary meristems in response to wounding or disease.

One of the main questions about cell division in eukaryotic organisms is, How is it regulated? This question is of more than passing interest. If biologists can determine the factors that control when cells divide, they may discover how those factors go awry in *cancer,* a disease in which cell division is uncontrolled. **EVOLUTION LINK** Biologists have found that certain regulatory molecules that control the cell cycle are common to all eukaryotes. Genetically programmed in the cell's nucleus, these regulatory molecules are parts of the *cell-cycle control system* found in organisms as diverse as plants, yeasts (a fungus), clams, frogs, and humans. These findings strongly suggest that the cell-cycle control system is an ancient legacy that evolved very early in the history of eukaryotes. ■

Interphase is the stage between successive cell divisions

A meristematic cell spends most of its life in **interphase,** a period of active growth and maintenance that precedes mitosis (•**Figure 12-2a**). Interphase ("between phases") is so named because it occurs between the phases of successive cell divisions. During interphase, the cell synthesizes needed materials and grows. Chromosomes undergo duplication during interphase, although the process is not readily visible. Then, during mitosis, they condense into visibly separate structures and are distributed to the two daughter nuclei.

Interphase is subdivided into three periods: G_1, S, and G_2 (see Figure 12-1). The first period, **G_1,** or the **first gap phase,** is the time between the end of the previous cell division and the beginning of DNA replication. The cell grows during the G_1 phase, which is typically the longest phase. Cells that are not actively dividing usually remain in this part of the cell cycle. Toward the end of G_1, the cell synthesizes certain enzymes used in DNA

replication. These activities make it possible for the cell to enter the S phase.

The second period of interphase, called the **S phase** or **synthesis phase,** involves the replication of DNA. Other materials, such as proteins that are components of chromosomes, are also synthesized at this time so that the cell can make duplicate copies of its chromosomes.

Following the completion of the S phase, the cell enters a **second gap phase,** or **G_2.** At this time, increased protein synthesis occurs as the cell prepares to divide. For many cells, the G_2 phase is short relative to the G_1 and S phases. The beginning of mitosis marks the completion of the G_2 phase.

To summarize, here is the sequence of interphase and M phase in the eukaryotic cell cycle:

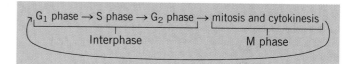

Mitosis

The completion of interphase is signaled by the beginning of mitosis, in which visible changes associated with the division of the nucleus take place. Most other cellular activities, such as protein synthesis, are suspended during mitosis, which is a relatively brief period of the cell's life. Mitosis is a continuous process, but for descriptive purposes, it is divided into four stages:

Prophase ⟶ metaphase ⟶ anaphase ⟶ telophase

HOMOLOGOUS CHROMOSOMES Members of a chromosome pair that are similar in size, shape, and genetic constitution.

DIPLOID (2N) The condition of having two sets of chromosomes per nucleus.

HAPLOID (N) The condition of having one set of chromosomes per nucleus.

CELL CYCLE The cyclic series of events in the life of a dividing eukaryotic cell.

MITOSIS The division of the cell nucleus resulting in two daughter nuclei, each with the same number of chromosomes as the parent nucleus.

CYTOKINESIS The stage of cell division in which the cytoplasm divides to form two daughter cells.

INTERPHASE The stage of the cell cycle between successive mitotic divisions.

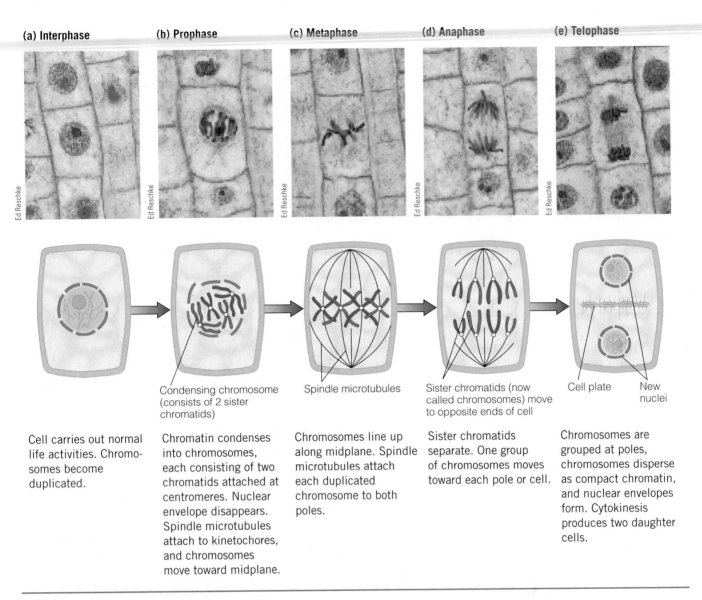

FIGURE 12-2 Interphase and the stages of mitosis.
The micrographs are onion (*Allium cepa*) root tip cells prepared with stains. The diploid number for the cells shown in the diagrams is four.

Duplicated chromosomes condense and become visible during prophase

The first stage of mitosis, **prophase**, begins when chromatin (the long, threadlike material of which chromosomes are composed) begins to condense and coil into visible chromosomes (•**Figure 12-2b**). In this form, the chromosomes easily move into position and eventually pass into the daughter cells with less likelihood of becoming tangled. Although each chromosome may contain several centimeters of DNA, at mitosis this DNA is condensed into a chromosome that is only 5 to 10 micrometers in length—a 10,000-fold shortening!

As prophase proceeds, the chromosomes become shorter and thicker and are individually visible under the light microscope. It is now apparent that each chromosome is actually a duplicated chromosome (recall that it was duplicated during the preceding S phase) that consists of two identical subunits called **sister chromatids** (•**Figure 12-3**).

Each chromatid includes a constricted region called the **centromere.** Sister chromatids are tightly associated in the vicinity of their centromeres. Attached to each centromere is a **kinetochore,** a protein complex to which **microtubules** (hollow, cylindrical fibers) can bind (see Chapter 3). These microtubules function in chromosome distribution during mitosis, in which one copy of each chromosome is delivered to each daughter cell.

During prophase, the nuclear envelope (the double membrane that encloses the nucleus) breaks apart. The parts of the fragmented membrane are temporar-

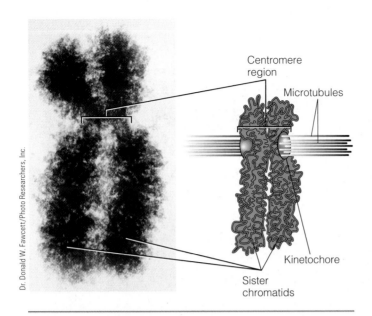

Centromere
region

Microtubules

Kinetochore

Sister
chromatids

Dr. Donald W. Fawcett/Photo Researchers, Inc.

FIGURE 12-3 A duplicated chromosome.
Each duplicated chromosome is composed of two identical
sister chromatids that are tightly associated at their centromere
regions. Associated with each centromere is a kinetochore,
which serves as a microtubule attachment site.

ily stored in *vesicles* (small, membrane-enclosed sacs) to be used later. The *nucleolus* (the structure in the nucleus where ribosomes are synthesized) shrinks and disappears.

A dividing cell can be described as a globe, with an equator that determines the cell's *midplane* and two opposite *poles*. This terminology is used for all cells regardless of their actual shape. During prophase, microtubules organize between the poles to form the mitotic **spindle**, a special structure that aids in the movement of chromosomes (see *Plants and People: The Autumn Crocus and Colchicine*).

Toward the end of prophase, the spindle microtubules grow and shrink as they move toward the center of the cell in a "search-and-capture" process. Their random movements give the appearance that they are "searching" for the chromosomes. If a microtubule comes near the centromere of a chromosome, the microtubule "captures" it. As the now-tethered chromosome continues moving toward the cell's midplane, a spindle microtubule from the cell's other pole attaches to the opposite side of the centromere.

During metaphase, duplicated chromosomes line up on the midplane

The short period during which the chromosomes are lined up along the midplane of the cell is **metaphase** (•**Figure 12-2c**). The mitotic spindle, now completely visible, is composed of numerous microtubules that extend from pole to pole. As already mentioned, each

of the two sister chromatids of each chromosome is attached by its kinetochore to microtubules from one pole, and its sister chromatid is attached by its kinetochore to microtubules from the opposite pole. During metaphase, each chromatid is quite condensed and appears thick and distinct. Because individual chromosomes are more distinct at metaphase than at any other time, they are typically photographed and studied during this stage.

Chromosomes move toward the poles during anaphase

Anaphase begins as the sister chromatids separate (•**Figure 12-2d**). Once the chromatids are no longer attached to their duplicates, each chromatid is considered an independent chromosome. The now-separated chromosomes move to opposite poles, using the spindle microtubules as tracks. The kinetochores, still attached to the spindle microtubules, lead the way, with the chromosome arms trailing behind. Anaphase ends when all the chromosomes have reached the poles.

Biologists are making progress in understanding the way chromosomes move apart in anaphase. Evidence indicates that microtubules are dynamic structures that shorten during anaphase. This shortening "pulls" the chromosomes toward the poles and is in part responsible for chromosome separation.

During telophase, two separate nuclei form

During the final stage of mitosis, **telophase,** the chromosomes arrive at the poles and return to their interphase condition. The chromosomes begin to elongate by uncoiling and then become invisible chromatin threads (•**Figure 12-2e**). A new nuclear envelope, made at least in part from small vesicles and other components of the old nuclear envelope, forms around each set of chromosomes. Nucleoli reorganize, and spindle microtubules disappear.

Cytokinesis forms two daughter cells

Cytokinesis, the division of the cytoplasm that usually accompanies mitosis, generally begins during telophase. In plant cells, cytokinesis occurs by the formation of a **cell plate,** a partition between the newly formed nuclei in the equatorial region of the cell. The cell plate grows laterally (sideways) toward the edges of the cell, forming two adjacent daughter cells (•**Figure 12-4**). The cell

SPINDLE The structure consisting mainly of microtubules that provides the framework for chromosome movement during cell division.

CELL PLATE The structure that forms during cytokinesis in plants, separating two daughter cells produced by mitosis.

PLANTS AND PEOPLE | The Autumn Crocus and Colchicine

■ **The autumn crocus (*Colchicum autumnale*). This small flowering plant produces the powerful drug colchicine, which blocks cell division.**

The autumn crocus (*Colchicum autumnale*) is a small perennial herb that produces light purple flowers in late summer and autumn (see figure). A member of the lily family, the autumn crocus is native to Europe, where it grows in subalpine meadows. It is also widely cultivated in temperate areas as an ornamental and for the chemical **colchicine,** which is obtained from its seeds and corms.

Colchicine blocks cell division by interfering with the normal function of the spindle, a special structure that aids in the movement of chromosomes during cell division. In cells treated with colchicine, chromosomes duplicate, but they cannot separate and move to the opposite ends of the cell. As a result, cell division stops, and the cell contains an extra set of chromosomes.

Colchicine is used in plant breeding to induce **polyploidy,** a condition in which cells have more than two sets of chromosomes. Polyploidy is also a naturally occurring phenomenon that is common in many plant species but rare in most animal species. In general, plants are relatively tolerant of extra chromosome sets—30 to 80 percent of all flowering plants are polyploids—whereas extra chromosome sets are often lethal in animals. Polyploid plants are often larger and more vigorous than diploid plants and may have other desirable qualities, such as larger fruits or extra flower petals.

Colchicine can be used to induce polyploidy in cells destined to become gametes—that is, in eggs or sperm cells. When gametes with extra chromosome sets unite, they give rise to a new generation of polyploid plants. Thus, plant breeders use colchicine to induce changes in the number of chromosomes in certain plants and to develop polyploid varieties of ornamental and agricultural plants. For example, colchicine was used to produce the polyploid condition in *triticale,* a wheat–rye hybrid that has wheat's high protein content and rye's ability to withstand cold and drought. Before treatment with colchicine, triticale plants were sterile; after colchicine treatment, which made them polyploids, they were fully fertile.

plate is destined to become two new plasma membranes and cell walls that will separate the daughter cells.

The Golgi bodies that gather near the midplane produce vesicles that form the cell plate. The vesicles contain materials to construct both a primary cell wall for each daughter cell and a middle lamella that will cement the primary cell walls together. The vesicle membranes fuse to become the plasma membrane of each daughter cell.

Sexual Reproduction

In most plants and animals, when certain cells in the reproductive organs divide, the result is not new somatic cells with a full (diploid) number of chromosomes but cells that become or give rise to reproductive cells, which have *half* the diploid number of chromosomes. These reproductive cells—eggs and sperm cells—are called **gametes.** Instead of becoming part of the body of the organism that produced them, gametes form a complete, new organism by uniting. If the female gamete (egg) unites with the male gamete (sperm cell), the result is a fertilized egg, or **zygote,** which is the first somatic cell of a completely new organism. This type of reproduction, involving the union of male and female gametes, is called **sexual reproduction.**

Gametes cannot have a full complement of chromosomes, or the zygote resulting from their union would have twice as many chromosomes as it should. A special type of reduction division, meiosis, reduces the number of chromosomes in reproductive cells by half. (The term

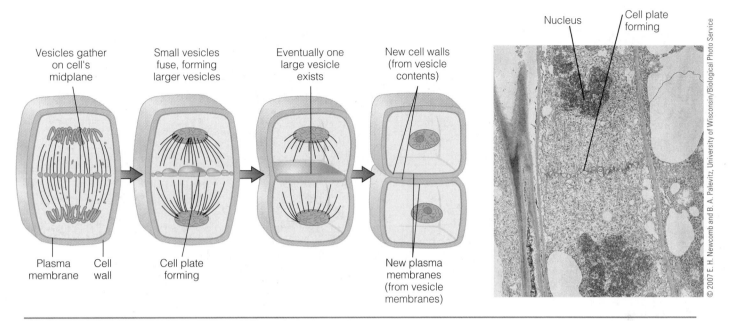

FIGURE 12-4 Cytokinesis.
Stages in the formation of plasma membranes and cell walls from the fusing vesicles of the cell plate. The electron micrograph shows cytokinesis in a maple (*Acer*) leaf cell. Note the cell plate, which consists of gathering vesicles.

meiosis means "to make smaller.") As a result of meiosis, each gamete has only one chromosome of each pair, resulting in an egg or sperm cell with a haploid, or *n*, number of chromosomes. It is important to note that haploid cells do not contain just any combination of chromosomes, but one member of each homologous pair. When two haploid gametes join in fertilization, the normal diploid number of chromosomes is restored. In this way, meiosis allows the chromosome number of a species to remain the same from generation to generation.

In animals and certain algae, meiosis usually gives rise to haploid eggs and sperm cells directly. In plants, however, meiosis results in haploid spores, which are reproductive cells that give rise to new individuals without first fusing with another cell.

Meiosis

Remember that in mitosis each daughter cell receives exactly the same number and kinds of chromosomes that the parent cell had. In contrast, in meiosis the number of chromosomes is reduced by half. Thus, each new cell that results from meiosis is a haploid cell; it has *n*, not *2n*, chromosomes.

Unlike mitosis, in which pairs of homologous chromosomes end up in *each* daughter cell, the process of meiosis *separates* the members of each homologous pair of chromosomes. As a result, any contrasting genetic

traits on each homologous pair of chromosomes are separated and distributed independently to different gametes or spores. The outcome of this distribution method is the likelihood that no two offspring of the same parents will be exactly alike.

Meiosis consists of two cell divisions, designated the first and second meiotic divisions, or simply meiosis I and meiosis II. Each includes prophase, metaphase, anaphase, and telophase stages. During **meiosis I,** the members of each homologous pair of chromosomes first join and then separate and move into different nuclei. The separation of homologous chromosomes that occurs during meiosis is like shuffling a deck of cards—one of each pair is randomly "dealt" to each new cell. These chromosomes were duplicated prior to meiosis I, so each consists of two chromatids. In **meiosis II,** the sister chromatids that make up each duplicated chromosome separate from each other and are distributed into two different nuclei. Thus, what began as one diploid cell gives rise after meiosis to four haploid cells.

Since it is easier to follow these events in an organism that has only a few chromosomes, we will discuss a plant with a diploid chromosome number of four (two homologous pairs).

MEIOSIS Process in which a *2n* cell undergoes successive nuclear divisions, potentially producing four *n* nuclei; leads to formation of spores in plants.

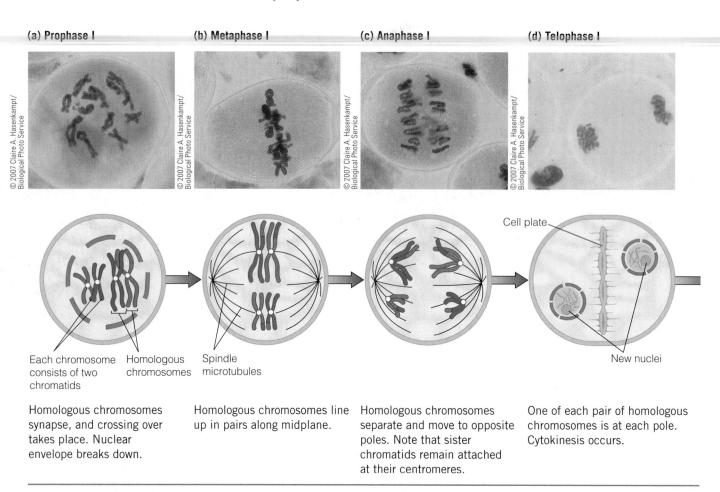

(a) Prophase I

(b) Metaphase I

(c) Anaphase I

(d) Telophase I

© 2007 Claire A. Hasenkampt/Biological Photo Service

Each chromosome consists of two chromatids

Homologous chromosomes

Spindle microtubules

Cell plate

New nuclei

Homologous chromosomes synapse, and crossing over takes place. Nuclear envelope breaks down.

Homologous chromosomes line up in pairs along midplane.

Homologous chromosomes separate and move to opposite poles. Note that sister chromatids remain attached at their centromeres.

One of each pair of homologous chromosomes is at each pole. Cytokinesis occurs.

FIGURE 12-5 The stages of meiosis.
The micrographs are Easter lily (*Lilium longiflorum*) flower cells prepared with stains and flattened on slides. The diploid number for the cells shown in the diagrams is four.

Homologous chromosomes separate into different daughter cells during meiosis I

As in mitosis, the chromosomes duplicate during the S phase of interphase before the complex movements of meiosis actually begin. Recall that each duplicated chromosome consists of two chromatids joined at their centromeres.

During **prophase I** (prophase of the first meiotic division), while the chromatids are still elongated and thin, the homologous chromosomes come together to lie side by side lengthwise (•**Figure 12-5a**). This pairing of homologous chromosomes is called **synapsis**. Since the diploid number in our example is four chromosomes, at synapsis we would see two homologous pairs. One of each pair, the *maternal homologue,* was originally inherited from the female parent during the formation of the zygote; the other member of a homologous pair, the *paternal homologue,* was inherited from the male parent. Because each chromosome duplicated during interphase and now consists of two chromatids, synapsis results in the association of *four* chromatids (2 homolo-

gous chromosomes × 2 chromatids per chromosome = 4 chromatids).

All the genes on a particular chromosome are *linked* and tend to be inherited together. However, this tendency for linked genes to stay together is not absolute. During synapsis, genetic material may be exchanged between chromatids of paired homologous chromosomes, a process called **crossing over.** During crossing over, pieces of the maternal and paternal chromatids break off and are then precisely rejoined to the opposite chromatid (•**Figure 12-6**). Crossing over produces new combinations of genes. The **genetic recombination** from crossing over greatly enhances the variety among sexually produced offspring. Some biologists think that recombination, with its resulting genetic diversity, is the main reason for sexual reproduction in eukaryotes.

While synapsis and crossing over are occurring, other events also take place. During prophase I, a spindle forms consisting of microtubules. The nuclear envelope and nucleolus disappear. In cells with large chromosomes, the association of the four chromatids of each

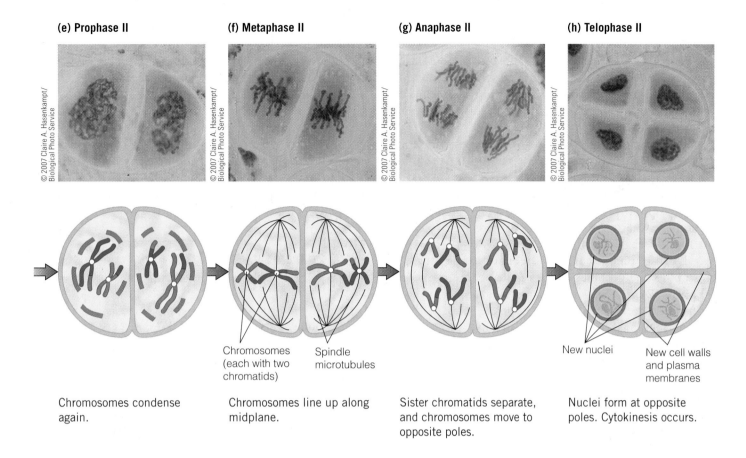

(e) Prophase II

Chromosomes condense again.

(f) Metaphase II

Chromosomes (each with two chromatids)

Spindle microtubules

Chromosomes line up along midplane.

(g) Anaphase II

Sister chromatids separate, and chromosomes move to opposite poles.

(h) Telophase II

New nuclei

New cell walls and plasma membranes

Nuclei form at opposite poles. Cytokinesis occurs.

pair of homologous chromosomes can be seen clearly with the microscope.

During **metaphase I,** homologous chromosomes line up *in pairs* along the midplane (●**Figure 12-5b**). Both kinetochores of one duplicated chromosome are attached by spindle microtubules to the same pole, and both kinetochores of the duplicated homologous chromosome are attached to the opposite pole.

During **anaphase I,** the paired homologous chromosomes separate, with one chromosome moving toward one pole and its homologue moving toward the other pole (●**Figure 12-5c**). Each pole receives a random combination of maternal and paternal chromosomes, but only one member of each homologous pair is present at each pole. The sister chromatids of each duplicated chromosome remain united at their centromere regions.

In **telophase I** in this example, there would be two duplicated chromosomes, one of each homologous pair, at each pole. During telophase I, the nuclei often reorganize, the chromatids generally elongate, and cytokinesis may take place (●**Figure 12-5d**). Note that the haploid number of chromosomes ($n = 2$) is now established, although each chromosome is still duplicated (that is, still consists of two chromatids).

During meiosis II, sister chromatids separate

An interphase-like stage usually follows meiosis I. Because it is not a true interphase—there is no S phase—no further DNA replication or chromosome duplication takes place. In most organisms this period is brief, and in some organisms it is absent.

Because the chromatids do not completely elongate between meiotic divisions, **prophase II** is also brief.

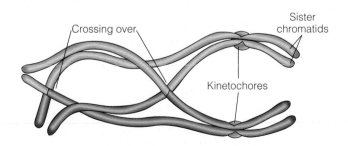

Crossing over

Sister chromatids

Kinetochores

FIGURE 12-6 A pair of homologous chromosomes during late prophase I of meiosis.
Note the four chromatids that make up the paired homologous chromosomes. Crossing over is visible at two sites.

SYNAPSIS The physical association of homologous chromosomes during prophase I of meiosis.

Prophase II is similar to mitotic prophase in many respects. In prophase II there is no pairing of homologous chromosomes (indeed, only one chromosome of each pair is in each nucleus), and no crossing over occurs (•**Figure 12-5e**). New spindle microtubules form, and the nuclear envelope, if it re-formed during telophase I, breaks down.

During **metaphase II,** the chromosomes, each consisting of two sister chromatids, line up on the midplanes of their cells (•**Figure 12-5f**). Metaphase I and II are distinguished from each other because in metaphase I there are four chromatids (of two homologous chromosomes), whereas in metaphase II there are two chromatids of a single chromosome.

During **anaphase II,** the sister chromatids, attached to spindle microtubules at their kinetochores, separate (•**Figure 12-5g**). The sister chromatids, each now considered a chromosome, move to opposite poles.

At **telophase II** there is one member of each homologous chromosome pair at each pole. Each chromosome is in an unduplicated state. Nuclear envelopes then re-form around each set of chromosomes, the chromo-

somes gradually elongate into threadlike chromatin, and cytokinesis occurs (•**Figure 12-5h**).

The two successive divisions of meiosis result in four haploid daughter cells, each containing one of each kind of chromosome. Each resulting haploid cell has a different combination of genes. This genetic variation has two sources: (1) DNA segments are exchanged between maternal and paternal homologues during crossing over. (2) During meiosis, the maternal and paternal chromosomes of homologous pairs separate independently so that each member of a pair is randomly distributed to one of the poles at anaphase.

A Comparison of Mitosis and Meiosis

Although the events of mitosis and meiosis are somewhat similar, there are several important differences (•**Figure 12-7**). Mitosis and meiosis are distinct in the number of divisions. In mitosis there is a single division, typically yielding two daughter cells. In meiosis there

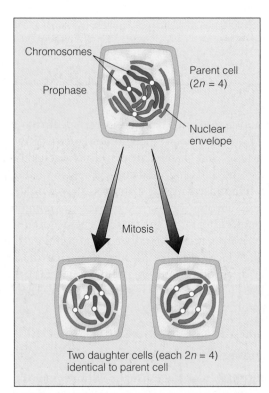

(a) Mitosis. Note that the two daughter cells have identical sets of four chromosomes (two pairs), which is the diploid number. (The chromosomes in the daughter cells are in an unduplicated state, whereas those in the original parent cell are duplicated.)

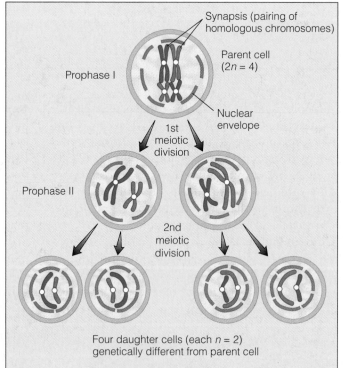

(b) Meiosis. Two divisions take place, giving rise to four daughter cells. Each daughter cell is haploid and has a total of two chromosomes, one of each pair.

FIGURE 12-7 Mitosis compared with meiosis.
The diploid number for each cell is four.

are two successive divisions, typically producing a total of four cells.

Each daughter cell produced in mitosis contains the diploid number of chromosomes. (An exception to this occurs in the plant life cycle, as we will discuss shortly.) Each of the four cells produced in meiosis contains the haploid number of chromosomes—that is, only one member of each homologous pair.

After mitosis, each daughter cell contains a set of chromosomes that is identical in every way to that of the parent cell. In contrast, the homologous chromosomes containing genetic information from each parent (that is, from egg and sperm cell) are thoroughly shuffled during meiosis, and one chromosome of each pair is randomly distributed to each new cell. The resulting haploid cells have new combinations of chromosomes and therefore unique combinations of genes.

In mitosis there is little opportunity for an exchange of genetic material between homologous chromosomes (that is, there is no crossing over) because homologous chromosomes do not associate physically at any time. Crossing over, however, is an important part of meiosis, which means that the genes originally located on one chromosome do not always stay together. Thus, crossing over further increases the genetic shuffling of meiosis.

Plant Life Cycles and Alternation of Generations

Plants and other organisms, such as certain algae and fungi, have a life cycle in which the diploid plant known as a sporophyte produces haploid spores by meiosis (•Figure 12-8). Each spore divides by mitosis to give rise to a multicellular plant known as a gametophyte, whose cells *all* contain the haploid number of chromosomes. The gametophyte then produces haploid gametes (eggs and sperm cells) by *mitosis*.

The gametes of plants are haploid, just as animal gametes are, but unlike animal gametes, they are not the

In each life cycle, the doubling of chromosomes that occurs during fertilization is compensated for by the reduction in chromosome number that occurs during meiosis.

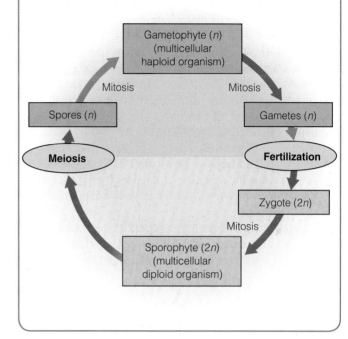

FIGURE 12-8 Plant life cycle.
The life cycle of plants involves an alternation of generations between diploid and haploid stages.

immediate products of meiosis. The gametes unite to form a diploid zygote, which divides mitotically to form a multicellular, diploid sporophyte. Chapters 22 to 25 provide further discussion of alternation of generations.

SPOROPHYTE The 2*n*, spore-producing stage in the life cycle of a plant.

GAMETOPHYTE The *n*, gamete-producing stage in the life cycle of a plant.

STUDY OUTLINE

❶ **Distinguish between a haploid cell and a diploid cell, and define *homologous chromosomes*.**
In the somatic cells of diploid organisms, chromosomes are present in pairs. Members of a chromosome pair that are similar in structure and genetic constitution are called **homologous chromosomes**. A **diploid (2*n*)** cell has two sets of chromosomes per nucleus. A **haploid (*n*)** cell has one set of chromosomes per nucleus.

❷ **Identify the phases in the cell cycle, and describe the main events of each phase.**
The **cell cycle** is the cyclic series of events in the life of a dividing eukaryotic cell. **Interphase** is the stage of the cell cycle between successive mitotic divisions. During interphase, the cell grows and prepares for the next division, and DNA replicates. Interphase is divided into the **first gap phase (G₁)**, the **synthesis phase (S)**, and the

second gap phase (G$_2$). **Mitosis** is the division of the cell nucleus resulting in two daughter nuclei, each with the same number of chromosomes as the parent nucleus. **Cytokinesis** is the stage of cell division in which the cytoplasm divides to form two daughter cells.

❸ **Describe the events that occur in each stage of mitosis.**
During **prophase,** chromatin condenses into chromosomes, the nucleolus disappears, the nuclear envelope breaks down, and the mitotic spindle begins to form. The **spindle** is a structure consisting mainly of **microtubules** that provides the framework for chromosome movement during cell division. At the end of prophase, each duplicated chromosome is composed of two sister **chromatids.** During **metaphase,** the duplicated chromosomes line up along the midplane of the cell. During **anaphase,** the sister chromatids separate and move to opposite poles of the cell; each chromatid is now considered a separate chromosome. During **telophase,** a nuclear envelope forms around each set of chromosomes, nucleoli reappear, the chromosomes lengthen and become chromatin, and the spindle disappears. Cytokinesis generally takes place in telophase. The **cell plate** is the structure that forms during cytokinesis in plants, separating two daughter cells produced by mitosis.

❹ **Explain why meiosis is needed at some point in the life cycle of every sexually reproducing organism, and describe the events that occur during meiosis I and meiosis II.**
Meiosis is the process in which a 2*n* cell undergoes successive nuclear divisions, potentially producing four *n* nuclei; meiosis leads to formation of **spores** in plants. Meiosis must occur at some time in the life of

a sexually reproducing organism if **gametes** are to be haploid. Meiosis consists of two cell divisions, meiosis I and meiosis II. During **meiosis I,** the members of each homologous pair of chromosomes separate and are distributed into separate nuclei in two daughter cells. These chromosomes were duplicated prior to meiosis I, so each consists of two chromatids. In **meiosis II,** the chromatids separate into individual chromosomes and are distributed into different haploid daughter cells; as a result of meiosis, four haploid cells form. **Synapsis,** the physical association of homologous chromosomes, and **crossing over,** the exchange of segments of homologous chromosomes, occur during prophase I of meiosis.

❺ **Compare and contrast mitosis and meiosis.**
Mitosis involves a single nuclear division in which the two daughter cells formed are genetically identical to each other and to the original cell; synapsis of homologous chromosomes does not occur during mitosis. Meiosis involves two successive nuclear divisions and forms four haploid cells, each with a different combination of genes; synapsis of homologous chromosomes occurs during prophase I of meiosis.

❻ **Define *alternation of generations.***
Plants alternate haploid and diploid generations. The **sporophyte** is the 2*n,* spore-producing stage in the life cycle of a plant. A diploid sporophyte plant forms haploid spores by meiosis. A spore divides mitotically to form a haploid gametophyte plant. The **gametophyte** is the *n,* gamete-producing stage in the life cycle of a plant. The gametophyte produces haploid gametes by mitosis. Two gametes then fuse to form a diploid zygote, which divides mitotically to produce a diploid sporophyte.

REVIEW QUESTIONS

1. Distinguish among chromatin, chromosome, and chromatid.
2. What is the relationship between genes and chromosomes?
3. Define the following terms: *diploid, haploid,* and *homologous chromosomes.*
4. Briefly summarize what occurs during the three phases of interphase (G$_1$, S, and G$_2$).
5. Draw and label the stages in mitosis.
6. Define *cytokinesis,* and distinguish between mitosis and cytokinesis.

7. Describe the stages in meiosis. Indicate at which stages synapsis, crossing over, and separation of homologous chromosomes occur.
8. What is the genetic consequence of crossing over?
9. How does meiosis differ from mitosis?
10. What is the *immediate* product of meiosis in plants? In animals?
11. What is alternation of generations?

THOUGHT QUESTIONS

1. When during the life of a cell are chromosomes visible? Why are they visible at this time?
2. Explain why two different organisms—for example, olives and humans—might have cells with the same number of chromosomes.
3. Why does mitosis take so much longer than cytokinesis?
4. How does genetic recombination aid in the long-term survival of a species?

5. The developing egg cell of a plant with a diploid chromosome number of 20 was treated with colchicine. A normal sperm cell subsequently fertilized this egg. How many chromosomes would you expect to find in the zygote?

Visit us on the web at http://www.thomsonedu.com/biology/berg for additional resources such as flashcards, tutorial quizzes, further readings, and web links.

Patterns of Inheritance

O ur modern knowledge of genetics, the branch of biology that deals with inheritance, is based on the work of a 19th-century Augustinian monk, Gregor Johann Mendel (1822–1884). Mendel lived in the town of Brünn, Austria, which is now Brno in the Czech Republic. In Mendel's time, inheritance was thought to be the result of a blending of two parents' characters, and there was no clear-cut understanding of the special role of reproductive cells in inheritance.

Mendel changed all that with a series of elegant experiments involving the garden pea (*Pisum sativum*). Native to the Mediterranean region, the garden pea is an annual plant cultivated during cool seasons for its edible seeds. The garden pea has a climbing stem, which grows as tall as 1.8 meters (6 feet) and bears fruits that are pods, each containing 2 to 10 seeds. There are many varieties of garden peas that exhibit differences in height, flower color, seed-coat color, and seed shape. Mendel selected the garden pea for his experiments, in part because it is easy to grow and because many varieties were commercially available.

Mendel bred his garden peas for many generations, carefully counting the number of plants that had each trait in each generation. His conclusions revolutionized the field of biology and laid the foundation for understanding inheritance in all sexually reproducing organisms, from garden peas to humans.

Mendel's work indicated that inheritance of traits was not due to a blending of genetic information but to the transmission of specific units of inheritance, which are now called *genes.* Mendel predicted not only the existence of genes but also the occurrence of genes in pairs. His results suggested that during the formation of gametes (haploid reproductive cells), the two members of a gene pair separate from each other so that each gamete contains only one of each pair. During sexual reproduction, the pairs are restored when an egg combines with a sperm cell. Thus, Mendel's research also predicted the existence of meiosis and of haploid and diploid cells.

Mendel's experiments with the garden pea (*Pisum sativum*) laid the foundations of heredity.

Inheritance in the Garden Pea

Before describing Gregor Mendel's experiments and findings, it is helpful to briefly review the process and structures involved in fertilization in flowering plants, which were discussed in Chapter 9. Like other flowering plants, garden peas produce pollen grains, each of which contains haploid male gametes called sperm cells. When a pollen grain lands on the stigma of a carpel (the female flower part), the pollen grain develops a long tube that grows through the carpel, eventually reaching the haploid egg. The egg and the sperm cell fuse, and an embryonic plant develops from the zygote (fertilized egg). As development continues, a seed forms, which contains a miniature plant and the nutrients it needs to get a start in life.

Unlike most flowering plants, garden peas normally self-fertilize—that is, the male and female gametes that fuse during sexual reproduction are from the same flower. Therefore, the surgical removal of the male flower parts makes the plant incapable of being pollinated and fertilized except by artificial means (•**Figure 13-1**). In this way, the cross-fertilization (or, more simply, *crossing*) of different varieties is closely controlled.

In his experiments Mendel studied several characters of garden peas: (1) seed shape (round or wrinkled); (2) seed color (yellow or green); (3) flower color (purple or white); (4) pod shape (inflated or pinched); (5) pod color (green or yellow); (6) position of flower and fruit on the stem (axial or terminal); and (7) stem length (tall or short) (•**Figure 13-2**). (*Characters* are the attributes for which heritable differences, or *traits,* are known. For

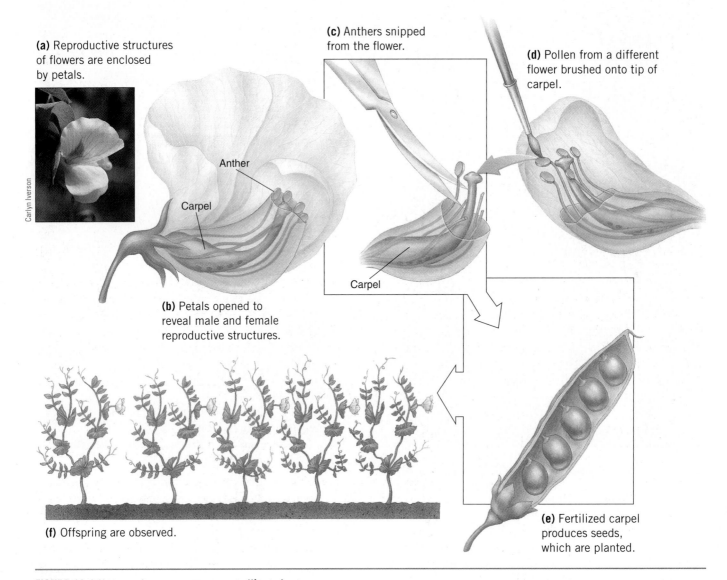

(a) Reproductive structures of flowers are enclosed by petals.

(c) Anthers snipped from the flower.

(d) Pollen from a different flower brushed onto tip of carpel.

Anther

Carpel

(b) Petals opened to reveal male and female reproductive structures.

Carpel

(e) Fertilized carpel produces seeds, which are planted.

(f) Offspring are observed.

Carlyn Iverson

FIGURE 13-1 How garden peas are cross-pollinated.
Since the petals completely enclose these parts, there is little chance of natural cross-pollination between separate flowers.

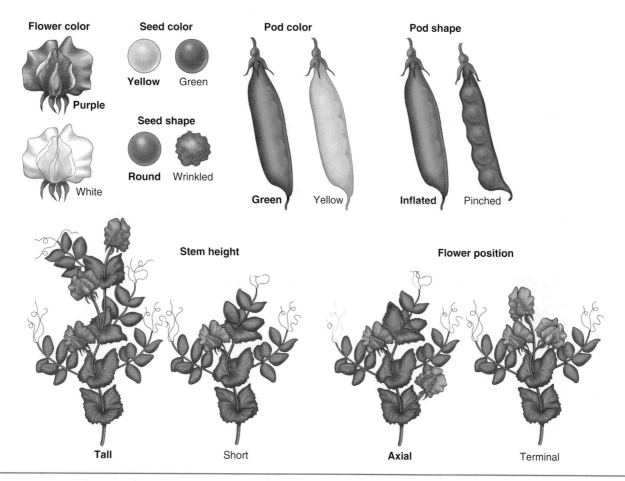

FIGURE 13-2 Seven characters in Mendel's study of pea plants.
Each character had two clearly distinguishable traits.

example, seed color is a character, and yellow seeds and green seeds are traits.)

In a typical experiment, Mendel crossed a tall variety of garden pea with a short, or bush, variety (●**Figure 13-3**). Both parent plants were genetically pure, or **true-breeding.** The offspring were not of some intermediate height but instead resembled the tall parent. When two of these offspring were crossed (or when one offspring was allowed to fertilize itself), some of the offspring of the second generation were short. The second generation occurred in a 3:1 ratio—that is, about three tall plants developed for every short one. No matter what character Mendel studied in garden peas, he obtained the same results. The first generation of offspring always resembled one of the parents, and in the second generation, plants with both traits always appeared in a 3:1 ratio.

The generation with which a particular genetic experiment is begun is called the **parental generation,** or **P.** Offspring of this generation are the **first filial generation,** or F_1. When two F_1 individuals are bred, or one self-fertilizes, the offspring constitute the **second filial generation,** or F_2.

Mendel's carefully recorded and repeated results indicated that some traits persist "silently" even though they are not expressed visibly. Mendel therefore concluded that an individual must have two sets of instructions for a particular character, such as height; that is, the instructions must occur in pairs. The 3:1 ratio in the second generation indicated to Mendel that these pairs of instructions combine and recombine in accordance with the rules of probability (probability being the likelihood that a given event will occur).

Mendel was active in the Natural History Society of Brünn and presented his findings in a series of research reports, which were published in the society's journal. His research was revolutionary, but its significance was not understood or appreciated until the early 20th century, when several biologists independently recognized most of the principles of inheritance and then rediscovered Mendel's papers describing them.

Because the behavior of chromosomes in meiosis and the role of deoxyribonucleic acid (DNA) in cells were not determined until many decades after Mendel's death, Mendel had no clear conception of the physical basis of inheritance. From his experimental results he

QUESTION: When the F₁ generation of tall plants is self-pollinated, what phenotypes appear in the F₂ generation?

HYPOTHESIS: Although only the "factor" (gene) for tall height is expressed in the F₁ generation, Mendel hypothesized that the factor for short height is not lost. He predicted that the short phenotype would reappear in the F₂ generation.

EXPERIMENT: Mendel crossed true-breeding tall pea plants with true-breeding short pea plants, which yielded only tall offspring in the F₁ generation. He then allowed these F₁ individuals to self-pollinate to yield the F₂ generation.

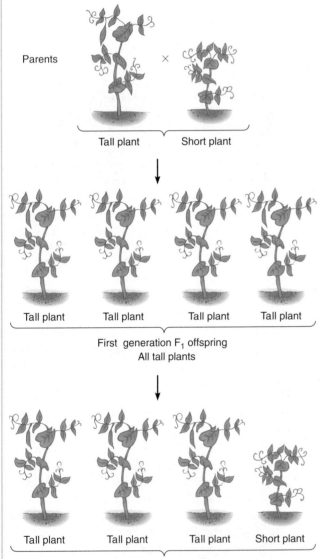

Parents

Tall plant   ×   Short plant

Tall plant   Tall plant   Tall plant   Tall plant

First generation F₁ offspring
All tall plants

Tall plant   Tall plant   Tall plant   Short plant

Second generation F₂ offspring
3 tall:1 short

RESULTS AND CONCLUSION: The F₂ generation included 787 tall and 277 short plants, which resulted in a ratio of about 3:1. Thus, Mendelian traits pass to successive generations in fixed ratios. ∎

FIGURE 13-3 One of Gregor Mendel's crosses using different varieties of garden pea.

inferred that units of inheritance existed, but he could not have said what they were or how they influenced an organism's characters. To Mendel, the unit of inheritance was a mathematical abstraction whose behavior could be described by equations. Perhaps the significance of Mendel's work was not appreciated because most biologists of his time lacked his mathematical background and were not used to thinking in quantitative terms.

Mendel's Principles of Inheritance

Mendel's breeding experiments led him to certain conclusions about the mechanisms of heredity. Later scholars restated some of those conclusions as **Mendel's principles of inheritance:**

1. Inherited characters are transmitted by genes, which occur in pairs called *alleles.*

2. Principle of dominance: When two alternative forms of the same gene (that is, two different alleles) are present in an individual, often only one—the *dominant* allele—is expressed.

3. Principle of segregation: When gametes form in meiosis, the two alleles of each gene, which are found on *homologous chromosomes,* segregate (separate) from each other, and each gamete receives only one allele for each gene (•**Figure 13-4**).

4. Principle of independent assortment: When two or more characters are examined in a single cross, each character is inherited without relation to the other characters. The reason is that the alleles for each trait assort into the gametes independently of one another. All possible combinations of genes thus occur in the gametes.

Genes occur in pairs and are inherited as parts of chromosomes

A typical gene is now viewed as a region of DNA that contains the information necessary to manufacture a specific polypeptide or protein. (Some DNA, particularly that involved in the control of genes, may not code for protein directly; this DNA usually affects the function of protein-producing genes.)

Let us illustrate the concept of gene expression with one of the characters Mendel studied in the garden pea, seed shape. Pea seeds are either round or wrinkled, and this character is related to the type of carbohydrate stored in the seed. Round seeds contain starch, and wrinkled seeds contain sugar. A high sugar concentration is thought to cause seeds to accumulate water during development. Then, as they mature and dry out, these seeds become wrinkled.

Mendel's principle of segregation is related to the events of meiosis: The separation of homologous chromosomes during meiosis results in the segregation of alleles.

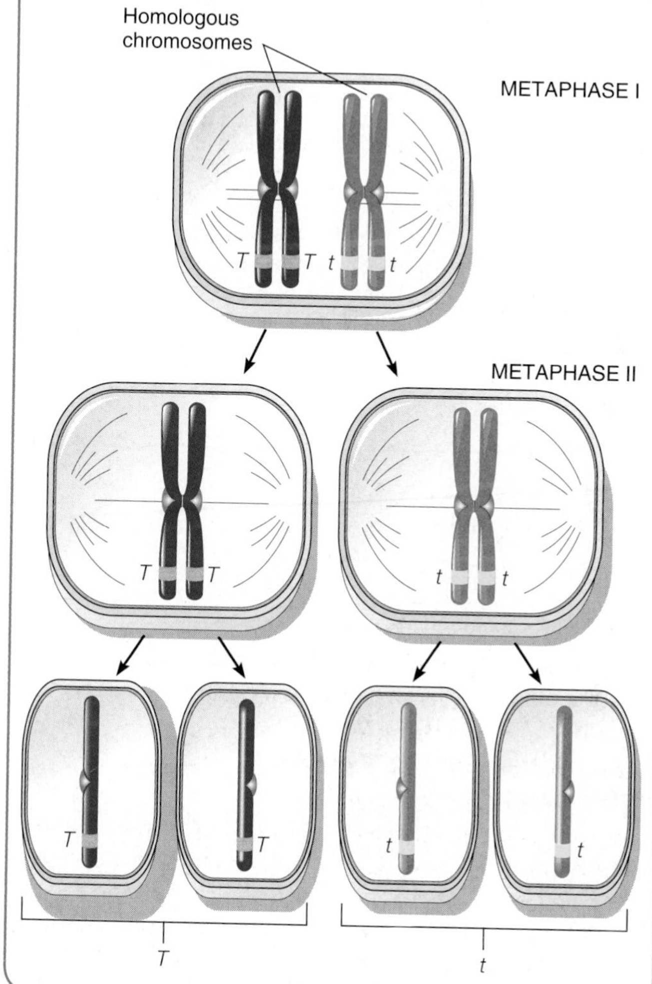

FIGURE 13-4 The chromosomal basis for segregation.
The separation of homologous chromosomes during meiosis results in the segregation of alleles. Note that half of the gametes will carry *T* and half will carry *t*.

The allele for round seed is the normal form of the gene. It codes for an enzyme that enables the seed to make starch from sugar. The allele for wrinkled seed codes for a defective enzyme, one that cannot make starch from sugar. Thus, the seed contains sugar and accumulates water, leading to its wrinkled state.

If an individual plant contained a defective wrinkled allele and a normal round allele, the seed would be round due to the presence of starch produced by the functional enzyme. Although the **genotype** (genetic makeup) of such a genetically mixed individual contains a wrinkled allele, one would not know this from the seed's appearance. The individual's **phenotype**, or observable features (the portion of the genotype that is actually expressed), is round.

A dominant allele masks the expression of a recessive allele

The two members of a pair of chromosomes are homologous and contain genes for similar characters arranged in similar order. The gene for each character occurs at a particular site in the chromosome called a **locus** (pl., *loci*). In garden peas, for example, if one homologous chromosome contains a gene for pod color at a particular locus, so will the other chromosome of that pair.

Now that we understand that alleles of a gene occur at the same site on homologous chromosomes, we can define *allele* more precisely. The alternative forms of a gene that govern the same character and that occupy corresponding loci on homologous chromosomes are **alleles** (●Figure 13-5).

The definition of *allele* implies that there are at least two alternative forms of the gene that can occupy a specific locus on homologous chromosomes. For book-keeping purposes, each of these forms is assigned a letter as its symbol.[1] It is customary to designate the **dominant** allele—the one that always manifests itself—with a capital letter, and the **recessive** allele—the allele that does not express itself in the presence of a dominant allele—with the same letter but lowercase. Thus, to specify the gene that determines height in garden peas, the letter *T* could be used to represent the allele for tall plant and the letter *t* the allele for short; the tall allele is dominant, whereas the short allele is recessive.

A monohybrid cross involves a single pair of alleles at a given locus

Let us illustrate genetic terms and some of the basic rules of genetics by performing a simple **monohybrid cross,** which is a cross between two individuals in which only one character at a single locus is studied. ●Figure 13-6a

---

[1]Early geneticists developed their own symbols to represent genes and alleles. Later, groups of scientists met and decided on specific symbols for a given research organism, but each research group had its own rules for assigning symbols. Universally accepted rules for assigning symbols for genes and alleles still do not exist.

GENOTYPE The genetic makeup of an individual.

PHENOTYPE The physical expression of an individual's genes.

LOCUS The location of a particular gene on a chromosome.

ALLELE One of two or more alternative forms of a gene.

DOMINANT Said of an allele that is always expressed when it is present.

RECESSIVE Said of an allele that is not expressed in the presence of a dominant allele.

Gene loci

(a) These chromosomes are nonhomologous. Each chromosome is made up of thousands of genes. A locus is a specific place on a chromosome where a gene is located.

A pair of alleles

These genes are not alleles

(b) These chromosomes are homologous. Alleles are members of a gene pair that occupy corresponding loci on homologous chromosomes.

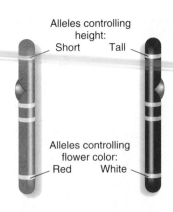

Alleles controlling height:
Short   Tall

Alleles controlling flower color:
Red   White

(c) Alleles govern the same character but do not necessarily contain the same information.

FIGURE 13-5 Homologous chromosomes, genes, and alleles.
Chromosomes occur in pairs in diploid cells. The members of a given pair correspond in shape, size, and type of genetic information and are referred to as homologous chromosomes. For purposes of illustration, each chromosome is shown in the unduplicated state.

shows the cross between a genetically pure tall garden pea (designated *TT*) and a genetically pure short garden pea (designated *tt*). The two *tt* alleles separate during meiosis, so each gamete produced by the short plant has only one *t* allele. In the formation of gametes in the tall plant, the *TT* alleles separate, so each gamete has only one *T* allele. The union of the *T*-bearing gamete with a *t*-bearing gamete during sexual reproduction results in offspring with the genotype *Tt*, that is, with one allele for tall and one allele for short. As you would expect, these offspring are all tall.

Suppose that two tall plants, each having alleles for both tall and short (*Tt*), are crossed (•**Figure 13-6b**). After meiosis occurs, half of the gametes produced by each plant have a single allele for tall (*T*), and the other half have a single allele for short (*t*).

The probable combinations of gametes are represented in a diagram called a **Punnett square** (see Figure 13-6). The Punnett square shows all possible combinations of gametes to form offspring. In a Punnett square, the types of gametes from one individual are written across the top of the grid, and the types of gametes from the other individual are written along the left side. The squares, which are filled in with the resulting

combinations of gametes, indicate the genotypes of all possible offspring from this cross.

A genotype is either homozygous or heterozygous

If both alleles specify short (*tt*), then the individual is **homozygous**—that is, it has identical alleles for the same gene. In another garden pea, both alleles may specify tall (*TT*). This second plant is also homozygous. It often happens, however, that one allele carries instructions for tall while the other carries instructions for short (*Tt*). In this case, the individual is **heterozygous** for height—it contains two different alleles for the same gene. In the heterozygous condition, only the dominant allele is expressed; thus, a garden pea with a genotype designated *Tt* is tall.

To summarize:

1. When an organism is homozygous for the dominant allele, its phenotype reflects the dominant allele. (A *TT* plant is tall.)

2. When an organism is homozygous for the recessive allele, its phenotype reflects the recessive allele. (A *tt* plant is short.)

3. When an organism is heterozygous, the dominant allele is expressed in the phenotype just as it would be if the organism were homozygous for the dominant allele. (A *Tt* plant is tall.) In such cases, one cannot readily distinguish a heterozygous individual (*Tt*) from one that is homozygous for the dominant allele (*TT*).

HOMOZYGOUS Possessing a pair of identical alleles for a particular gene.

HETEROZYGOUS Possessing a pair of unlike alleles for a particular gene.

Mendel inferred the existence of genes by observing the offspring of crosses between individuals with different phenotypes.

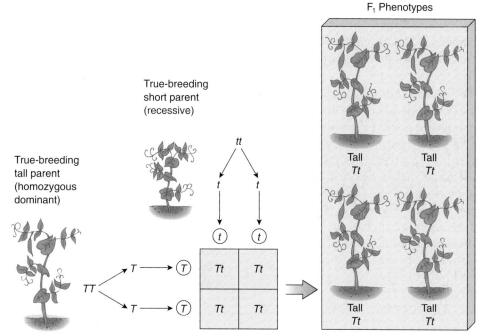

(a) When a genetically pure tall garden pea is crossed with a genetically pure short garden pea, all the offspring are tall.

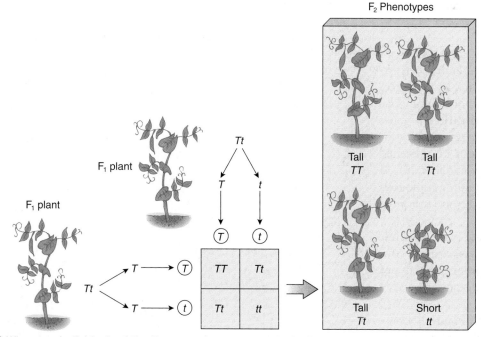

(b) When two individuals of the F$_1$ generation are crossed, the F$_2$ generation is produced. The phenotypic ratio of the F$_2$ is 3:1.

FIGURE 13-6 A monohybrid cross.
A Punnett square is used to determine all possible genotypes in the offspring.

All this information permits us to define *dominant* and *recessive* with greater precision. When one allele tends to dominate the other completely, so that it alone is expressed in the heterozygous condition, it is *dominant*. If an allele is expressed *only* when homozygous, it is *recessive*.

A test cross determines if an individual with a dominant phenotype is homozygous or heterozygous

There remains the practical question of how we determine the genotype of an individual that displays a dominant phenotype, because it could be homozygous for the dominant allele (*TT*) or heterozygous (*Tt*). One way to discover the answer is with a **test cross,** an experimental cross with an individual that is recessive (*tt*). Suppose, for example, that a tall garden pea is heterozygous (*Tt*). If one were to cross it with a short plant (which must be homozygous for the recessive short trait, or *tt*), at least some of the offspring should display the recessive trait and be short (•Figure 13-7). On the other hand, if the tall plant is homozygous for tall (*TT*), all offspring will be tall.

Genes located on different chromosomes are inherited independently

We now know that Mendel's principle of independent assortment applies only to characters carried on nonhomologous chromosomes. If different genes are carried on the same chromosome, they tend to be inherited together (discussed shortly). If genes occur on nonhomologous chromosomes, they are inherited independently.

To illustrate the principle of independent assortment, consider two different garden pea characters, seed shape and seed color. A single gene controls each character. A dominant allele determines round seed, whereas the wrinkled allele is recessive. The allele that results in yellow seed is dominant, and the allele for green seed is recessive.

A cross that involves individuals differing in *two* characters at two loci is called a **dihy-**

brid cross. If a plant homozygous for round seed (*RR*) is crossed with a plant homozygous for wrinkled seed (*rr*), all their offspring (*Rr*) are expected to have round seeds, since round is dominant and the offspring are heterozygous (*Rr*). Similarly (but quite unrelated), if one of the plants is homozygous for yellow seed (*YY*) and the other is homozygous for green seed (*yy*), the offspring will have yellow seeds (*Yy*). The offspring's genotype

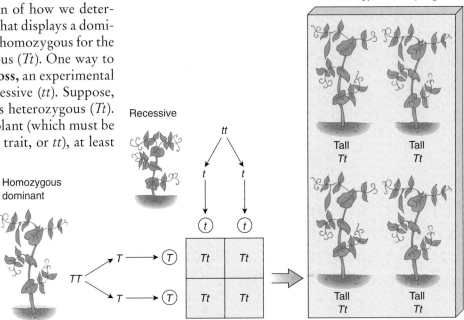

(a) If a homozygous tall garden pea is crossed with a short one, all the offspring are tall.

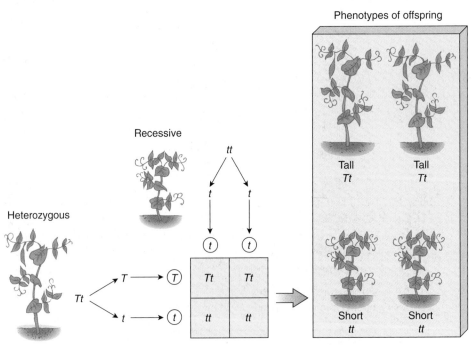

(b) If any of the offspring are short, the tall garden pea must be heterozygous.

FIGURE 13-7 A test cross.
This particular test cross would help determine the genotype of a tall garden pea.

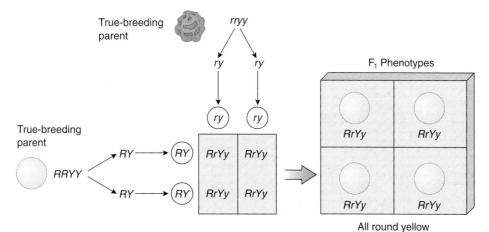

(a) When Mendel crossed a true-breeding plant having round and yellow seeds with a true-breeding plant having wrinkled and green seeds, the seeds produced by the F_1 plants were all round and yellow.

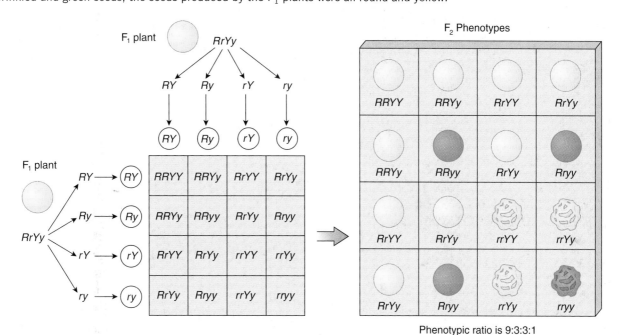

(b) When two heterozygous individuals are crossed, the ratio of phenotypes is 9:3:3:1 (9 round yellow:3 round green: 3 wrinkled yellow:1 wrinkled green).

FIGURE 13-8 A dihybrid cross.
In garden peas, the gene for round seed (*R*) is dominant over its allele for wrinkled (*r*), and the gene for yellow seed (*Y*) is dominant over its allele for green (*y*).

for both sets of characters is designated *RrYy* because the offspring are heterozygous for both traits (•**Figure 13-8a**). Since the two characters are the results of genes found on separate chromosome pairs, they are inherited (or assorted) independently of each other.

Let us suppose that some of these offspring self-fertilize (*RrYy* × *RrYy*). As you see in •**Figure 13-8b**, four kinds of gametes are possible because each gamete contains one allele for each gene. If, by chance, the chromosome bearing *R* is sorted during meiosis into the same gamete as the chromosome bearing *y*, the resulting gamete will have the genotype *Ry*. Similarly, if the chromosomes bearing *r* and *Y* are assorted together, the

gamete will have the genotype *rY*. Other gametes will contain *RY*, and still others *ry*. *There are no possible combinations other than these four.* Since alleles are always borne on homologous chromosomes that are *separated* from one another in meiosis, a gamete ordinarily has no more than one copy of each allele (for example, *R* or *r* but not both; *Y* or *y* but not both).

Because two parents produce four possible kinds of gametes, working out a cross of that kind requires a Punnett square of 16 boxes (16 different combinations of offspring are possible when these gametes come together randomly). Count the individuals exhibiting each of the four possible phenotypes (round yellow seeds,

FOCUS ON | Solving Genetics Problems

Some students intuitively understand the logic behind solving genetics problems, but many others have difficulty. The following discussion teaches a methodical approach to solving genetics problems.

Step 1. Read the problem. If it gives you information about the parents and asks you to determine the offspring, go to Step 2. If it gives you information about the offspring and asks you to work backward to determine the parents, go to Step 3.

Example 1. In garden peas, the allele for inflated pod is dominant over the allele for pinched pod, and green pod color is dominant over yellow pod. The two characters are inherited independently. Determine the genotypes and phenotypes of the offspring produced from a cross between an individual that is heterozygous for both characters and one that is recessive for both characters.

Example 2. In squash, the allele for disc-shaped fruit is dominant over

the allele for round fruit. Two squash plants are crossed, and of their 110 F_1 offspring, 54 have disc-shaped fruits and 56 have round fruits. What are the genotypes and phenotypes of each parent?

In Example 1, you are given information about the parents and asked to determine the offspring, so go to Step 2.

In Example 2, you are asked to work backward, so go to Step 3.

Step 2. To determine offspring when given information about the parents:

1. Make a key, using capital and lowercase letters to represent the various alleles.

   I = allele for inflated pod

   i = allele for pinched pod

   G = allele for green pod

   g = allele for yellow pod

2. Determine the genotypes of the parents. One parent is heterozygous for both characters—*IiGg;* the other parent is recessive for both characters—*iigg.* (Remember that each individual has two alleles for each character.) Therefore, the parents are *IiGg* × *iigg.*

3. Determine the possible gametes produced by each parent. Individual *IiGg* produces four different types of gametes: *IG, Ig, iG, ig.* Individual *iigg* produces only one type of gamete: *ig.* (Remember that gametes contain only one allele for each gene.)

4. Set up a Punnett square. Put all possible gametes for one parent across the top of the square and all possible gametes for the other parent down the side of the square (see below).

	IG	*Ig*	*iG*	*ig*
ig				

round green seeds, wrinkled yellow seeds, and wrinkled green seeds). The phenotypic ratio among the F_2 offspring is 9:3:3:1; that is, nine offspring produce round yellow seeds, three produce round green seeds, three produce wrinkled yellow seeds, and one produces wrinkled green seeds.

When comparing a monohybrid cross with a dihybrid cross, it becomes clear that as the number of characters being considered increases, the number of possible genotypes in the offspring multiplies rapidly. Imagine, then, how many different genotypes are possible among the offspring of two parents that differ in hundreds of characters! (*Focus On: Solving Genetics Problems* provides step-by-step directions for working genetics problems.)

PROCESS OF SCIENCE **Recognition of Mendel's work came during the early 20th century**

Mendel reported his findings at a meeting of the Natural History Society of Brünn; he published his results in the society's report in 1866. At that time biology was

largely a descriptive science, and biologists had little interest in applying quantitative and experimental methods such as Mendel had used. Other biologists of the time did not appreciate the importance of his results and his interpretations of those results. For 34 years his findings were largely neglected.

In 1900, Hugo DeVries in Holland, Carl Correns in Germany, and Erich von Tschermak in Austria each recognized Mendel's principles in their own experiments; they later discovered Mendel's paper and found that it explained their own individual research observations. By this time biologists had a much greater appreciation of the value of quantitative experimental methods. Correns gave credit to Mendel by naming the basic laws of inheritance after him.

Although gametes and fertilization were known at the time Mendel carried out his research, mitosis and meiosis had not yet been discovered. It is truly remarkable that Mendel formulated his ideas mainly on the basis of mathematical abstractions. Today, his principles are much easier to understand because we relate the transmission of genes to the behavior of chromosomes.

	IG	Ig	iG	ig
ig	IiGg Inflated green pods	Iigg Inflated yellow pods	iiGg Constricted green pods	iigg Constricted yellow pods

5. Perform the cross by filling in the boxes to represent the offspring. The four boxes (on right) show the genotypes of the offspring. (To keep things straight, place the *I*'s together and the *G*'s together, and put capital letters before lowercase letters.)

6. Finish the problem by listing the phenotypes for the offspring.

Step 3. To determine parents when given information about the offspring:

1. Make a key, using capital and lower-case letters to represent the various alleles.

   D = allele for disc-shaped fruit
   d = allele for round fruit

2. List what is known about the geno-types of the parents. Put a blank for each unknown allele. (Remember that each individual has two alleles for each character.)

   _ _ × _ _

3. Work backward by filling in the blanks. Because approximately half of the offspring exhibit the dominant trait, you know that at least one of the parents has at least one allele *D*.

   D _ × _ _

4. Continue this type of reasoning. Because half of the offspring exhibit the recessive trait, *each* parent must have at least one allele *d*. Fill in the blanks.

   Dd × _ *d*

5. Calculate a phenotypic ratio for the offspring if it is not given in the original problem. Do this by dividing each number of offspring by the largest number given.

   $$\frac{56}{56} = 1$$

   $$\frac{54}{56} = 0.96$$

   When rounded, the ratio is 1:1; that is, half of the offspring have disc-shaped fruits and half are round (1 round to 1 disc-shaped).

6. Determine which cross would give a 1:1 ratio. Because you already know that the parental genotypes are *Dd* × _ *d*, there are only two possibilities: *Dd* × *Dd* and *Dd* × *dd*. You may be able to determine the answer by thinking about the two possibilities. If not, then perform both crosses, and calculate the phenotypic ratio expected in each case. The phenotypic ratio for the offspring of *Dd* × *Dd* is 3:1; the phenotypic ratio for the offspring of *Dd* × *dd* is 1:1. Therefore, the geno-types of the parents are *Dd* × *dd*. Their phenotypes are disc-shaped fruit (for the *Dd* parent) and round fruit (for the *dd* parent).

Additional genetics problems (and their answers) are given at the end of the chapter.

The details of mitosis and meiosis were described during the late 19th century, and in 1902, American biologist Walter Sutton and German biologist Theodor Boveri independently pointed out the connection between Mendel's segregation of alleles and the separation of homologous chromosomes during meiosis. This connection developed into the **chromosome theory of inheritance,** which stated that inheritance can be explained by assuming that genes are linearly arranged in specific locations along the chromosomes.

The chromosome theory of inheritance was initially controversial, because at that time there was no direct evidence that genes are found on chromosomes. However, new research provided the findings necessary for wider acceptance and extension of these ideas and their implications. For example, the work of American geneticist Thomas Hunt Morgan in 1910 provided evidence for the location of a particular gene (white eye color) on a specific chromosome (the X chromosome) in fruit flies. Morgan and his graduate students also provided insight into the way genes are organized on chromosomes. ▪

Genetic Linkage

We have seen that genes that occur on nonhomologous chromosomes are assorted (and inherited) independently. The inheritance of multiple genes borne on a single pair of homologous chromosomes, however, is like that of a *single* gene, if the genes are located close together. The tendency for a group of genes on the same chromosome to be inherited together is known as **linkage.** Linkage occurs because genes that happen to occur on the same chromosome tend to remain with one another during meiosis.

Sometimes linked genes are not inherited together, however. By watching the behavior of chromosomes in meiosis, researchers determined that the explanation for the failure of linked genes to always stay together lay in the **crossing over** of segments of homologous chromosomes. In meiosis, homologous chromosomes, each consisting of two chromatids, come together during pro-

LINKAGE The grouping of genes on the same chromosome.

phase I and exchange pieces of themselves during crossing over (•**Figure 13-9**). Crossing over results in new combinations of genes in gametes. **EVOLUTION LINK** Greater variation in gametes gives rise to greater variation in traits among the offspring of this parent. The evolutionary significance of greater variation is that some of these combinations of genes may improve the chances of survival in the individuals that have them. (See *Plants and People: Protecting Genetic Diversity in Crop Plants*.) ∎

Genetic maps give the order and relative distances between all known genes on the chromosome

By observing a large number of crosses involving linked genes, one can determine how frequently linked genes stay together in their original combination and how frequently they cross over. This information is used to estimate the relative distances between linked genes: the farther from one another the loci of linked genes are on the chromosomes, the more frequently they tend to cross over. Thus, maps of chromosomes can be constructed that show the physical locations of their known genes.

Why is genetic mapping important? In humans, genetic mapping is helping to locate the genes responsible for genetic disorders. This knowledge will assist physicians in their diagnosis and treatment. Genetic mapping of plants is important because precise genetic maps facilitate the development of new varieties of agricultural plants by genetic engineering.[2]

Extensions of Mendelian Genetics

Some characters are not inherited by simple Mendelian rules. Geneticists know of many examples in which an organism's appearance (its phenotype) is the result of interactions among genes. These interactions include incomplete dominance and polygenes. In addition, interactions between genes and the environment affect phenotype.

Dominance can be incomplete

Not every gene consists of a dominant allele and a recessive allele. For example, when red-flowered (R^1R^1) and white-flowered (R^2R^2) four-o'clocks are crossed, the offspring do not have red or white flowers (•**Figure 13-10**). Instead, all the F_1 offspring have pink flowers. Does this

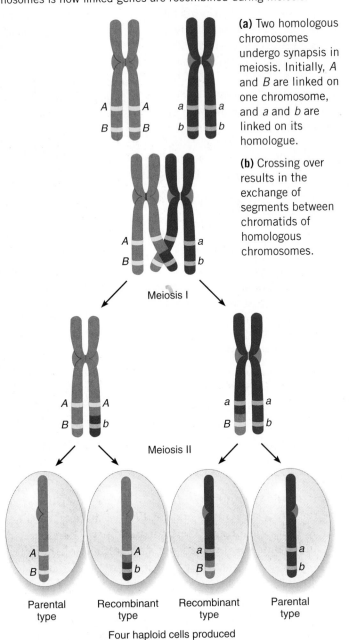

FIGURE IN FOCUS

The exchange of segments between chromatids of homologous chromosomes is how linked genes are recombined during meiosis.

(a) Two homologous chromosomes undergo synapsis in meiosis. Initially, *A* and *B* are linked on one chromosome, and *a* and *b* are linked on its homologue.

(b) Crossing over results in the exchange of segments between chromatids of homologous chromosomes.

Meiosis I

Meiosis II

Parental type   Recombinant type   Recombinant type   Parental type

Four haploid cells produced

(c) Crossing over permits the formation of new combinations of genes in the gametes. For example, note that genes *A* and *b* are now linked in one of the gametes, as are genes *a* and *B*. These linkages are not possible without crossing over.

FIGURE 13-9 Linkage and crossing over.

[2]*The* Arabidopsis *plant **genome** (the complete genetic material carried in an* Arabidopsis *cell) has been analyzed to determine the chromosome locations and structures of the thousands of genes of this important research plant. Some important agricultural plants, such as rice, have also been genetically analyzed.*

PLANTS AND PEOPLE | Protecting Genetic Diversity in Crop Plants

EVOLUTION LINK A plant that is cultivated is protected from natural competition with other plants and from plant-eating animals. The farmer selects the traits that are desirable in a cultivated plant, such as high yield, good flavor, or appealing color. He encourages the transmission of those traits to future generations by saving, and later planting, the seeds from those plants with desirable traits. Over time, such selection brings about genetic changes in the plants, and they may become so altered from their original ancestors that it is doubtful that they could survive and compete successfully in the wild. Such plants are **domesticated**.

Plants that exist in the wild often exhibit **genetic diversity**, which means that great genetic variety exists among individuals of a single species. Genetic diversity contributes to a species' long-term survival by providing the variation that enables a population to adapt to changing environmental conditions. When plants become domesticated, much of this genetic diversity is usually lost, because the farmer selects for propagation only those plants with certain marketable traits. At the same time, the farmer selects against other traits that may not be of obvious value in the domesticated plant. As a result, an individual crop variety generally contains a small fraction of the genetic variation inherent in the species.

Despite the fact that domestication reduces genetic diversity, the many farmers who have selected for specific traits worldwide have produced an enormous number of varieties of each domesticated plant. Each variety—which is the legacy of hundreds of farmers who have developed it over thousands of years—is adapted to the

■ Genetic diversity in corn. Some of these are heirloom corn varieties that are no longer widely planted. Each variety contains genetic variation that results in different colors, flavors, and textures.

David Cavagnaro/Peter Arnold, Inc.

specific climate where it was bred and contains a unique combination of traits conferred by its unique combination of genes (see figure).

The trend worldwide is to replace the many unique local varieties of a given crop with a few improved varieties. When local farmers stop planting their traditional varieties in favor of the more modern ones, the traditional varieties frequently become extinct. This represents a loss in genetic diversity, because each variety's unique combination of genes gives it a distinctive nutritional value, size, color, flavor, resistance to disease, and adaptability to a specific climate and soil type. The gene combinations in traditional varieties are potentially valuable to plant breeders, who use traditional breeding methods or genetic engineering to transfer these combinations to other varieties.

The conservation of plant genetic resources—both wild relatives and locally adapted varieties of important crop plants—is important because the world's food supply depends on it. Plant breeding today is an ongoing process, and breeders are continually developing new varieties of crops in response to changing conditions, such as new insect pests, rapidly evolving disease organisms, changing climate, and new agricultural techniques. Fortunately, agricultural research centers and seed banks are conserving many crop varieties (see *Plants and People: Seed Banks,* in Chapter 9). An international effort is also under way to set aside land reserves to preserve the wild relatives of some important crop species where they exist in the natural environment. ■

result negate Mendel's assumptions that inheritance is not a blending phenomenon and that some characters persist silently? Quite the contrary, for when two of these pink-flowered plants are crossed, the F_2 offspring have a ratio of one red-flowered to two pink-flowered to one white-flowered plant. If inheritance were a blending phenomenon, all offspring of two pink-flowered plants would be pink.

PROCESS OF SCIENCE In this instance, as in other aspects of science, results that differ from those predicted simply prompt scientists to reexamine and modify their assumptions to account for the new experimental results. The pink-flowered plants are clearly the heterozygous individuals in which neither the red allele nor the white allele is completely dominant. When the heterozygote has a phenotype that is intermediate between those of its two parents, the responsible alleles are said to show **incomplete dominance.** ■

Incomplete dominance is not unique to four-o'clocks. Red- and white-flowered snapdragons also produce pink-flowered plants when crossed. The reason is that the single "red" allele in these plants cannot code for the production of enough red pigment to make the petals look red.

Some characters are governed by more than one gene

Some characters are governed in their expression by many related genes, known as **polygenes.** Because each individual gene in polygenic inheritance makes a small contribution to the organism's phenotype, a character affected by polygenes exhibits a wide range of variability. The color of wheat kernels is an example of polygenic inheritance, as are height, fruit size, and several other characters in certain plants.

If a wheat plant with dark red kernels is crossed with a plant that produces white kernels, their offspring are light red—that is, intermediate in color. If that were all there was to it, this would be an obvious case of incomplete dominance. To test

In incomplete dominance, an F_1 heterozygote has a phenotype intermediate between those of its parents.

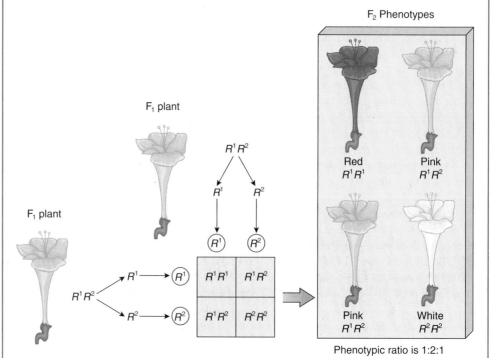

(a) Flower color in four-o'clocks is controlled by a single gene with two alleles, one for red (R^1) and one for white (R^2) flower color. Red is incompletely dominant to white. A plant with the genotype R^1R^2 has pink flowers. Note that there are alleles only for red and white flowers, not for pink.

(b) When two pink-flowered plants are crossed, approximately a fourth of the offspring have red flowers, half have pink flowers, and a fourth have white flowers.

FIGURE 13-10 Incomplete dominance.
The four-o'clock (*Mirabilis jalapa*) gets its common name from the fact that the flowers open in late afternoon.

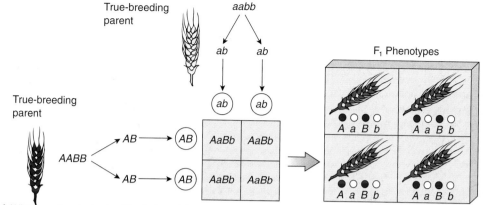

(a) When a wheat plant with dark red kernels is crossed with a plant with white kernels, the offspring are all intermediate in color.

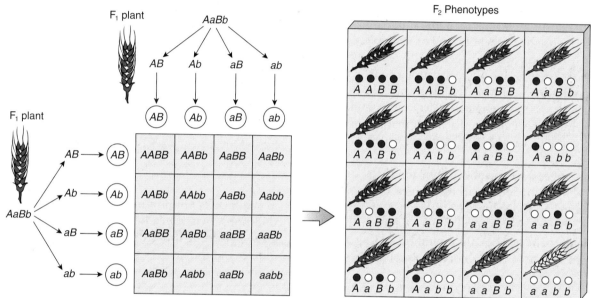

(b) The expected results in the F₂ generation. The dark dots (signifying the number of dark genes) are counted to determine the kernel colors in the F₂, as shown in Figure 13-12.

FIGURE 13-11 Polygenic inheritance.
In wheat, several genes affect a single character, kernel color. *A* and *B* represent the genes for dark red kernels, and *a* and *b* represent their respective light alleles.

whether incomplete dominance causes kernel color, you must examine F₂ plants that are the offspring of two plants with intermediate kernel color. If kernel color were the result of simple incomplete dominance, the F₂ plants would exhibit a 1:2:1 ratio among their offspring. In actuality, however, the offspring exhibit a complex distribution of kernel colors. Although the majority of offspring are intermediate in color, other offspring range from white to dark red, and the distribution of phenotypes is not easily expressed as a ratio. The explanation for this outcome is that at least three genes, each located on a different chromosome, govern kernel color in wheat.

To simplify this example, let us say that wheat-kernel color is governed by only two genes. Thus, a crossing between a plant with dark red kernels and one with white kernels may be represented as *AABB* × *aabb* (•**Figure 13-11**). The F₁ offspring are all *AaBb*. What

INCOMPLETE DOMINANCE A condition in which neither member of a pair of contrasting alleles is completely expressed when the other is present.

POLYGENES Two or more pairs of genes that affect the same character in additive fashion.

about the F_2? The genotypes of the F_2 are shown in the illustration. It is easy to determine their phenotypes: an individual with the genotype *AABB* has dark red kernels, whereas an individual with the genotype *aabb* has white kernels; intermediate kernel colors are predicted by counting the number of capital letters (represented visually as darkened circles in Figure 13-11) in each genotype (•**Figure 13-12**). For example, an *AABb* individual is darker than an *aaBB* individual.

Genes interact with the environment to shape phenotype

We have seen some of the ways in which genes interact to determine an organism's phenotype. In addition, the environment in which an individual lives exerts a strong influence over phenotype. For example, consider ear length in corn, which is genetically determined by many genes (that is, by polygenic inheritance). A corn plant may have a genotype for long ears, but if it does not have access to adequate light, water, and essential minerals (inorganic nutrients) during growth, it will not produce ears as long as those called for by its genotype.

Hydrangeas are shrubs grown for their attractive flowers. The color of certain hydrangea flowers ranges from blue to pink depending on the level of aluminum in the soil before the flowers begin to develop. It is easy to manipulate the level of aluminum by adding alum or aluminum sulfate to the soil or by lowering the soil pH

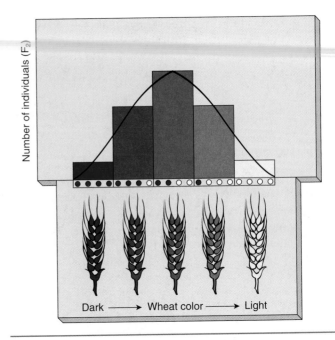

FIGURE 13-12 Continuous variation from polygenic inheritance. The data from Figure 13-11b produce a bell-shaped curve, which demonstrates continuous variation in wheat-kernel color, from dark to white, in the F_2. The bell-shaped curve is typical of the distribution of phenotypes in a character governed by polygenes.

(in acidic soils, aluminum is more soluble); under these conditions, hydrangea flowers are blue. In alkaline soils (made by adding limestone to the soil), aluminum is less soluble, and the flowers are pink (•**Figure 13-13a**).

(a) Hydrangea (*Hydrangea macrophylla*) inflorescences vary in color depending on availability of aluminum, which in turn is dependent on soil pH. Blue flowers develop in a soil pH of 5.5 or less, purple flowers in a soil pH of 5.6 to 6.4, and pink flowers in a soil pH of 6.5 to 7.0.

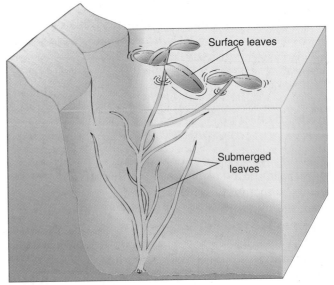

(b) Water starwort (*Callitriche heterophylla*) leaves vary depending on the environment to which they are exposed.

FIGURE 13-13 Interaction between genes and the environment.

Certain aquatic plants provide another example of how the external environment affects gene expression. The water starwort produces two kinds of leaves depending on whether they are underwater or aerial (•Figure 13-13b). The submerged leaves are long and narrow, whereas the leaves in air are broad. Thus, environmental conditions affect the expression of the genes that determine leaf shape.

STUDY OUTLINE

❶ **Define the following terms relating to genetic inheritance: *dominant* and *recessive*; *homozygous* and *heterozygous*; *genotype* and *phenotype*.**
A **dominant** allele is always expressed when it is present; a **recessive** allele is not expressed in the presence of a dominant allele. A **homozygous** individual possesses a pair of identical alleles for a particular gene; a **heterozygous** individual possesses a pair of unlike alleles for a particular gene. An individual's **genotype** is its genetic makeup; an individual's **phenotype** is the physical expression of its genes.

❷ **Distinguish among chromosomes, genes, and alleles.**
Genes occur in pairs and are inherited as parts of chromosomes. A **locus** is the location of a particular gene on a chromosome. An **allele** is one of two or more alternative forms of a gene. Alleles are carried on homologous chromosomes. Alleles may be alike (homozygous) or unlike (heterozygous).

❸ **Solve simple genetics problems involving monohybrid and dihybrid crosses.**
A **monohybrid cross** is a cross between two individuals in which only one character is being studied. A cross between two individuals, one homozygous dominant and the other recessive, yields F_1 offspring that are all heterozygous. When two heterozygous individuals are crossed, a 3:1 phenotypic ratio occurs in the F_2 offspring. A **dihybrid cross** is a cross between two individuals in which two characters are being studied. A cross between two individuals, one homozygous dominant for both characters and the other recessive for both characters, yields F_1 offspring that are heterozygous for both characters. When two heterozygous individuals are crossed, a 9:3:3:1 phenotypic ratio occurs in the F_2 offspring.

❹ **Explain how a test cross is used to determine the genotype of an individual exhibiting a dominant phenotype.**
A **test cross** is an experimental cross that determines if an individual with a dominant phenotype is homozygous or heterozygous. In a test cross the dominant individual is crossed with an individual that is recessive (for example, *tt*). If the dominant individual is heterozygous, some of the offspring should have the recessive phenotype.

❺ **Define *linkage*, and explain why linked genes are sometimes not inherited together.**
Linkage is the grouping of genes on the same chromosome. Two linked genes tend to be inherited together because genes that happen to occur on the same chromosome tend to remain with each other during meiosis. Sometimes linked genes are not inherited together; the failure of linked genes to stay together is the result of **crossing over** of segments of homologous chromosomes during prophase I of meiosis.

❻ **Contrast incomplete dominance and polygenic inheritance.**
Some characters are not inherited by simple Mendelian rules. **Incomplete dominance** is a condition in which neither member of a pair of contrasting alleles is completely expressed when the other is present; in incomplete dominance a heterozygote exhibits a phenotype intermediate between those of its two parents. **Polygenes** are two or more pairs of genes that affect the same character in additive fashion. In polygenic inheritance each individual gene makes a small contribution to the organism's phenotype.

REVIEW QUESTIONS

1. Distinguish between genes and alleles.
2. State Mendel's principle of dominance.
3. Why can an individual receive only one member of a homologous pair of chromosomes from each parent?
4. State Mendel's principle of segregation.
5. Distinguish between homozygous and heterozygous individuals.
6. Distinguish between dominance and recessiveness.
7. Distinguish between the genotype and phenotype of an individual.
8. State Mendel's principle of independent assortment.

9. Would you need to perform a test cross to determine the genotype of four-o'clocks that have pink flowers? Why or why not?

10. What are linked genes?
11. Define *incomplete dominance* and *polygenic inheritance.*

THOUGHT QUESTIONS

1. Mendel originated the principle of independent assortment on the basis of several crosses involving genes that today are known to be linked. Why did the linked genes that he studied appear to be inherited independently?

2. If you want to determine whether or not an organism exhibiting a dominant trait is homozygous, you should cross the unknown organism with an individual homozygous for the recessive allele. Why?

3. A new species of plant is discovered growing in the mountains of Tibet. Some individuals have blue flowers, and others have red. When a biologist crosses blue-flowered and red-flowered plants, the offspring all have purple flowers. What are the two most likely explanations? How would you determine which explanation is correct?

GENETICS PROBLEMS

1–3. Most of the individuals of a certain wildflower population have white flowers, although a few have purple flowers. Crosses have demonstrated that the allele for white flower color (W) is dominant over the allele for purple (w).

1. Give the genotype of a purple-flowered plant, and show the gametes that it would produce as a result of meiosis.
2. If two heterozygous white-flowered plants are crossed, what fraction of the offspring do you expect to be purple?
3. If two purple-flowered plants are crossed, what fraction of the offspring do you expect to be white?

4. In one study, Gregor Mendel crossed a yellow-seeded, tall garden pea with a green-seeded, short garden pea. The F_1 offspring were all yellow-seeded and tall. Assuming independent assortment of these two genes, what phenotypes and proportions did he find among the F_2 offspring when the F_1 garden peas were allowed to fertilize themselves?

5–6. Thorn apples produce either purple (P) or white (p) flowers, and their fruits are either spiny (S) or smooth (s). These two genes are located on nonhomologous chromosomes.

5. What is/are the phenotype(s) and genotype(s) of the offspring that result from the cross $PPss \times PpSs$?
6. If the cross $PpSS \times ppSs$ is made, which of the following is/are *not* represented in the offspring: (a) $PpSS$; (b) $PpSs$; (c) $ppSS$; (d) $ppSs$; (e) $PPSs$?

7–8. In garden peas, the gene for tall (T) is dominant over its allele for short (t), and round seed (R) is dominant over wrinkled seed (r). A homozygous tall, wrinkle-seeded plant is crossed with a homozygous short, round-seeded plant.

7. What is/are the phenotype(s) of the F_1? The genotype(s) of the F_1?
8. What is the phenotypic ratio of the F_2 generation?

9. In corn, a single gene controls height, with the allele for normal height (N) being dominant over the allele for short (n). A normal corn plant was crossed with a short plant, but the genotypes of the two plants were not known. Of the offspring, 150 were normal height and 153 were short. Based on the information given, determine the genotypes of the parent plants.

10. Sesame plants produce seed pods with one chamber (O) or three chambers (o), and leaves that are normal (N) or wrinkled (n). The two characters are inherited independently. Determine the genotypes and phenotypes of the two parents that produced the following offspring: 304 with one-chamber pods and normal leaves; 100 with one-chamber pods and wrinkled leaves; 298 with three-chamber pods and normal leaves; 103 with three-chamber pods and wrinkled leaves.

KEY TO GENETICS PROBLEMS

1. *ww;* all gametes would have *w*
2. 1/4 purple
3. None
4. 9:3:3:1 (9 yellow, tall:3 green, tall:3 yellow, short:1 green, short)
5. *PPSs* (purple, spiny), *PPss* (purple, smooth), *PpSs* (purple, spiny), and *Ppss* (purple, smooth)
6. (e) *PPSs*
7. All tall, round (all *TtRr*)
8. 9:3:3:1
9. *Nn* × *nn*
10. *OoNn* × *ooNn*

Visit us on the web at http://www.thomsonedu.com/biology/berg/ for additional resources, such as flashcards, tutorial quizzes, further readings, and web links.

The Molecular Basis of Inheritance

You may have seen the classic science fiction movie *Jurassic Park,* in which living dinosaurs are created from ancient dinosaur DNA (the genetic material of organisms) preserved in amber. Although amber deposits have provided interesting information about extinct animals and plants, it is not possible to grow dinosaurs from dinosaur DNA except in movies. Currently, humans cannot develop any animal, living or extinct, from a single cell or its DNA.

On the other hand, a complete plant *can* be developed from a single plant cell by using **tissue culture techniques,** methods of growing cells in a sterile, synthetic culture medium. These techniques make it possible to grow an entire plant from a single cell—a root cell, for example. This ability of a single cell to develop into an entire organism is called **totipotency** (from the Latin *toti* and *poten,* meaning "all powerful").

Although totipotency in plants was suggested in the early 1900s, it was not actually demonstrated until the late 1950s, when plant biologist F. C. Steward placed some cells from a carrot root into a culture medium containing sugar, minerals (inorganic nutrients), vitamins, and coconut milk, which was known to contain certain growth factors for plants. In this culture medium, the carrot cells grew and divided but did not *differentiate;* that is, they did not develop into the many types of cells found in carrot plants. This mass of disorganized, undifferentiated cells is called a **callus.** When the callus was transferred to a slightly different culture medium, however, differentiation occurred, and some of the cells formed tiny stems with leaves (see photograph). Roots later formed on those shoots, and the carrot plants were transplanted to soil.

Thus, Steward demonstrated that each carrot cell contained the necessary genetic information to grow and develop into an entire plant. Since Steward's pioneering work with carrots, plant tissue culture techniques have been used to generate many kinds of plants, from African violets to coastal redwoods. Tissue culture techniques have the potential to aid in the improvement of crop plants through genetic engineering and to preserve biological diversity by enabling scientists to grow numerous individuals of an endangered species.

A callus, or clump of disorganized, undifferentiated cells. When transferred to different culture media, callus grows tiny roots and shoots.

The instructions that tell a carrot—or any plant, for that matter—what to develop into, how large to grow, and what pigments to produce are contained in the nucleus of every cell of the plant. The cell's nucleus contains DNA, which stores genetic information. In this chapter we describe the molecular basis of inheritance, that is, the structure of the DNA molecule and how the genetic information in DNA is expressed.

Chromosomes

Each organism contains within it a set of hereditary instructions characteristic of its species—a blueprint for life—which is passed from generation to generation. If these instructions are to contain all the information necessary for an organism to live and grow and develop, the genetic material must be copied accurately and delivered to each new cell and to each offspring. The instructions embodied in the genetic material specify each cell's characteristic structure and functions. Through their control of cells, the instructions influence the entire organism.

In eukaryotes, these hereditary instructions are stored in the nucleus of every cell. In each nucleus are threadlike structures called **chromosomes** that contain **genes,** units of hereditary information. Most genes contain instructions for the manufacture of the proteins that make up the various structural and functional components of the cell.

Each chromosome consists of a single large *deoxyribonucleic acid (DNA)* molecule that contains hundreds or thousands of genes (●**Figure 14-1;** also see *Focus On: How Scientists Demonstrated That DNA Is the Genetic Material*). The DNA molecule is so long that it would become hopelessly tangled during cell division were it not organized in some fashion. The DNA molecule is wound around special protein molecules called **histones,** which help keep the DNA organized. The histones and their associated DNA are part of **chromatin,** the DNA and protein complex that makes up the chromosome.

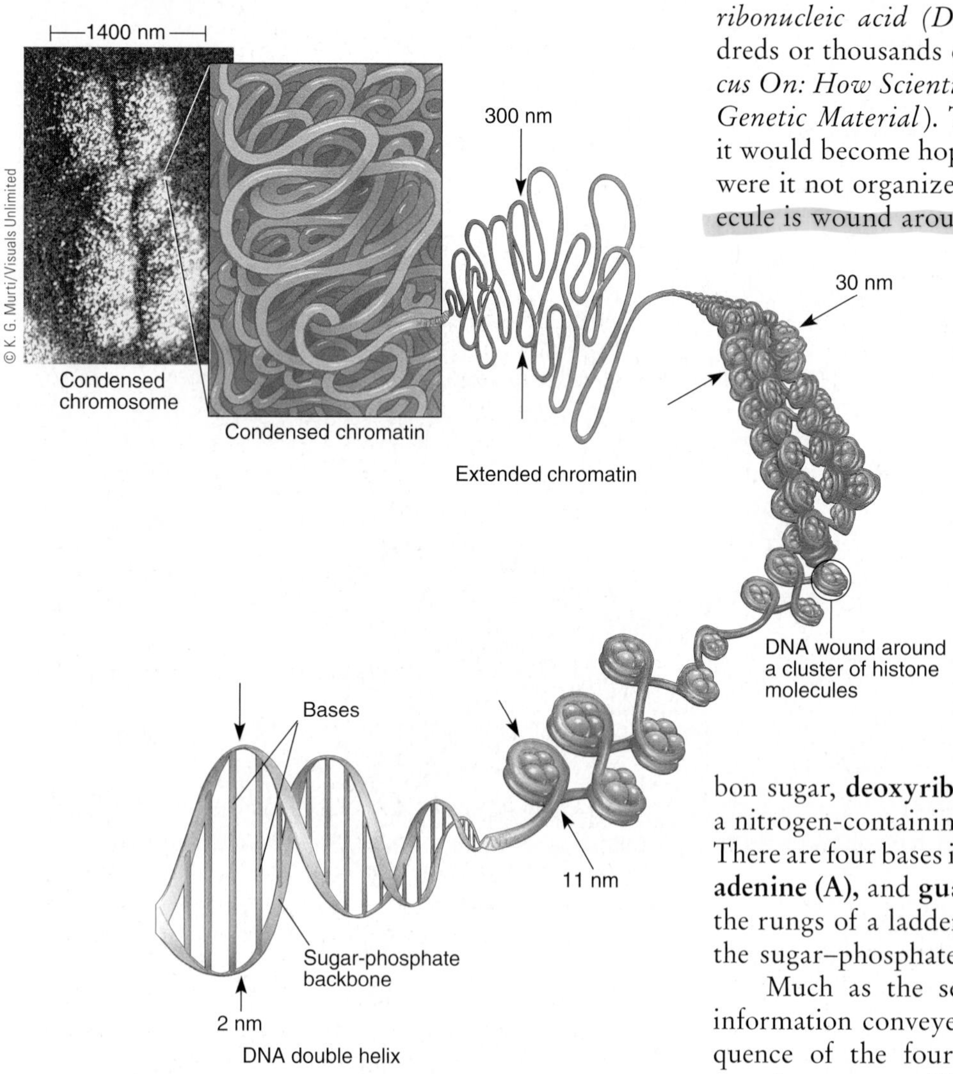

FIGURE 14-1 **Levels of organization in the eukaryotic chromosome.**

Structure of DNA

In Chapter 2 you learned that DNA is composed of molecular units called **nucleotides** (see Figure 2-18a). Each nucleotide consists of three parts: (1) a five-carbon sugar, **deoxyribose;** (2) a phosphate group; and (3) a nitrogen-containing organic compound called a base. There are four bases in DNA: **cytosine (C), thymine (T), adenine (A),** and **guanine (G).** These bases project, like the rungs of a ladder, more or less at right angles from the sugar–phosphate backbone of the DNA molecule.

Much as the sequence of letters determines the information conveyed by words and sentences, the sequence of the four nitrogen bases determines their genetic message. Thus, the sequence —AGGTAC— encodes a different set of information from that of the sequence —TAGTCC—.

FIGURE IN FOCUS

Base pairing and the sequence of bases in DNA provide a foundation for understanding both DNA replication and the inheritance of genetic information.

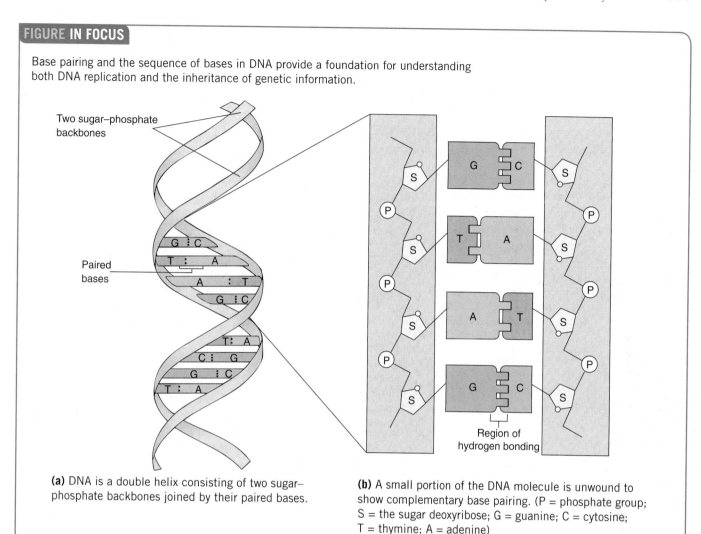

(a) DNA is a double helix consisting of two sugar–phosphate backbones joined by their paired bases.

(b) A small portion of the DNA molecule is unwound to show complementary base pairing. (P = phosphate group; S = the sugar deoxyribose; G = guanine; C = cytosine; T = thymine; A = adenine)

FIGURE 14-2 **DNA structure.**

The two nucleotide chains of DNA form a double helix

James D. Watson, a cell biologist, and Francis Crick, a biochemist, elucidated the structure of DNA in 1953. They determined that DNA consists of *two* strands attached to each other at their bases: adenine always pairs with thymine, and guanine always pairs with cytosine. The ladderlike double strand of DNA is twisted into a **double helix,** which superficially resembles a spiral staircase (•**Figure 14-2**).

The two strands of a double helix are *complementary* but not identical. So, if you know the base sequence of one of the two strands, you can predict the base sequence of the other, because adenine on one strand forms hydrogen bonds only with thymine on the other strand, and cytosine forms hydrogen bonds only with guanine. Under normal circumstances, no other pairing relationship is possible; for example, hydrogen bonds do not form properly between adenine and guanine or between adenine and cytosine. Thus, if a portion of one chain reads —AGGCTA—, then the corresponding portion of the other strand must have the sequence —TCCGAT—.

Replication of DNA

One of DNA's most distinctive (and essential!) properties is that it undergoes precise replication; that is, DNA makes an exact duplicate of itself. Before a cell divides,

CHROMOSOME One of several rod-shaped bodies in the cell nucleus that contain the hereditary units (genes).

GENE A discrete unit of hereditary information that usually specifies a polypeptide (protein).

PROCESS OF SCIENCE One of the earliest experiments suggesting that nucleic acid contains genetic information involved viruses. Viruses, a group of disease-producing agents, lack most of the traits of organisms; they are not composed of cells, nor do they metabolize, grow, or move by themselves. A virus can reproduce, however, with the aid of a living cellular host.

All organisms, including microscopic bacteria, are probably susceptible to viral infection. Viruses that attack bacteria are called **bacteriophages,** or **phages** for short. Like other viruses, a phage consists of a protein coat that surrounds a molecule of nucleic acid. A bacterium's infection begins when a phage first attaches to the bacterial cell wall. Part of the phage then penetrates the bacterial cell's outer coverings and takes over its metabolic machinery. The phage causes this machinery to replicate viral genetic material and form new phages like itself. In other words, the phage forces the bacterium host to follow the genetic directions the phage has injected into it. Eventually, what remains of the host cell bursts open, releasing many phage particles. Some of these particles may come in contact with another bacterial cell and repeat the process.

It was not known before 1952 whether the nucleic acid core or the protein coat (or perhaps both) carried the phage's genetic information into the bacterial cell, but in that year Alfred Hershey and Martha Chase settled the question. Hershey and Chase worked with the common intestinal bacterium *Escherichia coli* and a phage that attacked it. They knew that the phage's DNA core contained phosphorus but no sulfur. They also knew that the phage's protein coat contained sulfur but no phosphorus. They reasoned that if they could find some sulfur from a phage that had made its way into a bacterial cell, they would know that the phage's protein coat, not its DNA core, had entered the cell. On the other hand,

if they found phosphorus inside the bacterium, they would know that the phage's DNA core, not its protein coat, had entered the cell. Whichever substance entered the cell, Hershey and Chase reasoned, was responsible for replicating the virus and was therefore the material that transmitted traits to the next generation.

Hershey and Chase prepared a batch of phage whose protein coats contained radioactive sulfur ^{35}S. The sulfur would serve as a tracer that would enable them to detect even small quantities of the protein. They also prepared a separate batch of phage whose DNA core contained radioactive phosphorus ^{32}P. When the phage whose protein

was labeled with radioactive sulfur was added to a fresh culture of bacteria, the bacteria did not become radioactive. This showed that the protein coat did not enter the bacteria. When the phage whose DNA was labeled with radioactive phosphorus was added to a bacterial culture, however, the bacteria did become radioactive, indicating that viral DNA had entered the bacteria (see figure).

Since phages contain only protein and DNA, Hershey and Chase had investigated all possible genetic materials. It was clear that DNA had to be the material that directed the production of new viruses. DNA, not protein, was the genetic material. ■

PROCESS OF SCIENCE

QUESTION: Is DNA or protein the genetic material in bacterial viruses (phages)?

HYPOTHESIS: DNA (or protein) is the genetic material in bacterial viruses.

EXPERIMENT: Hershey and Chase produced phage populations with either radioactively labeled DNA or radioactively labeled protein. In both cases, they infected bacteria with the phages and then determined whether DNA or protein was injected into bacterial cells to direct the formation of new viral particles.

Bacterial viruses grown in ^{35}S to label protein coat or ^{32}P to label DNA

RESULTS AND CONCLUSION: New phages arising from phages with protein coats labeled with the radioactive isotope ^{35}S (*left side*) were not radioactive, indicating that protein was not involved in viral production. However, some new phages arising from phages with DNA labeled with the radioactive isotope ^{32}P (*right side*) were radioactive, indicating that viral DNA directs the formation of new particles. Thus, DNA is the genetic material in phages. ■

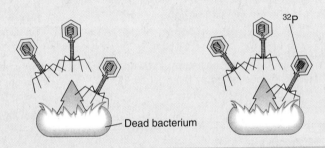

Dead bacterium

■ The Hershey–Chase experiments.

(a) Electron micrograph of DNA replication. During replication, two DNA molecules are synthesized from the original parent molecule. Replication is occurring at the Y-shaped structure, which is called a replication fork.

FIGURE 14-3 **The mechanism of DNA replication.**

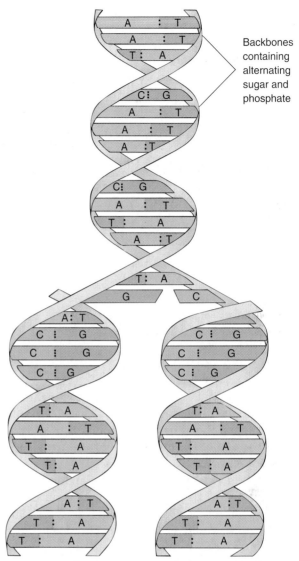

Backbones containing alternating sugar and phosphate

(b) The two strands are shown separating, and both are being copied. In the new strand, as in the old, adenine pairs with thymine, and cytosine pairs with guanine.

the two strands of DNA separate, and each one is used as a template, that is, a strand on which a new complementary strand is built. The result of replication is *two* double helix molecules, each identical in base sequence to the original double helix. Each "new" DNA molecule contains one of the original strands plus a newly synthesized strand (•**Figure 14-3**). Because the two original strands are not simply copied and then left behind but are *conserved*, or kept, as part of the two new double helices, this process is known as **semiconservative replication** (*semiconservative* here meaning "keeping a portion the same"). Each new double helix contains one old strand and one new strand.

Replication need not begin at the end of a double helix. Enzymes break the hydrogen bonds linking the two DNA strands at various sites along their length, causing them to separate. As the two strands of DNA separate, they form a Y-shaped region known as the **replication fork.** Many replication forks may occur simultaneously in a DNA molecule (•**Figure 14-4**). Because it occurs in many places at once, the replication of the enormous amount of information contained in DNA

takes as little as 20 minutes in rapidly dividing cells, such as those in a root tip.

After separating to form replication forks, both DNA strands act as templates for the assembly of new complementary strands. This assembly is the task of the enzyme **DNA polymerase,** which adds free nucleotides onto the "unzipped" DNA molecule. As the new

SEMICONSERVATIVE REPLICATION The type of replication characteristic of DNA, in which each new double-stranded molecule consists of one strand from the original DNA molecule and one strand of newly synthesized DNA.

DNA POLYMERASE An enzyme complex that catalyzes DNA replication by adding nucleotides to a growing strand of DNA.

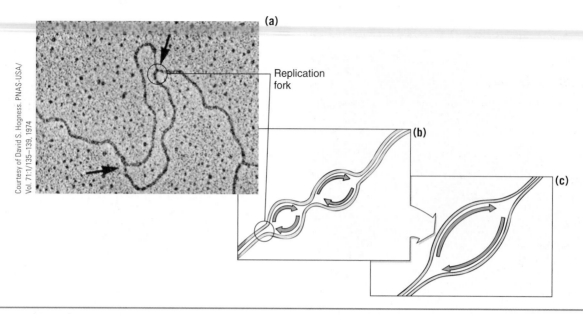

FIGURE 14-4 Bidirectional replication of DNA.
(a) Eukaryotic chromosomal DNA that is partially replicated. Note the replication forks
(*arrows*). **(b)** DNA synthesis proceeds in both directions until two replication forks meet
(c) and are joined.

complementary strands are produced, the original double helix separates further, and the newly synthesized strands continue to form hydrogen bonds with the old strands. At the end of the process, there are two identical double helices of DNA.

DNA polymerase is one of the most important enzymes in organisms, because DNA replication could not take place without it. An enzyme speeds up a chemical reaction that would ordinarily occur much more slowly, and enzymes act only on specific substances, or *substrates*. DNA polymerase is a complex of several enzymes that catalyzes the reaction in which nucleotides are joined in a precise way.

DNA polymerase accomplishes several remarkable tasks. It recognizes a specific base on an original strand of a replicating DNA molecule and then identifies the nucleotide base that is its complement (an A for a T, a C for a G, and so on). DNA polymerase links the complementary nucleotide to the proper end of the growing DNA strand; that is, it catalyzes the chemical reaction that fastens the nucleotide to the new strand. The enzyme then moves along to the next base on the existing strand and repeats the process. DNA polymerase continues this process at a rate of about 200 bases per second and does so for hundreds or thousands of bases every time the cell divides, and occasionally at other times as well!

Errors in replication do occur, but infrequently. Although exact base pairing of DNA is necessary for a regular double helix to form, improper pairing occasionally produces a distortion in the structure of the helix. However, there is a backup system that catches most of the infrequent errors that DNA polymerase makes. This backup system is a team of enzymes that snips out improperly paired bases and replaces them with the right ones.

DNA replication occurs in cells and involves making copies from a DNA template one at a time. Once biologists understood the process of DNA replication, they wondered if it was possible to amplify, or make multiple copies of, a specific DNA molecule. In 1985 biochemist Kary Mullis developed a way to make millions of copies of DNA in a test tube (see *Focus On: The Polymerase Chain Reaction*). In 1993 Mullis was awarded the Nobel Prize in Chemistry for his work.

Protein Synthesis

The messages in a plant's genes control everything about it by governing its production of protein. Both obvious traits, such as flower color and shape of leaves, and other aspects of a plant's structure and function that are difficult to observe are the direct result of the presence of protein molecules or of other molecules that are synthesized by protein enzymes.

Proteins are large, complex molecules manufactured from amino acids joined by peptide bonds. A long chain of amino acids is called a **polypeptide.** Some proteins consist of a single polypeptide chain, whereas others are composed of two or more polypeptide chains.

A typical cell produces—continuously and simultaneously—hundreds of kinds of proteins. The sequence

FOCUS ON | The Polymerase Chain Reaction

PROCESS OF SCIENCE The **polymerase chain reaction (PCR)** is a laboratory procedure that lets researchers produce millions of copies of DNA from a tiny sample in just a few hours. PCR has hundreds of applications, such as amplifying small amounts of DNA found at crime scenes, analyzing DNA from fossil leaves, and diagnosing disease, including HIV, the virus that causes AIDS.

The polymerase chain reaction involves subjecting DNA to alternating treatments of heat and enzymatic action (see figure). First, a double-stranded DNA molecule is heated to separate it into two single strands. Then the enzyme DNA polymerase assembles nucleotides along each strand, causing the formation of two double-stranded DNA molecules. The two double-stranded molecules are then heated, which causes them to separate into four single strands, and a second cycle of replication with DNA polymerase results in *four* double-stranded DNA molecules. After the next cycle of heating and replication, there are *eight* DNA molecules, and so on, with the number of DNA molecules doubling with each cycle.

Because the reaction can be carried out only if the DNA polymerase remains stable through many heating cycles, researchers use a heat-resistant DNA polymerase. They obtained this enzyme from a bacterium that lives in hot springs in Yellowstone National Park. Because the water in this environment is close to the boiling point, all enzymes in the bacterium have evolved to be stable at high temperatures.

Using PCR to Identify Mycorrhizae

PCR has many applications in basic biological research. For example, ecologists have used PCR to help survey the individual partners in symbiotic relationships (intimate associations between members of different species). One such relationship, known as

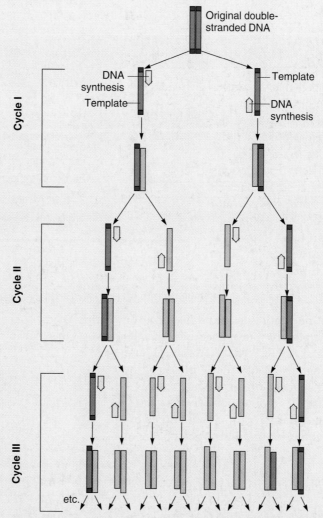

1 DNA is denatured (separated into single strands) by heat.

2 Each DNA strand acts as template for DNA synthesis catalyzed by DNA polymerase. The number of DNA molecules doubles each time cycle is repeated.

■ **Amplification of DNA by PCR. The initial reaction mixture includes a very small amount of double-stranded DNA, DNA precursors, and heat-resistant DNA polymerase. The number of DNA strands doubles each time the cycle is repeated.**

mycorrhizae, occurs between the roots of forest trees and fungi. Mycorrhizae are widespread in nature. The roots of most, if not all, trees are colonized by such fungi. This association is beneficial for both organisms: the fungus supplies the plant with essential minerals that it has absorbed from the soil, and the plant provides the fungus with carbohydrates and other organic molecules.

Identifying the exact species and strains of fungi found in mycorrhizal associations is often difficult. However, biologists successfully used PCR techniques to amplify DNA from a single mycorrhizal fungus colonizing jack pine roots. They demonstrated that the fungus was a different species from the one with which the jack pine had been thought to form an association.

Knowing exactly which species or strain of fungus forms mycorrhizae with which tree species has a practical value. Because mycorrhizal associations enhance tree growth, deliberately inoculating the roots of young seedlings with the specific fungus that colonizes them would cause forest and nursery trees to grow faster. ■

of bases in a cell's DNA contains the directions, or blueprints, for the production of each kind of protein molecule produced by a plant.

The genetic code is read as a series of triplets

In Chapter 2 you learned that 20 amino acids are commonly found in proteins. These amino acids can be considered the 20 letters of the protein alphabet. How can the four nucleotides in the DNA alphabet (A, C, T, G) specify instructions for the 20 amino acids? If single nucleotides were the instructions, DNA could specify the sequence of only 4 kinds of amino acids in any protein. If pairs of nucleotides were the instructions, there would be only 16 amino acids, because in a pool of four nucleotides there are only 16 unique pairs.

However, trios of nucleotides can and do form the genetic code, because a pool of four nucleotides permits 64 unique combinations (AAA, AAC, AAG, AAT, and so on), more than enough to specify the synthesis of the 20 amino acids. Each sequence of three nucleotides—more simply, three bases—is known as a **triplet**, and it is the basic unit of genetic information. With some exceptions, each triplet codes for one amino acid. Some of the 64 triplets are synonymous; that is, they code for the same amino acid. A few triplets do not code for amino acids but have other roles in protein synthesis.

Protein synthesis is a two-step process

Although the triplet sequence of bases in DNA specifies the order of amino acids in a polypeptide chain, DNA does not produce protein directly. DNA is found in the nucleus, and proteins are made by ribosomes in the cytoplasm outside the nucleus. Instead, a related nucleic acid, *ribonucleic acid (RNA)*, translates the information encoded in the DNA into a manufactured protein.

RNA differs from DNA in that each RNA nucleotide contains the sugar **ribose** rather than deoxyribose (•**Table 14-1** and •**Figure 14-5**). As in DNA, each RNA nucleotide contains one of four bases, although in RNA the base **uracil (U)** replaces the thymine (T) found in DNA. Uracil behaves chemically like thymine and readily forms base pairs with adenine. Another difference between RNA and DNA is that RNA molecules usually do not form double helices but function as single strands, although some RNA molecules are elaborately folded.

TABLE 14-1	A Comparison of DNA and RNA	
	DNA	**RNA**
Units	Nucleotides	Nucleotides
Sugar	Deoxyribose	Ribose
Phosphate	Yes	Yes
Bases	Adenine	Adenine
	Guanine	Guanine
	Cytosine	Cytosine
	Thymine	Uracil
Structure	Double helix	Single strand, sometimes folded elaborately
Function	Stores genetic information; used to make RNA	Used to make proteins
Location in cell	Part of chromosomes in nucleus	Made in nucleus; works in cytoplasm (ribosomes)
Structural forms	—	Three forms (messenger RNA, transfer RNA, ribosomal RNA)

In the first stage of protein synthesis, an RNA molecule is made on the basis of the information encoded in DNA. RNA synthesis is similar to DNA replication in that the DNA molecule unwinds and unzips. One of the DNA strands serves as a template on which RNA

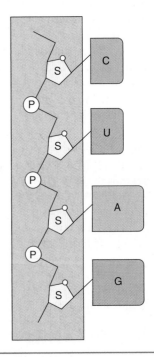

FIGURE 14-5 RNA structure.
RNA differs from DNA in that RNA is composed of a single strand rather than a double helix. The sugar in RNA is ribose, rather than deoxyribose. Three of RNA's four bases—adenine, guanine, and cytosine—are also found in DNA, but the fourth RNA base is uracil rather than thymine. (P = phosphate group; S = the sugar ribose; G = guanine; C = cytosine; U = uracil; A = adenine)

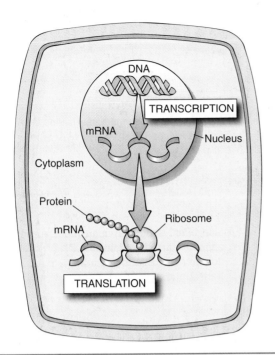

FIGURE 14-6 An overview of protein synthesis.
The master plans (DNA) are kept in the nucleus. During transcription, copies (mRNA) of the master plans are made and subsequently sent to the protein production area of the cell (the ribosomes). During translation, enzymes use the mRNA to assemble the correct sequence of amino acids into a protein.

nucleotides (A, U, C, G) are assembled in the proper order, based on complementary base pairing. The process of making RNA from a DNA template is called **transcription**, and it takes place in the nucleus in eukaryotic cells.[1]

When the RNA molecule is complete, it detaches from its DNA template, leaves the nucleus, and goes to a ribosome, a cellular structure that is the site of protein synthesis. (The two strands of the DNA molecule rezip and rewind to again form a double helix.) In this way the RNA molecule carries the instructions for the synthesis of a particular polypeptide from DNA in the nucleus to the ribosome in the cytoplasm.

In the second stage of protein synthesis, the ribosome reads the message encoded in the RNA molecule and uses it to assemble amino acids in the proper order to produce a specific polypeptide. The formation of a polypeptide at the ribosome on the basis of the information encoded in RNA is known as **translation.**

To summarize, protein synthesis involves two steps (●**Figure 14-6**). In step 1, transcription, the genetic message is copied from DNA to RNA. In step 2, translation, the genetic message is translated from the language of nucleic acids into the language of proteins.

Three kinds of RNA are involved in protein synthesis

There are three types of RNA molecules: messenger RNA (mRNA), ribosomal RNA (rRNA), and transfer RNA (tRNA). All three types are made in the nucleus from DNA by complementary base pairing, and each has a role in protein synthesis. **Messenger RNA (mRNA)** carries the coded instructions about protein structure from DNA to the ribosome. The ribosome assembles amino acids in the proper sequence into a polypeptide, guided by the information in the mRNA molecule. **Transfer RNA (tRNA)** carries specific amino acids to the ribosome during protein assembly. Ribosomes, the sites of protein synthesis, are composed of various proteins and **ribosomal RNA (rRNA).** Ribosomal RNA assists in the attachment of the ribosome to the mRNA molecule and in the assembly of amino acids in the proper order to make a polypeptide.

Details of Transcription and Translation

Now that we have presented an overview of information flow from DNA to RNA to protein, let us examine transcription and translation more closely.

Transcription is the synthesis of RNA from DNA

In transcription, a sequence of nucleotides on a DNA strand acts as the template for the synthesis of a complementary RNA strand. For example, where a nucleotide containing guanine occurs in the DNA strand, an RNA nucleotide containing cytosine pairs with it and becomes part of the new RNA strand.

TRANSCRIPTION Synthesis of RNA from a DNA template.

TRANSLATION Conversion of information provided by RNA to a specific sequence of amino acids in the production of a polypeptide chain.

MESSENGER RNA (mRNA) RNA that specifies the amino acid sequence of a polypeptide.

TRANSFER RNA (tRNA) RNA that transfers amino acids to the ribosome during protein synthesis.

RIBOSOMAL RNA (rRNA) An important part of the structure of ribosomes that also has catalytic functions needed during protein synthesis.

---

[1]*Some transcription also takes place in chloroplasts and mitochondria, organelles that contain their own DNA and ribosomes.*

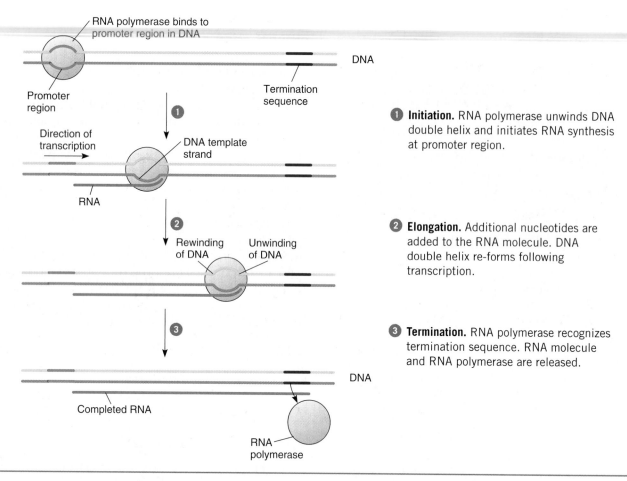

FIGURE 14-7 An overview of transcription: initiation, elongation, and termination.

An enzyme complex called **RNA polymerase** controls transcription. This remarkable enzyme responds to certain base sequences in the DNA molecule that tell the RNA polymerase which gene to transcribe, which of the two paired DNA strands it should copy, and where it should begin and end transcription.

In preparation for transcription, RNA polymerase unwinds the DNA double helix. Transcription begins when RNA polymerase binds to DNA at a specific base sequence it recognizes as a start signal (•**Figure 14-7**). RNA polymerase moves along a strand of DNA like a railroad locomotive on a track, assembling a single strand of RNA as it goes. Complementary RNA nucleotides are matched to the bases along the DNA template and added to the growing strand. The RNA strand separates from the DNA template as it forms. Transcription continues until the RNA polymerase comes to a specific nucleotide sequence in the DNA that it recognizes as a stop signal. The RNA molecule then separates completely from the DNA strand.

DNA replication and transcription differ in several important ways. First, in transcription only a single strand of DNA serves as the template, rather than both strands as in DNA replication. Second, RNA contains uracil rather than thymine; thus, during transcription, an RNA nucleotide containing uracil pairs with the DNA nucleotide containing adenine. Third, the RNA molecule detaches from the DNA template rather than forming a double helix with it.

Translation is the synthesis of a polypeptide specified by mRNA

Recall that the genetic code is based on triplets of nucleotide bases in DNA. This same information is encoded in mRNA as a sequence of three bases known as a **codon**. Each codon in mRNA is complementary to a DNA triplet and codes for a specific amino acid (•**Figure 14-8**). During translation, the order of the codons in the mRNA determines the order in which amino acids are added to a growing polypeptide chain.

Translation, which takes place at a ribosome, begins when a strand of mRNA becomes attached to one or more ribosomes. Each ribosome moves along the

First letter	Second letter: U	Second letter: C	Second letter: A	Second letter: G	Third letter
U	UUU Phe	UCU Ser	UAU Tyr	UGU Cys	U
U	UUC Phe	UCC Ser	UAC Tyr	UGC Cys	C
U	UUA Leu	UCA Ser	UAA Stop	UGA Stop	A
U	UUG Leu	UCG Ser	UAG Stop	UGG Trp	G
C	CUU Leu	CCU Pro	CAU His	CGU Arg	U
C	CUC Leu	CCC Pro	CAC His	CGC Arg	C
C	CUA Leu	CCA Pro	CAA Gln	CGA Arg	A
C	CUG Leu	CCG Pro	CAG Gln	CGG Arg	G
A	AUU Ile	ACU Thr	AAU Asn	AGU Ser	U
A	AUC Ile	ACC Thr	AAC Asn	AGC Ser	C
A	AUA Ile	ACA Thr	AAA Lys	AGA Arg	A
A	AUG Met or start	ACG Thr	AAG Lys	AGG Arg	G
G	GUU Val	GCU Ala	GAU Asp	GGU Gly	U
G	GUC Val	GCC Ala	GAC Asp	GGC Gly	C
G	GUA Val	GCA Ala	GAA Glu	GGA Gly	A
G	GUG Val	GCG Ala	GAG Glu	GGG Gly	G

= Stop codon

= Start codon

FIGURE 14-8 The genetic code.
The genetic code contains all possible combinations of the three bases that compose codons in mRNA. The abbreviations represent amino acids that the individual codons specify. For example, Phe stands for phenylalanine, and Leu for leucine. The codon AUG specifies the amino acid methionine and also signals the ribosome to initiate translation. Three codons (UAA, UGA, and UAG) do not specify amino acids; they terminate protein synthesis.

mRNA strand, and amino acids are added one by one to the growing polypeptide chain. The final step of translation occurs when the completed polypeptide chain is released. The polypeptide chain subsequently performs a specific cellular function; for example, it might function as an enzyme to catalyze a chemical reaction needed to produce a certain flower color.

Translation employs all three types of RNA. The rRNA and its associated enzymes (as part of the ribosome) play several important roles; the tRNA identifies and transports amino acids to the ribosome, the site of protein synthesis; and the mRNA specifies the order of amino acids in the polypeptide.

Ribosomal RNA is part of the ribosome

The ribosome is the cell's protein assembly factory. Each ribosome has a small subunit and a large subunit, each of which consists of rRNA and many protein molecules. Ribosomes provide everything needed for translation—mRNA, tRNAs with their associated amino acids, rRNA, and the enzymes necessary to join the amino acids with peptide bonds. The ribosome serves, in effect, as an enzymatic "matchmaker" that ensures not only that peptide bonds form but that they form between the correct amino acids in the order specified by mRNA.

Transfer RNA carries amino acids to the ribosome

The individual amino acids to be assembled into protein are carried to the ribosome by molecules of tRNA. Each tRNA molecule is elaborately folded and can form a temporary chemical bond with only one kind of amino acid (•**Figure 14-9**). The combination of a tRNA molecule with an amino acid is called an **aminoacyl-tRNA.** Enzymes, one for each kind of amino acid, ensure that each tRNA molecule links up with the correct amino acid.

Each tRNA contains a trio of bases, called an **anticodon,** that projects from the tRNA molecule. The ribosome recognizes the amino acid because the anticodon of its tRNA carrier matches with a complementary three-base sequence (the codon) on the mRNA strand.

Messenger RNA codons are the instructions the ribosome uses to construct a protein

After the mRNA binds to the ribosome, a special sequence of three bases (the codon AUG) on mRNA signals the start of translation. As the mRNA molecule passes through the ribosome, each codon on mRNA is read in turn. A tRNA molecule whose anticodon pairs with the codon on mRNA by complementary base pairing moves into the ribosome, carrying its specific amino acid. That amino acid is joined by a peptide bond to the amino acid at the end of the growing polypeptide chain (•**Figure 14-10**). The ribosome then moves along the mRNA strand so that a new codon can be read, and the process repeats.

Termination of protein synthesis is usually accomplished by special stop signals. These codons (UAA, UAG, UGA) do not code for amino acids but instead trigger the separation of the polypeptide chain and mRNA from the ribosome.

RNA POLYMERASE An enzyme that catalyzes the synthesis of RNA molecules from DNA templates.

CODON A sequence of three nucleotides in mRNA that specifies an individual amino acid or a start or stop signal.

ANTICODON A sequence of three nucleotides in tRNA that is complementary to a specific codon in mRNA.

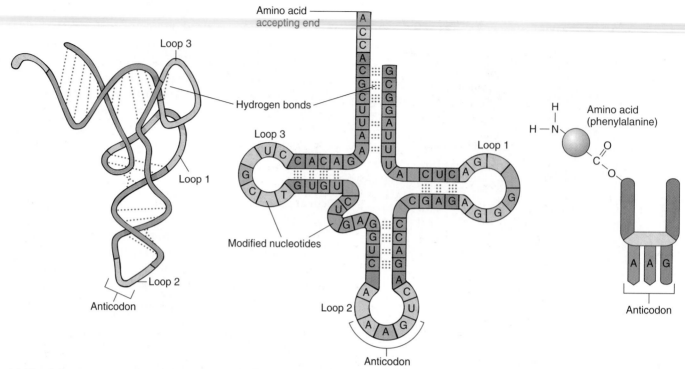

(a) The 3-D shape of a tRNA molecule is determined by hydrogen bonds formed between complementary bases.

(b) One loop contains the anticodon; these unpaired bases pair with a complementary mRNA codon. The amino acid attaches to the terminal nucleotide at the opposite end of the tRNA molecule.

(c) This stylized diagram of an aminoacyl-tRNA shows that the amino acid attaches to tRNA by its carboxyl group, leaving its amino group exposed for peptide bond formation.

FIGURE 14-9 Three representations of a tRNA molecule.

Each strand of mRNA may be used to make multiple copies of a particular protein. As many as 10 or 20 ribosomes usually bind to a single strand of mRNA, each one independently producing a separate copy of the same polypeptide. After the instructions in mRNA are used to produce the polypeptide chains that the cell requires, mRNA is degraded to its constituent nucleotides, which may be used to construct new RNA molecules.

Almost all organisms share the genetic code

DNA is the molecule of inheritance for all organisms except RNA viruses,[2] and the order of bases in a DNA molecule determines the characteristics of the traits for which DNA codes. Thus, species owe their individual traits to different sequences of bases in their DNA. **EVOLUTION LINK** The fact that DNA is the molecule of inheritance in virtually all organisms is evidence that all life is related. Moreover, when two species are closely related, their DNA has a greater proportion of identical sequences of nucleotide bases. Thus, a daisy and a tulip have significant stretches of base sequences that are identical because they are both flowering plants and they shared a common ancestor in the not-so-distant past. On the other hand, a yeast (a single-celled fungus) and a tulip do not have as many identical stretches of bases in their DNA. Yeasts and tulips evolved for hundreds of millions of years along different lines of descent. The fact that they share some identical base sequences, however, indicates that they had a common ancestor in the distant past.

An important aspect of the genetic code is that it is nearly universal throughout the biological world. The same DNA triplets code for the same amino acids in

---

[2]*RNA, rather than DNA, serves to store genetic information in some viruses. Of course, most biologists do not consider viruses to be living organisms.*

Initiation (steps ① and ②)

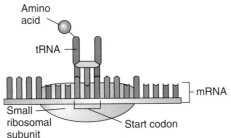
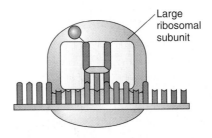

① The tRNA binds to the start codon by complementary base pairing between the tRNA's anticodon and the mRNA's codon.

② When the large ribosomal subunit binds to the small subunit, initiation is complete.

Elongation (steps ③ , ④ , and ⑤)

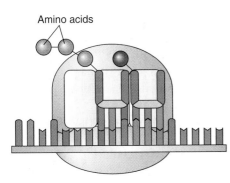
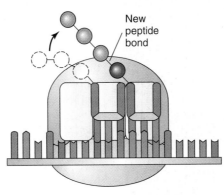
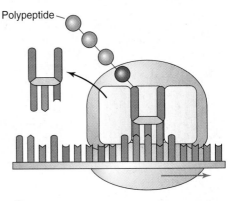

③ An aminoacyl-tRNA binds to mRNA by complementary base pairing between the anticodon and the codon.

④ The growing polypeptide chain becomes attached by a peptide bond to the amino acid that linked to mRNA in step ③.

⑤ The ribosome moves one codon to the right. The tRNA that gave up the growing polypeptide chain exits the ribosome. Elongation continues until a stop codon is reached.

Termination (steps ⑥ and ⑦)

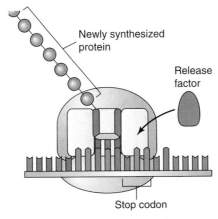
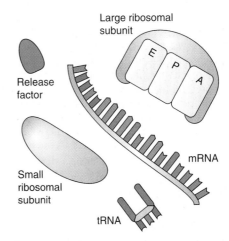

⑥ When the ribosome reaches a stop codon, a protein release factor attaches to the stop codon.

⑦ The polypeptide chain is released, and the remaining parts of the translation complex separate.

FIGURE 14-10 An overview of translation: initiation, elongation, and termination.

© Science VU/Keith Wood/Visuals Unlimited

FIGURE 14-11 Universality of the genetic code.
A gene from a firefly was genetically engineered into this tobacco (*Nicotiana tabacum*) plant. The gene codes for an enzyme that catalyzes the reaction in fireflies that produces light. The tobacco plant expressed the firefly gene and produced light—a dramatic example of the universality of the genetic code.

all organisms, with only a few minor variations (•**Figure 14-11**). The types of protein assembled vary from one species to another, and this variation accounts for the specific traits that make each species separate. The universality of the genetic code is compelling evidence that all organisms evolved from the same early life-forms. ■

Mutations

What would happen if during DNA replication, a base not complementary to the template base were placed in the newly forming strand? What if an entire sequence of bases from one DNA molecule moved and became part of another portion of the molecule or of a different DNA molecule? Such a change would be transmitted to all the DNA strands replicated from the strand in which it originally occurred. Thus, all the descendants of that cell would share its genetic difference.

Such accidental changes in genetic information, called **mutations,** do occur (•**Figure 14-12**). Sometimes a mutation causes no change in a protein that it codes for, but often a mutation alters the structure and function of that protein. Mutations are not necessarily harmful; some are harmless, and some are actually beneficial. These instances are rare, however, because of the already highly adapted nature of organisms. A random change in the genetic material is unlikely to improve the function of a cell or an organism.

Every organism ever investigated has been subject to mutation. The wonder is that we do not observe more mutations than we do. Fortunately, cells have enzymes whose function is to repair or remove damaged sections of DNA. As a result, most mutations are not passed on to other cells.

EVOLUTION LINK Mutations are important because they are the ultimate source of variation in the living world and provide the raw material for evolution. Mutations furnish the genetic variability on which natural selection acts, sometimes resulting in the evolution of new, advantageous traits. ■

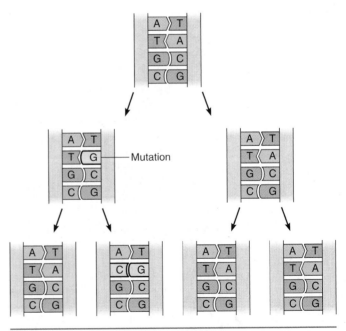

FIGURE 14-12 The perpetuation of a mutation.
The process of DNA replication can stabilize a mutation (*bright yellow*) so that it is transmitted to future generations.

Some mutations involve transposons

Some types of mutations result from a change in chromosome structure. These changes usually have a wide range of effects because they involve many genes. One type of mutation is caused by DNA sequences that "jump" into the middle of a gene. These movable sequences of DNA are known as **mobile genetic elements,** or **transposons.** Mobile genetic elements not only disrupt the functions of some genes but also inactivate some previously active genes.

Transposons were discovered in maize (corn) by the U.S. geneticist Barbara McClintock in the 1950s (●**Figure 14-13**). She observed that certain genes appeared to be turned off and on spontaneously. She deduced that the mechanism involved a segment of DNA that moved from one region of a chromosome to another, where it would inactivate a gene. However, biologists did not understand this phenomenon until the discovery of transposons in a wide variety of organisms. We now know that transposons are segments of DNA that range from a few hundred to several thousand bases. In recognition of her insightful findings, McClintock was awarded the Nobel Prize in Physiology or Medicine in 1983.

© Jerome Wexler/Visuals Unlimited

FIGURE 14-13 Transposons.
The white lines and streaks in these kernels are caused by transposons that have moved from one location to another, turning off pigment production in certain cells. The longer the transposon is in a location that turns off pigment production, the less pigment is produced. Some kernels are colorless (unpigmented) because the pigment-producing genes have been turned off for a long time.

MUTATION A change in the nucleotide sequence of DNA of an organism.

STUDY OUTLINE

❶ **Distinguish between a chromosome and a gene.**
A **chromosome** is one of several rod-shaped bodies in the cell nucleus that contain the hereditary units (genes). A **gene** is a discrete unit of hereditary information that usually specifies a **polypeptide** (protein).

❷ **Name the units that compose the DNA molecule, and identify the three parts of each unit.**
The deoxyribonucleic acid (DNA) molecule is a **double helix,** with each strand composed of **nucleotides.** Each nucleotide consists of **deoxyribose** (a sugar), a phosphate group, and an organic base. There are four bases: **cytosine, thymine, adenine,** and **guanine.**

❸ **Describe base pairing in DNA molecules. Given the base sequence of one strand of DNA, predict that of a complementary strand of DNA.**

Hydrogen bonds between complementary bases join the two strands of the DNA double helix. Adenine pairs (that is, forms hydrogen bonds) with thymine, and cytosine pairs with guanine.

❹ **Summarize the process of DNA replication, and explain what is meant by "semiconservative replication."**
DNA replication is the process of making an exact copy of a DNA molecule. DNA strands unwind during replication. DNA synthesis takes place at **replication forks,** Y-shaped regions where the two strands of DNA separate and where DNA synthesis occurs on both strands at once. **DNA polymerase** is an enzyme complex that catalyzes DNA replication by adding nucleotides to a growing strand of DNA. The type of replication characteristic of DNA is **semiconservative replication,** in which each new double-stranded molecule consists of one strand

from the original DNA molecule and one strand of newly synthesized DNA.

⑤ Compare the structures of DNA and RNA molecules, and identify the functions of the three types of RNA: messenger RNA, transfer RNA, and ribosomal RNA.
Ribonucleic acid (RNA) is formed from nucleotide subunits, each of which contains the sugar **ribose,** a base (cytosine, thymine, adenine and **uracil**), and a phosphate group. Unlike DNA, RNA is a single strand, although it may be elaborately folded. Three types of RNA have a role in protein synthesis. **Messenger RNA (mRNA)** specifies the amino acid sequence of a polypeptide. **Transfer RNA (tRNA)** transfers amino acids to the ribosome during photosynthesis. **Ribosomal RNA (rRNA)** is an important part of the structure of ribosomes that also has catalytic functions needed during protein synthesis.

⑥ Outline the flow of genetic information in cells from DNA to protein, and summarize the processes of transcription and translation.
DNA directs protein synthesis through an intermediary, RNA. Thus, the flow of genetic information is from DNA to RNA to proteins. **Transcription** is the synthesis of RNA from a DNA template. **RNA polymerase** is the enzyme that catalyzes the synthesis of RNA molecules from DNA templates. **Translation** is the conversion of information provided by RNA to a specific sequence of amino acids in the production of a polypeptide chain. Translation requires codons and anticodons. A **codon** is a sequence of three nucleotides in mRNA that specifies an individual amino acid or a start or stop signal. An **anticodon** is a sequence of three nucleotides in tRNA that is complementary to a specific codon in mRNA. During translation, each tRNA molecule attaches to a specific amino acid and carries it to the ribosome. The ribosome recognizes the anticodon of the tRNA molecule and allows it to base-pair with the mRNA codon. The amino acid brought in by tRNA forms a peptide bond with the growing polypeptide chain. The tRNA molecule is then released, and the process is repeated.

⑦ Explain the universality of the genetic code and its evolutionary significance.
The genetic code is nearly universal throughout the biological world: the same DNA triplets code for the same amino acids in all organisms, with only a few minor variations. The universality of the genetic code is compelling evidence that all organisms evolved from the same early life-forms.

⑧ Explain the effect of mutations on protein synthesis and on evolution.
A **mutation** is a change in the nucleotide sequence of DNA of an organism. Sometimes a mutation causes no change in a protein that it codes for, but often a mutation alters the structure and function of that protein. Mutation is the ultimate source of variation in the living world and provides the raw material for evolution. Mutations furnish the genetic variability on which natural selection acts, sometimes resulting in the evolution of new, advantageous traits.

REVIEW QUESTIONS

1. What are the informational units on chromosomes called? Of what chemical do these informational units consist?
2. Sketch a short section of DNA, and label the following parts: double helix, nucleotide, sugar–phosphate backbone, and base-pair rungs. Which base pairs are complementary in DNA?
3. If a segment of DNA is —CTAGTAAGC—, what is the complementary DNA strand?
4. How does the molecular structure of DNA enable it to form exact copies of itself during replication?
5. What is semiconservative replication?
6. Contrast DNA replication with transcription.
7. Compare and contrast the structures of DNA and RNA.
8. Summarize the roles of mRNA, tRNA, and rRNA in protein synthesis.
9. What is the relationship among triplets, codons, and anticodons?
10. What is the logical basis of the fact that the unit of the genetic code is a triplet of bases rather than a single base or pair of bases?
11. Explain the evolutionary significance of the fact that almost all organisms have the same genetic code.
12. What effect might a mutation have on protein synthesis?
13. Match these terms with the appropriate letter in the diagram: *translation, replication, transcription.*

$$(a) \; \text{DNA} \xrightarrow{(b)} \text{RNA} \xrightarrow{(c)} \text{protein}$$

THOUGHT QUESTIONS

1. What characteristics must a molecule have if it is to serve as genetic material?

2. How is the molecular structure of DNA uniquely adapted to its function as hereditary material?

3. A researcher analyzes the DNA in a dandelion and determines that guanine makes up 20 percent of the bases in the DNA. What is the percentage of cytosine? Of adenine? Of thymine?

4. What would happen to evolution if DNA were always transmitted precisely from generation to generation with no mutations?

5. The following sequence of triplets in DNA codes for which amino acids in a polypeptide chain? (*Hint:* You may use Figure 14-8 to help you answer the question, but you must do something *before* using the figure.)
   DNA: —TGT—TTT—TCA—GGT—CTA—

Visit us on the web at http://www.thomsonedu.com/biology/berg/ for additional resources such as flashcards, tutorial quizzes, further readings, and web links.

Genetic Frontiers

orn (*Zea mays*) is a grain crop first cultivated thousands of years ago somewhere in South or Central America. Today, corn is cultivated in many countries, including the United States, Canada, Russia, Chile, Australia, and South Africa. It is the world's third most important food crop; only rice and wheat are produced in greater quantities.

Selective breeding of corn over the centuries has resulted in many varieties. Most of the corn grown for livestock is a starchy feed type known as *dent corn* because a dent forms in the top of the grain when the kernel dries. Some varieties of corn that humans eat directly include sweet corn, in which the grains contain a high percentage of sugar, and popcorn, whose small, hard grains explode into soft white masses when heated.

The cultivation of corn removes many essential minerals (inorganic nutrients) from the soil, particularly nitrogen. For every 2500 kilograms (5500 pounds) of grain harvested from a field, 80 kilograms (175 pounds) of nitrogen are removed. To prevent a loss in soil fertility and an accompanying decline in crop productivity, nitrogen fertilizer must be added to the soil.

The expense of applying such large amounts of fertilizer is prohibitive, and farmers in poorer nations cannot afford it. In addition, water pollution is associated with applying fertilizer to farmlands. Fertilizer runoff from agricultural land into streams, rivers, and lakes causes excessive growth of algae and aquatic plants, which disrupts the natural balance between aquatic organisms and causes other environmental problems.

Given the need for fertilizing to grow corn and the problems associated with fertilizers, a variety of corn that made its own fertilizer would be a wonderful boon to humanity. Some food crops, including peas and soybeans, make use of bacteria to fertilize themselves. **Nitrogen-fixing bacteria** live in their roots and convert atmospheric nitrogen to a form that the plants can use. However, attempts to get nitrogen-fixing bacteria to colonize the roots of other crops, such as corn, have not been successful.

Research is under way in many laboratories to transfer genes from a plant such as soybean into corn, with the hope that the genes will be expressed in the corn plant and enable it to provide a suitable environment in its roots for the nitrogen-fixing bacteria. Other scientists are taking a different approach; they want to transfer certain genes from nitrogen-fixing bacteria into corn so that corn can make its own fertilizer directly. The technical challenges involved in such gene transfers are formidable and were thought insurmountable just a few years ago. However, new genetic techniques developed during the last three decades make the dream of self-fertilizing corn seem attainable. This chapter is concerned with these techniques.

A developing ear of corn (*Zea mays*). Because corn is such an important crop plant, it is the focus of research, including genetic engineering.

Genetic Modification of Organisms

Selective breeding, in which humans breed plants with desired traits, has been used for thousands of years to develop better food crops, such as corn, apples, and beans. The first farmer who ever decided which seeds to save for the following year's crop was practicing selective breeding. However, selective breeding requires a long time to produce results and does not always have the desired outcome. Corn with the protein content of soybeans, for example, is beyond the capability of selective breeding.

Using organisms to produce products that benefit humanity is known as **biotechnology.** Biotechnology does not necessarily refer to "technology" in the 21st-century sense: Selective breeding of plants is an example of biotechnology, and so is the use of yeast (a single-celled fungus) to cause bread to rise during baking. However, advances in genetics promise unprecedented control of inheritance. When people talk about biotechnology today, they are often using the word in a narrower sense, referring to the manipulation of the genetic material in cells so that organisms with new characteristics are constructed. Such modification of the DNA of an organism to produce new characteristics is also known as **genetic engineering** (•**Figure 15-1**).

Genetic engineering differs from selective breeding in several respects. One difference is that genetic engineering enables new strains of organisms to be developed in a fraction of the time required for selective breeding. With traditional breeding methods, incorporating genes for disease resistance into a particular crop plant might take 15 years. Genetic engineering has the potential to accomplish the same goal in a fraction of that time. Moreover, genetic engineering differs from traditional breeding methods in that it can make use of desirable genes from *any* organism, not just those from the species of plant being improved. If a gene for disease resistance in petunias would be beneficial in tomatoes, the genetic engineer can splice the petunia gene into the tomato plant's DNA to produce a disease-resistant tomato plant. This process could never be carried out by traditional breeding methods, because petunias and tomatoes are different species and cannot interbreed.

Recombinant DNA Technology

Recombinant DNA technology is a field of biology that started in the mid-1970s. It actually had its roots in the 1940s with genetic studies of bacteria and **bacteriophages,** the viruses that infect them (see *Focus On: How Scientists Demonstrated That DNA Is the Genetic Material,* in Chapter 14). After decades of basic research and the accumulation of extensive knowledge, recombinant DNA technology became feasible and available to the many scientists who now use these methods.

Recombinant DNA technology permits the formation of new combinations of genes by isolating genes from one organism and introducing them into either a similar or an unrelated organism. Using recombinant DNA technology, geneticists can insert DNA not only into unicellular bacteria but also into cells of plants and animals. Once inside the cell, this foreign DNA may be expressed. A foreign gene may alter the genetic composition of a bacterium, for example, so that it produces novel proteins, such as human growth hormone.

Scientists are not the only ones who practice recombinant DNA technology. Organisms have natural mechanisms for exchanging genes, and such gene transfers have apparently always occurred in nature. For example, bacteria sometimes transfer DNA through a special tube (a *pilus;* see Chapter 19) that extends from one cell to another. The transfer of genetic material from one bacterium to another by a bacterial virus is another example of natural genetic exchange. However, these natural "experiments" have occurred randomly, and certainly not to satisfy human goals and desires.

Inserting a foreign gene into a cell requires several steps. The gene that is being introduced into a different cell must be (1) isolated from its original cell; (2) used to

FIGURE 15-1 Genetic engineering.
A gene from a common soil bacterium (*Bacillus thuringiensis*) was engineered into cotton (*Gossypium*); this gene codes for the production of a protein that acts as a natural insecticide. In a field test, cotton plants were subjected to an insect infestation. The genetically engineered plant (*left*) received little damage and, as a result of being healthier, produced larger bolls of cotton. The unaltered plant (*right*) was weakened by insects and produced much smaller bolls.

Courtesy of Monsanto Company

construct a recombinant DNA molecule; (3) transferred into a new (that is, foreign) cell; (4) identified as having been taken up by the cell; and (5) expressed in that cell. We consider each of these steps in detail.

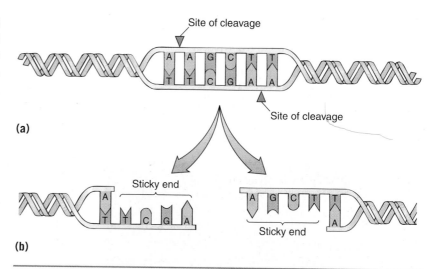

(a)

(b)

A specific region of DNA is isolated from the cell's DNA

To begin the process of genetic engineering, the investigator must isolate the gene of interest. The researcher breaks the cell's DNA into manageable fragments by using bacterial enzymes known as **restriction enzymes,** which cut DNA molecules at specific base sequences. Restriction enzymes generally recognize a specific DNA sequence four to six base pairs in length. One restriction enzyme recognizes and cuts DNA whenever it encounters the base sequence —AAGCTT—, whereas another cuts DNA at the sequence —GAATTC—. Discovering restriction enzymes was a major breakthrough in developing recombinant DNA technology. Today, many restriction enzymes, each with its own characteristics, are readily available to researchers.

Many base sequences recognized by restriction enzymes read the same from left to right and from right to left; that is, the base sequence of one strand reads the same as the bases on its complementary strand read in the opposite direction. For example, note that the DNA base-pair sequence —AAGCTT— reads the same as the reverse of its complementary sequence, —TTCGAA— (•**Figure 15-2**). When the particular restriction enzyme recognizes the —AAGCTT— sequence, it cleaves the double-stranded DNA between A and A on *each* strand. This action leaves one strand longer than the other on each DNA fragment; the single-stranded stub that extends from the end of each fragment of DNA is called a *sticky end*.

DNA segments from different organisms potentially can recombine with one another if they were broken apart by the same restriction enzyme. Each restriction enzyme always cuts the same base sequence, no matter what kinds of DNA it cleaves. For this reason, the sticky ends of the DNA are complementary and can be joined, or spliced, by an enzyme known as **ligase.**

Recombinant DNA molecules contain DNA from two sources

Restriction enzymes split the potential donor DNA into many sticky-ended fragments. The next step is to

FIGURE 15-2 Restriction enzymes.
(a) A portion of a DNA molecule (unwound for simplification) contains the recognition site for a restriction enzyme that recognizes the DNA base sequence —AAGCTT— and its complementary base pairs. Note that —AAGCTT— and its complement, —TTCGAA—, have the same base sequence in opposite directions. **(b)** The cleavage of DNA by this restriction enzyme results in two DNA fragments that have short single-stranded stubs—that is, sticky ends.

construct recombinant DNA molecules by combining these isolated DNA fragments with DNA from another source, which acts as a **vector,** or DNA carrier. The vector carries the gene from one organism into the cells of a second organism, where the gene is incorporated into that organism's existing DNA. The most common vector in recombinant DNA technology comes from bacteria.

In addition to having one large, circular DNA molecule that contains a bacterium's genetic information, some bacteria have small, circular DNA molecules, known as **plasmids,** that carry one or two genes (•**Figure 15-3**). When a bacterial cell divides, any plasmids that are present within the cell are replicated along with the bacterial chromosome, and the copies are distributed to daughter cells.

GENETIC ENGINEERING Manipulation of genes, often through recombinant DNA technology.

RECOMBINANT DNA TECHNOLOGY The techniques used to make DNA molecules by combining genes from different organisms.

RESTRICTION ENZYME An enzyme used in recombinant DNA technology to cleave DNA at specific base sequences.

VECTOR An agent, such as a plasmid or virus, that transfers DNA from one organism to another.

PLASMID A small, circular DNA molecule that carries genes separate from the main DNA of a bacterial cell.

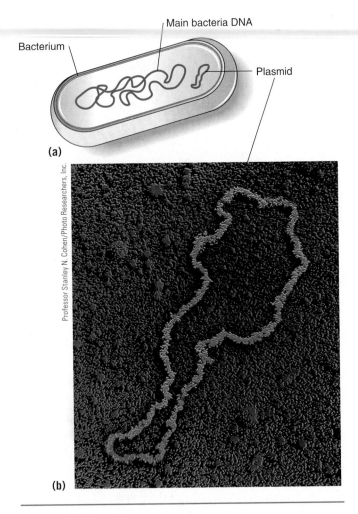

Bacterium

Main bacteria DNA

Plasmid

(a)

(b)

Professor Stanley N. Cohen/Photo Researchers, Inc.

FIGURE 15-3 Plasmid.
A plasmid is a small, circular DNA molecule that contains only a few genes. Plasmids are often used in genetic engineering to transfer desired genes into bacteria or other organisms. **(a)** The relative sizes of a plasmid and the main DNA of a bacterium. **(b)** An electron micrograph of a plasmid from the common intestinal bacterium *Escherichia coli.*

When plasmids are used as genetic vectors, desired genes are inserted into the isolated plasmids, which are then incorporated into bacterial cells. Each plasmid and its genes (its own genes plus the inserted gene) are then replicated every time the bacterial cells divide. As a result, the entire bacterial population descended from the original bacterial cells contains the novel genes in its plasmids.

To construct recombinant DNA molecules for which a plasmid is the vector, the investigator treats plasmids isolated from bacterial cells with a restriction enzyme (•**Figure 15-4**). Because the restriction enzyme cuts the plasmids at the base sequence it recognizes, the formerly circular plasmids now become linear molecules of DNA with sticky ends. These plasmid strands are then mixed with sticky DNA fragments from the donor cell. Then, the scientist uses ligase to join the sticky ends of the donor and plasmid DNA fragments into new circular plasmids.

The result is a mixture of recombinant plasmids, with each recombinant plasmid containing a different fragment of donor DNA. Although the vast majority of plasmids in the mixture do *not* contain the desired gene, a few will have exactly what is needed. To identify the plasmid that contains the gene of interest, the investigator *amplifies* all the plasmids; that is, many copies of each are made so that there will be millions of copies to work with. Amplification takes place inside bacterial cells as they divide.

Recombinant DNA molecules are transferred into a new host organism's cell

The recombinant DNA molecules (now existing as plasmids) are next taken into bacterial cells. Under normal conditions, bacteria do not take in plasmids; however, some bacteria receive plasmids under certain conditions, such as when the bacteria are treated with heat and calcium. Bacteria that take in plasmids are said to be **transformed.**

Because only a small proportion of the bacteria take up the recombinant plasmids, even under these special conditions, there has to be a way to determine which bacteria contain the genetically altered plasmids. Fortunately, plasmids often carry genetic information that makes the bacterium resistant to certain antibiotics, such as penicillin and tetracycline, that kill or inhibit bacteria and other microorganisms. If the plasmid that is used as a vector has a gene for resistance to a particular antibiotic, then investigators can grow, or *culture,* the bacterial population in the presence of that antibiotic. Only bacteria that have taken up the genetically altered plasmids resist the antibiotic and survive; the rest are killed by the antibiotic (•**Figure 15-5**).

To select the cells with the desired gene from among all other genetically altered cells, the researcher spreads a sample of the bacterial culture on plates of nutrient agar, a solid growth medium. If the sample is dilute enough, each bacterial cell is separated from other cells on the plate. As each cell reproduces, it gives rise to a **colony,** a cluster of millions of genetically identical cells; each bacterium in the colony contains the same recombinant plasmid. Because there may be as many as several thousand colonies, each containing a different portion of foreign DNA, researchers must then identify which colony contains the gene of interest, that is, the one they hope to transfer into a different organism.

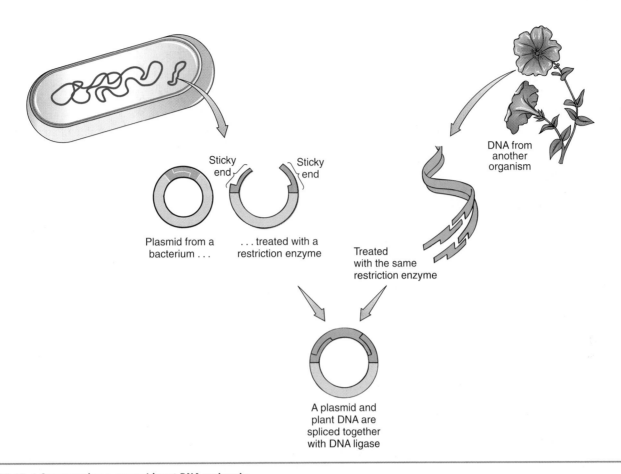

FIGURE 15-4 Constructing a recombinant DNA molecule.
DNA molecules from two sources are cut with the same restriction enzyme, which results in complementary sticky ends. In this example, one of the DNA molecules is a bacterial plasmid, and the other is a segment of DNA from a petunia. When mixed, their sticky ends form base pairs, and ligase splices the two together.

Labels in figure:
- Sticky end
- Sticky end
- Plasmid from a bacterium . . .
- . . . treated with a restriction enzyme
- Treated with the same restriction enzyme
- DNA from another organism
- A plasmid and plant DNA are spliced together with DNA ligase

Cells that have taken up the gene of interest are identified

Because a single gene is only a small part of the total DNA in an organism, isolating the piece of DNA containing that particular gene once the organism's DNA is spread among multiple recombinant plasmids is like finding a needle in a haystack. A needle could be found in a haystack of any size if one had a metal detector. The genetic equivalent of a metal detector can find just the desired DNA base sequence among the vast number of genes in the bacterium's chromosomes. This device, called a **genetic probe**, is a radioactively labeled segment of RNA or single-stranded DNA that is complementary to the target gene.

For example, suppose we want to identify the gene that codes for the protein nitrogenase, an enzyme involved in nitrogen fixation. Because we know the amino acid sequence of nitrogenase, we can synthesize the mRNA that produces it. This mRNA, which is the probe nucleic acid, attaches by complementary base pairing to the DNA that contains the base sequence needed for production of nitrogenase. Because the synthesized mRNA is radioactively labeled, it can be detected by X-ray film. If after this treatment, any radioactive DNA is detected in a particular colony of bacterial cells, that is the colony that contains the gene for the production of the desired protein—in this case, nitrogenase (•**Figure 15-6**).

The gene is expressed inside the new cell

Even though a gene is now known to be incorporated into certain bacteria, growing the bacterial strain would not necessarily produce the desired protein. The gene of interest cannot be transcribed (and subsequently trans-

GENETIC PROBE A single-stranded nucleic acid used to identify a complementary sequence by base pairing with it.

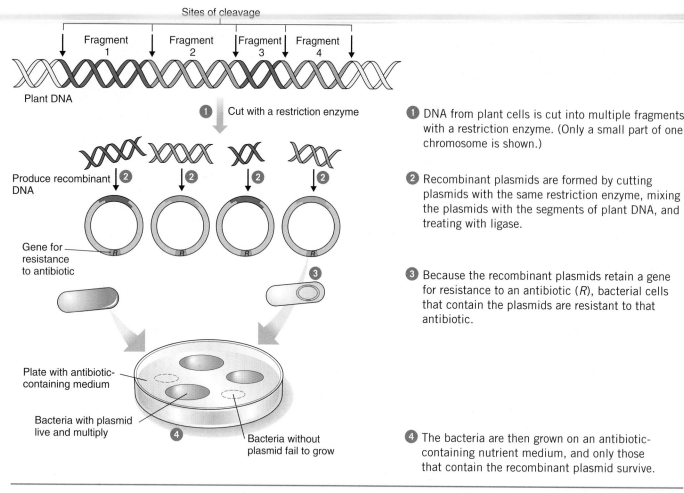

1. DNA from plant cells is cut into multiple fragments with a restriction enzyme. (Only a small part of one chromosome is shown.)

2. Recombinant plasmids are formed by cutting plasmids with the same restriction enzyme, mixing the plasmids with the segments of plant DNA, and treating with ligase.

3. Because the recombinant plasmids retain a gene for resistance to an antibiotic (*R*), bacterial cells that contain the plasmids are resistant to that antibiotic.

4. The bacteria are then grown on an antibiotic-containing nutrient medium, and only those that contain the recombinant plasmid survive.

FIGURE 15-5 Identifying bacteria that have taken up genetically altered plasmids.

lated into a protein) unless it is associated with a set of regulatory genes that, in effect, turn the gene "on." When these genes are added to plasmids, the bacterial cells can transcribe and translate the gene of interest.

The plasmid is replicated and distributed to its daughter cells during cell division. When a *recombinant plasmid*—one that has foreign DNA spliced into it—replicates in this way, many identical copies of the foreign DNA are made; in other words, the foreign DNA is cloned.

Gene insertion in plants is technically challenging

Although several vectors for the introduction of genes into plant cells are used, the most widely used vector is the crown gall bacterium (*Agrobacterium tumefaciens*), which produces tumors in a variety of plant species (●**Figure 15-7**). This bacterium causes disease

by inserting DNA from its plasmid into the plant cell; the plasmid DNA goes to the nucleus and is integrated into a plant chromosome. The inserted DNA instructs the plant cells to produce abnormal quantities of plant growth hormones. The result is repeated cell division, forming a tumor.

It is possible, however, to disarm the plasmid so that it does not induce tumor formation and to incorporate desirable genes in the *A. tumefaciens* plasmid for insertion into a plant host (●**Figure 15-8**). The plant cells into which such genetically altered plasmids are introduced are essentially normal except for the newly inserted genes. Genes placed in the plant cells in this fashion are transmitted to the next generation by sexual reproduction (that is, in seeds) or asexual reproduction.

There are several problems with using *A. tumefaciens* plasmids to introduce foreign genes into plants. First, only eudicots (the largest class of flowering plants) are susceptible to this bacterium; important monocots

FIGURE **IN FOCUS**

Genetic probes can be used to determine which cells have taken up the gene of interest. The pattern of spots on the X-ray film enables the researcher to identify which colonies from the original plate contain the correct plasmid.

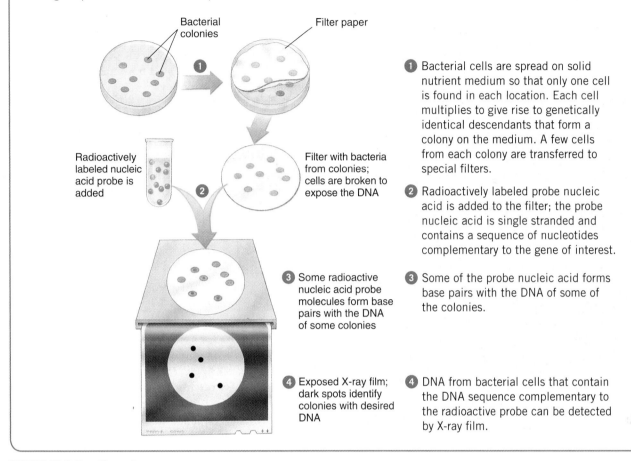

1 Bacterial cells are spread on solid nutrient medium so that only one cell is found in each location. Each cell multiplies to give rise to genetically identical descendants that form a colony on the medium. A few cells from each colony are transferred to special filters.

2 Radioactively labeled probe nucleic acid is added to the filter; the probe nucleic acid is single stranded and contains a sequence of nucleotides complementary to the gene of interest.

3 Some of the probe nucleic acid forms base pairs with the DNA of some of the colonies.

4 DNA from bacterial cells that contain the DNA sequence complementary to the radioactive probe can be detected by X-ray film.

(Labels in figure:)
Bacterial colonies
Filter paper
Radioactively labeled nucleic acid probe is added
Filter with bacteria from colonies; cells are broken to expose the DNA
3 Some radioactive nucleic acid probe molecules form base pairs with the DNA of some colonies
4 Exposed X-ray film; dark spots identify colonies with desired DNA

FIGURE 15-6 **Genetic probe.**

(the smaller class of flowering plants) such as corn, wheat, and rice are outside the host range of the crown gall bacterium.

Viruses are also used to introduce recombinant DNA into cells, and they are commonly used as vectors in mammalian cells, including human cells. These viruses are disabled so that they do not kill the cells they infect. Instead, the viral DNA, as well as any foreign DNA they carry, becomes incorporated into the cell's chromosomes after infection.

Other methods used to introduce foreign DNA into a cell do not involve a biological vector, that is, a bacterial plasmid or virus. In some cases, the researcher alters the cells chemically or by *electroporation*—delivering an electric shock—to make the plasma membrane permeable to the DNA molecules.

Unfortunately, not all plants take up DNA readily, particularly the cereal grains that are a major food source for humans. One useful approach is a genetic "shotgun." Researchers coat microscopic gold or tungsten fragments with DNA and then shoot them into plant cells so that they penetrate the cell walls. The gene gun was first used successfully to introduce genes into algal and yeast cells, but it is now used for plant and animal cells as well. For example, geneticists used such an approach to transfer a gene for resistance to a bacterial disease into cultivated rice from one of its wild relatives.

DNA CLONING The process of selectively amplifying DNA sequences so their structure and function can be studied.

FIGURE 15-7 **Crown gall tumors.**
A plasmid carried by the crown gall bacterium (*Agrobacterium tumefaciens*) induces the growth of tumors on the stem of a tomato (*Lycopersicon lycopersicum*) plant. The strain of *A. tumefaciens* used in genetic engineering is modified so that it contains a plasmid that does not cause tumors.

A further problem with gene insertion in plants is that researchers must use plant tissue culture to regenerate whole plants after individual cells have been genetically transformed. Tissue culture techniques do not work for all plants, although more than 1000 species have been regenerated in this way, and the number increases yearly. Recombinant DNA techniques cannot be used for plants in which tissue culture is a problem.

An additional complication of plant genetic engineering is that some important plant genes are located *in the chloroplasts* instead of the nucleus. Recall from Chapter 3 that chloroplasts are essential in photosynthesis, the process that is the basis of plant productivity. Chloroplast genetic engineering is currently the focus of much research interest.

Determining the Sequence of Nucleotides in DNA

Researchers have many applications for cloned DNA. They may clone a gene to obtain the encoded protein for some industrial or pharmaceutical process. Regardless of the particular application, before they engineer a gene, researchers must know a great deal about the gene and how it functions. The usual first step is determining the sequence of nucleotides.

PROCESS OF SCIENCE In the 1990s, the advent of automated **DNA-sequencing** machines connected to powerful computers let scientists sequence huge amounts of DNA quickly and reliably. Sequencing machines are powerful and decode about 1.5 million bases in a 24-hour period. These advances in sequencing technology made it possible for researchers to study the nucleotide sequences of entire **genomes** in a wide variety of organisms, both prokaryotic and eukaryotic. Much of this research received its initial impetus from the **Human Genome Project,** which began in 1990. The sequencing of the 3 billion base pairs of the human genome was essentially completed in 2001. The genomes of more than 100 species have been sequenced now, and several hundred more are in various stages of planning or completion. For example, scientists finished the rice genome in 2005. This was the second plant genome to be sequenced; the model research plant *Arabidopsis thaliana* was the first. We are in the middle of an extraordinary explosion of gene sequence data, largely due to automated sequencing methods. ■

DNA sequence information is now kept in large computer databases, many of which are accessible through the Internet. Examples are databases maintained by the National Center for Biotechnology Information and by the Human Genome Organization (HUGO). Geneticists use these databases to compare newly discovered sequences with those already known, to identify genes, and to access many other kinds of information. By searching for DNA sequences in a database, researchers gain insight into the function and structure of the gene's protein product, the evolutionary relationships among genes, and the variability among gene sequences within a population.

Genomics

Now that we have sequenced the genomes of humans, rice, and many other species, what do we do with that information? **Genomics** is the emerging field of biology that studies the entire DNA sequence of an organism's genome to identify all the genes, determine their RNA or protein products, and ascertain how the genes

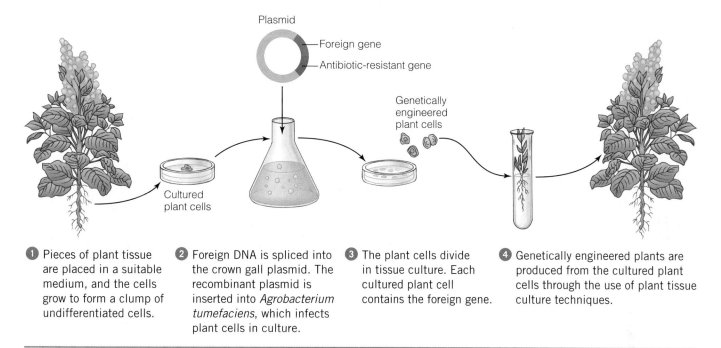

Plasmid
Foreign gene
Antibiotic-resistant gene

Genetically
engineered
plant cells

Cultured
plant cells

1 Pieces of plant tissue are placed in a suitable medium, and the cells grow to form a clump of undifferentiated cells.

2 Foreign DNA is spliced into the crown gall plasmid. The recombinant plasmid is inserted into *Agrobacterium tumefaciens*, which infects plant cells in culture.

3 The plant cells divide in tissue culture. Each cultured plant cell contains the foreign gene.

4 Genetically engineered plants are produced from the cultured plant cells through the use of plant tissue culture techniques.

FIGURE 15-8 **Genetic engineering.**

are regulated. As you will see, genomics has important practical applications in addition to answering scientific questions.

Most plant (and animal) genomes contain many DNA sequences that do not code for proteins. Although the entire genome is of interest to biologists, the protein-coding regions have the greatest potential for research and for medical applications. Researchers have compiled tens of thousands of short (25 to 30 nucleotides) DNA sequences, known as **expressed sequence tags (ESTs)**, that help identify protein-coding genes.

One way to study gene function is to silence genes one at a time. **RNA interference (RNAi)**, first discovered in the mid-1990s, can be used to quickly determine the function of a specific gene. RNAi is caused by small RNA molecules that interfere with the expression of genes or, more precisely, with their mRNA transcripts. After a protein-coding gene is identified, perhaps by using ESTs, the function of that gene is studied by using RNAi to shut the gene off. To do this, biologists synthesize a short stretch of RNA that is complementary to part of the DNA sequence of the gene being examined. After the gene is silenced, biologists observe any changes in the phenotype to help determine the function of the missing protein.

Several new fields have emerged as a result of the human genome project

We have already mentioned that the amount of genomic data now available requires that it be evaluated using powerful computers The discipline known as **bioinformatics,** or *biological computing,* includes the storage, retrieval, and comparison of DNA sequences within a given species and among different species. Bioinformatics uses powerful computers and sophisticated software to manage and analyze large amounts of data generated by sequencing and other technologies. For example, as researchers determine new DNA sequences, automated computer programs scan the sequences for patterns typically found in genes. Comparison of the databases of DNA sequences from different organisms provides insights about gene identification, gene function, and evolutionary relationships.

The new science of gene-based medicine known as **pharmacogenetics** has the goal to customize drugs to match a patient's genetic makeup (see *Focus On: DNA Microarrays*). Currently, a physician does not know in advance whether a particular medication will benefit a particular patient or cause severe side effects. An individual's genes, especially those that code for drug-

DNA SEQUENCING Procedure by which the sequence of nucleotides in DNA is determined.

GENOME All the genetic material contained in an individual.

GENOMICS The field of biology that studies the genomes of various organisms.

RNA INTERFERENCE Certain small RNA molecules that interfere with the expression of genes or their RNA transcripts.

FOCUS ON | DNA Microarrays

PROCESS OF SCIENCE DNA microarrays are a new technology that provide a way to study patterns of gene activity (expression) in an individual's cells. A DNA microarray consists of hundreds of different DNA molecules placed on a glass slide or chip (see figure). In ❶, a mechanical robot prepares the microarray that will be used for a medical test. The robot spots each location on the grid with thousands to millions of copies of a specific DNA strand. The single-stranded DNA molecules for each spot are amplified using the polymerase chain reaction (see *Focus On: The Polymerase Chain Reaction,* in Chapter 14).

In ❷, researchers isolate mRNA molecules from two cell populations—for example, liver cells treated with a newly developed drug and control liver cells not treated with the drug. Liver cells are often tested because the liver produces many enzymes that metabolize foreign molecules, including drug molecules. Drugs the liver cannot metabolize, or metabolizes weakly, may be too toxic to be used.

Researchers use isolated mRNA molecules to prepare DNA molecules for each cell population using a special enzyme called *reverse transcriptase.* (Recall from Chapter 14 that the process of transcription produces mRNA from DNA. Reverse transcriptase catalyzes the reverse of transcription—that is, it helps make DNA from mRNA.)

In ❸, the DNA molecules from each cell are then tagged with different-colored fluorescent dyes. The treated liver cells' DNA molecules might be labeled with red dye and the untreated liver cells' DNA molecules with green dye. In ❹, researchers add the two liver DNA populations to the array, and some of the liver DNA subsequently forms base pairs with the DNA on the array.

After washing the array to remove any DNA that has not formed base pairs, ❺ investigators scan the array with lasers to identify red and green fluorescence where base pairing has occurred. In ❻, computer analysis of the ratio of red to green fluorescence at each spot in the array produces a color-coded readout that researchers can further analyze. For example, a medical researcher might compare the experimental drug's overall pattern of gene activity with that of known drugs and toxins. If the gene activity for the treated cells matches that of a toxin that damages the liver, the drug will probably not go into clinical trials. ■

metabolizing enzymes, largely determine that person's response to a specific drug. Pharmacogenetics takes into account the subtle genetic differences among individuals. In as few as 10 years, patients may take routine genetic screening tests before a physician prescribes a drug.

The study of all the proteins encoded by an organism's genome and produced in its cells and tissues is **proteomics.** Scientists want to identify all the proteins made by a given kind of cell, but the process is much more complicated than sequencing an organism's genome. One reason is that some genes encode several different proteins. Also, every cell in a plant (or animal) has essentially the same genome, but cells in different tissues vary greatly in the kinds of proteins they produce. Protein expression patterns vary not only in different tissues but also at different stages in the development of a single cell. Scientists want to understand the role of each protein in a cell, how the various proteins interact, and the 3-D structure for each protein. While advancing biological knowledge, these goals also promise advances in medicine. If they know the shape of a protein associated with a certain type of cancer or other disease, pharmacologists may be able to develop drugs that bind to active sites on that protein to turn off its activity.

Applications of Genetic Engineering

Recombinant DNA technology has provided not only a new and unique set of tools for examining fundamental questions about cells but also new approaches to applied problems in many other fields. In some areas, the production of genetically engineered proteins and organisms has begun to have considerable impact on our lives, particularly in the fields of pharmacology and medicine.

DNA technology has revolutionized medicine and pharmacology

One of the first genetically engineered proteins approved for human use was human *insulin* produced by *Escherichia coli* bacteria. Before the use of recombinant DNA techniques to generate genetically altered bacteria capable of producing the human hormone, insulin was derived exclusively from other animals. Many diabetic patients become allergic to the insulin from animal sources because its amino acid sequence differs slightly from that of human insulin. The ability to produce the human hormone by recombinant DNA methods

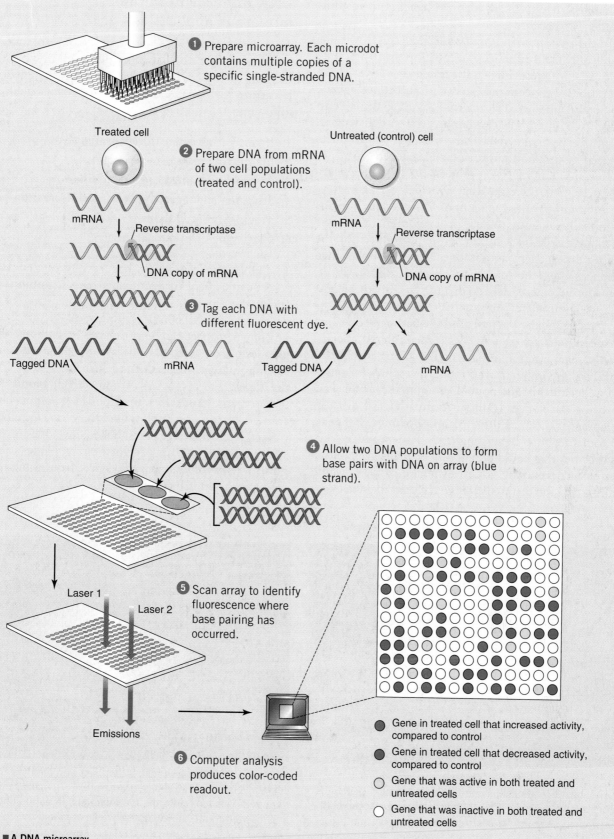

① Prepare microarray. Each microdot contains multiple copies of a specific single-stranded DNA.

Treated cell

Untreated (control) cell

② Prepare DNA from mRNA of two cell populations (treated and control).

mRNA

Reverse transcriptase

DNA copy of mRNA

mRNA

Reverse transcriptase

DNA copy of mRNA

③ Tag each DNA with different fluorescent dye.

Tagged DNA

mRNA

Tagged DNA

mRNA

④ Allow two DNA populations to form base pairs with DNA on array (blue strand).

⑤ Scan array to identify fluorescence where base pairing has occurred.

Laser 1

Laser 2

Emissions

⑥ Computer analysis produces color-coded readout.

● Gene in treated cell that increased activity, compared to control

● Gene in treated cell that decreased activity, compared to control

○ Gene that was active in both treated and untreated cells

○ Gene that was inactive in both treated and untreated cells

■ **A DNA microarray.**

has resulted in significant medical benefits to insulin-dependent diabetics.

In the past, human *growth hormone (GH)* was obtainable only from human cadavers. Only small amounts were available, and evidence suggested that some of the preparations were contaminated with infectious agents. Genetically engineered human GH is now available to children who need it to overcome growth deficiencies, specifically pituitary dwarfism. The list of products produced by genetic engineering continues to grow.

Recombinant DNA technology is increasingly used to produce vaccines that provide safe and effective immunity against infectious diseases in humans and animals. One way to develop a recombinant vaccine is to clone a gene for a surface protein produced by the disease-causing agent, or pathogen. The researcher then introduces the gene into a nondisease-causing vector. When the vaccine is delivered into the human or animal host, it stimulates an immune response. Because the immune system has "memory," if the pathogen carrying that specific surface protein is ever encountered, the immune system targets it for destruction. Human examples of antiviral recombinant vaccines are vaccines for influenza A, hepatitis B, and polio. Recombinant vaccines are also being developed against certain bacterial diseases and human cancers. Edible plants (potatoes, bananas, and tomatoes) are being engineered to produce low-cost vaccines that could potentially be made available to children in developing countries where proper vaccine handling and refrigerated storage are unavailable.

DNA fingerprinting has numerous applications

The analysis of DNA extracted from an individual, which is unique to that individual, is known as **DNA fingerprinting.** DNA fingerprinting has many applications in humans and other organisms, such as the following:

1. Investigating crime (forensic analysis)
2. Identifying mass disaster victims
3. Identifying human cancer cell lines
4. Studying endangered species in conservation biology
5 Tracking tainted foods
6. Clarifying disputed parentage

DNA fingerprinting has revolutionized law enforcement. The FBI established the Combined DNA Index System (CODIS) in 1990, consisting of DNA databases from all 50 states. Today, a DNA profile of an unknown suspect can be compared to millions of DNA profiles of convicted offenders in the database, often resulting in identification of the suspect. The DNA from the unknown suspect may come from blood, semen, bones, teeth, hair, saliva, urine, or feces left at the crime scene. Tiny amounts of human DNA have even been extracted from cigarette butts, licked envelopes or postage stamps, dandruff, fingerprints, razor blades, chewing gum, wristwatches, earwax, debris from under fingernails, and toothbrushes.

(a) The European corn borer, shown in the larval form, is the most destructive pest on corn in North America. Genetic engineers have designed *Bt* corn to control the European corn borer without heavy use of chemical insecticides.

(b) "Golden rice," shown here intermixed with regular white rice, contains high concentrations of β-carotene. To make golden rice, scientists took the gene for β-carotene in daffodil flowers and inserted it into the endosperm cells of rice.

FIGURE 15-9 Uses of transgenic plants.

PLANTS AND PEOPLE | Genetic Engineering and More Food for a Hungry World

The UN Food and Agriculture Organization estimates that more than 800 million people lack access to the food needed for healthy, productive lives. Most of these people live in rural areas of the poorest developing countries. Currently, there are 86 low-income, food-deficient countries that cannot produce enough food or afford to import enough food to feed the entire population. The two regions of the world with the greatest food insecurity are South Asia, with an estimated 270 million hungry people, and sub-Saharan Africa, with an estimated 175 million.

Producing enough food to feed the world's people is the largest single challenge in agriculture today, and the challenge grows more difficult each year as the human population expands. Since 1960, the human population has more than doubled, and science and industrialized agriculture have helped meet the food requirements of most of the population. We have expanded the area under cultivation and increased crop yields (output per unit of land area). But we have paid for these gains in food production with serious environmental problems, such as land degradation, depletion of groundwater, and diminishing surface water and wetland resources.

Genetic engineering has been touted by some as the high-technology answer to agriculture's problems. Although it is true that genetic engineering has the potential to revolutionize agriculture, the changes will not occur overnight. A great deal of basic research must be done before most of the envisioned benefits from genetic engineering are realized.

In addition, food security is influenced by many factors that genetic engineering cannot address quickly. For example, genetic uniformity has been bred into modern crop and livestock varieties for hundreds or even thousands of years. This loss of genetic diversity makes crops and livestock more vulnerable to diseases. Also, crops and livestock have been developed for specific climates and are therefore susceptible to climate change with its associated droughts, floods, and temperature shifts. Genetic engineering cannot provide quick answers to these kinds of challenges.

Despite its potential and real contributions to agriculture, genetic engineering is not the final solution to the challenge of producing enough food. Overshadowing all aspects of increased food production is one simple fact: The production of adequate food for the world's people will be an impossible goal until population growth is brought under control. The ultimate solution to world hunger is related to achieving a stable population in each nation at a level that it can sustainably support.

If applied properly, DNA fingerprinting has the power to identify the guilty and exonerate the innocent. Hundreds of convicted individuals have been released from incarceration, based on correlating DNA profiles with physical evidence from the crime scene. Such evidence is admissible in many court cases, although one limitation arises from the fact that the DNA samples are usually small and may have degraded. Obviously, great care must be taken to prevent contamination of the samples.

Transgenic plants are increasingly important in agriculture

Plants and animals in which foreign genes have been incorporated are referred to as **transgenic organisms.** The United States is currently the world's top producer of transgenic crops, also known as **genetically modified (GM) crops** (see *Plants and People: Genetic Engineering and More Food for a Hungry World*). Examples of GM crops are soybean, cotton, canola (used to make cooking oil), corn, papaya, and squash. In 2002, 74 percent of the U.S. soybean crop, 71 percent of its cotton, and 32 percent of its corn were GM crops. Farmers in Canada, Argentina, China, Brazil, Uruguay, and South Africa also grow a lot of GM crops. Most GM crops grown today are tolerant of herbicides (so herbicides can be sprayed to control weeds), resistant to insect pests (so insecticides do not have to be sprayed as much), or resistant to diseases.

Consider the European corn borer, which is the most damaging insect pest on corn in the United States and Canada (•**Figure 15-9a**). Efforts to control the European corn borer cost farmers more than $1 billion each year. Corn has been genetically modified to contain the *Bt* **gene,** a bacterial gene that codes for a protein with

TRANSGENIC ORGANISM A plant or other organism that has foreign DNA incorporated into its genome.

GENETICALLY MODIFIED (GM) CROP A crop plant that has had its genes intentionally manipulated; a transgenic crop plant.

PLANTS AND THE ENVIRONMENT | Field Testing of Genetically Engineered Plants

Much research focuses on determining the effects of introducing transgenic organisms into a natural environment. Carefully conducted tests have shown that transgenic organisms are not dangerous to the environment simply because they are transgenic. However, it is important to assess the risks of each new recombinant organism.

Scientists determine whether the organism has characteristics that might cause it to be environmentally hazardous under certain conditions. For example, if geneticists have engineered a transgenic crop plant, could those genes be transferred by pollen to that plant's relatives, generating "super-weeds" in a wild plant population? In 2003, ecologists at the University of Tennessee, Knoxville, announced the crossing of transgenic oilseed rape

plants that contained the *Bt* gene with its wild plant relative, which is a weed. They crossed the resulting hybrids with the wild relative again and then tested its ability to compete with other weeds in a field of wheat. The transgenic weed was a poor competitor and had less effect on wheat production than did its wild relatives in a control field. These results, although encouraging, must be interpreted with care. Scientists must evaluate each transgenic crop species individually to see if there is gene flow to wild relatives and, if so, the resulting effect.

Another concern is that nonpest species could be harmed. People paid a great deal of attention to the finding that monarch butterfly larvae raised in the laboratory are harmed if they are fed pollen from *Bt* corn plants.

Although more recent studies suggest that monarch larvae living in a natural environment do not consume enough pollen to cause damage, such concerns about effects on nontarget species persist and will have to be addressed individually.

EVOLUTION LINK Other concerns relate to plants engineered to produce pesticides, such as the *Bt* toxin. The future of the *Bt* toxin in transgenic crops is not secure, because low levels of the insecticide could potentially provide ideal conditions for selection for resistant individuals in the insect population. It appears certain that insects may evolve genetic resistance to the *Bt* toxin in transgenic plants in the same way they evolve genetic resistance to chemical insecticides. ■

insecticidal properties; *Bt* stands for the scientific name, *Bacillus thuringiensis. Bt* corn, introduced in the United States in 1996, does not need periodic sprays of chemical insecticides to control the European corn borer.

DNA technology also has the potential to develop crops that are more nutritious. For example, during the 1990s, geneticists engineered rice to produce high quantities of β-carotene, which the human body uses to make vitamin A (•**Figure 15-9b** on page 304). In developing countries, vitamin A deficiency is a leading cause of blindness in children. According to the World Health Organization, 275 million children are vitamin A deficient, and 250,000 to 500,000 become irreversibly blind each year. Vitamin A deficiency also makes children more susceptible to measles and other infectious diseases. Because rice is the staple diet in many countries with vitamin A deficiency, the widespread use of GM rice with β-carotene has the potential to prevent this deficiency in many of the world's children.

Certain transgenic plants can potentially be "pharmed" to produce large quantities of medically important proteins, such as human antibodies against the herpes virus, clot-busting drugs, and monoclonal antibodies used to treat colon cancer. Methods for de-

veloping transgenic plants are well established, but it is difficult to get plants to produce the desired protein in large enough quantities. To date, most transgenic plants developed for pharming have produced foreign proteins equal to only about 1 percent of the plant's total protein output (although this amount is not insignificant in soybeans, which contain 40 to 45 percent protein). The developers of this technology are working to increase the production of foreign proteins.

Some people are concerned about the health effects of consuming foods derived from GM crops and think that such foods should be restricted. For example, critics say some consumers may develop food allergies. Scientists also recognize this concern and routinely screen new GM crops for allergenicity.

DNA technology has raised safety concerns

When recombinant DNA technology was introduced in the early 1970s, many scientists considered the potential misuses at least as significant as the possible benefits. The possibility that an organism with undesirable environmental effects might be accidentally produced was of great concern, because new strains of bacteria or other organisms, with which the world has no previous experience, might be difficult to control. The scientists who initially developed the recombinant DNA methods insisted on stringent guidelines for making the new technology safe.

Experiments in thousands of university and industrial laboratories over more than 30 years have shown that recombinant DNA manipulations can be carried out safely. Geneticists have designed laboratory strains of *E. coli* to die in the outside world if they somehow escape from the lab. Researchers carry out experiments that might present unusual risks in facilities designed to hold pathogenic organisms; this precaution ensures that researchers work with them safely. So far no evidence suggests that researchers have accidentally cloned hazardous genes or have released dangerous organisms into the environment. However, malicious, *intentional* manipulation of dangerous genes certainly remains a possibility.

As the safety of the experiments has become established, scientists have relaxed many of the restrictive guidelines for using recombinant DNA. Stringent restrictions still exist, however, in certain areas of recombinant DNA research where there are known dangers or where questions about possible effects on the environment are still unanswered. These restrictions are most evident in research that proposes to introduce transgenic organisms into the wild, such as agricultural strains of plants whose seeds or pollen might spread in an uncontrolled manner (see *Plants and the Environment: Field Testing of Genetically Engineered Plants*).

To summarize, DNA technology in agriculture offers many potential benefits, including more nutritious foods, the reduced use of chemical pesticides, and higher yields, by providing disease resistance. However, like other kinds of technology, genetic engineering poses some risks, such as the risk that genetically modified plants and animals could pass their foreign genes to wild relatives, which could result in previously unknown environmental problems. The science of **risk assessment**, which uses statistical methods to quantify risks so they can be compared and contrasted, will help society decide whether to ignore, reduce, or eliminate specific risks of genetically engineered organisms.

STUDY OUTLINE

❶ Define *genetic engineering*, and outline the primary techniques used in recombinant DNA technology, including genetic probes and DNA cloning.

Genetic engineering is the manipulation of genes, often through recombinant DNA technology. **Recombinant DNA technology** consists of the techniques used to make DNA molecules by combining genes from different organisms. First, DNA molecules are cleaved at specific base sequences to break them into smaller fragments. Segments of DNA from different sources are then joined, forming a recombinant DNA molecule, and these molecules are taken into a host cell. The cells that take up the specific gene of interest are identified with a **genetic probe**, a single-stranded nucleic acid used to identify a complementary sequence by base pairing with it. The gene may be transcribed and translated within the host cell, leading to the production of a protein not previously produced by the host organism. **DNA cloning**, another aspect of recombinant DNA technology, is the process of selectively amplifying DNA sequences so their structure and function can be studied.

❷ Explain the actions and importance of restriction enzymes and ligase.

Restriction enzymes are used in recombinant DNA technology to cleave DNA at specific base sequences. That is, restriction enzymes break the DNA molecule into more manageable fragments. Segments of DNA from different sources are joined by the enzyme **ligase**.

❸ Identify the role of biological vectors, such as plasmids, in recombinant DNA technology, and describe a biological vector and a nonbiological method used to introduce genes into plant cells.

A **vector** is an agent, such as a plasmid or virus, that transfers DNA from one organism to another. A **plasmid** is a small, circular DNA molecule that carries genes separate from the main DNA of a bacterial cell. The plasmid of *Agrobacterium* is an effective vector for introducing genes into many plant cells. A nonbiological approach to introduce DNA into plant cells is a genetic "shotgun." Researchers coat microscopic gold or tungsten fragments with DNA and then shoot them into plant cells.

❹ Define *DNA sequencing*.
DNA sequencing is the procedure by which the sequence of nucleotides in DNA is determined. Automated DNA-sequencing machines connected to powerful computers let scientists sequence huge amounts of DNA quickly and reliably.

❺ Define *genome*, and briefly describe the emerging field of genomics.
A **genome** consists of all the genetic material contained in an individual. **Genomics** is the field of biology that studies the genomes of various organisms to identify all the genes, determine their RNA or protein products, and ascertain how the genes are regulated.

❻ Explain how RNA interference is used to study gene function.
RNA interference (RNAi) makes use of certain small RNA molecules that interfere with the expression of genes or their RNA transcripts. After a protein-coding gene is identified, the function of that gene can be studied using RNAi to shut the gene off. After the gene is silenced, biologists observe any changes in the phenotype to help determine the function of the missing protein.

❼ Describe at least one application of recombinant DNA technology in each of the following: medicine and pharmacology, DNA fingerprinting, and transgenic organisms, specifically genetically modified crops.
Escherichia coli have been genetically engineered to produce human insulin, which has resulted in significant medical benefits to insulin-dependent diabetics. The analysis of DNA extracted from an individual is known as **DNA fingerprinting.** DNA fingerprinting has many applications, such as investigating crime (forensic analysis), studying endangered species in conservation biology, and clarifying disputed parentage. A **transgenic organism** is a plant or other organism that has foreign DNA incorporated into its genome. A **genetically modified (GM) crop**—that is, a transgenic crop plant—has had its genes intentionally manipulated. Agricultural geneticists are developing GM plants that are resistant to insect pests, viral diseases, drought, heat, cold, herbicides, and salty or acidic soil.

❽ Discuss safety issues associated with recombinant DNA technology, and explain how these issues are being addressed.
Researchers were initially concerned that genetically engineered organisms would be dangerous if they escaped into the environment, so they carried out experiments that might present unusual risks in facilities designed to hold pathogenic organisms. So far no evidence suggests that researchers have accidentally cloned hazardous genes or have released dangerous organisms into the environment. As safety has become established, scientists have relaxed many of the restrictive guidelines for using recombinant DNA. Stringent restrictions still exist where questions about possible effects on the environment are unanswered, for example, in research that proposes to introduce transgenic organisms into the wild.

REVIEW QUESTIONS

1. What is genetic engineering?
2. What are restriction enzymes? Ligases? How are these enzymes employed in recombinant DNA research?
3. Explain how recombinant DNA molecules are usually constructed.
4. Describe how a foreign gene is implanted in a bacterial cell. What might be gained by this procedure?
5. Explain how a foreign gene might be inserted into a plant cell.
6. What is a genetic probe? How is it used to identify genetically modified cells that contain a specific gene?
7. How has the Human Genome Project triggered the DNA sequencing of many other organisms?
8. What is genomics?
9. What is RNA interference, and how is it used to study gene function?
10. What is a DNA microarray?
11. What are some of the benefits of plant genetic engineering?
12. Why does research involving recombinant DNA technology have safety guidelines?

THOUGHT QUESTIONS

1. Would genetic engineering be possible if restriction enzymes did not exist? Explain.
2. What are some of the environmental concerns regarding transgenic organisms? What kinds of information does society need to determine if these concerns are valid?
3. Think of at least two potential benefits of plant genetic engineering *not* discussed in this chapter.

Visit us on the web at http://www.thomsonedu.com/biology/berg/ for additional resources such as flashcards, tutorial quizzes, further readings, and web links.

Continuity through Evolution

Cabbage, brussels sprouts, broccoli, cauliflower, collard greens, kale, and kohlrabi are distinct vegetable crops, but they are all members of the same species, *Brassica oleracea,* which is thought to have originated in Europe. Cabbage, which forms a large head of overlapping leaves around a small terminal bud, has been cultivated for centuries. Brussels sprouts grow tiny cabbagelike axillary buds along a main unbranched stem. This vegetable is named for the city of Brussels, Belgium, where brussels sprouts first became popular. The edible parts of the broccoli plant are the highly nutritious green flower buds, which form dense clusters on the ends of succulent stems. The edible parts of the cauliflower plant are the dense clusters of short-stemmed, immature flower buds. Unlike broccoli flowers, cauliflower flowers are mostly sterile. Collards and kale are closely related leafy plants. Their edible leaves do not overlap to form dense heads as those of cabbage do. Kohlrabi is a lesser-known vegetable with a mild flavor reminiscent of turnips. It has an enlarged, fleshy, aerial storage stem that is eaten cooked or raw.

It is appropriate to begin our study of evolution with the vegetable crops of *B. oleracea.* All seven were produced by selective breeding of the colewort, or wild cabbage, a leafy plant native to Europe and Asia. In *selective breeding,* plants with desirable characteristics are selected for propagation. Beginning more than 4000 years ago, some breeders artificially selected wild cabbage plants that formed overlapping leaves; over time, these leaves were so emphasized that the plants, which resembled modern-day cabbages, were recognized as distinct from their wild cabbage ancestor. Other breeders emphasized different parts of the wild cabbage, giving rise to the other modifications. For example, kohlrabi developed after selection for an enlarged storage stem. Thus, humans are responsible for the evolution of *B. oleracea* into seven distinct vegetable crops.

Charles Darwin, the originator of much of the modern view of evolution, was impressed by the changes induced by selective breeding and hypothesized that a similar selective process occurred in nature. Darwin therefore used selective breeding as a model when he developed the concept of evolution by natural selection.

Selective breeding. Selective breeding has resulted in many varieties of the same species, *Brassica oleracea.* Shown are cabbage, broccoli, collards, cauliflower, brussels sprouts, and kohlrabi.

Pre-Darwinian Ideas about Evolution

All the organisms on our planet, from microscopic bacteria to coastal redwoods, from tropical tree frogs to desert cacti, evolved from one or a few simple kinds of organisms. This vast diversity of species developed from earlier species by a process Charles Darwin—a 19th-century naturalist widely considered one of the greatest biologists of all time—called "descent with modification," or evolution. In biology, **evolution** is defined as genetic change in a population of organisms. (A *population* is a group of individuals of the same species that live in a particular place at a specific time.) The word *evolution* refers not to the changes that occur to an individual organism during its lifetime but to the changes in populations over many generations.

Although Charles Darwin is universally associated with evolution, ideas concerning the gradual evolution of life predate Darwin by centuries. Aristotle (384–322 BCE) saw much evidence of design and purpose in nature and arranged all the organisms that he knew in a "Scale of Nature" that extended from the simple to the complex. He visualized organisms as being imperfect but moving toward a more perfect state. Some have interpreted this idea as a rudimentary concept of evolution, but Aristotle is vague on the nature of life's "movement toward the more perfect state" and certainly did not propose that natural processes drove the process of evolution. Furthermore, biologists now recognize that evolution does not move toward more "perfect" states, or even necessarily toward greater complexity.

Long before Darwin, fossils—fragments resembling bones, teeth, leaves, and shells—had been discovered embedded in rocks all over the world. Some of these fossils corresponded to parts of familiar living organisms, but others were strangely unlike any known form. Fossils of marine invertebrates, such as clams and corals, were found in rocks high on mountains. In the 15th century, Leonardo da Vinci correctly interpreted such finds as the remains of organisms that had existed in previous ages but had become extinct.

The French naturalist Jean Baptiste de Lamarck (1744–1829) was the first scientist to propose that organisms undergo change over time as a result of some natural phenomenon rather than divine intervention. According to Lamarck, a changing environment caused an organism to alter its behavior, thereby using some organs or body parts more and others less. Over several generations, a given organ or body part would increase in size if it was used a lot, or shrink and possibly disappear if it was used less. His hypothesis required that organisms pass traits they acquired during their lifetimes to their offspring.

As an example of this line of reasoning, Lamarck suggested that the long neck of the giraffe had evolved when a short-necked ancestor began browsing on the leaves of trees instead of grass. Lamarck reasoned that the ancestral giraffe, in reaching up, stretched and elongated its neck. Its offspring, after inheriting the longer neck, stretched still farther. The process, repeated over many generations, resulted in the long necks of modern giraffes.

Lamarck's proposed mechanism of evolution is quite different from the mechanism later proposed by Darwin. However, Lamarck's hypothesis remained a reasonable explanation for evolution until Mendel's basis of heredity was rediscovered at the beginning of the 20th century. At that time, Lamarck's ideas were largely discredited.

Darwin and Evolution

Charles Darwin (1809–1882), the son of a prominent physician, was sent at the age of 15 to study medicine at the University of Edinburgh. Finding himself unsuited for medicine, he transferred to Cambridge University to study theology. During that time, he became the protégé of the Reverend John Henslow, who was a professor of botany. Henslow encouraged Darwin's interest in the natural world.

Shortly after receiving his degree, the 22-year-old Darwin embarked on the HMS *Beagle,* a ship that was taking a 5-year exploratory cruise around the world to prepare navigation charts for the British Navy (•**Figure 16-1**). The *Beagle* sailed from Plymouth, England, in 1831, crossed the Atlantic, and cruised slowly down the east coast and up the west coast of South America. While other members of the company mapped the coasts and harbors, Darwin studied the animals, plants, fossils, and geologic formations of both coastal and inland regions, areas that had not been extensively explored. He observed firsthand the diversity of the plants and animals of these regions. During the *Beagle*'s voyage, Darwin collected and catalogued thousands of specimens of plants and animals and kept notes of his observations.

In 1832 the *Beagle* stopped for almost two months at the Galápagos Islands, 965 kilometers (600 miles) west of Ecuador. There Darwin continued his observations and collections (•**Figure 16-2**). He compared the plants and animals of the Galápagos with those of the South American mainland, which he had observed the previous year. He was particularly impressed by their similarities and wondered why, for example, the plants and animals of the Galápagos should resemble those from

EVOLUTION The accumulation of inherited changes within populations over time.

South America more than they resembled those from other islands, where the environmental conditions were more similar. Moreover, although there were similarities between Galápagos and South American species, there were also distinct differences. Birds and reptiles even differed from one island in the Galápagos to the next! After he returned home, Darwin pondered these observations and tried to develop a satisfactory explanation for the distribution of the plants and animals among the islands.

Europeans in the mid-1800s generally believed that organisms were the results of divine creation and therefore did not change significantly over time. They thought each species looked the same as it had the day it was created. True, there were some obvious exceptions to this premise. Breeders and farmers could develop many varieties of domesticated plants and animals in just a few generations. They did so by choosing certain traits and breeding only individuals with the desired traits, a procedure known as **selective breeding**, or *artificial selection*. Fossils discovered during the mid-19th century also contradicted the accepted views, because many fossils did not resemble any living organism.

The general notion in the mid-1800s was that Earth was too young for organisms to have changed significantly since they had first appeared. During the early 19th century, geologists advanced the idea that mountains, valleys, and other physical features of Earth's surface had not been created in their present forms. Instead, these features developed over long periods by the slow geologic processes of volcanic activity, uplift, erosion, and glaciation. Darwin took *Principles of Geology,* published by English geologist Charles Lyell in 1830, with him on his voyage and studied it carefully. Lyell provided an important concept for Darwin—that the slow pace of geologic processes, which still occur today, indicated that Earth was extremely old.

(a) Charles Darwin as a young man. This portrait was made shortly after Darwin returned to England from his voyage around the world.

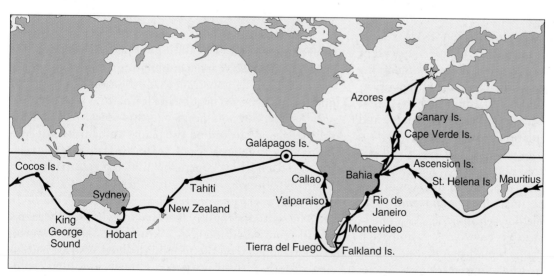

(b) The voyage of HMS *Beagle*. The 5-year voyage began in Plymouth, England, in 1831. Darwin's observations of the Galápagos Islands were the basis for his theory of evolution by natural selection.

FIGURE 16-1 Charles Darwin.

(a) Giant prickly pear trees (*Opuntia echios*). Other *Opuntia* species in the Galápagos are not trees.

(b) A lava cactus (*Brachycereus nesioticus*) that grows on recent lava flows is found only in the Galápagos Islands.

(c) A land iguana (*Conolophus subcristatus*) is also found nowhere in the world except the Galápagos.

(d) A red-footed booby (*Sula sula*) nests in shrubs and trees. Why its ducklike feet are bright red is not known.

FIGURE 16-2 The plants and animals of the Galápagos Islands.

The ideas of Thomas Malthus (1766–1834), a British clergyman and economist, were another important influence on Darwin. Malthus noted that populations increase in size until checked by factors in the environment. In the case of humans, Malthus suggested that wars, famine, and pestilence served as the inevitable and necessary brakes on population growth.

Darwin's years of observing the habits of plants and animals had introduced him to the struggle for existence described by Malthus. It occurred to Darwin that in this struggle, favorable variations would tend to be preserved and unfavorable ones eliminated. As a result, a population would adapt to the environment, and eventually, enough modifications would give rise to a new species. Time was all that was required for new species to evolve, and the geologists of the era, including Lyell, had supplied evidence that Earth was indeed old enough to provide an adequate amount of time for such changes to have occurred.

Darwin had at last conceived of a working theory of evolution—that of evolution by natural selection. He spent the next 20 years accumulating an immense body of evidence to demonstrate that evolution had occurred and formulating his arguments for natural selection.

As Darwin was pondering his ideas, Alfred Russel Wallace, who studied the plants and animals of Malaysia and Indonesia for 8 years, was similarly struck by the diversity of organisms and the peculiarities of their distribution. Wallace, too, arrived at the conclusion that evolution occurred by natural selection. In 1858 he sent a brief essay to Darwin, who was by then a world-renowned biologist, asking his opinion. Darwin recognized his own theory and realized Wallace had independently arrived at the same conclusion—that evolution occurs by natural selection.

Darwin's colleagues persuaded him to have Wallace's paper presented along with an abstract of his own views, which he had prepared and circulated to a few friends several years earlier. Both papers were presented at a meeting of the Linnaean Society in London in July 1858. Darwin's classic book *On the Origin of Species by Means of Natural Selection* was published in 1859.

Natural selection has four premises

Natural selection, the process in nature that causes evolution, is based on the tendency of organisms that have favorable **adaptations** to their environments to survive and become the parents of the next generation. Evolution by natural selection consists of four observations about the natural world: overproduction, inherited variation, limits on population growth, and differential reproductive success.

1. *Overproduction.* Natural populations have the reproductive potential to continuously increase their numbers over time. For example, each individual puffball (see Figure 21-12c) produces millions of spores, and if each of those spores germinated and, in turn, gave rise to millions of spores, the land would be covered with puffballs in a short time. Yet puffballs have not overrun the planet. Thus, in every generation, each species has the capacity to produce more offspring than will survive to reproductive maturity.

2. *Inherited variation.* The individuals in a population exhibit inherited variation. Each individual has a unique combination of traits, such as size, color, and ability to tolerate harsh environmental conditions. Some traits improve the chances of an individual's survival and reproductive success, whereas others do not. Although Darwin recognized the importance to evolution of inherited variation, he did not know the mechanism of inheritance.

3. *Limits on population growth.* Only so much food, water, light, space for growing, and other resources are available in the environment, and organisms must compete with one another for these limited resources in their struggle for existence. Because there are more individuals than the environment can support, not all will survive to reproduce. Other limits on population growth include predators, disease organisms, and unfavorable weather conditions.

4. *Differential reproductive success.* The individuals with the most favorable combination of characteristics (those that make individuals better adapted to their environment) are more likely to survive and reproduce. Offspring tend to resemble their parents, because the next generation inherits the parents' genetically based traits. Successful reproduction is the key to natural selection: The best-adapted individuals produce the most offspring, whereas individuals that are less well adapted die prematurely or produce fewer or inferior offspring.

Over time, enough changes may accumulate in a geographically separated population (living in a slightly different environment) to form a new species. The evolution of new species is considered in greater detail in Chapter 17.

The modern synthesis combines Darwin's theory with Mendelian genetics

One of the premises on which Darwin based his theory of evolution by natural selection is that individuals pass traits on to the next generation. However, Darwin could not explain *how* traits were passed from one generation to another or *why* individuals vary within a population. Darwin was a contemporary of Gregor Mendel, who worked out the basic patterns of inheritance. However, Darwin was apparently not acquainted with Mendel's work, which was not recognized by the scientific community until the early part of the 20th century.

Beginning in the 1930s and 1940s, biologists experienced a conceptual breakthrough when they combined the principles of Mendelian inheritance with Darwin's theory of natural selection. The result was a unified explanation of evolution known as the **modern synthesis,** or the **synthetic theory of evolution.** In this context, *synthesis* refers to combining parts of several previous theories to form a unified whole.

Today, the modern synthesis incorporates our expanding knowledge in genetics, systematics, paleontology, developmental biology, behavior, and ecology. It explains Darwin's observation of variation among offspring in terms of **mutation,** or changes in DNA, such as nucleotide substitutions. Mutations provide the genetic variability on which natural selection acts during evolution. The modern synthesis, which emphasizes the genetics of populations as the central focus of evolution, has held up well since it was developed. It has dominated the thinking and research of biologists working in many areas and has resulted in an enormous accumulation of new discoveries that validate evolution by natural selection.

PROCESS OF SCIENCE Although virtually all biologists accept the basic principles of the modern synthesis, they

NATURAL SELECTION The mechanism of evolution in which individuals with inherited characteristics well suited to the environment leave more offspring than individuals that are less suited to the environment do.

ADAPTATION An evolutionary modification that improves an organism's chances of survival and reproductive success.

MODERN SYNTHESIS A comprehensive, unified explanation of evolution based on combining previous theories, especially of Mendelian genetics, with Darwin's theory of evolution by natural selection.

may disagree over certain aspects of the evolutionary process. For example, what is the role of chance in determining the direction of evolution? How rapidly do new species evolve? These questions have arisen in part from a reevaluation of the fossil record and in part from discoveries about molecular aspects of inheritance. Debates are an integral part of the scientific process; they stimulate additional observation, experimentation, and reanalysis of older evidence. Science is a continuing process, and information obtained in the future may require us to modify certain parts of the modern synthesis. ■

Evidence for Evolution

Evolution by natural selection is supported by a vast body of scientific observations and experiments (see *Plants and People: Evolution and Bacterial Resistance to Antibiotics*). In this book we can report only a small fraction of this evidence, which comes from both living organisms and the fossil record. Taken together, this evidence confirms the theory that life unfolded on Earth by the process of evolution.

Fossils indicate that organisms evolved in the past

Perhaps the most direct evidence for evolution comes from the discovery, identification, and interpretation of fossils. **Fossils** (Latin *fossilis*, "something dug up") are

the remains or traces of organisms usually left in layers of rock (●Figure 16-3). Fossils provide a record of plants and animals that lived earlier, some understanding of where and when they lived, and an idea of the lifestyles they had. By studying the fossils of organisms of different geologic ages, we can trace the lines of evolution that gave rise to those organisms. Thus, the fossil record, although incomplete, provides a history of life on Earth.

The formation and preservation of a fossil require that the organism be buried under conditions that slow the process of decay. Slow decay is most likely if the remains are covered quickly by fine particles of soil suspended in water, which envelop the organism with sediment. Remains of aquatic organisms may be buried in bogs, mudflats, sandbars, or river deltas. Remains of terrestrial organisms may also be covered by waterborne sediments in a floodplain or by windblown sand in an arid region. Over time, as more sediments accumulate, the underlying sediments become cemented together to form sedimentary rock.

Because of the conditions required for preservation, the fossil record is not a random sample of all past life. The record is biased toward aquatic organisms and those living in the few terrestrial habitats that are conducive to fossil formation, such as deserts, areas with tar pits, and arctic regions. For example, relatively few fossils of tropical rainforest organisms have been found, because plant and animal remains decay rapidly on the forest floor, before fossils can form. Another reason for

Carlyn Iverson

(a) An impression fossil of a cottonwood (*Populus*) leaf that formed about 50 million years ago.

© Willard Clay/Jupiterimages

(b) Fossilized wood from the Petrified Forest National Park in Arizona.

FIGURE 16-3 Plant fossils.

PLANTS AND PEOPLE | Evolution and Bacterial Resistance to Antibiotics

EVOLUTION LINK In the late 1980s, the U.S. Centers for Disease Control and Prevention began to document an alarming increase in the incidence of tuberculosis (TB). In the 30 or so years before that time, the number of TB cases had declined in the United States, largely as a result of treating TB with antibiotics, which are drugs intended to harm or kill bacteria and other microorganisms. Many people are exposed to the bacteria that cause TB, but only people who are young, old, or weakened from some other disease usually exhibit symptoms.

Tuberculosis exhibits a disturbing trend in which drug-resistant strains of the bacteria that cause TB have evolved. These strains are resistant to one or more antibiotics that traditionally were used to treat TB. Drug-resistant TB is deadly: as many as 80 percent of the people infected with multidrug-resistant TB (MDR-TB) die within two months of diagnosis, even with medical care.

Bacterial resistance to antibiotics is an example of evolution. Bacteria are continually evolving, even inside the bodies of human and animal hosts. When an antibiotic is used to treat a bacterial infection, a few bacteria may survive because they are genetically resistant to the antibiotic,* and they pass these genes to future generations. As a result, the bacterial population contains a larger percentage of antibiotic-resistant bacteria than before.

Drug resistance is usually found in individuals who were previously treated for TB, and quite often human behavior is a factor in the evolution of drug resistance. A person infected with TB must take 3 to 10 pills of antibiotics each day for at least six months. After the first week or two of treatment, the individual usually feels better. Many individuals decide to quit taking their medication at this point. At this time, the TB bacteria still lurking in their bodies—those with a resistance to the antibiotic—rally. (If the individual continued taking the antibiotic, the body's immune system would have enough time to effectively eliminate the antibiotic-resistant bacteria, presumably because the individual would not have to fight any other infections during

*Bacteria obtain the genes for resistance to antibiotics from mutations, plasmids, or viruses (see Chapters 14 and 15).

that period.) The evolution of bacteria resistant to several antibiotics is a worst-case scenario, because MDR-TB is extremely difficult to treat effectively.

Industrialized agriculture has added a further complication to antibiotic resistance. Healthy pigs, chickens, and cattle that consume low doses of antibiotics in their feed typically gain more weight than animals that do not receive antibiotics (because they expend less energy fighting infections). Several studies have linked the indiscriminate use of antibiotics in humans and livestock to the increasing resistance of bacteria to antibiotics. In 2003, the World Health Organization recommended that routine use of antibiotics in livestock be eliminated, but the United States and many other countries continue the practice.

The tuberculosis bacterium is not the only bacterium to evolve resistance to antibiotics. The Worldwatch Institute reports that antibiotic resistance has been documented in more than 20 kinds of potentially harmful bacteria; some of these bacterial strains are resistant to every known antibiotic—more than 100 drugs. ∎

bias is that organisms with hard body parts, such as wood, bones, or shells, are more likely to form fossils than are those with soft tissues, such as flowers, mushrooms, or worms.

To be interpreted, the sedimentary layers containing fossils must be arranged in chronological order (see *Focus On: Dating Fossils*). The layers of sedimentary rock, if they have not been disturbed, occur in the sequence of their deposition, with the more recent layers on top of the older, earlier ones.

Biologists use fossils to study the pace of evolution

Biologists have long recognized that the fossil record lacks many transitional forms; that is, the starting points and end points are present, but the intermediate stages in the evolution from one species to another may be absent. The traditional Darwinian view of evolution is embodied in the *gradualism mode*, which reasons that evolution proceeds at a more or less steady rate, but it is not observed in the fossil record because the record is incomplete. Occasionally, a complete fossil record of transitional forms *is* discovered and is cited as a case for gradualism. The gradualism mode maintains that populations diverge from one another slowly by the gradual accumulation of adaptive characteristics within each population. These adaptive characteristics accumulate as a result of differing selective pressures on the populations living in different environments.

FOSSIL A part or trace of an ancient organism, usually preserved in sedimentary rock.

FOCUS ON | Dating Fossils

PROCESS OF SCIENCE To interpret fossil evidence of the history of life, investigators must date the rocks in which fossils are found. The radioactive isotopes that are present in a rock give us an accurate measure of its age. Recall from Chapter 2 that elements exist as different isotopes. Some isotopes (those with excess neutrons) are unstable and tend to break down, or decay, to a more stable isotope, usually of a different element. Such isotopes, termed **radioactive isotopes,** emit powerful, invisible radiation when they decay. As a radioactive isotope emits radiation, its atomic nucleus changes into the nucleus of a different element. For example, the nucleus of one radioactive isotope of uranium (U-235) decays over time into lead (Pb-207).

Each radioactive isotope has its own characteristic rate of decay. The time required for one-half of a radioactive isotope to change into a different material is known as its **half-life** (see figure). There is enormous variation in the half-lives of different radioactive isotopes. For example, the half-life of iodine (I-132) is only 2.4 hours, whereas the half-life of uranium (U-235) is 704 million years. The half-life of a particular radioactive isotope is constant and never varies; it is not influenced by temperature, pressure, or any other environmental factor.

Scientists estimate the age of a fossil by measuring the proportion of the original radioactive isotope and its decay product. For example, the half-life of a radioactive isotope of potassium (K-40) is 1.25 billion years, meaning that in 1.25 billion years half of the radioactive potassium in a rock sample will have decayed into its decay product, argon (Ar-40). If the ratio of potassium (K-40) to argon (Ar-40) in the rock being tested is 1:1, the rock is 1.25 billion years old. (The radioactive clock begins ticking when the rock solidifies. The rock initially contains some potassium but no argon.)

Several radioactive isotopes are used in dating fossils; a few examples include potassium (K-40; half-life 1.25 billion years), uranium (U-235; half-life 704 million years), and carbon (C-14; half-life 5730 years). C-14 is used to date the carbon remains of anything that was once living, such as wood, bones, and shells; the other

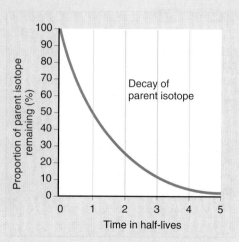

■ The decay of radioactive isotopes. At time zero, the radioactive clock begins ticking. At this point a sample is composed entirely of the radioactive isotope. After one half-life, only 50 percent of the original radioactive isotope remains.

isotopes are used to date the rock in which fossils are found. Because of its relatively short half-life, C-14 is useful for dating fossils and artifacts that are 50,000 years old or less. In contrast, K-40, with its long half-life, is used to date fossils that are hundreds of millions of years old. Wherever possible, the age of a fossil is verified independently, using two or more radioactive isotopes. ■

Beginning in the 1970s, many biologists began to question whether the fossil record really is incomplete. The *punctuated equilibrium mode* proposes that the fossil record reflects evolution accurately, with long periods of little or no evolutionary change punctuated, or interrupted, by short periods of rapid evolution. In punctuated equilibrium, evolution normally proceeds in relatively short "spurts" interspersed with long periods of inactivity. During each spurt, new species form, and many older species become extinct. Punctuated equilibrium accounts for the abrupt appearance of a new species in the fossil record with little or no record of intermediate forms. That is, there are few transitional forms in the fossil record because most species evolved so rapidly that they did not produce transitional forms.

The fact that there is abundant evidence in the fossil record of long periods with little or no change in a species seems to argue against gradualism. Gradualists, however, point out that little or no change in fossils is deceptive because fossils do not show all aspects of evolutionary change. Fossils typically show changes in hard body parts and external structures but do not reveal such characteristics as changes in cell physiology, behavior, and ecological roles, which also represent evolution.

Comparative anatomy of related species demonstrates similarities in their structures

Comparison of the structural details of any given organ found in different but related organisms reveals a basic similarity of form with some variation from one group to another. For example, a cactus spine and a pea tendril, although quite different in appearance, are similar in their underlying structure and development, because

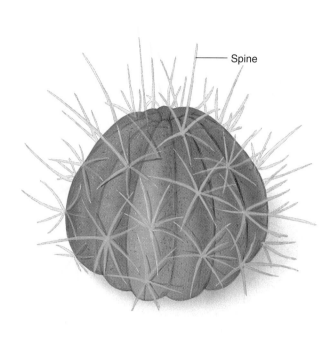

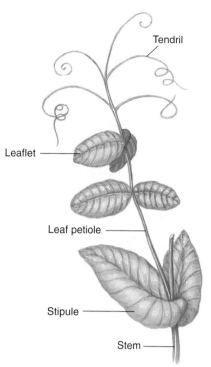

(a) The spines of the fishhook cactus (*Ferocactus wislizenii*) are modified leaves.

(b) The tendrils of the garden pea (*Pisum sativum*) are also modified leaves. Leaves of the garden pea are compound, and the terminal leaflets are modified into tendrils that are frequently branched. Note the leafy stipules at the base of the leaf; the stipules in the garden pea are often larger than the leaflets.

FIGURE 16-4 Homologous features.

both are modified leaves (•**Figure 16-4**). The spine protects the succulent stem tissue of the cactus, whereas the tendril, which winds around a small object once it makes contact, helps support the climbing stem of the pea plant. Darwin pointed out that such basic similarities in organs used in different ways are precisely the expected outcome of evolution. Superficially dissimilar organs or other structures that are similar in underlying form in different organisms because of a common evolutionary origin are termed **homologous features.** Spines and tendrils are homologous features.

Organs or other structures that are not homologous but simply have similar functions in different organisms are termed **homoplastic features.**[1] For example, spines, which are modified leaves, and thorns, which are modified stems, are homoplastic features that have evolved over time to meet, in different ways, the common need for protection (•**Figure 16-5**). Hawthorns and honey lo-

custs are examples of plants that produce thorns. Tendrils are an additional example of homoplastic features in plants. Grape and Boston ivy tendrils are modified shoots, and pea tendrils are modified leaves.

Like homologous features, homoplastic features offer crucial proof of evolution. Homoplastic features are of evolutionary interest because they show how unrelated groups may adapt in similar ways to common situations. The independent evolution of similarities in unrelated organisms as a result of adaptation to similar environmental conditions is known as **convergent evolution.** For example, convergent evolution has resulted

---

[1]*An older, less precise term that some biologists still use for nonhomologous features with similar functions is* analogous features.

HOMOLOGOUS FEATURES Dissimilar structures with an underlying similarity of form and development that occur in different species with a common ancestry.

HOMOPLASTIC FEATURES Structures in unrelated species that are similar in function and appearance but not in evolutionary origin.

CONVERGENT EVOLUTION The independent evolution of similar adaptations in unrelated species.

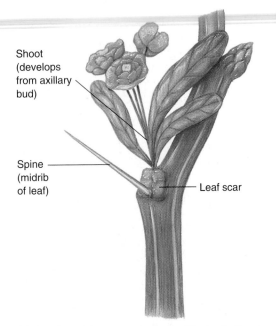

Shoot (develops from axillary bud)

Spine (midrib of leaf)

Leaf scar

Thorn (develops from axillary bud)

(a) A spine of Japanese barberry (*Berberis thunbergii*) is a modified leaf. (In this example, the spine is actually the midrib of the original leaf, which has been shed.)

(b) Thorns of downy hawthorn (*Crataegus mollis*) are modified stems that develop from axillary buds formed during the previous year's growth.

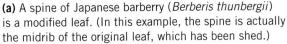

FIGURE 16-5 Homoplastic features.

in structural similarities in two unrelated plant families, the cactus family and the spurge family (●**Figure 16-6**). These plants, which evolved in similar desert environments in different parts of the world, are similar to one another in appearance even though they are not closely related.

Comparative anatomy also demonstrates the existence of **vestigial structures.** Many organisms contain organs or parts of organs that seem to serve no function. In plants, the greatly reduced petals of maple flowers are considered vestigial (●**Figure 16-7**).

Darwin was interested in vestigial structures because their presence conflicted with the prevailing view of creation. He wondered how organisms that were the products of a "perfect creation" could have useless parts. Evolution by natural selection offered an easy explanation. The occasional presence of a vestigial structure is to be expected as a species adapts to a different mode of life. Organs that become less important for an organism's survival may end up as vestiges. When an organ loses much or all of its function, it no longer has

any selective advantage. If its presence does not hurt the organism, however, the vestigial structure tends to remain.

Mimicry provides an adaptive advantage

Mimicry is a subtle example of how natural selection improves an organism's chances of surviving and producing offspring in a particular environment. **Mimicry** is the resemblance of one organism to another organism or to an inanimate object. The flowers of certain orchids mimic female bees. For example, the bee orchid looks like a female long-horned bee to the male bees (●**Figure 16-8a**). The highly specialized mimicry of bee orchids is not just visual. These plants also secrete a scent similar to that produced by female bees, and the males are irresistibly attracted to it. Note the two clumps of yellow pollen grains near what appears to be the head of the female bee. When the male bee attempts to copulate with the flower, the pollen clumps stick to his back. He carries them to another flower, which is pollinated when he tries to copulate again (a phenomenon called *pseudocopulation*). In this example, mimicry serves to attract pollinators.

Mimicry can also make an organism difficult to find. **Protective coloration,** coloration that permits an organism to blend into its surroundings, is a type of mimicry. Protective coloration can either benefit a prey

VESTIGIAL STRUCTURE An evolutionary remnant of a formerly functional structure.

BIOGEOGRAPHY The study of the geographic distribution of living organisms and fossils.

(a) *Euphorbia ingens*, a member of the spurge family, is native to Africa.

(b) *Cereus hankianus*, a member of the cactus family, is native to North America.

FIGURE 16-6 Convergent evolution.
Convergent evolution results in structural similarities in members of two unrelated plant families that evolved in similar desert environments in different parts of the world.

organism by screening it from its predators or benefit a predator by keeping the prey from noticing it until it is too late. Many examples of protective form and coloration involve animals resembling plants. Some caterpillars resemble twigs so closely that you would never guess they are animals—until they start to move (•**Figure 16-8b**). Natural selection has accentuated and preserved such protective coloration.

The distribution of plants and animals supports evolution

The study of the past and present distribution of plants and animals is called **biogeography.**

The geographic distribution of organisms affects their evolution. Darwin was interested in biogeography, and he considered why the species found on ocean islands tend to resemble species of the nearest mainland, even if the environment is different. He also observed that species on ocean islands do not tend to resemble species on islands with similar environments in other parts of the world. Darwin studied the plants and animals of

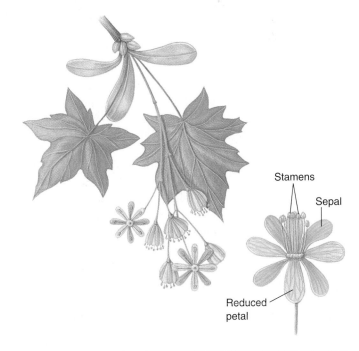

FIGURE 16-7 Vestigial structure.
The petals of wind-pollinated maple (*Acer*) flowers are reduced in size or totally absent. Because the petals do not serve a known function, they are interpreted as vestigial structures. A unisexual maple flower that contains stamens but not carpels is shown.

(a) Bee orchid (*Ophrys scolopax*) flowers resemble the shape, color, and smell of a female long-horned bee (*Eucera longicormis*).

(b) Geometrid larvae are caterpillars that resemble twigs when resting on a branch.

FIGURE 16-8 Mimicry.

two sets of arid islands—the Cape Verde Islands, nearly 640 kilometers (400 miles) off western Africa, and the Galápagos Islands, about 960 kilometers (600 miles) west of Ecuador, South America. On each group of islands, the plants and terrestrial animals were indigenous (native), but they did not resemble each other. Rather, those of the Cape Verde Islands resembled African species, and those of the Galápagos resembled South American species. The similarities of Galápagos species to South American species were particularly striking considering that the Galápagos Islands are dry and rocky and the nearest part of South America is humid and has a lush tropical rain forest. Darwin concluded that species from the neighboring continent migrated or were carried to the islands, where they subsequently adapted to the new environment and, in the process, evolved into new species.

One of the basic tenets of biogeography is that each species evolved only once. The particular place where a species originated is known as its **center of origin.** The center of origin is not a single point but the geographic range of the population when the new species formed. From its center of origin, each species spreads out until halted by a barrier of some kind. The barrier can be physical, such as an ocean, a desert, or a mountain; environmental, such as an unfavorable climate; or ecological, such as the presence of other species that compete for food or living space. (See *Plants and the Environment: Islands and the Problem of Invasive Species* for discussion of a serious environmental threat that is directly related to some of the principles of biogeography.)

Biogeography examines why species live where they do. Palm species, for example, are common in tropical rain forests of South America but almost nonexistent in comparable forests of central Africa. Similarly, South America has balsa trees (*Ochroma*), snakewood trees (*Cecropia*), sloths, and tapirs, whereas central Africa, with a similar climate and environmental conditions, has none of these. These organisms evolved in tropical America and could not expand their range into Africa

PLANTS AND THE ENVIRONMENT | Islands and the Problem of Invasive Species

EVOLUTION LINK **Biotic pollution,** the introduction of a foreign species into an area, often upsets the balance among the organisms living in that area. The foreign species may compete with native species for food or habitat or may prey on them. Generally, an introduced competitor or predator has a greater negative effect on local organisms than do native competitors or predators. Foreign species whose introduction causes economic or environmental harm are called **invasive species.**

Although invasive species may be introduced into new areas by natural means, humans are usually responsible for such introductions, either knowingly or accidentally. The water hyacinth, for example, was deliberately brought from South America to the United States because it has lovely flowers. Today, it has become a nuisance in waterways across the entire southeastern United States, clogging them so that boats cannot move easily and crowding out native aquatic plants (see figure).

Islands are particularly vulnerable to the introduction of invasive species. Many plant and animal species living on islands are *endemic*—that is, they are not found anywhere else in the world. Endemism is common on islands because they are so isolated. During an island's early history, only a few kinds of plant and animal colonists make it

■ The water hyacinth (*Eichhornia crassipes*), introduced into the southeastern United States from tropical South America, has become a pest.

there from across the water. Over many generations, the colonists' descendants adapt to the unique environmental characteristics of that island, gradually evolving into completely new species. Because they have evolved in isolation from competitors, predators, and disease organisms, island species have few defenses when such organisms are then introduced to their habitat. So, island species are particularly vulnerable to extinction.

For example, more than two-thirds of the plant species endemic to the

Galápagos Islands are at risk, largely due to the introduction of goats, pigs, and dogs. In Hawaii, the introduction of sheep has imperiled the mamane tree (because the sheep eat it) and the honeycreeper, an endemic bird (because it relies on the mamane tree for food).

In conclusion, islands were important centers of evolution in the past and are major centers of extinction today. Their isolation has resulted in a rich evolutionary history as well as a vulnerability to invasive species. ■

---

because the Atlantic Ocean was an impassable barrier. Likewise, central Africa has umbrella trees (*Musanga*), guapiruvu trees (*Schizolobium*), chimpanzees, and elephants, none of which are found in tropical America. Thus, the natural distribution of organisms on Earth is understandable only on the basis of where they evolved.

Molecular comparisons among organisms provide evidence for evolution

Similarities and differences in the biochemistry and molecular biology of various organisms provide evidence for evolutionary relationships. Lines of descent based

solely on biochemical and molecular data often resemble lines of descent based on structural and fossil evidence. Molecular evidence for evolution includes the universal genetic code and the identical sequences of amino acids (in proteins) and nucleotides (in DNA) that are found in different species.

Evidence that all life is related comes from the fact that all organisms use a genetic code that is virtually identical. The genetic code specifies a triplet (a sequence of three nucleotides in DNA) that codes for a particular codon (a sequence of three nucleotides in mRNA). The codon then codes for a particular amino acid in a polypeptide chain. For example, AAA in DNA codes for

PROCESS OF SCIENCE

QUESTION: Can a single-gene mutation in flower color affect animal pollinators?

HYPOTHESIS: If flower color is changed by a mutation, pollinator preferences for the resulting flower may change.

EXPERIMENT: *Mimulus lewisii* has violet-pink flowers and is pollinated by bumblebees; the closely related *M. cardinalis* has orange-red flowers and is pollinated by hummingbirds. Alleles at the same locus in both species affect flower color.

   Researchers transferred the allele for orange-red flowers from *M. cardinalis* to *M. lewisii*, which then produced yellow-orange petals. They also transferred the allele for pink flowers from *M. lewisii* to *M. cardinalis*, which then produced dark pink petals.

   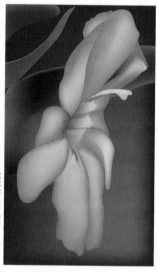

Wild-type *M. lewisii*   Wild-type *M. cardinalis*   Altered *M. lewisii*   Altered *M. cardinalis*

RESULTS AND CONCLUSION: The modified *M. lewisii* flowers were yellow-orange in color and had 68 times as many hummingbird visits as the unaltered *M. lewisii* flowers. The modified *M. cardinalis* flowers had dark pink petals and were visited by 74 times as many bumblebees as the unaltered *M. cardinalis* flowers. Thus, a single mutation in a gene for flower color resulted in a shift in animal pollinators.

FIGURE 16-9 Studying how pollinator preferences may have evolved.
Biologists H. D. Bradshaw Jr. from the University of Washington and Douglas Schemske from Michigan State University used monkey flowers to determine if a single gene mutation would have any effect on animal pollinators.

UUU in mRNA, which codes for the amino acid phenylalanine in organisms as diverse as shrimp, humans, bacteria, and tulips. In fact, AAA codes for phenylalanine in all organisms examined to date.

   The universality of the genetic code—no other code has been found in any organism—is compelling evidence that organisms evolved from a common ancestor. The genetic code has been maintained and passed along through all branches of the evolutionary tree since its origin in some extremely early (and successful) organism.

Proteins contain a record of evolutionary change

Organisms owe their characteristics to the types of proteins they have. Investigations of the sequences of amino acids in the proteins that play the same roles in different species have revealed both great similarities and specific differences. Even organisms that are related only remotely, such as humans, oaks, and bacteria, share some proteins, such as cytochrome *c*, which is part of the electron transport chain in aerobic respiration. To survive, all aerobic organisms need a respiratory protein with the same basic structure and function as the cytochrome *c* of their common ancestor. Any mutations that changed the amino acid sequence at structurally important sites of the cytochrome *c* molecule would have been harmful, and natural selection would have prevented such mutations from being passed to future generations. Consequently, not all amino acids that give cytochrome *c* its structural and functional features are free to change.

However, in the course of the long, independent evolution of different organisms, mutations have resulted in the substitution of many amino acids at less important locations in the cytochrome *c* molecule. The longer it has been since two species diverged (took separate evolutionary pathways), the greater are the differences in the amino acid sequences of their cytochrome *c* molecules.

DNA contains a record of evolutionary change

Because DNA codes for proteins, the differences in amino acid sequences indirectly demonstrate the nature and number of underlying DNA base-pair changes that must have occurred during evolution. Such molecular information is obtained directly by **DNA sequencing**—the determination of the order of nucleotide bases in DNA—of a gene shared by several organisms. Generally, the more closely species are thought to be related on the basis of other evidence, the greater is the percentage of nucleotide sequences that their DNA molecules have in common.

PROCESS OF SCIENCE **Evolutionary hypotheses are tested experimentally**

Increasingly, biologists are designing imaginative experiments to test evolutionary hypotheses. For example,

animal pollinators are known to affect the evolution of the plants they pollinate (see discussion of coevolution in Chapter 9). Botanists have wondered how plants evolve characteristics that attract different pollinators. Could a single gene mutation result in a shift in animal pollinators?

To address this question, botanists used two species of monkey flowers, both native to western North America, to determine if a single gene mutation would have any effect on which animals pollinate each species. The wild-type *Mimulus lewisii* has violet-pink flowers and is pollinated by bumblebees, whereas the closely related *M. cardinalis* has orange-red flowers and is pollinated by hummingbirds. Alleles at the same locus in both species affect flower color.

When the researchers transferred the allele for orange-red flowers from *M. cardinalis* to *M. lewisii*, the resulting *M. lewisii* flowers were yellow-orange instead of pink. These flowers were visited by 68 times as many hummingbirds as the (bee-pollinated) wild-type *M. lewisii* flowers (•**Figure 16-9**).

A shift in pollinators also occurred when *M. cardinalis* plants received the allele for pink flowers from *M. lewisii*. These *M. cardinalis* flowers were a dark pink in color and had 74 times as many bumblebee visits as the wild-type *M. cardinalis* flowers.

Biologists think that hummingbird-pollinated flowers evolved from insect-pollinated flowers many times during the course of flowering plant evolution. A shift in pollinators set the stage for future evolutionary divergence, ultimately resulting in new species. These data suggest that a single gene mutation affecting petal color could have been partly responsible for the pollinator shift that resulted in the evolution of two species, *M. lewisii* and *M. cardinalis*, from their common insect-pollinated ancestor. ■

STUDY OUTLINE

❶ Discuss the historical development of the theory of evolution.

Jean Baptiste de Lamarck was the first scientist to propose that organisms undergo change over time as a result of some natural phenomenon rather than divine intervention. Lamarck thought organisms could pass traits acquired during their lifetimes to their offspring. Charles Darwin's observations while voyaging on the HMS *Beagle* were the basis for his theory of evolution. Darwin tried to explain the similarities between animals and plants of the arid Galápagos Islands and the humid South American mainland.

❷ Define *evolution*, and explain the four premises of evolution by natural selection as proposed by Darwin.

Evolution is the accumulation of inherited changes within populations over time. An **adaptation** is an evolutionary modification that improves an organism's chances of survival and reproductive success. **Natural selection** is the mechanism of evolution in which individuals with inherited characteristics well suited to the environment leave more offspring than individuals that are less suited to the environment do. Darwin proposed four premises of evolution by natural selection: (1) Overproduction: Each species produces more offspring than will survive to maturity. (2) Inherited variation: The individuals in a population exhibit inheritable variation in their traits. (3) Limits on population growth: Organisms compete with one another for the food, space, water, light, and other resources needed for life. (4) Differential reproductive success: The offspring with the most favorable combination of characteristics are most likely to survive and reproduce, thus passing those genetic characteristics to the next generation.

❸ Compare the modern synthesis with Darwin's original theory of evolution.

One of the premises on which Darwin based his theory of evolution by natural selection is that individuals pass traits on to the next generation. However, Darwin could not explain how traits were passed from one generation to another or why individuals vary within a population. The **modern synthesis** is a comprehensive, unified explanation of evolution based on combining previous theories, especially of Mendelian genetics, with Darwin's theory of evolution by natural selection.

❹ Briefly discuss the evidence for evolution obtained from the fossil record.

A **fossil** is a part or trace of an ancient organism that is usually preserved in sedimentary rock. Fossils provide a record of plants and animals that lived earlier, some understanding of where and when they lived, and an idea of the lifestyles they had. By studying the fossils of organisms of different geologic ages, we can trace the lines of evolution that gave rise to those organisms.

❺ Summarize the evidence for evolution derived from comparative anatomy.

Homologous features are dissimilar structures with an underlying similarity of form and development that occur in different species that evolved from a common ancestor. **Homoplastic features** are structures in unrelated species that are similar in function and appearance but not in evolutionary origin. Homoplastic features, which do not indicate close evolutionary ties, are the result of **convergent evolution,** the process in which unrelated species living in similar environments evolve the same kinds of adaptations. A **vestigial structure** is an evolutionary remnant of a formerly functional structure. The occasional presence of a vestigial structure is to be expected as an ancestral species adapts to a different mode of life.

❻ Define *biogeography.*

Biogeography is the study of the geographic distribution of living organisms and fossils. The geographic distribution of organisms affects their evolution. Areas that have been isolated from the rest of the world for a long time have plants and animals unique to those areas.

❼ Describe how scientists make inferences about evolutionary relationships among organisms from the sequence of amino acids in specific proteins or the sequence of nucleotides in particular genes.

Biochemistry and molecular biology provide compelling evidence for evolution. The sequence of amino acids in common proteins, such as cytochrome *c,* reveals greater similarities in closely related species than in remotely related species. A greater proportion of the sequence of nucleotides in DNA is identical in closely related organisms than in remotely related organisms.

REVIEW QUESTIONS

1. Why is Aristotle linked to early evolutionary thought?
2. What were Jean Baptiste de Lamarck's ideas concerning evolution?
3. In what ways does Lamarck's idea of evolution not agree with present evidence?
4. Explain briefly the four premises of evolution by natural selection.
5. Why are only inherited variations important in the evolutionary process?
6. What part of his theory was Darwin unable to explain? How does the modern synthesis explain it?
7. What are fossils?
8. How do homologous and homoplastic features provide evidence of evolution?
9. Explain how cacti and spurges demonstrate convergent evolution.
10. Give an example of a plant that mimics an animal and of an animal that mimics a plant.
11. What is biogeography?
12. Explain why many island species are found nowhere else.
13. Summarize how DNA provides evidence for evolution.

THOUGHT QUESTIONS

1. In what ways are the studies of evolution and human history similar? How are they different?

2. How can you account for the fact that both Darwin and Wallace independently and almost simultaneously proposed essentially identical theories of evolution by natural selection?

3. Explain the following statement: "Natural selection picks from among the individuals in a population those that are best suited to current environmental conditions. It does not select based on some view of 'best design' or on 'possible future need.'"

4. During the past 50 or so years that insecticides have been widely used, more than 520 species of insects and mites have evolved resistance to certain insecticides. How is this process similar to the evolution of bacterial resistance to antibiotics?

Visit us on the web at http://www.thomsonedu.com/biology/berg/ for additional resources such as flashcards, tutorial quizzes, further readings, and web links.

The Evolution of Populations and Species

LEARNING OBJECTIVES

❶ Distinguish between the gene pool of a population and the genotype of an individual.

❷ Discuss the significance of the Hardy–Weinberg principle as it relates to evolution.

❸ Define *microevolution,* and explain how each of the following microevolutionary forces alters allele frequencies in populations: mutation, genetic drift, gene flow, natural selection.

❹ Define *biological species concept,* and describe some of the limitations of your definition.

❺ Distinguish among reproductive isolating mechanisms.

❻ Distinguish between allopatric and sympatric speciation.

❼ Discuss the evolutionary significance of adaptive radiation and extinction.

Primroses (*Primula*) are attractive perennial herbs that are popular as house and garden plants. They produce clusters of trumpet-shaped flowers at the end of a long stem; the leaves are generally located at the base of the stem. About 400 primrose species exist.

Each primrose species produces two kinds of flowers. Some primrose individuals have flowers that contain long ("male") stamens and short ("female") carpels, that is, carpels located deep inside the flower. Other individuals have flowers with short stamens and long carpels. These differing flower structures increase the likelihood of cross-pollination by insect pollinators such as bees and butterflies. Each type of insect enters primrose flowers in the same manner and penetrates the trumpet-shaped flowers the same distance each time. Thus, one kind of insect may carry pollen from short stamens of one flower to short carpels of another, and another insect may carry pollen from long stamens to long carpels in a different flower.

Primroses are a perfect plant with which to begin this chapter on evolution, because a new species of primrose, the kew primrose (*Primula kewensis*), evolved in 1898 at the Royal Botanic Gardens at Kew, England. The specific epithet *kewensis* was assigned in recognition of the plant's place of origin. Details about the evolution of the kew primrose are discussed later in this chapter.

This chapter will help you develop an understanding of the importance of genetic variation as the raw material for evolution and of the basic concepts of **population genetics,** the study of genetic variability within a population and of the forces that act on it. A **population** consists of all the individuals of the same species that live in a particular place at the same time. (Recall from Chapter 1 that a **species** is a group of similar organisms that interbreed in their natural environment.) First we examine genetic variation in populations, including how the relative abundances of *alleles* (alternative forms of a gene) change during the course of evolution. Then we consider how evolutionary changes within populations ultimately lead to the evolution of new species.

The kew primrose (*Primula kewensis*).

Genetic Variation in a Population

Individual organisms within a population exhibit genetic variation for the traits characteristic of the population. A field of garden peas may all have the same trait, such as a fruit that is a pod, but vary from one individual to another in pod color (green or yellow) or pod shape (inflated or constricted).

Evolution occurs in populations, not individuals (•**Table 17-1**). Although natural selection acts on individuals by determining which of them will survive to reproduce, individuals do not evolve during their lifetimes. Populations, however, undergo evolutionary change over many generations. Evolutionary change, which includes modifications in structure, physiology, ecology, and behavior, is inherited from one generation to the next. Although Darwin recognized that evolution occurs in populations, he did not understand how traits are passed to successive generations. One of the most significant advances in biology since Darwin's time is the demonstration of the genetic basis of evolution.

Each population has a **gene pool**, the total genetic material of all the individuals that make up that population at a given time (•**Figure 17-1**). An individual has only a small fraction of the alleles present in a population's gene pool. In diploid (*2n*) organisms, each somatic

TABLE 17-1	A Comparison of Individuals and Populations	
	INDIVIDUAL	**POPULATION**
Period of existence	One generation	Many generations
Genetic description	Genotype	Gene pool
Genetic variation	None	Substantial
Ability to evolve	No	Yes (change in allele frequencies)

(body) cell contains only two alleles for each gene, one on each of the homologous chromosomes (members of a chromosome pair). The genetic variation evident among individuals in a given population indicates that each individual has a different combination of the alleles in the gene pool.

If a population is not undergoing evolutionary change, the frequency of each allele in the gene pool remains constant from generation to generation. An **allele frequency** is the proportion of a specific allele (for example, of *T* or *t*) in a particular population. Changes in allele frequencies over successive generations indicate that evolution has occurred.

The Hardy–Weinberg Principle

Suppose you are studying a population of well-adapted plants growing in an environment that is relatively constant from year to year. If you were to compare the numbers of red-flowered plants and white-flowered plants in the population, for example, you might find 1820 plants with red petals and 180 with white petals, which is a ratio of 9 to 1. You might assume that after many generations, the dominant allele (red) would become more common in the population. You might also assume that the recessive allele (white) would eventually disappear altogether. These were common assumptions of many biologists early in the 20th century. However, these assumptions were incorrect, because the frequencies of alleles and genotypes (genetic makeup) do not change from generation to generation unless influenced by outside factors (discussed later).

Thus, if you came back to the same location the following summer, you might find a population essentially the same as the previous one, with roughly nine red-flowered plants to every white-flowered plant. If you continued the study for several generations and always got the same result, the population would clearly be in **genetic equilibrium**—that is, not undergoing evolutionary change with respect to the alleles being studied (in this case, for flower color). However, if allele frequencies change over successive generations, evolution is occurring.

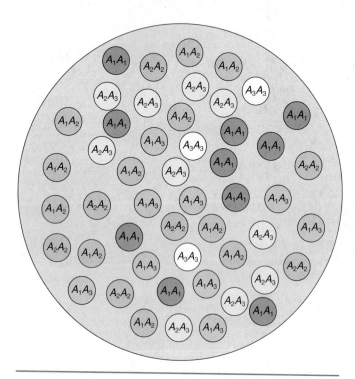

FIGURE 17-1 A gene pool.
This drawing depicts the individuals (*small circles*) in a hypothetical population (*large circle*). Only one gene (*A*) with three alleles (*A₁*, *A₂*, and *A₃*) is shown. Because each individual plant is diploid, it has only two alleles for each gene.

In 1908, G. H. Hardy, an English mathematician, and W. Weinberg, a German physician, each independently developed the explanation for this stability of successive generations in populations at genetic equilibrium. They pointed out that the expected frequencies of various genotypes in a population can be described mathematically. Each **genotype frequency** is the proportion of a particular genotype (for example, of *TT* or *Tt* or *tt*) in the population.

The principles of inheritance worked out by Mendel describe the genotype frequencies among offspring of a single mating pair (see Chapter 13). In contrast, the **Hardy–Weinberg principle** describes the frequencies of various alleles and genotypes of an *entire breeding population*. The Hardy–Weinberg principle shows that if the population is large, the process of inheritance by itself does not cause changes in allele frequencies. The Hardy–Weinberg principle represents an ideal situation that seldom occurs in the natural world. However, it is useful because it provides a model to help us understand the real world.

The Hardy–Weinberg principle of genetic equilibrium tells us what to expect when a sexually reproducing population is not evolving. The relative proportions of alleles and genotypes in successive generations will always be the same, provided the following five conditions are met:

1. *Random mating.* In unrestricted random mating, each individual in a population has an equal chance of mating with any individual of the opposite sex. The individuals represented by the genotypes *AA*, *Aa*, and *aa* must mate with one another at random and must not select their mates on the basis of genotype or any other factors that result in nonrandom mating.

2. *No mutations.* There must be no mutations that convert *A* into *a*, or vice versa. That is, the frequencies of *A* and *a* in the population must not change because of mutations. (Recall from Chapter 14 that a *mutation* is an alteration in a cell's genetic material, that is, its DNA.)

3. *Large population size.* Chance events do not alter the frequency of the alleles in a large population as they do in a small population. Allele frequencies in a small population are more likely to change by random fluctuations (that is, by genetic drift, which is discussed later) than are allele frequencies in a large population. Thus, the population must be large enough that the laws of probability function.

4. *No migration.* There can be no exchange of genes with other populations that might have different allele frequencies. In other words, there can be no movement of individuals (in animals) or of pollen or seeds (in plants) into or out of a population.

5. *No natural selection.* If natural selection is occurring, certain phenotypes (and their corresponding genotypes) are favored over others. Consequently, the allele frequencies would change, and the population would evolve.

Microevolution

Evolution represents a departure from the Hardy–Weinberg principle of genetic equilibrium. The degree of departure between the observed allele or genotype frequencies and those expected by the Hardy–Weinberg principle indicates the amount of evolutionary change. This type of evolution—generation-to-generation changes in allele or genotype frequencies *within* a population—is sometimes referred to as **microevolution,** because it often involves relatively small or minor changes, usually over a few generations.

Changes in the allele frequencies of a population result from four microevolutionary processes: mutation, genetic drift, gene flow, and natural selection. These microevolutionary processes are the opposite of the conditions that must be met if a population is in genetic equilibrium. When one or more of these processes act on a population, allele or genotype frequencies change from one generation to the next.

Mutation increases variation in the gene pool

Genetic variation is introduced into a gene pool through mutation. Mutations arise from (1) a change in the nucleotide base pairs of the gene, (2) a rearrangement of genes within chromosomes so that their interactions produce different effects, or (3) a change in the chromosome structure. Mutation is the source of all new alleles (●**Figure 17-2**). Mutations, which may or may not show a phenotypic effect, occur unpredictably and spontaneously.

Not all mutations pass from one generation to the next. Mutations occurring in somatic cells are not inherited. When an individual with a somatic mutation dies, the mutation is lost. Some mutations, however, oc-

GENE POOL All the alleles of all the genes in a freely interbreeding population.

HARDY–WEINBERG PRINCIPLE The mathematical prediction that allele and genotype frequencies do not change from generation to generation in the absence of microevolutionary processes.

MICROEVOLUTION Small-scale evolutionary changes caused by changes in allele or genotype frequencies in a population over a few generations.

(a) The leaf cactus (*Pereskia sacharosa*) still retains many of the features of its ancestors, including broad leaves and a "normal" stem. This plant's evolution was apparently the result of relatively few mutations.

(b) The fishhook cactus (*Ferocactus wislizenii*) is different from its ancestors. Mutations have resulted in much-reduced leaves (the spines) and a rounded, much-thickened succulent stem. These traits have enabled the cactus to exist in arid environments where its ancestors could not survive.

FIGURE 17-2 Cactus ancestors.
The ancestors of cacti were flowering trees and shrubs. Mutations provided the raw material that led to the evolution of the cactus family.

cur in reproductive cells, and these mutations may or may not affect the offspring. A mutation that occurs in the DNA that codes for a protein may still have little effect in terms of altering the structure or function of that protein. However, when the protein undergoes a change that affects its function, the mutation is usually harmful. For example, a mutation in a gene that codes for an enzyme involved in producing blue petals might result in a plant with nonblue petals that would not attract its normal insect pollinators.

By acting against seriously abnormal phenotypes, natural selection eliminates or reduces to low frequencies the most harmful mutations. Mutations with small phenotypic effects, even if slightly harmful, have a better chance of being incorporated into the population, where at some later time, under different environmental conditions, they may produce phenotypes that are useful or adaptive.

Mutations do not determine the *direction* of evolutionary change. Consider a plant population living in an environment that is becoming increasingly arid. A mutation that produces a new allele that helps an individual adapt to dry conditions is no more likely to occur than one for adapting to wet conditions or one that has no relationship to the changing environment. The production of new mutations simply increases the genetic variability that is acted on by natural selection. Mutation, therefore, increases the potential for **adaptation,** an evolutionary modification that improves an organism's chances of survival and reproductive success within a particular environment.

Mutation by itself causes small deviations in allele frequencies from those predicted by the Hardy–Weinberg principle. Although allele frequencies may be changed by mutation, these changes are typically several orders of magnitude smaller than changes caused by other evolutionary forces, such as genetic drift. As an evolutionary force, mutation is usually negligible, but it is important as the ultimate source of variation for evolution.

Genetic drift causes changes in allele frequencies through random events

The size of a population affects allele frequencies because random events, or chance, tend to cause changes

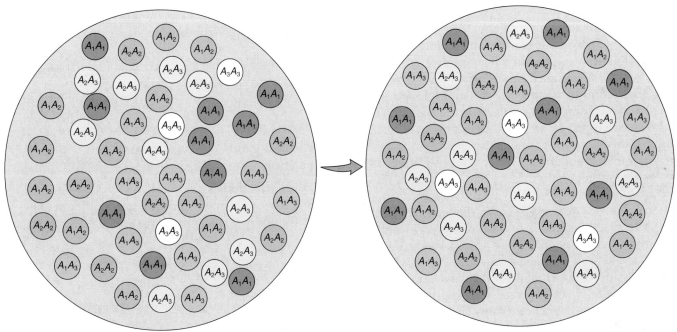

(a) Note how the allele frequencies in the large population vary only slightly due to chance.

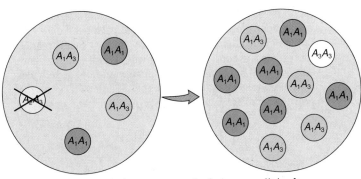

(b) In the small population, as a result of chance, allele A_2 disappears, resulting in a small population with quite different allele frequencies.

FIGURE 17-3 Genetic drift.
The frequency of three alleles for a given gene is shown in two hypothetical populations, one large and one small.

of relatively greater magnitude in a small population. If a population consists of only a few individuals, an allele present at a low frequency in the population could be completely lost by chance. Such an event would be unlikely in a large population.

For example, consider two populations, one with 10,000 individuals and one with 10 individuals. If an uncommon allele occurs at a frequency of 10 percent, or 0.10, then 1900 individuals in the larger population have the allele.[1] That same allele frequency, 10 percent,

[1]*Using Hardy–Weinberg mathematics:* $2(0.9)(0.1) + (0.1)^2 = 0.19$; $0.19 \times 10,000 = 1900$.

in the smaller population means that no more than 2 individuals have the allele.[2] From this example, you can see that there is a greater likelihood of losing the rare allele from the smaller population than from the larger one. Herbivores, for example, might eat the only two plants that have the uncommon allele in the smaller population, so these individuals would leave no offspring.

The production of random evolutionary changes in small breeding populations is known as **genetic drift**. Genetic drift results in changes in allele frequencies in a population from one generation to another. One allele may be eliminated from the population by chance, regardless of whether the allele is beneficial, harmful, or of no particular advantage or disadvantage (•Figure 17-3). Thus, genetic drift *decreases* genetic variation within a small population, although it tends to *increase* genetic differences among different populations.

Organisms in small populations are particularly vulnerable to genetic drift and its negative effects, because the loss of genetic variation reduces their ability to adapt to stressful changes in their environment. For many rare and endangered species whose populations are reduced in size, genetic drift tends to increase their chances of extinction.

---

[2]*Using Hardy–Weinberg mathematics:* $0.19 \times 10 = 1.9$.

GENETIC DRIFT A random change in allele frequencies in a small breeding population.

The founder effect occurs when a few individuals establish a new colony

When a few individuals from a large population establish, or found, a colony (as when wind carries a plant seed for a long distance), they bring with them only a small fraction of the genetic variation present in the original population. As a result, only the alleles that the colonizers happen to have will be represented in their descendants. Typically, the allele frequencies in the newly founded population differ markedly from those of the parent population. The genetic drift that results when a small number of individuals from a large population colonize a new area is called the **founder effect.**

The founder effect has been observed in populations of wild plants on islands off the Pacific coast of Canada. The Canadian mainland has several wild species of small, weedy annuals in the daisy family. These plants produce wind-dispersed seeds with fluffy parachutes similar to dandelion fruits. The seeds of mainland populations range in size from small to large, as do the fluffy parachutes. Sampling of these plants on 240 islands off the Canadian coast over a 10-year period revealed that the youngest island populations produced significantly smaller seeds than mainland populations. (Ages of island populations were easy to estimate because new colonizing occurred frequently.) This observation illustrates the founder effect, because only small seeds with large parachutes remain aloft to be blown by the wind to the nearby islands, and therefore only their alleles are available to later generations that grow on the islands.

The migration of pollen or seeds causes gene flow in plants

Members of a species tend to be found in local populations that are genetically isolated to some degree from other populations. Two populations of wind-pollinated plants, for example, are isolated genetically when the distance separating them is greater than the maximum distance their pollen is blown by the wind. Because each population is isolated to some extent from other populations, they have distinct genetic traits and gene pools.

When individuals of one population reproduce with members of another population, they contribute their alleles to that population's gene pool. Thus, migration causes a corresponding movement of alleles, or **gene flow,** between populations. Gene flow can have significant evolutionary consequences. As alleles "flow" from one population to another, they usually increase the amount of variability within the recipient population. If the gene flow between two populations is great enough, these populations become more similar genetically. Because gene flow reduces the amount of genetic variation between two populations, it tends to counteract the effects of natural selection (discussed shortly) and genetic drift, both of which cause individual populations to become increasingly distinct.

Natural selection changes allele frequencies in a way that increases adaptation

Natural selection is the mechanism of evolution first proposed by Darwin. Natural selection functions to preserve individuals with favorable phenotypes (and

---

PROCESS OF SCIENCE

QUESTION: Can we observe natural selection in action in polluted soil?

HYPOTHESIS: Excessive levels of copper in soil select for increased copper tolerance in plant populations growing near a copper mine.

EXPERIMENT: Bent grass (*Agrostis*) plants were taken from populations growing in areas known to have been contaminated by copper for 0, 4, 8, 14, and 70 years. Year 0 represents the control. All plants were grown in a nutrient solution containing a high level of copper. Root growth was measured to give an index of copper tolerance: 0 = no root growth (complete inhibition), and 100 = maximum root growth (no inhibition).

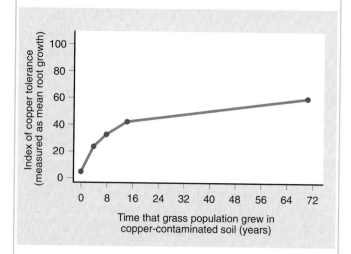

RESULTS AND CONCLUSION: The longer a grass population was exposed to copper-contaminated soil, the greater its tolerance for copper. This result clearly shows natural selection occurring in a direction that favors the population's survival in polluted soil. ■

FIGURE 17-4 Natural selection of genes for tolerance to high levels of copper.

TABLE 17-2   Processes of Microevolution

PROCESS	EFFECT ON GENE POOL	LEADS TO ADAPTATION (YES OR NO)
Mutation	Changes allele frequencies by providing inheritable variation	No
Genetic drift	Changes allele frequencies in small populations due to random events	No
Gene flow (migration)	Changes allele frequencies due to migration of individuals between populations	No
Natural selection	Changes allele frequencies due to differential survival and reproduction	Yes

therefore favorable genotypes) and to eliminate individuals with unfavorable genotypes. Individuals have a selective advantage if they survive and produce fertile offspring. Natural environmental pressures, such as competition for light or water or living space, select the individuals that survive to reproduce.

Over successive generations, the proportion of favorable alleles increases in the population. In contrast with other microevolutionary processes (mutation, genetic drift, and gene flow), natural selection leads to adaptive evolutionary change. Natural selection not only explains why organisms are well adapted to the environments in which they live but also helps account for the remarkable diversity of life. Natural selection enables populations to change, thereby adapting to different environments and different ways of life. In time, these evolving populations may become separate and distinct species.

The mechanism of natural selection does not develop a "perfect" organism. Rather, natural selection weeds out those individuals whose phenotypes are less adapted to environmental challenges, while allowing better-adapted individuals to survive and pass their alleles to their offspring. By reducing the frequency of alleles that result in the expression of less favorable traits, natural selection changes the composition of the gene pool in a direction that favors the population's survival (•**Figure 17-4**). These changes increase the probability that favorable alleles responsible for an adaptation will come together in the offspring.

Natural selection operates on an organism's phenotype

Natural selection does not act directly on an organism's genotype. Instead, it acts on the phenotype, which is, at least in part, an expression of the genotype. Rarely does a single gene have complete control over a single phenotypic trait, such as Mendel originally observed in garden peas. Much more common is the interaction of several genes for the expression of a single trait. Many plants are under this type of polygenic control (see the example of kernel color in Figure 13-11).

Three kinds of selection occur that cause changes in the normal distribution of phenotypes in a population: stabilizing, directional, and disruptive selection (•**Figure 17-5**). The process of natural selection associated with a population well adapted to its environment is known as *stabilizing selection*. Most populations are probably under the influence of stabilizing forces most of the time. Stabilizing selection selects against phenotypic extremes. In other words, individuals with an average, or intermediate, phenotype are favored.

If an environment changes over time, *directional selection* may favor phenotypes at one of the extremes of the normal distribution. Over successive generations, one phenotype gradually replaces another. For example, if greater size is advantageous in a new environment, larger individuals will become increasingly common in the population. Directional selection can occur, however, only if alleles favorable under the new circumstances are already present in the population.

Sometimes extreme changes in the environment may favor two or more different phenotypes at the expense of the mean. That is, more than one phenotype may be favored in the new environment. *Disruptive selection* is a special type of directional selection in which there is a trend in several directions rather than just one. It results in a divergence, or splitting apart, of distinct groups of individuals within a population. Disruptive selection selects against the average, or intermediate, phenotype.

•**Table 17-2** summarizes the four microevolutionary processes just discussed. We now shift our attention to how microevolution can contribute to the evolution of new species. But what exactly is a species?

GENE FLOW   The movement of alleles between local populations due to migration and subsequent interbreeding.

NATURAL SELECTION   The mechanism of evolution in which individuals with inherited characteristics well suited to the environment leave more offspring than individuals that are less suited to the environment do.

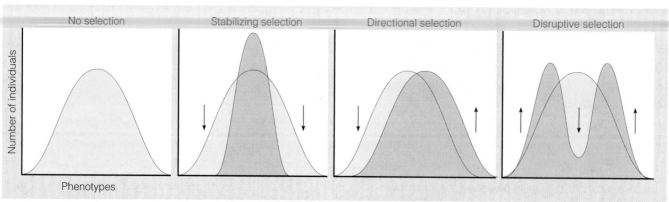

(a) A character under polygenic control exhibits a normal distribution of phenotypes in the absence of selection.

(b) As a result of stabilizing selection, which trims off extreme phenotypes, variation about the mean is reduced.

(c) Directional selection shifts the curve in one direction, changing the average value of the character.

(d) Disruptive selection, which trims off intermediate phenotypes, results in two or more peaks.

FIGURE 17-5 Modes of natural selection.
The lighter blue screen represents the distribution of individuals by phenotype in the original population. The darker blue screen represents the distribution of phenotypes in the evolved population. The arrows represent the pressure of natural selection on the phenotypes.

The Biological Concept of Species

The concept of distinct groups of organisms, known as species, is not new. However, every definition of exactly what constitutes a species has some sort of limitation. Linnaeus, the 18th-century botanist who is considered the founder of modern **taxonomy,** the science of describing, naming, and classifying organisms, classified plants into separate species based on their structural differences. This method, known as the *morphological species concept,* is still used to help characterize species, but structure alone is not adequate to explain what constitutes a species. For example, garden peas come in a wide variety of sizes and shapes, but all garden peas are clearly members of the same species (*Pisum sativum*).

Population genetics did much to clarify the concept of species. According to the **biological species concept,** a species consists of one or more populations whose members interbreed in nature to produce fertile offspring and do not interbreed with members of different species. In other words, species exhibit **reproductive isolation:** reproductive barriers restrict each species from interbreeding with other species.

One problem with the biological species concept is that it applies only to sexually reproducing organisms. Plants often reproduce asexually or by self-pollination, so the concept of reproductive isolation plays no role in defining them as species. The general concept of species is, however, still valid for these organisms; they are classified into species on the basis of structural and biochemical characteristics.

Another problem with the biological species concept is that two populations widely separated geographically may be so much alike that they are classified as the same species, but it is impossible to test whether they will interbreed in nature. For example, the quaking aspen (*Populus tremuloides*) has the largest range of any tree in North America, from northern Alaska and Canada to Mexico. It is not known whether quaking aspens from an extremely northern population could successfully interbreed with those from an extremely southern population. In the absence of this information, quaking aspens are considered a single species.

Conversely, organisms assigned to different species may interbreed if they are brought together in a greenhouse, zoo, aquarium, or laboratory. Therefore, we usually include in our definition of *species* that the members of a species do not normally interbreed with members of other species *in nature.*

To summarize, a species is a group of organisms that have a common gene pool and are capable of interbreeding with one another in nature but are reproductively isolated from other species. This definition is not perfect, and the biological concept of species has limitations.

(a) The black sage (*Salvia mellifera*) flowers in early spring.

(b) The white sage (*Salvia apiana*), photographed at the same time of year, has unopened flower buds.

FIGURE 17-6 Reproductive isolation due to timing of reproduction.

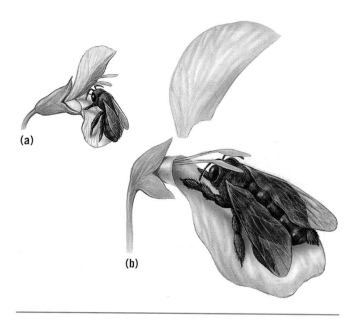

FIGURE 17-7 Reproductive isolation due to differences in floral structure.
Reproductive isolation occurs between black sage and white sage as a result of the differences in floral structure. Because the two species have different pollinators, they cannot interbreed. **(a)** The petal of the black sage is shaped as a landing platform for small bees. Larger bees cannot fit on this platform. **(b)** The larger landing platform and longer stamens of white sage allow pollination by larger California carpenter bees (a different species). If smaller bees land on white sage, their bodies do not brush against the pollen-bearing stamens. (The upper part of the white sage flower has been removed.)

Species have various mechanisms responsible for reproductive isolation

Certain mechanisms prevent interbreeding between species, thus maintaining reproductive isolation. These mechanisms preserve the integrity of the gene pool of each species by preventing gene flow between species. Sometimes timing prevents genetic exchange between two groups, as when they flower at different times of the day, season, or year. For example, black sage and white sage, two similar species, live in the same area of southern California but do not interbreed because black sage flowers in early spring and white sage blooms in late spring and early summer (•**Figure 17-6**).

Many flowering plants have physically distinct flower parts that help them maintain their reproductive isolation from one another. Black sage, which is pollinated by small bees, has a floral structure different from that of white sage, which is pollinated by large carpenter bees (•**Figure 17-7**). Presumably, the differences in floral structures prevent insects from cross-pollinating the two species should they happen to flower at the same time.

Sometimes, despite these reproductive isolating mechanisms, fertilization occurs between gametes of two species. However, other reproductive isolating mechanisms may still ensure reproductive failure. Often the embryo of such a union is aborted. A plant's development from fertilized egg to a complete multicellular

organism requires the precise interaction and coordination of many genes. Apparently, the genes from gametes of different species do not interact properly to regulate normal embryonic development. If an **interspecific hybrid** (a hybrid formed between two species) does live, it still may not reproduce, often because of chromosome differences. For example, if a radish (*Raphanus sativus*) is crossed with a cabbage (*Brassica oleraceae*), the resulting hybrid cannot form viable gametes because the chromosomes do not pair properly during meiosis.

BIOLOGICAL SPECIES CONCEPT The concept that a species consists of one or more populations whose members can interbreed to produce fertile offspring and cannot interbreed with individuals of other species.

REPRODUCTIVE ISOLATION The situation in which reproductive barriers prevent members of a species from successfully interbreeding with members of another species.

INTERSPECIFIC HYBRID The offspring of individuals belonging to different species.

Evolution of Reproductive Isolating Mechanisms

With our understanding of reproductive isolation and its role in defining what a species is, we are now ready to consider **speciation**, the formation of a new species. Speciation occurs when a population diverges from the rest of the species. A required step during speciation is the reproductive isolation of that population from other members of the species. When a population has become sufficiently different from its ancestral species that no genetic exchange occurs between them, even if the two populations meet, speciation has occurred. Such a situation is thought to arise in one of two ways, through allopatric or sympatric speciation.

Long geographic isolation and different selective pressures result in allopatric speciation

Speciation that occurs when one population becomes geographically separated from the rest of the species and subsequently evolves is known as **allopatric speciation** (*allo,* "different," and *patri,* "fatherland").

The geographic isolation required for allopatric speciation may occur in several ways. Earth's surface is in a constant state of change: rivers shift their courses; glaciers migrate; mountain ranges form; land bridges develop, separating previously united aquatic populations; and large lakes diminish into several smaller pools (•Figure 17-8).

What might be an imposing geographic barrier to one species may be of no consequence to another. For example, as a lake subsides into smaller pools, fish usually cannot cross the land barriers between the pools and become reproductively isolated. On the other hand, plants such as cattails, which disperse their fruits by air currents, would not be isolated by this barrier.

Allopatric speciation also occurs when a small population migrates and colonizes a new area away from the original species' range. This colony is geographically isolated from its parent species. The Galápagos Islands and the Hawaiian Islands were colonized by individuals of a few species of plants and animals. From these original colonizers, the distinctive species of plants and animals unique to each island evolved.

Clines may represent allopatric speciation in progress

You have seen that genetic differences often exist among different populations within the same species, a phenomenon known as *geographic variation.* One type of geographic variation is a **cline,** which is a gradual

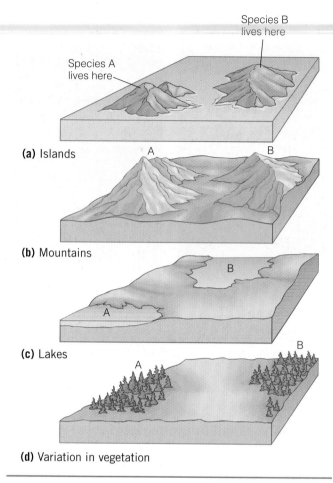

(a) Islands

(b) Mountains

(c) Lakes

(d) Variation in vegetation

FIGURE 17-8 Examples of geographic isolation.
New species may arise as a result of geographic isolation.

change in a species' phenotype and genotype frequencies through a series of geographically separate populations as a result of an environmental gradient. A cline exhibits variation in the expression of such attributes as color, size, shape, physiology, or behavior. Clines are common among species with continuous ranges over large geographic areas.

The common yarrow, a wildflower that grows in a variety of North American habitats from lowlands to mountain highlands, exhibits clinal variation in height in response to different climates at different elevations. Although substantial variation exists among individuals within each population, individuals in populations at higher elevations are, on average, shorter than those at lower elevations. The genetic basis of these clinal differences was demonstrated in a set of experiments in which series of populations from different geographic areas were grown in the same environment (•Figure 17-9). Despite being exposed to identical environmental conditions, each experimental population exhibited the height characteristic of the elevation from which it was collected.

PROCESS OF SCIENCE

QUESTION: Is clinal variation caused by genetic factors, or is it caused by environmental influences?

HYPOTHESIS: The differences in average height exhibited by populations of yarrow (*Achillea millefolium*) growing in their natural habitats at different altitudes (*represented in the figure*) are due to genetic differences.

EXPERIMENT: Seeds from widely dispersed populations in the Sierra Nevada of California and Nevada were collected and grown for several generations under identical conditions in the same test garden at Stanford, California.

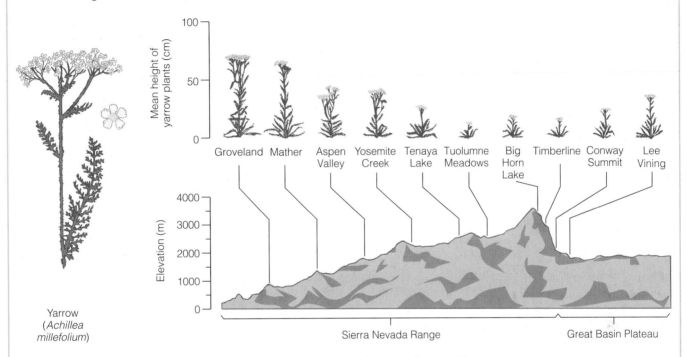

RESULTS AND CONCLUSION: The plants retained their distinctive heights, revealing genetic differences related to the elevation where the seeds were collected. ■

FIGURE 17-9 Clinal variation in yarrow.
(After J. Clausen, D. D. Keck, and W. M. Hiesey, "Experimental Studies on the Nature of Species: III, Environmental Responses of Climatic Races of *Achillea*," *Carnegie Institute Washington Publication*, Vol. 58, 1948)

New plant species may evolve in the same geographic region as the parent species

Although geographic isolation is an important factor in many cases of evolution, it is not an absolute requirement. When a population forms a new species within the same geographic region as its parent species, **sympatric speciation** (*sym*, "together," and *patri*, "fatherland") has occurred. Sympatric speciation is especially common in plants.

You have seen that interspecific hybrids, which form from the union of gametes from two species, rarely produce offspring and that, when they do, the offspring are usually sterile. The reason is that the chromosomes of an interspecific hybrid are not homologous and therefore cannot properly be parceled into gametes during meiosis. However, if the number of chromosomes doubles *before* meiosis takes place, then the chromosomes occur in pairs, and meiosis can occur successfully. This spontaneous doubling of chromosomes has been docu-

ALLOPATRIC SPECIATION Evolution of a new species that occurs when one population becomes geographically separated from the rest of the species.

SYMPATRIC SPECIATION Evolution of a new species that occurs within the parent species' geographic region.

mented many times in plants. When it occurs, the result is nuclei with multiple sets of chromosomes.

Polyploidy, the possession of more than two sets of chromosomes, is a major factor in plant evolution. When polyploidy occurs in conjunction with sexual reproduction between individuals of two species, known as **allopolyploidy,** it can produce a fertile interspecific hybrid (•**Figure 17-10**). Because allopolyploidy provides the

homologous chromosomes necessary for meiosis, gametes may be viable. Plants that are allopolyploids can reproduce with themselves (self-fertilize) or with similar individuals. However, they are reproductively isolated from both parent species because the allopolyploid has a different number of chromosomes from that of either of its parents.

Allopolyploidy was a significant factor in the evolution of flowering plants. As many as 80 percent of flowering plants are polyploids, and most of these are allopolyploids. Moreover, allopolyploidy provides a mechanism for extremely rapid speciation. A single generation is all that is needed to form a new, reproductively isolated species. Allopolyploidy helps explain the rapid appearance of flowering plants in the fossil record and the remarkable diversity (at least 300,000 species) in flowering plants today.

The kew primrose (*Primula kewensis*), discussed in the chapter introduction, is an example of sympatric speciation involving allopolyploidy (•**Figure 17-11**). The interspecific hybrid of two primrose species, *P. floribunda* (2n = 18) and *P. verticillata* (2n = 18), *P. kew-*

FIGURE IN FOCUS

Allopolyploidy is an important mechanism of sympatric speciation in plants.

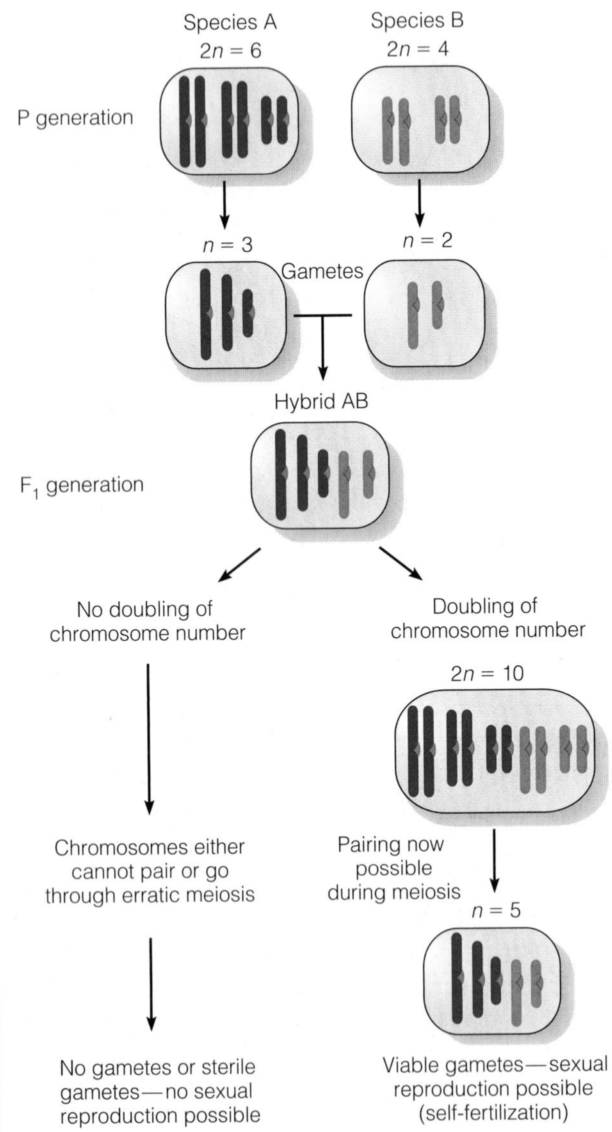

Species A
2n = 6

Species B
2n = 4

P generation

n = 3

n = 2

Gametes

Hybrid AB

F₁ generation

No doubling of chromosome number

Doubling of chromosome number

2n = 10

Chromosomes either cannot pair or go through erratic meiosis

Pairing now possible during meiosis

n = 5

No gametes or sterile gametes—no sexual reproduction possible

Viable gametes—sexual reproduction possible (self-fertilization)

FIGURE 17-10 How a fertile allopolyploid is formed.
When two species (designated the P generation) successfully interbreed, the interspecific hybrid offspring (the F₁ generation) are almost always sterile (*bottom left*). If the chromosomes double before meiosis occurs, the interspecific hybrid can undergo meiosis and is fertile (*bottom right*). Unduplicated chromosomes are shown for clarity.

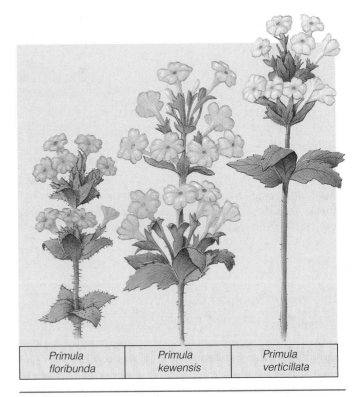

Primula floribunda	*Primula kewensis*	*Primula verticillata*

FIGURE 17-11 A documented case of sympatric speciation.
An allopolyploid primrose (*Primula kewensis*) arose during the early part of the 20th century in a greenhouse at the Royal Botanic Gardens at Kew, England. The F₁ hybrid of *P. floribunda* (2n = 18) and *P. verticillata* (2n = 18) was a sterile perennial with 18 chromosomes. Three separate times this sterile hybrid spontaneously formed a fertile branch, which was *P. kewensis,* a fertile allopolyploid (2n = 36) that produced seeds. Today, *P. kewensis* is a popular houseplant.

(a) The Haleakala silversword (*Argyroxiphium sandwicense*) is found only in the cinders on the upper slope of Haleakala Crater on the island of Maui. This plant is adapted to low precipitation and high levels of ultraviolet radiation.

(b) This tarweed species (*Wilkesia gymnoxiphium*), which superficially resembles a yucca, is found along the slopes of Waimea Canyon on the island of Kauai.

(c) *Dubautia platyphylla* is a large shrub found in moist ravines on the island of Maui. A close-up of a flowering shoot is shown.

FIGURE 17-12 Adaptive radiation in Hawaiian tarweeds.
The 28 tarweed species, which are classified in three closely related genera, live in a variety of habitats.

ensis had a chromosome number of 18 but was sterile. Then, at three separate times, it spontaneously formed a fertile branch, which was an allopolyploid ($2n = 36$) that produced viable seeds of *P. kewensis*.

Adaptive Radiation

In **adaptive radiation,** an ancestral organism evolves in a relatively short time into many new species that fill a variety of *ecological niches*—that is, roles or ways of living in a community. Only one type of organism can occupy each ecological niche. A newly evolved species can take over an ecological niche even if another species already occupies that niche, provided the new species has features that make it competitively superior to the original occupant. Alternatively, if a number of ecological niches are empty, adaptive radiation may exploit them.

Consider the Hawaiian tarweeds, 28 species of closely related plants found only on the Hawaiian Islands. When the tarweed ancestor, a California plant related to daisies, reached the Hawaiian Islands, there were many diverse environments, such as exposed lava flows, dry woodlands, moist forests, and bogs. The succeeding generations of tarweeds quickly diversified to occupy the many ecological niches available to them. The diversity in their leaves, which changed during the course of natural selection to enable different populations to adapt to various levels of light and moisture, is a particularly good illustration of adaptive radiation (•**Figure 17-12**). For example, leaves of tarweeds that are adapted to shady, moist forests are large, whereas those of tarweeds living in exposed dry areas are small. Leaves of tarweeds living on volcanic slopes, which receive intense ultraviolet radiation, are covered with dense silvery hairs that may reflect some of the radiation off the plant.

ALLOPOLYPLOIDY The situation in which an interspecific hybrid contains two or more sets of chromosomes from each of the parent species.

ADAPTIVE RADIATION The evolution of many related species from an ancestral species.

PLANTS AND THE ENVIRONMENT | Extinction Today—Endangered and Threatened Species

Currently, organisms are disappearing at an alarming rate. Conservation biologists estimate that species are becoming extinct at a rate approximately 10,000 times the normal rate of extinction. As much as one-fifth of all species may become extinct within the next 30 years. The Center for Plant Conservation estimates that more than 4000 native plant species are of conservation concern in the United States; this number represents about 20 percent of U.S. plant species (see figure).

A species is **endangered** when its numbers are so severely reduced that it is in danger of becoming extinct. When extinction is less imminent but the population of a particular species is still quite low, the species is said to be **threatened.** The genetic variability of an endangered or threatened species is severely diminished. Because long-term survival and evolution depend on genetic diversity, endangered and threatened species are at a much greater risk of extinction than are species with more genetic variability (see Chapter 27).

Endangered species share certain characteristics that make them more vulnerable to extinction than other species. These include (1) an extremely small, localized range; (2) an island habitat; and (3) low reproductive success, which is usually the result of a small population size.

Many endangered species have limited natural ranges, which makes them particularly prone to extinction if their habitats are altered. The Tiburon mariposa lily (*Calochortus tiburonensis*), for example, is found nowhere in nature except on a single hilltop near San Francisco. Similarly, the Florida stinking cedar (*Torreya taxifolia*) grows only on limestone banks along Florida's Apalachicola River. Development of either area would almost certainly cause the extinction of the species that lives there.

Many species that are endemic to certain islands (that is, are not found anywhere else in the world) are endangered. These organisms often have small populations that cannot be replaced should their numbers be destroyed. For example, by 1986 only two small populations of the endangered evergreen tree *Banara vanderbiltii* remained in Puerto Rico. The U.S. Fish and Wildlife Service has listed this tree as an endangered species and developed a plan for its recovery.

For a species to survive, its members must be present in large enough numbers in their range for successful reproduction. The minimum population density and size that ensure reproductive success vary from one type of organism to another. However, for all organisms, if the population density and size fall below a critical minimum level, the population decreases, making it susceptible to extinction.

Today, the most serious threat to the survival of many species is habitat disruption caused by human activities. We alter habitats when we build roads, parking lots, and buildings; clear forests to grow crops or graze domestic animals; or log forests for timber. We drain marshes to build on aquatic habitats, thus making them terrestrial, and we flood terrestrial habitats when we build dikes and dams, making them aquatic. Because most organisms are utterly dependent on a particular type of environment, habitat destruction reduces their biological range and compromises their ability to survive.

Even habitats left undisturbed and in their natural state are modified by human activities that produce acid precipitation and other forms of *pollution.* Acid precipitation is thought to have contributed to the decline of large stands of forest trees and to the biological death of many freshwater lakes. The production of other types of pollutants also adversely affects wildlife. Such pollutants include industrial and agricultural chemicals, organic pollutants from sewage, acid wastes seeping from mines, and greenhouse gas pollution, which contributes to global warming (see Chapter 27).

Adaptive radiation is apparently more common during periods of major environmental change, but it is difficult to determine whether such change actually triggers adaptive radiation. It is possible that major environmental change has an indirect effect on adaptive radiation by increasing the rate of extinction. Extinction produces empty ecological niches, which are then available for adaptive radiation. Mammals, for example, had evolved millions of years before they underwent adaptive radiation, which may have been triggered by the extinction of the dinosaurs. Originally, mammals were mostly small animals that ate insects. In a relatively short time after the dinosaurs' demise, mammals evolved to occupy and exploit a variety of roles that the dinosaurs had previously filled. Flying bats, running gazelles, burrowing moles, and swimming whales evolved from the ancestral mammals as a result of adaptive radiation.

Extinction and Evolution

Extinction occurs when the last individual of a species dies. Extinction is a permanent loss, because once a species is extinct, it never reappears. Extinctions have occurred continually since the origin of life. By one estimate, only 1 species is living today for every 2000 that have become extinct. Extinction is the eventual fate of

Monterrey manzanita

Green pitcher plant

Bladderpod

Arizona century plant

Cyanea

Texas snowbell

Large-fruited sand verbena

Nellie cory cactus

Macfarlane's four-o'clock

Harparella

Ashmeadows sunray

Florida stinking cedar

Scrub mint

Mead's milkweed

Na'u

Ko'oloa'ula

■ Sixteen of the more than 4000 native plant species of conservation concern in the United States.

all species, in the same way that death is the eventual fate of all individuals.

During the history of life, there have been two types of extinction, background extinction and mass extinction. The continual, low-level extinction of species that has occurred in response to gradual changes in the environment over time is called **background extinction.**

EXTINCTION The death of every member of a species.

Mass extinctions have occurred five or six times during Earth's history. At these times, numerous species and more inclusive taxonomic categories (such as genera and families) have gone extinct in both terrestrial and aquatic environments. Mass extinctions may have occurred over several million years, a relatively short time compared with the more than 3.5-billion-year history of life. Each episode of mass extinction was followed by a period during which the surviving species evolved rapidly to fill many of the ecological niches vacated by the extinct species.

The causes of extinction, particularly mass extinction, are not well understood, but environmental and biological factors are probably involved. Major changes in the climate, for example, would have adversely affected plants and animals that could not adapt. Marine organisms, in particular, are adapted to a steady, unchanging climate. If Earth's temperature were to decrease or increase overall by just a few degrees, numerous marine species would probably perish. Many biologists think that climate changes were responsible for some mass extinctions in the past.

It is also possible that mass extinctions were due to changes in the environment triggered by catastrophes.

MASS EXTINCTION The extinction of many species during a relatively short period of geologic time.

If Earth were bombarded by a large meteorite or comet, for example, the dust going into the atmosphere on impact could have blocked much of the sunlight from reaching Earth for at least several months. In addition to killing many plants, the atmospheric dust would have lowered Earth's temperature, leading to the deaths of many marine organisms.

Biological factors also trigger extinction. When a new species evolves, it may outcompete an older species, leading to the older species' extinction. Humans have had a particularly profound impact on the rate of extinction. Whenever humans move into new areas, their activities alter or destroy the habitats of many plant and animal species. Habitat destruction endangers an organism's survival and may cause the extinction of that species.

Some biologists fear that we have entered the greatest period of mass extinction in Earth's history, but the current mass extinction differs from previous ones in several respects. First, it is directly attributable to human activities. Second, it is occurring in a tremendously compressed period—just a few decades as opposed to millions of years. Perhaps even more serious, more plant species are becoming extinct than in previous mass extinctions. Since plants are the base of food webs—that is, animals eat plants or eat animals that eat plants—the extinction of animals may not be far behind (see *Plants and the Environment: Extinction Today—Endangered and Threatened Species* on pages 342–343).

STUDY OUTLINE

❶ Distinguish between the gene pool of a population and the genotype of an individual.

A **gene pool** consists of all the alleles of all the genes in a freely interbreeding population. An individual's genotype has only a small fraction of the alleles present in a population's gene pool. In diploid organisms, each somatic cell contains only two alleles for each gene.

❷ Discuss the significance of the Hardy–Weinberg principle as it relates to evolution.

The **Hardy–Weinberg principle** is the mathematical prediction that allele and genotype frequencies do not change from generation to generation in the absence of microevolutionary processes. The Hardy–Weinberg principle shows that if the population is large, the process of inheritance by itself does not cause changes in allele and genotype frequencies. This principle tells us what to expect when a sexually reproducing population is not evolving.

❸ Define *microevolution,* and explain how each of the following microevolutionary forces alters allele frequencies in populations: mutation, genetic drift, gene flow, natural selection.

Microevolution consists of small-scale evolutionary changes caused by changes in allele or genotype frequencies in a population over a few generations. Allele or genotype frequencies may be changed by mutation, genetic drift, gene flow, or natural selection. The source of new alleles in a gene pool is mutation, an alteration in a cell's genetic material. **Genetic drift** is a random change in allele frequencies in a small breeding population. **Gene flow** is the movement of alleles between local populations due to migration and subsequent interbreeding. **Natural selection** is the mechanism of evolution in which individuals with inherited characteristics well suited to the environment leave more offspring than individuals that are less suited to the environment do.

Natural selection causes changes in allele frequencies that lead to **adaptation** to the environment.

❹ **Define *biological species concept*, and describe some of the limitations of your definition.**
The **biological species concept** is the concept that a species consists of one or more populations whose members can interbreed to produce fertile offspring and cannot interbreed with individuals of other species. One problem with the biological species concept is that it applies only to sexually reproducing organisms. Another problem is that two widely separated populations may be so much alike that they are classified as the same species, but it is impossible to test whether they will interbreed in nature. Also, organisms assigned to different species may interbreed if they are brought together in a greenhouse, zoo, aquarium, or laboratory.

❺ **Distinguish among reproductive isolating mechanisms.**
Reproductive isolation is the result of reproductive barriers that prevent members of a species from successfully interbreeding with members of another species. Reproductive isolation occurs when two species reproduce at different times of the day, season, or year. Reproductive isolation also occurs as a result of structural differences in the reproductive organs of species. Reproductive failure is common even when fertilization has taken place between gametes of two species. The interspecific hybrid usually dies at an early stage of embryonic development, and if an interspecific hybrid survives to adulthood, it usually cannot reproduce successfully.

❻ **Distinguish between allopatric and sympatric speciation.**
Allopatric speciation is the evolution of a new species that occurs when one population becomes geographically separated from the rest of the species and subsequently evolves. **Sympatric speciation** is the evolution of a new species that occurs within the parent species' geographic region. In plants, sympatric speciation may occur as a result of allopolyploidy in an interspecific hybrid. An **interspecific hybrid** is the offspring of individuals belonging to different species. **Allopolyploidy** is the situation in which an interspecific hybrid contains two or more sets of chromosomes from each of the parent species. Allopolyploidy may enable the interspecific hybrid to reproduce successfully as a new species.

❼ **Discuss the evolutionary significance of adaptive radiation and extinction.**
Adaptive radiation is the evolution of many related species from an ancestral species. **Extinction** is the death of every member of a species. **Mass extinction** is the extinction of many species during a relatively short period of geologic time.

REVIEW QUESTIONS

1. Define *population*, *species*, and *gene pool*.
2. What does the Hardy–Weinberg principle explain?
3. Explain the effect of each of these microevolutionary processes on allele frequencies: (a) natural selection, (b) mutation, (c) gene flow, (d) genetic drift.
4. If a mutation occurs in a somatic cell, can the new allele become established in a population? Explain why or why not.
5. Why are mutations almost always neutral or harmful?
6. What is the biological species concept? How did population genetics help clarify the definition of a species?
7. Give several examples of reproductive isolation.
8. Describe why interspecific reproduction usually fails even when fertilization has taken place.
9. Identify at least five geographic barriers that may lead to allopatric speciation.
10. Why is allopatric speciation more likely to occur if the original isolated population is small?
11. Explain how allopolyploidy causes a new plant species to evolve in as little time as one generation.
12. What role does extinction play in evolution?

THOUGHT QUESTIONS

1. Given that the Hardy–Weinberg principle occurs only under conditions that populations in nature seldom, if ever, experience, why is it important?

2. Explain why we discuss evolution in terms of the selective advantage that a particular *genotype* confers on an individual, yet natural selection acts on an organism's *phenotype*.

3. Why is it necessary to understand the biological species concept to understand evolution?

4. How far apart must two populations of insect-pollinated plants be to be genetically isolated from each other?

5. Examine the maps, which show the continents (a) 240 million years ago, when they were joined, and (b) today. How might continental drift—the separation and drifting apart of once-joined landmasses—have affected the evolution of organisms on those landmasses? Organisms in the ocean?

6. Examine the figure on page 347.
   a. Are these variations in yarrows due to genetic differences among the plants or to the differences in environmental conditions? What kind of experiment would help answer this question?
   b. If you took seeds from yarrows growing at Big Horn Lake and planted them at Groveland, do you think the resulting plants would be tall or short? Explain your answer.

Visit us on the web at http://www.thomsonedu.com/biology/berg/ for additional resources such as flashcards, tutorial quizzes, further readings, and web links.

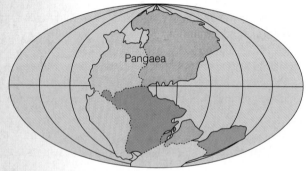

(a) 240 million years ago

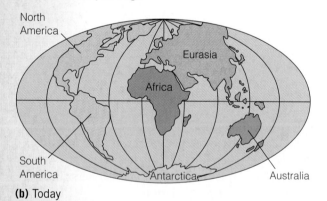

(b) Today

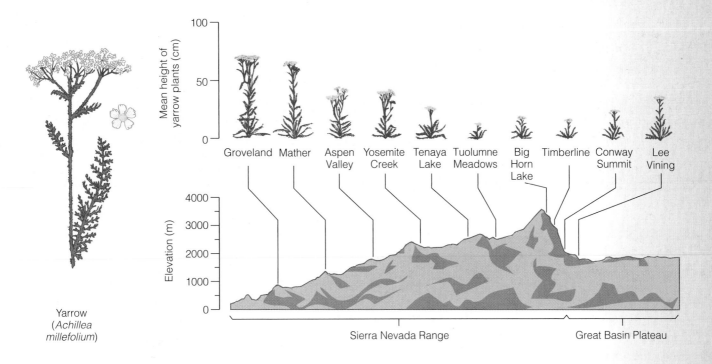

Yarrow
(*Achillea
millefolium*)

The Classification of Plants and Other Organisms

The European white water lily (*Nymphaea alba*), which is found in Europe, North Africa, and parts of Asia, is an herbaceous, aquatic flowering plant that thrives in shallow, still water. About 35 temperate and tropical species of water lilies belong to the genus *Nymphaea* worldwide. (A genus is a taxonomic group composed of related, similar species.) All have underground stems (either tubers or rhizomes) from which leaves arise. Each leaf consists of a long petiole and a large round or heart-shaped blade that usually floats on the water's surface. European white water lilies have large, attractive white flowers that also float on the water's surface, although the fruit that develops from the flower ripens underwater. Like some other water lilies, the European white water lily is widely cultivated in aquatic gardens.

Nymphaea, the genus of water lilies, is named after water nymphs, the beautiful fairylike goddesses in Greek mythology. Linnaeus, the 18th-century scientist who described and named these plants, was familiar with classical literature that depicted nymphs. Because water lilies live in ponds, lakes, and slow-moving waterways, they reminded Linnaeus of the mythical water nymphs.

The scientific name of the European white water lily, *Nymphaea alba,* consists of two words. One tells its genus, *Nymphaea,* which is common to all water lilies, and the other its species (or, more accurately, its specific epithet), *alba,* which is Latin for "white."

Although scientific names are often incomprehensible to lay people, most of the names, like *Nymphaea alba,* are full of meaning. Such names are also a scientific necessity. The European white water lily has more than 200 local, or common, names in at least four languages (English, French, German, and Dutch); in English it is called, among others, water lily, white water lily, European white water lily, water nymph, and platter dock. Clearly, biologists from different parts of the world who are interested in *Nymphaea alba* need to know they are all talking about the same species. Scientific nomenclature enables biologists of all nationalities to be precise when they discuss and share data on organisms.

European white water lily. The European white water lily has more than 200 common names but only one scientific name, *Nymphaea alba*.

Taxonomy

How would you use what you already know about plants if you wanted to assign them to groups? Would you place the red Christmas cactus, red rose, and red hibiscus in one group because they all have the same flower color? Or would you classify plants according to their uses, placing edible plants in one group and poisonous plants in another?

Each of these schemes might be valid, depending on your purpose. To study the diverse life-forms that share this planet and to effectively communicate findings, biologists must organize their knowledge. **Taxonomy** is the science of describing, naming, and classifying organisms. The term *classification* means arranging organisms into groups based on their similarities.

Classification methods have been used throughout history. Theophrastus, a Greek philosopher and biologist who lived during the third century BCE, classified several hundred plants into such groups as herbs, shrubs, and trees. His system of classification persisted for many centuries.

Dioscorides was a Roman military surgeon in the first century CE who traveled with the Roman army. Dioscorides wrote *Materia Medica*, a book that describes some 600 species of medicinal plants. *Materia Medica* was widely used as a medical reference for about 1500 years—until the end of the Middle Ages. Although Dioscorides did not attempt to classify the plants he wrote about, he often grouped similar plants together.

After the invention of a printing press with movable metal type in 1448, early botanical works known as **herbals** were printed. These volumes contained drawings and descriptions of plants, particularly those with medicinal uses (•**Figure 18-1**). Some herbals also contained attempts at classification. European explorers of the 16th and 17th centuries introduced to Europe hundreds of new plants gathered from other continents, compounding the need for a simple method of classification (see *Plants and People: Plant Exploration*).

The classification system designed in the mid-18th century by Carolus Linnaeus, a Swedish botanist, has survived, with some modification, to the present day (•**Figure 18-2**). In *Species Plantarum,* which was published in 1753, Linnaeus described all the plants known in his time—some 7300 species—and provided each with a binomial name. Linnaeus based his system of classi-

FIGURE 18-1 Drawing from an herbal.
This drawing of pot marigold (*Calendula*) was created in the 15th century from an illustration in *Materia Medica*. Scholarly books of plants, known as herbals, were printed during the 15th, 16th, and 17th centuries.

fication in *Species Plantarum* on visual observations of flower parts. He noted, for example, that all flowers of the same kind of plant contain the same number of stamens (pollen-producing structures).

Linnaeus probably intended to design an unchanging system of classification, for he carried out his work long before Darwin's theory of evolution made common ancestry the basis for classification. Neither Linnaeus nor his colleagues had any concept of the vast number of living and extinct species that would later be discovered. Yet it is remarkable how flexible and adaptable his system is to new biological knowledge. Few other 18th-century inventions survive today in a form that their originators would recognize. Linnaeus's system of classification provided the foundation on which modern biology was built.

TAXONOMY The science of describing, naming, and classifying organisms.

BINOMIAL NOMENCLATURE A system for giving each organism a two-word scientific name.

© Mary Evans Picture Library/Alamy

FIGURE 18-2 Carolus Linnaeus.
A Swedish botanist and physician, Linnaeus (1707–1778) is considered the father of modern taxonomy because of his contributions to the classification of plants and other organisms. He standardized scientific terminology by using binomial nomenclature to assign scientific names.

ganisms based on a unique two-part name for each. The first part of the name designates the genus, and the addition of the second part, the specific epithet, designates the species. The specific epithet is usually a word that describes some particular quality of the organism.

The generic (genus) name can be used alone to designate all species in the genus, but the specific epithet is never used alone; it is always preceded by the full or abbreviated generic name. In each scientific name, the generic name is first, and its first letter is always capitalized; the specific epithet is given second and is not capitalized. Both names must be underlined or italicized. Corn, for example, is assigned to the genus *Zea;* its specific epithet is *mays.* The proper scientific name for corn is *Zea mays.*

Although any given specific epithet is used for only one species within each genus, it can be used in more than one genus. For example, *Quercus alba* is the scientific name for the white oak, and no other oak species can have *alba* as its specific epithet. However, *Nymphaea alba* is the scientific name for the European white water lily, and *Salix alba* for the white willow. (Recall that *alba* is a Latin word meaning "white.") Both parts of an organism's scientific name must be used to accurately identify the species.

Scientific names are generally composed from Greek or Latin roots or from Latinized versions of the names of persons, places, or characteristics (•**Table 18-1** on page 354). *Hedera,* the generic name for the ivy *Hedera canariensis,* for example, is based on the Latin

Species are classified using binomial nomenclature

Prior to Linnaeus, scholars used Latin sentences up to 12 words long to describe each type of plant. For example, the spiderwort, a popular garden plant, was scientifically described as *Tradescantia ephemerum phalangoides tripetalum non repens Virginianum gramineum* (•**Figure 18-3**). Translated, this name means "the annual upright Tradescantia from Virginia with a grasslike habit, three petals, and stamens with hairs like spider legs." Linnaeus used this lengthy scientific description as well, but he added in the margin a simplified scientific name for this plant, *Tradescantia virginiana.*

Thus, Linnaeus's most significant contribution to biology is **binomial nomenclature,** a system of naming or-

Marion Lobstein

(a) Found in meadows and along roadsides and the edges of woods, spiderworts (*Tradescantia virginiana*) are also popular garden plants.

Marion Lobstein

(b) A close-up of the flower reveals six golden stamens with hairy filaments.

FIGURE 18-3 Spiderworts.

PLANTS AND PEOPLE | Plant Exploration

Humans have continually introduced wild plants into their gardens since they began cultivating food crops. The deliberate introduction of plants from other regions has occurred throughout history as people explored or invaded new areas. As a result of the Spanish invasion of South America, for example, the Spanish introduced potatoes, peppers, tomatoes, and many other plants from the Americas to Europe.

The first expeditions that were organized exclusively to collect plants are thought to have occurred during the 18th century, when plant explorers were typically employed by kings, wealthy men, or nurseries. Plant explorers faced many dangers, including hostile native people and deadly diseases. During the 19th century, for example, 7 of the 22 plant explorers hired by the Veitch nursery in England died overseas from disease, and one had to have his leg amputated.

Transporting the plants home frequently involved long sea journeys. Live plants that were kept on deck to obtain sunlight often succumbed to the salt spray. Plants that were transported as seeds, bulbs, corms, or tubers were more likely to survive. However, they had to be completely dried out to avoid mold, no easy feat if the collecting occurred in the humid tropics. The dried plant parts were packed in wooden cases that were filled with broken glass to keep rats on board from eating them. If a turbulent storm threatened

the ship, the plant material was often tossed overboard to lighten the load.

The experiences of two important plant explorers—Robert Fortune and David Fairchild—highlight the important contributions of the many plant explorers who introduced plants from the wild. Robert Fortune (1812–1880) was a Scottish plant explorer who collected plants in China during the 1840s and 1850s. Fortune, who worked for the British East India Company and the Horticultural Society of London, smuggled tea, along with knowledge of its cultivation and processing after harvest, out of China (see figure a). He discovered that green tea and black tea were produced by the same plant, *Camellia sinensis*. Black tea is made from "fermented" leaves, whereas green tea is "unfermented." Fortune also introduced winter jasmine, forsythia, Chinese holly, and many other beautiful ornamental plants to Europe.

Fortune smuggled plants out of China despite the fact that China was little known to outsiders and was politically unstable. To obtain access to areas declared off-limits to foreigners, Fortune sometimes disguised himself as Chinese. Once he even had to fight and kill pirates who were attempting to board his boat on the South China Sea.

David Fairchild (1869–1954) was a famous American plant explorer. Working for the U.S. Department of Agriculture, Fairchild searched all over the

world for economically useful plants. He caught typhoid in Ceylon (now Sri Lanka) but continued his travels after a brief recovery period. Fairchild introduced thousands of fruits, vegetables, grains, and ornamental plants to the United States. Some of the plants he collected include the water chestnut from China; cotton, grape, carob, barley, and dates from Egypt; pistachios from Greece; blood oranges from Malta; Jordan almonds from Spain; wheat from Russia; alfalfa from Peru; and rice and soybeans from Japan.

Plant exploration still goes on but only rarely involves the dangers of former times—although a botanist sent to Colombia by the Missouri Botanical Garden in the 1980s was beaten and held captive by indigenous Indians until the military intervened. Plant exploration today is generally focused on the collection of a particular plant—for example, cotton plants to breed with existing cultivars. This focus replaces the "Bring back whatever you find that is interesting" approach of past plant exploration.

Usually a team of two botanists is sent to a country to collect the plants, but first they do extensive research to determine where they should look for their target plants. For example, plant explorers first examine a collection of dried, pressed, and carefully labeled plant specimens collected in the past, which are stored in an **herbarium** (see figures b and c). Some herbarium

word for "ivy." The specific epithet *canariensis* tells us that this plant is native to the Canary Islands.

The use of Latin, a dead language that does not change, rather than a modern language in naming organisms is a carryover from the days when Latin was the language of European scholars. Why do we continue to use Latin rather than common names for plants and animals? Why call a sugar maple *Acer saccharum*? (*Acer* is derived from the Latin word meaning "maple," and *saccharum* from the Greek for "sugar.") The main

reason is to be precise and to avoid confusion, for in some parts of the United States this same tree is called hard maple or rock maple. The plant *Mirabilis jalapa* is generally called four-o'clock, but some people call it marvel-of-Peru, and others call it beauty-of-the-night. Because of the many instances of confusing common names, exact scientific names are important for the accurate identification of organisms. In addition, many scientifically important organisms lack common names. Since each organism has only one scientific name, a specimens more than 200 years old are still in use. The specimens brought back from the Lewis and Clark expedition (1804–1806), for example, are still preserved and are included in the Smithsonian National Herbarium. Often the labels on herbarium specimens tell plant explorers precisely where the plants were collected—for example, "in a field 15 km south of the city, along the western banks of the river."

Plant exploration occurs in many places worldwide. A team of Australian plant biologists might be sent to North Africa to collect plants to be screened as sources of new drugs to fight cancer. Or a team of U.S. botanists might be sent to the Middle East to collect seeds of forage grasses (grasses grown for animal fodder) that will be used in selective-breeding experiments to improve the rangeland grasses of western North America. Tropical areas of Central and South America and Asia are common sites for plant exploration, in part because many of the world's fruits and vegetables originated in the tropics. Locating the wild ancestors of important crops might enable us to introduce beneficial genes from these wild plants into existing crops.

The transport of plant material from one country or continent to another is more successful now than it once was, because the plants or seeds are wrapped in polyethylene, refrigerated, and transported quickly by air rather than ship. Plant exploration today relies on international cooperation, and special care is taken to avoid the accidental introduction of devastating pests along with the plant material.

(a) Robert Fortune was a 19th-century plant explorer from Scotland. This lithograph of tea plantations in China was published in his book, *A Journey to the Tea Countries of China,* in 1852.

(b) An herbarium specimen of shrubby althaea (*Hibiscus syriacus*), collected in 1992.

(c) A closeup of the label. Location 121 refers the researcher to a list that indicates exactly where this specimen was collected, on the University of Virginia's Blandy Experimental Farm in Boyce.

researcher in Canada reading a study published by a German scientist knows exactly which organisms were used.

A complete scientific name has a third part that cites the person who formally described the organism. Thus, the complete scientific name of the spiderwort is *Tradescantia virginiana* Linnaeus. (Note that the author citation is not italicized.) The author citation is often abbreviated, as in *Tradescantia virginiana* L. The *Draft Index for Author Abbreviations* and other sources explain the abbreviated names of author citations. The author citation is useful because it helps researchers find the original description of a particular plant.

The species is the basic unit of classification

Recall from Chapter 17 that a **species** consists of one or more populations whose members can interbreed to produce fertile offspring but are reproductively isolated

TABLE 18-1	Derivations of Selected Plant Genera Mentioned in This Chapter
GENUS	**DERIVATION**
Acer	Latin name for "maple"
Camellia	For George Kamel (Camellus was his Latinized name) (1661–1706), a Moravian Jesuit who studied Asian plants
Capsicum	From the Greek word *kapto,* meaning "to bite" (refers to the hot taste)
Gilia	Perhaps for Felipe Gil, an 18th-century Spanish botanist
Hedera	Latin name for "ivy"
Hibiscus	Greek name for "mallow" (a kind of plant)
Hymenocallis	From the Greek words for "membrane" and "beauty," in reference to the stamens, which are joined by a membrane
Mirabilis	From the Latin word for "wonderful"
Nymphaea	After Nymphe, a water nymph
Picea	Latin name for "pitch," a resin produced by spruce
Pyrus	From the Latin word for "pear"
Quercus	Latin name for "oak"
Salix	Latin name for "willow"
Tradescantia	For John Tradescant (1570–1638), English gardener and botanist
Zingiber	From *zingiberi,* named by Dioscorides from an Indian word

from other organisms.[1] Members of a species share a common evolutionary ancestry.

The species is the basic unit of classification but not the smallest taxonomic group in use. Geographically distinct populations within a species often display certain characteristics that distinguish them from other populations of the same species. If they can interbreed, however, they are not truly separate species but are known as **subspecies.**

Experts can usually distinguish subspecies from one another. However, subspecies may grade into one another at the borders of their geographic ranges, where there is opportunity to interbreed. Some of these subspecies are apparently in the process of becoming reproductively isolated and may, over time, become separate species (•**Figure 18-4**). Thus, some subspecies provide biologists with the opportunity to study evolution in progress.

Plants that are produced using techniques such as *selective breeding* (also called artificial selection) are called **cultivars.** The term *cultivar* is an abbreviated form of *cultivated variety.* A cultivar is not equivalent to a subspecies; a cultivar is produced while under cultivation, whereas a subspecies is a geographically distinct population that evolves by natural selection. Cultivar names may be used after scientific names. For example,

two cultivars of peaches are *Prunus persica* cv. Rosea and *Prunus persica* cv. Early Flame, which is a nectarine (a smooth-skinned cultivar of peaches).

Each taxonomic level is more general than the one below

Classification is hierarchical; the narrowest category in the Linnaean system is the species, and the broadest is the kingdom. Closely related species are grouped in the next higher level of classification, the **genus** (pl., *genera*). One or more related genera are assigned to the same **family,** and families are grouped into **orders,** orders into **classes,** classes into **phyla,** and phyla into **kingdoms** and/or **domains** (•**Table 18-2;** also see Table 1-1). These groupings can also be separated into subgroupings—for example, subphylum or subclass.

Each taxonomic level is broader (more inclusive) than the level below. For example, the family Zingiberaceae (the ginger family) contains about 1300 species in 49 genera, including genus *Zingiber* (the ginger genus), genus *Curcuma* (the turmeric genus), and genus *Elettaria* (the cardamom genus) (•**Figure 18-5**). The family Zingiberaceae, along with four other families, is placed in the order Zingiberales. Order Zingiberales and 18 other orders belong to class Monocotyledones (the monocots)—one of two main classes of flowering plants. The monocots produce seeds that contain one cotyledon (seed leaf) and possess flower parts (for example, petals, stamens, and carpels) in threes or multiples of three. Class Monocotyledones (monocots) and class Eudicotyledones (eudicots), along with several minor classes, are

---

[1]*Chapter 17 discussed some of the shortcomings of the definition of* species. *One shortcoming is that different plant species can sometimes interbreed to form fertile hybrids. Thus, reproductive isolation is not always characteristic of plant species.*

TABLE 18-2 Classification of Pear, Spider Lily, and Colorado Blue Spruce

	PEAR	SPIDER LILY	COLORADO BLUE SPRUCE
Kingdom	Plantae	Plantae	Plantae
Phylum	Anthophyta	Anthophyta	Coniferophyta
Class	Eudicotyledones	Monocotyledones	Coniferopsida
Order	Rosales	Liliales	Coniferales
Family	Rosaceae	Liliaceae	Pinaceae
Genus	*Pyrus*	*Hymenocallis*	*Picea*
Species	*Pyrus communis*	*Hymenocallis caroliniana*	*Picea pungens*

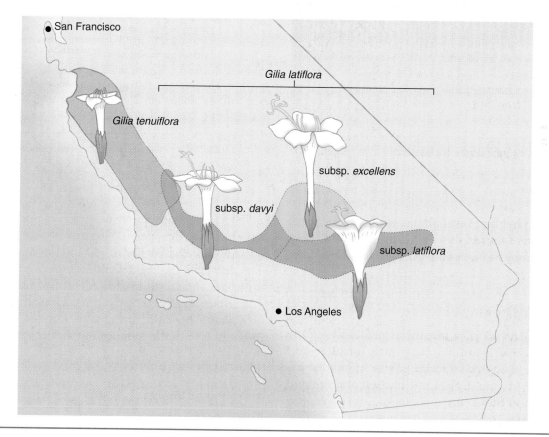

FIGURE 18-4 Ranges of several subspecies of the California wildflower *Gilia latiflora* and of a closely related species, *G. tenuiflora*.
From their similarities, it is probable that *G. tenuiflora* was originally a subspecies of *G. latiflora*. Because it now overlaps *G. latiflora* geographically without interbreeding, *G. tenuiflora* is considered a separate species. For simplicity, three other subspecies of *G. latiflora* have been omitted.

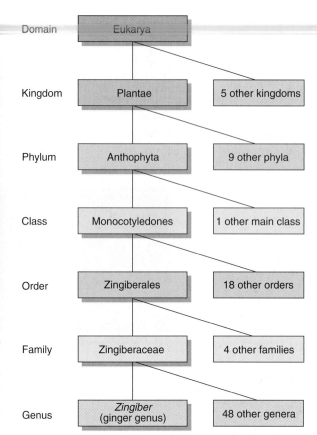

(a) A spike of ginger (*Zingiber*) flowers. The genus name is from *Zingiberi,* the name given this plant by Dioscorides; it is derived from an Indian word.

(b) Each taxonomic level is more inclusive than the one below it. For example, the order Zingiberales consists of 5 families. The family Zingiberaceae contains 49 genera and a total of about 1300 species.

FIGURE 18-5 **Hierarchical levels of classification for ginger.**

grouped into phylum Anthophyta, which, along with phylum Coniferophyta, phylum Gnetophyta, and several others, belongs to kingdom Plantae.

Taxonomists construct dichotomous keys to help identify plants

Most taxonomists identify organisms primarily on the basis of structural features. Using such features, taxonomists have developed special guides known as dichotomous keys to aid in the identification of unknown plants. A **dichotomous key** consists of a series of two contrasting statements made about plants. The person trying to identify the unknown plant chooses which of two statements is correct for the plant in question and is then directed to another set of paired, contrasting statements. By working through the statements in the key, the investigator eventually identifies and classifies the unknown organism.

To understand how such a key works, try to identify the plants depicted in •**Figure 18-6** by using the simplified, abbreviated key to 10 common woody plants. The entry in the right-hand column opposite the appropriate structural feature identifies the plant or tells you which number to go to next. Be aware that students often have difficulty using such keys because they are unfamiliar with the terminology or because they make a "wrong turn" in haste.

EVOLUTION LINK Classification Based on Evolutionary Relationships

The classification of plants and other organisms into groups determined by their evolutionary relationships is called **systematics.** A systematist seeks to reconstruct

(a)

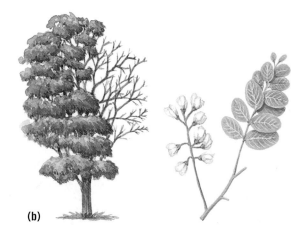

(b)

Key to 10 Common Woody Plants in the Eastern United States
1. a. Needles (elongated, thin leaves). . . . . Pine
   b. Leaves with broad, flat blades 2
2. a. Tree (one main stem) 3
   b. Shrub (multiple stems) or vine 8
3. a. Simple leaves (undivided leaves) 4
   b. Compound leaves (each leaf composed
      of two or more leaflets) 7
4. a. Palmate venation (veins radiating from
      a common point) 5
   b. Pinnate venation (veins arranged like
      a feather) 6
5. a. Opposite leaf arrangement on stem . . . . Maple
   b. Alternate leaf arrangement on stem . . . . Sweet gum

6. a. Narrow blade (1 centimeter or less) . . . . Willow
   b. Broad blade. . . . . Tulip poplar
7. a. Leaflets sharply tapered at tip . . . . Hickory
   b. Leaflets blunt at tip . . . . Black locust
8. a. Simple (undivided) leaves . . . . Mountain laurel
   b. Compound leaves (each leaf composed
      of two or more leaflets) 9
9. a. Leaves palmately compound (leaflets
      radiating from a common point) . . . . Virginia creeper
   b. Leaves pinnately compound (leaflets
      arranged like a feather) . . . . Sumac

FIGURE 18-6 Identifying plants using a dichotomous key. Use the key to identify the two plants depicted here.

the evolutionary history, or **phylogeny,** of organisms. Once these relationships are defined, the classification of organisms is based on common ancestry.

If all the plants within a taxonomic group share the same common ancestor, the group is referred to as **monophyletic** (one branch). Monophyletic groups are natural groupings, as they represent true evolutionary relationships, and they include all close relatives.

Many groups are **polyphyletic,** consisting of several evolutionary lines and not including a common ancestor. Polyphyletic groups may misrepresent evolutionary relationships. For this reason, taxonomists try to avoid constructing polyphyletic groups.

Cladistics emphasizes evolutionary relationships

An approach to systematics that has been widely adopted in the past several decades is **cladistics.** Cladistics emphasizes phylogeny by focusing on when evolutionary lineages (lines of descent) divide into two branches.

Cladists (taxonomists who practice cladistics) insist that taxa be monophyletic and use carefully defined objective criteria to determine branch points. Thus, common ancestry is the basis for classification. A cladist would classify daisies with flowering plants rather than with ferns because daisies and other flowering plants share a more recent common ancestor than do daisies and ferns. Cladists develop branching diagrams called **cladograms.**

SYSTEMATICS The scientific study of the diversity of organisms and their natural (evolutionary) relationships.

PHYLOGENY The evolutionary history of a species or other taxonomic group.

MONOPHYLETIC Said of a group consisting of organisms that evolved from a common ancestor.

CLADISTICS Classification of organisms based on recency of common ancestry rather than degree of structural similarity.

CLADOGRAM A diagram that illustrates evolutionary relationships based on the principles of cladistics.

The most difficult step in building a cladogram is to organize the characters in their correct evolutionary order. In this example, the character state "absent" for a particular character is the ancestral condition.

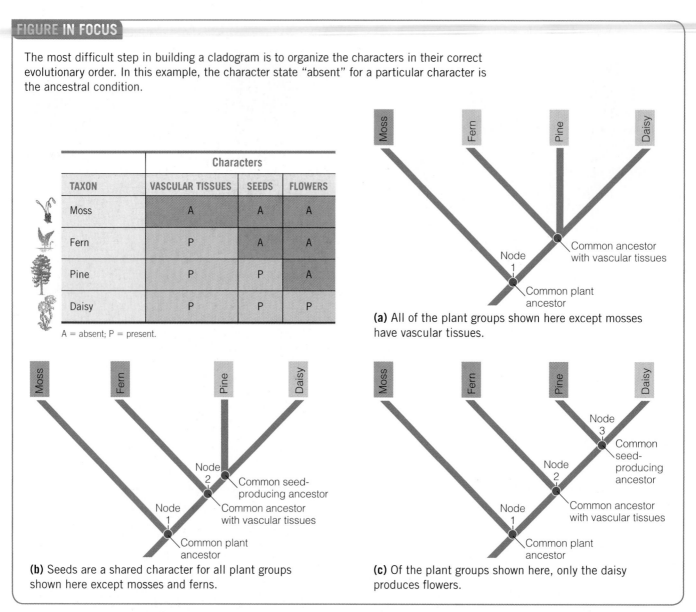

(a) All of the plant groups shown here except mosses have vascular tissues.

(b) Seeds are a shared character for all plant groups shown here except mosses and ferns.

(c) Of the plant groups shown here, only the daisy produces flowers.

FIGURE 18-7 Building a cladogram.
Refer to the table as you follow the process in (a) through (c). (Adapted from Dr. John Beneski, Department of Biology, West Chester University, West Chester, Pennsylvania)

The general goal of cladistics is to reconstruct phylogenies using an analysis of evolutionary changes in specific characters. Structural, physiological, or molecular characters can be used for the analysis. The characters must be homologous (identical by common descent) and must have evolved independently of one another.

The first step in constructing a cladogram is to select the groups. Here we use four plant groups: moss, fern, pine, and daisy. The next step is to select the homologous characters to be analyzed. In our example we use three characters: vascular tissue, seeds, and flowers (see table in •**Figure 18-7**). For each character, we must

define all of the conditions. For simplicity, we will consider our characters to have only two states—present and absent.

The last step in preparing the data is to organize the character states into their correct evolutionary order. The most common method of doing this is by *outgroup analysis*. An **outgroup** is a group that is considered to have diverged earlier than the other group under investigation and thus may represent an approximation of the ancestral condition. In our example, moss is the chosen outgroup, and its character states are considered ancestral. Therefore, the character state "absent" is the ancestral condition, and the character state "present" is the derived (evolved) condition for each of the characters.

Examine the table, and notice that all groups except the outgroup share a derived trait, vascular tissues. We may therefore conclude that these three groups—fern, pine, and daisy—are monophyletic, and using these data, we may construct a preliminary cladogram (see Figure 18-7a). The branch points (referred to as *nodes*) represent the divergence of the ancestral lineage into two lineages. Thus, node 1 represents the common plant ancestor as well as the divergence of the outgroup (moss) and the common ancestor of the three vascular plant groups.

Similarly, node 2 represents a subsequent divergence of the fern and the ancestor of plants with seeds (see Figure 18-7b). The branching process is continued with the data from the table (see Figure 18-7c). Note that pine and daisy are more similar to one another than to any other group. The presence of a common ancestor at node 3 indicates this relationship. The distance from the base of the tree indicates the relative time of divergence. For example, the farther a node is up the tree, the more recent the time of divergence. In our example, node 3 represents the most recent divergence, and node 1 represents the most ancient divergence. Thus, the daisy is more closely related to the pine and more distantly related to the fern. ■

PROCESS OF SCIENCE | Major Branches in the Tree of Life

Until relatively recently, the broadest level of taxonomic classification was the kingdom. The history of taxonomy at the kingdom level is an excellent example of how scientific knowledge expands and changes over time. From the time of Aristotle to the mid-19th century, biologists divided the living world into two kingdoms, plants and animals. With the development of microscopes, it became increasingly obvious that many organisms could not easily be assigned to either the plant or the animal kingdom. For example, the unicellular organism *Euglena*, which has been classified at various times in the plant kingdom and the animal kingdom, carries on photosynthesis in the light but in the dark uses its flagellum to move about in search of food (see Figure 20-3).

More than a century ago, a German biologist, Ernst Haeckel, suggested that a third kingdom be established, the kingdom Protista. Simple and ambiguous organisms, such as bacteria and most microorganisms, including *Euglena*, were classified in the kingdom Protista. Today, many biologists place algae (including multicellular forms), protozoa, water molds, and slime molds

in kingdom Protista. Thus, protists are a diverse group of mainly unicellular, mainly aquatic organisms.

In the 1930s, the French biologist Edouard Chatton suggested the term *procariotique* ("before nucleus") to describe bacteria, and the term *eucariotique* ("true nucleus") to describe all other cells. Virtually all biologists now accept this separation between prokaryotes and eukaryotes as a fundamental evolutionary divergence.

In the 1960s, advances in electron microscopy and biochemical techniques revealed basic cellular differences that inspired many new proposals for classifying organisms. In 1969, R. H. Whittaker proposed a five-kingdom classification (Monera, Protista, Plantae, Fungi, and Animalia) based mainly on cell structure and the ways that organisms obtain nutrition from their environment. Kingdom Monera was established to accommodate the bacteria, which are fundamentally different from all other organisms in that they are prokaryotic and lack distinct nuclei and other membranous organelles. Whittaker also suggested that fungi (such as mushrooms, molds, and yeasts) be classified as a separate kingdom, kingdom Fungi, rather than as part of the plant kingdom, because fungi are nonphotosynthetic and obtain energy by absorbing nutrients. Fungi also differ from plants in the composition of their cell walls, in body structure, and in reproduction.

Most biologists accepted the five kingdoms proposed by Whittaker until DNA studies by Carl Woese (pronounced "woes") and others beginning in the late 1970s revealed that there are two groups of prokaryotes. Woese proposed that the prokaryotes account for two of the three major branches of organisms. Woese's hypothesis continued to gain supporting evidence, and now most biologists think the two groups of prokaryotes, the more familiar Bacteria and the distinctive Archaea, each merit their own kingdom (see Chapter 19).

As a result, many biologists currently recognize a six-kingdom scheme that takes into account these evolutionary relationships as well as cell structure and ways of obtaining energy (•Figure 18-8 and •Table 18-3). The six kingdoms are Bacteria (most prokaryotes); Archaea (some prokaryotes found in extreme environments); Protista (algae, protozoa, water molds, and slime molds); Plantae (plants); Fungi (mushrooms and other fungi); and Animalia (animals).

Most biologists also use a level of classification above the kingdom, called a *domain*, based on fundamental molecular differences among the bacteria, archaea, and eukaryotes. They classify organisms in three domains: **Bacteria** (which corresponds to kingdom

DOMAIN BACTERIA The domain of metabolically diverse, unicellular, prokaryotic organisms.

Bacteria), **Archaea** (which corresponds to kingdom Archaea), and **Eukarya** (eukaryotes) (●**Figure 18-9**). Biologists have inferred that the three domains are the three main branches of the tree of life. Although you might think that the two domains of prokaryotic organisms would be closely related, the domain Archaea is apparently more closely related to the domain Eukarya than it is to the domain Bacteria.

Protists do not fit well in a single kingdom

Biologists use molecular data and ultrastructure to clarify evolutionary relationships among the protists. For example, biologists compare the nucleotide sequences for the same gene found in different protists. *Ultrastructure* consists of the fine details of cell structure revealed by electron microscopy.

Both molecular data and ultrastructure suggest that members of kingdom Protista are not a monophyletic group. Instead, most biologists regard the protist kingdom as a **paraphyletic group.** The common eukaryote ancestor that gave rise to the members of kingdom Protista also gave rise to plants, fungi, and animals, all of which are multicellular eukaryotes not classified in the protist kingdom.

Some biologists want to resolve this problem by splitting the organisms in the protist kingdom into several smaller kingdoms, each with its own common ancestor. However, most biologists (and students!) would rather deal with the limitations of a single, paraphyletic protist kingdom than with several additional monophy-

letic kingdoms. As you have seen, scientific knowledge changes as new information comes to light. It may be that kingdom Protista, despite its convenience, will be partitioned into several kingdoms in the future. We say more about the organisms in the protist kingdom in Chapter 20. ■

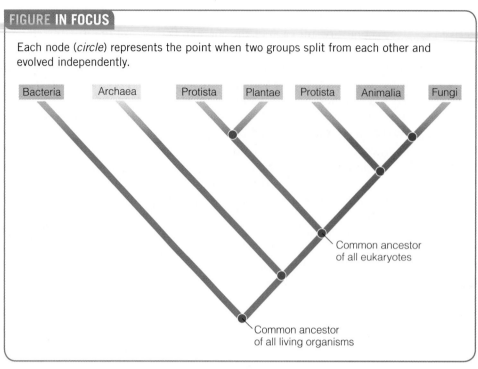

FIGURE IN FOCUS

Each node (*circle*) represents the point when two groups split from each other and evolved independently.

FIGURE 18-8 The six kingdoms.
This cladogram illustrates the evolutionary relationships among the six kingdoms. The branches portray ancestral populations of each group through time. Note that the Protista are a diverse group with two major branches. Systematists may eventually divide them into two or more kingdoms.

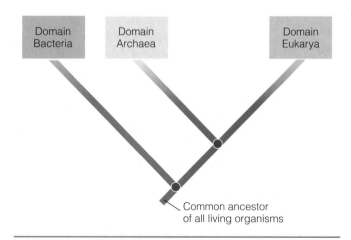

FIGURE 18-9 The three domains.
This diagram illustrates the evolutionary relationships among organisms in the three domains. Kingdoms Protista, Plantae, Fungi, and Animalia are assigned to domain Eukarya.

DOMAIN ARCHAEA The domain of unicellular, prokaryotic organisms adapted to extreme conditions, such as very hot or very salty environments.

DOMAIN EUKARYA The domain that includes all eukaryotic organisms (protists, plants, fungi, and animals).

PARAPHYLETIC Said of a group consisting of a common ancestor and some, but not all, of its descendants.

TABLE 18-3 The Six Kingdoms of Life

DOMAIN	KINGDOM	EXAMPLES	CHARACTERISTICS
Bacteria	Bacteria	Most bacteria belong to this group, including the nitrogen-fixing bacteria, cyanobacteria, lactic acid bacteria (in yogurt), and enterobacteria (in the intestines of humans and other animals)	Prokaryotic cells (lack distinct nuclei and other membranous organelles); single celled; microscopic; cell walls composed of peptidoglycan
Archaea	Archaea	Halophiles (live in salt ponds); thermoacidophiles (live in hot sulfur springs); methanogens (live in swamps and in the digestive tracts of humans and other animals)	Prokaryotic cells; single celled; microscopic; cell walls without peptidoglycan; very different biochemically from bacteria; adapted to extreme environments, such as hot springs and undersea thermal vents
Eukarya	Protista	Algae; slime molds; water molds; protozoa	Eukaryotic cells (possess distinct nuclei and other membranous organelles); single celled or simple multicellular; varied modes of nutrition—some are photosynthetic (like plants), some consume food (like animals), and some absorb nutrients (like fungi)
Eukarya	Fungi	Mushrooms; puffballs; mildews; yeasts	Eukaryotic cells; most have a threadlike, multicellular body; nonphotosynthetic; absorb nutrients; cell walls of chitin
Eukarya	Plantae	Mosses; ferns; pines; flowering plants	Eukaryotic cells; multicellular; photosynthetic; multicellular reproductive organs; cell walls of cellulose
Eukarya	Animalia	Jellyfish; clams; grasshoppers; frogs; mammals	Eukaryotic cells; multicellular; no cell walls; nonphotosynthetic; most obtain nutrients by consuming other organisms; most move about by muscular contraction; specialized nervous tissue to coordinate responses

STUDY OUTLINE

❶ Define *taxonomy,* and explain why the assignment of a scientific name to each species is important for biologists.
Taxonomy is the science of describing, naming, and classifying organisms. The main reason biologists use scientific names is to precisely identify organisms. Since each organism has only one scientific name, scientists avoid the confusion of one organism having many common names.

❷ Identify the biologist who originated the binomial system of nomenclature, and describe the general scheme of the system.
Binomial nomenclature is a system for giving each organism a two-word scientific name. Binomial nomen-

clature was first used consistently by Linnaeus. In this system, the basic unit of classification is the species. The scientific name of each species has two parts: the generic name (genus) and the specific epithet.

❸ List and describe the hierarchical groupings of classification.
The hierarchical system of classification includes the following groups, from most inclusive to least inclusive: **domain, kingdom, phylum, class, order, family, genus,** and **species.**

❹ Define *systematics,* and describe the cladistic approach to systematics.
Systematics is the scientific study of the diversity of organisms and their natural (evolutionary) relation-

ships. A systematist seeks to reconstruct **phylogeny,** the evolutionary history of a species or other taxonomic group. A group consisting of organisms that evolved from a common ancestor is said to be **monophyletic. Cladistics** is the classification of organisms based on recency of common ancestry rather than degree of structural similarity. Cladists emphasize phylogeny by focusing on when evolutionary lineages (lines of descent) divide into two branches. Cladists develop **cladograms,** diagrams that illustrate evolutionary relationships based on the principles of cladistics.

❺ **List and briefly describe the three domains and six kingdoms recognized by many biologists.**
The **domain Bacteria** consists of metabolically diverse, unicellular, prokaryotic organisms. The **domain Archaea** consists of unicellular, prokaryotic organisms adapted to extreme conditions, such as very hot or very salty environments. The **domain Eukarya** includes all eukaryotic organisms (protists, plants, fungi, and animals). The six-kingdom classification recognizes the kingdom Bacteria (which corresponds to domain Bacteria), kingdom Archaea (domain Archaea), kingdom Protista (domain Eukarya), kingdom Fungi (domain Eukarya), kingdom Plantae (domain Eukarya), and kingdom Animalia (domain Eukarya).

❻ **Summarize the scientific limitations of the kingdom Protista.**
Ideally, all members of a kingdom should have a common ancestor. Members of kingdom Protista are **paraphyletic** and consist of a common ancestor and some, but not all, of its descendants. Some biologists think the protists should not be grouped in a single kingdom.

REVIEW QUESTIONS

1. Distinguish between an organism's common names and its scientific name, and explain why scientific names are used in scientific work.
2. How did scientific names change as a result of Linnaeus's contributions?
3. What are the key features of binomial nomenclature?
4. The complete scientific name for white oak is *Quercus alba* L. What is the genus? The specific epithet? The author citation?
5. What is a species?
6. What is the difference between a subspecies and a cultivar?
7. List the taxonomic levels in the hierarchical system of classification, beginning with species.
8. Distinguish among systematics, taxonomy, and cladistics.
9. What is a cladogram? What does a node represent in a cladogram?
10. Why do many biologists recognize three domains and six kingdoms?
11. Explain why the kingdom Protista is not a natural kingdom, as compared with the other kingdoms.

THOUGHT QUESTIONS

1. Biologists think that there may be millions of species that are not yet scientifically described or named. Discuss why these organisms have not yet been studied, where most of them probably live, and to which kingdom or kingdoms they probably belong.
2. All fields of biology use taxonomy. List at least five disciplines within plant biology (see Chapter 1 for the disciplines), and tell why taxonomy would be important to each.
3. In which domain and kingdom would you classify each of the following: (a) oak, (b) *Euglena*, (c) *Streptococcus* (a prokaryote that can cause strep throat), (d) black bread mold, (e) tapeworm, (f) *Halobacterium* (a prokaryote that lives in extremely high concentrations of salt)?
4. How does a monophyletic group differ from a polyphyletic group?
5. Examine the figure on page 363.
   a. Which group of organisms in this cladogram was the first (that is, earliest) to diverge? What do cladists call such a group?
   b. Complete the following statement: Animals are _____ (closely or distantly) related

to bacteria. Explain your answer using the cladogram.

c. Which group of organisms shares the most recent common ancestor with fungi? Explain your answer using the cladogram.

Visit us on the web at http://www.thomsonedu.com/biology/berg/ for additional resources such as flashcards, tutorial quizzes, further readings, and web links.

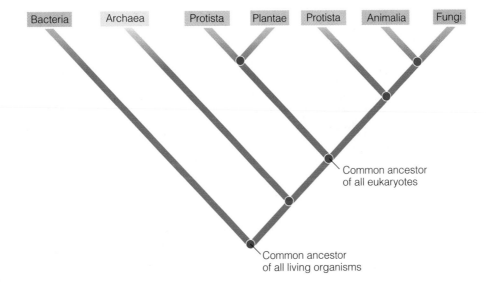

Viruses and Prokaryotes

A*nabaena* is the genus that contains several species of photosynthetic organisms whose bodies consist of filaments, or unbranched chains, of prokaryotic cells. *Anabaena* is a type of bacteria, known as *cyanobacteria,* found worldwide in fresh water, marine environments, and the soil.

Formerly, *Anabaena* species were considered a primitive type of alga, called blue-green algae, and were classified in the plant kingdom. Because cyanobacteria and other bacteria were traditionally considered plants, they are usually studied in plant biology courses. Despite their prokaryotic cell structure, *Anabaena* and other cyanobacteria resemble plants and algae in some ways. For example, photosynthesis is essentially identical in cyanobacteria, algae, and plants, and all contain the green pigment, chlorophyll *a.*

Anabaena and other cyanobacteria play crucial roles in the living world. Along with algae, cyanobacteria form the base of the food web in aquatic environments. Without producers such as *Anabaena,* aquatic animals could not exist because they would have nothing to eat. The oxygen that *Anabaena* and other cyanobacteria generate during photosynthesis also benefits aquatic organisms.

Anabaena and many other cyanobacteria are also ecologically important because they take nitrogen from the air and convert it to ammonia (NH_3), a form of nitrogen that plants use to synthesize proteins and other nitrogen-containing organic compounds. The conversion of atmospheric nitrogen to ammonia is called **nitrogen fixation.** Although nitrogen is an essential component of such biologically important molecules as proteins and nucleic acids, only a few prokaryotic organisms have the ability to fix nitrogen.

Anabaena sometimes forms a close association with an aquatic fern known as *Azolla.* The *Azolla–Anabaena* symbiosis, or partnership, is common in flooded rice paddies, where it provides needed nitrogen "fertilizer" to the rice. Although rice has been grown in Asia for thousands of years, the rice fields there do not exhibit a nitrogen deficiency. In contrast, crops grown in fields without populations of *Azolla–Anabaena* require the annual application of nitrogen fertilizer.

This chapter examines viruses and prokaryotes. Although they are not closely related organisms, we discuss them in a single chapter for convenience.

Anabaena, a filamentous cyanobacterium that fixes nitrogen. Nitrogen fixation occurs in the large, rounded cells called heterocysts.

Viruses

PROCESS OF SCIENCE During the late 1800s, botanists searched for the cause of tobacco mosaic disease, which stunts the growth of tobacco plants and gives the infected leaves a spotted, mosaic appearance. Botanists found they could transmit the disease to healthy plants by daubing their leaves with the sap of diseased plants. In 1892, Dmitri Ivanowsky, a Russian botanist, showed that the sap was still infective after it was passed through filters that removed particles the size of all known bacteria. A few years later, in 1898, his work was expanded by Martinus Beijerinck, a Dutch microbiologist. Apparently unaware of Ivanowsky's work, Beijerinck provided independent evidence that the agent that caused tobacco mosaic disease had many characteristics of a living organism. He hypothesized that the infective agent could reproduce only within a living cell and named it *virus* (the Latin word *virus* means "poison").

Early in the 20th century, scientists discovered infective agents, like those responsible for tobacco mosaic disease, that could cause disease in animals or kill bacteria. These pathogens (disease-causing agents) also passed through filters that removed known bacteria and were so small they were invisible under the light microscope. Curiously, they would not grow in laboratory cultures unless living cells were present. Today, we know these pathogens as viruses. ▪

A virus consists of nucleic acid surrounded by a protein coat

Viruses are not cellular, cannot move about on their own, and cannot carry on metabolic activities independently. All cellular forms of life, which are classified in the domains and kingdoms, contain both DNA and RNA, but a virus contains *either* DNA *or* RNA, not both. A typical virus consists of a core of nucleic acid (DNA or RNA) surrounded by a protein coat (●Figure 19-1a and b). Viruses lack ribosomes and the enzymes necessary for protein synthesis. They reproduce, but only within the living cells that they infect.

Because they are not cellular, viruses are not classified using the traditional Linnaean system into domains, kingdoms, or phyla. Today, the International Committee on Taxonomy of Viruses classifies viruses according to common characteristics, such as type of nucleic acid and presence or absence of a protein coat.

Where do viruses come from? The most widely held hypothesis is that viruses were originally fragments of DNA or RNA that "escaped" from cellular organisms. According to this *escaped gene hypothesis,* some viruses may trace their origins to animal cells, others to plant cells, and still others to bacterial cells. These multiple origins may explain why each virus usually infects only certain species—that is, a virus may infect a species that is the same as or closely related to the organism in which it originated. Support for this hypothesis includes the genetic similarity between some viruses and their host cells—a closer similarity than exists between one type of virus and another.

Another hypothesis suggests that viruses, because of their simplicity, represent a primitive form of life that arose even before the three domains diverged. However, viruses are parasites, dependent on their hosts. How could they have existed before their hosts evolved?

Bacteriophages are viruses that attack bacteria

Microbiologists called viruses that killed bacteria **bacteriophages** ("bacteria eaters"), or **phages.** Much of our knowledge of viruses has come from studying phages,

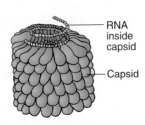

(a) Tobacco mosaic virus is an RNA-containing plant virus that is rod shaped. It consists of a core of RNA surrounded by a protein coat (capsid).

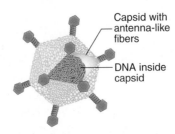

(b) Adenoviruses are DNA-containing animal viruses. An adenovirus has a coat composed of 252 oval subunits arranged into a 20-sided polyhedron.

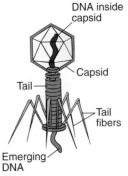

(c) Bacteriophages are DNA-containing bacterial viruses with a protein coat consisting of a polyhedral head and a helical tail. The DNA is normally coiled within the head.

FIGURE 19-1 Virus structure.

Viruses reproduce by seizing control of the metabolic machinery of a host cell. The host cell is destroyed in a lytic infection.

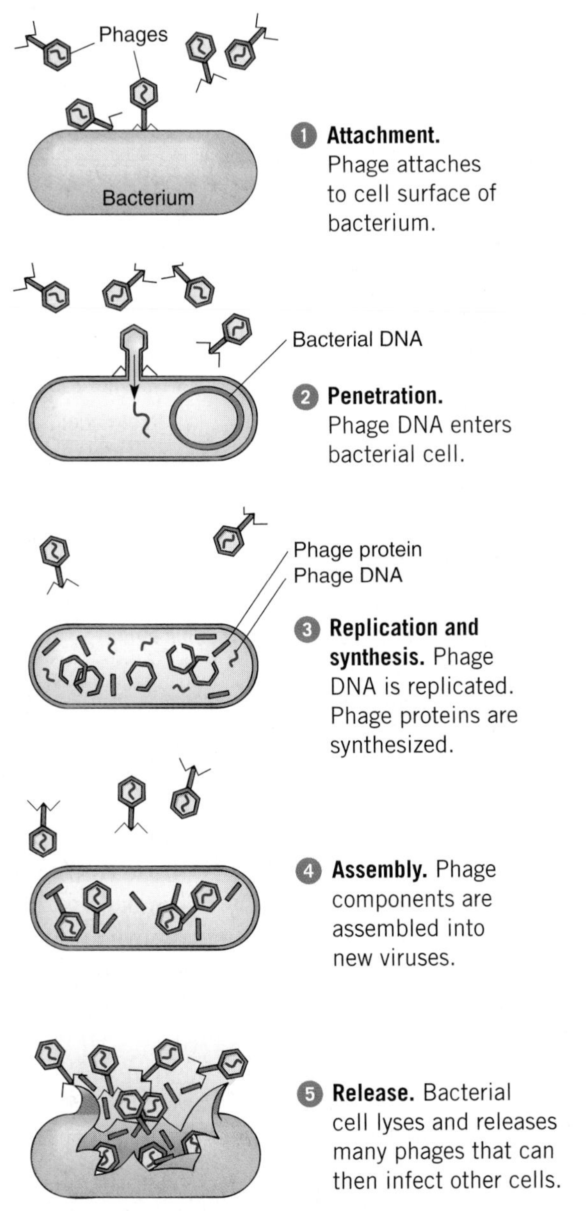

Phages

Bacterium

1 Attachment. Phage attaches to cell surface of bacterium.

Bacterial DNA

2 Penetration. Phage DNA enters bacterial cell.

Phage protein
Phage DNA

3 Replication and synthesis. Phage DNA is replicated. Phage proteins are synthesized.

4 Assembly. Phage components are assembled into new viruses.

5 Release. Bacterial cell lyses and releases many phages that can then infect other cells.

FIGURE 19-2 Sequence of events in a lytic cycle.

because they can be cultured easily within living bacteria in the laboratory. Phages are among the most complex viruses (•**Figure 19-1c**). Their most common structure consists of a long nucleic acid molecule (usually DNA) coiled within a polyhedral head. Many phages have a tail attached to the head. The phage may use fibers extending from the tail to attach to a bacterium. More than 2000 phages have been identified.

Before the age of sulfa drugs and antibiotics, phages were used clinically to treat infection. Then in the 1940s, they were abandoned (at least in Western countries) in favor of antibiotics, which were more dependable and easier to use. Now, with the widespread and growing problem of bacterial resistance to antibiotics (see *Plants and People: Evolution and Bacterial Resistance to Antibiotics,* in Chapter 16), these bacteria-killing viruses are once again the focus of research. With new knowledge of phages and sophisticated technology, several research groups are investigating phages to determine which ones kill which species of bacteria. Scientists are genetically engineering phages so that bacterial resistance to the phages will evolve more slowly.

Phages are also used to improve food safety. For example, certain phages kill deadly strains of *Escherichia coli* in cattle. Apparently, these *E. coli* bacteria do not make cattle ill but cause illness and death in people who eat infected, undercooked meat.

Viruses reproduce at the expense of their host cells

Viruses reproduce in two ways, the lytic and lysogenic cycles. In the **lytic cycle,** the virus lyses (destroys) the host cell. When the virus infects a susceptible host cell, it forces the host to use its metabolic machinery to replicate viral particles (•**Figure 19-2**). First the virus attaches to receptors on the host cell wall. Then the nucleic acid of the virus moves through the plasma membrane and into the cytoplasm of the host cell. The protein coat of a phage remains on the outside, but many viruses that infect animal cells enter the host cell intact.

Because the viral nucleic acid contains all the information necessary to produce new viruses, once the virus is inside the cell, it degrades the host cell nucleic acid and uses the host cell to synthesize the necessary components for the replication of the virus. After the newly synthesized viral components are assembled into new viruses, the viruses are released from the cell.

Phage release typically occurs all at once and results in rapid cell lysis, whereas animal viruses are often released slowly or bud off from the plasma membrane. The new viruses infect other cells, and the process begins anew. The time required for viral reproduction, from attachment to the cell to the release of new viruses, varies from less than 20 minutes to more than an hour.

VIRUS A tiny disease-causing agent consisting of a core of nucleic acid usually encased in protein.

LYTIC CYCLE The life cycle of a virus that kills the host cell by lysing it.

Some viruses integrate their DNA into the host DNA

Viruses do not always destroy their hosts. In a **lysogenic cycle** the viral nucleic acid becomes integrated into the host bacterial DNA and is then referred to as a **prophage.** When the bacterial DNA replicates, the prophage also replicates (•**Figure 19-3**). The viral genes that code for viral structural proteins may be repressed indefinitely. Bacterial cells carrying prophages are called *lysogenic cells.* Certain external conditions (such as ultraviolet light and X-rays) cause prophages to revert to a lytic cycle and then destroy their host. Sometimes prophages become lytic spontaneously.

Bacterial cells containing certain viruses in the lysogenic state may exhibit new properties, that is, the bacteria undergo **lysogenic conversion.** An interesting example involves the bacterium *Corynebacterium diphtheriae,* which causes diphtheria. Two strains of this species exist, one that produces a toxin (and causes diphtheria) and one that does not. The only difference between these two strains is that the toxin-producing bacteria contain a specific lysogenic phage. The phage DNA codes for the powerful toxin that causes the symptoms of diphtheria. Similarly, the bacterium *Clostridium botulinum,* which causes botulism, a serious form of food poisoning, is harmless unless it contains a prophage that induces synthesis of the toxin.

Some viruses infect animal cells

Hundreds of different viruses infect humans and other animals. Most viruses cannot survive very long outside a living host cell, so their survival depends on their being transmitted from animal to animal. The type of attachment proteins on the surface of a virus determines what type of cell it can infect. Some viruses have fibers that project from the protein coat and may help the virus adhere to the host cell. Other viruses, such as those that cause herpes, influenza, and rabies, are surrounded by an envelope of lipid and protein with projecting spikes that aid in attachment to a host cell.

After attachment to a host cell, some viruses fuse with the animal cell's plasma membrane. The viral protein coat and nucleic acid are both released into the animal cell. Other viruses enter the host cell by *endocytosis,* in which the plasma membrane of the animal cell invaginates to form a membrane-bounded vesicle that contains the virus.

In DNA animal viruses, the synthesis of viral DNA and protein is similar to the processes the host cell would normally perform during its own DNA and protein synthesis. In most RNA animal viruses, RNA replication and transcription take place with the help of an RNA polymerase enzyme.

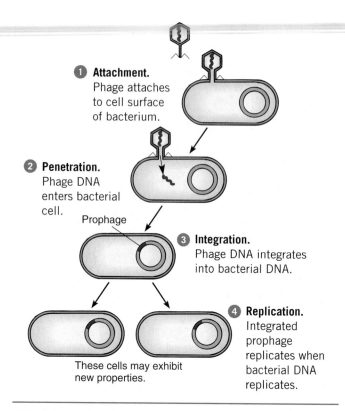

1 Attachment. Phage attaches to cell surface of bacterium.

2 Penetration. Phage DNA enters bacterial cell.

Prophage

3 Integration. Phage DNA integrates into bacterial DNA.

4 Replication. Integrated prophage replicates when bacterial DNA replicates.

These cells may exhibit new properties.

FIGURE 19-3 **Sequence of events in a lysogenic cycle.**

However, **retroviruses** are RNA viruses that have a DNA polymerase called **reverse transcriptase** used to transcribe the RNA genome into a DNA intermediate. This DNA intermediate is integrated into the host DNA. Copies of the viral RNA are synthesized as the incorporated DNA is transcribed by host RNA polymerases. The *human immunodeficiency virus (HIV)* that causes **acquired immunodeficiency syndrome (AIDS)** is a retrovirus. Certain cancer-causing viruses are also retroviruses.

After the viral genes are transcribed, the viral structural proteins are synthesized. The protein coat is produced, and new virus particles are assembled. Viruses exit the cell by lysis (the plasma membrane ruptures) or by budding from the host cell.

Animal viruses cause hog cholera, foot-and-mouth disease, canine distemper, swine influenza, and certain types of cancer (such as feline leukemia). Most humans suffer from two to six viral infections each year, including common colds. Viruses cause chickenpox, herpes simplex (one type causes genital herpes), mumps, rubella (German measles), rubeola (measles), rabies, warts, infectious mononucleosis, influenza, viral hepatitis, and AIDS.

Some viruses infect plants

Animals are not the only organisms infected by viruses. Viral diseases are spread among plants mostly by insects

FIGURE 19-4 Tobacco mosaic virus.
The virus produces a yellow and green mottling, or mosaic pattern, on a variety of plants, including these pepper leaves. The disease tends to reduce crop yields rather than kill the plants outright.

such as aphids and leafhoppers as they feed on plant tissues. Because of the thick plant cell wall, viruses cannot penetrate plant cells unless the cells are damaged. Plant viruses are also inherited by infected seeds or by asexual propagation. Once a plant is infected, the virus can spread throughout the plant body by passing through *plasmodesmata* (cytoplasmic connections) that penetrate the walls between adjacent cells (see Figure 3-12). Most plant viruses are RNA viruses.

Symptoms of viral infection include reduced plant size and spots, streaks, or mottled patterns on leaves, flowers, or fruits (•**Figure 19-4**). Plant viruses cause serious agricultural losses. Infected crops almost always produce lower yields. Because cures are not known for most viral diseases of plants, farmers commonly burn infected plants. Some agricultural scientists are focusing their efforts on developing virus-resistant strains of important crop plants.

Viroids and Prions

In 1961, an infective agent in potatoes was discovered that had a short RNA strand but no protein coat. The agent, named a **viroid**, is much smaller than a virus and has no protective protein coat and no associated proteins to assist in duplication. Each viroid serves as a template that is copied by host-cell RNA polymerases. These infective agents are generally found within the host-cell nucleus and may interfere with gene regulation. Viroids

are extremely hardy and resist heat and ultraviolet radiation because of the condensed folding of their RNA. Viroids cause a variety of plant diseases.

American biochemist Stanley Prusiner began studying **prions** in the early 1970s, motivated by the death of a patient from *Creutzfeldt–Jakob disease (CJD)* a degenerative brain disease. Prusiner discovered that the infective agent was not affected by radiation (which typically mutates nucleic acids), and he could not find DNA or RNA in the particles. He named the infective agent *prion* for "proteinaceous infectious particle."

Animals have a gene that encodes the prion protein, and this protein is normally harmless. However, it sometimes converts to a different shape that accumulates in the brains of patients with *transmissible spongiform encephalopathies (TSEs)*. This group of fatal degenerative brain diseases is found in birds and mammals. Bovine spongiform encephalopathy (BSE), a prion disease popularly referred to as "mad cow disease," became epidemic in cattle in the United Kingdom in the 1990s. More than 120 people in Europe have died from a human variety of BSE, providing evidence that the disease is transmissible from cow to human. The human disease is called vCJD because it is a variant of CJD. Human-to-human transmission has been associated with tissue and organ transplants and transfusion with contaminated blood. Recently, *chronic wasting disease,* a disease related to mad cow disease, has spread among deer and elk populations in the United States. Studies are under way to determine whether this disease can infect livestock and humans. To date, efforts to develop treatments for prion diseases have been unsuccessful.

Prokaryotes

In contrast to viruses, viroids, and prions, **prokaryotes** are cellular organisms. The cell structure of prokaryotes is fundamentally different from the cell structure of other living organisms, and as a result, all prokaryotes were grouped in a single kingdom until recently. However, molecular work verified that there are two groups of prokaryotes that differ from each other as much as they differ from the eukaryotes. Thus, microbiologists now assign prokaryotes to two domains, Bacteria and

LYSOGENIC CYCLE The life cycle of a virus in which viral DNA becomes incorporated into the host cell's genetic material.

VIROID A small, circular, infectious molecule of RNA that causes many plant diseases.

PRION An infectious agent composed only of protein.

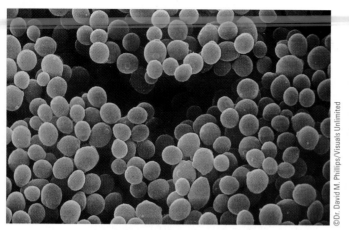

(a) *Micrococcus* are cocci (spherical) prokaryotes.

©Dr. David M. Phillips/Visuals Unlimited

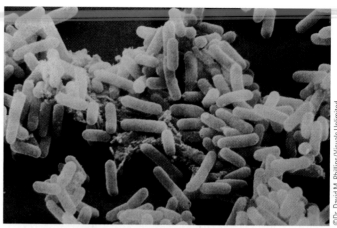

(b) *Escherichia coli* are bacilli (rod-shaped) prokaryotes.

©Dr. David M. Phillips/Visuals Unlimited

FIGURE 19-5 **Characteristic prokaryotic shapes.**

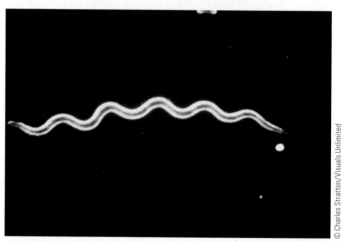

(c) *Borrelia* are spirilla (helical) prokaryotes.

© Charles Stratton/Visuals Unlimited

Archaea, which correspond to two kingdoms by the same names (see Figure 1-11).

Prokaryotes have three main shapes: sphere, rod, and helix (•**Figure 19-5**). Spherical bacteria, known as **cocci** (sing., *coccus*), occur singly in some species, in groups of two in others (diplococci), in long chains (streptococci), or in irregular clumps that look like bunches of grapes (staphylococci). Rod-shaped bacteria, called **bacilli** (sing., *bacillus*), may occur as single rods or as long chains of rods. Helical bacteria are known as **spirilla** (sing., *spirillum*).

Prokaryotic cell structure is simpler than eukaryotic cell structure

Prokaryotic cells lack the membrane-bounded organelles that are typical of eukaryotic cells: they have no nuclei, no mitochondria, no chloroplasts, no endoplasmic reticulum, and no Golgi apparatus. Most prokaryotes are unicellular organisms, but some form colonies or filaments that contain specialized cells. •**Figure 19-6** shows the structure of a typical prokaryote.

Almost all prokaryotic cells are tiny. Their cell volume is typically about one-thousandth that of a small eukaryotic cell, and their length only about one-tenth. Most prokaryotic cells vary from 1 to 10 micrometers

in length. (A prokaryotic cell 1 micrometer long is not visible without a microscope, because a micrometer is one-millionth of a meter, or one-thousandth of a millimeter.)

A **capsule**, or slime layer, covers some prokaryotes and may prevent the cells from drying out under adverse conditions. The capsule may also serve for defense by providing the cell with added protection against being engulfed by other microorganisms in the environment (or by white blood cells in the human body).

Most prokaryotic cells have a cell wall just inside the capsule. The composition of prokaryotic cell walls differs from that of eukaryotic cell walls in that most prokaryotic cell walls are composed of **peptidoglycan**, a macromolecule that confers strength and rigidity. The prokaryotic cell wall provides a rigid framework that supports the cell, maintains its shape, and keeps it from bursting. Normally, prokaryotes cannot survive without their cell walls.

Inside the cell wall is the prokaryote's plasma membrane, the active barrier between the cell and its external environment. The plasma membrane governs the

CAPSULE The outermost layer, usually composed of carbohydrate, of many prokaryotic cells.

PLASMID A small, circular DNA molecule that carries genes separate from the main DNA of a prokaryote.

PILUS One of numerous hairlike structures on the surface of a prokaryotic cell.

ENDOSPORE A highly resistant resting structure that forms within cells of certain prokaryotes.

passage of molecules into and out of the cell. The plasma membrane of a prokaryote carries out most of the functions performed by elaborate systems of internal membranes in eukaryotic cells.

Within the prokaryote's cytoplasm are the ribosomes, where protein synthesis occurs, and the genetic material (DNA). Prokaryotic DNA is found mainly in a single, long, circular molecule that lies in the cytoplasm and is not surrounded by a nuclear envelope. No proteins are associated with the prokaryotic DNA as there are with eukaryotic chromosomes, which have histones and other proteins as part of the chromosome structure. When stretched out to its full length, the genetic material is about 1000 times as long as the cell itself (•**Figure 19-7**). In addition to the circular DNA molecule, a small amount of genetic information may be present as smaller DNA loops, called **plasmids,** which replicate independently of the large DNA molecule. Plasmids often bear genes that provide resistance to antibiotics, a fact that makes plasmids useful in recombinant DNA technology (see Chapter 15).

Some prokaryotes have specialized structures

Some prokaryotes have flagella (•**Figure 19-8**), but their structure differs from that of eukaryotic flagella. Each prokaryotic flagellum has only a single filament, whereas each eukaryotic flagellum is composed of nine pairs of microtubules that surround a pair of microtubules running down its center. At the base of the prokaryotic flagellum is a structure that rotates the flagellum and pushes the cell forward much as a propeller pushes a ship.

Some prokaryotes have hundreds of straight, hairlike appendages known as **pili** (sing., *pilus*) that help the prokaryotes adhere to certain surfaces, such as the cells that they infect. Some pili are involved in the transmission of genetic material (discussed later in the chapter).

When the environment of a prokaryote becomes unfavorable—very dry, for example—many species become dormant. The cell loses water, shrinks slightly, and remains dormant until water is available again. Other species form extremely durable resting structures called **endospores** that survive when environmental conditions are extremely dry, hot, or cold or when food is scarce (•**Figure 19-9**). Some endospores are so resistant that they can survive an hour or more of

Prokaryotic cells are fundamentally different from eukaryotic cells. The organelles of prokaryotic cells are not enclosed by membranes.

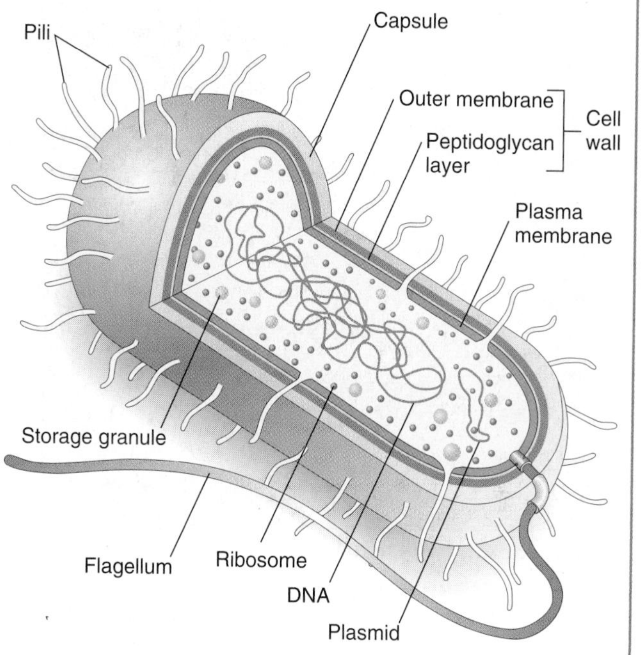

FIGURE 19-6 The structure of a typical prokaryote.
Shown is a gram-negative bacterium, which has an outer membrane exterior to the peptidoglycan layer (discussed later in the chapter). Other prokaryotes lack the outer membrane. Note the absence of a nuclear envelope surrounding the genetic material.

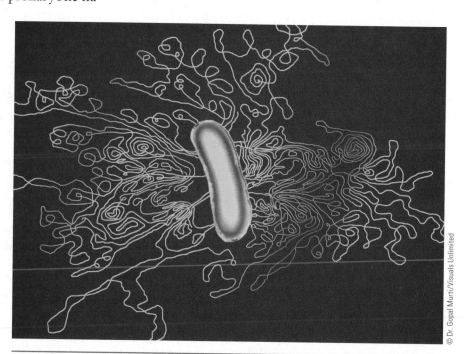

FIGURE 19-7 DNA spilling out of a ruptured *E. coli* cell.
The prokaryotic DNA molecule, which is about 1000 times the length of the cell, is tightly coiled inside the cell.

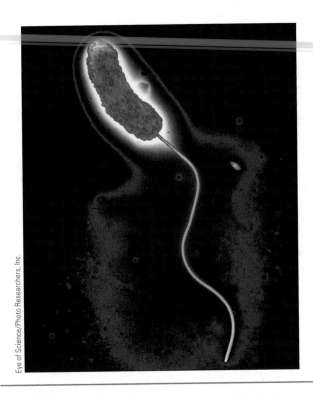

Eye of Science/Photo Researchers, Inc.

FIGURE 19-8 Prokaryotic flagellum.
This electron micrograph of *Vibrio cholerae,* the bacterium that causes cholera, shows its single flagellum. Other prokaryotes have a tuft of flagella at one end of the cell, and some have numerous flagella that project from many areas of the cell.

boiling or centuries of freezing. When environmental conditions are again suitable for growth, the endospore absorbs water, breaks out of its inner wall, and becomes an active, growing cell again.

Prokaryotes have diverse methods of obtaining organic molecules

Most prokaryotes are **heterotrophs** (from the Greek *hetero,* "other," and *tropho,* "nourishment") and must obtain organic molecules from other organisms. The majority of heterotrophic prokaryotes are free-living **decomposers,** organisms that get their nourishment from dead organic matter. Other heterotrophic prokaryotes live in or on other organisms. These prokaryotes are either **commensals,** which neither help nor harm their hosts, or **parasites,** which live at the expense of their hosts and can cause disease. Other

prokaryotes form **mutualistic** associations with another organism, in which both the prokaryote and its partner derive benefits from the association. For example, some bacteria inhabit the human intestine, where they obtain nutrients and produce vitamin K and some of the B vitamins, which are absorbed and used by the body.

Some prokaryotes are **autotrophs** (from the Greek *auto,* "self," and *tropho,* "nourishment") and manufacture their own organic molecules from carbon dioxide. Autotrophic prokaryotes are either photosynthetic or chemosynthetic. **Photosynthetic prokaryotes** obtain energy to manufacture organic compounds from light, whereas **chemosynthetic prokaryotes** obtain energy from chemical reactions, primarily the oxidation of (addition of oxygen to) inorganic compounds that contain ammonia, iron, or sulfur. Five groups of photosynthetic prokaryotes exist: cyanobacteria, green sulfur bacteria, purple sulfur bacteria, green nonsulfur bacteria, and purple nonsulfur bacteria.

As mentioned in the chapter introduction, cyanobacteria photosynthesize in the same way algae and plants do. Photosynthesis in other photosynthetic prokaryotes differs from "standard" photosynthesis in two important ways. First, prokaryotic chlorophyll, which is different from chlorophyll *a,* absorbs light most strongly in the near-infrared portion of the electromagnetic spectrum, which means that these prokaryotes carry on photosynthesis in red light at wavelengths that would not work well for plants, algae, or cyanobacteria. Second, photosynthesis by prokaryotes other than cyanobacteria does not produce oxygen, because water is not used as a hydrogen donor. Instead, hydrogen sul-

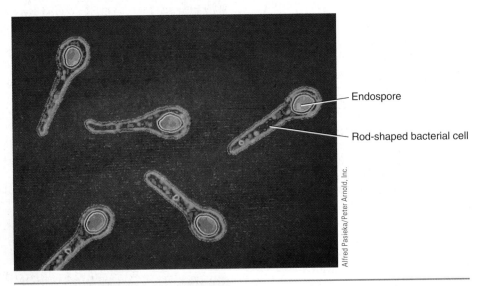

Endospore

Rod-shaped bacterial cell

Alfred Pasieka/Peter Arnold, Inc.

FIGURE 19-9 Endospore.
This endospore is inside a cell of *Clostridium tetani,* the bacterium that causes tetanus. Each cell contains only one endospore, which is a resistant, dehydrated remnant of the original cell.

fide (H₂S) is the hydrogen donor, and sulfur (S) is the by-product.

Prokaryotes differ in their needs for oxygen

Whether they are heterotrophs or autotrophs, most prokaryotes are **aerobes**; that is, they require atmospheric oxygen for cellular respiration. Plant and animal cells are also aerobic. Some prokaryotes are **facultative anaerobes**, meaning that they use oxygen for cellular respiration if it is available but will respire anaerobically when oxygen is absent. Other prokaryotes are **obligate anaerobes** and carry on cellular respiration only in the absence of oxygen. Some obligate anaerobes are actually poisoned by low concentrations of oxygen.

Prokaryotes reproduce by binary fission

Prokaryotes generally reproduce asexually by **binary fission,** in which one cell divides into two similar cells. After the circular DNA molecule replicates, the cell splits into two cells. A transverse wall forms between the two new cells by an ingrowth of both the plasma membrane and the cell wall. Binary fission occurs with remarkable speed; under ideal conditions, some species divide every 20 to 30 minutes! At this rate, if nothing interfered, one prokaryote could give rise to more than 130,000 prokaryotes within 6 hours. The speed of binary fission explains why the entrance of only a few pathogenic prokaryotes into a human can cause disease symptoms so quickly. Fortunately, prokaryotes cannot reproduce at this rate for long, because they are soon checked by a lack of food or the accumulation of waste products.

Although sexual reproduction involving the fusion of gametes does not occur in prokaryotes, genetic material is sometimes exchanged between individuals. In **conjugation,** two cells of different physiological mating types come together, and genetic material is transferred through pili from one cell (the donor) to the other (the recipient) (**•Figure 19-10**). In addition, fragments of DNA released by a broken cell are sometimes taken in by another prokaryotic cell (**transformation**). Alternatively, a virus may carry genetic material from one prokaryotic cell to another (**transduction**). These three methods of genetic exchange transfer genes for resistance to antibiotics from one prokaryote to another.

Many prokaryotes form biofilms

Many prokaryotes that inhabit watery environments do not live as unicellular organisms swimming about in a solitary existence. Instead, archaea and bacteria form dense films called **biofilms** that attach to solid surfaces.

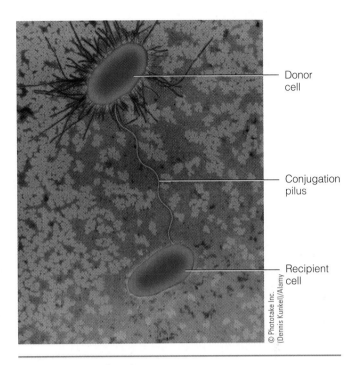

Donor cell

Conjugation pilus

Recipient cell

© Phototake Inc. (Dennis Kunkel)/Alamy

FIGURE 19-10 Conjugation.
This conjugation pilus connects two *Escherichia coli* cells. Plasmid DNA is transferred from donor cell to recipient cell during conjugation.

The prokaryotes in a biofilm secrete a slimy, gluelike substance rich in polysaccharides and become embedded in this matrix, which may be layered or clumped into various shapes.

Most biofilms are microbial communities that consist of many species of prokaryotes and may include other types of organisms, such as fungi and protozoa. The dental plaque that forms on teeth and can lead to tooth decay and gum disease is a familiar example of a biofilm. These films also form on the surfaces of contact lenses and catheters. They sometimes develop on surgical implants, such as pacemakers and joint replacements. According to the Centers for Disease Control and Prevention, biofilms cause as many as 70 percent of

HETEROTROPH An organism that obtains the carbon it needs from organic compounds in other organisms.

AUTOTROPH An organism that synthesizes all its complex organic compounds from carbon dioxide.

AEROBE An organism that grows and metabolizes only in the presence of molecular oxygen.

FACULTATIVE ANAEROBE An organism capable of growing and metabolizing in either the presence or absence of molecular oxygen.

OBLIGATE ANAEROBE An organism that grows and metabolizes only in the absence of molecular oxygen.

bacterial infections. The biofilm often protects the prokaryotes from white blood cells and antibodies of the body's immune system. Extremely high concentrations of antibiotics are required to effectively treat bacteria embedded in a biofilm.

Prokaryotes are classified into two domains, Archaea and Bacteria

EVOLUTION LINK Under a microscope, all prokaryotes appear similar. Their structural similarities, along with a scant fossil record, have made it difficult to determine evolutionary relationships among them. However, evidence from molecular biology has helped biologists conclude that ancient prokaryotes split into two lineages early in the history of life. The modern descendants of these two ancient lines of prokaryotes are the archaea, which include a few genera of prokaryotes that live in extreme environments, and the bacteria, which comprise all other groups of prokaryotes. The differences between archaea and bacteria are considered great enough to warrant their classification into separate domains and kingdoms. ∎

Many archaea survive in harsh environments

Biochemically, the 200 or so species of domain/kingdom **Archaea** are quite different from bacteria. One of their most distinguishing features is the absence of peptidoglycan in the cell wall. There are also other important (though quite technical) differences that set the archaea apart from bacteria.

The biochemical differences between the archaea and bacteria suggest that these groups may have diverged from each other long ago—relatively early in the history of life. Many of the extreme environments to which the modern archaea are adapted resemble conditions that were probably common on early Earth billions of years ago but that are somewhat rare today. Examples include hot springs and thermal vents at the bottom of the ocean, both of whose temperatures may exceed 100°C (212°F).

The archaea include three groups: the extreme halophiles, the methanogens, and the extreme thermophiles. The most common archaea are the **methanogens** (methane producers), which are obligate anaerobes that produce methane gas from simple carbon compounds. They inhabit sewage and swamp sediments and are common in the digestive tracts of humans and

Georg Gerster/Photo Researchers, Inc.

(a) Seawater evaporating ponds in Malinda, Kenya, are pink and other shades from the many extreme halophiles (salt-loving archaea) growing in them. (Algae also add color to the ponds.) The salt that remains after the water has evaporated has commercial value.

© Gustav Verderber/Visuals Unlimited

(b) Yellowstone National Park's Grand Prismatic Spring, the world's third largest hot spring, teems with thermophilic archaea. The rings around the perimeter, where the water is cooler, get their distinctive colors from the various kinds of prokaryotes living there.

FIGURE 19-11 Archaea.

other animals. Methanogens are important in recycling components of organic products of organisms that inhabit swamps. These archaea produce most of the methane in Earth's atmosphere.

The **extreme halophiles** live only in very salty environments, such as salt ponds, the Dead Sea, and Great Salt Lake (•**Figure 19-11a**). The extreme halophiles use aerobic respiration to make ATP, but they also carry out a form of photosynthesis in which they capture the energy of sunlight by using a purple pigment (bacteriorhodopsin). This pigment is similar to the pigment rhodopsin involved in animal vision.

The **extreme thermophiles** normally grow in hot (45°C to 110°C, or 113°F to 230°F), sometimes acidic, environments (•**Figure 19-11b**). For example, one species that is found in the hot sulfur springs of Yellowstone National Park flourishes at temperatures near 60°C (140°F) and pH values of 1 to 2 (the pH of concentrated sulfuric acid). Other extreme thermophiles inhabit volcanic areas under the sea. One species, found in hot, deep-sea vents on the seafloor, lives at temperatures from 80°C to 110°C (176°F to 230°F).

Members of the Archaea also inhabit less extreme conditions. For example, they are abundant in the soil and in the cold ocean surface waters near Antarctica. Evidence suggests that the Archaea are important in global biogeochemical cycles and in marine food chains. **EVOLUTION LINK** Interestingly, the archaea are more closely related to eukaryotes, such as plants and humans, than to the bacteria (see Figure 18-9). Evidence that supports this relationship comes from molecular data, including the fact that archaea and eukaryotes share certain genes that are not found in bacteria. The relationship between archaea and eukaryotes is a distant one, however; the line of descent leading to present-day archaea is thought to have diverged from the eukaryote line some 3 billion years ago! ■

The bacteria are the most familiar prokaryotes

About 3000 species of bacteria belong to domain/kingdom **Bacteria**. Early biologists quickly determined that bacteria are present almost universally—abundant in air, soil, and water and in and on the bodies of living and dead organisms. In fact, relatively few places in the world are completely devoid of bacteria; these organisms are found in fresh and salt water, as far down as several meters deep in the soil, in deep underground water supplies, in the ice of glaciers, and even in oil deposits far underground.

Staining properties of bacteria. In 1884 the Danish physician Hans Christian Gram developed the **Gram staining procedure**, which divides the bacteria into two groups. Bacteria that absorb and retain the primary stain during the procedure are referred to as **gram-positive**, whereas

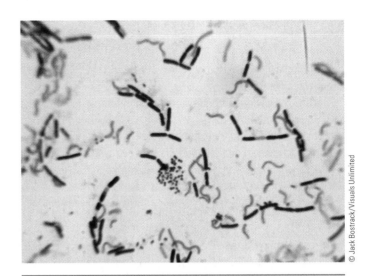

FIGURE 19-12 Gram-stained bacteria.
As viewed under a light microscope, gram-positive cells are purple, and gram-negative cells are pink.

those that do not retain the stain are **gram-negative** (•**Figure 19-12**). The cell walls of gram-positive bacteria are thick and consist primarily of peptidoglycan, which retains the Gram stain. The cell wall of a gram-negative bacterium consists of two layers, a thin peptidoglycan wall on the inside and an outer membrane that does not retain the stain.

The differences in composition of the cell walls of gram-positive and gram-negative bacteria are of great practical importance. The antibiotic penicillin, for example, interferes with peptidoglycan synthesis and ultimately results in a cell wall so fragile that it cannot protect the cell or keep it from bursting. Penicillin works most effectively against gram-positive bacteria because of the greater abundance of peptidoglycan in their cell walls.

Microbiologists recognize three main groups of bacteria on the basis of differences in their cell walls: the wall-less bacteria, the gram-negative bacteria, and the gram-positive bacteria (•**Table 19-1**).

Mycoplasmas. A **mycoplasma** is a tiny bacterium bounded by a plasma membrane but lacking a typical bacterial cell wall. Some mycoplasmas are so small that, like viruses, they pass through bacteriological filters. In fact, mycoplasmas are smaller than some viruses. Mycoplasmas may be the simplest form of cellular life. Some live in soil, some live in sewage, and others are parasitic on plants or animals. Many plant diseases originally attributed to viruses are now known to be caused by mycoplasmas (•**Figure 19-13**).

Gram-negative bacteria. The gram-negative bacteria exhibit a lot of variation in shape, structure, and metabo-

TABLE 19-1	Examples of Common Bacteria

Wall-less Bacteria

    Mycoplasmas—extremely small cells that lack cell walls

Gram-Negative Bacteria

    Nitrogen-fixing bacteria—aerobic bacteria that fix nitrogen

    Enterobacteria—a large group of diverse bacteria; facultatively anaerobic, heterotrophic

    Spirochetes—spiral shaped; flexible cell walls

    Cyanobacteria—photosynthetic autotrophs, occur mainly as colonial masses or filaments; some fix nitrogen

    Rickettsias—obligate parasites; a few cause diseases that are transmitted to humans by arthropods

    Chlamydias—obligate parasites; are not transmitted by arthropods

    Myxobacteria—unicellular bacilli; move by gliding

Gram-Positive Bacteria

    Lactic acid bacteria—fermenting bacteria that can ferment the sugar in milk

    Streptococci—fermenting bacteria

    Staphylococci—aerobic bacteria

    Clostridia—fermenting bacteria; anaerobic

lism. Among them are the **nitrogen-fixing bacteria,** which have the ability to fix atmospheric nitrogen into inorganic compounds plants can use. These gram-negative rods, most of which are flagellated, are common in soil and aquatic environments worldwide. Some, such as *Azotobacter,* are free-living, whereas others, such as *Rhizobium,* form symbiotic relationships with plants (see Chapter 26).

Many important pathogens are gram-negative bacteria. About one-half of the almost 200 species of bacteria that cause disease in plants are gram-negative rods in a single genus, *Pseudomonas. Pseudomonas* species cause a variety of diseases in plants, such as leaf spots, cankers (local decay of bark and wood), wilts, and seedling death. In humans, the gram-negative *Neisseria gonorrhoeae* causes the sexually transmitted disease gonorrhea.

Enterobacteria are a group of gram-negative bacteria that include free-living decomposers, plant pathogens, and some bacteria that inhabit humans. One enterobacterium, *E. coli,* lives in the intestines of humans and other animals as part of the normal microbial population (see *Plants and People: Problems Associated with Bacteria in Wastewater*). Some strains of the enterobacterium *Salmonella* cause food poisoning.

Spirochetes are corkscrew-shaped, gram-negative bacteria with flexible cell walls. Some spirochetes are free-living and inhabit freshwater and marine habitats, whereas some form close associations with other organisms, including a few that are parasitic. The spirochetes

of greatest medical importance cause Lyme disease, which is transmitted by ticks, and the sexually transmitted disease syphilis.

Cyanobacteria (such as *Anabaena,* mentioned in the chapter introduction) are gram-negative bacteria found in ponds, lakes, swimming pools, and moist soil as well as on logs and the bark of trees. Some cyanobacteria also occur in the ocean, and a few species inhabit hot springs. Several cyanobacteria are unicellular, but most species occur as round colonies or as long filaments (•**Figure 19-14**). Most cyanobacteria are photosynthetic autotrophs and contain chlorophyll *a,* which is also found in plants and algae. Cyanobacteria have several accessory pigments, including **phycocyanin** (a blue pigment) and **phycoerythrin** (a red pigment).

Rickettsias are gram-negative bacteria that must live within cells as parasites to survive. Most rickettsias parasitize arthropods, such as fleas, lice, ticks, and mites, without causing specific diseases in them. Diseases caused by the few rickettsias that are pathogenic to humans are transmitted by arthropods through bites or contact with their excretions. Among rickettsial diseases are typhus, transmitted by fleas and lice, and Rocky Mountain spotted fever, transmitted by ticks.

Chlamydias also must live within cells as parasites to survive; unlike rickettsias, however, chlamydias

FIGURE 19-13 Yellow blight in palm.
A mycoplasma causes yellow blight. Originally, a virus was thought to be the causative agent of this disease.

PLANTS AND PEOPLE | Problems Associated with Bacteria in Wastewater

Sewage is wastewater carried off from toilets, washing machines, and showers by drains or sewers. Because sewage contains human wastes, its release into bodies of water (for example, rivers, lakes, and oceans) causes two pollution problems that are related to bacteria. First, untreated municipal wastewater usually contains many disease-causing bacteria as well as other pathogenic microorganisms such as viruses, protozoa, and parasitic worms. Typhoid, cholera, bacterial dysentery, and enteritis are some of the more common bacterial diseases that are transmissible through contaminated water. Thus, sewage-polluted water poses a threat to public health.

The bacteria in sewage cause a second serious environmental problem in water: oxygen demand. The action of microorganisms, particularly bacteria, decomposes sewage and other organic materials into carbon dioxide, water, and similar inoffensive materials. This degradation process, known as cellular respiration, requires the presence of oxygen. Most organisms living in healthy aquatic ecosystems, including fish and aquatic plants, also use dissolved oxygen. But oxygen has a limited ability to dissolve in water, and when an aquatic ecosystem contains high levels of sewage or other organic material, the decomposing bacteria use up most of the dissolved oxygen, leaving little for fish or other aquatic organisms. At extremely low oxygen levels, fish die off in large numbers.

Sewage and other organic wastes are measured in terms of their **biochemical oxygen demand (BOD),** also

Courtesy of Millipore Corp.

(a) A water sample is first passed through a filtering apparatus. Then, for 24 hours, the filter disc is placed on a culture medium that supports coliform bacteria.

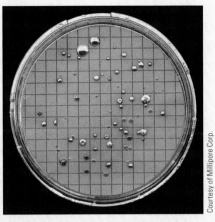

Courtesy of Millipore Corp.

(b) After incubation, the number of colonies is counted. Each colony arose from a single coliform bacterium in the original water sample.

■ The fecal coliform test indicates the likely presence of disease-causing agents in water.

called **biological oxygen demand**—the amount of oxygen needed by bacteria to decompose the wastes. A large amount of sewage in water creates a high BOD, which robs the water of dissolved oxygen. When dissolved oxygen levels are low, anaerobic bacteria also produce compounds that have unpleasant odors, further deteriorating water quality.

Because sewage-contaminated water is a threat to public health, periodic tests are conducted for the presence of sewage in our water supplies. Although many microorganisms thrive in sewage, the common intestinal bacterium *Escherichia coli* is typically used as an indication of the amount of sewage present in water and as an indirect measure of the presence of disease-causing agents. Although **coliform bacteria** (bacteria, such as *E. coli,* that inhabit

the large intestine) do not usually cause disease, their presence indicates the likely presence of disease-causing agents in water. *Escherichia coli* is perfect for monitoring sewage because it is not present in the environment except in human and animal feces, where it is found in large numbers.

The **fecal coliform test** is performed to test for the presence of *E. coli* in water (see figures). Safe drinking water should contain no more than 1 coliform bacterium per 100 milliliters (about $\frac{1}{2}$ cup) of water, safe swimming water should have no more than 200 per 100 milliliters of water, and general recreational water (for boating) should have no more than 2000 per 100 milliliters. In contrast, raw sewage may contain several million coliform bacteria per 100 milliliters of water.

do not depend on arthropods for transmission. Studies indicate that these gram-negative bacteria infect almost every species of bird and mammal. Perhaps 10 to 20 percent of the human population worldwide is infected, but interestingly, individuals may be infected for many years without apparent harm. However, chlamydias sometimes cause acute infectious diseases. For

example, a strain of *Chlamydia* causes trachoma, the leading cause of blindness in the world, and chlamydias cause an infection of the genitourinary tract that is a common sexually transmitted disease in the United States.

Myxobacteria are gram-negative bacteria that excrete slime. When they are cultured in a petri dish,

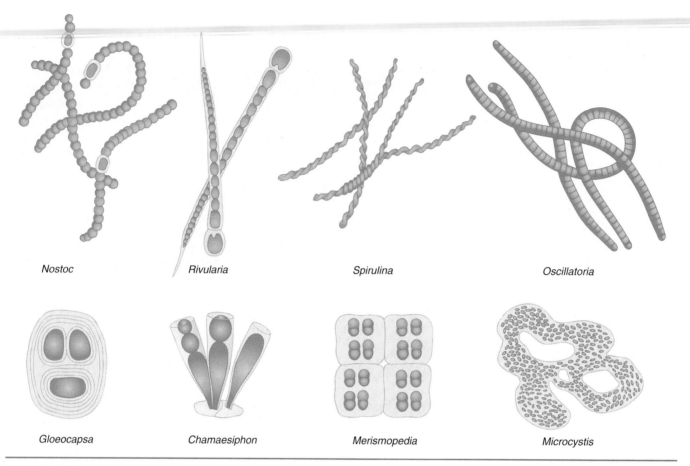

Nostoc   Rivularia   Spirulina   Oscillatoria

Gloeocapsa   Chamaesiphon   Merismopedia   Microcystis

FIGURE 19-14 Variation in cyanobacteria.

their growth is marked by a spreading layer of slime on which they glide or creep along. Most myxobacteria are decomposers that break down organic matter in the soil, manure, or rotting wood that is their habitat. A few myxobacteria prey on other prokaryotes.

Gram-positive bacteria. Gram-positive bacteria include the lactic acid bacteria, streptococci, staphylococci, and clostridia. **Lactic acid bacteria** are gram-positive bacteria that produce lactic acid as the main end product of their fermentation of sugars. Lactic acid bacteria are found in decomposing plant material and in milk, yogurt, and other dairy products. They are commonly present in animals and are among the normal inhabitants of the human mouth and vagina.

Streptococci, spherical gram-positive bacteria that occur in chains, are found in the human mouth as well as the digestive tract. Among the harmful species of streptococci are those that cause "strep throat," scarlet fever, dental caries, and a form of pneumonia.

Staphylococci, spherical gram-positive bacteria that occur in irregular clusters, normally live in the human nose and on the skin. They are opportunistic, which means that they cause disease when the immunity of the host is lowered. *Staphylococcus aureus* causes

boils and skin infections and may infect wounds. Certain strains of *S. aureus* cause food poisoning, and some are thought to cause toxic shock syndrome.

Clostridia are a notorious group of anaerobic gram-positive bacteria. One species causes tetanus, or lockjaw; another causes gas gangrene; and a third causes botulism, a potentially fatal type of food poisoning.

Prokaryotes are ecologically important

Although it is easy to overlook prokaryotes in the world of living things or to associate them exclusively with diseases, prokaryotes perform important environmental services. Indeed, prokaryotes are essential for all life-forms, and there are many more useful species than harmful ones. The contribution of some prokaryotes is to alter atmospheric nitrogen to a form that plants can use. Nitrogen fixation enables plants (and animals, because they eat plants) to manufacture essential nitrogen-containing compounds such as proteins and nucleic acids. The process of nitrogen fixation is extremely important to life. If there were no prokaryotic nitrogen fixers, there would be no biological way to convert atmospheric nitrogen to usable nitrogen, and life would cease to exist.

Other prokaryotes play an essential role in the biosphere as decomposers, which break down organic molecules into their simpler components. Prokaryotes, along with fungi, are nature's recyclers. Without prokaryotes and fungi, all available carbon, nitrogen, phosphorus, and sulfur would eventually be tied up in the wastes and dead bodies of plants and animals. Life would soon cease because of the lack of raw materials for the synthesis of new cellular components.

Prokaryotes are used to clean up hazardous waste sites. In a process known as **bioremediation,** a contaminated site is exposed to prokaryotes or other microorganisms that produce enzymes that break down the poisons, leaving behind harmless by-products such as carbon dioxide and chlorides. To date, more than 1000 species of prokaryotes and fungi have been used to clean up various forms of pollution.

STUDY OUTLINE

❶ Characterize viruses, and explain why they are not grouped in any of the three domains or six kingdoms.
A **virus** is a tiny disease-causing agent consisting of a core of nucleic acid usually encased in protein. Viruses are not cellular and therefore do not fall into any of the domains or kingdoms.

❷ Contrast lytic and lysogenic cycles.
A **lytic cycle** is the life cycle of a virus that kills the host cell by lysing it. A **lysogenic cycle** is the life cycle of a virus in which viral DNA becomes incorporated into the host cell's genetic material.

❸ Compare viroids and prions.
A **viroid** is a small, circular, infectious molecule of RNA that causes many plant diseases. A **prion** is an infectious agent composed only of protein.

❹ Describe the basic structure of a prokaryotic cell.
Prokaryotes lack the membrane-bounded organelles that are typical of eukaryotic cells: they have no nuclei, no mitochondria, no chloroplasts, no endoplasmic reticulum, and no Golgi apparatus. Most prokaryotes are unicellular organisms, but some form colonies or filaments that contain specialized cells.

❺ Define the following terms: *capsule, plasmid, pilus,* **and** *endospore.*
A **capsule** is the outermost layer, usually composed of carbohydrate, of many prokaryotic cells. A **plasmid** is a small, circular DNA molecule that carries genes separate from the main DNA of a prokaryote. A **pilus** is one of numerous straight, hairlike structures on the surface of a prokaryotic cell. An **endospore** is a highly resistant resting structure that forms within cells of certain prokaryotes.

❻ Characterize the metabolic diversity of prokaryotes using the following terms: *autotroph* **and** *heterotroph; aerobe, facultative anaerobe,* **and** *obligate anaerobe.*
Prokaryotes are either heterotrophs or autotrophs. A **heterotroph** obtains the carbon it needs from organic compounds in other organisms. An **autotroph** synthesizes all its complex organic compounds from carbon dioxide. Prokaryotes may be aerobes, facultative anaerobes, or obligate anaerobes. An **aerobe** grows and metabolizes only in the presence of molecular oxygen. A **facultative anaerobe** is capable of growing and metabolizing in either the presence or absence of molecular oxygen. An **obligate anaerobe** grows and metabolizes only in the absence of molecular oxygen.

❼ Distinguish between archaea and bacteria.
Members of domain/kingdom **Archaea** have cell walls of unusual chemical composition and are often adapted to harsh conditions. Three groups of archaea are halophiles, methanogens, and thermophiles. **Methanogens** are obligate anaerobes that produce methane gas from simple carbon compounds. **Extreme halophiles** live in very salty environments; they use aerobic respiration to make ATP but carry out photosynthesis by using a purple pigment to capture the energy of sunlight. **Extreme thermophiles** grow in hot, sometimes acidic, environments. Walled members of domain/kingdom **Bacteria** have cell walls composed of **peptidoglycan,** a macromolecule that confers strength and rigidity. Bacteria are divided into three groups on the basis of cell-wall composition. **Mycoplasmas** lack cell walls. **Gram-negative bacteria** have thin cell walls of peptidoglycan surrounded by an outer membrane. **Gram-positive** bacteria have thick cell walls of peptidoglycan.

❽ Explain why prokaryotes are so crucial to the environment.
Prokaryotes play an important role in the biosphere, because many of them are **decomposers** and break down dead organic materials, so they can be recycled in the biosphere. Some prokaryotes are important in **nitrogen fixation,** the conversion of atmospheric nitrogen to a form that other organisms can use.

REVIEW QUESTIONS

1. Why don't biologists assign viruses to one of the three domains or six kingdoms?
2. What is a lytic cycle? A lysogenic cycle?
3. Distinguish among viruses, viroids, and prions.
4. How does a prokaryotic cell differ from a eukaryotic cell?
5. Draw and label the three major shapes of prokaryotic cells.
6. Describe a capsule, a plasmid, and an endospore.
7. Distinguish between autotrophs and heterotrophs. What is a chemosynthetic autotroph?
8. Describe one way in which prokaryotes might exchange genetic material.
9. What are the differences between archaea and bacteria? To which group are plants and other eukaryotes more closely related?
10. How does the cell wall of a gram-positive bacterium differ from that of a gram-negative bacterium?
11. Explain how prokaryotes that are decomposers help recycle materials in the environment.
12. When bacteriophages enter a bacterial cell, they may take one of two pathways, lytic or lysogenic. Identify each of these pathways in the figures on page 381. Using the figures, describe the sequence of events in each pathway. (Refer to Figures 19-2 and 19-3 to check your answers.)

13. Label the following diagram of a prokaryotic cell. (Refer to Figure 19-6 to check your answers.)

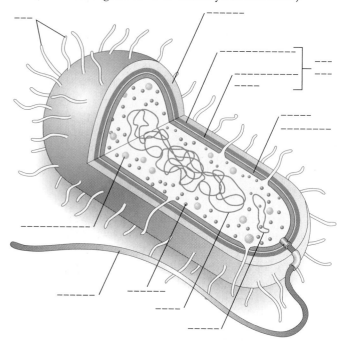

THOUGHT QUESTIONS

1. Imagine that you discover a new microorganism. After careful study, you decide to classify it as a cyanobacterium in kingdom Bacteria. What characteristics might lead you to this classification?
2. The bacterium *Clostridium tetani*, which causes tetanus (lockjaw), is an obligate anaerobe. Which type of wound is more likely to develop tetanus, a surface cut or a deep puncture or gunshot wound? Why?

3. Why are prokaryotes classified primarily on the basis of chemical and metabolic features rather than their shapes?

Visit us on the web at http://www.thomsonedu.com/biology/berg/ for additional resources such as flashcards, tutorial quizzes, further readings, and web links.

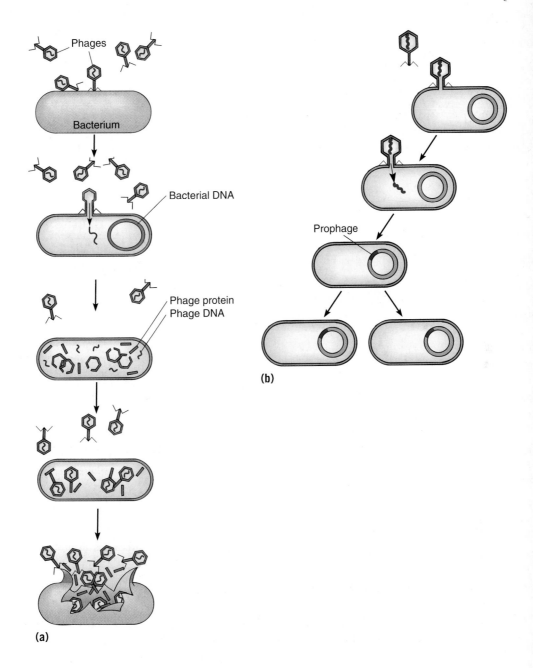

(a)

(b)

20

Kingdom Protista

LEARNING OBJECTIVES

❶ Briefly describe the features common to the members of the kingdom Protista.

❷ Discuss in general terms the diversity in the protist kingdom, including means of locomotion, modes of nutrition, interactions with other organisms, and modes of reproduction.

❸ Describe the kinds of data biologists use to classify eukaryotes.

❹ Briefly describe and compare euglenoids and dinoflagellates.

❺ Briefly describe and compare water molds, diatoms, golden algae, and brown algae.

❻ Explain why red algae, green algae, and land plants are considered a monophyletic group.

❼ Briefly describe and compare plasmodial slime molds and cellular slime molds.

A large area in the middle of the North Atlantic Ocean—between Africa and the West Indies (between 20° and 35° north latitude)—is known as the Sargasso Sea. Surface ocean currents, which rotate clockwise around the margins of the North Atlantic, do not have an appreciable effect on this central expanse of warm water. Early European explorers feared sailing into the Sargasso Sea, because their ships would drift aimlessly through the comparatively still water for days or even weeks.

The Sargasso Sea gets its name from *Sargassum,* a seaweed that floats there in great abundance. (The genus name *Sargassum* is derived from *sargazo,* the Spanish word for "seaweed.") Although most species of sargassum grow attached to rocks in shallow ocean waters along rocky coastlines, sargassum in the Sargasso Sea has adapted to the open ocean. Sargassum looks like a plant instead of an *alga;* it grows a meter or more in length and has stemlike and leaflike parts that produce round, gas-filled bladders, which keep it buoyant.

The floating meadows of sargassum offer shelter from the dangers of the open ocean and supply food to a diverse community in the middle of the Atlantic Ocean. Many fishes and other animals spend their entire lives among the tangled branches of sargassum, hiding from larger predators that lurk nearby and preying on smaller fish and invertebrates that consume sargassum. Other fishes, including dolphinfish, flying fish, and eels, journey to the Sargasso Sea to breed. Eels migrate spectacular distances from European and American rivers to reproduce in the Sargasso Sea; the young return to the rivers to live out their adult lives.

Sargassum is one of many plantlike organisms placed in the eukaryotic kingdom Protista. Many protists are usually studied in plant biology classes because they were traditionally considered plants. Other animal-like protists are studied in zoology classes.

Sargassum seaweed and small fishes.

Introduction to the Protists

When Robert Whittaker proposed the five-kingdom system of classification in 1969, only unicellular eukaryotic organisms were placed in kingdom Protista (from the Greek *protos,* "the very first"). The boundaries of this kingdom have expanded since that time, although there is no universal agreement among biologists about what constitutes a protist. Today, most biologists interpret the protist kingdom broadly, to include not only unicellular eukaryotes but eukaryotes of multicellular yet relatively simple organization. Kingdom Protista encompasses animal-like protists (protozoa), plantlike protists (algae, including seaweeds), and funguslike protists (slime molds and water molds). Although these groups superficially resemble animals, plants, or fungi in certain respects, they are not considered merely simple kinds of animals, plants, or fungi.

EVOLUTION LINK Kingdom Protista is not a natural grouping of organisms, and organisms are placed within it for taxonomic convenience. If natural relationships based on a shared evolutionary history were the sole means of classifying organisms into kingdoms, we would have many more than six kingdoms. Some biologists think as many as 50 phyla and 5 eukaryote kingdoms are needed to recognize natural relationships within the protists alone. ■

The kingdom Protista consists of a vast assortment of eukaryotic organisms whose diversity makes them difficult to characterize (•Figure 20-1). Biologists estimate that there are as many as 200,000 species of protists living today. Their salient feature, eukaryotic cell structure characterized by the presence of nuclei and other membrane-bounded organelles, is shared with organisms from three other kingdoms—animals, plants, and fungi. There is, however, a distinct separation between eukaryotic protists and the prokaryotes in kingdoms Bacteria and Archaea.

Size varies considerably within the protist kingdom, from microscopic algae and protozoa to giant kelps, brown algae that reach 75 meters (about 250 feet) in length. Although most protists are unicellular, some form **colonies** (loosely connected groups of cells), some are **coenocytic** (a multinucleate mass of cytoplasm), and some are multicellular. Multicellular protists have relatively simple body forms without specialized tissues.

Methods of obtaining nutrients in the kingdom Protista are varied. Most of the algae are *autotrophic* and photosynthesize as plants do. Some *heterotrophic* protists obtain their nutrients by absorption, as fungi do, whereas others resemble animals and ingest food. Some protists switch their mode of nutrition and are autotrophic at certain times and heterotrophic at others.

Although many protists are free-living, others form symbiotic relationships with various organisms. These

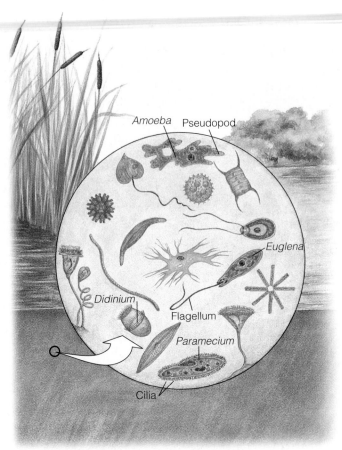

(a) Various protists in a drop of pond water.

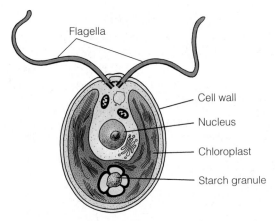

(b) *Chlamydomonas*, a unicellular, photosynthetic protist, is a motile organism with two flagella and a cup-shaped chloroplast.

FIGURE 20-1 Protists.

intimate associations range from mutualism, a more-or-less equal partnership in which both partners benefit, to parasitism, in which one partner lives on or in another and is metabolically dependent on it. Some parasitic protists are important pathogens (disease-causing agents) of plants or animals. Specific examples of symbiotic associations are described throughout this chapter.

Most protists are aquatic and live in oceans or freshwater ponds, lakes, and streams. They make up most of the **plankton,** the floating, often microscopic organisms that are the base of the food web in aquatic ecosystems. Other aquatic protists attach to rocks and other surfaces in the water. Terrestrial (land-dwelling) protists are restricted to damp places such as soil, bark, and leaf litter. Even the parasitic protists are aquatic because they live in the wet environments of other organisms' body fluids.

Reproduction is varied in the kingdom Protista. Almost all protists reproduce asexually, and many also reproduce sexually, often by **syngamy,** the union of gametes. Most protists, however, do not develop multicellular reproductive organs, nor do they form embryos the way more complex eukaryotes do.

Protists, most of which are motile at some point in their life cycles, have various means of locomotion. They may move by pushing out cytoplasmic extensions as an amoeba does, by flexing individual cells, by waving *cilia* (short, hairlike structures), or by lashing *flagella* (long, whiplike structures) (see Figure 20-1b). Some protists use a combination of two or more means of locomotion.

Evolution of the Eukaryotes

EVOLUTION LINK Biologists hypothesize that protists were the first eukaryotic cells to evolve from ancestral prokaryotes. They may have appeared in the fossil record as early as 2.2 billion years ago. The first eukaryotes were probably *zooflagellates,* heterotrophic unicellular protists that have flagella. However, other than a few protists with hard shells, such as diatoms, most ancient protists did not leave many fossils, because their bodies were too soft to leave permanent traces. Evolutionary studies of protists focus primarily on molecular and structural comparisons of present-day organisms, which contain many clues about their evolutionary history. ■

A consensus is emerging in eukaryote classification

PROCESS OF SCIENCE Scientists reevaluate evolutionary relationships among the eukaryotes as additional evidence becomes available. Two types of modern research, molecular data and ultrastructure, have contributed substantially to scientific understanding of the phylogenetic relationships among protists. Molecular data were initially obtained for the gene that codes for the small subunit ribosomal RNA in different eukaryotes. More recently, biologists have compared other nuclear genes, many of which code for proteins, in various protists.

Ultrastructure is the fine details of cell structure revealed by electron microscopy. In many cases, ultrastructure data complement molecular data. Electron microscopy reveals similar structural patterns among those protists that comparative molecular evidence suggests are **monophyletic**—that is, evolved from a common ancestor. For example, molecular and ultrastructure data suggest that water molds, diatoms, golden algae, and brown algae—protists that at first glance seem to share few characteristics—are a monophyletic group.

Given the diversity in protist structures and molecules, most biologists regard the protist kingdom as a **paraphyletic** group—that is, the protist kingdom contains some, but not all, of the descendants of a common eukaryote ancestor. Molecular and ultrastructural analyses continue to help biologists clarify relationships among the various protist phyla and among protists and the other eukaryotic kingdoms (•**Figure 20-2**).

Biologists also use these data to develop various proposals that split organisms in kingdom Protista into several kingdoms. One proposal, for example, removes red algae and green algae from the protists and classifies them within the plant kingdom. Biologists who support this reclassification base it on similarities in the inner and outer chloroplast membranes of red algae, green algae, and land plants. Figure 20-2 supports such a classification change, because molecular evidence also suggests these groups share a common ancestor (see node A). ■

Representative Protists

In this chapter we discuss 10 representative phyla of the kingdom Protista: euglenoids, dinoflagellates, water molds, diatoms, golden algae, brown algae, red algae, green algae, plasmodial slime molds, and cellular slime molds. They are summarized in •**Table 20-1.** Other protists, including protozoa such as paramecia and amoebas, were traditionally classified as animals and are not usually studied in botany.

Most euglenoids are freshwater unicellular flagellates

There are about 900 species of **euglenoids** (phylum Euglenophyta). Most are unicellular flagellates that have two flagella, one long and one so short that it does not protrude outside the cell (•**Figure 20-3** on page 388). A euglenoid changes shape continually as it moves

EUGLENOID A mostly freshwater, flagellated, unicellular protist that moves by an anterior flagellum and is usually photosynthetic.

Unraveling evolutionary relationships among the protists has been difficult, but progress is occurring.

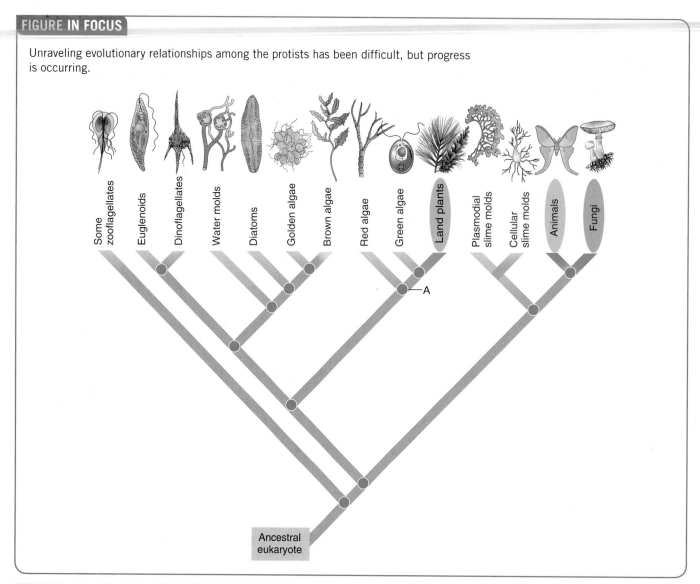

FIGURE 20-2 Evolutionary relationships among eukaryotes.
Relationships among eukaryotes are diverse and debated; this cladogram presents one interpretation. Groups within ovals represent separate eukaryotic kingdoms in the six-kingdom system used in this book. All other eukaryotes (without the ovals) are in kingdom Protista. Node A indicates where the red algae branched off from the green algae and land plants. Only the zooflagellate and animal branches shown in the figure are not discussed in this text.

through the water, because its outer covering, called a *pellicle,* is flexible rather than rigid. The red to orange *eyespot,* a carotenoid-containing region of the cell, is thought to help the euglenoid perceive the direction of light. Euglenoids reproduce asexually by cell division; none has ever been observed to reproduce sexually.

Most euglenoids contain chloroplasts and photosynthesize. They have chlorophyll *a,* chlorophyll *b,* and carotenoids, the same pigments found in green algae and plants. (Although the euglenoids have the same pigmen-

tation as the green algae and plants, they are not closely related to either group.) Energy reserves are stored as a polysaccharide called paramylon. Some photosynthetic euglenoids lose their chlorophyll when grown in the dark and turn into heterotrophs that obtain their nutrients by ingesting organic matter. Some species of euglenoids are always colorless and heterotrophic.

Euglenoids are both plantlike and animal-like and are often used as an example of one of the dilemmas of classification. When biologists recognized only two

TABLE 20-1 Comparison of Representative Phyla of the Kingdom Protista

PHYLUM	BODY FORM	LOCOMOTION	PHOTOSYNTHETIC PIGMENTS	SPECIAL FEATURES
Euglenophyta (euglenoids)	Unicellular	Two flagella (one of them very short)	Chlorophylls *a* and *b*; carotenoids	Flexible outer covering; store paramylon
Dinophyta (dinoflagellates)	Unicellular, some colonial	Two flagella	Chlorophylls *a* and *c*; carotenoids, including fucoxanthin	Many covered with cellulose plates; store oils and polysaccharides
Oomycota (water molds)	Coenocytic mycelium	Two flagella on zoospores	None	Cellulose and/or chitin in cell walls
Bacillariophyta (diatoms)	Unicellular, some colonial	Most nonmotile; some move by gliding over secreted slime	Chlorophylls *a* and *c*; carotenoids, including fucoxanthin	Silica in shell; store oils and carbohydrates
Chrysophyta (golden algae)	Unicellular or colonial	Two flagella or none	Chlorophylls *a* and *c*; carotenoids, including fucoxanthin	May be covered by calcium carbonate or silica scales; store oils and carbohydrates
Phaeophyta (brown algae)	Multicellular	Two flagella on reproductive cells	Chlorophylls *a* and *c*; carotenoids, including fucoxanthin	Differentiation of body into blade, stipe, and holdfast; store laminarin
Rhodophyta (red algae)	Most multicellular, some unicellular	Nonmotile	Chlorophyll *a*; carotenoids; phycocyanin; phycoerythrin	Some reef builders; store floridean starch
Chlorophyta (green algae)	Unicellular, colonial, siphonous, multicellular	Most flagellated at some stage in life; some nonmotile	Chlorophylls *a* and *b*; carotenoids	Reproduction highly variable; store starch
Myxomycota (plasmodial slime molds)	Multinucleate plasmodium	Streaming cytoplasm; flagellated or amoeboid reproductive cells	None	Reproduce by spores formed in sporangia
Dictyosteliomycota (cellular slime molds)	Vegetative form—single cell; reproductive form—multicellular slug	Amoeboid (for single cells); cytoplasmic streaming (for multicellular)	None	Aggregation of cells triggered by chemical signal

kingdoms—plants and animals—euglenoids were classified in the plant kingdom (with the algae) because many are photosynthetic. At the same time, other biologists classified them in the animal kingdom because they move, and some of them feed.

Most euglenoids inhabit freshwater ponds and puddles, particularly those with large amounts of organic material. For that reason their numbers are used to detect organic pollution. If a body of water has a large population of euglenoids, it is probably polluted with excess organic matter. Some euglenoids are found in marine waters and mudflats.

Most dinoflagellates are a part of marine plankton

Some 2000 to 4000 species of **dinoflagellates** (phylum Dinophyta) are known. Most of these unusual organisms are unicellular, although a few are colonies. Their cells are often covered with shells of interlocking cellu-

lose plates, some of which contain *silica* (silicon dioxide) that impart strength (●**Figure 20-4**). Each dinoflagellate is *biflagellated* (has two flagella); one is wrapped like a belt around a transverse groove in the center of the cell, and the other is located in a longitudinal groove (at a right angle to the transverse groove) and projects behind the cell. The undulation of these flagella propels the dinoflagellate through the water like a spinning top. Indeed, the dinoflagellates' name is derived from the Greek *dinos*, "whirling."

Most dinoflagellates are photosynthetic and have the pigments chlorophyll *a*, chlorophyll *c*, and carotenoids. A special yellow-brown carotenoid, **fucoxanthin,** is found only in dinoflagellates and a few other

DINOFLAGELLATE A unicellular, biflagellated, typically marine protist that is usually photosynthetic, contains the brown pigment fucoxanthin, and is an important component of plankton.

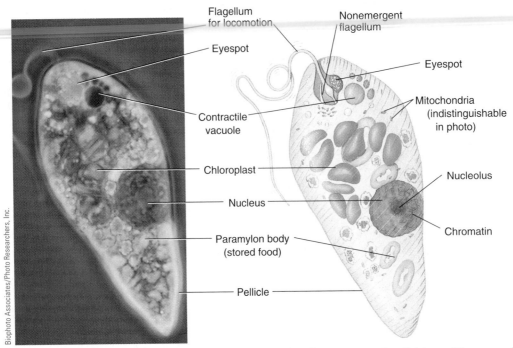

Flagellum for locomotion

Nonemergent flagellum

Eyespot

Eyespot

Contractile vacuole

Mitochondria (indistinguishable in photo)

Chloroplast

Nucleus

Nucleolus

Chromatin

Paramylon body (stored food)

Pellicle

Biophoto Associates/Photo Researchers, Inc.

(a) A living species of *Euglena*. Euglenoids are unicellular, flagellated algae that have at various times been classified in the plant kingdom (with the algae) and in the animal kingdom (when protozoa were considered animals).

(b) *Euglena* has an eyespot, a light-sensitive organelle that helps it react to light. Euglena's outer covering, called a pellicle, is flexible and enables the organism to change shape easily.

FIGURE 20-3 Euglenoids.

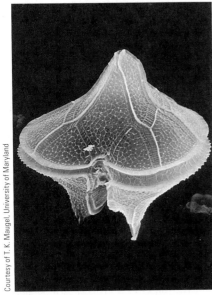

Courtesy of T. K. Maugel, University of Maryland

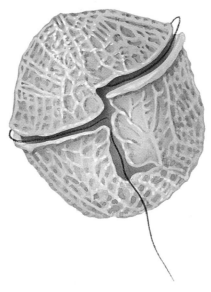

Kevin Schafer/Peter Arnold, Inc.

(a) A scanning electron micrograph of *Protoperidium*. Note the plates that encase the unicellular body.

(b) A species of *Gonyaulax*, showing the two flagella located in grooves.

(c) A red tide in Mexico. Billions of dinoflagellates produce the orange cloudiness in the water.

FIGURE 20-4 Dinoflagellates.

groups of algae (the diatoms and brown algae). However, other dinoflagellates are colorless (and therefore nonphotosynthetic) and ingest other microorganisms for food. Dinoflagellates usually store energy reserves as oils or polysaccharides.

Many dinoflagellates reside as **endosymbionts** in the bodies of marine invertebrates such as jellyfish, corals, and mollusks. These symbiotic dinoflagellates lack cellulose plates and flagella and are called **zooxanthellae.** Zooxanthellae photosynthesize and provide carbohydrates for their invertebrate partners. The contribution of zooxanthellae to the productivity of coral reefs is substantial. Other dinoflagellates that reside in invertebrates lack pigments and do not photosynthesize; these heterotrophs are parasites that live off their hosts.

Reproduction in the dinoflagellates is primarily asexual, by cell division, although a few species reproduce sexually. The dinoflagellate nucleus is unusual because the chromosomes are permanently condensed and always evident. Meiosis and mitosis are unique because the nuclear envelope remains intact throughout cell division and the spindle is located *outside* the nucleus.

In terms of ecological contributions, the dinoflagellates are one of the most important groups of producers in marine ecosystems. A few dinoflagellates are known to have occasional population explosions, or **blooms.** These blooms, known as **red tides** (see Figure 20-4c), frequently color the water orange, red, or brown. It is not known what environmental conditions initiate dinoflagellate blooms, but they are more common in the warm waters of late summer and early fall. The number of red tides has been increasing in recent years, and many experts think that human-produced coastal pollution triggers them, presumably by providing nutrients to the dinoflagellates. Some of the dinoflagellate species that form red tides produce a toxin that attacks the nervous systems of fishes, leading to massive fish kills. When airborne, this toxin irritates the eyes and makes breathing difficult for coastal residents and beachgoers.

Humans sometimes get paralytic shellfish poisoning by eating oysters, mussels, or clams that have fed on certain dinoflagellates. Paralytic shellfish poisoning induces respiratory failure and can result in death. (The dinoflagellates do not appear to harm the shellfish.)

The water molds produce flagellated zoospores

The 700 species of **water molds** (phylum Oomycota) were once classified as fungi because of their superficial structural resemblance. Both a water mold and a fungus have a body, termed a **mycelium,** that grows over a food source, digests it with enzymes that it secretes, and then absorbs the predigested nutrients (●**Figure 20-5a**). The threadlike *hyphae* that make up the mycelium in a water mold are coenocytic; that is, the water mold body consists of one giant multinucleate "cell." The cell wall of a water mold may contain cellulose (as in a plant), chitin (as in a fungus), or both cellulose and chitin. Because a water mold produces flagellated cells at some point in its life cycle, whereas fungi never produce motile cells, most biologists classify the water molds as protists rather than fungi.

When food is plentiful and environmental conditions are favorable, water molds reproduce asexually (●**Figure 20-5b**). A hyphal tip swells and a cross wall is formed, separating the hyphal tip from the rest of the mycelium. Within this structure, called a *zoosporangium,* tiny biflagellate **zoospores** form. Each zoospore has the potential to develop into a new mycelium. When environmental conditions worsen, water molds initiate sexual reproduction. After fusion of male and female nuclei, a thick-walled **oospore** develops from the zygote. Water molds often survive low temperatures in the winter as oospores.

Some water molds have played infamous roles in human history. For example, the water mold that causes a disease in potatoes called *late blight* precipitated the Irish potato famine of the 19th century. As a result of several rainy, cool summers in Ireland during the 1840s, the late blight water mold, which thrives under such conditions, multiplied unchecked. Potato tubers rotted in the fields, and because potatoes were the staple of the Irish peasants' diet, many people starved to death. Estimates of the number of deaths that resulted from the outbreak of this plant disease range from 250,000 to more than 1 million. The famine prompted a mass migration out of Ireland to other countries, such as the United States.

Late blight is still a problem today. New strains of the late blight water mold have appeared in the United States, Canada, northern Europe, Russia, South America, Japan, South Korea, the Middle East, and Africa. These strains resist the chemicals usually used to control the disease. Because today most people eat a varied diet, late blight does not cause famine, but it costs potato growers millions of dollars annually in lost crops.

A close relative of the late blight water mold causes *sudden oak death,* which is killing oak forests in California, Oregon, and Washington (see *Plants and the Environment: Sudden Oak Death* on page 392).

WATER MOLD A protist with a body consisting of a coenocytic mycelium and with asexual reproduction by motile zoospores and sexual reproduction by oospores.

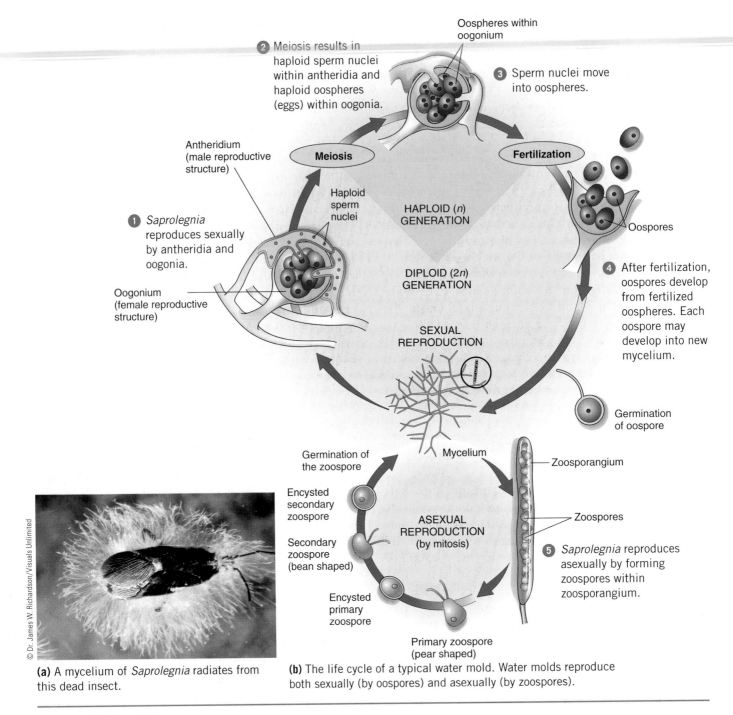

2 Meiosis results in haploid sperm nuclei within antheridia and haploid oospheres (eggs) within oogonia.

Oospheres within oogonium

3 Sperm nuclei move into oospheres.

Meiosis

Fertilization

Antheridium (male reproductive structure)

Haploid sperm nuclei

HAPLOID (*n*) GENERATION

1 *Saprolegnia* reproduces sexually by antheridia and oogonia.

DIPLOID (2*n*) GENERATION

Oospores

4 After fertilization, oospores develop from fertilized oospheres. Each oospore may develop into new mycelium.

Oogonium (female reproductive structure)

SEXUAL REPRODUCTION

Germination of oospore

Mycelium

Zoosporangium

Germination of the zoospore

Encysted secondary zoospore

Zoospores

ASEXUAL REPRODUCTION (by mitosis)

Secondary zoospore (bean shaped)

5 *Saprolegnia* reproduces asexually by forming zoospores within zoosporangium.

Encysted primary zoospore

Primary zoospore (pear shaped)

© Dr. James W. Richardson/Visuals Unlimited

(a) A mycelium of *Saprolegnia* radiates from this dead insect.

(b) The life cycle of a typical water mold. Water molds reproduce both sexually (by oospores) and asexually (by zoospores).

FIGURE 20-5 Water molds.

Diatoms are unflagellated and have silica in their cell walls

There are at least 100,000 species of **diatoms** classified in the phylum Bacillariophyta. Most are unicellular, although a few exist as filaments or colonies. Diatoms are protected by a shell that is composed of two halves that overlap where they fit together, much like the two halves of a petri dish. The shells are impregnated with silica, a glasslike material laid down in striking, intricate patterns, which are used to classify the species (•**Figure 20-6a**).

Diatom cells have one of two shapes: *radial symmetry* (wheel shaped) and *bilateral symmetry* (boat shaped or needle shaped). Although most diatoms are part of the floating plankton, some diatoms live on rocks and other surfaces, where they move by gliding. The secre-

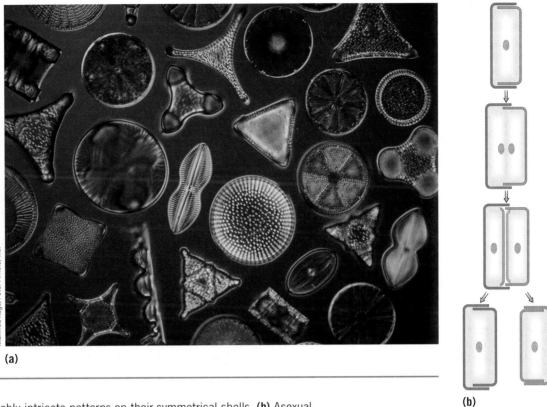

Manfred Kage./Peter Arnold, Inc.

(a)

(b)

FIGURE 20-6 Diatoms.

(a) Diatoms have remarkably intricate patterns on their symmetrical shells. **(b)** Asexual reproduction in diatoms. As cell division occurs, each new cell retains half of the original shell. The other half of the shell is synthesized, always to fit inside the original half. As a result, one of the new cells is smaller than the other.

tion of a slimy material from a small groove along the shell facilitates this gliding movement.

Most diatoms are photosynthetic and contain the pigments chlorophyll *a,* chlorophyll *c,* and carotenoids, including the yellow-brown fucoxanthin; their pigment composition gives them a yellow or brown color. Energy reserves are stored as oils or carbohydrates.

Diatoms most often reproduce asexually by cell division (•**Figure 20-6b**). When a diatom divides, the two halves of its shell separate, and each becomes the larger half of the shell for a new diatom cell. Therefore, the diatom cells that receive the smaller halves of the shells get progressively smaller with each succeeding generation. When diatom descendants reach a certain fraction of the original size of the ancestral individual, sexual reproduction is triggered, and the small diatoms produce shell-less gametes. Sexual reproduction restores the diatoms to the original size, because each resulting zygote (fertilized egg) grows substantially before producing a new shell.

Diatoms are common in fresh water and ocean water, but they are especially abundant in cooler marine waters. They are major producers in aquatic ecosystems because of their extremely large numbers. By some es-

timates, diatoms are responsible for about 20 percent of all the photosynthesis that takes place in the world, which means these organisms are of major importance in the global carbon cycle (see Chapter 26).

When diatoms die, their shells, which do not decompose, trickle down and accumulate in layers of what eventually becomes sedimentary rock. After millions of years, geologic upheaval exposes some of these deposits, called **diatomaceous earth,** on land. For example, huge deposits of diatomaceous earth are found in the White Hills of Lompoc, California. Diatomaceous earth is mined and used as a filtering, insulating, and soundproofing material. As a filtering agent, it is used to refine raw sugar and to process vegetable oils. Because of its abrasive properties, diatomaceous earth is a common ingredient in scouring powders and metal polishes. It is no longer added to most toothpastes because it is too abrasive for tooth enamel.

DIATOM Usually a unicellular protist, covered by an ornate siliceous shell and containing the brown pigment fucoxanthin, that is an important component of plankton in both marine and fresh water.

PLANTS AND THE ENVIRONMENT | Sudden Oak Death

The water molds, which cause potato late blight, are also responsible for **sudden oak death,** a devastating disease that affects oaks and many other plants. On coast live oak, California black oak, and canyon live oak, the disease causes "bleeding" bark cankers on the trunk and stems. These dead areas spread and usually kill the plant; as the infected stems die, the leaves wilt and turn brown (see figure). The disease in other host species, such as California bay laurel, big leaf maple, and California buckeye, causes leaf spots and twig cankers that weaken and sometimes kill the plant. Young coast redwood and Douglas fir, two species of particular importance to Pacific coastal states, are also susceptible to leaf spots and twig cankers caused by sudden oak death. The disease also affects shrubs such as rhododendron and azalea and herbs such as western starflower.

Botanists do not know where *Phytophthora ramorum,* the recently described water mold that causes sudden oak death, originated; some plant pathologists suspect it came from Asia. The disease has also been identified in Europe, but there it is caused by a

■ The coast live oak with brown foliage has died from sudden oak death. Photographed in California.

genetic strain of *P. ramorum* different from the one in North America.

Sudden oak death appeared in oak woodlands in California during the mid-1990s and has spread to many states since then, probably on diseased nursery plants. In 2005, federal regulations went into effect requiring inspection and certification of nurseries in California, Oregon, and Washington before they can ship plants out of state. It is hoped that this effort will stop the spread of sudden oak death before it

damages the oak forests of eastern North America.

Another line of defense against sudden oak death may be controlled burns. Fire ecologists compared California data on the location of past forest fires with the location of sudden oak death outbreaks; they found that the disease was almost nonexistent in areas where fires had occurred. More research will be required to determine if controlled burns will help check the spread of *P. ramorum.*

Most golden algae are unicellular biflagellates

About 1000 known species of **golden algae,** classified in the phylum Chrysophyta, are found in both freshwater and marine environments. Most species are biflagellate, unicellular organisms, although some are colonial (**•Figure 20-7**). A few golden algae lack flagella and are similar to amoebas in appearance except that they contain chloroplasts. Tiny scales of either silica or calcium carbonate may cover the cells. Reproduction in golden algae is primarily asexual and involves the production of biflagellate zoospores.

Most golden algae are photosynthetic and produce the same pigments as diatoms: chlorophylls *a* and *c* and carotenoids, including fucoxanthin. The pigment composition of golden algae gives them a golden or golden brown color. As in diatoms, energy reserves are stored

FIGURE 20-7 **Golden algae.**
LM of a colonial golden alga (*Synura*) found in freshwater lakes and ponds.

(a) *Laminaria* is a typical brown alga. Note the blade, stipe, and holdfast.

Blade

Stipe

Holdfast

© Blickwinkel/Alamy

(b) A giant kelp (*Macrocystis pyrifera*) bed off the coast of California is ecologically important to aquatic organisms, including the sea lion shown here.

Gregory Ochocki/Photo Researchers, Inc.

FIGURE 20-8 Brown algae.

as oils or carbohydrates. A few species ingest bacteria and other particles of food. Ecologically, golden algae are important producers in marine environments. They compose a significant portion of the ocean's **nanoplankton,** extremely minute algae that are major producers because of their great abundance.

Brown algae are the giants of the protist kingdom

About 1500 species of **brown algae** (phylum Phaeophyta) exist. All brown algae are multicellular. Their bodies, which are tufts, ropes, or thick, flattened branches, range in size from several centimeters to 75 meters (about 250 feet) in length. The largest brown algae, called kelps, are tough and leathery in appearance and exhibit considerable differentiation into leaflike **blades,** stemlike **stipes,** and anchoring **holdfasts** (•**Figure 20-8a**). They often have gas-filled floats to increase buoyancy (recall sargassum in the chapter introduc-

(c) Close-up of *M. pyrifera*, showing blades attached to a stipe. Note the gas-filled float at the base of each blade.

Lawrence Naylor/Photo Researchers, Inc.

GOLDEN ALGA A protist that is biflagellated, unicellular, and photosynthetic and contains the brown pigment fucoxanthin.

BROWN ALGA A predominantly marine, photosynthetic protist that is multicellular and contains the brown pigment fucoxanthin.

tion). It is important to remember that the blades, stipes, and holdfasts of brown algae are not homologous with the leaves, stems, and roots of plants. Brown algae and plants are thought to have evolved from different unicellular ancestors (see Figure 20-2).

Brown algae are photosynthetic and have chlorophyll *a*, chlorophyll *c*, and carotenoids, including the yellow-brown fucoxanthin, in their chloroplasts. The main energy reserve in brown algae is a carbohydrate called laminarin.

Brown algae are commercially important for several reasons. Their cell walls contain a polysaccharide, **algin,** which possibly helps cement the cell walls of adjacent cells. Algin is used as a thickening and stabilizing agent in ice creams, marshmallows, toothpastes, shaving creams, hair sprays, and hand lotions. Humans, particularly in East Asian countries, eat brown algae, which are a rich source of certain vitamins and of minerals such as iodine. Brown algae are one source of the antiseptic tincture of iodine.

Reproduction is varied and complex in the brown algae. They reproduce sexually, and most spend a portion of their lives as haploid organisms and a portion as diploid organisms. Their reproductive cells, both asexual zoospores and sexual gametes, are biflagellated.

Brown algae are common in cooler marine waters, especially along rocky coastlines, where they are found mainly in the intertidal zone or in relatively shallow waters. Kelps form extensive underwater "forests" and are essential in that ecosystem as the primary producer of food (•**Figure 20-8b and c**). Kelp beds also provide habitats for many marine invertebrates, fishes, and mammals such as sea otters. The diversity of life in these underwater forests of brown algae is remarkable.

Most red algae are multicellular organisms

There are 4000 to 6000 species of **red algae** (phylum Rhodophyta). The vast majority are multicellular organisms, although a few unicellular species exist. The multicellular body of a red alga is commonly composed of complex, interwoven filaments. Red algae are sometimes delicate and feathery, although a few species are flattened sheets of cells (•**Figure 20-9**). Most multicellular red algae attach to rocks or other solid materials with an anchoring *holdfast*.

The chloroplasts of red algae contain **phycoerythrin,** a red pigment, and **phycocyanin,** a blue pigment, in addition to chlorophyll *a* and carotenoids. The cyanobacteria are the only other organisms that have the same pigment composition as the red algae. Red algae store their energy reserves as a polysaccharide (floridean starch) similar to glycogen.

The cell walls of red algae often contain thick, sticky polysaccharides that are of commercial value. For example, **agar** is a polysaccharide extracted from certain red algae and used as a food thickener and culture medium (a substance on which to grow microorganisms and propagate some plants, such as orchids). Another polysaccharide extracted from red algae, **carrageenan,** is a food additive used to stabilize chocolate milk and to provide a thick, creamy texture to ice creams and other soft, processed foods. (A stabilizer keeps the texture uniform by preventing various ingredients from settling out.) Carrageenan is also used to stabilize cosmetics and paints. Red algae are a source of vitamins (particularly A and C) and minerals for humans, particularly in Japan and other East Asian countries where they are eaten fresh, dried, or toasted in such traditional foods as sushi and nori.

Nori is the Japanese name for edible red algae of the genus *Porphyra,* which has been cultivated in shallow bays for centuries. After it is harvested, nori is processed into thin, dry, paperlike sheets that are used to wrap sushi and rice balls. Nori is also added to soups, noodle dishes, and sauces as a flavoring.

Reproduction in the red algae has been studied in detail for only a few species, but it is remarkably complex, with an alternation of sexual and asexual stages. Although sexual reproduction is common, at no stage in the life history of red algae are there any flagellated cells.

The red algae are primarily found in warm tropical oceans, although a few species occur in fresh water and in soil. Some red algae incorporate calcium carbonate into their cell walls from the ocean waters (see Figure 20-9b). These coralline red algae are central in building "coral" reefs and are possibly more important than coral animals in this process.

The green algae exhibit a great deal of diversity

If one had to pick a single word to describe the 17,000 species of **green algae** (phylum Chlorophyta), it would have to be *variety*. Green algae exhibit many body forms and methods of reproduction. Their body forms range from unicells to colonies to coenocytic, **siphonous** (tubular) algae to multicellular filaments and sheets (•**Figure 20-10**). The multicellular forms do not have cells differentiated into specialized tissues, however. Most green algae are flagellated during at least part of their life history, although there are a few that are totally nonmotile.

Although the green algae are structurally diverse, they are biochemically uniform. Green algae are photosynthetic, with chlorophyll *a*, chlorophyll *b*, and ca-

(a) *Rhodymenia*, common in the Northern Hemisphere, has a flattened, branching body.

(b) *Bosiella,* found in the Pacific Ocean, is a coralline red alga with a hard, brittle body.

(c) *Polysiphonia*, widely distributed throughout the world, has a highly branched body.

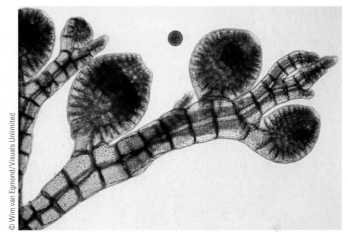

(d) Light micrograph of reproductive structures of *Polysiphonia*, which has a remarkably complex life cycle.

FIGURE 20-9 Red algae.
Most red algae are multicellular, many with highly branched, filamentous bodies.

rotenoids present in chloroplasts of a wide variety of shapes. Starch (a polysaccharide) is the main energy reserve. Most green algae have cell walls that contain cellulose, although some lack walls and some are covered with scales. Green algae have a significant number of characteristics in common with plants, including pigments, storage products (energy reserves), and cell-wall composition. Because of these and other similarities, it is generally accepted that plants evolved from green algalike ancestors.

Reproduction in the green algae is as varied as structure, and both sexual and asexual reproduction occur. Asexual reproduction is by cell division in unicellular forms or by fragmentation in multicellular forms; in fragmentation, pieces that break off from the original alga give rise to new individuals. Many green algae reproduce asexually by forming reproductive cells called **spores** by mitosis; if these spores are flagellated and motile, they are called *zoospores.* Each spore is capable of developing into a new individual without fusing with another cell as gametes do.

Sexual reproduction in the green algae involves the formation of gametes in unicellular gametangia. Three

RED ALGA One of a diverse group of photosynthetic protists that contain the pigments phycocyanin and phycoerythrin.

GREEN ALGA One of a diverse group of protists that contain the same pigments as land plants (chlorophylls *a* and *b* and carotenoids).

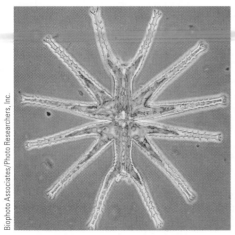

(a) Many green algae, such as the intricately beautiful *Micrasterias,* are unicells.

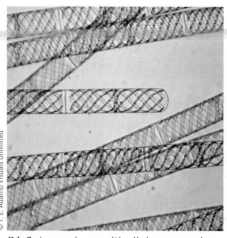

(b) *Spirogyra* is a multicellular green alga with a filamentous body form.

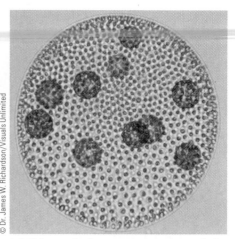

(c) *Volvox* colonies are each composed of up to 50,000 cells. New colonies are inside the parental colony, which eventually breaks apart.

(d) Some multicellular green algae are sheetlike. The thin, leaflike form has given *Ulva* its common name, "sea lettuce."

(e) Siphonous green algae such as dead man's fingers (*Codium fragile*) are coenocytic, which means that each individual's body is composed of one giant cell with multiple nuclei.

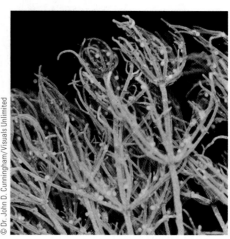

(f) *Chara,* a green alga commonly called a stonewort, is closely related to land plants. *Chara* is widely distributed in fresh water.

FIGURE 20-10 Green algae.

types of sexual reproduction are recognized in green algae: isogamy, anisogamy, and oogamy (•**Figure 20-11**). If the two gametes that fuse are identical in size and appearance, sexual reproduction is **isogamous** (•**Figure 20-12**). **Anisogamous** sexual reproduction involves the fusion of two gametes that differ in size and/or motility (•**Figure 20-13**). Some green algae are **oogamous** and produce large, nonmotile female gametes (eggs) and small, flagellated male gametes (sperm cells).

Instead of sexual reproduction by the fusion of gametes, some green algae join temporarily and exchange genetic information through **conjugation,** a pro-

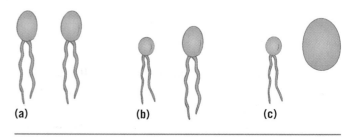

(a)   **(b)**   **(c)**

FIGURE 20-11 Types of sexual reproduction.
(a) In isogamy, the gametes are similar in size and form. **(b)** In anisogamy, the gametes are similar in form, but one of them is larger than the other. **(c)** In oogamy, the female gamete is larger and nonmotile, whereas the male gamete is smaller and usually motile.

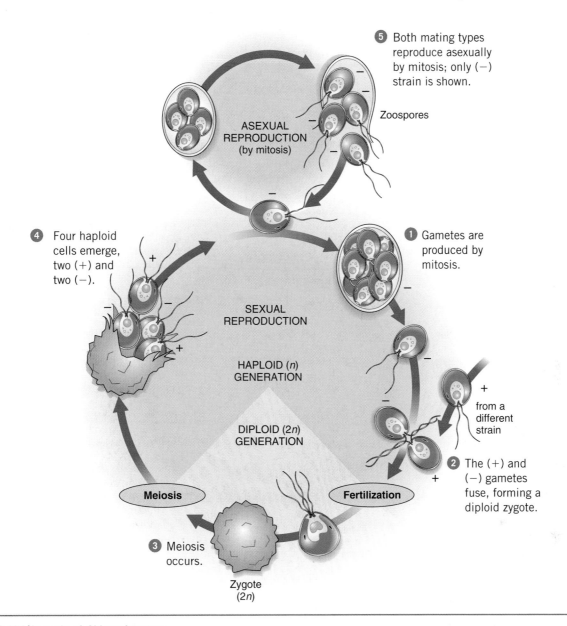

5 Both mating types reproduce asexually by mitosis; only (−) strain is shown.

Zoospores

ASEXUAL REPRODUCTION (by mitosis)

4 Four haploid cells emerge, two (+) and two (−).

1 Gametes are produced by mitosis.

SEXUAL REPRODUCTION

HAPLOID (*n*) GENERATION

DIPLOID (2*n*) GENERATION

from a different strain

Meiosis

Fertilization

2 The (+) and (−) gametes fuse, forming a diploid zygote.

3 Meiosis occurs.

Zygote (2*n*)

FIGURE 20-12 Life cycle of *Chlamydomonas.*
This haploid green alga has two indistinguishable strains, (+) and (−). Both strains reproduce asexually by mitosis. During sexual reproduction, a (+) gamete fuses with a (−) gamete to form a diploid zygote. Meiosis occurs, and four haploid cells emerge, two (+) and two (−).

cess in which the genetic material of one cell passes into and fuses with the genetic material of a recipient cell (•**Figure 20-14**).

Both aquatic and terrestrial green algae exist. Aquatic green algae primarily inhabit fresh water, although marine species also occur. Green algae that inhabit the land are restricted to damp soil, cracks in tree bark, and other moist places. Some green algae live as endosymbionts in the bodies of animals such as freshwater sponges, mollusks, and flatworms; others grow with fungi as a "dual organism" called a lichen (see Chapter 21).

Regardless of where they live, green algae are ecologically important. Because of their photosynthetic activity, they are at the base of many food webs, particularly in freshwater habitats. Green algae also help oxygenate the water during daylight hours (recall that oxygen is a by-product of photosynthesis). Thus, aquatic animals depend on green algae to provide them with both food and oxygen.

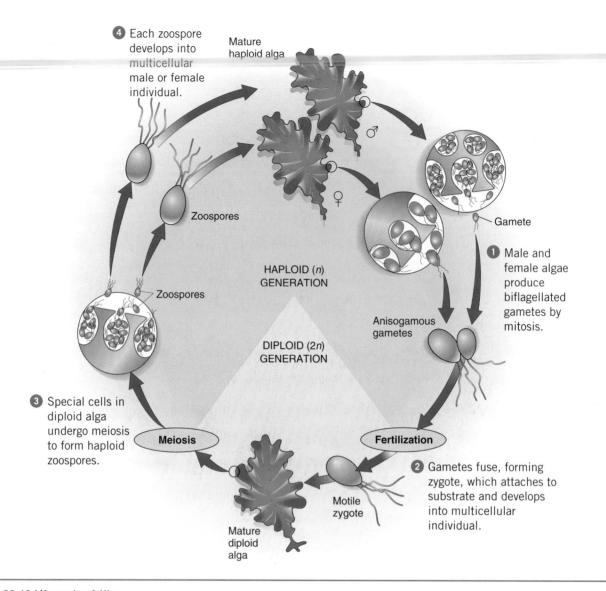

④ Each zoospore develops into multicellular male or female individual.

Mature haploid alga

Zoospores

Zoospores

HAPLOID (*n*) GENERATION

DIPLOID (2*n*) GENERATION

Meiosis

③ Special cells in diploid alga undergo meiosis to form haploid zoospores.

Mature diploid alga

Motile zygote

Fertilization

② Gametes fuse, forming zygote, which attaches to substrate and develops into multicellular individual.

① Male and female algae produce biflagellated gametes by mitosis.

Gamete

Anisogamous gametes

FIGURE 20-13 Life cycle of *Ulva*.
This green alga alternates between haploid and diploid multicellular phases, which are identical in overall appearance. The male and female haploid algae give rise to anisogamous gametes that fuse and subsequently develop into the diploid alga. Special cells in the diploid alga undergo meiosis to form haploid zoospores that develop directly into haploid algae, and the cycle continues.

The plasmodial slime molds have an unusual feeding stage

There are about 700 species of **plasmodial slime molds** (phylum Myxomycota). In the feeding stage of the plasmodial slime mold life cycle, a **plasmodium** streams over damp, decaying logs and leaf litter, ingesting bacteria, yeasts, spores, and decaying organic matter (•**Figure 20-15a**). The plasmodium, a wall-less amoeba-like mass that contains many diploid nuclei in its cytoplasm, changes shape as it moves, much as an amoeba does. Often brightly colored, it forms a network of channels to cover a larger surface area as it creeps along.

When the food supply dwindles or there is insufficient moisture, the plasmodium flows to a surface exposed to the air and initiates reproduction. Intricate stalked structures usually form from the drying plasmodium (•**Figure 20-15b**). Within these structures, called **sporangia,** meiosis occurs to produce haploid nuclei. Later, a wall of cellulose and/or chitin develops around each nucleus to form a spore. These spores are extremely resistant to adverse environmental conditions, such as drought and extremes of temperature. When conditions are favorable, however, the spores crack open, and a haploid reproductive cell emerges from each. This haploid

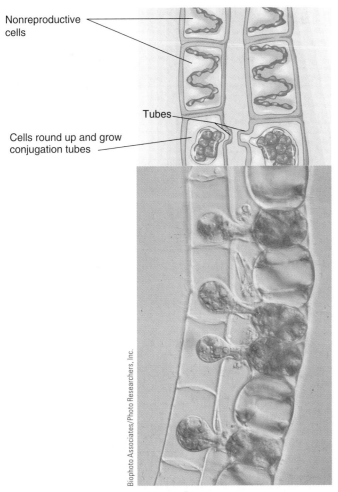

Nonreproductive cells

Tubes

Cells round up and grow conjugation tubes

Biophoto Associates/Photo Researchers, Inc.

FIGURE 20-14 Conjugation.
Spirogyra is a haploid organism that undergoes a sexual phenomenon known as conjugation. (*Top*) Filaments of two strains line up, and conjugation tubes form between cells of the two filaments. (*Micrograph on bottom*) The contents of one cell pass into the other through the conjugation tube. The two cells then fuse and form a diploid zygote. Following a period of dormancy, the zygote undergoes meiosis, restoring the haploid condition.

© Robert Calentine/Visuals Unlimited

(a) The plasmodium of *Physarum polycephalum* is colored bright yellow. This naked mass of protoplasm, shown on a rotting log, is multinucleate and feeds on bacteria and other microorganisms.

© Robert Calentine/Visuals Unlimited

(b) The reproductive structures of *P. polycephalum* are stalked sporangia.

FIGURE 20-15 Plasmodial slime molds.
Plasmodial slime molds are characterized by a creeping plasmodium and the production of spores.

cell is either flagellated or amoeba-like, depending on how wet the environment is; flagellated cells form in wet conditions. These two reproductive cells, the flagellated **swarm cell** and the amoeba-like **myxamoeba,** act as gametes. Eventually, two gametes fuse to form a zygote with a diploid nucleus. The resultant diploid nucleus divides by mitosis, but the cytoplasm does not divide, so the result in this case is a multinucleate plasmodium.

The cellular slime molds aggregate into a "slug"

The 50 or so species in the phylum Dictyosteliomycota are called **cellular slime molds.** Cellular slime molds

have close affinities with the amoebas and plasmodial slime molds. Each cellular slime mold is a small, individual, amoeba-like cell that behaves as a separate, solitary organism; it moves over rotting logs and soil, ingesting bacteria and other particles of food. Each cell has a haploid nucleus.

PLASMODIAL SLIME MOLD A protist whose feeding stage consists of a multinucleate, amoeboid plasmodium.

CELLULAR SLIME MOLD A protist whose feeding stage consists of a unicellular, amoeboid organism that aggregates to form a pseudoplasmodium during reproduction.

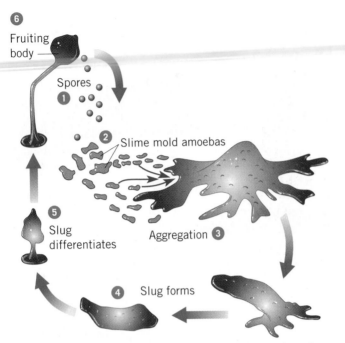

(a) The fruiting body releases spores, each of which gives rise to an amoeba-like organism. Slime mold amoebas ingest food, grow, and reproduce by cell division. After their food is depleted, the cells aggregate and organize into a sluglike pseudoplasmodium that migrates for a period before forming a stalked fruiting body, and the cycle continues.

FIGURE 20-16 **Cellular slime molds.**
The life cycle of the cellular slime mold, *Dictyostelium discoideum,* includes a unicellular stage and a multicellular pseudoplasmodium.

When moisture or food becomes inadequate, the individual haploid cells send out a chemical signal that causes them to aggregate by the hundreds or thousands. During this stage the cells migrate as a single multicellular "organism," called a **pseudoplasmodium** or "slug." Each cell of the slug retains its plasma membrane and individual identity. Eventually, the slug settles and develops into a stalked structure. The cells in the front third of the slug form the stalk, and the rear portion of the slug forms a rounded structure at the top of the stalk, within which spores differentiate. Each spore grows into an individual amoeba-like cell, and the cycle

(b) A developing fruiting body.

(c) The sluglike pseudoplasmodium.

repeats itself (●**Figure 20-16**). This reproductive cycle is generally asexual, although sexual reproduction has been observed occasionally.

STUDY OUTLINE

❶ **Briefly describe the features common to the members of the kingdom Protista.**
The kingdom Protista is composed of "simple" eukaryotic organisms, most of which are aquatic. Protists exhibit a remarkable diversity and range in size from microscopic unicells to large multicellular organisms.

❷ **Discuss in general terms the diversity in the protist kingdom, including means of locomotion, modes of nutrition, interactions with other organisms, and modes of reproduction.**
Although some protists are nonmotile, most have some means of locomotion, at least for part of the life cycle.

Some protists move by pushing out cytoplasmic extensions, some flex individual cells, some wave cilia, and some lash flagella. Protists obtain their nutrients photosynthetically or heterotrophically. Protists may be free-living or form symbiotic relationships, which range from mutualism to parasitism. Almost all protists reproduce asexually; many reproduce both sexually and asexually.

❸ Describe the kinds of data biologists use to classify eukaryotes.
Molecular data and ultrastructure have contributed to our understanding of phylogenetic relationships among protists. Molecular data were initially obtained for the gene that codes for small subunit ribosomal RNA in different eukaryotes. Biologists have also compared other nuclear genes, many of which code for proteins, in various protists. **Ultrastructure** is the fine details of cell structure revealed by electron microscopy. Ultrastructure is often similar among protists that comparative molecular evidence suggests are **monophyletic** and evolved from a common ancestor.

❹ Briefly describe and compare euglenoids and dinoflagellates.
Euglenoids are mostly freshwater, flagellated, unicellular protists that move by an anterior flagellum and are usually photosynthetic. **Dinoflagellates** are unicellular, biflagellated, typically marine protists that are usually photosynthetic, contain the brown pigment fucoxanthin, and are an important component of plankton.

❺ Briefly describe and compare water molds, diatoms, golden algae, and brown algae.

Water molds are protists with a body consisting of a coenocytic mycelium and with asexual reproduction by motile zoospores and sexual reproduction by oospores. **Diatoms** are usually unicellular protists, covered by ornate siliceous shells and containing the brown pigment fucoxanthin, that are an important component of plankton in both marine and fresh water. **Golden algae** are protists that are biflagellated, unicellular, and photosynthetic and contain the brown pigment fucoxanthin. **Brown algae** are predominantly marine, photosynthetic protists that are multicellular and contain the brown pigment fucoxanthin.

❻ Explain why red algae, green algae, and land plants are considered a monophyletic group.
Red algae are a diverse group of photosynthetic protists that contain the pigments phycocyanin and phycoerythrin. **Green algae** are a diverse group of protists that contain the same pigments as land plants (chlorophylls *a* and *b* and carotenoids). Ultrastructure—specifically, similarities in the inner and outer chloroplast membranes—and molecular data support the hypothesis that red algae, green algae, and land plants are monophyletic.

❼ Briefly describe and compare plasmodial slime molds and cellular slime molds.
Plasmodial slime molds are protists whose feeding stage consists of a multinucleate, amoeboid plasmodium. **Cellular slime molds** are protists whose feeding stage consists of unicellular, amoeboid organisms that aggregate to form a pseudoplasmodium during reproduction.

REVIEW QU

1. What are the ch
   Why are protists s
2. In what way does a unice...m a bacterium?
3. How do protists vary in their means of obtaining nutrients?
4. What are some of the ways protists interact with other organisms?
5. What are the two main types of data biologists use to help classify the protists?
6. How are euglenoids and dinoflagellates alike? How are they different?
7. Water molds are considered close relatives of brown algae. What feature(s) does/do these two groups of organisms share?
8. Why do some biologists classify the red and green algae as plants? Why do other biologists classify these algae as protists rather than plants?
9. Which group of organisms discussed in this chapter is closely related to the plasmodial slime molds? How do they differ?
10. How are the protists important to humans?
11. How are the protists important ecologically?
12. What evidence supports the hypothesis that the protist kingdom is a paraphyletic group?

13. Label the parts of the *Chlamydomonas* cell. (Refer to Figure 20-1b to check your answers.)

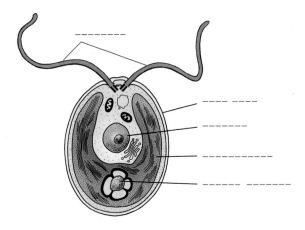

14. Use the following diagram to explain the life cycle of *Chlamydomonas*. Make sure you include both the sexual and asexual parts. (Refer to Figure 20-12 to check your answers.)

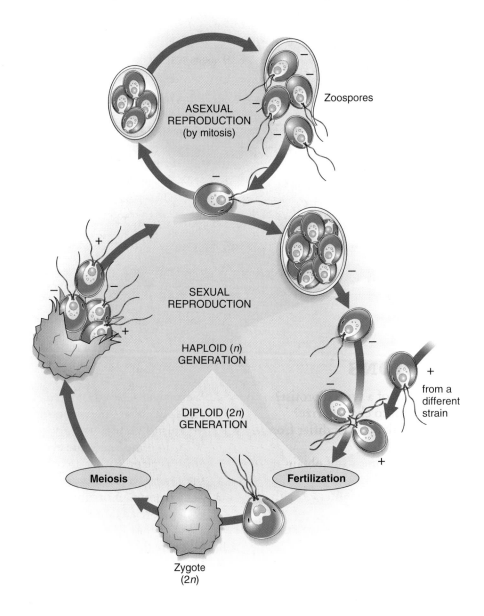

THOUGHT QUESTIONS

1. How might *Sargassum* have first colonized the Sargasso Sea?

2. *Euglena* is a freshwater alga that lacks a cell wall. Why, then, is the presence of a contractile vacuole, which expels excess water, so important to *Euglena*?

3. Plasmodial slime molds reproduce sexually, whereas cellular slime molds do not. What advantages might there be for each organism to have the type of reproduction it does?

4. When food is plentiful and environmental conditions are favorable, water molds reproduce asexually; when environmental conditions worsen, water molds reproduce sexually. The same is true for many other protists. Explain why it might be advantageous to reproduce asexually in favorable environments but sexually in unfavorable, changing environments.

Visit us on the web at http://www.thomsonedu.com/biology/berg/ for additional resources such as flashcards, tutorial quizzes, further readings, and web links.

Kingdom Fungi

Since ancient times edible mushrooms have been considered a delicacy. Although mushrooms contain some vitamins and minerals (inorganic nutrients), most are not particularly nutritious and do not contain much in the way of proteins, carbohydrates, or lipids (fats). Nonetheless, they are an important part of the world's major cuisines because they add distinctive flavors and textures to food.

In the United States, more than 850 million pounds of mushrooms are cultivated annually. The common mushroom (*Agaricus bisporus*) is the only fungal species grown extensively for food, although several mushrooms (such as oyster, shiitake, portobello, and straw mushrooms) are becoming increasingly popular.

Cultivating mushrooms requires exacting conditions. First the microscopic spores germinate in a laboratory under sterile conditions and grow into a network of delicate threads called a *mycelium*. Tiny plugs of the mycelium are then placed on sterile millet seeds, which they invade and digest. The seeds are then sent to a mushroom farm.

In windowless buildings at the mushroom farm, the farmer prepares beds of compost formed from the decomposed remains of straw and horse, cow, or chicken manure. The mycelium-containing millet seeds are then scattered across the compost, and the temperature and humidity are carefully regulated. After 2 to 3 weeks, the mycelium, which has formed a mat across the compost, is covered by peat moss. The temperature of the room is lowered to 10°C to 15°C (50°F to 60°F) and the humidity raised to 95 percent; these sudden changes in environmental conditions cause the mycelium to form tiny "buttons," immature mushrooms, in a week or two. Mushrooms are harvested from the bed over the next 3 to 7 months.

This chapter examines the biology and significance of mushrooms and other organisms classified in kingdom Fungi. Although fungi are not plants, they are usually studied in plant biology because they were traditionally considered plants when all organisms were classified into two or three kingdoms. The economic and medical importance of fungi to humans, as well as their ecological relationships with plants, justifies their coverage in general botany.

A mushroom bed at a commercial mushroom farm.

Characteristics of Fungi

Mushrooms, morels, and truffles—the delights of gourmet cooks around the world—have much in common with the black mold that grows on stale bread and the mildew that appears on damp shower curtains. All of these organisms belong to the kingdom Fungi, a diverse group of more than 70,000 known species, most of which are terrestrial. Fungi are eukaryotes; their cells contain membrane-bounded organelles such as nuclei and mitochondria. Fungi, which vary strikingly in size and shape, were originally classified in the plant kingdom, but biologists today recognize that they are not plants. Because they are distinct from plants and other eukaryotes in many ways, fungi are accorded a separate kingdom.

Fungal cells, like plant cells, are enclosed by cell walls. However, fungal cell walls have a different chemical composition from that of plant cell walls. In most fungi the cell wall is composed in part of **chitin,** which is also a component of the external skeletons of insects and other arthropods. Chitin, a polysaccharide that consists of subunits of a nitrogen-containing sugar, is far more resistant to breakdown by microorganisms than is the cellulose that makes up plant cell walls.

Fungi lack chlorophyll and chloroplasts and therefore are nonphotosynthetic. Although they are **heterotrophs**—that is, they cannot synthesize their own organic materials from simple inorganic raw material—fungi do not ingest food as animals do. Instead, fungi secrete digestive enzymes onto living or dead organic material and then *absorb* small organic molecules of the predigested food through their cell walls and plasma membranes. Some fungi are decomposers, and others are parasites. As **decomposers,** fungi obtain their nutrients from dead organic matter (animal carcasses, leaves, garbage, wood, and wastes); as **parasites,** they obtain nutrients from living organisms.

Fungi grow best in moist habitats but are found wherever organic material is available. They require moisture to grow and obtain water from the atmosphere as well as from the organic material on which they live. When the environment becomes dry, fungi survive by going into a resting stage or by producing spores that are resistant to desiccation (drying out). Although the optimum pH for most species is slightly acidic (about 5.6), some fungi can tolerate and grow in environments where the pH ranges from 2 (very acidic) to 9 (alkaline). Many fungi are less sensitive than bacteria to high osmotic pressures and therefore can grow in concentrated salt or sugar solutions, such as jelly, that discourage or prevent bacterial growth. Because fungi thrive over a wide temperature range, even refrigerated food is subject to spoilage by fungi.

Most fungi have a filamentous body plan

The body structures of fungi range from the unicellular *yeasts* to the multicellular, filamentous *molds,* a term used loosely to include mildews, rusts and smuts, mushrooms, and many other fungi. **Yeasts** are unicellular fungi that reproduce asexually mainly by **budding,** in which a small bulge (bud) grows and eventually separates from the parent cell. Each bud that separates from the parent cell is a new yeast cell that can grow and bud again. Yeasts also reproduce asexually by fission (the equal division of one cell into two cells) and sexually through spore formation (discussed shortly).

Most fungi are filamentous **molds.** A mold consists of long, branched threads (filaments) of cells called **hyphae** (sing., *hypha*). Hyphae form a tangled mass or tissuelike aggregation known as a **mycelium** (pl., *mycelia*) (●**Figure 21-1**). The cobweblike mold sometimes seen on bread is the mycelium of a fungus. What is not seen is the extensive part of the mycelium that grows down into the substance of the bread.

Some hyphae are **coenocytic;** that is, they are not divided into individual cells but are like an elongated, multinucleated giant cell. Other hyphae are divided by cross walls, called **septa** (sing., *septum*), into individual cells that each contain one or more nuclei. The septa of septate fungi often contain large pores that permit cytoplasm and sometimes nuclei to flow from cell to cell, providing a system of internal transport.

Fungi reproduce by spores

Fungi reproduce by microscopic **spores,** nonmotile (nonflagellated) reproductive cells dispersed by wind or by animals. Spores are usually produced on aerial hyphae (hyphae that project up into the air). This arrangement permits air currents to carry the spores to new areas. In some fungi, such as mushrooms, the aerial hyphae form large reproductive structures, called **fruiting bodies,** in which spores are produced. The familiar part of a mushroom is a large fruiting body. We do not normally see the nearly invisible network of hyphae buried out of sight in the rotting material on which the mushroom grows.

Fungi produce spores either asexually (by mitosis) or sexually (by meiosis). Fungal cells usually contain haploid nuclei. In sexual reproduction, the hyphae of two genetically distinct mating types come together and their cytoplasm fuses, a process known as **plasmogamy.** The resulting cell has two haploid nuclei, one from each fungus. This cell gives rise by mitosis to other cells with two nuclei. At some point, the two haploid nuclei fuse. This process, called **karyogamy,** results in a cell containing a diploid *zygote nucleus.* Meiosis then occurs, and haploid spores form.

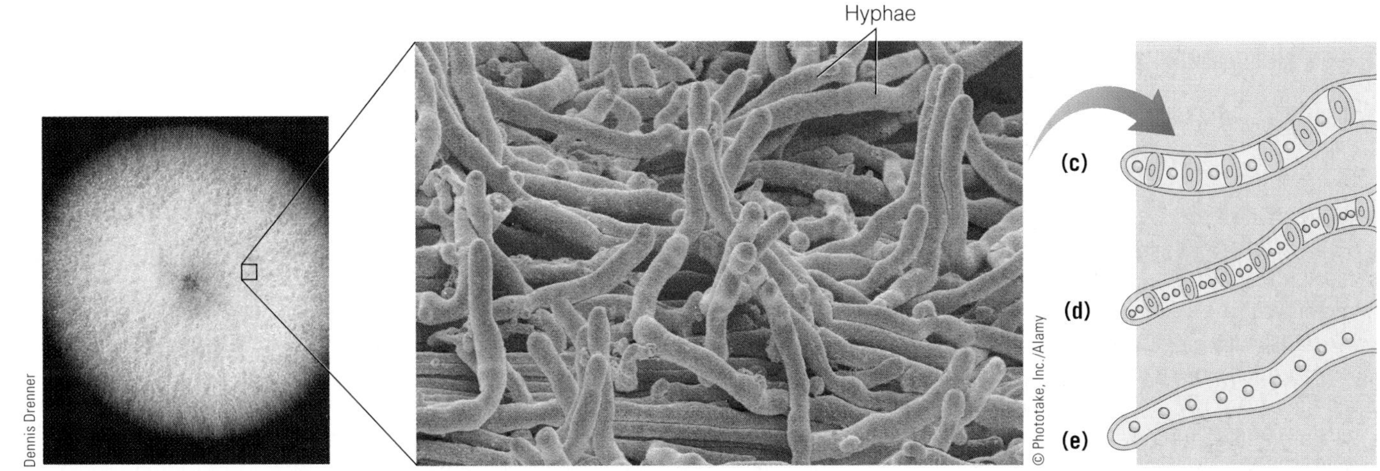

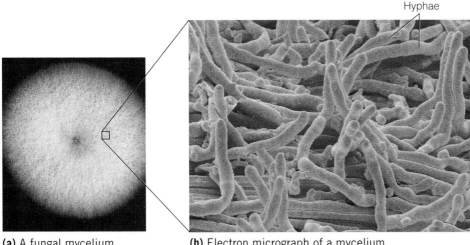

Hyphae

(a) A fungal mycelium growing on agar in a culture dish. In nature, fungal mycelia are rarely so symmetrical.

(b) Electron micrograph of a mycelium.

(c) A hypha divided into cells by septa; each cell is monokaryotic. In some fungi the septa are perforated (as shown).

(d) A septate hypha in which each cell is dikaryotic (has two nuclei).

(e) A coenocytic hypha.

FIGURE 21-1 **Filamentous fungi.**

During sexual reproduction in the two largest fungal phyla (the ascomycetes and basidiomycetes), plasmogamy occurs (the hyphae fuse), but karyogamy (the fusion of the two different nuclei) does not occur immediately. For a time, the nuclei remain separate within the fungal cytoplasm. Hyphae that contain two genetically distinct nuclei within each cell are referred to as **dikaryotic,** which is described as *n + n* rather than *2n* (see Figure 21-1d). Hyphae that contain only one nucleus per cell are **monokaryotic** (see Figure 21-1c). Fungal reproduction is discussed in greater detail later in the chapter.

When a fungal spore contacts an appropriate food source, perhaps an overripe peach that has fallen to the ground or has lain in the fruit basket in the kitchen too long, it germinates and begins to grow (•**Figure 21-2**). A threadlike hypha emerges from the tiny spore and, as it grows into the peach, branches frequently. Soon a tangled mat of hyphae infiltrates the peach and secretes digestive enzymes, degrading the peach's organic compounds to small molecules that the fungus absorbs. Later, other hyphae extend upward into the air to bear spores.

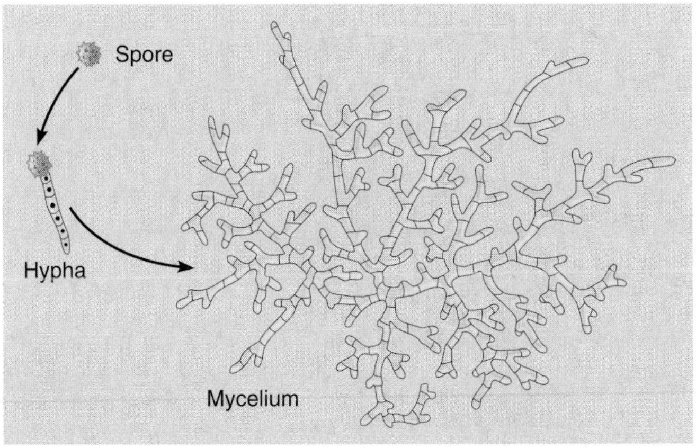

Fungal Diversity

Historically, fungi were classified mainly on the characteristics of their sexual spores and fruiting bodies. More recently, molecular data, such as comparative DNA and RNA sequences, have clarified relationships among fungal groups. Biologists currently assign fungi to five main phyla: Chytridiomycota, Zygomycota, Glomeromy-

Spore

Hypha

Mycelium

FIGURE 21-2 **Germination of a spore to form a mycelium.**

CHITIN A nitrogen-containing polysaccharide that forms the cell walls of many fungi.

HYPHA One of the threadlike filaments composing the mycelium of a fungus.

MYCELIUM The vegetative (nonreproductive) body of most fungi, consisting of a branched network of hyphae.

SPORE A reproductive cell that gives rise to individual offspring in fungi and certain other organisms.

TABLE 21-1 Phyla of Kingdom Fungi

PHYLUM AND COMMON TYPES	ASEXUAL REPRODUCTION	SEXUAL REPRODUCTION	OTHER KEY CHARACTERS
Chytridiomycota (chytrids or chytridiomycetes) *Chytridium*	Flagellated zoospores produced in zoosporangia	Flagellated gametes in some species	
Zygomycota (zygomycetes) Black bread mold Microsporidia are classified with the zygomycetes	Nonmotile, haploid spores produced in sporangia	Zygospores	Important decomposers; some are insect parasites; microsporidia are opportunistic pathogens that infect animals
Glomeromycota (glomeromycetes)	Large, multinucleate blastospores	Has not been observed	Form arbuscular mycorrhizae with plant roots
Ascomycota (ascomycetes or sac fungi) Yeasts, powdery mildews, molds, morels, truffles	Conidia pinch off from conidiophores	Ascospores	Have a dikaryotic stage; form important symbiotic relationships as lichens and mycorrhizae
Basidiomycota (basidiomycetes or club fungi) Mushrooms, bracket fungi, puffballs, rusts, smuts	Uncommon	Basidiospores	Have a dikaryotic stage; many form mycorrhizae with tree roots

cota, Ascomycota, and Basidiomycota. Microsporidia, a group of intracellular parasites, are classified in this text with the zygomycetes (•Figure 21-3 and •Table 21-1).

Until recently, biologists grouped about 25,000 species of fungi that did not fit into the major groups as *deuteromycetes* (phylum Deuteromycota). This was a **polyphyletic group** (they did not share a common ancestor) of species assigned together simply as a matter of convenience. Mycologists classified fungi as deuteromycetes if no sexual stage had been observed for them at any point in their life cycle. Some of these fungi had lost the ability to reproduce sexually, whereas others reproduced sexually only rarely. Biologists have now identified relationships between deuteromycetes and their sexually reproducing relatives, based on DNA comparisons among various species. Most of the deuteromycetes have been reassigned to phylum Ascomycota, and a few reassigned to phylum Basidiomycota.

Chytrids have flagellated spores

At one time biologists thought the **chytrids,** also known as **chytridiomycetes** (phylum Chytridiomycota), were funguslike protists, similar in many respects to the water molds. However, several lines of evidence indicate that the approximately 790 species of chytrids are members of kingdom Fungi. Like fungi, their cell walls con-

tain chitin. Molecular comparisons of DNA and RNA sequences also provide evidence that the chytrids are indeed fungi.

Chytrids are small, relatively simple fungi that inhabit ponds and damp soil; a few species live in salt water. Most chytrids are decomposers that degrade organic material. However, a few species cause disease in

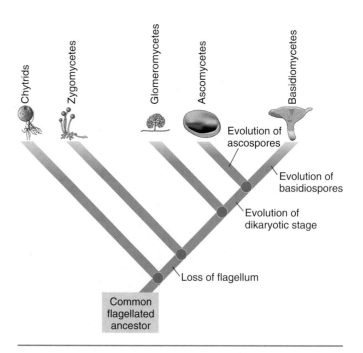

FIGURE 21-3 Major phyla of fungi.
This cladogram shows evolutionary relationships among the fungi, based on comparisons of DNA sequence data for many species. The chytrids branched off first during fungal evolution.

CHYTRID A fungus characterized by the production of flagellated cells at some stage in its life history.

ZYGOMYCETE A fungus characterized by the production of nonmotile, asexual spores and sexual zygospores.

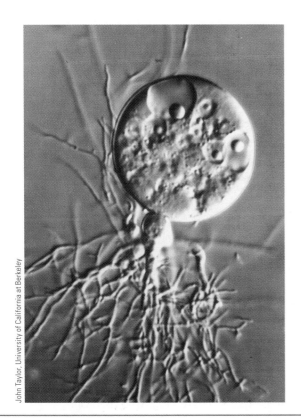

John Taylor, University of California at Berkeley

FIGURE 21-4 Chytrid.
Micrograph of a common chytrid (*Chytridium convervae*). Many chytrids have a microscopic body form consisting of a rounded, coenocytic thallus and branched rhizoids that superficially resemble roots.

has flagellated cells. At some point in their evolutionary history, other fungal groups apparently lost the ability to produce motile cells, perhaps during the transition from aquatic to terrestrial habitats. ■

Zygomycetes reproduce sexually by forming zygospores

The approximately 1060 species of phylum Zygomycota are referred to as **zygomycetes.** They produce **zygospores,** sexual spores that survive unfavorable environments by remaining dormant. Their hyphae are coenocytic—that is, they lack septa; however, septa do form to separate the hyphae from reproductive structures. Most zygomycetes are decomposers that live in the soil on decaying plant or animal matter (•**Figure 21-5**). A few zygomycetes are parasites of plants and animals.

Perhaps the best-known zygomycete is the black bread mold *Rhizopus nigricans,* a decomposer that breaks down bread and other foods. Before preservatives were added to food, bread left at room temperature often became covered with a black, fuzzy growth in a few days. Bread becomes moldy when a spore falls on it and then germinates and grows into a tangled mass of hyphae, known as a mycelium. Hyphae penetrate the bread and absorb nutrients. Eventually, certain hyphae grow upward and develop **sporangia** (spore sacs) at

plants and animals. A parasitic chytrid is partly responsible for declining amphibian populations dating back to the 1970s. Infected frogs have been identified in many parts of the world.

Most chytrids are unicellular or composed of a few cells that form a simple body, called a **thallus** (pl., *thalli*). The thallus may have slender extensions, called **rhizoids,** that anchor it to a food source and absorb food (•**Figure 21-4**). Chytrids are the only fungi that have flagellated cells. Their spores bear a single, posterior flagellum. Sexual reproduction has not been observed in most chytrids, but those species that do reproduce sexually have flagellated gametes. **EVOLUTION LINK** Molecular evidence suggests that chytrids were probably the earliest fungal group to evolve. Scientists hypothesize that the common ancestor of all fungi was a flagellated ancient protist. No other fungal group

© Peter Arnold, Inc. (Darlyne A. Murawski)/Alamy

FIGURE 21-5 *Pilobolus*, a zygomycete that grows in animal dung.
Stalked sporangia of *Pilobolus* protrude from a pile of dung, which contains an extensive mycelium of the fungus. The stalked sporangia, which are 5 to 10 millimeters tall, act like shotguns and forcibly discharge the sporangia (*black tips*) away from the dung onto nearby grass. When cattle or horses eat the grass, the spores pass unharmed through the digestive tract and are deposited in a fresh pile of dung.

FIGURE IN FOCUS

Like most fungi, most zygomycetes reproduce both asexually and sexually.

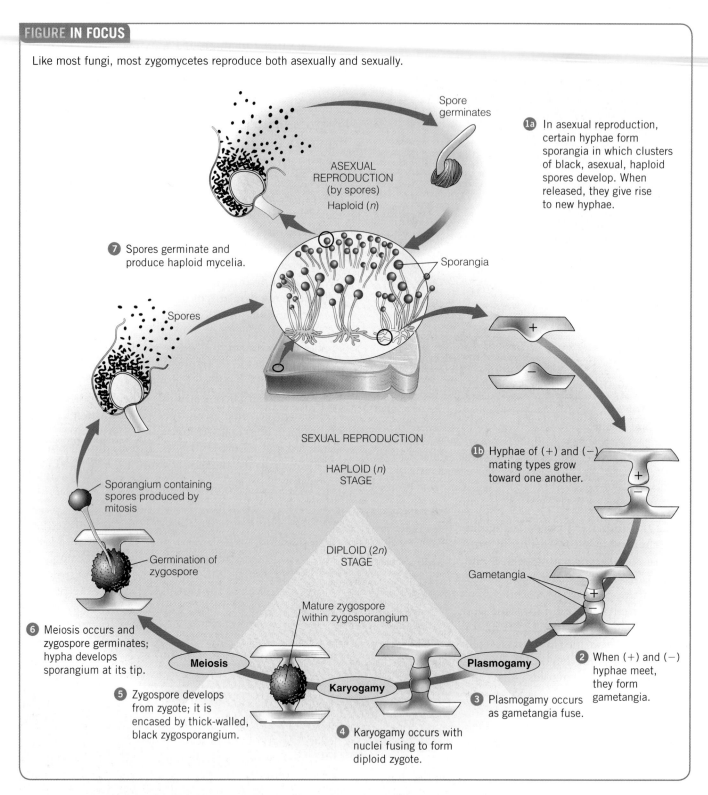

1a In asexual reproduction, certain hyphae form sporangia in which clusters of black, asexual, haploid spores develop. When released, they give rise to new hyphae.

Spore germinates

ASEXUAL REPRODUCTION
(by spores)
Haploid (*n*)

7 Spores germinate and produce haploid mycelia.

Spores

Sporangia

1b Hyphae of (+) and (−) mating types grow toward one another.

SEXUAL REPRODUCTION

HAPLOID (*n*) STAGE

DIPLOID (2*n*) STAGE

Gametangia

Sporangium containing spores produced by mitosis

Germination of zygospore

6 Meiosis occurs and zygospore germinates; hypha develops sporangium at its tip.

Meiosis

Mature zygospore within zygosporangium

Karyogamy

Plasmogamy

2 When (+) and (−) hyphae meet, they form gametangia.

3 Plasmogamy occurs as gametangia fuse.

5 Zygospore develops from zygote; it is encased by thick-walled, black zygosporangium.

4 Karyogamy occurs with nuclei fusing to form diploid zygote.

FIGURE 21-6 The life cycle of the black bread mold.
(*Top*) Asexual reproduction involves the formation of haploid spores. (*Bottom*) Sexual reproduction in the black bread mold (*Rhizopus nigricans*) takes place only between genetically distinct mating types, which are designated + and −.

their tips. Clusters of black asexual spores that develop within each sporangium are released when the delicate sporangium ruptures. The black spores give the black bread mold its characteristic color.

Sexual reproduction in the black bread mold occurs when the hyphae of two genetically distinct mating types, designated plus (+) and minus (−), grow into contact with one another (•**Figure 21-6**). An individual

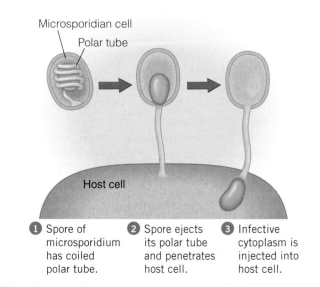

1 Spore of microsporidium has coiled polar tube.

2 Spore ejects its polar tube and penetrates host cell.

3 Infective cytoplasm is injected into host cell.

FIGURE 21-7 **Infection by a microsporidium.**

hypha mates only with a hypha of a different mating type; that is, sexual reproduction occurs only between a hypha of a plus (+) strain and a hypha of a minus (−) strain. Because there is no obvious sexual differentiation between the two mating types, it is not appropriate to refer to them as "male" and "female."

When hyphae of opposite mating types are near one another, hormones are produced that cause the hyphae to come together. Plus and minus nuclei then fuse to form a diploid nucleus, the zygote. A zygospore develops with a thick protective covering. The zygospore may lie dormant for several months and can survive desiccation and extreme temperatures. Meiosis probably occurs during or just before germination of the zygospore. When the zygospore germinates, an aerial hypha develops with a sporangium at the top. Haploid spores form within the sporangium by mitosis. When these spores are released, they germinate to form new hyphae with either + or − nuclei. Only the zygote of a black bread mold is diploid; its hyphae and asexual spores are haploid.

Molecular data suggest that microsporidia are zygomycetes

Microsporidia are small, unicellular parasites that infect eukaryotic cells. Although microbiologists estimate there may be more than 1 million species of microsporidia, only about 1500 species are currently named. Molecular studies have provided evidence that microsporidia are fungi or are closely related to fungi. At present, these common pathogens (disease-causing agents) are classified with the zygomycetes. Microsporidia are opportunistic pathogens that infect animals. For example, microsporidia infect people with compromised immune

systems, such as those with HIV/AIDS. Microsporidia cause a variety of diseases involving many organ systems, and some species cause lethal infections.

Microsporidia have two developmental stages inside their host: a feeding stage and a reproductive stage. The spores, which have thick protective walls, pass from cell to cell inside the host or are excreted in urine or through the skin. The spores, the only stage with distinct characters, are used to identify groups. Each spore is equipped with a unique structure with a long, thread-like **polar tube.** The spore, when it enters the gut of a new host, discharges its polar tube and penetrates the lining of the gut. Acting as a hypodermic needle, the polar tube injects the contents of the spore into the host cell (•**Figure 21-7**).

Glomeromycetes are symbionts with plant roots

Glomeromycetes (phylum Glomeromycota) are symbionts that form intracellular associations with the roots of most trees and herbaceous plants. (A *symbiotic association* is an intimate relationship between individuals of different species.) Glomeromycetes were previously considered zygomycetes, but biologists have concluded that they form a separate group based on molecular comparisons. Glomeromycetes have coenocytic (no septa) hyphae. They reproduce asexually with large, multinucleate spores called *blastospores.* Sexual reproduction has not been observed.

The symbiotic relationships between fungi and the roots of plants are called **mycorrhizae** (from Greek words meaning "fungus roots"). The roots supply the fungus with sugars, amino acids, and other organic substances. The mycorrhizal fungus benefits the plant by extending the reach of its roots. The slender mycelia are far thinner than roots and extend into narrow spaces, absorbing minerals that the plant could not obtain on its own. With the help of the mycorrhizal fungus, the plant takes in more water and minerals such as phosphorus and nitrogen.

Glomeromycetes extend their hyphae through cell walls of root cells but do not penetrate the plasma membrane. As the hyphae push forward, the root plasma membrane surrounds them. Thus, the hyphae are like fingers pushing into a glove formed by the plasma membrane. Because they are intracellular, these fungi are referred to as **endomycorrhizal fungi.** The most widespread endomycorrhizal fungi are called **arbuscular my-**

GLOMEROMYCETE A fungus that forms a distinctive branching form (arbuscular mycorrhizae) of endomycorrhizae.

ENDOMYCORRHIZAL FUNGI Fungi that form mycorrhizae that extend into plant roots.

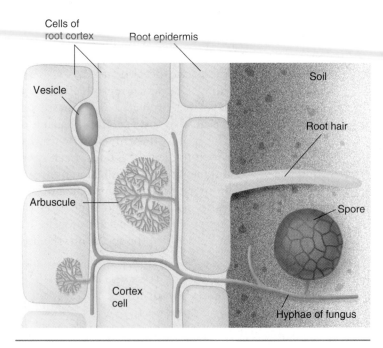

FIGURE 21-8 Glomeromycetes.
These fungi form arbuscular mycorrhizae. Note how the mycelium has grown into the root, and its hyphae branch between the cells of the root. Hyphae have penetrated through the cell walls of two root cells and have branched extensively to form arbuscules. The tip of one hypha between root cells has enlarged and serves as a vesicle that stores food. The tip of a hypha in the soil has enlarged, forming a spore. The spaces between the root cells have been magnified.

corrhizae because the hyphae inside the root cells form branched, tree-shaped structures known as **arbuscules** (•**Figure 21-8**). The arbuscules are the sites of nutrient exchange between the plant and the fungus. Arbuscular mycorrhizae live entirely underground.

EVOLUTION LINK Scientists have discovered mycorrhizae fungi within ancient plant fossils in rocks that are about 400 million years old. These findings suggest that when plants moved onto the land, their fungal partners moved with them. In fact, botanists have suggested that fungal hyphae may have provided plants with water and minerals before their own root systems evolved. ■

Ascomycetes reproduce sexually by forming ascospores

Phylum Ascomycota is a large group of fungi, the **ascomycetes**, consisting of about 32,300 species. The ascomycetes are sometimes referred to as **sac fungi** because their sexual spores are produced in little sacs called **asci** (sing., *ascus*). Their hyphae usually have septa, but these cross walls are perforated so that cytoplasm and even nuclei can move from one cell to another.

Ascomycetes include most yeasts; the powdery mildews; most of the blue-green, pink, and brown molds

that cause food to spoil; cup fungi; and the edible morels and truffles. Some ascomycetes cause serious plant diseases such as Dutch elm disease, chestnut blight, ergot disease on rye, and powdery mildew on fruits and ornamental plants (see *Plants and the Environment: Fighting the Chestnut Blight*).

In most ascomycetes, asexual reproduction involves the production of spores called **conidia,** which are pinched off at the tips of certain specialized hyphae known as **conidiophores** (conidia bearers) (•**Figure 21-9**). Sometimes called "summer spores," conidia are a means of rapidly propagating new mycelia when environmental conditions are favorable; each conidium that lands on a suitable food source rapidly grows into a new mycelium, which in turn produces conidia. Conidia occur in various shapes, sizes, and colors in different species; the color of the conidia is what gives many of these molds their characteristic brown, blue-green, pink, and other tints.

Some ascomycetes have genetically distinct mating strains; others are self-fertile, that is, they have the ability to mate with themselves. Sexual reproduction takes place after two hyphae grow together

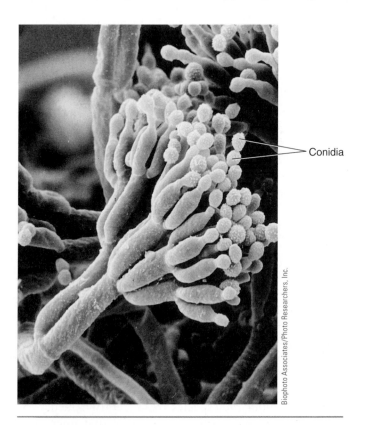

Conidia

Biophoto Associates/Photo Researchers, Inc.

FIGURE 21-9 Conidia.
Conidia are asexual reproductive cells produced by ascomycetes and a few basidiomycetes. Shown is an electron micrograph of *Penicillium* conidiophores, which resemble paintbrushes. Note the conidia pinching off the tips of the "brushes."

PLANTS AND THE ENVIRONMENT | Fighting the Chestnut Blight

Prior to the 20th century, the American chestnut (*Castanea dentata*) was an important tree, both ecologically and economically, in eastern forests. Throughout its natural range, which extended

from Maine to Georgia and west to Illinois, the chestnut tree provided food and shelter for forest animals, including squirrels, birds, and insects. The tree grew rapidly, forming a straight, tall trunk that provided valuable, decay-resistant wood for house foundations and interior trim, furniture, railroad ties, and fence posts. The wood was also an important source of tannin, a chemical used to convert animal hides to leather. The nuts of the American chestnut were harvested, roasted, and eaten directly or ground into flour.

Early in this century, the chestnut blight fungus (*Cryphonectria parasitica*), an ascomycete, was accidentally imported on diseased oriental chestnuts and quickly attacked native chestnuts, which had no resistance to it. First identified in New York in 1904, the blight killed or damaged several bil-

lion mature North American chestnut trees—almost every specimen throughout the tree's entire natural range—by the late 1940s (see figure).

A spore of the fungus enters a chestnut tree through a wound or crack in the bark, and the hyphae grow rapidly, attacking the inner bark. When the trunk is girdled (completely encircled) by the fungus, the portion of the plant above the injury dies. The fungus does not kill the roots, however, and they periodically sprout branches, only to be killed by the blight once again. Wind, water, and animals carry the spores from tree to tree.

Although the American chestnut is on the brink of extinction, there are hopes for its eventual recovery. Some biologists have bred the American chestnut with the Chinese chestnut, a related species that is resistant to chestnut blight. With careful selective breeding over many years, it may be possible to develop a disease-resistant variety of American chestnut.

Other biologists are trying to isolate strains of the chestnut blight fungus that are less virulent (less deadly). Such a strain was first identified in Europe in the 1950s; some of the trees there have recovered from the disease. More recently, a less virulent strain of the chestnut blight fungus has appeared in the United States. Biologists determined that a certain virus causes the reduced virulence of the chestnut blight fungus. They genetically engineered a synthetic form of the virus and introduced it into the virulent strain of the fungus, which subsequently lost its virulence. However, the virally infected blight fungus does not spread as rapidly as the healthy blight fungus. More research will be needed if the American chestnut is to make a comeback.

■ (*Top*) The massive trunks of American chestnuts photographed in the late 1800s in the Great Smoky Mountains. These trees and almost all other American chestnuts were killed by the chestnut blight fungus during the first part of this century. (*Bottom*) Chestnuts that sprout from the roots do not grow very large before they are killed by the chestnut blight.

and their cytoplasm mingles (•**Figure 21-10**). Within this fused structure the two nuclei come together, but they do not fuse. New hyphae develop from the fused structure, and the cells of these hyphae are dikaryotic.

ASCOMYCETE A fungus characterized by the production of nonmotile, asexual conidia and sexual ascospores.

Ascomycetes produce asexual spores called conidia and sexual spores called ascospores.

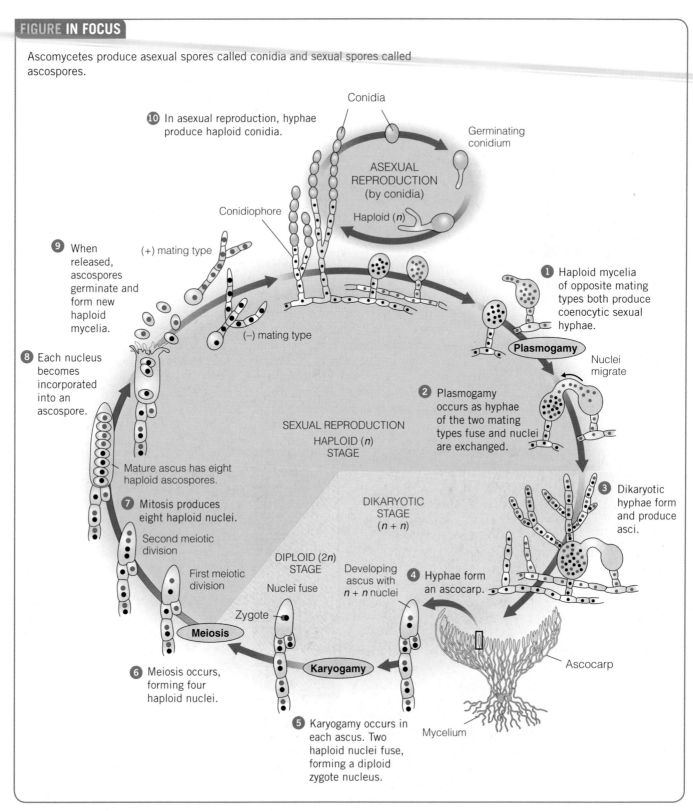

Conidia

10 In asexual reproduction, hyphae produce haploid conidia.

Germinating conidium

ASEXUAL REPRODUCTION (by conidia)

Haploid (*n*)

Conidiophore

9 When released, ascospores germinate and form new haploid mycelia.

(+) mating type

(−) mating type

1 Haploid mycelia of opposite mating types both produce coenocytic sexual hyphae.

Plasmogamy

Nuclei migrate

2 Plasmogamy occurs as hyphae of the two mating types fuse and nuclei are exchanged.

SEXUAL REPRODUCTION HAPLOID (*n*) STAGE

8 Each nucleus becomes incorporated into an ascospore.

Mature ascus has eight haploid ascospores.

7 Mitosis produces eight haploid nuclei.

Second meiotic division

First meiotic division

DIKARYOTIC STAGE (*n* + *n*)

3 Dikaryotic hyphae form and produce asci.

DIPLOID (2*n*) STAGE

Nuclei fuse

Developing ascus with *n* + *n* nuclei

4 Hyphae form an ascocarp.

Zygote

Meiosis

6 Meiosis occurs, forming four haploid nuclei.

Karyogamy

5 Karyogamy occurs in each ascus. Two haploid nuclei fuse, forming a diploid zygote nucleus.

Ascocarp

Mycelium

FIGURE 21-10 The life cycle of an ascomycete.
(*Top*) Asexual reproduction involves the formation of haploid conidia. (*Bottom*) Sexual reproduction involves the fusion of two haploid hyphae to form a dikaryotic (*n* + *n*) structure. Asci form from this structure; in each ascus the *n* + *n* nuclei fuse, followed by meiosis and mitosis to produce eight ascospores. The asci form the inner layer of a fruiting body known as an ascocarp.

(a) The ascocarp (fruiting body) of the common brown cup (*Peziza badio-confusa*) is shaped like a bowl and is 3 to 10 centimeters (1 to 4 inches) wide. It grows on damp soil in woods throughout North America. Photographed in Muskegon, Michigan.

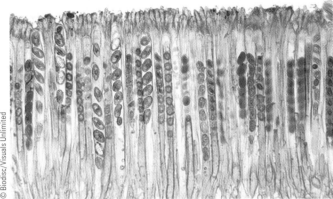

(b) Asci, each containing eight ascospores, line the inner portion of the ascocarp.

FIGURE 21-11 **Sexual reproduction in the ascomycetes.**

These *n* + *n* hyphae form a fruiting body, known as an **ascocarp,** that is characteristic of the species (•**Figure 21-11**). The ascocarp is where the asci develop.

Within a cell that develops into an ascus, the two nuclei fuse and form a diploid nucleus, the zygote nucleus. Each zygote then undergoes meiosis to form four haploid nuclei. This process is usually followed by one mitotic division of each of the four nuclei, resulting in the formation of eight haploid nuclei. Each haploid nucleus (surrounded by cytoplasm) develops into an **ascospore,** so there are usually eight haploid ascospores within the ascus. The ascospores are released when the tip of the ascus breaks open. Air currents carry individual ascospores, often for long distances. If an ascospore lands in a suitable location, it germinates and forms a new mycelium.

Yeasts are unicellular ascomycetes that reproduce asexually by budding and sexually by forming ascospores. During sexual reproduction, two haploid yeasts fuse and form a diploid zygote. The zygote undergoes meiosis, and the resulting haploid spores remain enclosed for a time within the original diploid cell wall. This sac of spores corresponds to an ascus and ascospores. Yeasts are essential in making bread and in fermenting alcoholic beverages (discussed later in this chapter).

Basidiomycetes reproduce sexually by forming basidiospores

The approximately 22,300 species that make up the phylum Basidiomycota include the most familiar of the fungi—mushrooms, bracket fungi, and puffballs—as well as some destructive plant parasites of important crops, such as wheat rust and corn smut (•**Figure 21-12**).

Basidiomycetes, or **club fungi,** derive their name from the fact that they develop a **basidium** (pl., *basidia*), a structure comparable in function to the ascus of ascomycetes. Each basidium is an enlarged, club-shaped hyphal cell, at the tip of which four **basidiospores** develop (•**Figure 21-13**). Note that basidiospores develop on the *outside* of the basidium, whereas ascospores develop *within* the ascus. The mature basidiospores are released, and when they come in contact with the proper environment, each develops into a new mycelium.

The mycelium of a basidiomycete, such as the cultivated mushroom introduced at the beginning of the chapter, consists of a mass of white, branching, threadlike hyphae that occur mostly underground. Septa divide the hyphae into cells, but the septa are perforated and allow cytoplasmic streaming between cells.

Compact masses of hyphae called buttons develop along the mycelium. Each button grows into the structure that we ordinarily call a mushroom. More formally, the mushroom, which consists of a stalk and cap, is referred to as a **basidiocarp.** The lower surface of the cap usually consists of many **gills,** thin plates that extend radially from the stalk to the edge of the cap. The basidia develop on the surfaces of these gills.

BASIDIOMYCETE A fungus characterized by the production of sexual basidiospores.

Dennis Drenner

Jeff Lepore/Photo Researchers, Inc.

(a) Basidia line the gills of the jack-o'-lantern mushroom (*Omphalotus olearius*), a poisonous species whose gills glow green in the dark.

(b) Rounded earthstars (*Geastrum saccatum*) release puffs of microscopic spores after the sac wall is hit by a raindrop or brushed by animals. This fungus is common in leaf litter under trees.

© Wally Eberhart/Jupiterimages

© Richard D. Poe/Visuals Unlimited

Richard H. Gross

(c) A giant puffball (*Calvatia gigantea*), with a woman's hand to show scale. At maturity, a dried-out puffball often has a pore through which the basidiospores are discharged as a puff of dust.

(d) The stinkhorn (*Phallus ravenelii*) has a foul smell that attracts flies, which help disperse the slimy mass of basidiospores at the tip.

(e) Bracket fungi (*Fomes*) grow on both dead and living trees, producing shelflike fruiting bodies. Basidiospores are produced in pores located beneath each shelf.

FIGURE 21-12 Diversity in basidiomycete fruiting bodies.

Each individual fungus produces millions of basidiospores, and each basidiospore has the potential, should it be transported to an appropriate environment, to give rise to a new **primary mycelium** (•Figure 21-14). Hyphae of a primary mycelium consist of monokaryotic cells, each of which contains a single haploid nucleus. When, in the course of its growth, such a hypha encounters another hypha of a genetically distinct mating type,

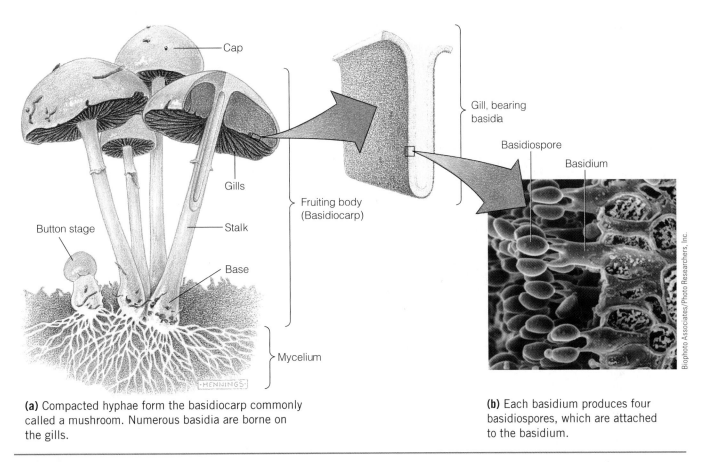

(a) Compacted hyphae form the basidiocarp commonly called a mushroom. Numerous basidia are borne on the gills.

(b) Each basidium produces four basidiospores, which are attached to the basidium.

FIGURE 21-13 Sexual reproduction in the basidiomycetes.

the two fuse. As in the ascomycetes, the two haploid nuclei remain separate within each cell. In this way a **secondary mycelium** with dikaryotic hyphae is produced, in which each cell contains two separate haploid nuclei.

The *n + n* hyphae of the secondary mycelium grow extensively and eventually form compact masses, which are the mushrooms, or basidiocarps. Each basidiocarp actually consists of a mat of intertwined hyphae. On the gills of the mushroom the dikaryotic nuclei fuse and form diploid zygotes. These are the only diploid cells that form in the life history of a basidiomycete. Each zygote undergoes meiosis, resulting in four haploid nuclei. These nuclei move to the outer edge of the basidium. Fingerlike extensions of the basidium develop, and the nuclei and some cytoplasm move into them; each of these extensions becomes a basidiospore, which is attached to the rest of the basidium by a delicate stalk. When the stalk breaks, the basidiospore is released.

Ecological Importance of Fungi

Fungi make important contributions to the ecological balance of our world. Like bacteria, most fungi are free-living decomposers that absorb nutrients from organic wastes and dead organisms. For example, many fungal decomposers degrade cellulose and lignin, the main components of plant cell walls. When fungi degrade wastes and dead organisms, carbon (as CO_2) and the mineral components of organic compounds are released into the environment, where they become available to plants and other organisms. Without this continuous decomposition, essential minerals would remain locked up in huge mounds of dead animals, feces, branches, logs, and leaves. The minerals would be unavailable for use by new generations of organisms, and life would cease.

Fungi form many important symbiotic relationships with animals, plants, bacteria, and protists. These relationships have major effects on ecosystems.

Fungi form symbiotic relationships with some animals

Because animals do not have the enzymes necessary to digest cellulose and lignin, cattle and other grazing animals cannot, by themselves, obtain nutrients from the plant material they eat. Their survival depends on fungi that inhabit their guts, because fungi, like many other microorganisms, do have the enzymes that break down these organic compounds. The fungi benefit by living in a nutrient-rich environment.

FIGURE IN FOCUS

Basidiomycetes produce sexual basidiospores on the gills of basidiocarps (fruiting bodies).

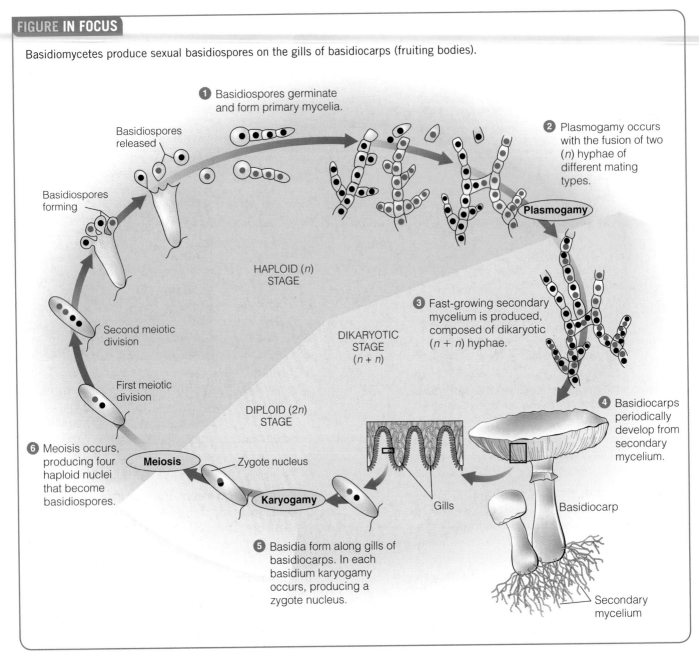

FIGURE 21-14 Life cycle of a typical basidiomycete.
Note the dikaryotic stage and the separation of plasmogamy and karyogamy. Asexual reproduction is uncommon in this group.

Fungi also form symbiotic relationships with ants and termites. More than 200 species of ants farm fungi. Leaf-cutting ants bring leaves to their fungi and protect them from competitors and predators. The ants also disperse the fungi to new locations. In exchange, the fungi digest the leaves, thereby providing nutrients for the ants. This symbiosis can involve other organisms. Fungal parasites sometimes infest the farmed fungi. In response, the ants culture bacteria (actinomycetes) that produce antibiotics to control these parasites. These symbiotic relationships, among the most complex known, are the product of perhaps 50 million years of coevolution.

Mycorrhizae are symbiotic relationships between fungi and plant roots

Mycorrhizae occur in about 80 percent of plants (and in more than 90 percent of all plant families). As discussed in the section on glomeromycetes, mycorrhizal fungi decompose organic material in the soil and increase the surface area of a plant's roots so that the plant can ab-

PROCESS OF SCIENCE

QUESTION: How do mycorrhizae affect their plant symbionts?

HYPOTHESIS: Mycorrhizal fungi enhance the growth of their plant partners.

EXPERIMENT: Control western red cedar (*Thuja plicata*) seedlings were grown in the absence of the mycorrhizal fungus (*see below*).

The experiment group was identical in all respects to the control group, except the experiment plants had roots that formed mycorrhizal associations (*see below*).

RESULTS AND CONCLUSION: Mycorrhizae enhance growth in western red cedars. ▪

FIGURE 21-15 Western red cedar experiment with mycorrhizae.

sorb more water and minerals. In exchange, the roots supply the fungus with organic nutrients.

The importance of mycorrhizae first became evident when horticulturalists observed that orchids do not grow unless an appropriate fungus lives with them. Similarly, biologists have shown that many forest trees

decline and die from mineral deficiencies when transplanted to mineral-rich grassland soils that lack the appropriate mycorrhizal fungi. When forest soil that contains the appropriate fungi or their spores is added to the soil around these trees, they quickly resume normal growth. Studies performed with various types of plants, including western red cedar, have confirmed the role of mycorrhizae in plant growth (•**Figure 21-15**).

As we have discussed, glomeromycetes form *endomycorrhizal* connections; they infiltrate the cells of plant roots (see Figure 6-14b). At least 500 species of ascomycetes and basidiomycetes also form mycorrhizal connections, but their hyphae coat the plant root rather than penetrate its cells. They are referred to as **ectomycorrhizal fungi** (see Figure 6-14a). Interestingly, researchers have shown that some mycorrhizal fungi harbor bacteria in their cytoplasm. Although the role of the bacteria is not yet clear, their presence suggests that they may be members of a three-way partnership: fungus, plant, and bacteria.

Mycorrhizal fungi connect plants, allowing nutrient transfer among them. Scientists have measured the movement of organic materials from one tree species to another through shared mycorrhizal connections. Mycorrhizal fungi also release chemicals that protect the plant against herbivores and pathogens.

Lichens are symbiotic relationships between a fungus and a photosynthetic organism

Although a **lichen** looks like a single organism, it is actually a symbiotic association between a photosynthetic organism and a fungus (•**Figure 21-16a**). The photosynthetic partner is usually a green alga or a cyanobacterium, and the fungus is most often an ascomycete. In some lichens in tropical regions, the fungal partner is a basidiomycete. The algae or cyanobacteria found in lichens are also found as free-living species in nature, but the fungal components of lichens are generally found only as part of lichens. Scientists have identified more than 13,200 species of lichen-forming fungi.

In the laboratory the fungal and photosynthetic partners of a lichen can be separated and grown separately in appropriate culture media. The alga or cyanobacterium grows more rapidly when separated, whereas the fungus grows slowly and requires a culture medium that provides many complex carbohydrates. Gener-

ECTOMYCORRHIZAL FUNGI Fungi that form mycorrhizae consisting of a dense sheath over the root's surface.

LICHEN A compound organism consisting of a symbiotic fungus and an alga or cyanobacterium.

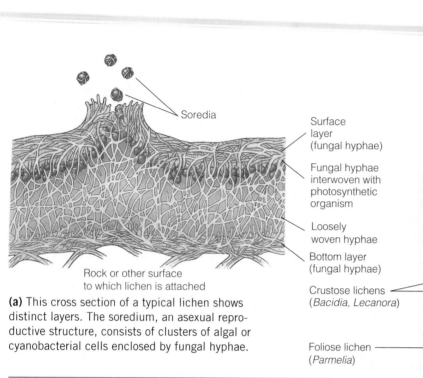

(a) This cross section of a typical lichen shows distinct layers. The soredium, an asexual reproductive structure, consists of clusters of algal or cyanobacterial cells enclosed by fungal hyphae.

FIGURE 21-16 **Lichens.**

(b) Lichens vary in color, shape, and overall appearance. Three growth forms—crustose, foliose, and fruticose—are shown on a maple branch in Washington State.

ally, the fungus does not produce fruiting bodies when separated from its photosynthetic partner; as part of a lichen, however, the fungus does produce fruiting bodies. The photosynthetic organism and the fungus can be reassembled as a lichen, but only if they are placed together in a culture medium under conditions that do not support either of them independently.

What is the nature of this partnership? In the past the lichen was considered a definitive example of *mutualism,* a symbiotic relationship equally beneficial to both species. The photosynthetic partner carries on photosynthesis, producing carbohydrate molecules for itself and the fungus, and the fungus obtains water and minerals for the photosynthetic partner as well as protects it against desiccation. Some biologists think, however, that the lichen partnership is not a true case of mutualism but one of controlled parasitism of the photosynthetic partner by the fungus; microscopic examination reveals that fungal hyphae penetrate and destroy some of the algal cells.

Lichens typically have one of three general growth forms—crustose, foliose, and fruticose (•**Figure 21-16b**). **Crustose** lichens are flat and grow tightly attached to the rock or whatever they are growing on; **foliose** lichens are also flat, but they have leaflike lobes and are

not so tightly attached; **fruticose** lichens grow erect and are branched and shrublike.

Able to tolerate extremes of temperature and moisture, lichens grow everywhere except in heavily polluted cities. For example, there are 350 lichen species that exist in the arctic region as compared with only 2 species of flowering plants.[1] Many other lichen species are at home in steaming equatorial rain forests. They grow on tree trunks, mountain peaks, and bare rock. In fact, lichens are often the first organisms to colonize bare rocky areas and play a role in the formation of soil. Lichens gradually make tiny cracks in the rocks to which they cling, facilitating weathering of the rocks by wind and rain.

The "reindeer mosses" of arctic regions are not mosses but lichens that serve as the main source of food for the caribou (reindeer) herds of the region. Some li-

---

[1]*Although lichens are composed of two organisms, lichen thalli are distinctive enough that biologists designate them as species.*

chens produce colored pigments; one pigment, orchil, is used to dye woolens, and another, litmus, is widely used in chemistry laboratories as an acid–base (pH) indicator.

Lichens vary greatly in size. Some are almost invisible, whereas others, such as the reindeer mosses, may cover kilometers of land with a growth that is ankle deep. Lichens grow slowly, and the radius of a crustose lichen may increase by less than a millimeter each year. Some mature crustose lichens are thousands of years old.

Lichens absorb minerals mainly from the air and from rainwater, but they also absorb them directly from the surface on which they grow. Lichens cannot excrete or sequester into vacuoles the elements that they absorb, and perhaps for this reason they are sensitive to toxic compounds in the environment. A reduction in lichen growth is an indication of air pollution, especially sulfur dioxide. Absorption of such toxic compounds results in damage to the chlorophyll of the photosynthetic partner. The return of lichens to an area indicates an improvement in air quality in that vicinity.

Lichens reproduce mainly by asexual means, usually fragmentation, in which bits of the lichen break off and, if they land in a suitable place, establish themselves as new lichens. Some lichens release special dispersal units called **soredia** (sing., *soredium*) that contain cells of both partners (see Figure 21-16a). In others, the alga reproduces asexually by mitosis, whereas the fungus produces ascospores. Wind disperses the ascospores, which may find an appropriate photosynthetic partner only by chance.

Economic and Medical Impact of Fungi

The same powerful digestive enzymes that enable fungi to decompose wastes and dead organisms also enable fungi to reduce wood, fiber, and food to their basic components with great efficiency. Many species of basidiomycetes have enzymes that break down lignin in wood. From the human perspective, various fungi cause incalculable damage to stored goods and building materials each year. Bracket fungi, for example, cause enormous losses by decaying wood—both stored lumber and the wood of living trees.

Fungi are more destructive than any other disease-causing organism to plants. Their activities cost billions of dollars in agricultural damage annually. Some fungi cause diseases in humans and other animals. Yet fungi also contribute to our quality of life. They are responsible for economic gains as well as losses. People eat them

and grow them to make various chemicals, such as citric acid and other industrial chemicals.

Fungi provide beverages and food

To make wine and beer and bake bread, humans exploit the ability of yeasts to ferment sugars, which produces ethyl alcohol and carbon dioxide. Wine is produced when yeasts ferment fruit sugars; beer is made when yeasts ferment grain, usually barley. During the process of making bread, carbon dioxide produced by the yeast becomes trapped in the dough as bubbles, which cause the dough to rise and give leavened bread its light texture. Both the carbon dioxide and the alcohol produced by the yeast evaporate during baking (•**Figure 21-17**).

The unique flavors and smells of cheeses such as Roquefort and Camembert are produced by the action of the mold *Penicillium* (see Figure 21-17a). *Penicillium roquefortii*, for example, is found in caves near the French village of Roquefort, and only cheeses produced in that area can be called Roquefort cheese.

Many cultures have used fungi to improve the nutrient quality of the diet. For example, in East Asian countries, the mold *Aspergillus tamarii* and other microorganisms (bacteria and yeasts) are used to produce soy sauce by fermentation of soybeans. Soy sauce enhances other foods with more than its special flavor; to the low-protein rice diet it adds vital amino acids from both the soybeans and the fungi themselves.

Among the basidiomycetes there are some 200 species of edible mushrooms and about 70 species of poisonous ones, sometimes called toadstools. Some edible mushrooms are cultivated commercially (recall *Agaricus bisporus* in the chapter introduction). Morels, which superficially resemble mushrooms, and truffles, which produce underground fruiting bodies, are ascomycetes (•**Figure 21-18**). These gourmet delights are now cultivated from mycorrhizae on the roots of tree seedlings, although many professional mushroom pickers still harvest morels and truffles in the wild.

Edible and poisonous mushrooms can look alike and may even belong to the same genus. There is no simple way to distinguish edible mushrooms from poisonous ones; they must be identified by an expert. Some of the most poisonous mushrooms belong to the genus *Amanita*. Toxic species of this genus are appropriately called such names as "destroying angel" (*A. virosa*) (•**Figure 21-19**) and "death cap" (*A. phalloides*). Eating part of a single cap of either of these mushrooms can be fatal.

Ingestion of certain species of mushrooms causes intoxication and hallucinations. Central American Indians still use the sacred mushrooms of the Aztecs,

(a) Wine, beer, bread, and distinctive cheeses are produced in part by fungi. The bluish splotches in the cheese are patches of conidia.

Raymond Tschoepe

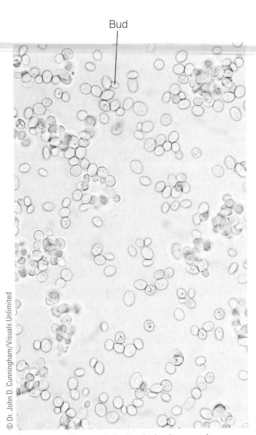

Bud

© Dr. John D. Cunningham/Visuals Unlimited

(b) Yeasts ferment fruits (wine) or grains (beer), producing ethyl alcohol. The same process produces the carbon dioxide bubbles responsible for making bread rise.

FIGURE 21-17 **Fungal food products.**

© Gary Meszaros/Visuals Unlimited

(a) The common morel (*Morchella esculenta*), which grows 6 to 10 cm (2.5 to 4 in) tall, is found throughout North America.

© Dr. John D. Cunningham/Visuals Unlimited

(b) The Oregon white truffle (*Tuber gibbosum*), which is found underground near Douglas firs and possibly oak trees in British Columbia and northern California, is 1 to 5 cm (0.4 to 2 in) wide. People find these subterranean ascocarps with the help of trained dogs or pigs. Here, truffles are shown whole and sectioned.

FIGURE 21-18 **Edible ascomycetes.**

the process of determining the functions of the proteins encoded by its genes. Researchers use this yeast to study such processes as the correlation between cell age and cancer. *Saccharomyces cerevisiae* is also being used to study the mechanism of action of antifungal drugs and resistance to these drugs.

Researchers are investigating certain species of microsporidia for the biological control of pathogens and insect pests. Some of these species are already being used to parasitize insect pests. In some cases, they interfere with reproduction in their insect host. Other microsporidia take over the metabolism and reproduction of the host. Researchers are studying the use of microsporidia in controlling the spread of malaria. A recent study showed that when female *Anopheles* mosquitoes are infected with microsporidia, their blood feeding is decreased. Fungal infection of the mosquito also interferes with the development of the protist that causes malaria.

Fungi produce useful drugs and chemicals. Discovered in 1928 by the British bacteriologist Alexander Fleming, penicillin, produced by the mold *Penicillium notatum*, is still one of the most widely used and most effective antibiotics. Other drugs derived from fungi include the antibiotic griseofulvin (used clinically to inhibit the growth of athlete's foot and ringworm fungi); cyclosporine (used to suppress immune responses in patients receiving organ transplants); statins (used to lower blood cholesterol levels); and fumagillin (which inhibits the formation of new blood vessels and may become an effective anticancer agent).

The ascomycete *Claviceps purpurea* infects the flowers of rye plants and other cereals. It produces a structure called an **ergot** where a grain would normally form (•**Figure 21-20**). When livestock eat ergot-

FIGURE 21-19 The destroying angel.
This extremely poisonous mushroom (*Amanita virosa*), found in grass or near trees throughout North America, is recognizable by the ring of tissue around its stalk and by the underground cup from which the stalk protrudes. About 50 grams (2 ounces) of this mushroom could kill an adult man.

Conocybe and *Psilocybe*, in religious ceremonies, for their hallucinogenic properties. The chemical ingredient psilocybin, chemically related to lysergic acid diethylamide (LSD), is responsible for the trancelike state and colorful visions experienced by those who eat these mushrooms. Ingestion of psychoactive mushrooms is not recommended, because negative reactions vary considerably—from mild indigestion to death. In addition, the possession and use of such mushrooms are illegal in most states.

Fungi are important to modern biology and medicine

The yeast *Saccharomyces cerevisiae* (baker's yeast), an ascomycete, has served as a model eukaryotic cell. It was the first eukaryote whose genome was sequenced; its 6000 genes make up the smallest genome of any eukaryotic model organism. Molecular biologists are in

FIGURE 21-20 *Claviceps purpurea* infecting rye flowers.
This fungus produces a brownish black structure called an ergot (*left*) where a seed would normally form in the grain head. A healthy grain head is shown for comparison (*right*).

contaminated grain or when humans eat bread made from ergot-contaminated rye flour, they may be poisoned by the toxic substances in the ergot. These substances may cause a condition called ergotism, which involves nervous spasms, convulsions, psychotic delusions, and constriction of blood vessels that, if severe enough, can lead to gangrene. During the Middle Ages, ergotism occurred often and was known as St. Anthony's Fire. In the year 994, for example, an epidemic of St. Anthony's Fire caused more than 40,000 deaths. In 1722 the cavalry of Czar Peter the Great was felled by ergotism on the eve of the battle for the conquest of Turkey. This was one of several times in recorded history that a fungus changed the course of human affairs. Some of the compounds produced by ergot are now used clinically in small quantities as drugs to induce labor, to stop uterine bleeding (recall that it constricts blood vessels), to treat high blood pressure, and to relieve one type of migraine headache.

Fungi cause many important plant diseases

Fungi are responsible for about 70 percent of all major crop diseases, including epidemic diseases that spread rapidly and often result in complete crop failure. Damage may be localized in certain tissues or structures of the plant, or the disease may be systemic and spread throughout the entire plant. Fungal infections may cause stunting of plant parts or of the entire plant, may cause growths similar to warts, or may kill the plant.

A plant often becomes infected after hyphae enter through stomata (pores) in the leaf or stem or through wounds in the plant body. Alternatively, the fungus may produce an enzyme that dissolves the plant's cuticle, after which it easily invades the plant's tissues. As the fungal mycelium grows, it may remain mainly between the plant cells or it may penetrate the cells. Parasitic fungi often produce special hyphal branches, **haustoria**, that penetrate host cells and obtain nourishment from the cytoplasm (●**Figure 21-21**).

Ascomycetes cause important plant diseases, including powdery mildews, chestnut blight, Dutch elm disease, apple scab, and brown rot, which attacks cherries, plums, apricots, peaches, and nectarines (●**Figure 21-22a**). Diseases caused by basidiomycetes include smuts and rusts that attack a variety of plants, including soybeans and the cereals (corn, wheat, oats, and other grains) (●**Figure 21-22b**). Some of these parasites, such as wheat rust, have complex life cycles that involve two or more plant species, during which several kinds of spores are produced. For example, the wheat rust must infect an American barberry plant to complete the sexual stage of its life cycle, and the eradication of American barber-

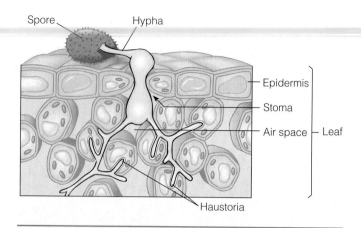

FIGURE 21-21 How a fungus parasitizes a plant.
In this example, the hypha enters the leaf through a stoma. As the hypha grows, it branches extensively through the internal air spaces and penetrates plant cells with specialized hyphal extensions called haustoria.

ries in wheat-growing regions has reduced infection with wheat rust. Wheat rust has not been eliminated by eradication of the barberry, however, because the fungus overwinters on wheat at the southern end of the Grain Belt and forms asexual spores. During the spring, wind blows these spores for hundreds of kilometers, reinfecting northern areas of the United States and Canada.

Some fungi cause animal diseases

Some fungi cause superficial infections in which only the skin, hair, or nails are infected. Ringworm, athlete's foot, and jock itch are examples of superficial fungal infections. Because these fungi infect dead layers of skin that are not fed by capillaries, the immune system cannot launch an effective response.

Many pathogenic fungi are opportunists that cause infections only when the body's immune system is compromised, for example, in individuals with HIV. Cancer patients and organ transplant recipients who are given medications to suppress their immune systems are also at risk. *Candida* is a fungus that inhabits the human mouth and vagina. The immune system and the normal bacteria of these regions usually prevent this yeast from causing infection. However, when the immune system is compromised, *Candida* multiplies, causing thrush, a painful yeast infection of the mouth, throat, and vagina.

Other fungi infect internal tissues and organs and may spread through the body. Histoplasmosis, for example, is an infection of the lungs caused by inhaling the spores of a fungus abundant in bird droppings. Most people in the eastern and midwestern parts of the United States have been exposed to this fungus at one time or another, and an estimated 40 million Americans

(a) Brown rot of nectarines is caused by *Monilinia fruiticola,* an ascomycete.

FIGURE 21-22 **Representative fungal plant pathogens.**

(b) Corn smut on an ear of sweet corn. *Ustilago maydis* is the basidiomycete that causes corn smut.

have had mild infections. Fortunately, the infection is usually confined to the lungs and is of short duration. If the infection spreads into the bloodstream, however, it can be serious and sometimes fatal.

Some fungi produce poisonous compounds collectively called **mycotoxins.** A few species of *Aspergillus,* for example, produce potent mycotoxins called **aflatoxins** that harm the liver and are known cancer-causing agents. Foods on which aflatoxin-producing fungi commonly grow include peanuts, pecans, corn, and other grains. Other foods that may contain traces of aflatoxins include animal products such as milk, eggs, and meat (from animals that consumed feed contaminated by aflatoxin). Avoiding aflatoxin in the diet is impossible, but exposure should be minimized as much as possible. Any human food or animal forage product that has become moldy should be suspected of aflatoxin contamination and discarded.

Fungi contribute to *sick building syndrome,* a situation in which the occupants of a building experience adverse health effects linked to the time spent in that building. Mold-related insurance claims amount to hundreds of millions of dollars each year. When conditions are moist, molds can grow on carpet, leather, cloth, wood, insulation, and food. Mold spores, fragments, and mold products make their way into the air, and people are exposed through inhalation as well as by skin contact. The most common responses to mold exposure are allergic reactions that range from mild to severe illnesses, including hay fever, sinusitis, asthma, and dermatitis.

STUDY OUTLINE

❶ **Describe the distinguishing characteristics of the kingdom Fungi.**
Fungi are eukaryotes that lack chlorophyll and are heterotrophic; they absorb predigested food through the cell wall and plasma membrane. Fungal cell walls contain **chitin,** a nitrogen-containing polysaccharide. The vegetative (nonreproductive) body of most fungi is a **mycelium** that consists of a branched network of

threadlike filaments called **hyphae.** Other fungi—the **yeasts**—are unicellular.

❷ **Explain the fate of a fungal spore that lands on an appropriate food source.**
Fungi reproduce by **spores,** reproductive cells that give rise to individual offspring. Spores may be produced sexually or asexually. When a fungal spore comes into contact with an appropriate food source, the spore germinates and begins to grow a mycelium.

❸ **List distinguishing characteristics and give examples of each of the following fungal groups: chytridiomycetes, zygomycetes, glomeromycetes, ascomycetes, and basidiomycetes.**
Chytrids (chytridiomycetes) are fungi characterized by the production of flagellated cells at some stage in their life history; a parasitic chytrid is partly responsible for declining amphibian populations. **Zygomycetes** are fungi characterized by the production of nonmotile asexual spores and sexual zygospores; the black bread mold is a zygomycete. **Glomeromycetes** are symbionts that form intracellular associations (**arbuscular mycorrhizae**) with the roots of most trees and herbaceous plants. **Ascomycetes** are fungi characterized by the production of nonmotile asexual conidia and sexual ascospores; ascomycetes include yeasts, cup fungi, morels, truffles, and pink and green molds. **Basidiomycetes** are fungi characterized by the production of sexual basidiospores; basidiomycetes include mushrooms, puffballs, rusts, and smuts.

❹ **Explain the ecological significance of fungi as decomposers.**
Most fungi are **decomposers** that break down organic compounds in dead organisms, leaves, garbage, and wastes into simpler materials that can be recycled.

Without this continuous decomposition, essential minerals would be unavailable for use by new generations of organisms, and life would cease.

❺ **Describe the important ecological role of mycorrhizae.**
Mycorrhizae are mutualistic relationships between fungi and the roots of plants. The fungus supplies minerals to the plant, and the plant secretes organic compounds needed by the fungus. **Endomycorrhizal fungi** form mycorrhizae that extend into plant roots, and **ectomycorrhizal fungi** form mycorrhizae consisting of a dense sheath over the root's surface.

❻ **Characterize the unique nature of a lichen.**
A **lichen** is a compound organism consisting of a symbiotic fungus and an alga or cyanobacterium. Lichens have three main growth forms: **crustose, foliose,** and **fruticose.**

❼ **Summarize some of the ways that fungi impact humans economically.**
Mushrooms, morels, and truffles are foods; yeasts are vital in the production of beer, wine, and bread; certain fungi are used to produce cheeses and soy sauce. Fungi are also used to make citric acid and other industrial chemicals.

❽ **Summarize the importance of fungi to biology and medicine.**
Fungi are used to make medications, including penicillin and other antibiotics. Fungi cause many important plant diseases, including wheat rust, Dutch elm disease, and chestnut blight. Fungi are opportunistic pathogens in humans. They cause human diseases, including ringworm, athlete's foot, candidiasis, and histoplasmosis. Some fungi produce **mycotoxins,** such as **aflatoxins,** which can cause liver damage and cancer.

REVIEW QUESTIONS

1. What characteristics distinguish fungi from other organisms?
2. How does the body plan of a yeast differ from that of a mold?
3. What happens when a fungal spore lands on a suitable food source?
4. What is the difference between a hypha and a mycelium? Between a fruiting body and a mycelium?
5. Describe the ecological significance of each of the following: decomposer fungi, lichens, and mycorrhizae.
6. Distinguish between fungi that are decomposers and those that are parasites.
7. Give several characteristic features of each of the following fungal groups: chytrids, zygomycetes, glomeromycetes, ascomycetes, and basidiomycetes.
8. Describe the life cycle of a typical mushroom.
9. Explain the economic importance of yeasts.
10. Briefly describe three important fungal diseases of plants and three fungal diseases of humans.

11. Label the following diagrams. (Refer to Figure 21-13 to check your answers.)

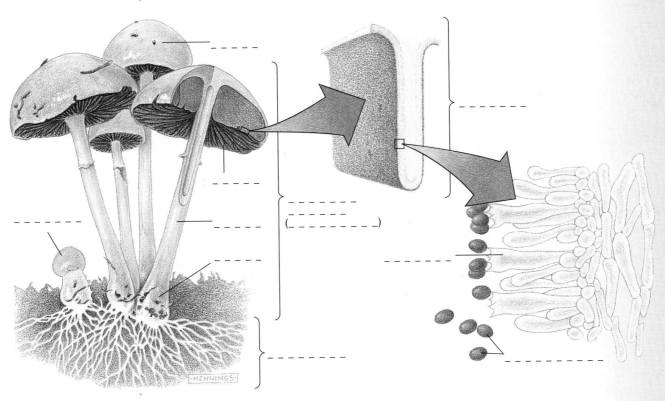

THOUGHT QUESTIONS

1. If you do not see any mushrooms in your lawn, can you conclude that there are no fungi living there? Why or why not?

2. Why are mushrooms and puffballs usually located aboveground, whereas the fungal mycelium is underground?

3. Under what kinds of environmental conditions might it be more advantageous for a fungus to reproduce asexually? Sexually?

4. What measures can you suggest to prevent bread from becoming moldy?

5. Biologists have determined that many mycorrhizal fungi are sensitive to a low pH. What human-caused environmental problem might prove catastrophic for these fungi? How might this problem affect their plant partners?

Visit us on the web at http://www.thomsonedu.com/ biology/berg/ for additional resources such as flashcards, tutorial quizzes, further readings, and web links.

Introduction to the Plant Kingdom: Bryophytes

Several hundred species of peat mosses (*Sphagnum*) grow in bogs, swamps, and other wet places throughout the world. Also known as sphagnum moss and bog moss, peat moss is a "giant" among mosses because it is capable of growing 30 centimeters (1 foot) or more in length; most mosses are less than 5 centimeters (2 inches) long. *Sphagnum* often forms "quaking bogs," which are spongy mats that grow over the surfaces of bodies of water.

Commercially, peat mosses are the most important mosses. One of the distinctive features of *Sphagnum* is the presence of large, empty cells in the whitish green "leaves," which absorb and hold water. These water storage cells are also found in the "stems." The capacity of dried peat moss to absorb water makes it a useful packing material for shipping live plants as well as a good soil conditioner. Peat moss in soil soaks up and retains moisture, and its acidic nature makes it a particularly good soil conditioner for acid-loving plants such as azaleas and rhododendrons.

The acidic and the anaerobic conditions of the bog retard the growth of bacterial and fungal decomposers. As a result, dead peat mosses accumulate as thick deposits—some several meters in depth—under the growing mat of living peat mosses. Over time, the organic material compresses to form *peat*. In some countries such as Ireland, peat is cut into blocks, dried, and burned as fuel. Peat is also burned to produce smoke, which gives the barley malt used to make Scotch whiskeys its characteristic flavor.

In some bogs, the lowest layers of peat are thousands of years old. When the pollen and spores found in various layers are carefully sampled and accurately identified, they provide a history of the changing plant communities in a given area. This information provides indirect evidence of long-term climate changes.

Occasionally the remains of humans have been uncovered during excavations of old peat bogs in Ireland and other parts of Europe. In such cases the clothing and features are remarkably well preserved, because the acidic conditions of bogs inhibit decay. In a peat bog in western Denmark, a man's body, estimated at 2000 years old, was found with a noose about its neck. Archaeologists think that other bodies discovered in peat bogs were sacrificed during religious ceremonies.

Carlyn Iverson

Peat mosses (*Sphagnum*) grow along the edges of ponds, slow-moving streams, and bogs.

Evolution of Plants

About 445 million years ago, planet Earth would have seemed a most inhospitable place because, although life abounded in the oceans, it did not yet exist in abundance on land. The oceans were filled with vast numbers of fishes, mollusks, and crustaceans as well as countless microscopic algae, and the water along rocky coastlines was home to large seaweeds. Occasionally, perhaps, an animal would crawl out of the water onto land, but it never stayed there permanently because there was little to eat on land—not a single blade of grass, no fruit, and no seeds.

During the next 30 million years, a time corresponding roughly to the Silurian period in geologic time, plants appeared in abundance and colonized the land. Where did they come from? Although plants living today exhibit great diversity in size, form, and habitat, botanists hypothesize that they all evolved from a common ancestor that was an ancient green alga (see Figure 20-2). Biologists infer this common ancestor because modern green algae share many biochemical and metabolic traits with modern plants. Both green algae and plants contain the same photosynthetic pigments: chlorophylls *a* and *b* and carotenoids, including **carotenes** (orange pigments) and **xanthophylls** (yellow pigments). Both store excess carbohydrates as starch and have cellulose as a major component of their cell walls. In addition, plants and some green algae share certain details of cell division, including the formation of a cell plate during cytokinesis (see Figure 12-4).

Recent molecular and structural data indicate that land plants probably descended from a group of green algae called **charophytes,** or **stoneworts** (see Figure 20-10f). Molecular comparisons, particularly of DNA and RNA sequences, provide compelling evidence that charophytes are closely allied to land plants. These comparisons among plants and various green algae include sequences of chloroplast DNA, certain nuclear genes, and ribosomal RNA. In each case, the closest match occurs between charophytes and plants, indicating that modern charophytes and plants probably share a recent common ancestor. ∎

(a) The giant sequoias (*Sequoiadendron giganteum*) of California are the world's largest plants, growing to 76 meters (250 feet) in height with a massive trunk many meters in diameter.

(b) Duckweeds (*Spirodela*) are tiny floating aquatic plants about 1 centimeter (3/8 inch) long. If you look closely at the frog's body, you will see minute green dots. Each is a water-meal (*Wolffia*) plant. These tiny floating herbs, about 1.5 millimeters (1/16 inch) long, are the smallest known flowering plants.

FIGURE 22-1 Size variation in plants.

The Plant Kingdom

The plant kingdom comprises hundreds of thousands of species that live in varied habitats, from the frozen arctic tundra to lush tropical rain forests to harsh deserts to moist streambeds. Land plants are complex multicellular organisms that range in size from minute, almost microscopic duckweeds and water-meal to massive giant sequoias, some of the largest organisms that have ever lived (•**Figure 22-1**).

What are some features of plants that have let them colonize so many types of environments? One important difference between plants and algae is that a waxy **cuticle** covers the aerial portions of a plant. Essential for existence on land, the cuticle helps prevent desiccation, or drying out, of plant tissues by evaporation. Plants are rooted in the ground and, unlike animals, cannot move to wetter areas during dry periods; therefore, a cuticle is critical to a land plant's survival.

Plants obtain the carbon they require for photosynthesis from the atmosphere as carbon dioxide (CO_2). To be fixed into organic molecules such as sugar, CO_2 must first diffuse into the chloroplasts that are inside green plant cells. Because a waxy cuticle covers the external surfaces of stems and leaves, however, gas exchange through the cuticle between the atmosphere and the interior of cells is negligible. Tiny pores called **stomata** (sing., *stoma*), which dot the surfaces of leaves and stems, facilitate gas exchange; algae lack stomata.

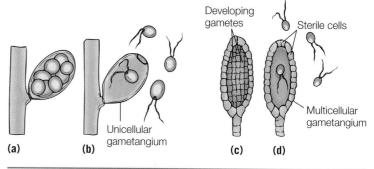

(a) **(b)** Unicellular gametangium **(c)** **(d)**

Developing gametes — Sterile cells — Multicellular gametangium

FIGURE 22-2 The reproductive structures of plants and algae differ. (a,b) In algae, gametangia are generally unicellular. When the gametes are released, only the wall of the original cell remains. **(c,d)** In plants, gametangia are multicellular, but only the inner cells become gametes. A protective layer of sterile (nonreproductive) cells surrounds the gametangium.

The sex organs, or **gametangia** (sing., *gametangium*), of most plants are multicellular, whereas the gametangia of algae are unicellular (•**Figure 22-2**). Each gametangium has a layer of sterile (nonreproductive) cells that surrounds the gametes (eggs or sperm cells). This outer jacket of sterile cells protects the delicate gametes from desiccation. In plants, the fertilized egg develops into a multicellular **embryo** (young plant) within the female gametangium. Thus, the embryo is protected during its development. In algae, the fertilized egg develops away from its gametangium. In some algae the gametes are released into the water before fertilization, whereas in others the fertilized egg is released.

The plant life cycle alternates between haploid and diploid generations

Plants have clearly defined **alternation of generations** in which they spend part of their lives in a multicellular haploid stage and part in a multicellular diploid stage (•**Figure 22-3**).[1] The haploid portion of the life cycle is

[1] For convenience we limit our discussion to plants that are not polyploid, although polyploidy is common in the plant kingdom. We therefore use the terms diploid and 2n (and haploid and n) interchangeably, although these terms are not actually synonymous.

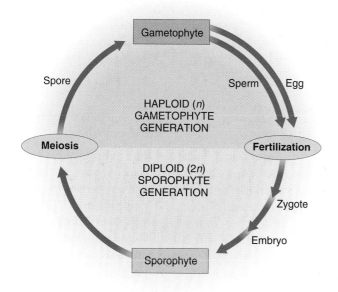

FIGURE 22-3 The basic plant life cycle. Plants undergo an alternation of generations, spending part of the cycle in a haploid gametophyte stage and part in a diploid sporophyte stage. Depending on the plant group, the haploid or the diploid stage may be greatly reduced.

CUTICLE A noncellular, waxy covering over the epidermis of the aerial plant parts that reduces water loss.

STOMA A small pore in the plant epidermis that provides for gas exchange for photosynthesis.

ALTERNATION OF GENERATIONS A type of life cycle characteristic of plants and a few algae and fungi in which they spend part of their life in a multicellular *n* gametophyte stage and part in a multicellular *2n* sporophyte stage.

called the **gametophyte generation** because it gives rise to haploid gametes by mitosis. When two gametes fuse, the diploid portion of the life cycle, called the **sporophyte generation,** begins. The sporophyte generation produces haploid **spores** by the process of meiosis; these spores represent the first stage in the gametophyte generation.

Let us examine alternation of generations more closely. The haploid gametophyte plant produces male gametangia, called **antheridia** (sing., *antheridium*), in which numerous sperm cells form (•**Figure 22-4a**). The gametophyte plant also forms female gametangia, known as **archegonia** (sing., *archegonium*), each bearing a single egg (•**Figure 22-4b**). Sperm cells reach the archegonium in a variety of ways, such as by water, animals, and wind. They swim down the neck of the archegonium, and one sperm cell fuses with the egg. This process, known as **fertilization,** results in a fertilized egg, or **zygote.**

The diploid zygote is the first stage in the sporophyte generation. The zygote divides by mitosis and develops into a multicellular embryo, the young sporophyte plant. Embryo development takes place within the archegonium; thus, the embryo is supported, nurtured, and protected as it develops. Eventually, the embryo grows into a mature sporophyte plant. The mature sporophyte has special spore-producing cells that are capable of dividing by meiosis. Each of these **spore mother cells** divides by meiosis to form four haploid spores.

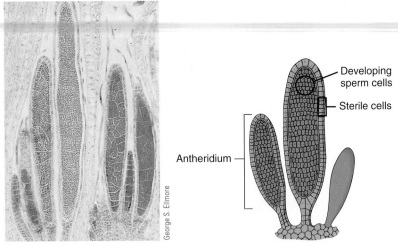

(a) Each antheridium, the male gametangium, produces numerous sperm cells.

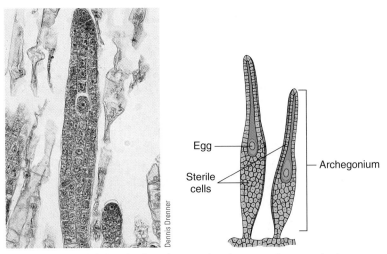

(b) Each archegonium, the female gametangium, produces a single egg.

FIGURE 22-4 **Plant gametangia.**
Shown are generalized moss gametangia.

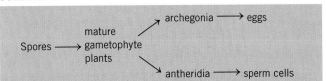

Fertilization of egg by sperm cell ⟶ zygote ⟶ embryo ⟶ mature sporophyte plant ⟶ spore mother cells ⟶ meiosis ⟶ spores

All plants produce spores by meiosis, in contrast with algae and fungi, which may produce spores by meiosis or mitosis. The spores represent the first stage in the gametophyte generation. Each spore divides by mitosis to form a multicellular gametophyte, and the cycle continues.

Spores ⟶ mature gametophyte plants ⟶ archegonia ⟶ eggs
⟶ antheridia ⟶ sperm cells

Thus, plants alternate between a haploid gametophyte generation and a diploid sporophyte generation.

Four major groups of plants evolved

The plant kingdom consists of four main groups: bryophytes, seedless vascular plants, and two groups of seeded vascular plants: gymnosperms and angiosperms (flowering plants) (•**Table 22-1** and •**Figure 22-5**). The mosses and other *bryophytes* are small **nonvascular plants**—they lack a specialized vascular, or conducting, system to transport dissolved nutrients, water, and essential inorganic minerals throughout the plant body. In the absence of such a system, bryophytes rely on diffusion and osmosis to obtain required materials. This reliance means that bryophytes are restricted in size; if they were much larger, some of their cells could not obtain enough necessary materials. Bryophytes do not form seeds, the reproductive structures that will be discussed

| TABLE 22-1 | **The Plant Kingdom** |

NONVASCULAR PLANTS

I. Nonvascular plants with a dominant gametophyte generation (bryophytes)
  Phylum Bryophyta (mosses)
  Phylum Hepatophyta (liverworts)
  Phylum Anthocerophyta (hornworts)

VASCULAR PLANTS

II. Vascular plants with a dominant sporophyte generation
  A. Seedless plants
      Phylum Lycopodiophyta (club mosses)
      Phylum Pteridophyta (ferns, whisk ferns, and horsetails)

  B. Seed plants
    1. Plants with naked seeds (gymnosperms)
        Phylum Coniferophyta (conifers)
        Phylum Cycadophyta (cycads)
        Phylum Ginkgophyta (ginkgoes)
        Phylum Gnetophyta (gnetophytes)

    2. Plants with seeds enclosed within a fruit
        Phylum Anthophyta (angiosperms or flowering plants)
          Class Eudicotyledones (eudicots)
          Class Monocotyledones (monocots)

in Chapter 24; bryophytes reproduce and disperse primarily via haploid spores.

The other three groups of plants—seedless vascular plants, gymnosperms, and flowering plants—have vascular tissues and thus are known as **vascular plants.** The two vascular tissues are **xylem** for conducting water and dissolved minerals and **phloem** for conducting dissolved organic molecules such as sugar. A key step in the evolution of vascular plants was the ability to produce **lignin,** a strengthening polymer found in the walls of cells that function for support and conduction. The stiffening property of lignin enabled plants to grow tall (which let them maximize light interception). The successful occupation of the land by plants, in turn, made the evolution of terrestrial (land-dwelling) animals possible by providing them with both habitat and food (see

Plants and the Environment: The Environmental Challenges of Living on Land on page 436).

EVOLUTION LINK Club mosses and ferns (which include whisk ferns and horsetails) are seedless vascular plants that, like the bryophytes, reproduce and disperse primarily via spores. Seedless vascular plants arose and diversified during the Silurian and Devonian periods of the Paleozoic era, between 444 million years ago and

SPORE A reproductive cell that gives rise to individual offspring in plants, fungi, and certain algae and protozoa.

ANTHERIDIUM A multicellular male gametangium that produces sperm cells.

ARCHEGONIUM A multicellular female gametangium that produces an egg.

359 million years ago. Club mosses and ferns extend back more than 420 million years and were of considerable importance as Earth's dominant plants in past ages. Fossil evidence indicates that many species of these plants were the size of immense trees. Most living representatives of club mosses and ferns are small.

The gymnosperms are vascular plants that reproduce by forming seeds. Gymnosperms produce seeds borne exposed (unprotected) on a stem or in a cone. Plants with seeds as their primary means of reproduction and dispersal first appeared about 359 million years ago, at the end of the Devonian period. These early seed plants diversified into varied species of gymnosperms.

The most recent plant group to appear is the flowering plants, or angiosperms, which arose during the early Cretaceous period of the Mesozoic era, about 130 million years ago. Like gymnosperms, flowering plants reproduce by forming seeds. Flowering plants, however, produce seeds enclosed within a fruit. ■

Use care when comparing one plant group to another

In comparing groups of plants, it is convenient to use terms such as *lower* and *higher*, *simple* and *complex*, and *primitive* and *advanced*. Do not take such terms to imply, however, that plants labeled higher, complex, or advanced are better or more nearly perfect than others. Rather, these labels are used in a comparative sense to describe hypothesized evolutionary relationships. For example, the terms *higher* and *lower* usually refer to the level at which a particular group diverged from a main line of evolution. It is customary, for instance, to refer to bryophytes as lower plants because they are thought to have originated first—that is, near the base of the cladogram depicting evolution of the plant kingdom (see Figure 22-5). However, bryophytes are not primitive in all structural or physiological characteristics. Each species has become highly specialized to its own lifestyle.

MOSS A member of a phylum of spore-producing nonvascular plants in which the dominant *n* gametophyte alternates with a 2*n* sporophyte that remains attached to the gametophyte.

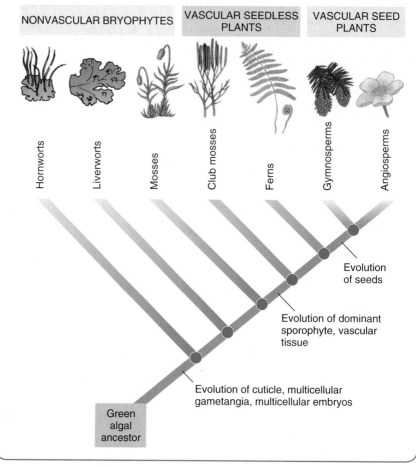

FIGURE IN FOCUS

This cladogram shows hypothetical evolutionary relationships among living plants, based on current evidence. The four major groups of plants are bryophytes, seedless vascular plants, and two groups of seed plants—gymnosperms and angiosperms.

NONVASCULAR BRYOPHYTES

VASCULAR SEEDLESS PLANTS

VASCULAR SEED PLANTS

Hornworts

Liverworts

Mosses

Club mosses

Ferns

Gymnosperms

Angiosperms

Evolution of seeds

Evolution of dominant sporophyte, vascular tissue

Evolution of cuticle, multicellular gametangia, multicellular embryos

Green algal ancestor

FIGURE 22-5 Plant evolution.
Cladograms such as this one represent an emerging consensus that is open to change as new discoveries are made. Although the arrangement of nonvascular, seedless vascular, and seed plant groupings is widely recognized, the order in which the hornworts, liverworts, and mosses evolved is not yet resolved.

Bryophytes

The **bryophytes** (from the Greek words meaning "moss plant") consist of about 16,000 species of mosses, liverworts, and hornworts; bryophytes are the only living nonvascular plants (•**Table 22-2**). Because they have no means for extensive internal transport of water, sugar, and essential minerals, bryophytes are typically small. Although some bryophytes have a cuticle, others do not and instead absorb water directly through the surfaces of their leafy shoots. They generally require a moist environment for active growth and reproduction, but some bryophytes tolerate dry areas.

The bryophytes are divided into three distinctive phyla: mosses (phylum Bryophyta), liverworts (phy-

TABLE 22-2	A Comparison of Major Groups of Bryophytes		
PLANT GROUP	**DOMINANT STAGE OF LIFE CYCLE**	**REPRESENTATIVE GENERA**	
Mosses (phylum Bryophyta)	Gametophyte: leafy plant	*Polytrichum, Sphagnum, Physcomitrella*	
Liverworts (phylum Hepatophyta)	Gametophyte: thalloid or leafy plant	*Marchantia, Bazzania*	
Hornworts (phylum Anthocerophyta)	Gametophyte: thalloid plant	*Anthoceros*	

lum Hepatophyta), and hornworts (phylum Anthocerophyta). These three groups differ in many ways and may or may not be closely related. They are usually studied together because they lack vascular tissues and have similar life cycles.

Moss gametophytes are differentiated into "leaves" and "stems"

Mosses (phylum Bryophyta), which include about 9900 species, usually live in dense colonies or beds on moist soil, rocks, or tree bark (•**Figure 22-6**). Each individual plant has tiny, hairlike, absorptive structures called **rhizoids** and an upright, stemlike structure that bears leaflike blades, each normally consisting of a single layer of undifferentiated cells except at the midrib. Because mosses lack vascular tissues, they do not have true roots, stems, or leaves; the moss structures are not homologous to roots, stems, or leaves in vascular plants. Some moss species have water-conducting cells and sugar-conducting cells, although these cells are not as specialized or as effective as the conducting cells of vascular plants.

Alternation of generations is clear in the life cycle of mosses (•**Figure 22-7**). The moss gametophyte is a leafy, green, usually perennial plant that often bears its gametangia at the top of the plant (see Figure 22-7 ❶). Many moss species have separate sexes: male plants that bear antheridia and female plants that bear archegonia.

(a) Mosses cover several rocks in a dry creek bed.

(b) A close-up of moss gametophytes. Mosses grow in dense clusters.

Capsule

Seta

Foot

(c) The sporophytes, each consisting of a foot, seta, and capsule, grow out of the top of the gametophytes following sexual reproduction.

FIGURE 22-6 Mosses.

PLANTS AND THE ENVIRONMENT | The Environmental Challenges of Living on Land

Life began in the oceans, but many organisms have since adapted to terrestrial life in a "sea" of air. Every organism living on land has to meet the same environmental challenges: obtaining enough water; preventing excessive water loss; getting enough energy; and in temperate and polar regions, tolerating widely varying temperatures. How those challenges are met varies from one organism to another and in large part explains the diversity of life encountered on land today.

Let us compare how plants and vertebrates (animals with backbones) meet several terrestrial challenges.

Obtaining Enough Water Animals are motile and walk, slither, fly, run, or crawl to water sources; this mobility requires not only the ability to move (with the use of skeletal and muscular systems) but the ability to sense the water's presence (with the nervous system).

Plants adapted differently to the challenge of obtaining water: they have roots that both anchor them in the soil and absorb water and dissolved minerals. Some plants have shallow, spreading root systems that obtain moisture from precipitation as it drains into the soil. Other plants have root systems that penetrate deeply into the soil to reach groundwater.

Preventing Excessive Water Loss The outer layers of terrestrial vertebrates and plants protect the moist inner tissues from drying out. Vertebrates that are adapted to living on land have accumulations of a water-insoluble protein called keratin in their epithelial cells.

Keratin is particularly thick in reptiles, where it forms scales that greatly reduce water loss by evaporation.

Plants have a water-insoluble waxy coating, the cuticle, over their epidermal cells. Plants that are adapted to moister habitats (such as water lilies) may have a thin cuticle, whereas those adapted to drier environments (such as cacti) often have a thick, crusty cuticle. Many desert plants also have a reduced surface area, particularly of leaves, which minimizes water loss.

Getting Enough Energy Animals are heterotrophs and eat plants or other animals that eat plants. This means that animals could not become permanent colonizers of land until plants were established.

Almost all plants are autotrophs and must absorb enough sunlight for effective photosynthesis. Plants that are relatively tall avoid the shade of competitors; the evolution of lignin-reinforced supporting cells such as fibers facilitated such growth; plants, of course, lack skeletal systems for support. Other plants are adapted to lower light intensities and grow in the shade of larger plants, albeit more slowly.

Tolerating High Temperatures Air temperature on land varies greatly, particularly in temperate regions. Many animals avoid the heat by resting in the shade or by burrowing in the ground during the day; these animals become active at night when it is cooler. Mammalian skin has sweat glands that secrete a watery solution (sweat) when the air temperature is high. The

evaporation of sweat off the skin cools the body.

Plants also rely on evaporative cooling, although they do not produce sweat. Plants lose large quantities of water through their stomata in a process known as transpiration. As this water evaporates, it carries heat with it.

Tolerating Low Temperatures Vertebrates deal with the low temperatures of winter in several ways. Mammalian hair and bird feathers provide insulation and enable the body to conserve heat by trapping air next to the skin's surface. In addition, some animals avoid cold weather by migrating to warmer climates or by passing the winter in a dormant state called hibernation.

Many plants (biennials and perennials) also overwinter in a dormant state. The aerial parts of some plants die during the winter, but the underground parts remain alive and dormant; the following spring, these underground parts resume metabolic activity and develop new aerial shoots. Likewise, the seeds of certain plants remain dormant and do not germinate until they are exposed to near-freezing temperatures for a certain period. (What would happen to these plants if the seeds germinated immediately?) Many trees are deciduous and shed their leaves, remaining bare for the duration of their dormancy. By shedding its leaves, a plant reduces water loss during the cold winter months, when obtaining water from the soil is difficult, because roots cannot absorb water from ground that is cold or frozen. Thus, the shedding of leaves is actually an adaptation to the "dryness" of winter.

Other mosses produce antheridia and archegonia on the same plant.

Minute, flagellated sperm cells are released from antheridia during rainy weather, and splashing raindrops transport the sperm cells to archegonia. A raindrop lands on the top of a male gametophyte plant, and

sperm cells are discharged into it from the antheridia. When another raindrop lands on the male plant, the first sperm-laden droplet may splash into the air and onto the top of a nearby female plant. Alternatively, insects may touch the sperm-laden fluid and inadvertently carry it for a considerable distance. Once in a film of water on

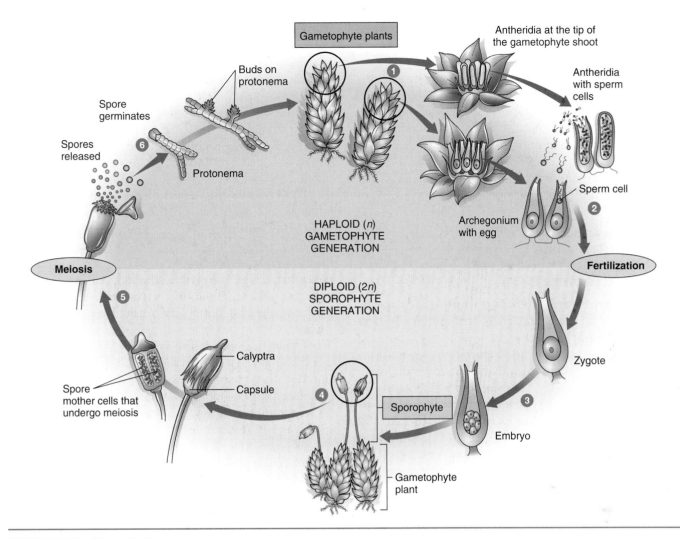

FIGURE 22-7 The life cycle of mosses.
The dominant generation in the moss life cycle is the gametophyte, represented by the leafy green plants. The sporophyte generation grows out of the gametophyte. Mosses require water as a transport medium for sperm cells during fertilization. See text for a detailed description.

the female moss plant, a sperm cell swims down the neck of the archegonium, which secretes chemicals to attract and guide the sperm cells ❷. The requirement for water as a transport medium for sperm cells is considered a primitive characteristic, inherited from green algal ancestors, that mosses have retained.

Fertilization occurs when one of the sperm cells fuses with the egg within the archegonium. The diploid zygote, formed as a result of fertilization, grows by a series of mitotic divisions into a multicellular embryo ❸ that develops into a moss sporophyte. This sporophyte grows out of the top of the female gametophyte and remains attached and nutritionally dependent on the gametophyte throughout its existence (see Figure 22-6c). The sporophyte is initially green and photosynthetic but turns golden brown at maturity. A moss sporophyte plant is composed of three parts: a **foot,** which anchors the sporophyte to the gametophyte and absorbs minerals and nutrients from it; a **seta,** or stalk; and a **capsule,**

the sporangium (spore case) that contains spore mother cells (see Figure 22-7 ❹). A caplike structure, the *calyptra,* which is derived from the archegonium, covers the capsule of some species.

The spore mother cells undergo meiosis to form haploid spores ❺. When the spores are mature, the capsule bursts open from the pressure that has developed as it dried out, and it releases the spores. Wind or rain then carries the spores to other places. If a moss spore lands in a favorable spot, it germinates and grows into a branching, filamentous thread of green cells called a **protonema** ❻. The protonema, which superficially resembles a filamentous green alga, forms buds, each of

CAPSULE The portion of the bryophyte sporophyte in which spores are produced.

PROTONEMA In mosses, a filament of *n* cells that grows from a spore and develops into leafy moss gametophytes.

which grows into a leafy green gametophyte plant; and the life cycle continues.

Botanists consider the haploid gametophyte generation the dominant generation in mosses because it is larger, more persistent, and nutritionally independent of the sporophyte. In contrast, the moss sporophyte, which is relatively short-lived, is attached to and nutritionally dependent on the gametophyte plant.

The name "moss" is often misused to refer to plants that have no biological relationship to the mosses. For example, reindeer "moss" is a lichen that is a dominant form of vegetation in the arctic tundra, Spanish "moss" is a flowering plant, and club "moss" is a relative of ferns.

Mosses are a significant part of their environment

Although mosses are typically inconspicuous in the environment, they are important ecologically. By frequently colonizing rock that was previously colonized by lichens, mosses play an important role in forming soil. Mosses, which form mats that cover the rock, eventually die, forming a thin soil in which grasses and other plants can grow. Because they grow in dense colonies, mosses hold the soil in place and help prevent soil erosion. At the same time, they retain moisture that they and other organisms need. Waxwings and other birds use moss, along with twigs and grass, as nesting material.

Liverworts are either thalloid or leafy

Liverworts (phylum Hepatophyta) consist of about 6000 species of nonvascular plants with a dominant ga-

metophyte generation. Liverworts are small, generally inconspicuous plants that grow on moist soil, rocks, old stumps, and tree bark. They are common, particularly in moist coniferous forests. Unlike mosses, hornworts, and other plants, liverworts lack stomata (although some liverworts have surface pores thought to be analogous to stomata).

The liverwort body form is often a flattened, lobed structure called a **thallus** (pl., *thalli*) that is not differentiated into leaves, stems, or roots. In fact, liverworts are so named because the lobes of their thalli resemble the lobes of the liver; *wort* is derived from the Old English word *wyrt*, meaning "plant." The common liverwort, *Marchantia polymorpha,* is thalloid (•**Figure 22-8a**). On the underside of the liverwort thallus are hairlike rhizoids that anchor the plant to the soil. Although *Marchantia* and other thalloid liverworts are distinctive in appearance, they are not typical of the liverwort group as a whole.

Other liverworts, known as *leafy liverworts,* have a branching, leafy form rather than a lobed thallus and superficially resemble mosses. Leafy liverworts have prostrate, leafy "shoots" and rhizoids (•**Figure 22-8b**). As in the mosses, leafy liverwort "leaves" consist of a single layer of undifferentiated cells.

The life cycle of liverworts is basically the same as that of mosses, although some of the structures look quite different. The haploid gametophyte plant is considered the dominant generation, and the sporophyte is relatively short-lived. Sexual reproduction of liverworts involves production of archegonia and antheridia on the haploid gametophyte. The common liverwort (*Marchantia*) has archegonia and antheridia on separate plants

(a) The common liverwort (*Marchantia polymorpha*) gametophyte consists of flattened, ribbonlike lobes. The thallus produces gemmae cups containing reproductive bodies, called gemmae.

(b) This leafy liverwort (*Bazzania trilobata*) is often found on rotting logs or tree stumps. Its generic name indicates that the "leaves" are divided into three lobes (which you need a hand lens to discern).

FIGURE 22-8 Liverworts.

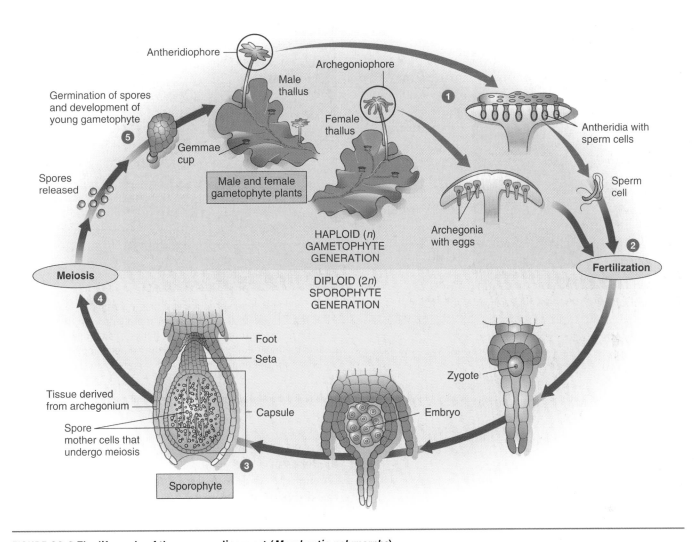

FIGURE 22-9 The life cycle of the common liverwort (*Marchantia polymorpha*).
The dominant generation is the gametophyte, represented by the separate ribbonlike male and female thalli. The stalked structures are the antheridiophores, with antheridia that produce sperm cells, and the archegoniophores, with archegonia that each bear an egg cell. See text for a detailed description.

(•Figure 22-9 ❶). In some liverworts, including *Marchantia*, the gametangia are borne on stalked structures called *antheridiophores*, which bear antheridia, and *archegoniophores*, which bear archegonia (•Figure 22-10a and **b**).

Like mosses, liverworts require water as a transport medium for flagellated sperm cells. Splashing raindrops transport sperm cells to the archegonia, where fertilization takes place (see Figure 22-9 ❷). The resulting zygote develops into a multicellular embryo that becomes a mature sporophyte ❸. The liverwort sporophyte, which is usually somewhat spherical, is attached to and nutritionally dependent on the gametophyte plant, as in mosses (•Figure 22-10c). It is only a few millimeters long and, like moss sporophytes, consists of a foot, a seta, and a capsule. Spore mother cells in the capsule of the sporophyte undergo meiosis, producing haploid

spores (see Figure 22-9 ❹). Each spore has the potential to develop into a green gametophyte ❺, and the cycle continues.

Some liverworts reproduce asexually by forming small balls of tissue called **gemmae** (sing., *gemma*), which are borne in a saucer-shaped structure, the **gemmae cup**, directly on the liverwort thallus (see Figure 22-8a). Splashing raindrops and small animals help

LIVERWORT A member of a phylum of spore-producing, nonvascular, thalloid or leafy plants with a life cycle similar to that of mosses.

THALLUS A body that lacks roots, stems, or leaves.

GEMMA A small body of tissue that becomes detached from a parent liverwort and is capable of developing into a new organism.

© David Sieren/Visuals Unlimited

Antheridiophore

Gametophyte thallus

(a) Antheridiophores produce sperm-bearing antheridia.

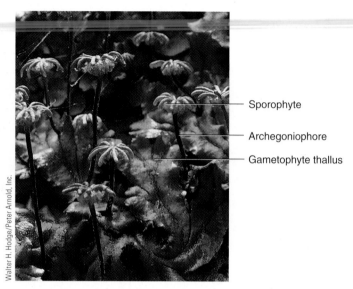

Walter H. Hodge/Peter Arnold, Inc.

Sporophyte

Archegoniophore

Gametophyte thallus

(b) Archegoniophores produce egg-bearing archegonia. After fertilization of the egg within an archegonium, the sporophyte generation develops. It hangs upside down under the fingerlike projections of the archegoniophore and resembles a miniature coconut hanging on a palm tree.

FIGURE 22-10 *Marchantia* gametangia and sporophytes.

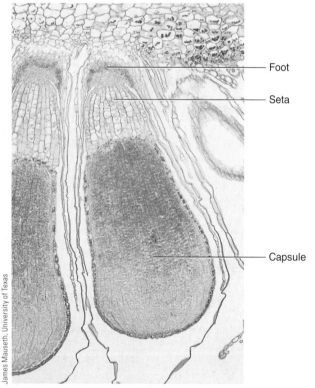

James Mauseth, University of Texas

Foot

Seta

Capsule

(c) The liverwort sporophyte, which is always attached to and dependent on the gametophyte plant, consists of a foot, seta, and capsule. Meiosis occurs in the capsule, and produces haploid spores.

disperse gemmae. When a gemma lands in a suitable place, it grows into a new liverwort thallus. Liverworts may also reproduce asexually by thallus branching and growth. The individual thallus lobes elongate, and each becomes a separate plant when the older part of the thallus that originally connected the individual lobes dies. Both gemmae and thallus branching are asexual because they do not involve the fusion of gametes or meiosis.

Liverwort thalli superficially resemble the lobes of a liver

During the Middle Ages, many Europeans believed in the **doctrine of signatures,** which held that each disease or ailment could be treated by a specific plant and that each plant's "signature," or symbolic physical feature, provided a clue to its use. Because their form suggested the lobes of an animal liver, thalloid liverworts were thought to have medicinal value for the treatment of liver ailments. Modern medical research has not supported the doctrine of signatures, and liverworts do not help cure any diseases of the liver.

Hornworts are inconspicuous thalloid plants

Hornworts (phylum Anthocerophyta) are a small group of about 100 species of bryophytes whose gametophytes superficially resemble those of the thalloid liverworts (•**Figure 22-11**). Hornworts live in disturbed habitats, such as fallow fields and roadsides.

Hornworts may or may not be closely related to other bryophytes. For example, their cell structure, particularly the presence of a single large chloroplast in each cell, resembles that of certain algal cells more than that of plant cells. In contrast, mosses, liverworts, and other plants have many disc-shaped chloroplasts per cell.

In the common hornwort (*Anthoceros natans*), the leafy green thallus, which is 1 to 2 centimeters (less than 1 inch) in diameter, is the gametophyte generation. Ar-

Robert A. Ross

(a) The gametophyte with mature sporophytes of the common hornwort (*Anthoceros natans*).

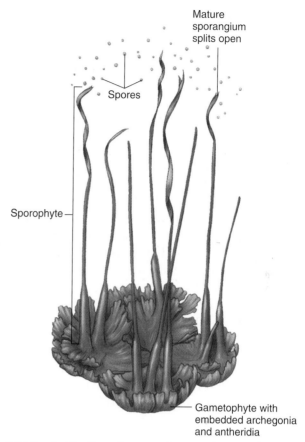

Mature sporangium splits open

Spores

Sporophyte

Gametophyte with embedded archegonia and antheridia

(b) After fertilization, the sporophytes project up out of the gametophyte thallus, forming "horns."

FIGURE 22-11 Hornwort.

chegonia and antheridia are embedded in the thallus rather than on archegoniophores and antheridiophores. After fertilization and development, the needlelike sporophyte projects out of the gametophyte thallus, forming a spike, or "horn"—hence the name "hornwort." A single gametophyte often produces multiple sporophytes. Meiosis occurs, during which spores form within each sporangium. The sporangium splits open from the top to release the spores; each spore has the potential to give rise to a new gametophyte thallus. A unique feature of hornworts is that the sporophytes, unlike those of mosses and liverworts, continue to grow from their bases for the remainder of the gametophyte's life.

PROCESS OF SCIENCE Bryophytes are used for experimental studies

Botanists use certain bryophytes as experimental models to study many fundamental aspects of plant biology, including genetics, growth and development, plant ecology, plant hormones, and *photoperiodism,* which is plant responses to varying periods of night and day length. The moss *Physcomitrella patens* is a particularly important research organism for studying plant evolution, because its genes and structural and physiological features can be compared to those of flowering plants. In this regard, *Physcomitrella* is the nonvascular plant equivalent of *Arabidopsis thaliana,* which is an important model organism for studies of flowering plant inheritance, development, and evolution (see Figure 9-1). As experimental organisms, *Physcomitrella* and other bryophytes are easy to grow on artificial media and do not require much space because they are so small (•**Figure 22-12**). ▪

HORNWORT A member of a phylum of spore-producing, nonvascular thalloid plants with a life cycle similar to that of mosses.

Evolution of Bryophytes

EVOLUTION LINK

Plants are a **monophyletic group**—that is, all plants are thought to have evolved from a common ancestral green alga. Fossil evidence indicates the bryophytes are ancient plants, probably the first group of plants to arise from the most recent common plant ancestor. The fossil record of ancient bryophytes is incomplete, consisting mostly of spores and small tissue fragments, and can be interpreted in different ways. As a result, it does not provide a definite answer on bryophyte evolution.

The oldest known recognizable plant spores are dated at about 470 million years old, whereas the oldest known plant fossils are dated at about 425 million years old. These fossils resemble modern liverworts in many respects, and their spores are very similar to those in the 470-million-year-old rocks. This evidence suggests that liverwort-like plants may have been the earliest plants to colonize land.

Structural and molecular evidence, however, supports the hypothesis that hornworts may be the most ancient group of plants alive today. A 1998 cladistic analysis using structural data was the first to strongly support this hypothesis, which was reinforced in 2000 by DNA evidence. The study of nuclear, chloroplast, and mitochondrial genes from numerous plant species,

Courtesy of David J. Cove, University of Leeds

FIGURE 22-12 Mosses as research organisms.
Researchers inoculated this petri dish with offspring from a genetic cross in the moss *Physcomitrella patens*. They will test the cultures for their phenotypes with respect to vitamin requirements. (The culture medium in which the mosses are currently growing is supplemented with all the vitamins required by the parental strains.)

from bryophytes to flowering plants, suggests an early plant ancestor gave rise to two lineages, one from which the hornworts descended and the other from which liverworts, mosses, and all other plants descended. ∎

STUDY OUTLINE

❶ **Discuss some environmental challenges of living on land, and describe how several plant adaptations meet these challenges.**
The colonization of land by plants required the evolution of structural, physiological, and reproductive adaptations. Plants produce gametes in multicellular **gametangia** that contain a protective layer of sterile cells. A **cuticle** is a noncellular, waxy covering over the epidermis of the aerial plant parts that reduces water loss. **Stomata** are small pores in the plant epidermis that provide for gas exchange for photosynthesis.

❷ **Name the algal group from which plants are hypothesized to have descended, and describe supporting evidence.**
Plants probably evolved from green algal ancestors. Plants and green algae have similar biochemical characteristics: the same pigments (chlorophylls *a* and *b*, **carotenes**, and **xanthophylls**), cell-wall components (cellulose), and carbohydrate storage material (starch). Plants and green algae share similarities in certain fundamental processes, such as cell division. Recent molecular and structural data indicate that land plants

probably descended from a group of green algae called **charophytes,** or **stoneworts.**

❸ **Explain what is meant by alternation of generations, and diagram a generalized plant life cycle.**
Alternation of generations is a type of life cycle in which plants spend part of their life in a multicellular *n* **gametophyte generation** and part in a multicellular 2*n* **sporophyte generation.** The gametophyte generation produces haploid gametes by mitosis. An **antheridium** is a multicellular male gametangium that produces sperm cells; an **archegonium** is a multicellular female gametangium that produces an egg. Gametes fuse to form a diploid **zygote** in a process known as **fertilization.** The first stage in the sporophyte generation is the zygote, which develops into an embryo protected and nourished by the gametophyte plant. The mature sporophyte plant has **spore mother cells** that undergo meiosis to produce haploid spores that are the first stage in the gametophyte generation. A **spore** is a reproductive cell that gives rise to an individual offspring.

❹ **Summarize the features that distinguish bryophytes from other plants.**

Bryophytes are small, fairly simple plants. Unlike other plants, bryophytes are nonvascular. The gametophyte is the dominant generation in the bryophyte life cycle; that is, it grows independently of the sporophyte and is usually perennial.

❺ **Name and briefly describe the three phyla of bryophytes.** **Mosses** are a phylum of spore-producing, nonvascular plants in which the dominant *n* gametophyte alternates with a 2*n* sporophyte that remains attached to the gametophyte. Leafy moss gametophytes develop from a **protonema**, a filament of *n* cells that grows from a spore. A moss sporophyte consists of a **capsule**, the portion of the bryophyte sporophyte in which spores are produced; a **seta**; and a **foot**. **Liverworts** are a phylum of spore-producing, nonvascular, thalloid or leafy plants with a life cycle similar to that of mosses. A **thallus** is a body that lacks roots, stems, or leaves. **Gemmae** are small bodies of tissue that become detached from a parent liverwort and are capable of developing into a new organism. **Hornworts** are a phylum of spore-producing, nonvascular thalloid plants with a life cycle similar to that of mosses. Hornwort gametophytes are thalloid, and their sporophytes form hornlike projections out of the gametophyte thallus.

❻ **Describe the ecological significance of the mosses.** By colonizing rock that was previously colonized by lichens, mosses play a role in forming a thin soil in which grasses and other plants can grow. Because they grow in dense colonies, mosses hold the soil in place and help prevent soil erosion.

REVIEW QUESTIONS

1. What are the most important environmental challenges that plants experience on land, and what adaptations do plants have to meet these challenges?
2. What features distinguish plants from algae?
3. Which group of algae is probably the ancestor of plants? Cite evidence for your answer.
4. What features distinguish bryophytes from other plants?
5. Define *alternation of generations,* and distinguish between gametophyte and sporophyte generations.
6. What limits the size of bryophytes? Why do most bryophytes live in moist environments?
7. Which of the following are parts of the gametophyte generation in mosses: antheridia, zygote, embryo, capsule, archegonia, sperm cells, egg cell, spores, and protonema?
8. List and briefly describe the three phyla of bryophytes.
9. How are mosses, liverworts, and hornworts similar? How is each group distinctive?
10. How are mosses ecologically important?

11. Label the following diagram. (Refer to Figure 22-6c to check your answers.)

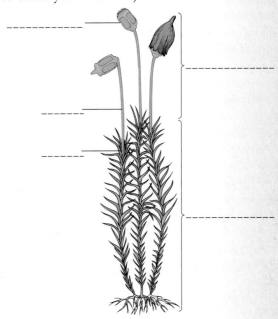

THOUGHT QUESTIONS

1. Which organisms probably colonized the land first, plants or animals?
2. Fossils of trees are easier to find than fossils of bryophytes. Offer a possible explanation.
3. Suppose you are given an unknown plant. How could you determine whether or not it is a bryophyte?
4. Develop a hypothesis to explain why mosses grow in clusters or mats. How would you test your hypothesis?

Visit us on the web at http://www.thomsonedu.com/biology/berg/ for additional resources such as flashcards, tutorial quizzes, further readings, and web links.

The Plant Kingdom: Seedless Vascular Plants

One of the best ways to learn about plants is to grow them. With a little care and attention, you can maintain a delicate maidenhair fern (*Adiantum*) indoors even if you live in a small apartment or a dormitory (see photograph). Most of the approximately 200 species of maidenhair ferns are native to tropical American forests. Maidenhair ferns are popular as houseplants because they are attractive and easy to grow.

The leaves, or fronds, of maidenhair ferns typically grow 30 to 38 centimeters (12 to 15 inches) long and consist of shiny, black stalks that bear delicate blades divided into small, fanlike leaflets. (The maidenhair fern is so named because the slender leaf stalks resemble human hair.) The fronds of the maidenhair fern are borne on a horizontal, fast-spreading underground stem called a *rhizome,* which grows just beneath the surface of the soil.

Like all ferns, the maidenhair reproduces by forming spores. When mature, the outer edges of fertile leaves (leaves that bear spores) typically fold over the spore cases, which are produced on the undersides of the leaf margins. Although it is possible to grow a new maidenhair fern from a spore, a lot of patience and special care are required. For that reason, most amateur plant growers purchase maidenhair ferns from a florist or plant nursery.

Like all houseplants, the maidenhair fern is a fascinating organism that reacts to the care (or lack of care) it receives. It prefers bright, filtered light, but the delicate leaves will scorch if placed in direct sunlight. Normal room temperature is generally fine for maidenhair ferns, although it is best to place them above (not in) a tray or saucer of wet pebbles to maintain a high humidity if the air temperature rises above 24°C (75°F). Like all plants, maidenhair ferns must be watered more frequently when the air is warmer, but avoid overwatering. Add enough water so that the soil is slightly moist but not waterlogged. Moisten (but do not soak) the soil, and then water again only after the top inch of the soil has dried out.

Maidenhair fern is just one of many examples of ferns. We now examine other ferns, as well as other spore-forming vascular plants, in detail.

Maidenhair ferns (*Adiantum*) thrive indoors with a minimum of care.

Spore-Forming Vascular Plants

In Chapter 22 we learned that the bryophytes are plants with dominant gametophytes whose sporophytes are attached to and dependent on the gametophyte plants. The bryophytes are also characterized by something they do *not* have: they are the only plants that lack vascular tissues. In contrast, virtually all plants other than bryophytes have vascular tissues and long-lived, dominant sporophytes.

This chapter introduces vascular plants that reproduce by forming spores rather than seeds (•**Table 23-1**). The most important adaptation found in seedless vascular plants, as compared with algae and bryophytes, is the presence of specialized vascular tissues—xylem and phloem—for support and conduction. Recall from Chapter 10 that **xylem** conducts water and dissolved minerals (inorganic nutrients) and **phloem** conducts carbohydrates, predominantly sucrose; both tissues also provide structural support. This system of internal conduction enables vascular plants to grow larger than bryophytes because water, minerals, and sugar are transported to all parts of the plant. Although seedless vascular plants in temperate environments are relatively small, tree ferns in the tropics may grow to heights of 18 meters (60 feet) (•**Figure 23-1**). All seedless vascular plants have true stems with vascular tissues, and most also have true roots and leaves.

There are two main groups of seedless vascular plants: the *ferns* and the *lycophytes*. The lycophytes consist of three groups of seedless vascular plants—the club mosses, spike mosses, and quillworts. Relatively few lycophyte species exist today, but these plants dominated the landscape for more than 100 million years. We will

FIGURE 23-1 Tasmanian tree fern (*Dicksonia antarctica*).
Tree ferns, most of which are native to tropical rain forests, are found in Australia, New Zealand, South Africa, and South America.

TABLE 23-1	A Comparison of Major Groups of Seedless Vascular Plants	
PLANT GROUP	**DOMINANT STAGE OF LIFE CYCLE**	**REPRESENTATIVE GENERA**
Phylum Pteridophyta	Sporophyte	
Ferns	Roots; rhizomes; and leaves (megaphylls)	*Adiantum, Azolla, Dicksonia, Platycerium, Polystichum,* and *Pteridium*
Whisk ferns	Rhizomes and erect stems; no true roots or leaves	*Psilotum*
Horsetails	Roots; rhizomes; erect stems; and leaves (reduced megaphylls)	*Equisetum*
Phylum Lycopodiophyta	Sporophyte	
Club mosses	Roots; rhizomes; erect stems; and leaves (microphylls)	*Lycopodium*
Spike mosses	Roots; erect or prostrate stems; and leaves (microphylls)	*Selaginella*
Quillworts	Roots; short, erect underground stem; and quill-like leaves (microphylls)	*Isoetes*

consider their long evolutionary history later in the chapter.

Ferns are an ancient group of plants that are still successful today, as evidenced by their considerable diversity. Biologists originally considered whisk ferns and horsetails distinct enough to classify in separate phyla. However, many kinds of evidence, such as DNA comparisons and similarities in sperm structure, have resulted in their being reclassified as ferns. As shown in •**Figure 23-2**, ferns, including horsetails and whisk ferns, are a monophyletic group and the closest living relatives of seed plants.

The leaf evolved as the main organ of photosynthesis

Biologists have extensively studied the evolution of the leaf as the main organ of photosynthesis. The two basic types of true leaves, microphylls and megaphylls, evolved independently of each other (•**Figure 23-3**). The **microphyll** is usually small and has a single vascular strand (that is, a single vein). Microphylls probably evolved from small, projecting extensions of stem tissue (*enations*). Lycophytes are the only group of living plants with microphylls.

In contrast, **megaphylls** probably evolved from stem branches that gradually filled in with additional tissue (*webbing*) to form most leaves as we know them today. Megaphylls have a vein system consisting of more than one vascular strand, as we would expect if they evolved from branch systems. Ferns (with the exception of whisk ferns, discussed later in the chapter) and seed plants (gymnosperms and flowering plants) have megaphylls.

Evidence suggests that megaphylls evolved over a 40-million-year period in the Late Paleozoic era in response to a gradual decline in the level of atmospheric CO_2. As CO_2 declined, plants developed a flattened blade with more stomata for gas exchange. The greater number of stomata allowed cells inside the leaf to get enough CO_2 in spite of the atmospheric change.

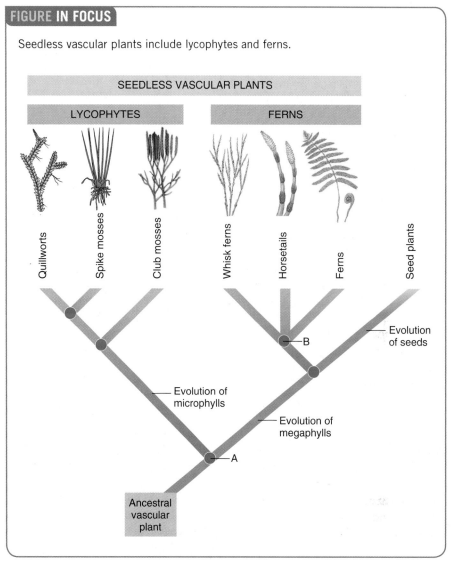

FIGURE IN FOCUS

Seedless vascular plants include lycophytes and ferns.

FIGURE 23-2 Evolutionary relationships among seedless vascular plants. These relationships are based on structural and molecular comparisons. Node A indicates where the lycophytes branched off from other vascular plants. Node B has multiple branches emerging from a single point because the exact relationships among these closely related organisms are still unclear.

XYLEM The vascular tissue that conducts water and dissolved minerals.

PHLOEM The vascular tissue that conducts dissolved sugar and other organic compounds.

MICROPHYLL Type of leaf found in lycophytes; contains one vascular strand.

MEGAPHYLL Type of leaf found in virtually all vascular plants except lycophytes; contains multiple vascular strands.

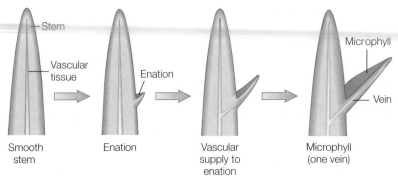

(a) Microphyll evolution

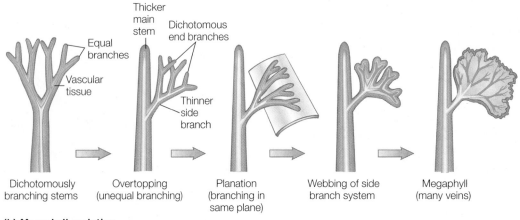

(b) Megaphyll evolution

FIGURE 23-3 Evolution of leaves.
(a) Microphylls probably originated as outgrowths (enations) of stem tissue that developed a single vascular strand later. Club mosses have microphylls. **(b)** Megaphylls, which have multiple veins, probably evolved from the evolutionary modification of side branches. Webbing is the evolutionary process in which the spaces between close branches are filled with chlorophyll-containing cells. Ferns, horsetails, gymnosperms, and flowering plants have megaphylls.

Ferns

Most of the 11,000 species of **ferns** (phylum Pteridophyta) are terrestrial, although a few have adapted to aquatic habitats (•**Figure 23-4**). Ferns range from the tropics to the Arctic Circle, with the most species living in tropical rain forests, where they perch high in the branches of trees. In temperate regions, ferns commonly inhabit swamps, marshes, moist woodlands, and stream banks. Some species grow in fields, rocky crevices on cliffs or mountains, or even deserts. The most common fern species throughout the world is bracken (*Pteridium aquilinum*), a rugged, coarse, "weedy" plant that grows well on poor soil and is uncommon in the moist habitats favored by other ferns.

Ferns have a dominant sporophyte generation

The life cycle of ferns involves a clearly defined alternation of generations (•**Figure 23-5**). The ferns grown as houseplants (such as Boston fern, maidenhair fern, and staghorn fern) represent the larger, more conspicuous sporophyte generation (see Figure 23-5 ❶). A fern sporophyte consists of a horizontal underground stem, called a **rhizome**, that bears thin, wiry roots and rela-

FERN One of a phylum of seedless vascular plants that reproduce by spores produced in sporangia; undergoes an alternation of generations between a dominant sporophyte and a gametophyte (prothallus).

tively large, conspicuous leaves (megaphylls) called **fronds.** The fronds have two functions, photosynthesis and reproduction (discussed shortly). Fern sporophytes are **perennials;** that is, they live for a number of years. In geographic areas with pronounced winters, the leaves die each autumn, and the underground rhizome produces new leaves each spring.

When a young frond first emerges from the ground, it is tightly coiled and resembles the top of a violin, hence the name **fiddlehead** (•**Figure 23-6**). As a fiddlehead grows, it unrolls from the tip and expands to form a frond. Fern fronds are usually compound; that is, the blade is divided into several leaflets, or *pinnae,* that branch from the axis, or *rachis.* The pinnae form beautifully complex leaves. Fern fronds, roots, and rhizomes are considered true plant organs because each contains vascular tissues.

The conspicuous plant body of the fern—the sporophyte generation—forms spores by meiosis (see Figure 23-5 ❷). Spore production usually occurs on the undersides of the fronds, where fertile areas develop **sporangia** (sing., *sporangium*), or spore cases. Many species bear the sporangia in clusters called **sori** (sing., *sorus*) on the fronds (•**Figure 23-7**). The sori of certain ferns are covered by a protective membranous covering, the *indusium.* In some species, sporangia are borne on separate, fertile stalks instead of on the fronds. Within the sporangia, spore mother cells undergo meiosis to form haploid spores.

The sporangium of many ferns is characterized by an *annulus,* a row of cells with unevenly thickened cell walls. As the relative humidity changes, tensions develop that break the sporan-

(a) The Christmas fern (*Polystichum acrostichoides*) is green at Christmas, making it a popular holiday decoration.

(b) Most spleenwort ferns (*Asplenium*) are tropical. Their common name refers to their purported medicinal properties.

(c) The staghorn fern (*Platycerium bifurcatum*), native to Australian rain forests, is widely cultivated. In nature, it is an epiphyte and grows attached to tree trunks.

(d) The mosquito fern (*Azolla caroliniana*) is a tiny, free-floating aquatic fern that sometimes grows so densely that it reportedly smothers mosquito larvae.

FIGURE 23-4 Fern diversity.

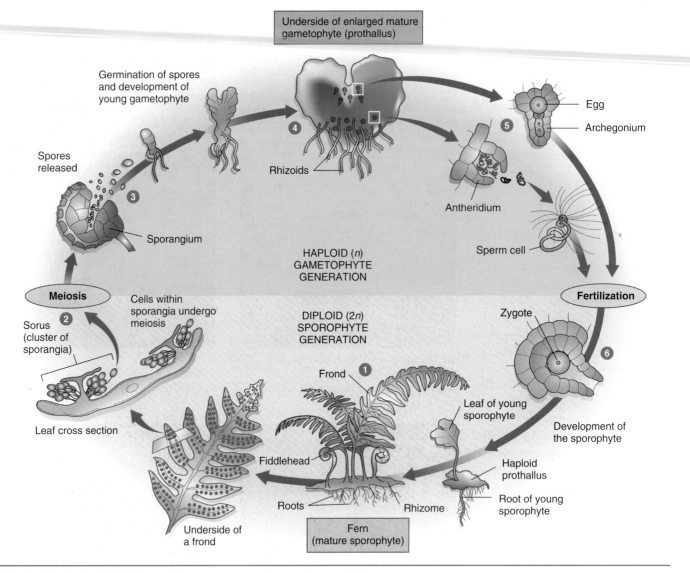

Underside of enlarged mature gametophyte (prothallus)

Germination of spores and development of young gametophyte

Spores released

Rhizoids

Egg

Archegonium

Antheridium

Sperm cell

Sporangium

HAPLOID (*n*) GAMETOPHYTE GENERATION

Meiosis

Cells within sporangia undergo meiosis

Fertilization

Sorus (cluster of sporangia)

DIPLOID (*2n*) SPOROPHYTE GENERATION

Zygote

Frond

Leaf of young sporophyte

Leaf cross section

Development of the sporophyte

Fiddlehead

Haploid prothallus

Roots

Rhizome

Root of young sporophyte

Underside of a frond

Fern (mature sporophyte)

FIGURE 23-5 The fern life cycle.
Note the clearly defined alternation of generations between the gametophyte (prothallus) and sporophyte (leafy plant) stages. See text for a detailed description.

gium open along the thin-walled cells, thrusting the spores into the air (see Figure 23-5 ❸). Fern spores are disseminated in air currents, often over great distances. When the spores land in suitable places, such as moist soil or cracks in rocks, they may germinate and grow by mitosis into gametophytes.

The gametophyte generation of ferns, which bears no resemblance to the sporophyte generation, is a tiny (less than half the size of one of your fingernails), green, often heart-shaped structure that grows flat against the ground (●**Figure 23-8**). Called a **prothallus** (pl., *prothalli*), the fern gametophyte lacks vascular tissues and has tiny, hairlike rhizoids that anchor it (see Figure 23-5 ❹).

The prothallus usually produces both female gametangia (archegonia) and male gametangia (antheridia) on its underside (see Figure 23-5 ❺). The flask-shaped archegonia, located in the central region of the prothallus near the "notch," each contain a single nonmotile egg. Numerous sperm cells are produced in the spherical antheridia, which are found scattered among the rhizoids. The sperm cells are shaped like tiny corkscrews, each with many flagella. Both eggs and sperm cells are produced by mitosis of haploid gametophyte cells.

Like the bryophytes, ferns have retained the primitive requirement of water to accomplish fertilization. Any thin film of water on the ground beneath the prothallus provides the transport medium in which the

Marion Lobstein

FIGURE 23-6 Fiddleheads.

John Arnaldi

FIGURE 23-7 Sori.

These round sori of rabbit's foot fern (*Polypodium aureum*) are arranged in two prominent rows on the leaf's underside. Considerable variation exists in arrangement of sori on fern leaves.

flagellated sperm cells swim to the archegonium. The neck of the archegonium provides a passageway from the archegonium surface to the egg. After a sperm cell fertilizes the egg, a diploid zygote ❻ grows by mitosis into a multicellular embryo (an immature sporophyte). At this stage, the sporophyte embryo is attached to and dependent on the gametophyte. (Generally, only one sporophyte plant develops from each prothallus.) As the embryo matures, the prothallus withers and dies, and the sporophyte becomes free-living.

The fern life cycle alternates between the diploid sporophyte, with its rhizome, roots, and fronds, and the haploid gametophyte (prothallus). The sporophyte generation is dominant not only because it is larger than the gametophyte but because it persists for an extended period (most fern sporophytes are perennials), whereas the gametophyte dies soon after reproducing.

Whisk ferns are classified as reduced ferns

Only about 12 species of **whisk ferns** exist today. Whisk ferns, which live mainly in the tropics and subtropics, are relatively simple in structure and lack true roots and leaves but have vascularized stems. *Psilotum nudum,* a representative whisk fern, has both a horizontal underground rhizome, bearing hairlike rhizoids, and verti-

cal aerial stems that grow to 61 centimeters (2 feet) in height (●**Figure 23-9a**). The aerial stem, which branches extensively, resembles a whisk broom—hence the common name "whisk fern." The rhizome with its attached rhizoids performs the functions of a root—that is, it absorbs water and dissolved minerals—whereas the aerial stems are green and photosynthetic. Small, scalelike

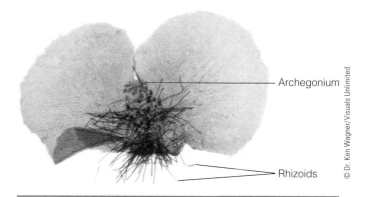

Archegonium

Rhizoids

© Dr. Ken Wagner/Visuals Unlimited

FIGURE 23-8 Prothallus.

The prothallus is the gametophyte generation of a fern. The dark spots near the notch of the "heart" are archegonia; no antheridia are visible. Many prothalli produce archegonia and antheridia at different times.

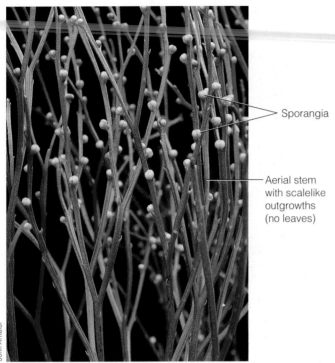

John Arnaldi

(a) The sporophyte of the whisk fern *Psilotum nudum*. The stem is the main photosynthetic organ of this rootless, leafless, vascular plant. Sporangia, which are initially green but turn yellow as they mature, are borne on short lateral branches directly on the stems. *Psilotum nudum* has been cultivated for centuries in Japan, where numerous varieties are highly prized.

Sporangia

Aerial stem with scalelike outgrowths (no leaves)

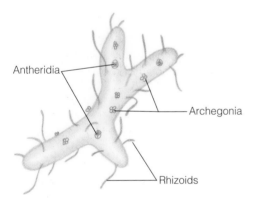

Antheridia

Archegonia

Rhizoids

(b) The gametophyte of *P. nudum* lives underground, nourished by mycorrhizae.

FIGURE 23-9 The whisk fern.

FIGURE 23-10 *Calamites.*

Calamites was an ancient horsetail the size of a small tree. Like modern-day horsetails, *Calamites* had an underground rhizome where roots and aerial shoots originated. (Redrawn from L. Emberger, *Les Plantes Fossiles,* Masson et Cie, Paris, 1968.)

projections extend from the stem, but these are not considered leaves because they lack vascular tissues.

Whenever the stem forks or branches, it always divides into two equal halves. Botanists call this forking **dichotomous branching.** In contrast, when most plant stems branch, one stem is more vigorous and becomes the main trunk, whereas the other becomes a lateral branch.

The upright stems of *Psilotum* are green and are the main organs of photosynthesis. Small, round sporangia, borne directly on the erect, aerial stems, contain spore mother cells that undergo meiosis to form haploid spores. After being dispersed by air currents, the spores, which require darkness to germinate, develop into haploid prothalli.

The prothalli of whisk ferns are difficult to study because they are minute (about 2 millimeters in diam-

Strobilus

Reproductive shoot

Vegetative shoot

John Arnaldi

(a) In some horsetail (*Equisetum*) species, both fertile shoots, which bear conelike strobili, and vegetative (nonreproductive) shoots are unbranched.

S. Solum/PhotoLink/Getty Images

(b) Some *Equisetum* species have unbranched fertile shoots and highly branched vegetative shoots (*shown*) with whorls of branches that resemble a bushy horse's tail.

FIGURE 23-11 Representative horsetails.

eter) and grow underground (•**Figure 23-9b**). They are nonphotosynthetic and apparently have a symbiotic relationship with mycorrhizal fungi, which provide them with sugar and essential minerals (see Chapter 21).

Botanists have carefully studied whisk ferns in recent years. Molecular data, including comparisons of nucleotide sequences of ribosomal RNA, chloroplast DNA, and mitochondrial DNA in living species, support the hypothesis that whisk ferns are reduced ferns rather than surviving relatives of extinct vascular plants (see discussion of rhyniophytes later in this chapter). A *reduced* organism is one that has become simplified during the course of evolution. Presumably, the ancestral ferns that gave rise to *Psilotum* had both roots and leaves.

Horsetails are an evolutionary line of ferns

About 300 million years ago (mya), **horsetails** were among Earth's dominant plants and grew as large as

modern trees (•**Figure 23-10**). Because they contributed to the formation of Earth's vast coal deposits, these ancient horsetails, like other ancient ferns and lycophytes, are still significant to us today. (See *Plants and People: Ancient Plants and Coal Formation* on page 454, and *Plants and the Environment: The Environmental Effects of Mining Coal* on page 458.)

The few surviving horsetails, about 15 species in the single genus *Equisetum,* grow mostly in wet, marshy habitats and are small (less than 1.3 meters, or 4 feet, tall) but distinctive plants (•**Figure 23-11**). They are widely distributed on every continent except Australia. Traditionally classified in their own phylum, horsetails are now grouped with ferns. This reclassification is based on molecular similarities between horsetails and other ferns.

Horsetails have true roots, stems (both rhizomes and erect aerial stems), and small leaves. The hollow, jointed stems are impregnated with silica, which gives them a gritty texture. The erect stems arise from perennial rhizomes that also bear wiry roots. Small leaves,

PLANTS AND PEOPLE | Ancient Plants and Coal Formation

The industrial society in which we live depends on energy from fossil fuels, which formed from the remains of ancient organisms. One of our most important fossil fuels is coal, which is burned to produce electricity, heat homes, and manufacture items made of steel and iron. Although coal is mined from the ground as minerals are, it is not a mineral like gold or aluminum but is an organic material, formed from ancient plants.

Much of the coal we use today formed from the prehistoric remains of ancient vascular plants, particularly those of the Carboniferous period, which occurred approximately 300 million years ago (see figure). Of the five main groups of plants that contributed to coal formation, three were seedless vascular plants: the club mosses, horsetails, and ferns. The other two important groups of coal formers were seed plants: the seed ferns (now extinct) and early gymnosperms.

It is hard to imagine that the small, relatively inconspicuous club mosses, ferns, and horsetails of today were so significant in forming the vast beds of coal in the ground. However, many of the members of these groups that existed during the Carboniferous period were giants by comparison and formed vast forests. Some giant club mosses, for example, grew almost 40 meters (130 feet) tall.

The climate during the Carboniferous period was warm and mild, and plants grew year-round because of the favorable weather conditions. Forests occurred in low-lying areas that were periodically flooded. When the water level receded, the plants would become established again.

When these large plants died or blew over during storms, they decomposed incompletely because they were covered by the swampy water. The anaerobic (oxygen-free) conditions of the water prevented wood-rotting fungi from decomposing the plants, and anaerobic bacteria do not decompose wood rapidly. Thus, over time the partially decomposed plant material accumulated and consolidated.

When the water level rose and flooded the low-lying swamps, layers of sediment formed over the plant material. Over time, heat and pressure built up in these accumulated layers and converted the plant material to coal and the sediment layers to sedimentary rock. Much later, geologic upheavals raised the layers of coal and sedimentary rock.

Coal is usually found in underground layers, called seams, that range from 2.5 centimeters (1 inch) to more than 30 meters (100 feet) in thickness. Various grades of coal (lignite, bituminous, and anthracite) formed as a result of different temperatures and pressures. The largest coal deposits are in North America, Russia, and China, but deposits are also found in the arctic islands, western Europe, India, South Africa, Australia, and eastern South America.

■ The plants of the Carboniferous period included giant ferns, horsetails, and club mosses.

interpreted as reduced megaphylls, are fused in whorls at each node (the area on the stem where leaves attach). The green stem is the main organ of photosynthesis.

The word *Equisetum* comes from the Latin words *equus*, "horse," and *saeta*, "bristle." Horsetails are so named because certain vegetative (nonreproductive) stems have whorls of branches that give the appearance of a bushy horse's tail. In the past, horsetails were called "scouring rushes" and were used to scrub out pots and pans along stream banks.

The aerial stems of horsetails are either vegetative (sterile) or reproductive (fertile). Each reproductive branch of a horsetail bears a terminal, conelike **strobilus** (pl., *strobili*). The strobilus consists of several

KEY

1–13	Club mosses
14–16	Seed ferns
17–19	Ferns
20–21	Horsetails
22	Early gymnosperm
23	Primitive insect
24	Early dragonfly
25,26	Early roaches

stalked, umbrella-like structures, each of which bears 5 to 10 sporangia in a circle around a common axis. Horsetail spores have appendages, or *elaters,* that uncoil as the spores dry out, helping to toss them out of the sporangium (•**Figure 23-12**).

The horsetail life cycle is similar in many respects to the fern life cycle. In horsetails, as in ferns, the spo-

rophyte is the conspicuous plant, whereas the gametophyte is a minute, lobed thallus ranging in width from the size of a pinhead to about 1 centimeter (less than 1/2 inch) across. The sporophyte and gametophyte are both photosynthetic and nutritionally independent at maturity. Like ferns, horsetails require water as a medium for flagellated sperm cells to swim to the egg.

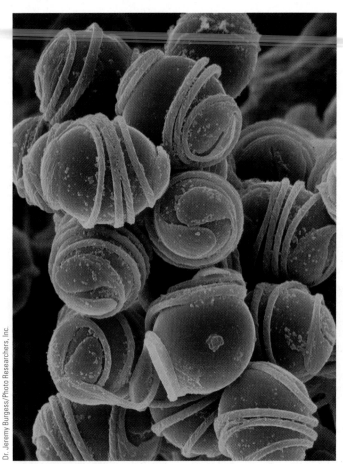

Dr. Jeremy Burgess/Photo Researchers, Inc.

(a) Scanning electron micrograph of spores of *Equisetum arvense.*

FIGURE 23-12 Horsetail spores.

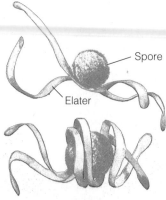

(b) When the spores dry out, the elaters folded about the spores expand, aiding in dispersal from the sporangium.

Lycophytes

Like horsetails, **lycophytes** (phylum Lycopodiophyta) were important plants millions of years ago, when species that are now extinct often reached great sizes (●**Figure 23-13**). These large, treelike plants, like the ancient horsetails, were major contributors to our present-day coal deposits. The approximately 1200 species of living lycophytes include three groups: club mosses, spike mosses, and quillworts. All lycophytes, living and extinct, are characterized by leaves that are microphylls.

Club mosses are small lycophytes with rhizomes and short, aerial stems

Club mosses, such as *Lycopodium,* are small (less than 25 centimeters, or 10 inches, tall), attractive plants com-

LYCOPHYTE One of a phylum of seedless vascular plants, some of which are heterosporous.

mon in temperate woodlands (●**Figure 23-14**). That common names are sometimes misleading in biology is vividly evident in this group of plants. The most common names are "club mosses" (because they superficially resemble large moss gametophytes) and "ground pines" (because some that are common in pine forests superficially resemble miniature trees). Yet these plants are neither mosses, which are nonvascular, nor pines, which are seed plants.

Club mosses have true roots; rhizomes and erect or trailing aerial stems; and small, scalelike, single-veined leaves (microphylls). Sporangia are borne on fertile leaves that are either clustered in conelike strobili at the tips of stems or scattered in reproductive areas along the stem. Club mosses are evergreen and often fashioned into Christmas wreaths and other decorations. In some areas they are endangered from overharvesting.

The life cycle of *Lycopodium* is similar in many respects to that of the ferns. As in ferns, both sporophyte and gametophyte are independent plants at maturity. In *Lycopodium,* the sporophyte is the conspicuous generation, whereas the gametophyte is a small thallus ranging from microscopic size to about 2.5 centimeters (1 inch) in length. The gametophytes are either photosynthetic or subterranean (and therefore nonphotosynthetic). Like whisk fern gametophytes, the subterranean gametophytes of club mosses apparently form a symbiotic relationship with mycorrhizal fungi that provide them with nourishment. Like ferns, *Lycopodium* requires water as a medium in which flagellated sperm cells swim to the egg in the archegonium.

Spike mosses produce male and female gametophytes

Spike mosses have a variety of growth forms; some are creeping and flat, some are cushiony, and others are erect

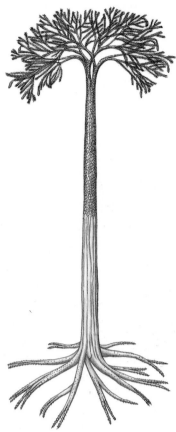

(a) This ancient club moss was a large tree—to 40 meters (about 130 feet). Numerous fossils of *Lepidodendron* were preserved in coal deposits, particularly in Great Britain and the central United States.

(b) Close-up of fossil *Lepidodendron* bark, showing scars where leaves were once attached. Each leaf scar is about 1.9 centimeters (0.75 inch) wide.

FIGURE 23-13 *Lepidodendron.*
(**a,** Redrawn from M. Hirmer, *Handbuch der Paläobotanik,* R. Olderbourg, Munich, 1927.)

Strobilus

Leaves
(microphylls)

(a) The sporophyte of *Lycopodium obscurum* has small, scalelike, evergreen leaves (microphylls). Spores are produced in sporangia on fertile leaves clustered in a conelike strobilus.

(b) Vegetative growth of *Lycopodium complanatum*, which has a creeping growth form. Like most club mosses, *L. complanatum* prefers moist, shady woodlands.

FIGURE 23-14 Representative club mosses.

PLANTS AND THE ENVIRONMENT | The Environmental Effects of Mining Coal

Coal mining, especially surface mining (also called strip mining), has substantial effects on the environment. In surface mining, vegetation and topsoil are completely removed by bulldozers, giant power shovels, and wheel excavators to expose the coal seam (see figure). The coal is then scraped out of the ground and loaded into railroad cars or trucks. Approximately 60 percent of the coal mined in the United States is obtained by surface mining, which destroys plant and animal habitat and increases soil erosion and water pollution.

Prior to the passage of the 1977 Surface Mining Control and Reclamation Act (SMCRA), abandoned surface mines were usually left as large open pits. Cliffs of excavated rock called highwalls, some more than 30 meters (100 feet) high, were left exposed. Acid and toxic mineral drainage from such mines and the burial of topsoil or its removal by erosion prevented most plants from recolonizing the land naturally. Streams were polluted with sediment and acid mine drainage. Dangerous landslides occurred on land that was unstable due to the lack of vegetation.

Surface-mined land can be restored to prevent such degradation and to make the land productive for other purposes, but restoration is expensive. Before the existence of laws requiring

■ Strip mining at a lignite coal mine in eastern Texas.

Department of the Interior, Office of Surface Mining/Chuck Meyers

coal companies to reclaim the land, few restorations of surface-mined land took place. Since 1977 the SMCRA has required coal companies to restore surface-mined areas.

Sometimes this restored land is used for wildlife habitat—for example, rangelands for wild grazing animals in some western states, and wetlands in midwestern and eastern states. The Eastern Kentucky Regional Airport is situated on a reclaimed mining site, as is a tree farm in Garrett County,

Maryland. Other postmining land uses include camping areas, golf courses, farmlands, sanitary landfills, cemeteries, and housing developments.

Land that was strip-mined prior to 1977 is gradually being restored as well, with money from a tax that coal companies pay on currently mined coal. However, because so many abandoned coal mines exist, it is doubtful that they will all be restored, and only those that are most dangerous from a health and safety viewpoint are reclaimed.

(●**Figure 23-15**). These lycophytes have long, creeping rhizomes and typically branch dichotomously. Their roots are produced at branchings along the length of the

stems; like the stems, the roots branch dichotomously. The overlapping, scalelike leaves, which are microphylls, have a single unbranched vein; leaves are arranged in four vertical rows in many species. Each scalelike leaf produces a small, inconspicuous outgrowth on the upper surface near the base of the leaf; this structure is called a **liguIè**.

Living spike mosses are classified in a single genus (*Selaginella*). They are most common in shady, moist tropics and subtropics, but some grow in temperate climates. Although most species grow on moist soil, some are adapted to arid environments. The resurrection plant (*S. lepidophylla*), for example, is native to prairies

HOMOSPORY Production of one type of *n* spore that gives rise to a bisexual gametophyte.

HETEROSPORY Production of two types of *n* spores, microspores and megaspores.

MICROSPORE The *n* spore in heterosporous plants that gives rise to a male gametophyte.

MEGASPORE The *n* spore in heterosporous plants that gives rise to a female gametophyte.

FIGURE 23-15 Spike moss.
Selaginella is a small, mosslike plant that is widely distributed in the tropics and subtropics, with a few species in temperate areas. *Selaginella* is heterosporous and produces both microspores (which develop into male gametophytes) and megaspores (which develop into female gametophytes).

FIGURE 23-16 Resurrection plant.
The plant on the left was recently watered, and the plant on the right was left dry.

and deserts of Texas and Mexico. When conditions are dry, this species curls up into a brown ball and becomes dormant, but when moisture is again available, it is "resurrected"—that is, it expands and becomes green and photosynthetic (•**Figure 23-16**).

Spike mosses, quillworts, and some ferns are heterosporous

In the life cycles examined thus far, plants produce only one type of spore as a result of meiosis. This condition, known as **homospory,** is characteristic of bryophytes (see Chapter 22), horsetails, whisk ferns, club mosses, and most ferns. However, spike mosses, quillworts (discussed in the next section), and some ferns are **heterosporous,** in which they produce two types of spores: microspores and megaspores. •**Figure 23-17** illustrates the generalized life cycle of a heterosporous plant.

Spike moss (*Selaginella*) is an example of a heterosporous plant (•**Figure 23-18**). The sporophyte plant produces sporangia within a conelike strobilus. Each strobilus usually bears two kinds of sporangia, microsporangia and megasporangia ❶. **Microsporangia** are sporangia that produce **microspore mother cells,** which undergo meiosis to form microscopic haploid **microspores** ❷. Each microspore develops into a male gametophyte that produces sperm cells within antheridia ❸.

Megasporangia in the *Selaginella* strobilus produce **megaspore mother cells.** When megaspore mother cells undergo meiosis, they form haploid **megaspores** ❹,

each of which develops into a female gametophyte that produces eggs within archegonia ❺.

In *Selaginella,* the development of male gametophytes from microspores and of female gametophytes

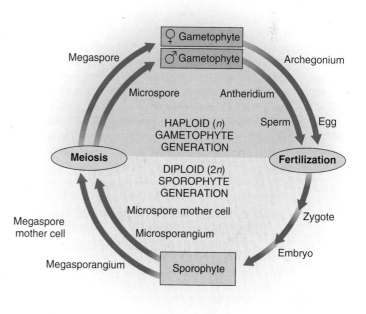

FIGURE 23-17 The basic life cycle of heterosporous plants.
The sporophyte of heterosporous plants produces two types of spores, microspores and megaspores.

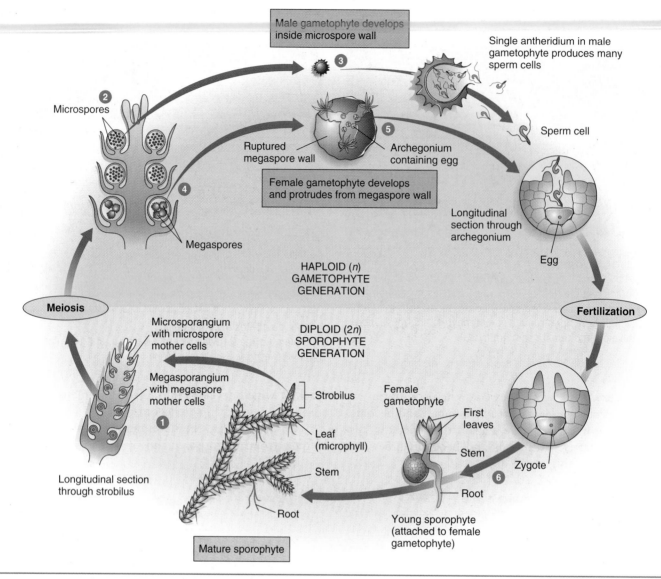

FIGURE 23-18 The life cycle of *Selaginella*.
Spike moss is heterosporous, which results in the formation of unisexual gametophytes.

from megaspores occurs within their respective spore walls, using stored food provided by the sporophyte. As a result, the male and female gametophytes are not truly free-living, unlike the gametophytes of other seedless vascular plants. Fertilization is followed by the development of a new sporophyte ❻.

Heterospory evolved several times during the history of land plants. It was a significant development in plant evolution because it was the forerunner of the evolution of seeds. Heterospory characterizes the two most successful groups of plants existing today, the gymno-

sperms and the flowering plants, both of which produce seeds.

Quillworts have quill-like leaves arising from a short underground stem

Quillworts (*Isoetes*) are perennial aquatic or semi-aquatic herbs that are widely distributed in temperate, freshwater habitats (●**Figure 23-19**). Quillworts are distinctive in their growth form: they have a *corm,* a short, swollen underground stem. Cylindrical, quill-like leaves (microphylls) grow in a clump from the base of the corm; roots that arise from the corm divide dichotomously. Aerial stems are absent.

Like other lycophytes, quillworts are spore-bearing vascular plants. Like the spike mosses, quillworts are heterosporous and produce two kinds of spores. Megasporangia that produce a few large megaspores are located at the bases of the outer leaves, whereas microsporangia that produce many smaller microspores are at the bases of inner leaves.

FIGURE 23-19 Quillworts.
Quillworts are classified into species based on markings (such as ridges and pits) on the surfaces of their megaspores.

PROCESS OF SCIENCE Seedless Vascular Plants as Research Organisms

Botanists use many seedless vascular plants as experimental models to study certain aspects of plant biology, such as physiology, growth, and development. Ferns and other seedless vascular plants are useful in studying how apical meristems give rise to plant tissues. An **apical meristem** is the area at the tip (apex) of a root or shoot where growth—cell division, elongation, and differentiation—occurs. Ferns and other seedless vascular plants have a single, large *apical cell* located at the center tip of the apical meristem. This apical cell is the source, by mitosis, of all the cells that eventually make up the root or shoot. The apical cell divides in an orderly fashion, and the smaller daughter cells produced by the apical cell, in turn, divide, giving rise to different parts of the root or shoot. It is possible to trace mature cells in the root or shoot back to their origin from the single apical cell.

Ferns are interesting research plants for studies in genetics because they are **polyploids** and have multiple sets of chromosomes. (Many ferns have hundreds of chromosomes.) However, gene expression in ferns is exactly what one would expect of a *diploid* plant. Apparently, genes in the extra sets of chromosomes are gradually silenced and therefore not expressed. ■

EVOLUTION LINK Evolution of Vascular Plants

Currently, the oldest known megafossils of early vascular plants are from Silurian (420 mya) deposits in Europe. (Plant *megafossils* are fossilized roots, stems, leaves, and reproductive structures.) Megafossils of several kinds of

small, seedless vascular plants were also discovered in Silurian deposits in Bolivia, Australia, and northwestern China. Microscopic spores of early vascular plants appear in the fossil record earlier than megafossils, suggesting that even older megafossils of simple vascular plants may be discovered.

Botanists assign the oldest known vascular plants to phylum Rhyniophyta, which, according to the fossil record, arose some 420 mya and became extinct about 380 mya. The rhyniophytes are so named because many fossils of these extinct plants were found in fossil beds near Rhynie, Scotland. *Rhynia gwynne-vaughanii* is an example of an early vascular plant that superficially resembled whisk ferns in that it consisted of leafless, upright stems that branched dichotomously from an underground rhizome (•**Figure 23-20**). *Rhynia* lacked roots, although it had absorptive rhizoids. Sporangia formed at the ends of short branches. The internal structure of its rhizome contained a central core of xylem cells for conducting water and minerals.

PROCESS OF SCIENCE For many years, botanists considered *R. major,* a plant that grew about 50 centimeters (20 inches) tall and probably lived in marshes, a classic example of a rhyniophyte. Fossils indicate that this plant had rhizoids, dichotomously branching rhizomes, and upright stems that terminated in sporangia. However, recent microscopic studies of fossil rhizomes indicate

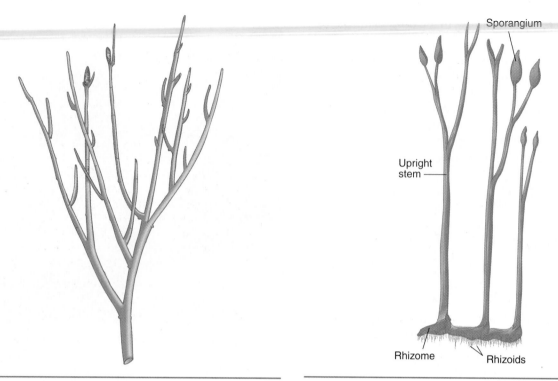

FIGURE 23-20 Reconstruction of *Rhynia gwynne-vaughanii.*
This leafless plant, one of Earth's earliest vascular plants, is now extinct. It grew about 18 centimeters (7 inches) tall. (Redrawn from D. Edwards, "Evidence for the Sporophytic Status of the Lower Devonian Plant *Rhynia gwynne-vaughanii*," *Review of Palaeobotany and Palynology,* Vol. 29, 1980.)

FIGURE 23-21 Reconstruction of *Aglaophyton major.*
Recent evidence indicates that this plant, although superficially similar to other early vascular plants, lacked conducting tissues that are characteristic of vascular plants. For that reason, it was reclassified into a new genus and is no longer considered a rhyniophyte. (Redrawn from J. D. Mauseth, *Botany: An Introduction to Plant Biology,* 2nd ed., Saunders College Publishing, Philadelphia, 1995.)

that the central core of tissue lacked the xylem cells characteristic of vascular plants. For that reason, *R. major* was reclassified into a new genus, *Aglaophyton,* and is no longer considered a rhyniophyte (•**Figure 23-21**). Science is an ongoing enterprise, and over time, existing knowledge is reevaluated in light of newly discovered evidence. *Aglaophyton major* is an excellent example of the self-correcting nature of science, which is in a perpetually dynamic state and changes in response to newly available techniques and data. ■ ■ ■

Ecological and Economic Importance of Seedless Vascular Plants

Ferns and lycophytes are ecologically important, in part because, like other plants, they help form soil. Their branching underground rhizomes and roots or rhizoids help hold the soil in place, thereby preventing erosion. Ferns are sometimes present in early stages of *ecological succession,* the orderly sequence of changes in the kinds of plants living in a particular area. Some animals, such as muskrats, eat seedless vascular plants that grow along stream banks.

Ferns are widely cultivated for their aesthetic appeal. Because of the beauty and variety of their fronds, ferns such as Boston fern, maidenhair fern, and staghorn fern are cultivated in homes, gardens, and conservatories, and florists often add fern leaves to floral arrangements.

The fiddleheads of some fern species, such as the ostrich fern (*Matteuccia struthiopteris*), are harvested in early spring, boiled or steamed, and eaten. Fiddleheads are sold fresh (as a seasonal item), frozen, and canned and are particularly popular in New England and the Canadian Maritime provinces.

The most economically important seedless vascular plants are the ones that died about 300 mya. The coal deposits that formed from the remains of these ancient ferns, club mosses, and horsetails are of great relevance to humans. Coal powered the steam engine and supplied the energy needed for the Industrial Revolution of the 19th century. Today, coal is used primarily by utility companies to produce electricity and, to a lesser extent, by heavy industries such as steelmaking.

STUDY OUTLINE

❶ Summarize the features that distinguish seedless vascular plants from bryophytes.

Seedless vascular plants have several adaptations that bryophytes lack, including vascular tissues and a dominant sporophyte generation. **Xylem** is the vascular tissue that conducts water and dissolved minerals, and **phloem** is the vascular tissue that conducts dissolved sugar and other organic compounds. As in bryophytes, reproduction in seedless vascular plants depends on water as a transport medium for motile sperm cells.

❷ Contrast microphylls and megaphylls.

The two basic types of true leaves, microphylls and megaphylls, evolved independently of each other. The **microphyll**, found in lycophytes, contains one vascular strand. The **megaphyll**, found in virtually all vascular plants except lycophytes, contains multiple vascular strands.

❸ Distinguish between the two phyla of seedless vascular plants.

Ferns (phylum Pteridophyta) are members of a phylum of seedless vascular plants that reproduce by spores produced in sporangia. Ferns, which undergo an alternation of generations between a dominant sporophyte and a gametophyte (prothallus), are the largest and most diverse group of seedless vascular plants. Whisk ferns and horsetails are currently classified as ferns. **Lycophytes** (phylum Lycopodiophyta) are members of a phylum of seedless vascular plants, some of which are heterosporous. Lycophytes include club mosses, spike mosses, and quillworts.

❹ Name and briefly describe ferns, and explain why whisk ferns and horsetails are currently classified as ferns.

The fern sporophyte consists of a **rhizome** that bears **fronds** (megaphylls) and true roots. Sporophytes of **whisk ferns** have **dichotomously branching** rhizomes and erect stems; they lack true roots and leaves. **Horsetail** sporophytes have roots, rhizomes, aerial stems that are hollow and jointed, and leaves that are reduced megaphylls. Biologists originally considered whisk ferns and horsetails distinct enough to be classified in separate phyla. However, many kinds of evidence, such as DNA compari-

sons and similarities in sperm structure, have resulted in their being reclassified as ferns.

❺ Describe the life cycle of ferns, and compare their sporophyte and gametophyte generations.

Fern fronds bear sporangia in clusters called **sori.** Meiosis in sporangia produces haploid spores. The fern gametophyte, called a **prothallus**, develops from a haploid spore and bears both archegonia and antheridia. Each archegonium contains a single, nonmotile egg, whereas each antheridium produces numerous sperm cells. Following fertilization, a diploid zygote grows by mitosis into a multicellular embryo (an immature sporophyte).

❻ Name and briefly describe the three groups of lycophytes.

Living lycophytes include three groups: club mosses, spike mosses, and quillworts. The **club moss** sporophyte has true roots; rhizomes and erect or trailing aerial stems; and small, scalelike microphylls. The **spike moss** sporophyte has long, creeping rhizomes that typically branch dichotomously; roots that branch dichotomously; and overlapping, scalelike microphylls. The **quillwort** sporophyte consists of an underground corm; cylindrical, quill-like microphylls; and roots.

❼ Compare the generalized life cycle of a homosporous plant with that of a heterosporous plant.

Homospory is production of one type of *n* spore that gives rise to a bisexual gametophyte. **Heterospory** is production of two types of *n* spores, microspores and megaspores. The **microspore** gives rise to a male gametophyte that produces sperm cells within antheridia. The **megaspore** gives rise to a female gametophyte that produces eggs within archegonia.

❽ Name and describe one of Earth's earliest vascular plants.

Botanists assign the oldest known vascular plants to phylum Rhyniophyta, which, according to the fossil record, arose some 420 mya and became extinct about 380 mya. *Rhynia gwynne-vaughanii* was an early vascular plant that superficially resembled whisk ferns in that it consisted of leafless upright stems that branched dichotomously from an underground rhizome. *Rhynia* lacked roots, although it had absorptive rhizoids. Sporangia formed at the ends of short branches. The internal structure of its rhizome contained a central core of xylem cells for conducting water and minerals.

❾ Describe the ecological and economic significance of the ferns and lycophytes.
Like other plants, ferns and lycophytes help form soil. Their branching underground rhizomes and roots or rhi-

zoids hold the soil in place, thereby preventing erosion. Living seedless vascular plants are of limited economic importance, although many are cultivated for their aesthetic appeal. The coal deposits that formed from the remains of ancient ferns, club mosses, and horsetails are of great relevance to humans. Coal powered the Industrial Revolution of the 19th century and is used today to produce electricity.

REVIEW QUESTIONS

1. Explain why bryophytes and seedless vascular plants are classified in different phyla.
2. Distinguish between microphylls and megaphylls.
3. In terms of the number of species, which group of seedless vascular plants is most diverse?
4. Which features do club mosses and ferns share?
5. How can whisk ferns be distinguished from other seedless vascular plants? Why are whisk ferns classified as ferns?
6. Why are certain horsetails called "scouring rushes"?
7. How is the common name "club moss" misleading?

8. How can spike mosses be distinguished from club mosses?
9. Are quillworts more closely related to ferns or club mosses? Explain your answer.
10. Describe the specific stages of alternation of generations in the ferns.
11. How is coal related to ancient ferns and lycophytes?
12. Define *heterospory*, and explain how heterospory modifies the plant life cycle.
13. What are rhyniophytes?
14. Label the following diagram. (Refer to Figure 23-5 [bottom] to check your answers.)

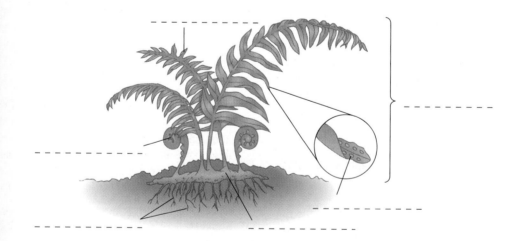

THOUGHT QUESTIONS

1. Which of the living plants studied in this chapter would you consider the most primitive? Why?

2. Suppose you are given an unknown seedless vascular plant. How could you determine whether it is a fern, a whisk fern, a horsetail, or a club moss?

3. How may the following trends in plant evolution be adaptive to living on land?
   a. Dependence on water for fertilization ⟶ no need for water as a transport medium
   b. Dominant gametophyte generation ⟶ dominant sporophyte generation
   c. Homospory ⟶ heterospory

Visit us on the web at http://www.thomsonedu.com/biology/berg/ for additional resources such as flashcards, tutorial quizzes, further readings, and web links.

The Plant Kingdom: Gymnosperms

LEARNING OBJECTIVES

❶ Compare the features of seeds with those of spores, and discuss the adaptive advantages of plants that reproduce by seeds.

❷ Distinguish between a seed and an ovule, and define *integuments.*

❸ Summarize the features that distinguish gymnosperms from seedless vascular plants.

❹ Name and briefly describe the four phyla of gymnosperms.

❺ Contrast monoecious plants and dioecious plants.

❻ Trace the steps in the life cycle of pine, and compare the sporophyte and gametophyte generations.

❼ Describe the ecological and economic significance of gymnosperms.

❽ Trace the evolution of gymnosperms from seedless vascular plants.

I n 1994, David Noble, an officer of the Australian National Park and Wildlife Service, was hiking through a rugged wilderness area in Wollemi National Park in the Blue Mountains, near Sydney, Australia. He found a small stand of fewer than 100 trees that he did not recognize, so he took a fallen branch back with him and asked botanists to identify it.

The botanists were amazed, because this plant was known only from the fossil record, extending back to about 150 million years ago (mya). It was presumed to have gone extinct millions of years ago. Botanists called it the Wollemi pine in recognition of where it was first discovered. After the Wollemi pine was studied and recognized to have properties not found in similar conifers, botanists assigned it to a new genus, *Wollemia,* and named it *Wollemia nobilis.* Its specific epithet, *nobilis,* honors the man who discovered it.

The Wollemi pine is a large, fast-growing tree that can grow as tall as 40 meters (130 feet) and have a trunk diameter of 1 meter (more than 3 feet) or more. Its leaves are pendulous and occur in two rows along the branches. Its unusual bark is dark brown and knobby.

The Wollemi pine is closely related to several gymnosperm species native to the Southern Hemisphere, such as the Norfolk Island pine (native to Norfolk Island in the South Pacific) and the monkey-puzzle tree (native to Chile). Fossil evidence suggests that the plant family to which these plants are assigned (the Araucariaceae) is 200 million years old—dating back to the time when dinosaurs dominated the planet.

Because there are so few Wollemi pines in the wild, their location is kept secret to protect them from unscrupulous collectors and vandals. Meanwhile, botanists are cultivating the plants by seeds and stem cuttings (Wollemi pines form sprouts at the base of the trunk) without harming the wild plants. These plants are available to botanical gardens and individual collectors; proceeds are used to protect the Wollemi pines and other endangered species in Australia.

This chapter focuses on pines and other gymnosperms, one of the two groups of seed plants. The other group—the flowering plants—is discussed in Chapter 25.

Wollemi pines (*Wollemia nobilis*). This wild population exists in a secret location near Sydney, Australia.

Seed Plants

So far, our discussion of plant diversity has covered bryophytes and seedless vascular plants, all of which reproduce by means of *spores*—haploid reproductive units that give rise to gametophytes. Although the most successful and widespread plants also produce spores, their primary means of reproduction and dispersal is by **seeds,** which represent an important adaptation for life on land. Each seed consists of an embryonic sporophyte plant, nutritive tissue, and a protective coat (•**Figure 24-1a**). Seeds develop from the fertilized egg cell, the female gametophyte, and its associated tissues. The two groups of seed plants, gymnosperms and angiosperms (flowering plants), exhibit the greatest evolutionary complexity in the plant kingdom and are the dominant plants in most terrestrial environments.

Seeds are reproductively superior to spores for three main reasons. First, a seed contains a well-developed multicellular young plant with embryonic root, stem, and leaves already formed, whereas a spore is a single cell. The parent plant protects and supports the young plant in the seed during its development; spores do not receive such protection.

Second, a seed contains an abundant food supply. After *germination* (in which the young plant begins to grow and establish itself as an independent plant), food stored in the seed nourishes the plant embryo until it becomes self-sufficient. Because a spore is a single cell, minimal food reserves exist to sustain the plant that develops from a germinating spore.

Third, a multicellular **seed coat** surrounds and protects a seed. Seeds can survive for extended periods at reduced rates of metabolism and germinate when conditions become favorable.

Seeds and seed plants are intimately connected with the development of human civilization. From pre-

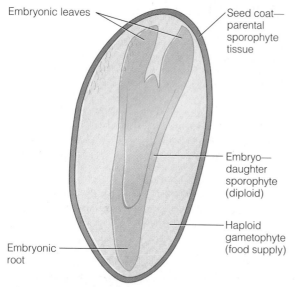

Embryonic leaves

Seed coat—parental sporophyte tissue

Embryo—daughter sporophyte (diploid)

Haploid gametophyte (food supply)

Embryonic root

(a) Cross section through a pine seed.

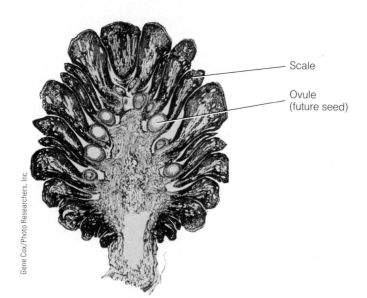

Gene Cox/Photo Researchers, Inc.

Scale

Ovule (future seed)

(b) Gymnosperm seed. Longitudinal section through a female pine cone, showing the ovules (which develop into seeds) borne on scales. Note the absence of an ovary wall.

C Squared Studios/Photodisc/Getty Images

Fruit (ovary wall)

Seed

(c) Angiosperm seed. Longitudinal section through an avocado fruit, showing the seed surrounded by ovary tissue of the maternal sporophyte.

FIGURE 24-1 Seeds.

historic times, early humans collected and used seeds for food. The food stored in seeds is a concentrated source of proteins, oils, carbohydrates, and vitamins, which are nourishing for humans as well as for germinating plants. Seeds are easy to store (if kept dry), so humans can collect them during times of plenty to save for times of need. Few other foods are stored as conveniently or for as long. Although flowering plants produce most seeds that humans consume, the seeds of certain gymnosperms—the pinyon pine (*Pinus edulis*), for example—are edible. A staple in the diets of many native Americans, pinyon pine seeds are now sold as "pine nuts."

Gymnosperms and flowering plants bear seeds

In Chapter 23, you learned that some seedless vascular plants are **heterosporous.** However, *all* seed plants are heterosporous and produce two types of spores: microspores and megaspores. In fact, heterospory is a requirement of seed production.

Following fertilization in seed plants, an **ovule,** which consists of a *megasporangium* and its enclosed structures, develops into a seed. Seed plants also have **integuments,** one or two layers of sporophyte tissue that surround and enclose the megasporangium. The integument has a **micropyle,** a tiny opening at one end through which the pollen tube enters. After fertilization takes place, the seed coat develops from the integuments.

Botanists divide seed plants into two groups based on whether or not an ovary wall surrounds their ovules (an *ovary* is a structure that contains one or more ovules) (•**Figure 24-1b and c**). The two groups of seed plants are the **gymnosperms** and the **angiosperms,** or flowering plants. The word *gymnosperm* is derived from the Greek for "naked seed." Gymnosperms produce seeds that are totally exposed or borne on the scales of female strobili (cones). In other words, there is no ovary wall surrounding the ovules of gymnosperms. Pine, spruce, fir, hemlock, and ginkgo trees are examples of gymnosperms.

The term *angiosperm* is derived from the Greek expression that means "seed enclosed in a vessel or case." Flowering plants produce their seeds within a *fruit* (a mature ovary). Thus, the ovules of angiosperms are protected. Flowering plants, which are extremely diverse, include corn, grasses, oaks, water lilies, maples, roses, cacti, apples, palms, and buttercups.

Both gymnosperms and flowering plants have vascular tissues: **xylem** for conducting water and dissolved minerals (inorganic nutrients), and **phloem** for conducting dissolved sugar. Both have life cycles with an **alternation of generations;** that is, they spend part of their lives in the multicellular diploid sporophyte stage and part in the multicellular haploid gametophyte stage. The sporophyte generation is the dominant stage in each group, and the gametophyte generation is significantly reduced in size and entirely dependent on the sporophyte generation. Unlike the plants we have considered so far (bryophytes and seedless vascular plants), gymnosperms and flowering plants do not have free-living gametophytes. Instead, the female gametophyte is attached to and nutritionally dependent on the sporophyte generation.

Gymnosperms

The gymnosperms include some of the most interesting members of the plant kingdom. For example, one of the world's most massive organisms is the General Sherman tree, a giant sequoia (*Sequoia giganteum*) in Sequoia National Park, California. It is 82 meters (267 feet) tall and has a girth of 23.7 meters (77 feet) measured 1.5 meters (5 feet) above ground level. A coast redwood (*S. sempervirens*) known as the Mendocino tree is among the world's tallest trees, measuring 112 meters (364 feet). Botanists used tree-ring analysis to determine that one of the oldest living trees, a bristlecone pine (*Pinus aristata*) in the White Mountains of California, is 4900 years old (•**Figure 24-2**).

The four phyla into which gymnosperms are classified represent four evolutionary lines (•**Figure 24-3** and •**Table 24-1**). Most gymnosperms belong to the phylum Coniferophyta and are commonly called conifers. Two phyla of gymnosperms, Ginkgophyta (the ginkgo) and Cycadophyta (the cycads), are evolutionary remnants of groups that were more significant in the past. The fourth phylum, Gnetophyta (gnetophytes), is a collection of unusual plants that share certain traits not found in the other gymnosperms.

SEED A reproductive body consisting of a young, multicellular plant and food reserves, enclosed by a seed coat.

OVULE The structure in seed plants that develops into a seed following fertilization.

INTEGUMENT The outer layer of an ovule that develops into a seed coat following fertilization.

GYMNOSPERM Any of a group of seed plants in which the seeds are not enclosed in an ovary.

FIGURE 24-2 The bristlecone pine.
Bristlecone pines (*Pinus aristata*), found in mountainous parts of California, Nevada, and Utah, are the world's longest-lived trees. Bristlecone pines are conifers, gymnosperms that produce their seeds in cones.

Seed plants include four gymnosperm phyla and one phylum of flowering plants (angiosperms).

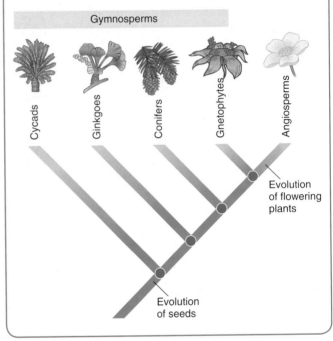

FIGURE 24-3 Gymnosperm evolution.
This cladogram shows one hypothesis of phylogenetic relationships among living seed plants. It is based on structural evidence and molecular comparisons. The arrangement of the gymnosperm groups shown here may change as future analyses help clarify relationships.

TABLE 24-1 A Comparison of Living Gymnosperm Phyla

PLANT GROUP	REPRESENTATIVE GENERA	REPRODUCTIVE PARTS	LEAVES	XYLEM
Phylum Coniferophyta (conifers)	*Abies, Picea, Pinus, Sequoia, Wollemia*	Most are monoecious (separate male and female cones in different locations on the same plant); seeds in cones	Needles or scalelike leaves	Tracheids
Phylum Cycadophyta (cycads)	*Encephalartos, Zamia*	Dioecious (male and female cones on separate plants); seeds in cones	Compound, palmlike leaves	Tracheids
Phylum Gingkophyta (ginkgo)	Only one genus: *Ginkgo*	Dioecious (ovules and male strobili on separate plants); exposed, fleshy seeds	Fan-shaped leaves	Tracheids
Phylum Gnetophyta (gnetophytes)	*Ephedra, Gnetum, Welwitschia*	Most are dioecious (male and female cones on separate plants); fleshy seeds	Varies depending on genus	Tracheids and vessel elements

Dennis Drenner

(a) Colorado blue spruce (*Picea pungens*).

David Cavagnaro/Peter Arnold, Inc.

(b) White fir (*Abies concolor*).

FIGURE 24-4 **Conifers.**

Conifers

There are about 630 species of **conifers** (phylum Coniferophyta). Pines, hemlocks, spruces, and firs are examples of conifers, which are the most familiar group of gymnosperms (•**Figure 24-4**). These woody trees or shrubs produce annual additions of secondary tissues (wood and bark); there are no herbaceous (nonwoody) conifers. The wood (*secondary xylem*) consists of tracheids—long, tapering cells with pits through which water and dissolved minerals move from one cell to another.

Many conifers produce **resin,** a viscous, clear or translucent substance consisting of several organic compounds that may protect the plant from fungal or insect attack. The resin collects in **resin ducts,** tubelike cavities that extend throughout the roots, stems, and leaves. Cells lining the resin ducts produce and secrete resin.

Most conifers have leaves (*megaphylls*) called **needles** that are commonly long and narrow, tough, and leathery (•**Figure 24-5**). Pines bear clusters of two to five needles, depending on the species. In a few conifers, such as American arborvitae (*Thuja occidentalis*), the leaves are scalelike and cover the stem.

Most conifers are **evergreen** and bear their leaves (needles) throughout the year. Pine needles, which are evergreen, have a thick, waxy cuticle and stomata that are sunken in depressions (see Figure 8-11). These features are water-conserving adaptations that enable the pine tree to retain its needles throughout the winter, when low temperatures inhibit the roots from absorbing water. Only a few conifers, such as the dawn redwood, larch, and bald cypress, are **deciduous** and shed their needles at the end of the growing season.

Most conifers are **monoecious;** they have separate male and female reproductive parts in different locations on the same plant. These reproductive parts are generally borne in *strobili* (commonly called **cones**), hence the name *conifer,* which means "cone bearing."

Conifers occupy extensive areas, ranging from the Arctic to the tropics, and are the dominant vegetation in *boreal forests,* the forested northern regions of Alaska,

CONIFER Any of a large phylum of gymnosperms that are woody trees and shrubs with needlelike, mostly evergreen, leaves and seeds in cones.

MONOECIOUS Having male and female reproductive parts in separate flowers or cones on the same plant.

(a) In white pine (*Pinus strobus*), leaves are long, slender needles that occur in clusters of five.

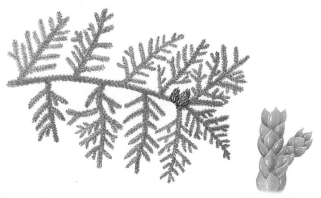

(b) In American arborvitae (*Thuja occidentalis*), leaves are small and scalelike (*see inset*).

FIGURE 24-5 **Leaf variation in conifers.**

Canada, Europe, and Siberia. In addition, they are important in the Southern Hemisphere, particularly in wet, mountainous areas of temperate and tropical regions in South America, Australia, New Zealand, and Malaysia. Southwestern China, with more than 60 species of conifers, has the greatest regional diversity of conifer species in the world. California, New Caledonia (an island east of Australia), southeastern China, and Japan also have considerable diversity of conifer species.

Pines represent a typical conifer life cycle

The genus *Pinus,* by far the largest genus in the conifers, consists of about 100 species. A pine tree is a mature sporophyte. Pine is heterosporous and therefore pro-

duces two kinds of spores, microspores and megaspores, in separate cones on the sporophyte (•**Figure 24-6 ❶**). Male cones, usually 1 centimeter or less in length, are smaller than female cones and generally occur in dense clusters on the ends of lower branches each spring (•**Figure 24-7**). The more familiar, woody female cones, which are on the tree year-round, are usually found on the upper branches of the tree and bear seeds after reproducing (see Figure 24-7). Female cones vary considerably in size among different species. The sugar pine (*P. lambertiana*) of California produces the world's longest female cones, which reach lengths of 60 centimeters (2 feet).

Each male cone (also called a *pollen cone*) consists of **sporophylls,** leaflike scales that bear sporangia on the underside (•**Figure 24-8**). At the base of each sporophyll are two **microsporangia,** which contain numerous *microspore mother cells.* Each microspore mother cell undergoes meiosis to form four haploid **microspores** (see Figure 24-6 ❷). Microspores then develop into extremely reduced *male gametophytes.* Each immature male gametophyte, also called a **pollen grain,** consists of four cells, two of which—a *generative cell* and a *tube cell*—are involved in reproduction (see Figure 24-8 inset). The other two cells soon degenerate. Two large air sacs on each pollen grain provide buoyancy for wind dissemination. Male cones shed pollen grains in great numbers, and wind currents carry some to the immature female cones (see Figure 24-6 ❸).

Many botanists think that the female cones (also called *seed cones*) are modified branch systems. Each cone scale bears two ovules, or **megasporangia,** on its upper surface ❹ (also see Figure 24-1b). Within each megasporangium, meiosis of a *megaspore mother cell* produces four haploid **megaspores.** One of these divides mitotically, developing into a *female gametophyte,* which produces several *archegonia,* each containing one egg. The other three megaspores are nonfunctional and soon degenerate.

When the ovule is ready to receive pollen, it produces a sticky droplet by the opening where the pollen grains land. **Pollination,** the transfer of pollen from the male cones to the female cones, occurs in the spring for a week or 10 days, after which the pollen cones wither and drop off the tree. One of the many pollen grains that adhere to the sticky female cone grows a **pollen tube,** an outgrowth that digests its way through the megasporangium to the egg within an archegonium. The germinated pollen grain with its pollen tube is the mature male gametophyte. The tube cell and the generative cell enter the pollen tube.

The tube cell is involved in the growth of the pollen tube. Within the pollen tube, the generative cell divides and forms a *stalk cell* and a *body cell;* the body cell later

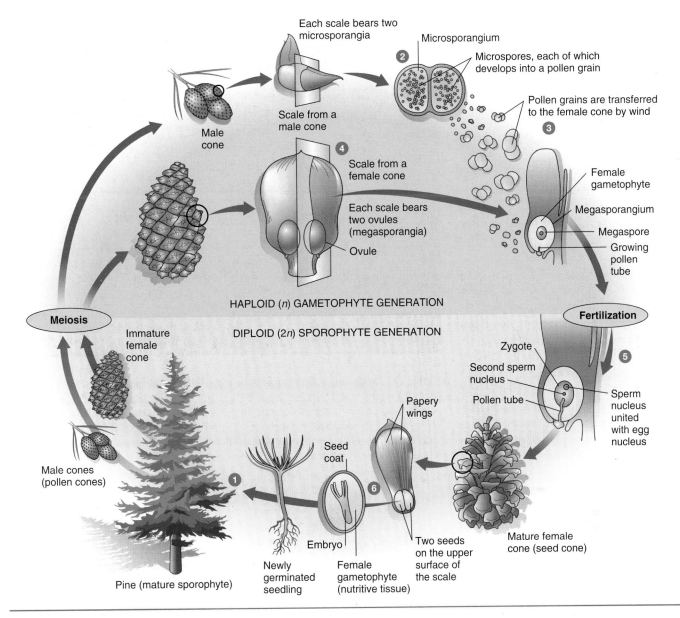

FIGURE 24-6 The life cycle of pine.
Pine and other gymnosperms produce wind-borne pollen grains. See text for a detailed description of the life cycle.

divides and forms two nonmotile (unflagellated) sperm cells. The pollen tube, when it reaches the female gametophyte, discharges the two sperm cells near the egg. One of these sperm cells fuses with the egg, in the process of **fertilization,** to form a zygote, or fertilized egg (❺), which subsequently grows into a young pine embryo in the seed (❻). The other sperm cell degenerates.

The developing embryo consists of an embryonic root and an embryonic shoot with several cotyledons (embryonic leaves). The embryo is embedded in haploid female gametophyte tissue that becomes the nutritive tissue in the mature pine seed. A tough protective seed coat surrounds the embryo and nutritive tissue. The

seed coat forms a thin, papery wing at one end that enables dispersal by air currents.

A long time elapses between the appearance of female cones on a pine tree and the maturation of seeds

SPOROPHYLL A leaflike structure that bears spores within a sporangium (or sporangia).

POLLEN GRAIN The structure in seed plants that develops from a microspore into a male gametophyte.

POLLEN TUBE In seed plants, a tube that forms after the germination of a pollen grain and through which male gametes (sperm cells) pass into the ovule.

in those cones. When pollination occurs during the first spring, the female cone is still immature, and meiosis of the megaspore mother cells has not yet occurred. After the megaspore has formed, eggs take more than a year to form within archegonia. Meanwhile, the pollen tube grows slowly through the megasporangium to the archegonia. About 15 months after pollination, fertilization occurs and the embryo begins to develop. Seed maturation takes several additional months, although some seeds remain within the female cones for several years before being shed.

In the pine life cycle, the sporophyte generation is dominant, and the gametophyte generation is restricted in size to microscopic structures in the cones. Although the female gametophyte produces archegonia, the male gametophyte is so reduced that it does not produce antheridia (male gametangia). The gametophyte generation in pines, as in all seed plants, depends totally on the sporophyte generation for nourishment.

A major adaptation in the pine life cycle is elimination of the requirement for external water as a sperm transport medium. Instead, air currents carry pine pollen grains to female cones, and sperm cells accomplish fertilization by moving through a pollen tube to the egg. Pine and other conifers are plants whose reproduction is totally adapted for life on land.

Some conifers produce fleshy, cuplike structures

Some conifers do not produce woody female cones. As shown in •**Figure 24-9**, each yew (*Taxus*) seed is almost completely surrounded by a fleshy, cuplike covering that is an outgrowth from the base of the seed. The seed coverings are red and attract birds, which eat them and disperse the seeds. Although the yew is an attractive ornamental and is planted extensively, it is not widely known that its bark, leaves, and seeds are poisonous. Chil-

FIGURE 24-7 Male and female cones in jack pine (*Pinus banksiana*).
(*Left*) A mature, woody female cone has opened to shed its seeds. (*Right*) A cluster of golden male cones produces copious amounts of pollen in the spring. The location of young female cones above male cones facilitates cross pollination between different individuals, because it is unlikely that wind will blow pollen grains directly upward to female cones on the same tree.

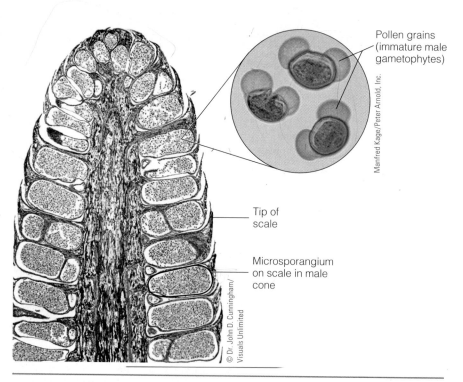

Pollen grains (immature male gametophytes)

Tip of scale

Microsporangium on scale in male cone

FIGURE 24-8 Pollen cone.
Longitudinal section through a male pine cone showing microsporangia containing microspores. Each microspore develops into a pollen grain (*see inset*).

Edgar E. Webber

FIGURE 24-9 Yew seeds.
The seeds of yews (*Taxus*) are borne in fleshy, cuplike structures. Birds and other animals eat the fleshy structures and disperse the seeds.

dren should be warned not to eat the brightly colored seed coverings so that they do not accidentally ingest the seeds.

Juniper (*Juniperus*) bears its seeds in fleshy cones that resemble berries. The fleshy scales of the cones are fused and completely envelop the seeds. At maturity, these cones are colored orange, red, brown, purple, or black, depending on the species. Birds eat the cones and swallow the seeds, which are dispersed in their droppings. Gin makers dry the fleshy cones of the common juniper (*Juniperus communis*) and use them to flavor the gin.

Cycads

The **cycads** (phylum Cycadophyta) were important during the Triassic period, which began about 251 mya and is sometimes referred to as the "Age of Cycads." Most species of cycads are now extinct, and the few surviving cycads—about 140 species—are tropical and subtropical plants with stout, trunklike stems and compound leaves that resemble those of palms or tree ferns (•**Figure 24-10**).

Reproduction in cycads is similar to that in pines except that cycads are **dioecious** and therefore have seed cones on female plants and pollen cones on male plants. Their seed structure is most like that of the earliest seeds found in the fossil record. Cycads have also retained motile sperm cells, each of which has many

DIOECIOUS Having male and female reproductive structures on separate plants.

Walter H. Hodge/Peter Arnold, Inc.

(a) This cycad (*Encephalartos transvenosus*) in South Africa grows to approximately 9.2 meters (30 feet) and resembles a palm.

John Arnaldi

(b) A female sago cycas (*Zamia integrifolia*) in Florida is shown with several seed cones. Like *Zamia*, most cycads are short plants (under 2 meters tall).

FIGURE 24-10 Cycads.

short, hairlike flagella. Motile sperm cells are a vestige retained from the ancestors of cycads, in which sperm cells swam from antheridia to archegonia. In cycads, air or insects carry pollen grains to the female plants and their cones; there the pollen grain germinates and grows a pollen tube. The sperm cells are released at the top of this tube and swim through the tube to the egg. In other words, despite having motile sperm cells, cycads do not require water as a transport medium for fertilization.

Many cycads are endangered species, primarily because they are popular as ornamentals and are gathered from the wild and sold to collectors. One species (*Encephalartos woodii*) consists of a single male individual native to South Africa; all of the female "woodiis" have apparently become extinct. Offshoots from the single male woodii have been cultivated in botanical gardens around the world, but the species cannot reproduce sexually because no females exist.

Ginkgo

There is only one living species in the phylum Ginkgophyta—the **ginkgo**, or maidenhair tree (*Ginkgo biloba;* •**Figure 24-11**). Ginkgo is so named because of the resemblance of its fan-shaped leaves to those of the maidenhair fern, discussed in the Chapter 23 introduction. Native to eastern China, ginkgo has been cultivated for centuries. It is found growing wild in only two locations in China, and it probably would have become extinct had it not been cultivated for its edible seeds in China and Japan.

Ginkgo is the oldest genus (and species) of living trees. In China, scientists have discovered fossil ginkgo leaves and wood that are 170 million years old and structurally similar to the modern ginkgo. Fossil ginkgo ovules more than 120 million years old also bear a remarkable resemblance to ginkgo ovules of the present day.

Ginkgo is widely cultivated in North America and Europe today, particularly in parks and along city streets, where it is planted frequently because it is hardy and somewhat resistant to air pollution and disease. Its leaves are deciduous and turn a beautiful yellow before being shed in the fall.

(a) A young male ginkgo, or maidenhair (*Ginkgo biloba*), tree.

(b) A young branch bearing male strobili. As the leaves age, they become a darker green.

(c) Close-up of a branch from a female ginkgo tree, showing the exposed seeds.

FIGURE 24-11 The ginkgo.

© Dr. John D. Cunningham/Visuals Unlimited

(a) The leaves of *Gnetum gnemon* resemble those of flowering plants. The exposed seeds are yellow to red when ripe.

© David Wall/Alamy

(c) *Welwitschia* is native to deserts in southwestern Africa. It survives on moisture-laden fogs that drift inland from the ocean. Photographed in the Namib Desert, Namibia.

FIGURE 24-12 **Gnetophytes.**

Like cycads, ginkgoes are dioecious, with separate male and female trees. They have flagellated sperm cells, an evolutionary vestige that is not required, because the ginkgo, like the cycads, produces airborne pollen grains. Ginkgo seeds are completely exposed rather than contained within cones. Only female trees produce seeds, a fact you should remember if you ever wish to plant ginkgoes. As the seeds mature, their outer fleshy coverings give off a foul odor, which makes the female trees undesirable. Some cities and towns have ordinances

© Bob Gibbons/Alamy

(b) A male joint fir (*Ephedra*) has pollen cones clustered at the nodes. In the 19th century, European pioneers used species native to the American Southwest to make a beverage, Mormon tea.

prohibiting the planting of female ginkgoes. In China and Japan, where people eat the seeds, the female trees are more common.

Ginkgo has been an important medicinal plant for centuries. Extracts from the leaves may enhance neurological functioning by increasing blood flow to the brain. Several studies are under way to determine if ginkgo improves memory in elderly people.

Gnetophytes

The **gnetophytes** (phylum Gnetophyta) is composed of about 70 species in three diverse genera (*Gnetum*, *Ephedra*, and *Welwitschia*). Gnetophytes share certain features that make them unique among the gymnosperms. For example, gnetophytes have more efficient water-conducting cells, called *vessel elements*, in their xylem. Flowering plants also have vessel elements in their xylem, but of the gymnosperms, only the gnetophytes do. Also, the cone clusters that some gnetophytes produce resemble flower clusters, and certain details in their life cycles resemble those of flowering plants.

The genus *Gnetum* contains tropical vines (called *lianas*) and a few shrubs and trees. *Gnetum* has simple, broad leaves with an opposite leaf arrangement on the stem. Its seeds have a fleshy, brightly colored outer envelope that superficially resembles a fruit of flowering plants (•**Figure 24-12a**).

Species in the genus *Ephedra* include many shrubs and vines found in deserts and other dry temperate and tropical regions (•**Figure 24-12b**). Most species of *Ephedra* are dioecious and have separate male and female plants. Some *Ephedra* species resemble horsetails in that they have jointed green stems with tiny leaves. Commonly called joint fir, *Ephedra* has been used medicinally for centuries. An Asiatic species (*E. distachya*) is the source of ephedrine, which stimulates the heart and raises blood pressure. Ephedrine at one time was sold over the counter in weight-control medications and herbal energy-boosters. Several deaths from chronic use or overdose of products containing ephedrine caused the U.S. Food and Drug Administration to take ephedrine off the market in 2004. (The combination of caffeine with ephedrine caused some users to have cardiac arrthymias, heart attacks, or strokes.)

The gnetophyte genus, *Welwitschia*, contains a single species found in deserts of southwestern Africa (•**Figure 24-12c**). Most of *Welwitschia*'s body—a long taproot—grows underground. Its short, wide stem forms a shallow disc, up to 0.9 meter (3 feet) in diameter, from which two ribbonlike leaves extend. These two leaves continue to grow throughout the plant's life, but their ends are usually broken and torn by the wind, giving the appearance of numerous leaves. Each leaf grows to about 2 meters (6.5 feet) in length. When *Welwitschia* reproduces, cones form around the edge of its disclike stem.

EVOLUTION LINK The Evolution of Seed Plants

One group of plants that evolved from ancestral seedless vascular plants was the **progymnosperms,** all now extinct. Progymnosperms had two derived features absent in their immediate ancestors: leaves with branching veins (megaphylls) and woody tissue (secondary xylem) similar to that of modern gymnosperms. Progymnosperms, however, reproduced by spores. *Archaeopteris,* a progymnosperm that lived about 370 mya, is the earliest known tree with "modern" woody tissue (•**Figure 24-13a**).

Scientists have discovered fossils of several progymnosperms with reproductive structures intermediate between those of spore plants and seed plants. For example, the evolution of microspores into pollen grains and of megasporangia into ovules (seed-producing structures) can be traced in fossil progymnosperms. Seed-producing plants appeared during the Late Devonian period, more than 359 mya. The fossil record indicates that different groups of seed plants apparently arose independently several times.

As mentioned previously, fossilized remains of ginkgo are found in 170-million-year-old rocks, and other groups of gymnosperms were well established by 160 mya to 100 mya. Although the gymnosperms are an ancient group, some questions persist about the exact pathways of gymnosperm evolution. The fossil record indicates that progymnosperms probably gave rise to conifers and to another group of extinct plants called **seed ferns,** which were seed-bearing woody plants with fernlike leaves (•**Figure 24-13b**). In turn, the seed ferns probably gave rise to cycads and possibly ginkgoes, as well as to several plant groups now extinct. The origin of gnetophytes remains unclear, although molecular data suggest that they are closely related to conifers. ■

(a) Progymnosperm. *Archaeopteris,* which existed about 370 mya, had some features in common with modern seed plants but did not produce seeds.

(b) Seed fern. *Emplectopteris* produced seeds on fernlike leaves. Seed ferns existed from about 360 mya to 250 mya.

FIGURE 24-13 Evolution of seed plants.
(**a,** Redrawn from C. B. Beck, "Reconstructions of *Archaeopteris* and Further Consideration of Its Phylogenetic Position," *American Journal of Botany,* Vol. 49, 1962; **b,** Redrawn from H. N. Andrews, *Ancient Plants and the World They Lived In,* Comstock, New York, 1947.)

PLANTS AND PEOPLE | Old-Growth Forests of the Pacific Northwest

The most commercially valuable public forest land in the United States is in northern California, western Oregon, and western Washington. A drama involving people's jobs and the environment unfolded there during the late 1980s and 1990s. At stake were thousands of jobs and the future of large tracts of old-growth (virgin) coniferous forest, along with the existence of organisms that depend on the forest. One of these forest animals, the northern spotted owl, came to symbolize the confrontation (see figure).

Biologists regard the few remaining old-growth forests of Douglas fir, western hemlock, and spruce in the Pacific Northwest as a living laboratory that demonstrates the complexity of one of the few natural ecosystems not extensively altered by humans. To environmentalists, the old-growth forests, with their 2000-year-old trees, are a national treasure. These stable forest ecosystems provide biological habitats for many species, including the northern spotted owl and 40 other endangered or threatened species. Provisions of the *Endangered Species Act* require the government to protect the habitat of endangered species so that their numbers increase. To enforce this law in the Pacific Northwest, in 1991 a court ordered the suspension of logging in about 1.2 million hectares (3 million acres) of federal forest, where the owl lives.

The timber industry opposed the moratorium, stating that thousands of jobs would be lost if the northern spotted owl habitat was set aside. The situation was more complicated than simply jobs versus the environment, however. The timber industry was already declining in terms of its ability to support people in the Pacific Northwest. During the decade between 1977 and 1987, logging in Oregon's national forests had increased by more

■ **The northern spotted owl. This endangered species is found primarily in old-growth forests in the Pacific Northwest.**

than 15 percent, whereas employment for loggers, sawmill operators, and the like dropped by an estimated 12,000 jobs during the same period. The main cause of this decline in jobs was automation of the timber industry. Also, the timber industry in that region was not operating *sustainably:* trees were being removed faster than the forest could regenerate. If the industry had continued to log at its 1980s rates, most of the remaining old-growth forest would have disappeared within 20 years.

Fortunately, timber is not as important to the economy of the Pacific Northwest as it used to be. This change started long before the northern spotted owl controversy. The economies of northwestern states have become more diversified; by the late 1980s, the timber industry's share of the economy in Oregon and Washington was less than 4 percent.

The 1994 Northwest Forest Plan provided federal aid to the area to retrain some timber workers for other

jobs. State programs also helped reduce unemployment. Logging resumed in federal forests in Washington, Oregon, and northern California, but at a fraction of the level permitted in the 1980s. About 75 percent of federal forests were reserved to safeguard watersheds and provide habitat for several hundred species.

As with many compromises, neither environmental groups nor timber-cutting interests were happy with the plan. Timber-cutting interests thought the plan was too restrictive, and they challenged the forestry laws and the Endangered Species Act on which the plan was based. In 1996, the U.S. Supreme Court upheld the legislation that protects the northern spotted owl and other endangered species. In 1999, certain environmental groups successfully sued the U.S. Forest Service and the Bureau of Land Management for not adequately carrying out provisions of the Northwest Forest Plan that protect endangered species.

Ecological and Economic Significance of Gymnosperms

The conifers are the most important phylum of gymnosperms, both ecologically and commercially, because they are far more common than the other, relatively obscure gymnosperms. Conifers are the predominant trees in about 35 percent of the world's forests. In North America, western and northern forests are primarily coniferous, as are those in the southeastern United States.

Coniferous forests play an essential role in the environment. Their roots hold vast tracts of soil in place, thereby reducing soil erosion. Forests are important *watersheds* (areas of land drained by river systems) because they absorb, hold, and slowly release water. Watersheds provide a well-regulated flow of water to river systems, even during dry periods, and help control floods by absorbing floodwaters. In addition, forests provide food and shelter to animals, fungi, and many other organisms (see *Plants and People: Old-Growth Forests of the Pacific Northwest* on page 479).

The conifers are also of considerable economic importance for both recreation and timber. Recreational uses of forests include such activities as camping, backpacking, picnicking, and observing nature. These activities, which require equipment, supplies, and guides, support a large recreation industry.

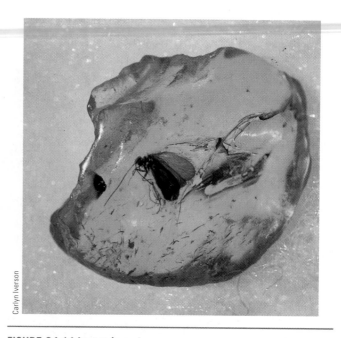

FIGURE 24-14 Insect in amber.
A prehistoric caddis fly embedded in Baltic amber has been preserved almost perfectly for 35 million years.

(a) Douglas fir (*Pseudotsuga menziesii*) grows along the Pacific coast and in the Rocky Mountains.

(b) Red spruce (*Picea rubens*), found in eastern Canada and the northeastern United States, also extends southward to the Great Smoky Mountains.

(c) Loblolly pine (*Pinus taeda*) is widely distributed through the southeastern United States.

FIGURE 24-15 Three commercially important conifers.

In the United States, about 80 percent of our timber crop comes from conifers. Humans use conifers for lumber (especially for building materials and paper products), medicinal products (such as the anticancer drug Taxol from the Pacific yew), and turpentine and resins, which are ingredients in varnishes and certain plastics. Because of their attractive appearance, conifers are grown commercially for landscape design and for Christmas trees. Amber, the fossilized resin of prehistoric conifers, is often cut and polished to make jewelry. Some amber contains fossilized insects, spiders, small lizards, and plant fragments (•**Figure 24-14**).

Three trees are highlighted in this discussion as representative conifers with commercial value: the Douglas fir from western forests, the red spruce from northern forests, and the loblolly pine from southeastern forests.

Douglas fir (*Pseudotsuga menziesii*), the state tree of Oregon, is a major timber tree in both the Rocky Mountain and Pacific forests of the West (•**Figure 24-15a**). Its wood, which is strong, durable, and attractive when stained, is used for many products, including telephone poles, railroad ties, fences, pulp for paper, furniture, and plywood. More plywood is made from Douglas fir than from any other tree. Douglas firs are also grown as ornamentals.

Red spruce (*Picea rubens*), the provincial tree of Nova Scotia, is an important timber tree in northern coniferous forests (•**Figure 24-15b**). Its primary range extends from southeastern Canada to New England, and it is found in the Appalachian Mountains.[1] The wood of red spruce is relatively lightweight but strong and is used to make such products as oars and ladders as well as musical instruments, such as violins and pianos. Much of the red spruce that is cut today is used to make paper, paperboard (for containers), and fiberboard. Small red spruces are harvested and sold as Christmas trees.

Loblolly pine (*Pinus taeda*), one of the southern yellow pines, is the most abundant timber tree in the southern coniferous forest (•**Figure 24-15c**). It is also planted extensively on tree plantations. Loblolly pine is used primarily to make paper, paperboard, and fiberboard. It is also valued as a construction material and is used to make turpentine.

---

[1]*Red spruce is declining and dying throughout much of its range in the Appalachians. Environmental pollution, including acid rain, is thought to be a contributing factor.*

STUDY OUTLINE

❶ Compare the features of seeds with those of spores, and discuss the adaptive advantages of plants that reproduce by seeds.

A **seed** is a reproductive body consisting of a young, multicellular plant and food reserves, enclosed by a seed coat. Seeds are reproductively superior to spores; a spore is a single cell with minimal food reserves to sustain the plant that develops from a germinating spore.

❷ Distinguish between a seed and an ovule, and define *integuments*.

An **ovule** is the structure in seed plants that develops into a seed following fertilization. An **integument** is the outer layer of an ovule that develops into a seed coat following fertilization.

❸ Summarize the features that distinguish gymnosperms from seedless vascular plants.

A **gymnosperm** is any of a group of seed plants in which the seeds are not enclosed in an ovary; that is, gymnosperm seeds are either totally exposed or are borne on the scales of cones. Unlike seedless vascular plants, gymnosperms produce seeds. Gymnosperms also produce wind-borne **pollen grains,** which develop from microspores of seed plants into male gametophytes; seedless vascular plants do not produce pollen grains.

❹ Name and briefly describe the four phyla of gymnosperms.

A **conifer** is any of a large phylum of gymnosperms that are woody trees and shrubs with needlelike, mostly **evergreen,** leaves and seeds in cones. **Cycads** are palmlike or fernlike in appearance and reproduce with pollen and seeds in conelike structures; there are relatively few surviving members of this once-large phylum. *Ginkgo biloba*, the only surviving species in its phylum, is a **deciduous** tree; female **ginkgoes** produce fleshy seeds directly on branches. **Gnetophytes** share a number of traits with angiosperms, including the presence of more efficient water-conducting cells, called vessel elements, in their xylem.

⑤ Contrast monoecious plants and dioecious plants.
Monoecious plants have male and female reproductive parts in separate flowers or cones on the same plant. **Dioecious** plants have male and female reproductive structures on separate plants. Most conifers are monoecious, whereas cycads, ginkgo, and most gnetophytes are dioecious.

⑥ Trace the steps in the life cycle of pine, and compare the sporophyte and gametophyte generations.
A pine tree is a mature sporophyte; pine gametophytes are extremely small and nutritionally dependent on the sporophyte generation. Pine is **heterosporous** and produces microspores and megaspores in separate cones. Each cone has **sporophylls**, leaflike structures that bear spores within a sporangium (or sporangia). Male cones produce **microspores** that develop into pollen grains that are carried by air currents to female cones. Female cones produce **megaspores**. One of the four megaspores produced by meiosis develops into a female gametophyte within an ovule (**megasporangium**). After **pollination**, the transfer of pollen to the female cones, a **pollen tube** grows through the megasporangium to the egg within the archegonium. Male gametes (sperm cells) pass through the pollen tube into the ovule. After **fertilization**, the zy-gote develops into an embryo encased in a seed adapted for wind dispersal.

⑦ Describe the ecological and economic significance of gymnosperms.
Conifers are the predominant trees in about 35 percent of the world's forests. Their roots hold soil in place, thereby reducing soil erosion. Conifer forests are important watersheds and provide habitat for many organisms. Recreational uses of forests include camping, backpacking, picnicking, and observing nature. Humans use conifers for lumber, medicinal products, and turpentine and resins. Conifers are also grown commercially for landscape design and for Christmas trees.

⑧ Trace the evolution of gymnosperms from seedless vascular plants.
Seed plants evolved from seedless vascular plants. **Progymnosperms** were seedless vascular plants that had megaphylls and "modern" woody tissue. Progymnosperms probably gave rise to conifers as well as to **seed ferns,** which in turn likely gave rise to cycads and possibly ginkgo. The evolution of gnetophytes is unclear, although molecular data indicate that they are closely related to conifers.

REVIEW QUESTIONS

1. Why are seeds such a significant evolutionary development?
2. What is an ovule? How are ovules and seeds related?
3. Which features distinguish gymnosperms from seedless vascular plants?
4. What are the four groups of gymnosperms? Briefly describe each.
5. Are pines monoecious or dioecious? Explain your answer.
6. What is the dominant generation in the pine life cycle?
7. Are pine spores homosporous or heterosporous? Explain your answer.
8. How does pollination occur in the gymnosperms? Fertilization?
9. List several ways in which conifers are advanced in relation to ferns.
10. Which features do cycads, ginkgoes, and gnetophytes share with conifers?
11. Which features distinguish cycads from ginkgo? From gnetophytes?
12. How are conifers ecologically important?
13. What are some of the commercial uses of conifers?
14. What were progymnosperms?
15. Label the four gymnosperm branches on the following cladogram. Refer to Figure 24-3 to check your answers.

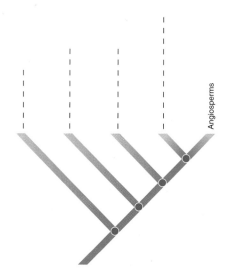

THOUGHT QUESTIONS

1. Which of the following gymnosperms could be successfully grown on your campus grounds: (a) Douglas fir, (b) cycad, (c) *Ginkgo,* (d) red spruce? Explain your answers.

2. If you were given the wood of an unknown plant and asked to identify it as either a conifer or a gnetophyte, what would you look for? How could you identify the unknown plant if you were given a few of its leaves?

3. Describe the evolutionary changes that had to take place as ancient seedless vascular plants evolved into gymnosperms.

4. Explain how a bristlecone pine can live for thousands of years yet remain "physiologically young." (*Hint:* The answer to this question is related to secondary growth.)

Visit us on the web at http://www.thomsonedu.com/biology.berg for additional resources such as flashcards, tutorial quizzes, further readings, and web links.

The Plant Kingdom: Flowering Plants

LEARNING OBJECTIVES

❶ Summarize the features that distinguish flowering plants from gymnosperms.

❷ Describe the ecological and economic significance of the flowering plants.

❸ Distinguish between monocots and eudicots, the two largest classes of flowering plants, and give specific examples of each class.

❹ Briefly explain the life cycle of a flowering plant, and describe double fertilization.

❺ Discuss some of the evolutionary adaptations of flowering plants.

❻ Trace the evolution of flowering plants from gymnosperms.

❼ Distinguish between basal angiosperms and core angiosperms.

❽ Briefly describe the distinguishing characteristics and give an example or two of each of the following flowering plant families: magnolia, walnut, cactus, mustard, rose, pea, potato, pumpkin, sunflower, grass, orchid, and agave.

The common foxglove (*Digitalis purpurea*), a member of the snap-dragon family, is an herbaceous biennial native to Europe and parts of Asia. Having been introduced widely to other parts of the world, including North America, foxglove is a popular garden plant. Besides being an ornamental, foxglove is the source of *digitalis,* an important medicine for the treatment of heart disease. Drugs produced from digitalis make the heart muscle contract strongly, increasing the flow of blood from the heart and improving the body's circulation. Digitalis also helps restore regular heartbeats in people suffering from congestive heart failure.

The scientific investigation that led to the widespread use of digitalis to treat heart problems occurred during the 18th century. An English physician, William Withering, noted that a woman dying from *dropsy,* a buildup of fluid in the tissues, recovered after drinking a tea made from the leaves and other parts of the common foxglove plant. (*Edema,* the modern term for dropsy, is a symptom of heart disease.) Withering determined that the plant's leaves contained the most active ingredients. He carefully tested increasing strengths of the dried, powdered leaves to determine the safest, most effective dosage. The painstaking care with which Withering carried out his research enabled physicians to administer digitalis safely and thereby save or prolong thousands of lives during the past 200 years. Had Withering been less precise in his studies, he almost certainly would have caused many deaths, because at higher doses digitalis is a powerful poison.

The common foxglove is an appropriate introduction to this chapter because foxglove shows how humans rely on flowering plants. Many earlier chapters of this book were devoted exclusively or primarily to the flowering plants. Chapters 5 (tissues and the multicellular plant body), 6 (roots), 7 (stems), 8 (leaves), and 9 (flowers, fruits, and seeds) examined the structures of flowering plants. This chapter emphasizes their life cycle and their economic significance.

Foxglove. The leaves of common foxglove (*Digitalis purpurea*) have medicinal value.

Flowering Plants

Flowering plants, or **angiosperms,** are classified in the phylum Anthophyta (derived from the Greek *anthus,* "flower," and *phyt,* "plant").[1] Flowering plants are the most successful and abundant plants today, surpassing even the gymnosperms in importance. The flowering plants are remarkably diverse and have adapted to almost every habitat. Consisting of at least 300,000 species, they are Earth's dominant plants. Flowering plants come in a wide variety of sizes and forms, from herbaceous violets to ivy vines to massive eucalyptus trees. Some flowering plants—tulips and roses, for example—

---

[1]*Some botanists prefer to classify flowering plants in phylum Magnoliophyta instead of phylum Anthophyta. The International Botanical Congress has not yet standardized phylum names, although the matter is under study.*

have large, conspicuous flowers; others, such as grasses and oaks, produce small, inconspicuous flowers.

Flowering plants are vascular plants that reproduce sexually by forming flowers and, after a unique double fertilization process (discussed later in the chapter), seeds within fruits. The main way in which flowering plants differ from gymnosperms, the other group of seed plants, is that the **ovules,** the structures that contain egg cells and that develop into **seeds** following fertilization, are *enclosed* within an **ovary,** which later becomes a **fruit** (•**Figure 25-1** and •**Table 25-1**). The ovary is the enlarged base of a **carpel** or the enlarged bases of a group of fused carpels. As discussed in Chapter 9, the female part of the flower is also referred to as a **pistil** and may be *simple,* consisting of a single carpel, or *compound,* consisting of several to many fused carpels (see Figure 9-4). The fruit protects the developing seeds and often aids in their dispersal.

(a) Orange flower (*Citrus sinensis*).

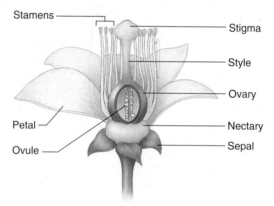

(b) Diagram of an orange flower. Note the organ that secretes nectar, known as a nectary.

(c) Developing ovaries. The stigma and style have dropped off.

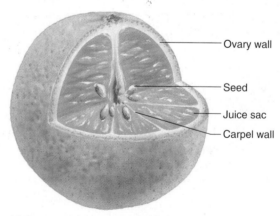

(d) A cross section through an orange. Each section of an orange is a carpel; the ovary that develops into the fruit consists of several fused carpels.

FIGURE 25-1 Development of orange fruits.

TABLE 25-1 A Comparison of Gymnosperms and Flowering Plants

CHARACTERISTIC	GYMNOSPERMS	FLOWERING PLANTS
Growth habit	Woody trees and shrubs	Woody or herbaceous
Conducting cells in xylem	Tracheids	Vessel elements and tracheids
Reproductive structures	Cones (usually)	Flowers
Pollen grain transfer	Wind	Animals, wind, or water
Fertilization	Egg and sperm cell $\longrightarrow$ zygote	Double fertilization:
		Egg and sperm cell $\longrightarrow$ zygote
		2 Polar nuclei and sperm cell $\longrightarrow$ endosperm
Seeds	Exposed or borne on scales of cones	Enclosed within fruit

Flowering plants are extremely important to humans; in fact, our survival as a species depends on them. All our major food crops are flowering plants, including the cereal crops rice, wheat, corn, and barley. Woody flowering plants such as oak, cherry, and walnut provide us with valuable lumber. Flowering plants give us fibers such as cotton and linen and medicines such as digitalis and codeine (see *Plants and People: The Medicinal Value of Flowering Plants*). Products as diverse as cork, rubber, tobacco, coffee, chocolate, and aromatic oils for perfumes come from flowering plants. **Economic botany** is the subdiscipline of botany that deals with plants of economic importance; most of these are flowering plants.

The two largest classes of flowering plants are the monocots and eudicots

Phylum Anthophyta is divided into two large classes and several small classes with only a few members each. The two large classes are the monocots (class Monocotyledones) and the eudicots (class Eudicotyledones) (•**Figure 25-2**).[2] The smaller classes will be discussed later in the chapter in the context of their evolutionary significance; for now, we restrict our discussion of flowering plants to the monocots and eudicots, which collectively represent about 97 percent of all flowering plant species. Eudicots are more diverse and include many more species (at least 200,000) than the monocots (at least 90,000). •**Table 25-2** provides a comparison of some of the general features of the two classes.

Monocots include grasses, orchids, irises, onions, lilies, and palms. Monocots are mostly herbaceous plants with long, narrow leaves that have parallel venation (the main leaf veins run parallel to one another). The parts of monocot flowers usually occur in threes. For example, a flower may have three sepals, three petals, six stamens, and a compound pistil consisting of three fused carpels. Monocot pollen grains have a single pore or furrow as a germination opening. Monocot seeds have a single **cotyledon**, or embryonic seed leaf; **endosperm**, a nutritive tissue, is usually present in the mature seed.

Eudicots include oaks, roses, mustards, cacti, blueberries, and sunflowers. Eudicots are either herbaceous (such as a tomato plant) or woody (such as a hickory tree). Their leaves vary in shape but are usually broader than monocot leaves, with netted venation (branched veins resembling a net). Flower parts usually occur in fours or fives or multiples thereof. Eudicot pollen grains have three pores or furrows as germination openings. Two cotyledons are present in eudicot seeds, and endosperm is usually absent in the mature seed, having been absorbed by the two cotyledons during seed development.

---

[2]*Botanists who prefer to classify flowering plants in phylum Magnoliophyta usually prefer class Magnoliopsida instead of class Eudicotyledones, and class Liliopsida instead of class Monocotyledones. Like phylum names, class names have not yet been standardized by the International Botanical Congress.*

ANGIOSPERM The traditional name for flowering plants, a large, diverse phylum of plants that form flowers for sexual reproduction and produce seeds enclosed in fruits.

OVULE The structure in the ovary that contains a female gametophyte and develops into a seed after fertilization.

OVARY The base of a carpel or fused carpels that contains ovules and develops into a fruit after fertilization.

MONOCOT One of two main classes of flowering plants; monocot seeds contain a single cotyledon.

COTYLEDON The seed leaf of a plant embryo, which may contain food stored for germination.

ENDOSPERM The $3n$ nutritive tissue that is formed at some point in the development of all angiosperm seeds.

EUDICOT One of two main classes of flowering plants; eudicot seeds contain two cotyledons.

TABLE 25-2 Distinguishing Features of Monocots and Eudicots

FEATURE	MONOCOTS	EUDICOTS
Flower parts	Usually in threes	Usually in fours or fives
Pollen grains	One furrow or pore	Three furrows or pores
Leaf venation	Usually parallel	Usually netted
Vascular bundles in stem cross section	Usually scattered or more complex arrangement	Arranged in a circle (ring)
Roots	Usually fibrous root system	Usually taproot system
Seeds	Embryo with one cotyledon	Embryo with two cotyledons
Secondary growth (wood and bark)	Absent	Often present

Lily (*Lilium*)

Bluets (*Hedyotis*)

- Petal
- Sepal
- Anther of stamen
- Stigmas of pistil

(a) Monocots, such as this nodding trillium (*Trillium cernuum*), have their floral parts in threes. Note the three green sepals, three white petals, six stamens, and three stigmas (the compound pistil consists of three fused carpels).

- Petal
- Anther of stamen
- Pistils

(b) Most eudicots such as this *Tacitus bellus* have floral parts in fours or fives. Note the five petals, 10 stamens, and five separate pistils. Five sepals are also present but barely visible against the background.

FIGURE 25-2 The two largest classes of flowering plants.

PLANTS AND PEOPLE | The Medicinal Value of Flowering Plants

From extracts of cherry and horehound for cough medicines to chemical compounds in periwinkle and autumn crocus for cancer therapy, derivatives of roots, stems, leaves, and flowers play important roles in the treatment of human illness and disease. About 25 percent of all prescription medicines contain one or more active ingredients extracted from plants.

Many of the plant-produced chemicals with medicinal properties are *alkaloids,* bitter-tasting organic compounds that contain nitrogen. For example, the rosy periwinkle produces two alkaloids, vinblastine and vincristine, which are used to treat two kinds of cancer—Hodgkin's disease, which generally afflicts young adults, and childhood leukemia. Other important alkaloids with medicinal value include quinine, morphine, and reserpine. Quinine, an antimalarial drug, comes from the bark of yellow cinchona. Morphine, used medically to relieve pain, is extracted from the opium poppy. The Indian snakeroot is the source of reserpine, an alkaloid used to treat hypertension and some psychiatric disorders.

Various medicinal plants that we employ today have been used for centuries in folk medicine. The modern medical establishment has learned to respect reports about medicinal plants in the traditions of indigenous peoples, because they provide clues about which plants to test. Studies show that plants identified as useful by shamans (native healers) and traditional plant users are up to 60 percent more likely to have medicinal value than plants that are randomly collected. **Ethnobotany,** the study of the traditional uses of plants by indigenous people, helps pharmaceutical companies identify medicinal plants (see figure). Unfortunately, much of this knowledge is disappearing as indigenous peoples are exposed to modern ways and their traditions are forgotten.

Only a fraction of the more than 300,000 species of flowering plants have been investigated for their medici-

■ An ethnobotanist consults with a Tirio Indian in Suriname about the uses of a rain-forest plant.

nal value. It is likely that the untested species contain other valuable alkaloids with medicinal properties. Unfortunately, we may never have the opportunity to find out, because many of the wild plant species that have medicinal potential are themselves threatened as human activities destroy wildlife habitats.

The Life Cycle of Flowering Plants

Flowering plants undergo alternation of generations in which the sporophyte generation is larger and nutritionally independent. The gametophyte generation in flowering plants is microscopic and nutritionally dependent on the sporophyte. Like gymnosperms and certain seedless vascular plants, flowering plants are heterosporous and produce two kinds of spores, microspores and megaspores.

Sexual reproduction occurs in the flower, which usually consists of sepals, petals, stamens, and pistil (●**Figure 25-3** ❶). The *stamen,* the male reproductive organ, consists of a thin stalk (the filament) and an anther, where pollen forms. The *pistil* consists of a stigma (where the pollen lands), a style (the neck through which the pollen tube must grow), and an ovary, which contains one or more ovules.

As in all seed plants, the ovules of flowering plants are enclosed by one or two layers called *integuments.*

A small opening in the integument, called a *micropyle,* permits passage of the pollen tube into the ovule.

Each young ovule within an ovary contains a megaspore mother cell that undergoes meiosis to produce four haploid megaspores ❷. Three of these usually disintegrate, and one divides mitotically and develops into a mature female gametophyte, also called an **embryo sac.** The most widely studied type of embryo sac contains seven cells with eight haploid nuclei. Six of these cells, including the egg cell, contain a single nucleus each, and a seventh, the *central cell,* has two nuclei, called **polar nuclei** ❸. The egg and the central cell with two polar nuclei are directly involved in fertilization; the other five cells in the embryo sac apparently have no direct role in the fertilization process and later disintegrate. The *synergids* (the two cells closely associated with the egg), however, release chemicals that may affect the direction of pollen tube growth.

Each pollen sac, or microsporangium, of the anther contains numerous microspore mother cells, each of which undergoes meiosis to form four haploid micro-

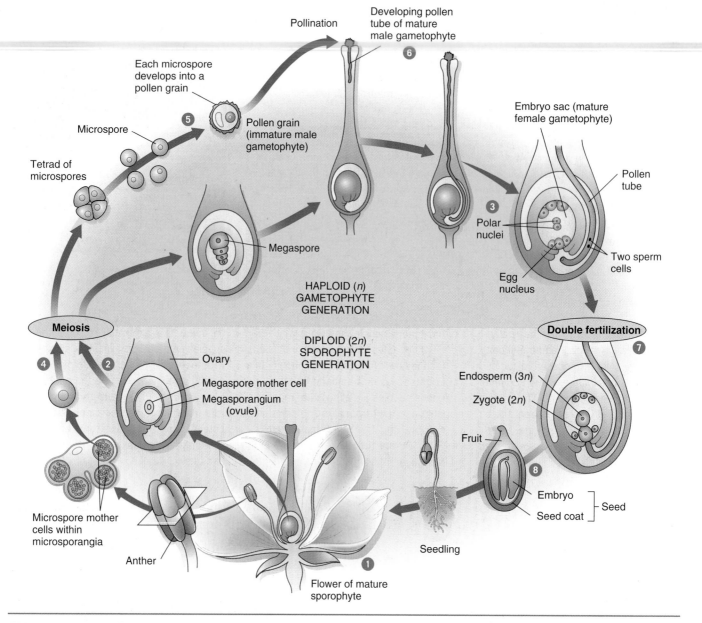

FIGURE 25-3 A generalized life cycle of flowering plants.
The most distinctive feature of the flowering plant life cycle is double fertilization, in which one sperm cell unites with the egg to form a zygote, and the other sperm cell unites with the two polar nuclei to form a 3n cell that gives rise to endosperm. See text for a detailed description.

spores ❹. Every microspore develops into an immature male gametophyte, also called a *pollen grain* (see Figure 25-3 ❺; •**Figure 25-4**). Pollen grains are extremely small; each consists of two cells, the tube cell, or vegetative cell, and the smaller generative cell.

When the pollen is mature, the anthers split open and begin to shed pollen grains. A variety of agents, including wind, water, insects, and other animal pollinators, transfer pollen grains to the stigma. If the pollen grain lands on the stigma of a suitable flower, it germinates; that is, the tube cell forms a pollen tube that grows

down the style and into the ovary (see Figure 25-3 ❻). The germinated pollen grain with its pollen tube is the mature male gametophyte. Next, the generative cell divides to form two nonflagellated male gametes called sperm cells. The sperm cells move down the pollen tube and are discharged into the embryo sac. Both sperm cells are involved in fertilization.

Something happens during sexual reproduction in flowering plants that does not occur anywhere else in the living world. When the two sperm cells enter the embryo sac, both participate in fertilization. One sperm

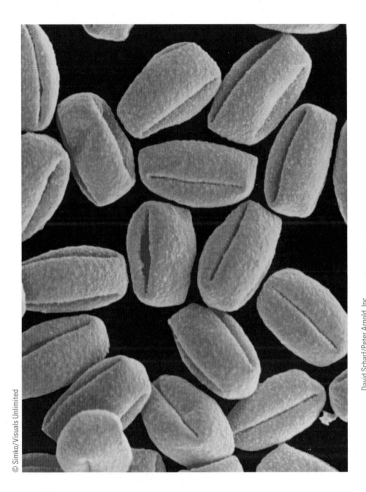

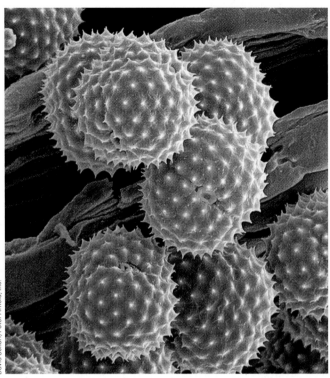

FIGURE 25-4 Scanning electron micrographs of pollen grains. Many pollen grains are so distinctive that they can be identified at the species level. **(a)** Oak (*Quercus*) pollen. **(b)** Ragweed (*Ambrosia bidentata*) pollen.

cell fuses with the egg to form a zygote that grows by mitosis and develops into a multicellular embryo in the seed. The second sperm cell fuses with the two haploid polar nuclei of the central cell to form a *triploid* (3*n*) cell that grows by mitosis and develops into the *endosperm,* a nutrient tissue rich in lipids, proteins, and carbohydrates that nourishes the growing embryo. This fertilization process, which involves two separate nuclear fusions, is called **double fertilization** and is, with two exceptions, unique to flowering plants ❼. (Double fertilization has been reported in the gymnosperms *Ephedra nevadensis* and *Gnetum gnemon.* This process differs from double fertilization in flowering plants in that an additional zygote, rather than endosperm, is produced. The second zygote later disintegrates.)

Seeds and fruits develop after fertilization

As a result of double fertilization and subsequent growth and development, each seed contains a young plant embryo and nutritive tissue, both of which are surrounded by a protective seed coat ❽. In monocots the endosperm persists and is the main source of food in the mature

seed. In most eudicots the endosperm nourishes the developing embryo, which subsequently stores food in its cotyledons.

As a seed develops from an ovule following fertilization, the integuments develop into a seed coat. The ovary wall surrounding the ovule(s) enlarges dramatically and develops into a fruit. In some instances, other tissues associated with the ovary also enlarge to form the fruit. Fruits serve two purposes: to protect the developing seeds from desiccation as they grow and mature and to aid in the dispersal of seeds. For example, dandelion fruits have feathery plumes that are lifted and carried by air currents. Animals often swallow edible fruits, including the seeds, which are dispersed in their droppings. Once a seed lands in a suitable place, it may germinate and develop into a mature sporophyte that produces flowers, and the life cycle continues as described.

DOUBLE FERTILIZATION A process in the flowering plant life cycle in which there are two fertilizations; one fertilization results in formation of a zygote, whereas the second results in formation of endosperm.

Apomixis is the production of seeds without the sexual process

Sometimes flowering plants produce embryos in seeds without the fusion of gametes. This process is known as **apomixis.** Seed production by apomixis is a form of asexual reproduction, and because there is no fusion of gametes, the embryo is genetically similar to the original parent. (The embryo may not be *identical* to the parent, because mutations may occur.) Apomixis has an advantage over other methods of asexual reproduction in that the seeds produced by apomixis can be dispersed by methods associated with sexual reproduction. Examples of plants that reproduce by apomixis include dandelions, citrus trees, blackberries, garlic, and certain grasses.

EVOLUTION LINK
Successful Adaptations of Flowering Plants

The evolutionary adaptations of flowering plants account for their success in terms of their ecological dominance and their great number of species. Seed production as the primary means of reproduction and dispersal, an adaptation shared with the gymnosperms, is clearly significant and provides a definite advantage over seedless vascular plants. Closed carpels, which give rise to fruits surrounding the seeds, and the process of double fertilization, with the consequent development of embryo-nourishing endosperm, increase the likelihood of reproductive success in the flowering plants. The evolution of a variety of interdependencies with many types of insects, birds, and bats, which disperse pollen from one flower to another of the same species, is another reason for angiosperm success (recall the discussion of coevolution in Chapter 9). Pollen transfer results in cross-fertilization, which mixes the genetic material and promotes genetic variation among the offspring.

Several distinctive features have contributed to the success of flowering plants in addition to their highly successful reproduction involving flowers, fruits, and seeds. Recall that most flowering plants have in their xylem efficient water-conducting cells, called **vessel elements,** in addition to tracheids. In contrast, the xylem of almost all seedless vascular plants and gymnosperms consists exclusively of tracheids. Most flowering plants also have efficient sugar-conducting cells, called **sieve-tube elements,** in their phloem. Vascular plants other than flowering plants and gnetophytes lack vessel elements and sieve-tube elements.

The leaves of flowering plants, with their broad, expanded blades, are efficient at absorbing light for photosynthesis. Abscission (shedding) of these leaves during cold or dry periods reduces water loss and has enabled some flowering plants to expand into habitats that would otherwise be too harsh for survival. The stems and roots of flowering plants are often modified for food or water storage, another feature that helps flowering plants survive in severe environments.

Probably most crucial to the evolutionary success of flowering plants, however, is the overall adaptability of the sporophyte generation. As a group, flowering plants readily adapt to new habitats and changing environments. This adaptability is evident in the great diversity exhibited by the various species of flowering plants. For example, the cactus is remarkably well adapted for desert environments (•**Figure 25-5a**). Its stem stores water; its leaves (spines) have a reduced surface area for *transpiration* (loss of water vapor) and may also protect against thirsty herbivorous animals; and its thick, waxy

(a) The barrel cactus (*Ferocactus acanthodes*) is adapted for desert environments.

(b) The water lily (*Nymphaea odorata*) is adapted for wet environments.

FIGURE 25-5 Adaptability of flowering plants.

cuticle reduces water loss. In contrast, the water lily is well adapted for wet environments, in part because it has air channels that provide adequate oxygen to stems and roots living in oxygen-deficient water and mud (•**Figure 25-5b**). Thus, flowering plants are remarkably adapted to their environments. ■

EVOLUTION LINK The Evolution of Flowering Plants

The flowering plants are the most recent group of plants to evolve. In evolution, new structures or organs often originate by modification of previously existing structures or organs. Much evidence supports the classic interpretation that the four organs of a flower—sepals, petals, stamens, and carpels—arose from highly modified leaves. This evidence includes comparisons of the arrangement of vascular tissues in both flowers and leafy stems and of the developmental stages of floral parts and leaves.

Sepals are the most leaflike of the four floral organs, and botanists generally agree that sepals are specialized leaves. Although petals of many flowering plant species are leaflike, botanists generally view petals as modified stamens that later became sterile and leaflike. Cultivated roses and camellias provide evidence supporting this hypothesis; in some varieties the stamens have been transformed by selective breeding into petals, forming showy flowers with large numbers of petals.

The remarkably leaflike stamens and carpels of certain tropical trees and other species support the hypothesis that stamens and carpels originated from leaves or leaflike organs. Consider, for example, the carpel of *Drimys,* a genus of flowering trees and shrubs native to Southeast Asia, Australia, and South America. This carpel resembles a leaf that is folded inward along the midrib, thereby enclosing the ovules, and joined along the entire length of the leaf's margin (•**Figure 25-6**).

PROCESS OF SCIENCE The fundamental question is whether these leaflike stamens and carpels are early organs that were conserved (retained) during the course of evolution or are highly specialized organs that do not resemble early stamens and carpels. In other words, do the leaflike stamens and carpels indicate that all stamens and carpels are derived from leaves or is their being leaflike just a coincidence unrelated to their evolution? Many botanists who have studied this question have concluded that stamens and carpels are probably derived from leaves. Not all botanists accept the origin of stamens and carpels from highly modified leaves, however. As we have noted throughout the text, uncertainty and debate are part of the scientific process, and scientists can never claim to know a final answer. ■

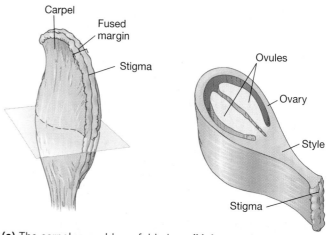

(a) The carpel resembles a folded leaf in which the ovules borne on its upper surface are enclosed.

(b) A cross section of the carpel, cut along the dashed line in **(a)**.

FIGURE 25-6 Carpel of *Drimys piperita.*

During the course of more than 130 million years of angiosperm evolution, flower structure diversified as floral organs fused or became reduced in size or number. These changes led to greater complexity in floral structure in some species and to greater simplicity in other species. Interpreting the floral structures of so many angiosperm species is sometimes difficult, but it is important, because correct interpretations are essential for devising a classification scheme based on evolutionary history.

Flowering plants probably descended from gymnosperms

The fossil record, although incomplete, suggests that flowering plants descended from gymnosperms. By the middle of the Jurassic period, approximately 180 million years ago (mya), several gymnosperm lines existed, with some features resembling those of flowering plants. Among other traits, these derived gymnosperms had leaves with broad, expanded blades and the first modified seed-bearing leaves, which nearly enclosed the ovules. Beetles were evidently visiting these plants and possibly transferring pollen from one plant to another; biologists have suggested that perhaps this relationship was the beginning of **coevolution**, a mutual adaptation between plants and their animal pollinators.

PROCESS OF SCIENCE One important task facing **paleobotanists** (biologists who study fossil plants) is determining which of the ancient gymnosperms are in the direct line

APOMIXIS A type of reproduction in which fruits and seeds are formed asexually.

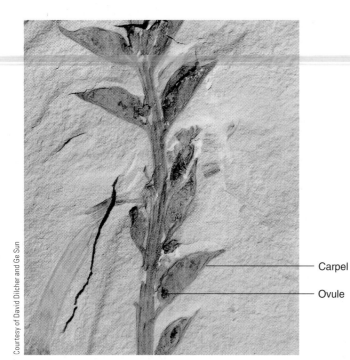

Courtesy of David Dilcher and Ge Sun

Carpel

Ovule

(a) The oldest known fossil angiosperm. This fossil of the extinct plant *Archaefructus* shows a carpel-bearing stem. It was discovered in northeastern China and is about 125 million years old.

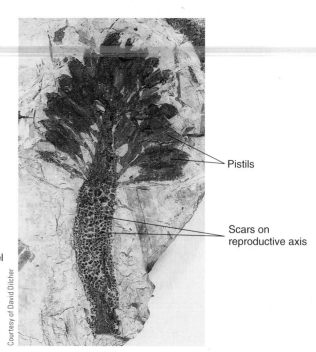

Courtesy of David Dilcher

Pistils

Scars on reproductive axis

(b) The fossilized flower of the extinct plant *Archae-anthus linnenbergeri*, which lived about 100 mya. The scars on the reproductive axis (receptacle) may show where stamens, petals, and sepals were originally attached but abscised (fell off). Many spirally arranged pistils were still attached at the time this flower was fossilized.

FIGURE 25-7 **Fossil angiosperms.**

of evolution leading to the flowering plants. Most botanists hypothesize, based on structural data, that flowering plants arose only once—that is, that there is only one line of evolution from the gymnosperms to the flowering plants. The gnetophytes are the gymnosperm group some botanists consider the closest living relatives of flowering plants; both structural similarities and certain comparative molecular data support this conclusion. Like angiosperms, gnetophytes have vessels, lack archegonia, have flowerlike compound strobili, and undergo double fertilization. However, some molecular data refute a close link between the gnetophytes and flowering plants. It is hoped additional studies will clarify the relationships among the various gymnosperm phyla and the flowering plants. ▪

The oldest definitive trace of flowering plants in the fossil record consists of ovules enclosed in tiny podlike fruits interpreted as carpels in Jurassic and Lower Cretaceous rocks some 125 million to 145 million years old (•**Figure 25-7a**). The oldest fossilized flowers are about 118 million to 120 million years old. Based on the fossil

record as well as both structural and molecular data of living angiosperms, the first flowering plants were probably small, weedy shrubs or herbaceous plants adapted to disturbed habitats. They may have been fragile plants that were not easily preserved, which could explain why there are few fossils of early angiosperms—the environment in which they evolved was repeatedly disturbed and not favorable for their preservation as fossils.

Botanists hypothesize that the rapid diversification of angiosperms did not occur until early flowering plants had invaded lowland regions. By 90 mya, during the Cretaceous period, flowering plants had diversified and had begun to replace gymnosperms as Earth's dominant plants. Fossils of flowering plant leaves, stems, flowers, fruits, and seeds are numerous and diverse. They outnumber fossils of gymnosperms and ferns in Late Cretaceous deposits, indicating the rapid success of flowering plants once they appeared (•**Figure 25-7b**). Many angiosperm species apparently arose from changes in chromosome number (see discussion of sympatric speciation in Chapter 17).

FIGURE IN FOCUS

There are six groups of angiosperms: three of basal angiosperms and three of core angiosperms. The monocot and eudicot groups have the greatest number of species.

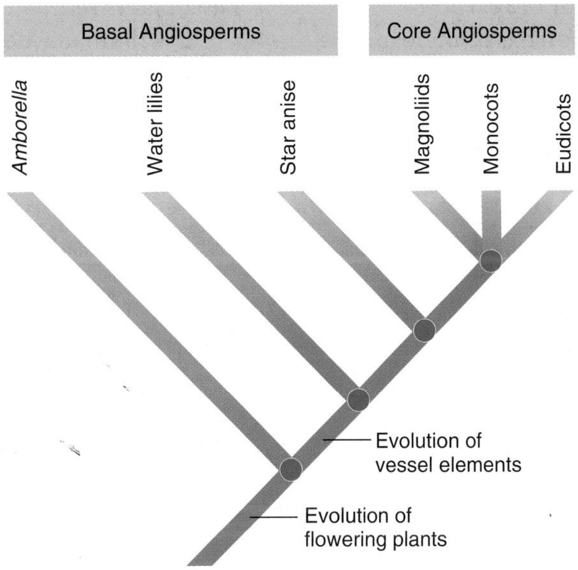

(a) One hypothesis of relationships among the flowering plants, based on fossil and molecular evidence. *Amborella,* water lilies, and star anise are living plants whose ancestors apparently branched off the angiosperm family tree early. These early groups were followed by the magnoliids, the monocot branch, and the eudicots.

Stephen McCabe, Arboretum University of California Santa Cruz

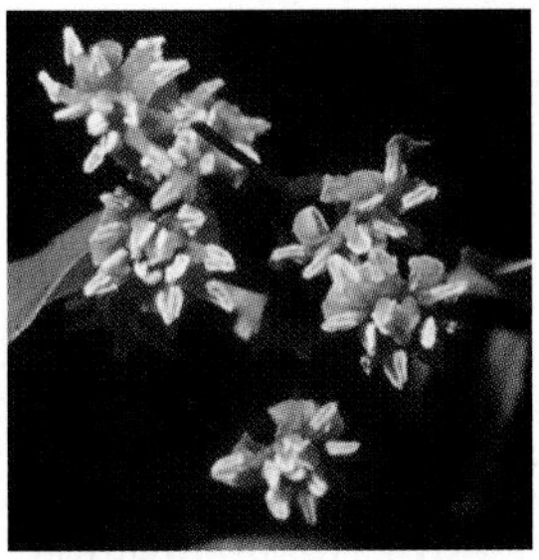

(b) *Amborella trichopoda,* a basal angiosperm.

Rod Planck/Photo Researchers, Inc.

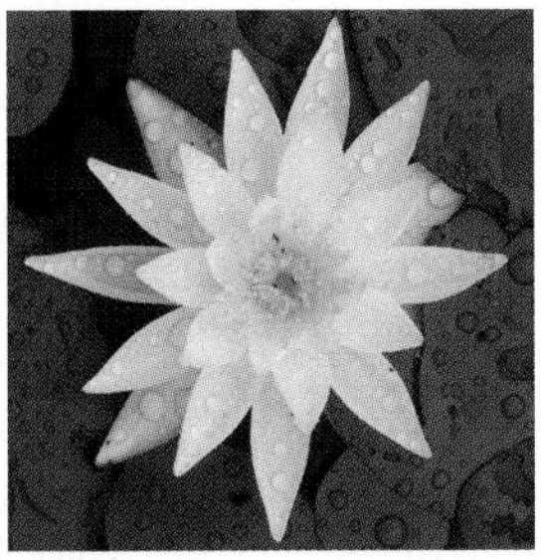

(c) Water lily (*Nymphaea*), a basal angiosperm.

© Robert and Ann Simpson/Visuals Unlimited

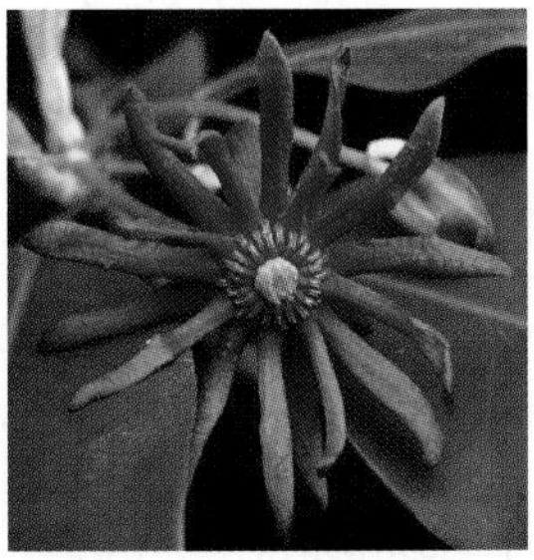

(d) Star anise (*Illicium verum*), a basal angiosperm.

Sinclair Stammers/Photo Researchers, Inc.

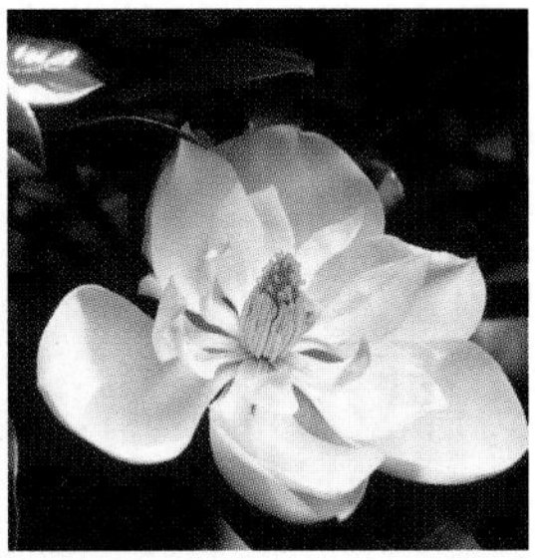

(e) *Magnolia grandiflora,* a core angiosperm.

FIGURE 25-8 Evolution of flowering plants.

The basal angiosperms comprise three groups

Recent analyses of structural features and molecular comparisons have helped clarify relationships among the angiosperm classes. The evidence indicates that three groups of basal angiosperms evolved before the divergence of core angiosperms (•**Figure 25-8a**). **Basal angiosperms** consist of about 170 species thought to be ancestral to all other flowering plants.

Basal angiosperms have features considered more primitive than those of most flowering plants. Some basal angiosperms lack vessels in their xylem, probably because vessels evolved after the evolutionary lines to the basal angiosperms branched off. The flowers of

BASAL ANGIOSPERM One of three groups of angiosperms that are thought to be ancestral to all other flowering plants.

basal angiosperms have many parts arranged spirally (as opposed to fewer parts with a whorled arrangement). Each carpel of basal angiosperms is shaped like a tube, and the stigma, instead of being located on the top of the carpel, extends down the side of the carpel.

The oldest surviving group of basal angiosperms is represented by a single living species, *Amborella trichopoda* (●**Figure 25-8b**). A shrub native to New Caledonia, an island in the South Pacific, *Amborella* may be the nearest living relative to the ancestor of all flowering plants. The water lilies and related families compose the second group of basal angiosperms (●**Figure 25-8c**). This group, which contains about 70 species of aquatic or wetland herbs, may be the second-oldest surviving lineage. Star anise and relatives—the third group of basal angiosperms—consist of about 100 species of vines, trees, and shrubs found mostly in warmer climates. Star anise is important economically because it is a source of star anise oil, used in sweets and in cough drops (●**Figure 25-8d**).

Magnoliids, monocots, and eudicots compose the core angiosperms

Most angiosperm species belong to a group of **core angiosperms,** which is divided into three subgroups: magnoliids, monocots, and eudicots. **Magnoliids** include species in the magnolia, laurel, and black pepper families, along with several related families (●**Figure 25-8e**). Although magnoliids were traditionally classified as "dicots," molecular evidence such as DNA sequence comparisons indicate that the magnoliids are neither eudicots nor monocots. Native to tropical or warm temperate regions, magnoliids include several economically important plants, such as avocado, black pepper, nutmeg, and bay laurel.

Note in Figure 25-8a that the magnoliids, monocots, and eudicots are all depicted as arising from a single branch point. Exact relationships among these three groups, each of which is probably **monophyletic,** remain to be clarified. In other words, at this time we cannot say with certainty which of the three groups was the first to branch off from the other two. ∎

A Survey of 12 Flowering Plant Families

The flowering plants, as the principal photosynthetic organisms on land, provide other organisms with a continuous supply of energy that they must have to survive.

All animals require flowering plants directly or indirectly. A deer browsing on magnolia twigs during the winter is using plants directly. A garter snake that consumes a toad that ate grasshoppers that consumed leaves is using plants indirectly.

Early humans depended on flowering plants for their survival—for the food they ate, the shelters they built, the clothing they wore, and the fires they burned. Even today, humans who live in highly industrialized societies still depend on flowering plants and use them in a variety of ways. Twelve of the more than 300 families of flowering plants are highlighted here to show their importance to humans and to the biosphere. One of these—the magnolia family—is a magnoliid; eight are eudicots—the walnut, cactus, mustard, rose, pea, potato, pumpkin, and sunflower families; and three are monocots—the grass, orchid, and agave families.

The magnolia family includes ornamentals and timber trees

About 220 species of trees and shrubs native to temperate and tropical areas of Asia and the Americas are classified in the magnolia family (Magnoliaceae). Members of the magnolia family are easy to recognize because they have simple, alternate leaves, and large, conspicuous flowers that contain numerous stamens and pistils. The fruit is conelike. Many biologists regard the magnolia family as one of the more primitive families of flowering plants alive today, based on its floral structure.

Two representative members of the magnolia family are the southern magnolia and the tuliptree (●**Figure 25-9**). The southern magnolia, with its shiny evergreen leaves and fragrant creamy white flowers, is a popular ornamental that is planted widely in parks and gardens. Many other species of the magnolia family are also cultivated extensively as ornamentals.

The tuliptree, also known as the yellow poplar, is the tallest of the eastern hardwoods. It is a fast-growing tree of some commercial importance, particularly in the eastern part of the United States. The soft, fine-grained wood of this forest tree is used for a variety of products from furniture to toys. The tuliptree is also planted as an ornamental. Songbirds and squirrels eat its seeds.

The walnut family provides nuts and wood for making furniture

About 50 to 60 species of walnuts, hickories, and pecans belong to the walnut family (Juglandaceae). All are deciduous trees native to the temperate and subtropi-

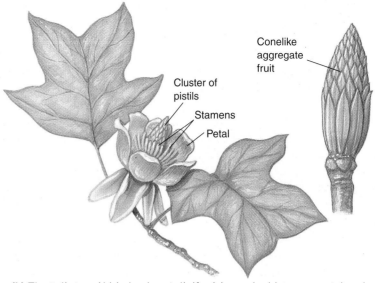

(a) Southern magnolia (*Magnolia grandiflora*) is highly prized for its beauty.

(b) The tuliptree (*Liriodendron tulipifera*) is a valuable ornamental and is also harvested for its wood.

FIGURE 25-9 The magnolia family.

cal areas of Asia and North and South America (•**Figure 25-10**). The leaves are pinnately compound and arranged alternately on the stem. Members of the walnut family are **monoecious** and bear separate male and female flowers on the same plant. The male flowers, which occur in **catkins,** clusters of tiny flowers borne on a pendulous stalk, produce pollen that is carried by the wind to female flowers. The female flowers are small, lack petals, and usually occur individually on small, erect stalks.

The walnut family is ecologically and economically important for its edible nuts and its wood. In forests, wildlife ranging from opossums to wild turkeys feed on the nuts. English walnuts—native to Asia, not England as their common name implies—and pecans, native to North America, are widely cultivated for their nuts. The edible portions of the "nut" are the greatly enlarged cotyledons of the embryo within the seed. In addition to being eaten directly, walnut and pecan seeds are pressed to yield oil, which is an important ingredient in many soaps and cosmetics.

Walnut and pecan trees produce a strong, rich, fine-grained wood that is used to make furniture, interior woodwork, and even the interiors of Rolls Royces. The wood of black walnut, native to eastern North America, is among the most valuable hardwood trees in North America; as a result of its commercial value, black walnut has been cut extensively.

The cactus family is important for its ornamentals

The cactus family (Cactaceae) consists of more than 2000 species of perennial herbs, vines, shrubs, and small trees that are native to North and South America, although some were introduced from the Americas and subsequently became naturalized in China, India, and the Mediterranean. The greatest number and variety of cacti occur in Mexico; in the United States cacti are most abundant in Texas, Arizona, and New Mexico.

Cacti usually live in deserts and are succulent plants—that is, they store water in their stems or leaves. Most cacti have **spines,** modified leaves borne on succulent stem.[3] The stem functions not only for water storage—to enable the plant to survive periods of

---

[3]*Other flowering plant families also have succulent members. Some of these, like the euphorbias, closely resemble cacti in appearance and are often confused with them.*

CORE ANGIOSPERM The group to which most angiosperm species belong; core angiosperms are divided into three subgroups: magnoliids, monocots, and eudicots.

MAGNOLIID One of the groups of flowering plants; magnoliids are core angiosperms that were once classified as "dicots," but molecular evidence indicates that they are neither eudicots nor monocots.

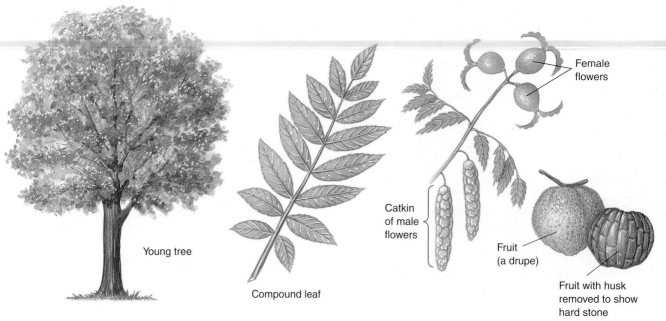

(a) The pinnately compound leaves and green fruits of black walnut (*Juglans nigra*) make the tree easy to recognize.

(b) Cutaway view of inner layer of the walnut fruit wall (the stone), showing the seed with its two lumpy cotyledons.

(c) Immature pecan fruits ripening on the tree (*Carya illinoinensis*). The pecan "nut" is actually a seed.

FIGURE 25-10 **The walnut family.**

drought—but as the main organ of photosynthesis (•**Figure 25-11a**). In certain cacti the succulent stems are prominently ribbed during periods of drought, but during rainy periods the stems swell and the ribs fill in, much like an extended accordion. Cacti roots vary depending on the growth habits of the particular species. Many cacti have an extensive fibrous root system that covers a large area horizontally but may extend

(a) The giant saguaro (*Carnegiea gigantea*) has ribbed stems that grow to 18.5 meters (60 feet) tall and 0.6 meter (2 feet) thick. This plant, which does not grow well in cultivation, thrives in the Sonoran Desert located in parts of Mexico, California, and Arizona.

(b) The Christmas cactus (*Schlumbergera bridgesii*) is popular as a houseplant because of its magenta flowers, which bloom near the Christmas holiday. Native to Brazil where it grows as an epiphyte in rain forests, the Christmas cactus has numerous jointed, spineless stems. The individual stem segments resemble leaves.

FIGURE 25-11 The cactus family.

vertically only a few inches; such a root system quickly absorbs any water that becomes available during the infrequent showers characteristic of arid environments. Not all cacti live in deserts, however. A few, such as the Christmas cactus of Brazil, live as epiphytes on trees in tropical rain forests (•**Figure 25-11b**).

Each cactus flower contains numerous sepals, petals, and stamens and a compound pistil composed of two to many fused carpels (•**Figure 25-11c**); after fertilization, berries containing many tiny black or brown seeds develop. **EVOLUTION LINK** Because of the fascinating adaptations of cacti that enable them to survive in deserts, botanists are interested in how cacti evolved. It is thought that the ancestors of cacti may have been vinelike plants with large leaves; such plants would have required a moist environment. As the climate changed and became much drier, many species of plants may not have been able to adapt and undoubtedly be-

(c) Cactus flowers contain numerous sepals, petals, and stamens. The compound pistil consists of two to many fused carpels. In this photo of a prickly pear cactus (*Opuntia humifusa*), the petals, stamens, and stigma are evident.

Stamens

Stigma

Petals

came extinct. The cacti probably survived this drying trend through a gradual reduction in leaf size and the evolution of succulent stems. ■

Cacti are of economic importance primarily as ornamentals, although some (for example, prickly pear) are cultivated for their fleshy, edible fruits. Cacti have wide appeal as houseplants, because they have attractive flowers and varied spine and rib formations and they can tolerate a lot of neglect (such as forgetting to water them). Some cactus species are in danger of extinction because they have been overcollected for "desert landscaping."

The mustard family contains many important food crops

The mustard family (Brassicaceae or, alternatively, Cruciferae) consists of about 3000 species, many of which are important food crops, ornamentals, and weeds. Members of the mustard family are found worldwide, although the greatest number of species are native to temperate regions of Asia and the Mediterranean.

Most species are annual or perennial herbs, although a few are shrublike. The leaves are simple and arranged alternately on the stem. The flowers are usually arranged in an **inflorescence.** Each flower almost always contains four sepals, four petals, six stamens, and a compound pistil composed of two fused carpels; the petals are arranged in the shape of a cross, or cruciform—hence the family name Cruciferae (•**Figure 25-12a**).

The mustard family also includes many ornamentals, such as sweet alyssum (•**Figure 25-12b**) and honesty. Many mustard species are food crops of considerable economic importance. One of the most important members of the mustard family is *Brassica oleracea,* which includes cabbage (•**Figure 25-12c**), cauliflower, broccoli, kale, kohlrabi, collards, and brussels sprouts (see the introduction to Chapter 16). Turnips, radishes (•**Figure 25-12d**), watercress, and Chinese cabbage are also in the family, and the condiments mustard and horseradish are made from members of the mustard family. Oil is extracted from the seeds of other species, such as the oilseed rape plant, which yields canola oil. Shepherd's purse and weedy mustards are bothersome weeds.

The rose family is important for its fruits and ornamentals

The rose family (Rosaceae), which includes apples, cherries, strawberries, and other fruits as well as roses and many other ornamentals, is a large and important family of about 3370 species. It includes woody trees, shrubs,

and herbaceous plants distributed worldwide, although most members of the rose family are concentrated in northern temperate regions.

Leaves of members of the rose family are alternate and either simple or compound. The flowers, which are pollinated by insects, are large and showy and typically have five sepals, five petals, numerous stamens, and one to many pistils (•**Figure 25-13a and b**). Wild roses always have five petals, and the numerous petals of cultivated roses adhere to the fivefold rule discussed earlier in this chapter; that is, the petal number is some multiple of five.

Many important fruiting plants in the temperate region are members of the rose family, including apples, pears, peaches, nectarines, cherries, apricots, plums, almonds, strawberries, raspberries, and blackberries (•**Figure 25-13c**). The apple, one of the most commercially important fruits in the world, is eaten raw, baked, roasted, sautéed, and candied. Apples are also used to make applesauce, apple butter, jelly, vinegar, and beverages such as apple cider and apple juice.

Many members of the rose family are cultivated as ornamentals because of the beauty and fragrance of their flowers; flowering apples, cherries, and plums are important ornamentals, as is the world's most widely cultivated flower, the rose. Thought to have been the first flower that humans cultivated, roses may have been grown in ancient Greek gardens as long ago as the fifth century BCE. Roses were also important to the Romans, the early Chinese, and the Moors. The rose was a symbol of the early Christian church and the emblem of a number of European kings. It was cultivated for its medicinal value during the Middle Ages and is still grown in parts of Europe and Asia for its fragrant oil, from which perfumes are made. Throughout history, humans have glorified the rose in literature, music, and art.

The pea family provides important food crops

The pea family (Fabaceae or, alternatively, Leguminosae), which includes as many as 17,000 species throughout the world, contains food and forage crops second only to the grass family in importance. Pea family members are herbs, vines, shrubs, or trees whose leaves are usually pinnately compound with an alternate leaf arrangement. The roots of most members of the pea family contain nodules in which nitrogen-fixing bacteria convert atmospheric nitrogen to nitrogenous compounds that the plant uses. The presence of these nodules enables members of the pea family to thrive in relatively poor soil.

Flowers occur in erect or pendulous clusters. The five sepals and five petals in each flower sometimes vary in size and degree of fusion with one another, making the flower distinctly irregular in shape (●**Figure 25-14a and b**). Each flower usually contains 10 stamens and a simple pistil. The fruit that forms after reproduction occurs is a legume, a dry fruit that splits open along two seams (●**Figure 25-14c**). (The term *legume* refers both to a fruit type and to any of the members of the pea family, all of which produce this fruit type.)

The most important members of the pea family are food and forage crops. The seeds and pods of many species—such as garden peas, chick peas, green beans and kidney beans, broad beans, lima beans, soybeans, lentils, and peanuts—are particularly nutritious because they are rich in proteins and oil (●**Figure 25-14d**). Every major civilization since the dawn of agriculture has incorporated a grain–legume combination into its cuisine for balanced

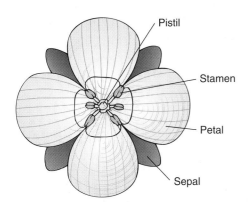

(a) A typical mustard flower, showing the four petals arranged in the shape of a cross.

Labels: Pistil, Stamen, Petal, Sepal

(b) Sweet alyssum (*Lobularia maritima*) is a popular garden plant used as a ground cover.

© Wally Eberhart/Visuals Unlimited

C. McIntyre/PhotoLink/Getty Images

(c) Cabbage (*Brassica oleracea*), one of the oldest cultivated vegetables, comes in a wide variety of colors and is eaten raw, fermented in brine (sauerkraut), and cooked.

S. Solum/PhotoLink/Getty Images

(d) Radish (*Raphanus sativus*) cultivars include red (*shown*), black, and white roots. Radishes are eaten raw or pickled in brine.

FIGURE 25-12 The mustard family.

(a) Plum (*Prunus*) flowers have five sepals, five petals, numerous stamens, and a single pistil.

(b) Strawberry (*Fragaria*) flowers have five sepals, five petals, numerous stamens, and numerous pistils.

FIGURE 25-13 **The rose family.**

(c) Apples, apricots, blackberries, cherries, nectarines, peaches, plums, raspberries, and strawberries are all members of the rose family.

nutrition. Examples of such pairings are rice and soybeans, corn and beans, and barley and lentils. Each of these combinations provides the proper complement of essential amino acids that human cells require yet cannot synthesize.

Forage crops—that is, crops on which livestock such as beef and dairy cattle feed—include alfalfa and crimson clover, both members of the pea family (•**Figure 25-14e**). The importance of forage crops cannot be overestimated; in the United States, more land is used to grow forage crops than to grow all other crops combined. Members of the pea family that are forage crops also produce nectar, which bees use to make honey.

Important ornamentals in the pea family include sweet peas, lupines, and wisteria. Other species within the pea family are sources of timber, fibers, dyes, tannins, and many other products. A notable example is brazilwood, or pernambuco (*Caesalpinia echinata*), a rainforest tree whose wood is used to make concert-quality bows for violins, violas, cellos, and basses.

The potato family is important for its food crops and medicines

The 2000 to 3000 species in the potato family (Solanaceae) are found worldwide, with many species centered in Central and South America and in Australia. The potato family consists primarily of herbs, although a few members are shrubs or small trees. The potato family is of great importance to humans because it contains many food crops, ornamentals, and medicinal and poisonous plants.

Leaves of members of the potato family are simple although highly varied, with alternate leaf attachment. The flowers are usually regular and typically contain five sepals, five petals, five stamens, and a compound

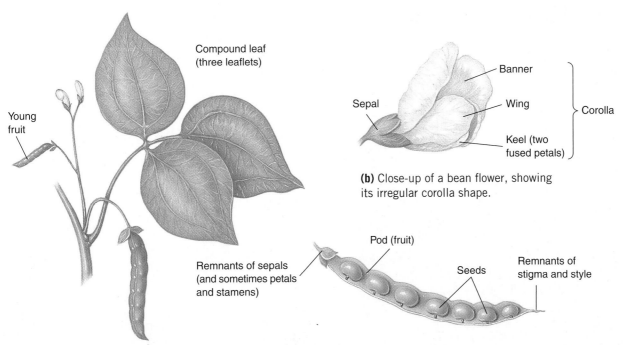

(a) Part of a common bean (*Phaseolus vulgaris*) plant, showing flowers and developing fruits.

Compound leaf (three leaflets)

Young fruit

(b) Close-up of a bean flower, showing its irregular corolla shape.

Banner

Sepal

Wing

Corolla

Keel (two fused petals)

Pod (fruit)

Remnants of sepals (and sometimes petals and stamens)

Seeds

Remnants of stigma and style

(c) The bean fruit, a legume, is opened to show the seeds.

(d) Peanuts (*Arachis hypogaea*) are seeds that develop underground after the aerial flowers are pollinated and grow downward into the soil.

(e) Crimson clover (*Trifolium incarnatum*) is grown for forage and in crop rotations to improve the soil.

FIGURE 25-14 **The pea family.**

Edgar E. Webber

(a) Potato (*Solanum tuberosum*) flowers. Note the five fused petals. The yellow stamens and stigma/style are also visible.

C Squared Studios/Getty Images

(b) Important food crops in the potato family include white potatoes, tomatoes, peppers, and eggplants.

FIGURE 25-15 **The potato family.**

pistil composed of two fused carpels (•**Figure 25-15a**). The fruit that develops from the ovary is either a berry or a capsule and contains a large number of seeds.

Some of the most important fruit and vegetable crops—such as white potatoes, tomatoes, bell peppers, eggplants, chili peppers, pimentos, and paprika—occur in the potato family (•**Figure 25-15b**). Notable ornamentals include the petunia, grown for its attractive flowers, and the Chinese lantern plant, grown for its ornamental fruits.

Many members of the potato family contain toxic *alkaloids* that have medicinal value at low concentrations. Deadly nightshade, also called belladonna, is the source of the alkaloid *atropine,* a substance used to treat poisoning from nerve gas and certain pesticides. In addition, drugs known as *tropane alkaloids* are extracted from belladonna and used to treat heart irregularities. Jimsonweed, which contains the hallucinogenic alkaloid *scopolamine,* was ingested by Native Americans to produce visions during important rituals such as puberty rites, but its overall toxicity prevented its widespread use; chewing the seeds can be fatal. In the introduction to Chapter 8, we discussed the toxicity of tobacco, a

member of the potato family that is widely cultivated for its leaves, which are smoked, chewed, or snuffed.

The pumpkin family includes major food plants

The pumpkin family (Cucurbitaceae) contains more than 700 species of pumpkins, melons, squashes, cucumbers, and gourds. The pumpkin genus (*Cucurbita*), from which the plant family takes its name, includes such plants as pumpkins, squashes, gourds, and vegetable marrow. Another important genus in this family is *Cucumis,* which includes cantaloupe, honeydew, muskmelon, and cucumber. Other important food sources are the watermelon and citron melon (both *Citrullus*). Gourds, which have hard rinds, have been cultivated for centuries to provide scoops and containers. Other members of the pumpkin family provide fibers—for example, the loofah sponge (*Luffa*).

Most species in the pumpkin family are annuals or perennials native primarily to tropical and subtropical areas, such as the South American rain forest and the savanna and bush of Africa. Many members of the family

Petals fused into bell shape

Stigma

Style

Ovary

© Robert Calentine/Visuals Unlimited

(a) Cutaway view of a female squash flower. Note the inferior ovary located beneath the point of attachment of the sepals and petals.

Anthers

© Robert Calentine/Visuals Unlimited

(b) Cutaway view of a male squash flower. Note the united anthers.

FIGURE 25-16 **The pumpkin family.**

are herbaceous or woody vines with tendrils and simple leaves with palmate venation. Their flowers are unisexual and occur either on the same or different plants. The female flowers have inferior ovaries with three (usually) fused carpels (●**Figure 25-16a**). An **inferior ovary** is one that is located below the point at which the other floral organs are attached. The stamens in the male flowers are fused to some degree (●**Figure 25-16b**). Fruits in the pumpkin family are berries, capsules, or **pepos,** which are modified berries with leathery rinds (●**Figure 25-16c**). The fruits often enclose numerous large, flattened seeds.

The sunflower family is one of the largest families of flowering plants

The sunflower family (Compositae or, alternatively, Asteraceae) is a large, complex family that consists of about 25,000 species found around the world. Most are annuals or perennial herbs or shrubs, although some are biennials. The sunflower family includes chrysanthemums, marigolds, dahlias, sunflowers, and daisies (●**Figure 25-17a and b**). Some are food plants, such as lettuce,

C Squared Studios/Getty Images

(c) Botanically, the squash fruit is a pepo that contains numerous large, flattened seeds.

(a) Head of a sunflower (*Helianthus annuus*).

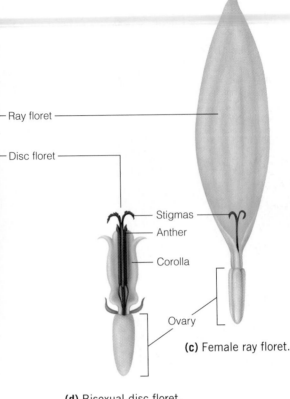

Ray floret

Disc floret

Stigmas

Anther

Corolla

Ovary

(c) Female ray floret.

(d) Bisexual disc floret.

(b) Flower heads of gerbera daisies (*Gerbera*).

FIGURE 25-17 The sunflower family.

endive, sunflower, Jerusalem artichokes, and globe artichokes; others, such as safflower and sunflower, are oil crops. Some members of the sunflower family produce

chemicals used as insecticides, dyes, or medicines; chrysanthemum flowers, for example, are the source of the insecticide pyrethrin. A few members of the sunflower family, such as dandelion and thistles, are weeds.

Plants belonging to the sunflower family have a dense, headlike inflorescence, called a **capitulum,** that consists of an aggregation of few to hundreds of flowers crowded together on a receptacle. The entire structure appears to be a single flower to people who are not botanists. The capitulum usually contains two types of flowers, or *florets:* the large, outer *ray florets,* which appear to be "petals"; and the small, inner *disc florets* (•**Figure 25-17c and d**). The fruits each contain a single seed and are dry and do not split open at maturity (see Figure 9-15c).

The grass family is the most important family of flowering plants

The grass family (Poaceae or, alternatively, Gramineae), which contains about 9000 species distributed throughout the world, is the most important family of flowering plants because it includes the cereal grains such as

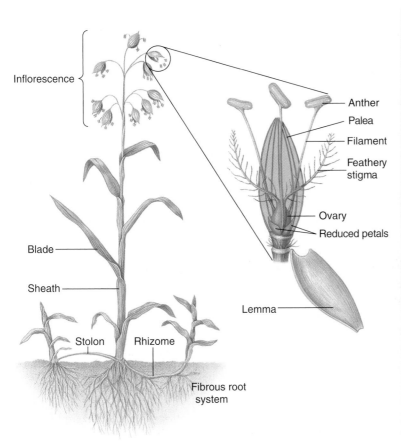

(a) The growth habit of a representative grass plant. Grass flowers, which are wind pollinated, are highly modified. Note, for example, the long, feathery stigmas, adapted to efficiently catch wind-borne pollen.

(b) Close-up of bread wheat (*Triticum aestivum*). More than 20,000 cultivars of bread wheat exist with variation in such characters as size, color, and shape of the grains and heads.

rice (see Chapter 4 introduction), wheat, oats, and corn. Grasses are monocots and are most often annual or perennial herbs; bamboos, which are somewhat "woody," are a notable exception.[4] Each grass leaf has parallel venation and consists of a long, narrow blade and a sheath that wraps around the stem above the nodes (•**Figure 25-18a**). The stem may be erect or creeping and is usually hollow except at the nodes, where it is always solid. Grasses have fibrous root systems.

Flowers of grasses occur in inflorescences and are highly modified. Petals are quite reduced or absent (because grasses are wind pollinated, they have little need for petals), and

[4] *The hardness of bamboo is due to an abundance of fibers, not to the production of secondary tissues as in woody eudicots and gymnosperms.*

(c) Sugarcane (*Saccharum officinarum*) supplies most of the world's sugar. Sugarcane plants photosynthesize efficiently and produce more calories (of carbohydrates) per acre than any other crop.

FIGURE 25-18 The grass family.

each flower typically contains three stamens and a pistil with two feathery stigmas and an ovary containing one ovule (see Figure 25-18a inset). The fruit is usually a seedlike grain, or **caryopsis** (see Figure 9-15a).

Grasses, which grow in virtually all of Earth's major terrestrial ecosystems, are of major ecological significance. In the great grasslands of the world, such as prairies and savannas, grasses are the principal vegetation. Many wild animals, from large hoofed mammals such as bison to tiny rodents, depend on grasses for food. Grasses are also important forage crops for livestock animals.

The grass family supplies humans with many important foods (●**Figure 25-18b and c**). In addition to corn, wheat, and rice, grasses provide other important grains, such as barley, rye, and oats. Sugarcane, also a member of the grass family, supplies much of the world's table sugar. Grasses also provide forage and fodder for domesticated animals, thereby feeding humans indirectly. In many parts of the world, grasses provide shelter, notably bamboo for buildings and thatch (grass leaves) for roofing. Ryegrasses, Bermuda grass, bluegrasses, and fescues are some of the grasses that are widely cultivated as turf for lawns and golf courses; other grasses are grown as ornamentals in flower gardens. Some grasses are significant weeds that are difficult to control.

The orchid family exhibits great variation in flower structure

The orchid family (Orchidaceae), which occurs worldwide, contains at least 18,000 species and is one of the largest families of flowering plants. Orchids are monocots and are perennial herbs. About half of these species are terrestrial and grow in soil; the other half are epiphytic and grow attached to other plants or rocky surfaces. One-fourth to one-third of all orchid species are in danger of extinction as a result of the destruction of their natural habitats and overcollection by orchid hobbyists.

The roots of terrestrial orchids are like those of other plants, whereas aerial roots of epiphytic orchids are thick and covered with **velamen,** a layer of dead cells that absorbs and holds water. The aerial roots of many orchids are unusual in that they are green and photosynthesize (●**Figure 25-19a**).

Some orchids have a single upright stem that bears leaves and flowers. Other orchids have **pseudobulbs** (false bulbs), which are thickened stems at or above the

soil level that arise from a horizontal rhizome and serve as a storage organ for food and water (●**Figure 25-19b**). Each pseudobulb bears one or more leaves and eventually a flower stalk; after it flowers, the pseudobulb slowly dies. Orchid leaves, which are alternate and often occur in two rows, are usually fleshy. Each leaf is simple and encircles the stem as a sheath.

Orchid flowers vary extensively, occurring in many colors, shapes, and sizes, although their overall pattern is unmistakable. Each flower consists of three petal-like sepals and three petals, the third petal forming a lip that differs in color and shape from the rest of the sepals and petals (●**Figure 25-19c**). The reproductive structures are borne in the center of the flower, with both male and female parts fused into a single structure called a *column*.

In orchids the pollen is aggregated into masses called **pollinia** (sing., *pollinium*) that are carried intact to other flowers by insects, hummingbirds, bats, and sometimes even frogs, depending on the species. Some orchids produce a quick-drying glue that adheres the pollinium to the body of the pollinator long enough for it to be transferred to another flower. After fertilization occurs, numerous tiny seeds develop within the fruit, a podlike *capsule*.

Orchids, which are monocots, are widely cultivated by florists and orchid fanciers for their fascinating flowers, and breeders have developed many thousands of hybrids by crossing different species. Vanilla flavoring is an extract from the immature seed pods of an orchid native to tropical America (●**Figure 25-19d**). The pods are fermented for several months, and the flavoring is extracted by soaking the chopped, fermented pods with ethyl alcohol.

The agave family is best known for its ornamentals

The agave family (Agavaceae) contains about 600 species and is found throughout tropical and semitropical areas, particularly in arid regions. Agaves are monocots that are perennial herbs or woody shrubs and trees. The leaves of plants in the agave family are crowded into a rosette at the base of the stem and are stiff, fleshy, narrow, and sharp pointed (●**Figure 25-20**). Many agaves are succulents.

Agaves bear their flowers in inflorescences. Each flower consists of three petal-like sepals and three petals, joined to form a tube; six stamens; and a compound pistil composed of three fused carpels. After fertilization, numerous seeds develop in the fruit (a capsule or berry).

John Arnaldi

Aerial roots

(a) The moth orchid (*Phalaenopsis hybrid*) has photosynthetic aerial roots.

Carlyn Iverson

Pseudobulbs

(b) A *Cymbidium* hybrid has pseudobulbs that function as storage organs.

Carlyn Iverson

Petal

Sepal

Lip (modified petal)

(c) Orchid flowers, such as *Dendrobium*, are distinctive in that the third petal forms a lip.

L. Hobbs/PhotoLink/Getty Images

(d) Vanilla flavoring comes from the immature seed pods of the vanilla (*Vanilla planifolia*) plant.

FIGURE 25-19 The orchid family.

Many agaves are cultivated as ornamentals, both outdoors in semiarid regions and indoors as houseplants. The century plant, quite popular as an outdoor ornamental, is so named because it was mistakenly thought to flower only once in a century. Actually, the century plant flowers once it is about 10 years old, after which it dies. At maturity some century plants, such as *Agave franzosinii*, are quite large; each leaf may reach 2.4 meters (8 feet) in length.

In addition to serving as important ornamentals, some species in the agave family—for example, bowstring hemp and sisal hemp—produce leaf fibers from which twine, ropes, mats, and fishing nets are made. Several species are used to make alcoholic beverages; in Mexico and Central America, the sweet sap of plants in the *Agave* genus is collected and fermented to make pulque, which can be drunk directly or distilled to make tequila.

© Dr. Robert Gustafson/Visuals Unlimited

FIGURE 25-20 The agave family.
The century plant (*Agave*) is a succulent with sword-shaped leaves arranged as a rosette around a short stem. Note the floral stalk. Shown is *A. shawii*.

STUDY OUTLINE

❶ **Summarize the features that distinguish flowering plants from gymnosperms.**
Angiosperm is the traditional name for flowering plants, a large, diverse phylum of plants that form flowers for sexual reproduction and produce seeds enclosed in fruits. Flowering plants have vascular tissues, as do ferns and gymnosperms, and produce seeds, as do gymnosperms. Unlike those of gymnosperms, the ovules of flowering plants are enclosed within an ovary. An **ovule** is the structure in the ovary that contains a female gametophyte and develops into a seed after fertilization. The **ovary** is the base of a carpel or fused carpels that contains ovules and develops into a fruit after fertilization.

❷ **Describe the ecological and economic significance of the flowering plants.**
Flowering plants are extremely important to humans because our survival as a species depends on them. All our major food crops are flowering plants. Woody flowering plants provide us with valuable lumber. Flowering plants give us fibers and medicines. Products as diverse as cork, rubber, tobacco, coffee, chocolate, and aromatic oils for perfumes come from flowering plants. **Economic botany** is the subdiscipline of botany that deals with plants of economic importance; most of these are flowering plants.

❸ Distinguish between monocots and eudicots, the two largest classes of flowering plants, and give specific examples of each class.

Monocots and **eudicots** are the two main classes of flowering plants. Monocot seeds contain a single cotyledon, whereas eudicot seeds contain two cotyledons; a **cotyledon** is the seed leaf of a plant embryo, which may contain food stored for germination. Monocots have floral parts in threes, whereas eudicots have floral parts in fours or fives. Monocots include grasses, orchids, irises, onions, lilies, and palms; eudicots include oaks, roses, mustards, cacti, blueberries, and sunflowers.

❹ Briefly explain the life cycle of a flowering plant, and describe double fertilization.

Flowering plants undergo an alternation of generations in which the sporophyte generation is larger and nutritionally independent, and the gametophyte generation is reduced to only a few microscopic cells. **Double fertilization** is a process in the flowering plant life cycle in which there are two fertilizations; one of the fertilizations results in the formation of a zygote, whereas the second results in the formation of endosperm. **Endosperm** is the 3*n* nutritive tissue that is formed at some point in the development of all angiosperm seeds.

❺ Discuss some of the evolutionary adaptations of flowering plants.

Flowering plants reproduce sexually by forming flowers. After double fertilization, seeds are formed within fruits. Flowering plants have efficient water-conducting **vessel elements** in their xylem and efficient carbohydrate-conducting **sieve-tube elements** in their phloem. Wind, water, insects, or other animals transfer pollen grains in various flowering plants. **Apomixis** is a type of reproduction in which fruits and seeds are formed asexually.

❻ Trace the evolution of flowering plants from gymnosperms.

Flowering plants probably descended from ancient gymnosperms that had specialized features, such as leaves with broad, expanded blades and closed carpels. Flowering plants probably arose only once.

❼ Distinguish between basal angiosperms and core angiosperms.

Basal angiosperms consist of three groups of angiosperms thought to be ancestral to all other flowering plants. **Core angiosperms** are the group to which most angiosperm species belong; core angiosperms are divided into three subgroups: magnoliids, monocots, and eudicots. **Magnoliids** include species in the magnolia, laurel, and black pepper families, along with several related families; magnoliids were once classified as "dicots," but molecular evidence indicates that they are neither eudicots nor monocots.

❽ Briefly describe the distinguishing characteristics and give an example or two of each of the following flowering plant families: magnolia, walnut, cactus, mustard, rose, pea, potato, pumpkin, sunflower, grass, orchid, and agave.

Flowering plants have been classified into more than 300 families. The magnolia family is important as ornamentals and as a source of timber; examples include the southern magnolia and the tuliptree. The walnut family provides nuts for food and wood for making furniture; examples include English walnut, black walnut, and pecan. The cactus family is important as ornamentals; examples include prickly pear and Christmas cactus. The mustard family contains many important food crops; examples include cabbage, broccoli, cauliflower, turnip, and mustard. The rose family is commercially important for its fruits and ornamentals; examples are apple, pear, plum, cherry, apricot, peach, strawberry, raspberry, and rose. The pea family includes important food crops; examples are garden pea, chick pea, green bean, soybean, lima bean, peanut, red clover, and alfalfa. The potato family is important for its food crops and chemicals used as drugs; examples include potato, tomato, green pepper, eggplant, petunia, and deadly nightshade (belladonna). The pumpkin family contains food crops such as pumpkins, melons, squashes, cucumbers, cantaloupe, honeydew, muskmelon, cucumber, and watermelon. The sunflower family, one of the largest families of flowering plants, includes chrysanthemums, marigolds, sunflowers, daisies, and some food plants such as lettuce and globe artichokes. The grass family is the most important family of flowering plants from a human standpoint; examples include rice, wheat, corn, oats, barley, rye, sugarcane, and bamboo. The orchid family is one of the largest families of flowering plants and contains a greater variety of flowers than any other family; an example is the vanilla orchid. The agave family is best known for its ornamentals; examples include the century plant, sisal hemp, and bowstring hemp.

REVIEW QUESTIONS

1. Name at least three features that distinguish flowering plants from other plants.
2. Explain how reproduction in flowering plants differs from that in gymnosperms, including how the plants are fertilized, whether the seeds are exposed or enclosed, and whether endosperm is produced.
3. How do the nonreproductive adaptations of flowering plants differ from those of gymnosperms?
4. How does pollination occur in flowering plants? Fertilization?
5. What special advantage does asexual reproduction by apomixis have over other kinds of asexual reproduction (such as corms and bulbs, discussed in Chapter 7)?
6. What are the two largest classes of flowering plants, and how can you distinguish between them?
7. Describe the evolutionary changes that had to take place as ancient gymnosperms evolved into flowering plants.
8. Describe the significant features of the oldest known fossil angiosperms.
9. Are monocots considered basal or core angiosperms? Explain your answer.
10. How are flowering plants ecologically important?
11. How are flowering plants important to humans?
12. Specify the family in which each of the following flowering plants is classified: raspberry, pecan, cucumber, barrel cactus, red clover, jimsonweed, century plant, wheat, tuliptree, turnip, daisy, garden pea, tomato, sisal hemp.
13. Fill in the blanks on the diagram on page 513. Use Figure 25-3 to check your answers.

THOUGHT QUESTIONS

1. Why are conifers and flowering plants placed in separate phyla?
2. How are cones and flowers alike? How are they different? (*Hint:* Your answer should consider microspores/megaspores and seeds.)
3. In contrast to the cones of gymnosperms, which are either male or female, most flowers of flowering plants contain both male and female reproductive structures. Explain how bisexual flowers might be advantageous for flowering plants.
4. Water lilies are adapted to wet environments, whereas cacti and agaves are adapted to dry environments. For each plant, specify at least two adaptations that might have been selected for (or against) during the course of its evolution.
5. You are given a plant that you have never seen before (see figure). Determine whether it is a eudicot or a monocot. What are the features that helped you make this determination?

Carlyn Iverson

Visit us on the web at http://www.thomsonedu.com/biology/berg for additional resources such as flashcards, tutorial quizzes, further readings, and web links.

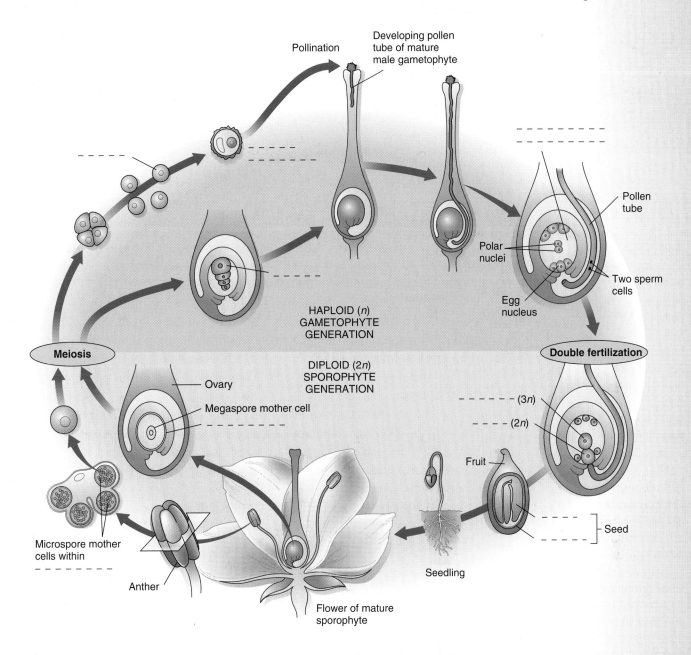

Pollination

Developing pollen
tube of mature
male gametophyte

Pollen
tube

Polar
nuclei

Two sperm
cells

Egg
nucleus

HAPLOID (*n*)
GAMETOPHYTE
GENERATION

Meiosis

Double fertilization

DIPLOID (2*n*)
SPOROPHYTE
GENERATION

Ovary

Megaspore mother cell

(3*n*)

(2*n*)

Fruit

Seed

Microspore mother
cells within

Anther

Seedling

Flower of mature
sporophyte

Ecosystems

LEARNING OBJECTIVES

❶ Define *ecology,* and distinguish among population, community, ecosystem, and biosphere.

❷ Explain the difference between J-shaped and S-shaped population growth curves.

❸ Summarize the three main types of survivorship curves.

❹ Characterize producers, consumers, and decomposers.

❺ Describe what is meant by an organism's ecological niche.

❻ Define *competition, predation,* and *symbiosis,* and distinguish among mutualism, commensalism, and parasitism.

❼ Define *ecological succession,* and distinguish between primary succession and secondary succession.

❽ Summarize the concept of energy flow through a food web.

❾ Describe the main steps in the carbon, nitrogen, and hydrologic cycles.

❿ Distinguish between bottom-up and top-down processes.

$\mathbf{C}$hesapeake Bay on the eastern coast of the United States is the world's richest *estuary,* a semi-enclosed body of water where fresh water drains into the ocean. Biological diversity and productivity abound in Chesapeake Bay, as they do wherever fresh water and salt water form a gradient from unsalty fresh water to salty ocean water. Chesapeake Bay has three distinct marsh communities: freshwater marshes at the head of the bay, brackish (moderately salty) marshes in the middle bay region, and salt marshes at the mouth of the bay. Each community has its own characteristic plants and other organisms.

A salt marsh presents a monotonous view—mile after mile of flooded meadows of cordgrass (*Spartina*). Cordgrass is a perennial grass with a horizontal rhizome and a stout, hollow stem. Its tough, flat leaves taper to a narrow point. Cordgrass has short, intermediate, and tall forms; the tall form grows in deeper water, whereas the short form is found in shallow water.

High salinity (although not as high as that of ocean water) and twice-daily tidal inundations create a challenging environment to which only a few plants, such as cordgrass, have adapted. Those that have adapted to the salt-marsh environment thrive—in part because nitrates and phosphates, which drain into the marsh from the land, promote their rapid growth. Cordgrass and the microscopic algae and cyanobacteria that live on and around it capture energy from the sun and use it to manufacture carbohydrates by photosynthesis.

Many small aquatic animals depend on the cordgrass for food, shelter, and hiding places (to avoid being eaten). Cordgrass is eaten directly by some animals, and when the cordgrass dies, its remains (called detritus) provide nutrients for many other inhabitants of the salt marsh and the bay.

Cordgrass is an appropriate plant with which to begin our discussion of ecology because, like other plants, it is essential to the survival of many organisms. Cordgrass plays a pivotal role in the salt-marsh community because it contributes to the flow of energy and the cycling of materials in this ecosystem.

Salt marsh in autumn. Cordgrass (*Spartina alterniflora*) is the dominant vegetation in a Chesapeake Bay salt marsh.

Introduction to Ecology

Ecology is the study of how living organisms and the physical environment interact in an immense and intricate web of relationships. The concept of ecology was first developed in the 19th century by Ernst Haeckel, who also created its name—*eco* from the Greek word for "house" and *logy* from the Greek word for "study." Thus, ecology literally means the study of one's house.

Biologists call environmental factors involving living organisms **biotic factors,** and those involving the nonliving, physical environment, **abiotic factors.** Abiotic factors include precipitation, temperature, pH, wind, and chemical nutrients. Ecologists develop hypotheses to explain such phenomena as the distribution and abundance of life, the ecological role of specific species, and the interactions among species. Ecologists then test these hypotheses.

The focus of ecology is local or global, specific or generalized, depending on what questions the scientist is trying to answer. Thus, one ecologist might study the organisms living in a rotting log (•**Figure 26-1**), another might study all the organisms that live in a forest where the rotting log is found, and another might examine how matter cycles between the forest and surrounding communities.

How does ecology fit into the organization of the biological world? Most ecologists are interested in the levels of biological organization including and above the level of the individual organism. Individual organisms are arranged into **populations,** members of the same species that live in a prescribed area at the same time. The boundaries of the area are defined by the ecologist performing a particular study. A population ecologist, for example, might study a population of pines or a population of cordgrass like that discussed in the chapter introduction.

Populations are organized into **communities,** which are all the populations of different organisms that live and interact within an area. A community ecologist might study how organisms interact with one another—including who eats whom—in a deciduous forest community or in an alpine meadow community. **Ecosystem,** a more inclusive term than *community,* encompasses a community and its environment. Thus, an ecosystem includes not only all the interactions among the organisms of a community but the interactions between organisms and their physical environment. An ecosystem ecologist, for example, might examine how energy and materials move through the organisms living in a salt-marsh community or a rainforest community.

Earth's ecosystems are organized into the **biosphere,** which includes all of Earth's organisms. In addition to the biosphere, there are divisions of Earth's physical environment on which the biosphere depends:

Michael P. Gadomski/Photo Researchers, Inc.

FIGURE 26-1 Community in a rotting log.
Fallen logs, called "nurse logs," shelter plants and other organisms and enrich the soil as they decay.

the atmosphere, hydrosphere, and lithosphere. The *atmosphere* is the gaseous envelope surrounding Earth, the *hydrosphere* is Earth's supply of water (liquid and frozen, fresh and salty), and the *lithosphere* is the soil and rock of Earth's crust. Ecologists who study the biosphere examine the interrelationships among Earth's atmosphere, land, water, and organisms.

Ecology, the broadest field within the biological sciences, has links to every other biological discipline and to other disciplines—geology, chemistry, and earth science, for example—that are not traditionally part of biology. Because humans are biological organisms, all human activities, including economics and politics, have profound ecological implications.

Population Ecology

Population ecology considers both the number of individuals of a particular species that are found in an area and the dynamics, or changes, in populations—how and why population numbers increase or decrease over time. Population ecologists try to determine the processes common to all populations. They study how predation, disease, and other environmental pressures affect a population; they also examine how a population interacts with its environment, such as how individuals in a population compete for sunlight or other resources. (Used in this context, a *resource* is anything from the environment that meets the needs of a particular species.) Because of competition, predation, and other environmental pressures, population growth, whether of bacteria, oaks, or humans, does not increase indefinitely.

Additional aspects of populations that interest ecologists are their reproductive success or failure (extinction), their evolution, their genetics, and the way they affect the normal functioning of communities and ecosystems. Biologists in applied disciplines, such as forestry and agronomy (crop science), must understand population ecology to manage populations of economic importance. Also, understanding the population dynamics of endangered and threatened species plays a key role in efforts to prevent their slide to extinction.

Spacing is an important feature of populations

Like other organisms, the individuals in a plant population often exhibit characteristic patterns of spacing, or *dispersion,* relative to one another. Individuals may be spaced in a random, clumped, or uniform dispersion. **Random dispersion** occurs when individuals in a population are spaced throughout an area in a way unrelated to the presence of others (•**Figure 26-2a**). Of the three major types of dispersion, random dispersion is least common and hardest to observe in nature, leading some ecologists to question its existence. Trees of the same species sometimes seem to be distributed randomly in a tropical rain forest, but in an extensive study of 1000 species, ecologists determined that most were not randomly dispersed.

Perhaps the most common spacing is **clumped dispersion,** or **patchiness,** which occurs when individuals

are concentrated in specific parts of the habitat (•**Figure 26-2b**). Clumped dispersion often results from the patchy distribution of resources in the environment (discussed later in the chapter). Clumped dispersion may also occur in plants because of limited seed dispersal or asexual reproduction. An entire grove of aspen trees, for example, may be clumped together because they originate asexually from a single plant.

Uniform dispersion occurs when individuals are more evenly spaced than would be expected from a random occupation of a given habitat (•**Figure 26-2c**). Uniform dispersion may occur when competition among individuals is severe or when plant roots or abscised leaves produce toxic substances that inhibit the growth of nearby plants.

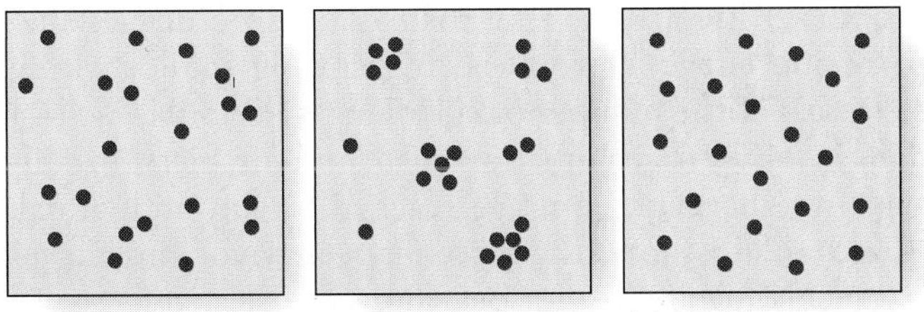

Population increase is limited by environmental resources

Population size, whether of sunflowers or redwoods, changes over time. The maximum rate at which a population increases per unit time under ideal conditions, that is, when resources are abundant, is known as its **intrinsic rate of increase.** Different species have different intrinsic rates of increase, depending on such factors as the age at which reproduction begins, the number of reproductive periods per lifetime, and the number of offspring the individual is capable of producing. These factors determine whether a particular species has a large or small intrinsic rate of increase.

If we plot the population size versus time, under optimal conditions, the graph has a J shape that is characteristic of **exponential population growth,** the accelerating population growth rate that occurs when optimal conditions allow a constant growth rate (•**Figure 26-3**). When a population grows exponentially, the larger the population gets, the faster it grows. Regardless of which species we are considering, plant or animal, whenever a population is growing at its intrinsic rate of increase, population size plotted versus time always gives a curve of the same shape.

(a) Random dispersion **(b)** Clumped dispersion **(c)** Uniform dispersion

FIGURE 26-2 Dispersion of individuals in a population.

ECOLOGY A discipline of biology that studies the interrelations between living things and their environments.

POPULATION A group of organisms of the same species that live in a defined geographic area at the same time.

COMMUNITY An association of populations of different species living in a defined habitat with some degree of interdependence.

ECOSYSTEM The interacting system that encompasses a community and its nonliving, physical environment.

BIOSPHERE All of Earth's living organisms, collectively.

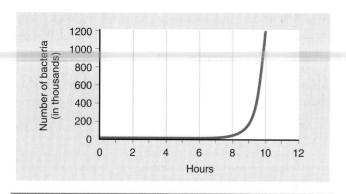

FIGURE 26-3 Exponential population growth.
When bacteria divide every 20 minutes, their numbers increase exponentially. The curve of exponential growth has a characteristic J shape. The ideal conditions under which bacteria or other organisms reproduce exponentially rarely occur in nature, and when these conditions do occur, they are of short duration.

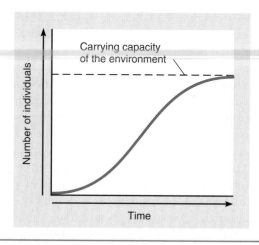

FIGURE 26-4 Carrying capacity and logistic population growth.
In many laboratory studies, exponential population growth slows as the carrying capacity of the environment is approached. The logistic model of population growth, when graphed, has a characteristic S-shaped curve.

No population can grow exponentially indefinitely

Certain populations may grow exponentially for brief periods. However, organisms cannot reproduce indefinitely at their intrinsic rate of increase, because the environment sets limits. These limits include such unfavorable environmental conditions as limited availability of sunlight, water, and other essential resources (resulting in increased competition) as well as limits imposed by disease and predators.

Over longer periods, the rate of population growth may decrease to nearly zero. This leveling out occurs at or near the limits of the environment to support the population. The **carrying capacity** represents the largest population that can be maintained for an indefinite period in a particular environment, assuming there are no changes in the environment. In nature, the carrying capacity is dynamic and changes in response to environmental changes. An extended drought, for example, could lower the carrying capacity for plants in that environment.

When a population regulated by environmental limits is graphed over longer periods, the curve has a characteristic S shape. The curve shows the population's initial exponential increase (note the curve's J shape at the start, when environmental limits are few), followed by a leveling out as the carrying capacity is approached (•**Figure 26-4**). The S-shaped growth curve, also called **logistic population growth,** describes a population increasing from a small number of individuals to a larger number of individuals, in which population growth is ultimately limited by the environment. Although the S curve is an oversimplification of how most populations change over time, it does fit some populations studied in the laboratory, as well as a few studied in nature.

Each species has specific life history traits

Each species is uniquely suited to its *life history*—its lifetime patterns of growth and reproduction. Many years pass before a young magnolia tree flowers and produces seeds, whereas a poppy plant grows from seed, flowers, and dies in a single season. Plant ecologists try to understand the adaptive consequences of various **life history traits,** characteristics such as reproductive patterns, potential capacity to produce offspring, life span, and survivorship.

Some species expend their energy in a single, immense reproductive effort. Agaves, for example, are commonly called century plants because it was thought that they flower only once in a century. However, an agave flowers after it is about 10 years old and then dies (see Figure 25-20). Other plant species—such as perennial herbaceous plants, shrubs, and trees—do not flower once and then die; instead, these plants exhibit repeated reproductive cycles. Pines, for example, produce male and female cones every spring once the trees reach a certain size.

Survivorship, the probability that a given individual in a population will survive to a particular age, is another life history trait that ecologists study. A survivorship curve is produced by plotting the logarithm (base 10) of the number of surviving individuals against age, from seed germination in plants (or birth in animals) to the maximum age reached by any individual. •**Figure 26-5** shows the three main survivorship curves that ecologists recognize.

In Type I survivorship, as exemplified by humans, the young and those at reproductive age have a high probability of surviving. The probability of surviving decreases more rapidly with increasing age; mortality (death) is concentrated later in life. •**Figure 26-6** shows a survivorship study for a natural population of Drummond phlox, an annual native to East Texas that became widely distributed in the southeastern United States after it escaped from cultivation. Because most Drummond phlox seedlings survive to reproduce after germination, the plant exhibits a Type I survivorship that is typical of annuals.

In Type III survivorship, the probability of mortality is greatest early in life, and those individuals that avoid early death subsequently have a high probability of survival. Type III survivorship is characteristic of many tree species, because many newly germinated seedlings die; those trees that become established in the canopy exhibit low mortality for many years.

In Type II survivorship, which is intermediate between Types I and III, the probability of surviving does not change with age. The probability of death is equally likely across all age groups, resulting in a linear decline in survivorship. Annual plant populations crowded in a limited amount of space often exhibit Type II survivorship. As these plants compete for available sunlight, water, and other resources, individuals die at a more-or-less constant rate (a process called self-thinning) before they reproduce.

Many plant species exist as metapopulations

The natural environment is a heterogeneous **landscape** consisting of interacting ecosystems that provide a variety of habitat patches. Landscapes, which are typi-

cally several to many square kilometers in area, cover larger land areas than individual ecosystems. Consider a forest, for example. The forest landscape is a mosaic of various elevations, temperatures, levels of precipitation, soil moisture, soil types, and other properties. Because each species has its own habitat requirements, this heterogeneity in physical properties is reflected in the different organisms that occupy the various patches in the landscape (•**Figure 26-7**). Some species occur in narrow habitat ranges, whereas others have wider habitat ranges.

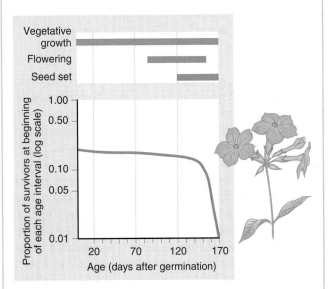

PROCESS OF SCIENCE

QUESTION: What type of survivorship does the annual plant Drummond phlox exhibit?

HYPOTHESIS: After seed germination, Drummond phlox has Type I survivorship, typical of other annual plants.

EXPERIMENT: A population of Drummond phlox was counted at varying periods after the plants germinated to determine survivorship. Data were collected at Nixon, Texas, in 1974 and 1975.

RESULTS AND CONCLUSION: Results were plotted (*shown above*) and compared with Type I, Type II, and Type III survivorship curves. On the *y*-axis, survivorship begins at 0.296 instead of 1.0 because the study took into account death during the seed dormancy period prior to germination (*not shown*). Bars above the graph indicate the various stages in the Drummond phlox life history, from germination to death. Drummond phlox exhibits a Type I survivorship after germination of the seeds. (Adapted from W. J. Leverich and D. A. Levin, "Age-Specific Survivorship and Reproduction in *Phlox drummondii,*" *American Naturalist* 113 (6): 1148, 1979.) ∎

FIGURE 26-6 Survivorship curve for a Drummond phlox population.

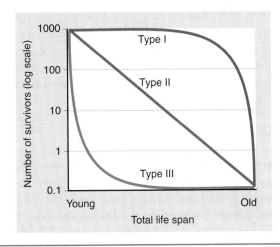

FIGURE 26-5 Survivorship curves.
These curves represent the survivorships of species in which mortality is greatest in old age (Type I), spread evenly across all age groups (Type II), and greatest among the young (Type III). The survivorships of most organisms are compared to these curves.

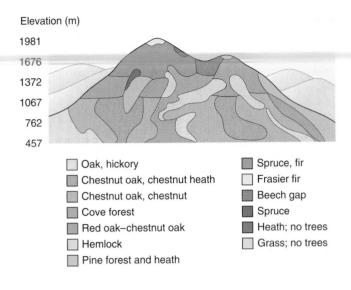

Elevation (m)

1981
1676
1372
1067
762
457

☐ Oak, hickory
☐ Chestnut oak, chestnut heath
☐ Chestnut oak, chestnut
☐ Cove forest
☐ Red oak–chestnut oak
☐ Hemlock
☐ Pine forest and heath

☐ Spruce, fir
☐ Frasier fir
☐ Beech gap
☐ Spruce
☐ Heath; no trees
☐ Grass; no trees

FIGURE 26-7 The mosaic nature of landscapes.
An evaluation of the distribution of vegetation on a typical west-facing slope in Great Smoky Mountains National Park reveals that the landscape consists of patches. Chestnut oak is a species of oak (*Quercus prinus*); chestnut heath is an area within chestnut oak forest where the trees are widely scattered and the slopes underneath are covered by a thick growth of laurel (*Kalmia*) shrubs; cove forest is a mixed stand of deciduous trees. (Adapted from R. H. Whittaker, "Vegetation of the Great Smoky Mountains," *Ecological Monographs*, Vol. 26, 1956.)

Population ecologists have discovered that many species are not distributed as one large population across the landscape. Instead, many species exist as a series of local populations distributed in distinct habitat patches. A population that is divided into several local populations among which individuals occasionally disperse (emigrate and immigrate) is known as a **metapopulation.** For example, note the various local populations of red oak on the mountain slope in Figure 26-7. The spatial distribution of a species occurs because different habitats vary in suitability, from unacceptable to preferred. The preferred sites are more productive habitats that increase the likelihood of survival and reproductive success for the individuals living there.

Community Ecology

Communities exhibit characteristic properties that populations lack. These properties, known collectively as *community structure* and *community functioning*, include the number and types of species present, the relative abundance of each species, the interactions among species, and community resilience to disturbances. **Community ecology** is the description and analysis of patterns and processes within the community. Find-

ing common patterns and processes in a wide variety of communities—for example, a pond community, a pine forest community, and a sagebrush desert community—helps ecologists understand community structure and functioning.

Communities are exceedingly difficult to study because a large number of individuals of many species interact with one another and are interdependent in a variety of ways. Species compete with one another for food, water, living space, and other resources. Some organisms kill and eat other organisms. Some species form intimate associations with one another, whereas other species are only distantly connected. Certain species interact in positive ways, in a process known as **facilitation,** which modifies and enhances the local environment for other species. For example, alpine plants in harsh mountain environments grow faster and reproduce more successfully when other plants are growing nearby. Unraveling the many positive and negative, direct and indirect interactions of organisms living as a community is one of the goals of community ecologists.

Communities contain producers, consumers, and decomposers

On the basis of how they obtain nourishment, each organism plays one of three main roles in community life: producer, consumer, or decomposer. Most communities contain representatives of all three groups, which interact extensively with one another.

Sunlight is the source of energy that powers almost all life processes. **Producers,** also called **autotrophs,** perform photosynthesis, the biological process in which light energy is captured and transformed into the chemical energy of organic molecules, such as carbohydrates, which are manufactured from carbon dioxide and water. By incorporating the organic molecules that they manufacture into their own cells, producers make their bodies or body parts potential food sources for other organisms. Plants are the most significant producers on land; in aquatic environments algae and certain prokaryotes are important producers.

Animals are **consumers;** that is, they are **heterotrophs** that use the tissues of other plant and animal organisms as a source of food energy and body-building materials. Consumers that eat producers are called **primary consumers,** which usually means that they are exclusively *herbivores* (plant eaters). Cattle and deer are examples of primary consumers. **Secondary consumers** eat primary consumers and include flesh-eating *carnivores*, which consume other animals exclusively. Lions and tigers are examples of carnivores. Other consumers, called *omnivores,* eat a variety of plant and animal organisms. Bears, pigs, and humans are examples of omnivores.

Some consumers, called **detritus feeders,** or *detritivores,* ingest **detritus,** which is dead organic matter that includes animal carcasses, leaf litter, and feces. Detritivores, such as snails, crabs, clams, and worms, are especially abundant in aquatic habitats, where they burrow in the bottom muck and consume the organic matter that collects there. Earthworms are terrestrial (land-dwelling) detritus feeders, as are termites, beetles, snails, and millipedes.

Many consumers do not fit readily into a single category of herbivore, carnivore, omnivore, or detritivore. To some degree, these organisms modify their food preferences as the need arises. For example, in the salt-marsh community, marsh crabs are both predators and detritus feeders. Furthermore, some organisms change their food preferences over their lifetimes. Tadpoles are primary consumers, but adult frogs are secondary consumers.

Decomposers are microbial heterotrophs that break down organic material and use the decomposition products as a source of energy. Decomposers, which complete the work of detritus feeders in consuming dead organisms and waste products, typically release simple inorganic molecules, such as carbon dioxide and minerals (inorganic nutrients), that can then be reused by producers. Bacteria and fungi are important decomposers. For example, sugar-metabolizing fungi are first to invade dead wood; they consume the wood's simple carbohydrates. When they exhaust those, other fungi, often aided by termites and bacteria, complete the decomposition of the wood by digesting cellulose, a complex carbohydrate that is the main constituent of wood.

Natural communities contain a balanced representation of producers, consumers, and decomposers. Producers and decomposers are necessary for the long-term survival of any community. Through photosynthesis, producers provide both food and oxygen for the rest of the community. Decomposers are also indispensable, because without them, dead organisms and waste products would accumulate indefinitely. Without decomposers, essential elements such as potassium, nitrogen, and phosphorus would remain permanently in dead organisms and organic compounds and therefore would be unavailable for use by new generations of organisms. Consumers also play an important role in communities by maintaining a balance between producers and decomposers.

An organism's role in the community is its ecological niche

Every organism has its own **ecological niche,** or simply **niche,** which is its role within the structure and function of a community. Although the concept of ecological niche has been used in ecology since early in the 20th century, Yale ecologist G. E. Hutchinson first described in 1957 the multidimensional nature of the niche that is accepted today.

A complete description of an organism's ecological niche involves all dimensions and aspects of the physical, chemical, and biological factors that the organism requires to survive, to remain healthy, and to reproduce. Among other things, the niche includes the local environment in which an organism lives (its **habitat**). A niche also encompasses how an organism obtains nourishment, what organisms eat it, what organisms it competes with, and how it interacts with and is influenced by the abiotic components of its environment, such as light, temperature, and moisture. Because a complete description of an organism's ecological niche involves many dimensions and is difficult to define precisely, ecologists usually confine their studies to one or a few niche variables, such as ability to tolerate temperature extremes.

Competition for the same resource occurs between organisms

Competition occurs when two or more individuals simultaneously require a single resource, which is usually in limited supply. The resources for which plants commonly compete include water, light, soil minerals, and growing space. Because resources are often in limited supply in the environment, their use by one individual decreases the amount available to others (•**Table 26-1**). If a tree in a dense forest grows taller than surrounding trees, for example, it absorbs more of the incoming sunlight. Less sunlight is therefore available for nearby trees shaded by the taller tree.

Competition occurs among individuals within a population or between species. Competition among individuals within a population (that is, within the same

PRODUCER An organism that synthesizes organic compounds from simple inorganic raw materials.

CONSUMER An organism that cannot synthesize its own food from inorganic raw materials and therefore must obtain energy and body-building materials from other organisms.

DECOMPOSER A microorganism that breaks down dead organic material and uses the decomposition products as a source of energy.

ECOLOGICAL NICHE The totality of an organism's adaptations, its use of resources, its interactions with other organisms, and the lifestyle to which it is fitted in its community.

COMPETITION The interaction among two or more individuals that attempt to use the same essential resource, such as food, water, sunlight, or living space.

TABLE 26-1 Ecological Interactions among Species

INTERACTION	EFFECT ON SPECIES 1	EFFECT ON SPECIES 2
Competition between species 1 and species 2	Harmful	Harmful
Predation of species 2 (prey) by species 1 (predator)	Beneficial	Harmful
Symbiosis		
Mutualism of species 1 and species 2	Beneficial	Beneficial
Commensalism of species 1 with species 2	Beneficial	No effect
Parasitism by species 1 (parasite) on species 2 (host)	Beneficial	Harmful

species), known as **intraspecific competition,** is a major factor that limits population size. Competition between two or more individuals of different species, known as **interspecific competition,** also affects population size and sometimes results in the local extinction of one or more competing species.

Ecologists traditionally assumed that competition is the most important determinant of both the number of species found in a community and the size of each population. Today, ecologists recognize that competition is only one of many interacting biotic and abiotic factors that affect community structure.

EVOLUTION LINK Competition has adverse effects on species that use a limited resource. Over time, natural selection favors individuals of each species that avoid or reduce competition for environmental resources. Reduced competition among coexisting species as a result of each species' niche differing from the others in one or more ways is called **resource partitioning.** Difference in root depth is an example of resource partitioning in plants. For example, three common annuals found in certain abandoned fields are smartweed, Indian mallow, and bristly foxtail. Smartweed roots extend deep into the soil, Indian mallow roots grow to a medium depth, and bristly foxtail roots are shallow. This difference reduces competition for the same soil resources—water and minerals—by allowing the plants to exploit different portions of the soil. ∎

EVOLUTION LINK Natural selection shapes both predator and prey

Predation is the consumption of one species, the *prey,* by another, the *predator* (see Table 26-1). Predation includes both herbivores eating plants and carnivores eating other animals. Predation has resulted in an evolutionary "arms race," with the evolution of predator strategies (more efficient ways to catch prey) and prey

strategies (better ways to escape the predator). A predator that is more efficient at catching prey exerts a strong selective force on that prey by consuming those individuals with less effective prey strategies. Over time, the prey may evolve some sort of countermeasure. In turn, the countermeasure acquired by the prey acts as a strong selective force on the predator. This interdependent evolution between two interacting species, as represented by predator-prey relationships, is an example of **coevolution.**

Plants cannot escape herbivores by fleeing, but they have a number of adaptations that protect them from being eaten. The presence of spines, thorns, tough leathery leaves, or even thick wax on leaves discourages foraging herbivores from grazing. Other plant adaptations involve an array of protective chemicals that are unpalatable or even toxic to herbivores. The active ingredients in such plants as marijuana, opium poppy, and tobacco, for example, may discourage the foraging of herbivores.

Milkweeds are an excellent example of the biochemical coevolution between plants and herbivores.

FIGURE 26-8 Plant defense against herbivores.
The common milkweed (*Asclepias syriaca*) is protected by its toxic chemicals. Its leaves are poisonous to most herbivores except monarch caterpillars (*shown*) and a few other insects.

FIGURE 26-9 Mutualism.
An ant guards the nectar of a passionflower (*Passiflora*) blossom against nectar-robbing thieves. The nectar is thus reserved for its intended recipients, pollen-transporting hummingbirds. The ant benefits from this association because it is rewarded with nectar produced by special glands in the leaves.

Milkweeds produce alkaloids and cardiac glycosides, chemicals that are poisonous to all animals except a small group of insects (•**Figure 26-8**). During the course of evolution, these insects acquired the ability to either tolerate or metabolize the milkweed toxins. As a result, they eat milkweeds without being poisoned. They avoid competition from other herbivorous insects, since few other insects eat milkweed leaves. Predators also learn to avoid milkweed-tolerant insects, which accumulate the toxins in their tissues and are usually brightly colored to announce that fact. The black-, white-, and yellow-banded caterpillar of the monarch butterfly is an example of a milkweed feeder. ■

Symbiosis is a close association between species

Symbiosis is an intimate relationship or association between individuals of two or more species. The partners of a symbiotic relationship, called **symbionts,** may either benefit from, be unaffected by, or be harmed by the relationship (see Table 26-1). The thousands, or even millions, of symbiotic associations in nature are all products of coevolution, and they may be classified into several types: mutualism, commensalism, and parasitism.

Benefits are shared in mutualism
Mutualism is a symbiotic relationship in which both partners benefit. For example, *mycorrhizae* are mutualistic associations between fungi and the roots of almost all plants. The fungus absorbs phosphorus and other essential minerals from the soil and provides them to

the plant, and the plant provides the fungus with carbohydrates produced by photosynthesis. Plants not only grow more vigorously in the presence of mycorrhizae but better tolerate environmental stresses such as drought and high soil temperatures (see Figure 21-15).

A remarkable example of mutualism occurs in the rain forests of South and Central America between ants or wasps and flowering vines. Vines that produce bright red flowers are usually pollinated by hummingbirds, which obtain a reward of nectar from the blossoms. Some insects, however, try to rob the flower of its nectar without performing the important task of pollination. Ants or wasps come into the picture at this point. These insects patrol the plant and attack any would-be nectar robbing insect (•**Figure 26-9**). In exchange for their vigilance, the ant guards are rewarded with nectar, which is produced in the plant's leaves.

Commensalism is taking without harming
Commensalism is a type of symbiosis in which one organism benefits and the other is neither harmed nor helped. One example of commensalism is the relationship between a rainforest tree and **epiphytes,** smaller plants that grow on other plants (•**Figure 26-10**). Spanish moss, orchids, and bromeliads are examples of

FIGURE 26-10 Commensalism.
Epiphytes are small plants that grow on larger plants. They are particularly common in the humid tropics, although a few epiphytes even occur in deserts.

PREDATION A relationship in which one organism (the predator) devours another organism (the prey).

SYMBIOSIS An intimate relationship between two or more organisms of different species.

epiphytes. Most epiphytes anchor themselves onto tree branches but do not obtain nutrients or water directly from the trees. Their location lets them obtain adequate light, rainwater dripping down the branches, and required minerals, which are washed out of the tree's leaves by rainfall. Thus, the epiphyte benefits from the association while the tree remains largely unaffected. (It should be noted that some epiphytes are parasitic and harm the plant on which they grow; see Figure 6-12.)

Parasitism is taking at another's expense

Parasitism is a symbiotic relationship in which one member, the *parasite,* benefits and the other, the *host,* is adversely affected. The parasite obtains nourishment from its host, and although a parasite may weaken its host, it rarely kills it. Many parasites do not cause disease, but some do. When a parasite causes disease and sometimes the death of a host, the parasite is called a *pathogen.*

The mistletoes are some of the best-known parasitic seed plants, mainly because of their popularity at Christmastime. About 1300 species of mistletoes are known to parasitize conifer and hardwood trees, particularly in the tropics. Mistletoes anchor themselves to their hosts by means of adventitious roots that penetrate and branch through the hosts' tissues (see Figure 6-12). Mistletoes are "water robbers"; that is, they take water and minerals from their hosts, but not carbohydrates produced by photosynthesis.

An interesting example of parasitism involves the fungus *Puccinia monoica*, which parasitizes rock cress (*Arabis holboellii*), a plant widely distributed across the northern part of North America. When *Puccinia* infects rock cress, it does not kill the plant immediately, but it changes the plant's growth pattern. An infected plant grows much taller than normal and produces a cluster of leaves at its top. The fungal mycelium, which is bright yellow, grows on and eventually covers these leaves, giving them the appearance of buttercup flowers (•**Figure 26-11**). The fungus also secretes a sugary solution and produces a strong scent, imitating the nectar and aroma of flowers. This elaborate mimicry increases the chances of successful fungal reproduction by attracting insects to the fungal "flowers." Sexual reproduction in *Puccinia* requires the union of nuclei from two mating types. Insects are attracted to the "flowers," where they eat the sugary syrup. As the insects feed, pieces of the fungus cling to their bodies. Flying from one *Puccinia* "flower" to another, the insects distribute complementary mating types from fungus to fungus over a broad area. Rock cress pays for this parasitic relationship with altered growth and an early death.

FIGURE 26-11 Parasitism.
In this example of parasitism, a fungus (*Puccinia monoica*) parasitizes a North American plant called rock cress (*Arabis holboellii*), causing it to change its growth and produce, with the fungus's help, fake flowers. The "flowers" are so realistic that they attract insect pollinators, which help the fungus (instead of the plant) reproduce.

Keystone species and dominant species affect the character of a community

Certain species, called **keystone species,** are crucial in determining the nature of the entire community. Other species of a community depend on or are greatly affected by the keystone species. Keystone species are usually not the most abundant species in the community. Although present in relatively small numbers, the individuals of a keystone species profoundly influence the entire community because they often affect the amount of available food, water, or some other resource. Thus, the impact of keystone species is greatly disproportionate to their abundance. Identifying and protecting keystone species are crucial goals of conservation biology, because if a keystone species disappears from a community, many other species in that community may become more common, rarer, or even disappear.

One problem with the concept of keystone species is that it is difficult to measure all the direct and indirect impacts of a keystone species on other organisms in a community. Consequently, most evidence for the existence of keystone species is based on indirect observations rather than on experimental manipulations. For example, consider the fig tree. Because fig trees produce a continuous crop of fruits, they may be keystone species in tropical rain forests of Central and South America. Fruit-eating monkeys, birds, bats, and other fruit-eating vertebrates of the forest do not normally consume large quantities of figs in their diets. During that time of the year when other fruits are less plentiful, however, fig trees become important in sustaining fruit-eating vertebrates. It is therefore assumed that, should the fig trees disappear, most of the fruit-eating vertebrates would also disappear. If, in turn, the fruit eaters disappear, the spatial distribution of other fruit-bearing plants would become more limited because the fruit eaters help disperse their seeds. Thus, protecting fig trees in tropical rain forests probably increases the likelihood that monkeys, birds, bats, and many other tree species will survive. The question is whether this anecdotal evidence of fig trees as keystone species is strong enough for policymakers to grant special protection to fig trees.

In contrast to keystone species, which have a large impact out of proportion to their abundance, **dominant species** greatly affect the community because they are very common. Trees, the dominant species of forests, change the local environment. Trees provide shade, which changes both the light and moisture availability on the forest floor. Trees provide numerous habitats and *microhabitats* (such as a hole in a tree trunk) for other species. Forest trees also provide food for many organisms and therefore play a large role in *energy flow* through the forest ecosystem. Similarly, cordgrass is the dominant species in salt marshes, prairie grass in grasslands, and kelp in kelp beds. Typically, a community has one or a few dominant species, and most other species are relatively rare.

Ecological succession is community change over time

A new community does not spring into existence overnight but develops gradually through a series of stages, each dominated by different species. The process of community development over time, which involves species in one stage being replaced by other species, is called **ecological succession,** or simply **succession.** An area is initially colonized by certain early-successional species that give way over time to mid-successional species, which in turn give way much later to late-successional species.

Ecological succession is usually described in terms of the changes in the plant composition of an area, although each successional stage also has its charac-

teristic kinds of animals and other organisms. The time involved in ecological succession is tens, hundreds, or thousands of years. Ecologists distinguish between two types of succession: primary and secondary.

In primary succession, a community develops on land not previously inhabited by plants

Primary succession is the change in species composition over time in an area not previously inhabited by plants or other organisms; no soil exists when primary succession begins. Bare rock surfaces, such as recently formed volcanic lava and rock scraped clean by glacial action, are examples of sites where primary succession might occur.

Exposed rock is an inhospitable environment that few organisms can tolerate. The temperature of bare rock may be quite high in the sunlight, and unless rain is falling, the rock may be totally devoid of moisture. Whatever minerals are present in the rock are locked up in its hard, crystalline structure, so are unavailable to organisms.

In primary succession on bare rock, the details vary from one site to another, but usually the first community that is observed consists of lichens (•**Figure 26-12**). Because lichens are the first organisms to colonize bare rock, they are called **pioneers.** Lichens live on the surfaces of many rocks and beneath the surfaces of porous rocks, in a sheltered and somewhat moister habitat. Lichens are resistant to desiccation (drying out). They cease to grow when water is unavailable but quickly resume active growth when moisture returns; they can absorb their own weight in water within moments of moistening.

With the passage of time, several important cumulative changes occur. The *biomass* (amount of organic material) of the lichen community increases. Lichens secrete acids that help break the rock apart, and as a result, fine particles of rock become detached from the rock's surface or even within the rock itself. As lichens die, their decomposing remains mix with the rock particles to form a rudimentary soil. Water is absorbed and retained in the tissues of the lichens and in the new, thin soil layer for longer periods than before. As these changes occur, an increasing number of small organisms move into the area and make their homes in the lichens and soil.

These changes—increased biomass, soil development, water retention, and an increased number of organisms—moderate the harsh conditions under which

ECOLOGICAL SUCCESSION The sequence of changes in the species composition of a community over time.

the pioneer community lives, making it possible for mosses to grow there. In fact, since mosses grow faster than lichens, they tend to replace the lichens. Their greater productivity leads to a greater accumulation of biomass and ultimately of soil. This leads to further habitat change.

Over time the moss community may be replaced by drought-resistant ferns, followed in turn by tough grasses and herbs. Once sufficient soil has accumulated, grasses and herbs are replaced by low shrubs, which in turn may be replaced by forest trees in several distinct stages. Primary succession from a pioneer community on bare rock to a forest community may take hundreds or thousands of years.

The Indonesian island of Krakatoa has provided scientists with a perfect long-term study of primary succession in a tropical rainforest environment. In 1883, a volcanic eruption destroyed all life on the island. Ecologists have surveyed the island in the more than 100 years since the devastation to document the return of life-forms. As of the 1990s, ecologists had found that the progress of primary succession to a diverse tropical rainforest community was slow, in part because of Krakatoa's isolation. Many species are limited in their ability to disperse over water. Krakatoa's forest, for example, may have only one-tenth the tree-species richness of undisturbed tropical rain forest of nearby islands. The lack of plant diversity has, in turn, limited the number of colonizing animal species. In a forested area of Krakatoa where zoologists would expect more than 100 butterfly species, for example, there are only 2 species.

In secondary succession, a community develops after removal of an earlier community

Secondary succession is the change in species composition over time in an area already substantially modified by a preexisting community; soil is already present at these sites. An abandoned field and an area cleared by a forest fire are common examples of sites where secondary succession occurs.

Since the late 1930s, many ecologists have studied secondary succession on abandoned farmland. Although all the stages of secondary succession may take more than 100 years to occur at a given site, it is possible for a single researcher to study old-field succession in its entirety by observing sites in the same area that were abandoned at different times. Often the ecologist accurately determines when each field was abandoned by examining court records.

One heavily studied example of secondary succession is the regrowth of abandoned farmland in North

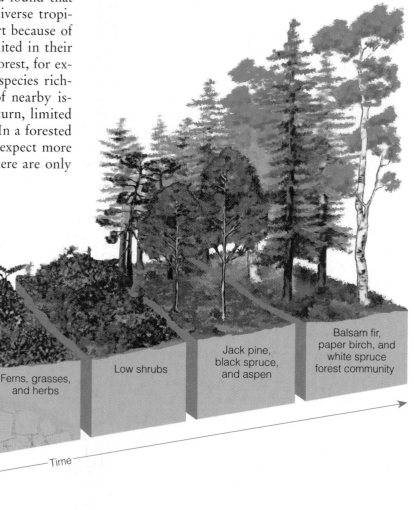

Exposed rocks

Lichens and mosses

Ferns, grasses, and herbs

Low shrubs

Jack pine, black spruce, and aspen

Balsam fir, paper birch, and white spruce forest community

Time

FIGURE 26-12 Primary succession on bare rock exposed by a retreating glacier.

Carolina. Abandoned farmland in North Carolina is colonized by a predictable sequence of plant communities. The first year after cultivation ceases, annual and perennial herbs grow in abundance, but crabgrass is by far the most common species. In the second year, horseweed, a tall, coarse annual, is the main species. Horseweed does not remain dominant more than 1 year, however, because decaying horseweed roots inhibit the growth of young horseweed seedlings. In addition, horseweed does not compete well with plants that become established in the third year, including broomsedge. Typically, broomsedge outcompetes other plants because it is more tolerant of droughts. Broomsedge continues to be dominant for the next few years.

In years 5 to 15, the dominant plants in the abandoned farmland are pines, such as shortleaf pine on drier sites or loblolly pine on moister sites (●**Figure 26-13**). Through shading and the buildup of litter (pine needles and branches) on the soil, the growing pines produce conditions that cause the earlier dominant plants to decline in importance.

Over time, pines give up their dominance to hardwoods such as oaks and hickories. This stage of secondary succession depends primarily on the environmental changes produced by the pines. The pine litter causes soil changes, such as an increase in water-holding capacity, that are necessary for young hardwood seedlings to become established. In addition, young pine seedlings do not thrive in the shade of older trees, whereas hardwood seedlings do.

During the summer of 1988, wildfires burned approximately one-third of Yellowstone National Park. This natural disaster provided a valuable chance for ecologists to study secondary succession in areas that had been forests. After the conflagration, gray ash covered the forest floor, and most of the trees, although standing, were charred and dead. Secondary succession in Yellowstone has occurred rapidly since 1988. In the following spring, trout lily and other herbs sprouted and covered much of the ground. Ten years after the fires, a young forest of knee-high to shoulder-high lodgepole pines dominated the area, and Douglas fir seedlings began growing. Ecologists continue to monitor the changes in Yellowstone as secondary succession unfolds.

Disturbance influences succession

Early studies suggested that succession inevitably progressed to a stable and persistent community, known as a *climax community*, which was determined solely by climate. Periodic disturbances, such as fires or floods, were not thought to exert much influence on climax communities. If the climax community was disturbed

FIGURE 26-13 **Secondary succession in an abandoned field.** Shown is a young loblolly pine (*Pinus taeda*) becoming established in a field of broomsedge (*Andropogon virginicus*).

in any way, it would return to a self-sustaining, stable equilibrium in time.

This traditional view of stability has fallen out of favor. The apparent end-point stability of species composition in a "climax" forest is probably the result of how long trees live relative to the human life span. It is now recognized that forest communities never reach a state of permanent equilibrium but instead exist in a state of continual disturbance. The species composition and relative abundance of each species vary in a mature community over a range of environmental gradients, although the community retains a relatively uniform appearance overall (recall the patches of local populations seen in Figure 26-7).

Ecosystem Ecology

Life would not be possible without the abiotic (nonliving) environment. As the sun warms the planet, it powers the hydrologic cycle (causes precipitation), drives ocean currents and atmospheric circulation patterns, and produces much of the climate to which organisms have adapted (•**Figure 26-14**). The sun also supplies the energy that almost all organisms use to carry on life processes.

Individual communities and their abiotic environments are ecosystems, which are the basic units of ecology. An ecosystem encompasses all the interactions among organisms living in a particular place, and among those organisms and their abiotic environment. Ecosystem interactions are complex because each organism responds not only to other organisms but to conditions in the atmosphere, soil, and water. In turn, organisms exert an effect on the abiotic environment. Like communities, ecosystems vary in size, lack precise boundaries, and are nested within larger ecosystems. The branch of ecology that analyzes the movement of energy and matter through large numbers of organisms and the abiotic environment is known as **ecosystem ecology.**

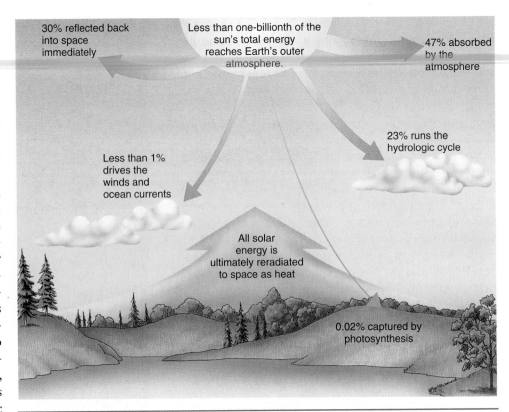

30% reflected back into space immediately

Less than one-billionth of the sun's total energy reaches Earth's outer atmosphere.

47% absorbed by the atmosphere

23% runs the hydrologic cycle

Less than 1% drives the winds and ocean currents

All solar energy is ultimately reradiated to space as heat

0.02% captured by photosynthesis

FIGURE 26-14 The fate of solar radiation on Earth.
Most of the energy released by the sun never reaches Earth. The solar energy that does reach Earth warms the planet's surface, drives the hydrologic cycle and other cycles of matter, produces the climate, and powers almost all life through the process of photosynthesis.

Energy flows through ecosystems in a one-way direction

The passage of energy in a linear, or one-way, direction through an ecosystem is known as **energy flow.** Energy enters an ecosystem as the radiant energy of sunlight, some of which is trapped by plants during the process of photosynthesis. This energy, now in chemical form, is stored in the bonds of organic molecules such as glucose. When the molecules are broken apart by cellular respiration, the energy becomes available to do work such as tissue repair, production of body heat, or reproduction. As the work is accomplished, the energy escapes the organism and disperses into the environment as low-quality heat. Ultimately, this heat energy radiates into space. Thus, once organisms use energy, it becomes unavailable for reuse (see the discussion of the second law of thermodynamics in Chapter 2).

In an ecosystem, energy flow can be traced through **food chains,** in which chemical energy passes from one organism to the next in a sequence (•**Figure 26-15**). Producers form the beginning of the food chain by capturing the sun's energy through photosynthesis. Herbivores and omnivores eat plants to obtain the chemical energy of the producers' molecules as well as building materials from which to construct tissues. In turn, carnivores and omnivores consume herbivores and reap the energy stored in the herbivores' molecules. At the end of a food chain are decomposers; through cellular respiration, they break down organic molecules in the carcasses and body wastes of all other members of the food chain.

The most important thing to remember about energy flow in food chains is that it is linear, or unidirectional. That is, stored chemical energy moves from one organism to the next *as long as it is not used.* After an individual uses that energy to do work, however, it is unavailable for any other organism because it has changed into heat energy that disperses in the environment.

Each level in a food chain is called a **trophic level.** The first trophic level is made up of producers (organisms that photosynthesize); the second trophic level,

Ecologists gain insights into how ecosystems function by examining energy flow and the energy content of each trophic level.

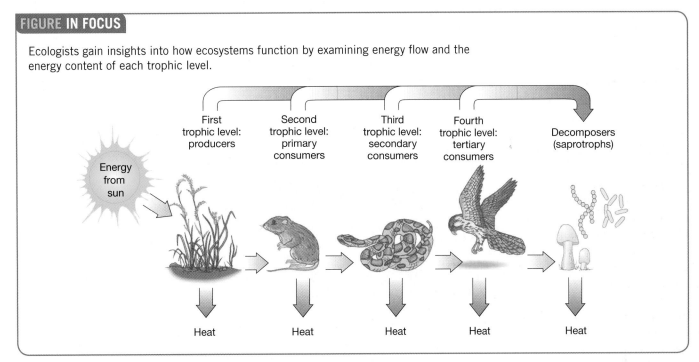

FIGURE 26-15 Trophic levels in a food chain.
Energy from an external source (the sun) enters a food chain and exits as heat loss. Most of the energy acquired by a given trophic level is used for metabolic purposes (and released as heat) and is therefore unavailable to the next trophic level.

primary consumers (herbivores); the third trophic level, secondary consumers (carnivores); and so on.

Food chains as simple as the one just described rarely occur in nature, because few organisms eat just one other kind of organism. More typically, the flow of energy and materials through ecosystems takes place with a range of food choices for each organism. In an ecosystem of average complexity, numerous alternative pathways are possible. Thus, a **food web,** an aggregate of interconnected food chains in an ecosystem, is a more realistic model of the flow of energy through ecosystems (•**Figure 26-16**).

Pyramids of energy illustrate how ecosystems work

An important feature of energy flow is that most energy dissipates into the environment when going from one trophic level to another. The relative energy value at each trophic level can be graphically represented by a **pyramid of energy** that indicates the energy contents (usually expressed in kilocalories) of the biomass of the various trophic levels (•**Figure 26-17**). Pyramids of energy help us understand that most food chains are short because of the dramatic reduction in energy content that occurs at each trophic level. Pyramids of energy also

demonstrate the environmental effects of a human vegetarian diet versus a meat-based diet: fewer plants (and less land on which to grow them) are required to support vegetarians than to support meat eaters.

Matter cycles through ecosystems

Matter, the material of which organisms are composed, cycles from the living world to the nonliving physical environment and back in what are called **biogeochemical cycles.** Three biogeochemical cycles—carbon, nitrogen, and water—are particularly important to organisms.

Carbon dioxide is the pivotal molecule of the carbon cycle

Carbon must be available to organisms because proteins, carbohydrates, and other organic molecules that

TROPHIC LEVEL Each sequential step in a food chain or food web, from producer to primary, secondary, or tertiary consumers.

BIOGEOCHEMICAL CYCLE The process by which matter cycles from the living world to the nonliving, physical environment and back again.

are essential to life contain carbon. Carbon is present in the atmosphere as a gas, carbon dioxide (CO_2), which makes up approximately 0.04 percent of the atmosphere. Carbon is present in the ocean in a dissolved form and is also found in rocks such as limestone. Carbon cycles between the abiotic environment, including the atmosphere, and organisms (•**Figure 26-18**). Refer to the figure as you read the following description.

During photosynthesis, plants remove CO_2 from the air and fix, or incorporate, it into organic molecules such as sugar (❶). Thus, photosynthesis incorporates carbon from the abiotic environment into the biological compounds that make up an ecosystem's producers. These compounds are usually used as fuel for cellular respiration (the process that releases energy from organic compounds) by the producer that made them, by a consumer that eats the producer, or by a decomposer that breaks down the remains of the producer or consumer. Because the end products of aerobic respiration are CO_2, water, and energy for biological work, CO_2 is returned to the atmosphere by the process of respiration (❷).

Sometimes the carbon in biological molecules is not cycled back to the abiotic environment for some time. For example, a lot of carbon is stored in the wood of trees, where it may stay for several hundred years. In addition, millions of years ago, vast coal beds formed from the bodies of ancient trees that did not decay fully before they were buried (❸). Similarly, the carbon-containing remains of unicellular marine organisms that accumulated in the geologic past probably gave rise to the underground deposits of oil and natural gas (❹).

FIGURE IN FOCUS

Food webs in all but the simplest ecosystems are too complex to depict all the species and links actually present. Rarely do food-web diagrams take into account that some links are strong and others are weak. Moreover, food webs change over time, with additions and deletions of links.

FIGURE 26-16 A food web at the edge of a deciduous forest.
This food web is greatly simplified compared to what actually happens in nature.

Coal, oil, and natural gas, called **fossil fuels** because they formed from the remains of ancient organisms, are vast depositories of carbon compounds that were the end products of photosynthesis millions of years ago. The carbon in coal, oil, natural gas, and wood may be returned to the atmosphere by the process of burning, or *combustion*. In combustion, organic molecules are

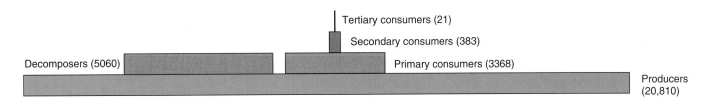

FIGURE 26-17 A pyramid of energy.
Energy values for this pyramid of energy for Silver Springs, Florida, are given in kilocalories per square meter per year. Representative organisms include tape grass (producers), snails (primary consumers), young river turtles (secondary consumers), gar fish (tertiary consumers), and bacteria and fungi (decomposers). Note the substantial loss of usable energy from one trophic level to the next. (Based on H. T. Odum, "Trophic Structure and Productivity of Silver Springs, Florida," *Ecological Monographs,* Vol. 27, 1957.)

rapidly oxidized (combined with oxygen) and converted to CO_2 and water with an accompanying release of heat and light (❺).

An even greater amount of carbon may be stored for millions of years once it is incorporated into the shells of marine organisms (❻). When those organisms die, their shells sink to the ocean floor and are covered by sediments. Such shells accumulate to form seabed deposits hundreds of meters thick, which eventually cement together to form the sedimentary rock *limestone.* Earth's crust is dynamically active, and over millions of years, sedimentary rock on the bottom of the seafloor may be lifted and become land surfaces. For

example, some cliffs in the Canadian Rockies are composed of limestone, indicating that the region was once beneath the ocean. When limestone is exposed by the process of geologic uplift, it slowly erodes (wears away) through chemical and physical weathering processes (❼), and its carbon returns to the water and atmosphere, where it is available to participate in the carbon cycle again.

In the carbon cycle, then, photosynthesis removes carbon from the abiotic environment and incorporates it into biological molecules; respiration, combustion, and erosion return carbon to the water and atmosphere of the abiotic environment.

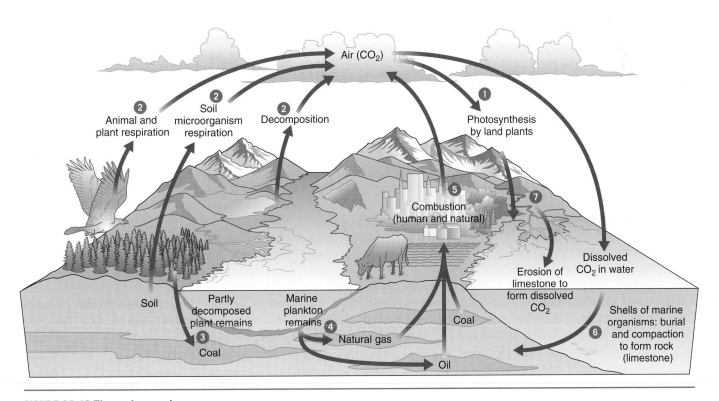

FIGURE 26-18 The carbon cycle.

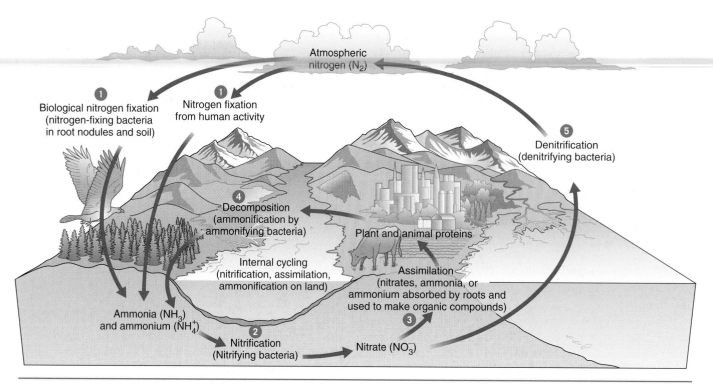

FIGURE 26-19 The nitrogen cycle.

Bacteria are essential to the nitrogen cycle

Nitrogen is crucial to life because it is an essential part of proteins, which are important structural components of cells, including enzymes, and nucleic acids. At first glance, a shortage of nitrogen for organisms seems impossible: Earth's atmosphere is about 80 percent nitrogen gas (N_2), a two-atom molecule. However, molecular nitrogen is so stable that it does not readily combine with other elements. Therefore, organisms cannot use nitrogen gas from the atmosphere to manufacture their proteins and nucleic acids. Nitrogen gas must be broken apart into individual nitrogen atoms before it combines with other elements to form molecules.

Five steps occur in the nitrogen cycle: nitrogen fixation, nitrification, assimilation, ammonification, and denitrification (•**Figure 26-19**). Bacteria perform all these steps except assimilation. Refer to the figure as you read the following description.

Nitrogen fixation involves the conversion of gaseous nitrogen (N_2) to ammonia (NH_3) (❶). This process is called nitrogen fixation because nitrogen is fixed (incorporated) into a form that organisms can use. Although considerable nitrogen is also fixed by combustion, volcanic action, and lightning discharges as well as by industrial processes, most nitrogen fixation is biological. Biological nitrogen fixation is carried out by

nitrogen-fixing bacteria, including cyanobacteria, in soil and aquatic environments.

Nitrogen-fixing bacteria employ an enzyme called **nitrogenase** to break up molecular nitrogen and combine it with hydrogen. Because nitrogenase functions only in the absence of oxygen, the bacteria that use nitrogenase must insulate the enzyme from oxygen by some means. Some nitrogen-fixing bacteria live beneath layers of oxygen-excluding slime on the roots of certain plants. But the most important nitrogen-fixing bacteria, *Rhizobium*, live in special swellings, or **nodules,** on the roots of legumes, such as beans or peas, and some woody plants (•**Figure 26-20a**). The relationship between *Rhizobium* and its host plants is mutualistic: the bacteria receive carbohydrates from the plant, and the plant receives nitrogen in a form that it can use to make proteins and nucleic acids. The host plants are not the only plants to benefit from this relationship. When the host plants die and are decomposed, the nitrogen becomes available for other plants.

In aquatic habitats, most nitrogen fixation is performed by cyanobacteria. Some filamentous cyanobacteria have special oxygen-excluding cells called **heterocysts** that fix nitrogen (•**Figure 26-20b**). Some water ferns have cavities in which cyanobacteria live, somewhat as *Rhizobium* lives in the root nodules of legumes.

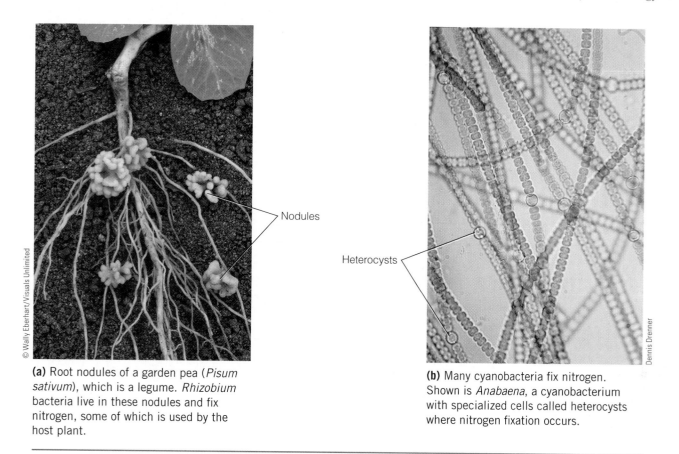

(a) Root nodules of a garden pea (*Pisum sativum*), which is a legume. *Rhizobium* bacteria live in these nodules and fix nitrogen, some of which is used by the host plant.

(b) Many cyanobacteria fix nitrogen. Shown is *Anabaena*, a cyanobacterium with specialized cells called heterocysts where nitrogen fixation occurs.

FIGURE 26-20 Organisms and nitrogen fixation.

Other cyanobacteria fix nitrogen in symbiotic association with certain plants or as the photosynthetic partners of certain lichens.

Nitrification is the conversion of ammonia (NH_3) to nitrate (NO_3^-) (Figure 26-19 ❷). Nitrification is a two-step process accomplished by aerobic soil bacteria. First the soil bacteria *Nitrosomonas* and *Nitrococcus* convert ammonia to nitrite (NO_2^-). Then the soil bacterium *Nitrobacter* oxidizes nitrite to nitrate. The process of nitrification furnishes these bacteria, called nitrifying bacteria, with energy.

In **assimilation,** roots absorb either nitrate (NO_3^-) or ammonia (NH_3) that was formed by nitrogen fixation and nitrification and then incorporate the nitrogen in these molecules into plant proteins and nucleic acids (❸). When animals consume plant tissues, they assimilate nitrogen as well, by ingesting the nitrogen compounds of plants and converting them to animal compounds. As a matter of fact, *all* organisms assimilate nitrogen.

Ammonification is the conversion of biological nitrogen compounds to ammonia (❹). Ammonification begins when organisms produce nitrogen-containing waste products. Land animals such as mammals and amphibians produce urea in urine. Other land animals, such as reptiles, birds, and insects, produce uric acid in their wastes. These substances, along with the nitrogen compounds in dead organisms, are decomposed, thereby releasing the nitrogen into the abiotic environment as ammonia (NH_3). The bacteria that perform ammonification in both the soil and aquatic environments are called ammonifying bacteria. The ammonia produced by ammonification enters the nitrogen cycle and is available for the processes of nitrification and assimilation. Most available nitrogen in the soil derives from the recycling of organic nitrogen by ammonification.

Denitrification is the reduction of nitrate (NO_3^-) to gaseous nitrogen (N_2) (❺). Denitrifying bacteria reverse the actions of nitrogen-fixing and nitrifying bacteria; that is, denitrifying bacteria return nitrogen to the atmosphere as nitrogen gas. Denitrifying bacteria are anaerobic: they prefer to live and grow where there is little or no free oxygen. For example, some denitrifying bacteria are found deep in the soil near the water table,

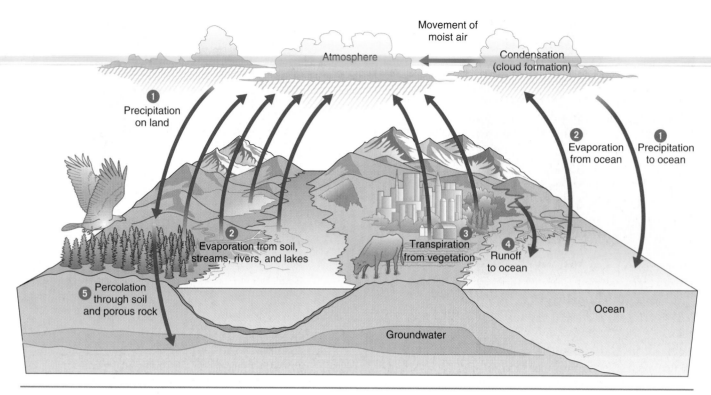

FIGURE 26-21 The hydrologic cycle.

an environment that is nearly oxygen-free. The nitrogen gas that is released works its way through the soil to the atmosphere.

Water circulates in the hydrologic cycle

In the hydrologic cycle, water continuously circulates from the ocean to the atmosphere to the land and back to the ocean (•**Figure 26-21**). Living organisms, from bacteria to plants and animals, are also a part of the hydrologic cycle: water makes up much of the mass of most organisms and serves as a medium for chemical reactions as well as for the transport of materials within and among cells. Refer to the figure as you read the following description. Water moves from the atmosphere to the land and ocean in the form of precipitation (rain, sleet, snow, or hail) (❶). When water evaporates from the ocean surface and from soil, streams, rivers, and lakes, it eventually condenses and forms clouds in the atmosphere (❷). In addition, *transpiration*, the loss of water vapor from land plants, adds a considerable amount of water vapor to the atmosphere (❸). Roughly 97 percent of the water a plant absorbs from the soil is transported to the leaves, where it is lost by transpiration.

Water may evaporate from land and reenter the atmosphere directly. Alternatively, it may flow in rivers and streams to coastal *estuaries*, where fresh water

meets the ocean (❹). The movement of surface water from land to ocean is called **runoff**, and the area of land drained by runoff is called a *watershed*. Water also percolates (seeps) downward in the soil to become **groundwater**, where it is trapped and held for a time (❺). The underground caverns and porous layers of rock in which groundwater is stored are called *aquifers*. Groundwater may reside in the ground for hundreds to many thousands of years, but eventually it supplies water to the soil, streams and rivers, plants, and the ocean. The human removal of more groundwater than precipitation or melting snow recharges, called *aquifer depletion*, eliminates groundwater as a water resource.

Regardless of its physical form (solid, liquid, or vapor) or location, every molecule of water eventually moves through the hydrologic cycle. Tremendous amounts of water cycle annually between Earth and its atmosphere. The volume of water entering the atmosphere from the ocean each year is estimated at about 425,000 cubic kilometers, an amount that is impossible to visualize. Approximately 90 percent of this water reenters the ocean directly as precipitation over water; the remainder falls on land. As is true of the other cycles, water (in the form of glaciers, polar ice caps, and certain groundwater) can be lost from the cycle for thousands of years.

Ecosystems may be regulated from the bottom up or the top down

One question that ecologists have recently considered is which process—bottom up or top down—is more significant in the regulation of various ecosystems. Both energy flow and cycles of matter are involved in bottom-up and top-down processes. **Bottom-up processes** are based on food webs that, as you know, always have producers at the first (lowest) trophic level (●**Figure 26-22a**). In a sense, the biogeochemical cycles that regenerate nutrients such as nitrates and phosphates for producers to assimilate are located "under" the first trophic level. Thus, bottom-up processes regulate ecosystem function by nutrient cycling and other aspects of the abiotic environment. If bottom-up processes dominate an ecosystem, the availability of resources such as water or soil minerals controls the number of producers, which in turn controls the number of herbivores, which controls the number of carnivores.

Bottom-up processes apparently predominate in certain aquatic ecosystems in which nitrogen or phosphorus is limiting. An experiment in which phosphorus was added to a phosphorus-deficient river (the Kaparuk River in Alaska) resulted in an increase in algae, followed over time by increased populations of aquatic insects, other invertebrates, and fish.

In contrast, **top-down processes** regulate ecosystem function by trophic interactions, particularly from the highest trophic level (●**Figure 26-22b**). Ecosystem regulation by top-down processes occurs because carnivores eat herbivores, which in turn eat producers, which in turn affect levels of nutrients. If top-down processes dominate an ecosystem, the effects of an increase in the population of top predators cascade down the food web through the herbivores and producers. Top-down processes are also known as a *trophic cascade*.

A change in the feeding preferences of killer whales off the coast of Alaska provides an excellent example of a trophic cascade. A few decades ago, killer whales began preying on sea otters, causing a sharp decline in the otter population. As sea otters have declined, the number of sea urchins, which sea otters eat, has increased.

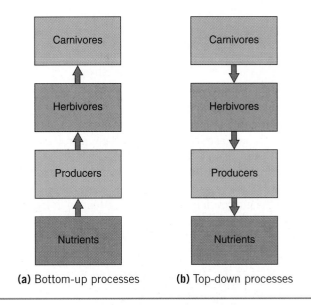

(a) Bottom-up processes   **(b)** Top-down processes

FIGURE 26-22 Bottom-up and top-down processes.

Sea urchins eat kelp, the producers at the base of the food web; the increase in the number of sea urchins has caused a decline in kelp populations. Top-down regulation may predominate in ecosystems with few trophic levels and low species richness. Such ecosystems may have only one or a few species of dominant herbivores, but those species have a strong impact on the producer populations.

It may be that top-down and bottom-up processes are not mutually exclusive. In a study of a thorn-scrub community in north-central Chile, ecologists demonstrated that top-down regulation predominates in certain small desert mammals and plant species. However, during periodic El Niño events,[1] the increase in precipitation resulted in bottom-up increases in producers and consumers. Thus, in this ecosystem, both top-down and bottom-up processes appear important in trophic dynamics over an extended period.

---

[1] *El Niño is a periodic, large-scale warming of surface waters of the tropical eastern Pacific Ocean that temporarily alters both ocean and atmospheric circulation patterns.*

STUDY OUTLINE

❶ **Define *ecology*, and distinguish among population, community, ecosystem, and biosphere.**
Ecology is a discipline of biology that studies the interrelations between living things and their environments; ecologists study populations, communities, ecosystems,

and the biosphere. A **population** is a group of organisms of the same species that live in a defined geographic area at the same time. A **community** is an association of populations of different species living in a defined habitat with some degree of interdependence. An **ecosystem** is the interacting system that encompasses a community and its nonliving, physical environment. The **biosphere** is all of Earth's living organisms, collectively.

② **Explain the difference between J-shaped and S-shaped population growth curves.**
Intrinsic rate of increase is the maximum rate at which a population could increase in number under ideal conditions. Although certain populations exhibit an accelerated pattern of growth known as **exponential population growth** for a limited period (the J-shaped curve), eventually the growth rate decreases, as exhibited in **logistic population growth**. Logistic population growth shows a characteristic S-shaped curve. Seldom do natural populations follow the logistic growth curve closely.

③ **Summarize the three main types of survivorship curves.**
Survivorship is the probability that a given individual in a population will survive to a particular age. There are three general survivorship curves: Type I survivorship, in which mortality is greatest in old age; Type II survivorship, in which mortality is spread evenly across all age groups; and Type III survivorship, in which mortality is greatest among the young.

④ **Characterize producers, consumers, and decomposers.**
A **producer** is an organism that synthesizes organic compounds from simple inorganic raw materials. A **consumer** is an organism that cannot synthesize its own food from inorganic raw materials and therefore must obtain energy and body-building materials from other organisms. A **decomposer** is a microorganism that breaks down dead organic material and uses the decomposition products as a source of energy.

⑤ **Describe what is meant by an organism's ecological niche.**
An organism's **ecological niche** includes the totality of an organism's adaptations, its use of resources, its interactions with other organisms, and the lifestyle to which it is fitted in its community.

⑥ **Define *competition, predation,* and *symbiosis,* and distinguish among mutualism, commensalism, and parasitism.**
Competition is the interaction among two or more individuals that attempt to use the same essential resource, such as food, water, sunlight, or living space. **Predation** is a relationship in which one organism (the predator) kills and devours another organism (the prey). **Symbiosis** is an intimate relationship between two or more organisms of different species. Symbiosis includes mutualism, commensalism, and parasitism. In **mutualism,** both partners benefit. In **commensalism,** one organism benefits and the other is unaffected. In **parasitism,** one organism (the parasite) benefits and the other (the host) is harmed.

⑦ **Define *ecological succession,* and distinguish between primary succession and secondary succession.**
Ecological succession is the sequence of changes in the species composition of a community over time. **Primary succession** begins in an area not previously inhabited (for example, bare rock). **Secondary succession** begins in an area where there was a preexisting community and a well-formed soil (for example, abandoned farmland).

⑧ **Summarize the concept of energy flow through a food web.**
Energy flows through an ecosystem in one direction, from the sun to producer to consumer to decomposer. A **trophic level** is one of each sequential step in a **food chain** or **food web,** from producer to primary, secondary, or tertiary consumers. Much of the energy acquired by a given trophic level is used for metabolic purposes at that level and is therefore unavailable to the next trophic level.

⑨ **Describe the main steps in the carbon, nitrogen, and hydrologic cycles.**
A **biogeochemical cycle** is the process by which matter cycles from the living world to the nonliving, physical environment and back again. Carbon dioxide is the important gas of the carbon cycle; carbon enters plants, algae, and cyanobacteria as CO_2, which photosynthesis incorporates into organic molecules; cellular respiration, combustion, and erosion of limestone return CO_2 to the water and atmosphere, making it available to producers again. The nitrogen cycle has five steps: **nitrogen fixation** is the conversion of nitrogen gas to ammonia; **nitrification** is the conversion of ammonia or ammonium to nitrate; **assimilation** is the conversion of nitrates, ammonia, or ammonium to proteins, chlorophyll, and other nitrogen-containing compounds by plants; **ammonification** is the conversion of organic nitrogen to ammonia and ammonium ions; **denitrification** is the conversion of nitrate to nitrogen gas. The hydrologic cycle involves an exchange of water between the land, ocean, atmosphere, and organisms; water enters the atmosphere by evaporation and transpiration and leaves the atmosphere as precipitation; on land, water filters through the ground or runs off to lakes, rivers, and the ocean.

⑩ **Distinguish between bottom-up and top-down processes.**
If **bottom-up processes** dominate an ecosystem, the availability of resources such as minerals controls the number of producers (that is, the lowest trophic level), which in turn controls the number of herbivores, which in turn controls the number of carnivores. **Top-down processes** regulate ecosystems from the highest trophic level—by consumers eating producers. If top-down processes dominate an ecosystem, an increase in the number of top predators cascades down the food web through the herbivores and producers.

REVIEW QUESTIONS

1. How is a community different from an ecosystem?
2. How does a J-shaped population growth curve relate to intrinsic rate of increase? How does an S-shaped curve relate to carrying capacity?
3. What is survivorship? How do Type I, Type II, and Type III survivorship curves differ?
4. How might one distinguish between a producer and a decomposer? A consumer and a decomposer?
5. Distinguish between an organism's habitat and its ecological niche.
6. Distinguish between intraspecific competition and interspecific competition.
7. How is mutualism different from commensalism? How do parasitism and predation differ?
8. Define *ecological succession,* and distinguish between primary and secondary succession.

9. Why is the concept of a food web generally preferable to that of a food chain?
10. Why is the cycling of matter essential to the continuance of life?
11. Describe the carbon cycle, including the following processes: photosynthesis, respiration, and combustion.
12. List and describe the five steps in the nitrogen cycle.
13. Biologists think the reintroduction of wolves to Yellowstone National Park, which began in 1995, will ultimately result in a more varied and lush plant composition. Is this an example of bottom-up or top-down processes? Why?

THOUGHT QUESTIONS

1. John Muir once said, "When one tugs at a single thing in nature, he finds it attached to the rest of the world." What did he mean?
2. Could a balanced ecosystem be constructed that contained only producers and consumers? Only consumers and decomposers? Only producers and decomposers? In each case, explain the reason for your answer.
3. Describe the ecological niche of humans. Do you think that our niche has changed during the past thousand years? Why or why not?
4. Is it possible to have an inverted (∇) pyramid of energy? Why or why not?
5. What kinds of environmental conditions trigger primary succession? Secondary succession?

6. Sometimes epiphytes are so abundant in tropical rain forests that they cover the entire surfaces of leaves and interfere with a larger plant's ability to photosynthesize. In this situation, would the epiphytes be an example of commensalism? If not, what type of interaction would they exemplify?
7. In Hawaii, there are no native grazing mammals. Based on what you have learned in this chapter, do you think Hawaiian plants possess or lack thorns and toxic chemicals? Explain your answer.

Visit us on the web at http://www.thomsonedu.com/biology/berg/for additional resources such as flashcards, tutorial quizzes, further readings, and web links.

Global Ecology and Human Impacts

The common dandelion (*Taraxacum officinale*) is a flowering plant native to Europe and Asia but found worldwide in temperate climates. Dandelions are common in meadows and fields and on roadsides and lawns. Bees use the nectar to make honey, and birds eat the seeds. Although this yellow-flowered plant is sometimes grown for salad greens, particularly in the Northeastern United States, dandelions are generally considered bothersome weeds.

The dandelion body consists of a long taproot topped by a stem that bears leaves in a basal rosette. The flower head, borne on a hollow stem, consists of a cluster of small yellow flowers. Although the flowers contain both male and female structures, the seeds commonly develop asexually—without pollination or fertilization—by *apomixis.* The seed balls, which are white, fluffy, and spherical, consist of many parachute-like fruits covered with silky hairs; the wind easily carries the fruits to new locations.

Unlike most plants, the dandelions' geographic range is quite extensive. Only a few plants, such as junipers and daisies, have ranges as extensive as that of dandelions. Most organisms have more limited distributions. For example, coconut palms grow only in the tropics, such as the sandy beaches of warm Pacific islands.

It is clear that species are not distributed uniformly, but what governs their distribution? Basically, organisms live only in the areas to which they are adapted. The greater the physical differences among habitats, the greater the differences among the organisms that inhabit them: cacti could not survive in a salt marsh, and marsh grasses could not survive in the desert.

In this chapter we examine nine major biomes, including their climates and soils, which largely determine which plants are characteristic of each. We also consider two serious environmental issues that affect biomes: declining biological diversity and global warming.

The common dandelion. Dandelions (*Taraxacum officinale*), found in North America from Alaska to Florida, thrive in disturbed sites such as fields, lawns, and roadsides.

Biomes

A **biome** is a large, relatively distinct terrestrial region characterized by a particular combination of climate, soil, plants, and animals that is approximately the same regardless of where it occurs in the world (•**Figure 27-1**). Because it covers such a large geographic area, a biome encompasses many interacting landscapes. Recall from Chapter 26 that a **landscape** is a large land area (several to many square kilometers) composed of interacting ecosystems.

Biomes largely correspond to major climate zones, with average temperature and precipitation being most important (•**Figure 27-2**). Temperature is generally the overriding climate factor near the poles, whereas precipitation becomes more significant than temperature in temperate and tropical regions. Other abiotic factors to which biomes are sensitive include temperature extremes as well as rapid temperature changes, floods, droughts, strong winds, and fires.

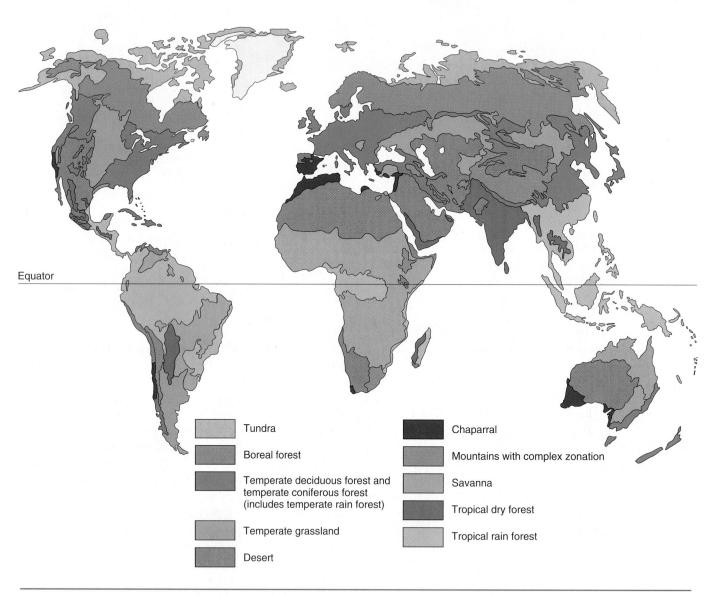

Equator

	Tundra		Chaparral
	Boreal forest		Mountains with complex zonation
	Temperate deciduous forest and temperate coniferous forest (includes temperate rain forest)		Savanna
	Temperate grassland		Tropical dry forest
	Desert		Tropical rain forest

FIGURE 27-1 The world's major biomes.
This simplified diagram shows sharp boundaries between biomes. Biomes actually intergrade at their boundaries, sometimes over large areas. Note that forested mountains such as the Appalachian Mountains are keyed according to their predominant vegetation type, whereas mountains with variable vegetation are keyed as "Mountains with complex zonation." (Based on data from the World Wildlife Fund.)

Our discussion covers nine major biomes: tundra, boreal forest, temperate rain forest, temperate deciduous forest, temperate grassland, chaparral, desert, savanna, and tropical rain forest.

Tundra is the northernmost biome

Cold, boggy plains called **tundra** (also called **arctic tundra**) exist in the extreme northern latitudes. The Southern Hemisphere has no equivalent of the arctic tundra, because no land exists in the proper latitudes. In North America, tundra is found in northern Canada and Alaska.

Arctic tundra has long, harsh winters and short summers. Although the growing season, with its warmer temperatures, is as short as 50 days, the days are long. Due to the tilt of Earth on its axis, above the Arctic Circle, the sun does not set at all for many days in midsummer, although the amount of light at midnight is one-tenth that at noon. Much of the tundra receives little precipitation (10 to 25 centimeters, or 4 to 10 inches, per year), and most of it falls during the summer months.

Tundra soils tend to be geologically young because most were formed only after the last Ice Age (glacier ice began retreating about 17,000 years ago). These soils are usually nutrient poor and have little organic litter (dead leaves and stems, animal droppings, and the remains of organisms) in the uppermost layer of soil. Although the soil surface melts during the summer, tundra has a layer of permanently frozen ground called **permafrost** that varies in depth and thickness. Because permafrost interferes with drainage, the thawed upper zone of soil is usually waterlogged during the summer. Permafrost also prevents the roots of larger plants from becoming established. The limited precipitation of the tundra, in combination with low temperatures, flat topography (surface features), and permafrost, produces a landscape of broad, shallow lakes, sluggish streams, and bogs (•Figure 27-3a).

Few plant species are found in the tundra, but the individual species present often exist in great numbers. The dominant producers are mosses, lichens such as reindeer moss, grasses, and grasslike sedges. Most of these short plants are herbaceous perennials that live 20 to 100 years. No readily recognizable trees or shrubs grow in the tundra except in sheltered locations, although dwarf willows, dwarf birches, and other dwarf trees and shrubs are common. Tundra trees and shrubs seldom grow taller than 30 centimeters (12 inches), largely because root growth is limited by the permafrost (•Figure 27-3b).

The distribution of the world's biomes is largely the result of climate patterns, which determine an area's water (measured as precipitation) and energy (measured as temperature).

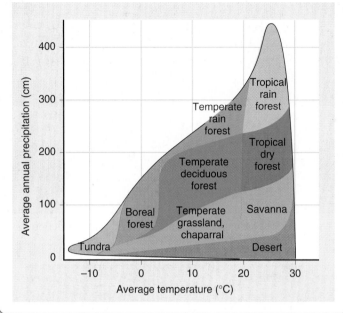

FIGURE 27-2 Using precipitation and temperature to identify biomes. Factors such as soil type, fire, and seasonality of precipitation affect whether temperate grassland or chaparral develops. (Adapted from R. H. Whittaker, *Communities and Ecosystems,* 2nd ed., Macmillan, New York, 1975.)

The year-round animal life of the tundra includes brown lemmings, weasels, arctic foxes, snowshoe hares, musk oxen, ptarmigans, and snowy owls (which sometimes migrate south when food supplies are low). In the summer, large herds of caribou migrate north into the tundra to graze on sedges, grasses, and dwarf willow. Dozens of bird species also migrate north in summer to nest and feed on the abundant insects. Mosquitoes, blackflies, deerflies, and other insects survive the winter as eggs or pupae and occur in great numbers during summer.

Tundra regenerates slowly after it has been disturbed, and even casual use by hikers can damage it. On portions of the arctic tundra, oil exploration and military use have inflicted long-lasting damage that is likely to persist for hundreds of years.

BIOME A large, relatively distinct terrestrial region characterized by a similar climate, soil, plants, and animals.

(a) The flat plains of Alaskan tundra are marshy and wet during the short summer months. This photo, taken in autumn, also shows Mount McKinley in the background.

(b) Arctic willow (*Salix arctica*) is a creeping shrub that rarely exceeds 10 centimeters (4 inches) in height.

FIGURE 27-3 **Arctic tundra.**

Boreal forest is dominated by conifers

Just south of the tundra is the **boreal forest,** or **taiga,** a huge evergreen forest biome that stretches across northern portions of North America and Eurasia (•**Figure 27-4**). Boreal forest is the world's largest biome, covering approximately 11 percent of Earth's land. In North America, boreal forest occurs throughout most of Canada and Alaska as well as in parts of New England and states around the Great Lakes. No biome comparable to the boreal forest occurs in the Southern Hemisphere, because there is no land at the corresponding latitudes.

Winters are extremely cold and severe in the boreal forest, although not as harsh as in the tundra. Boreal forest receives as little as 50 centimeters (20 inches) of precipitation per year. Its soil is typically acidic, low in minerals (inorganic nutrients), and characterized by a deep layer of partly decomposed conifer needles at the surface. Permafrost is either deep beneath the soil's surface (in the northernmost boreal forest) or absent (in the southern boreal forest). Boreal forest has numerous ponds and lakes where depressions were carved into the surface by grinding ice sheets that covered that area during the last Ice Age.

Black and white spruces, balsam fir, eastern larch, and other conifers dominate the boreal forest, although deciduous flowering trees, such as aspen or birch, form striking stands. Conifers have many drought-resistant adaptations, such as needlelike leaves with a minimal surface area to reduce water loss. Such adaptations enable conifers to withstand the "drought" of the northern winter months, when roots cannot absorb water because the ground is frozen.

Animal life of the boreal forest includes some larger species, such as wolf, bear, moose, and caribou. (The caribou migrate from the tundra to the boreal forest for winter.) However, most animal life is medium-sized to small, including rodents, rabbits, and fur-bearing predators such as lynx, sable, and mink. Birds are abundant in the boreal forest during the summer, but most species migrate to warmer climates in the winter. Insects are also abundant.

Most of the boreal forest is not well suited to agriculture because of its short growing season and mineral-poor soil. The boreal forest, which is harvested primarily by clear-cutting, is currently the primary source of the world's industrial wood and wood fiber.

Temperate rain forest has cool weather, dense fog, and high precipitation

Coniferous **temperate rain forest** occurs on the northwestern coast of North America (from northern California to Alaska), in southeastern Australia, and in southwestern South America. Annual precipitation in this biome is high, from about 200 to 380 centimeters (80 to 152 inches) and is augmented by condensation of water from dense coastal fogs. The proximity of temperate rain forest to the coastline moderates the temperature so that there is a narrow seasonal fluctuation; winters are mild, and summers are cool. The soil in temperate rain forest is relatively nutrient poor, although its organic content may be high. Cool temperatures slow the activity of bacterial and fungal decomposers. Thus, needles and large fallen branches and trunks accumulate on the ground as litter that takes many years to decay and release minerals to the soil.

The dominant vegetation in the North American temperate rain forest is large coniferous evergreen trees, such as western hemlock, Douglas fir, Sitka spruce, and western red cedar (•**Figure 27-5**). The temperate rain forest is also rich in epiphytic vegetation, which consists of small plants such as mosses, club mosses, lichens, and ferns that grow nonparasitically on the trunks and branches of large trees; mosses, lichens, and ferns also carpet the ground. Squirrels, wood rats, mule deer, elk, and numerous bird species are among the animals common in the temperate rain forest.

Temperate rain forest is one of the richest wood producers in the world and supplies us with lumber and pulpwood. It is also one of the most complex ecosystems in terms of number of species. Care must be taken to avoid overharvesting original old-growth forest, because such an ecosystem takes hundreds of years to develop. When the logging industry harvests old-growth forest, it typically replants the area with a *monoculture* (a single species) of trees that it can harvest in 40- to 60-year cycles. Thus, the old-growth forest ecosystem never has a chance to redevelop. A small fraction of the original, old-growth temperate rain forest in Washington, Oregon, and northern California remains untouched. These stable forest ecosystems provide biological habitat for many organisms, including 40 endangered and threatened species (see *Plants and People: Old-Growth Forests of the Pacific Northwest* in Chapter 24).

Temperate deciduous forest has a canopy of broadleaf trees

Summers are hot, winters are cold, and annual precipitation ranges from about

FIGURE 27-4 The boreal forest.
These coniferous forests occur in cold regions of the Northern Hemisphere adjacent to the tundra.

FIGURE 27-5 Temperate rain forest.
An elk peers out of temperate rain forest in Olympic National Park, Washington. Trees of the temperate rain forest include Douglas fir, western hemlock, and western red cedar. Moisture-loving ferns, mosses, and lichens grow on the trees as well as the ground.

Bruce Heinemann/Photodisc/Getty Images

FIGURE 27-6 Temperate deciduous forest.
This biome exhibits dramatic seasonal changes. The onset of autumn turns the leaves into a variety of golds, oranges, reds, and browns. A variety of hardwoods grow in temperate deciduous forest, including oak, hickory, beech, tuliptree, sweetgum, and dogwood.

other uses. Although these returning forests do not have the diversity found in virgin stands, many forest organisms have successfully become reestablished.

Worldwide, temperate deciduous forests were among the first biomes to be converted to agricultural use. In Europe and Asia, many soils that originally supported temperate deciduous forests have been cultivated by traditional agricultural methods for thousands of years without a substantial loss in fertility. During the 20th century, however, intensive agricultural practices were widely adopted; these, along with overgrazing and **deforestation** (clearance of large expanses of forest), have contributed to the degradation of some lands that were formerly forested.

75 to 125 centimeters (30 to 50 inches) in **temperate deciduous forest.** The soil of a temperate deciduous forest typically consists of a topsoil rich in organic material and a deep, clay-rich lower layer. As organic materials decay, mineral ions are released. If roots of the living trees do not immediately absorb these ions, they leach into the clay, which may retain them.

Temperate deciduous forest occupies most of the eastern United States, southeastern Canada, Europe (except for the northernmost and southernmost regions), and eastern Asia. Broadleaf hardwood trees, such as oak, hickory, maple, ash, and beech, that lose their foliage annually dominate temperate deciduous forest (•**Figure 27-6**). The trees form a dense canopy that overlies saplings and shrubs. In the southern reaches of the temperate deciduous forest, the number of broadleaf evergreen trees, such as magnolia and some oak species, increases.

Temperate deciduous forest originally contained a variety of large mammals, such as puma, wolf, deer, bison, and bear, plus many small mammals and birds. Reptiles and amphibians abounded, along with a denser and more varied insect life than exists today.

In Europe and North America, logging and land clearing for farms, tree plantations, and cities have removed much of the original temperate deciduous forest. Where it has regenerated, these forests are often in a seminatural state—that is, highly modified by humans for recreation, livestock foraging, timber harvest, and

Temperate grasslands occur in areas of moderate precipitation

Grasses are the dominant vegetation of **temperate grasslands,** which typically occur in the drier continental interiors of North America (*prairie*) and Eurasia (*steppe*). In the Southern Hemisphere, temperate grasslands are found in East and South Africa (*veld*) and Argentina (*pampas*).

Summers are hot, winters are cold, rainfall is often uncertain, and fires help shape the landscape in temperate grasslands (see *Plants and the Environment: Fire and Terrestrial Ecosystems*). Annual precipitation averages 25 to 75 centimeters (10 to 30 inches). Grassland soils are some of the best agricultural soils in the world. As a result of low precipitation, minerals in grassland soils tend to accumulate just below the topsoil instead of leaching to deeper ground. Grassland soil also has considerable organic material, because the aerial parts of many grasses die off each winter and contribute to the organic content of the soil. The roots and rhizomes survive the winter underground and send up new growth the following spring.

Moist temperate grasslands, also known as *tallgrass prairies,* occur in the United States in Iowa, western Minnesota, eastern Nebraska, and parts of other midwestern states and across Canada's prairie provinces. Few trees grow there except near rivers and streams, and perennial grasses grow in great profusion in the thick, fertile soil. Prairie violets, purple prairie clover, tall gold-

PLANTS AND THE ENVIRONMENT | Fire and Terrestrial Ecosystems

Fires started by lightning are an important environmental force in many geographic areas. The areas most prone to fires have wet seasons followed by dry seasons. Vegetation that grows and accumulates during the wet season dries out enough during the dry season to burn easily. When lightning hits the ground, it ignites the dry organic material, and a fire spreads through the area.

Fires have several effects on the environment. First, combustion frees the minerals that were locked in dry organic matter. The ashes remaining after a fire are rich in potassium, phosphorus, calcium, and other minerals essential for plant growth. Thus, vegetation flourishes after a fire. Second, fire removes plant cover and exposes the soil, which stimulates the germination of seeds that require bare soil and encourages the growth of shade-intolerant plants. Third, fire can cause increased soil erosion because it removes plant cover, leaving the soil more vulnerable to wind and water.

The evolutionary effects of fire on plants have been studied extensively. Grasses adapted to fire have underground stems and buds that are unaffected by the blaze sweeping over them; after the fire kills the aerial parts, these underground parts send up new shoots. Fire-adapted trees such as bur oak and ponderosa pine have thick bark that is resistant to fire. (In contrast, fire-sensitive trees such as many hardwoods have thin bark.) Certain pines, such as jack pine and lodgepole pine, depend on fire for successful reproduction, because the heat of the fire opens the cones and releases the seeds.

Fire was part of the natural environment long before humans, and many terrestrial ecosystems have adapted to fire. African savanna, California chaparral, North American grasslands, and pine forests of the southern United States are some of the fire-adapted biomes. Fire helps maintain grasses as the dominant vegetation in grasslands—for example, by removing fire-sensitive hardwood trees and shrubs.

The influence of fire on plants became even more pronounced once humans appeared, because humans set fires both deliberately and accidentally, making fire a more common occurrence. Humans purposely set fires for several reasons: to facilitate the growth of grasses and shrubs that many game animals require; to clear the land for agriculture and human development; and, in times of war, to reduce enemy cover.

Humans also try to prevent fires, and sometimes that effort can have its own disastrous consequences. When fire is excluded from a fire-adapted ecosystem, organic litter accumulates. As a result, when a fire does occur, it is much more destructive. To avoid this situation, humans sometimes conduct *controlled burns,* a tool of ecological management in which the organic litter is deliberately burned before it has accumulated to levels that can fuel dangerous, destructive fires. Prevention of fire also converts grassland to woody vegetation and facilitates the invasion of fire-sensitive trees into fire-adapted conifer forests. In this instance, controlled burns can be used to suppress oaks and other hardwoods, thereby maintaining the natural fire-adapted ecosystem (see figure).

■ Fire, a necessary tool of ecological management. Here, a controlled burn helps maintain a ponderosa pine (*Pinus ponderosa*) stand in Oregon.

enrod, scarlet paintbrush, and many other herbaceous flowering plants grow among the grasses. Before most of this area was converted to arable land, certain grass species grew as tall as a person on horseback, and the land was covered with large herds of grazing animals, particularly bison. The principal predators were wolves, although in sparser, drier grasslands coyotes took their place. Smaller animals included prairie dogs and their predators (foxes, black-footed ferrets, and various birds of prey), grouse, reptiles, and great numbers of insects.

Shortgrass prairies, in which the dominant grasses are less than 0.5 meter (1.6 feet) tall, are temperate grasslands with less precipitation than the moister grasslands just described but with greater precipitation than deserts (•**Figure 27-7**). In the United States, shortgrass prairies occur in the eastern half of Montana, the western half of South Dakota, and parts of other midwestern states, as well as western Alberta in Canada. The plants grow in less abundance than in the moister grasslands, and occasionally some bare soil is exposed.

Temperate grasslands are so well suited to agriculture that few remnants remain. The American Midwest, the Ukraine, and other moist temperate grasslands have become the "breadbaskets" of the world because they provide ideal growing conditions for crops such as corn and wheat, both of which are grasses. Little of the North American tallgrass prairie remains; more than 90 percent has vanished under the plow, and the remainder is so fragmented that almost nowhere can we see even an approximation of what European settlers saw when they settled in the Midwest. Today, the tallgrass prairie is considered North America's rarest biome.

FIGURE 27-7 **Temperate grasslands: shortgrass prairie.** This biome is mostly treeless but contains a profusion of grasses and other herbaceous flowering plants.

Chaparral is a thicket of evergreen shrubs and small trees

Some hilly temperate environments have mild winters with abundant rainfall, combined with extremely dry summers. Such Mediterranean climates, as they are called, occur not only around the Mediterranean Sea but also in California, western Australia, portions of Chile, and South Africa. In the North American Southwest, this Mediterranean-type community is known as **chaparral.** This vegetation type is also known as *maquis* in the Mediterranean region, *mallee scrub* in Australia, *matorral* in Chile, and *Cape scrub* in Africa. Chaparral soils are thin and relatively infertile. Frequent fires occur naturally in this habitat, particularly in late summer and autumn, and native plants are adapted to the fires.

Chaparral vegetation looks strikingly similar in different areas of the world, even though the individual species are different. A dense growth of evergreen shrubs, often of drought-resistant pine and scrub oak trees, dominates chaparral (•**Figure 27-8**). During the rainy winter season the landscape may be lush and green, but during the hot, dry summer, the plants lie dormant. Trees and shrubs often have **sclerophyllous leaves**—hard, small, leathery leaves that resist water loss. Some chaparral plants, such as sagebrush, are noted for **allelopathy,** in which their roots or leaves secrete toxic substances that inhibit the establishment of competing plants nearby. Many plants are also specifically fire adapted and grow best in the months following a fire. Such growth is possible because fire releases the minerals that were tied up in the plants that burned. The underground parts are not destroyed by fire, and with the new availability of essential minerals, the plants sprout vigorously during the winter rains. Mule deer, wood rats, chipmunks, skinks, and many species of birds are common animals of the chaparral.

The fires that occur at irregular intervals in California are often quite costly because they consume expensive homes built on the hilly chaparral landscape. Efforts to control the naturally occurring fires sometimes backfire. Denser, thicker vegetation tends to accumulate when periodic fires are prevented; consequently, when a fire does occur, it is much more severe. Removing the chaparral vegetation, whose roots hold the soil in place, also causes problems—for example, mudslides sometimes occur during winter rains in these areas.

Desert occurs where little precipitation falls

Deserts are dry areas in temperate (*cold deserts*) and subtropical or tropical regions (*warm deserts*). Although deserts occur on every continent, most are found along the Tropic of Cancer (between 15 and 30 degrees north of the equator) or the Tropic of Capricorn (between 15 and 30 degrees south of the equator). The Sahara in North Africa, approximately the size of the United States, is the world's largest desert. In the world's driest desert, the Atacama Desert in Chile, precipitation is barely measurable.

North America has four distinct deserts. The Great Basin Desert in Nevada, Utah, and neighboring states is a cold desert dominated by sagebrush. The Chihuahuan Desert, home of century plants (agaves; see Figure 25-20), is a warm desert found in Texas and New Mexico as well as Mexico. The warm Sonoran Desert, with its many species of cacti, is found in Arizona, California, and Mexico (•**Figure 27-9a**). The Mojave Desert in Nevada and California is a warm desert known for its Joshua trees (•**Figure 27-9b**).

© Dr. John Cunningham/Visuals Unlimited

FIGURE 27-8 Chaparral.
Drought-resistant evergreen shrubs and small trees are the main vegetation in the chaparral biome. Chaparral develops where hot, dry summers alternate with mild, rainy winters. Photographed in the Santa Monica Mountains, California.

Jules Frazier/Photodisc/Getty Images

(a) Large cacti, such as these saguaros (*Carnegiea gigantea*), are found in the Sonoran Desert. Photosynthesis is carried out in the stem, which also serves to store water. Leaves are modified into spines, which discourage herbivores from eating cactus tissue.

FIGURE 27-9 Desert.

© Gerald and Buff Corsi/Visuals Unlimited

(b) The Joshua tree (*Yucca brevifolia*) is one of the characteristic flowering plants of the Mojave Desert. Its evergreen leaves, which occur in dense clusters at the ends of branches, are stiff and narrow and have sharply pointed tips.

The dryness of the desert atmosphere leads to wide daily temperature extremes of heat and cold. Deserts vary greatly depending on the amount of precipitation they receive, which is generally less than 25 centimeters (10 inches) per year. Desert soils are low in organic material but often high in mineral content, such as the salts sodium chloride (NaCl) and calcium carbonate ($CaCO_3$). In some regions the concentration of certain minerals in the soil reaches toxic levels.

Vegetation, both perennial and annual, is sparse in deserts, and much of the desert soil is exposed. Perennial desert plants, such as the creosote bush, may have reduced leaves or no leaves, an adaptation that enables them to conserve water. Other desert plants, such as mesquite, shed their leaves for most of the year and grow only during the brief moist season. Wide spacing of plants allows each one to obtain enough of the available soil moisture around it to survive. Many desert plants have defensive spines, thorns, or toxins to prevent desert animals from feeding on them.

Desert animals tend to be small. In the heat of the day they remain under cover or periodically return to shelter, and at night they come out to forage or hunt. In addition to desert-adapted insects, there are many specialized desert reptiles, including lizards, tortoises, and snakes. Mammals include rodents such as the American kangaroo rat, which does not have to drink water but can subsist solely on the water content of its food plus metabolically generated water. American deserts are habitat for jackrabbits, and Australian deserts, for kangaroos. Desert carnivores such as the African fennec fox and some birds of prey, especially owls, live on the rodents and rabbits.

Humans alter deserts in several ways. Off-road vehicles damage desert vegetation, which sometimes takes years to recover. When the top layer of desert soil is disturbed, erosion occurs more readily, and less vegetation grows to support native animals. Another problem is that certain cacti have become rare as a result of illegal collecting. Irrigation of desert soils for agriculture often

Carlyn Iverson

FIGURE 27-10 Savanna.
African savanna supports large herds of grazing animals and their predators, although much savanna is vanishing under pressure from pastoral and agricultural land use.

causes them to become salty and unfit for crops or native vegetation.

Savanna is a tropical grassland with scattered trees

The **savanna** biome is a tropical grassland with widely scattered clumps of low trees. The world's largest savanna occupies about one-third of the African continent. Savanna also occurs in parts of South America (northern Argentina), northern Australia, and India. The Serengeti Plain of East Africa is the best-known savanna.

Savanna is found in warm, tropical areas of relatively low rainfall or seasonal rainfall with prolonged dry periods. Temperatures in savannas vary little throughout the year, and precipitation, not temperature, regulates seasons. Annual precipitation is 85 to 150 centimeters (34 to 60 inches). Savanna soil is low in essential minerals, but is often rich in aluminum; in places the aluminum reaches levels that are toxic to many plants.

Wide expanses of grasses interrupted by occasional trees characterize savanna (•**Figure 27-10**). Trees such as *Acacia* bristle with thorns that provide protection against herbivores. Both trees and grasses have fire-adapted features, such as extensive underground root systems, that enable them to survive the seasonal droughts as well as periodic fires that sweep across the savanna.

The African savanna contains the greatest assemblage of hoofed mammals in the world—great herds of wildebeest, antelope, giraffe, zebra, and the like. Large predators, such as lions and hyenas, kill and scavenge the herds. In areas of seasonally varying rainfall, the herds and their predators follow annual migration routes.

Humans are rapidly converting savannas to rangeland for cattle and other livestock, which are replacing the big herds of game animals. Severe overgrazing in places has converted some savannas to deserts (see *Plants and the Environment: Desertification,* in Chapter 6).

There are two basic types of tropical forests

There are many kinds of tropical forests, but ecologists generally classify them as one of two types: tropical dry forests or tropical rain forests. **Tropical dry forests** occur in regions with a wet season and a dry season (usually 2 to 3 months each year). Annual precipitation is 150 to 200 centimeters (60 to 80 inches). During the dry season, many tropical trees shed their leaves and remain dormant, much as temperate trees do during the winter. India, Brazil, Thailand, and Mexico are some of the countries that have tropical dry forests. Tropical dry forests intergrade with savanna on their dry edges and with tropical rain forests on their wet edges. Logging and overgrazing by domestic animals have fragmented and degraded many tropical dry forests.

Tropical rain forests occur where the temperature is warm throughout the year and precipitation occurs almost daily. The annual precipitation of tropical rain forests is 200 to 450 centimeters (80 to 180 inches). Much of this precipitation comes from locally recycled water that enters the atmosphere by transpiration from the forest's own trees. Tropical rain forests are often found in areas with ancient, highly weathered soil that has been extensively leached by heavy precipitation, producing a soil that is poor in minerals. Little organic matter accumulates in such soils; since the climate is warm and moist, decay organisms and detritus-feeding ants and termites decompose organic litter rapidly. Vast networks of roots and mycorrhizae quickly absorb minerals from decomposing materials. Thus, most of the minerals of tropical rain forests are tied up in the vegetation rather than the soil. Despite the scarcity of minerals in the soil, tropical rainforest plants grow rapidly, stimulated by the abundant solar energy and precipitation.

Tropical rain forests are found near the equator in Central and South America, Africa, and Southeast Asia. The world's largest tropical rain forest is in the Amazon River basin of northern South America (•**Figure 27-11a**). Of all the biomes, the tropical rain forest is unrivaled in types of species. No single organism dominates the tropical rain forest; one can travel for 0.4 kilometer (0.25 mile) without encountering two individuals of the same species of tree.

Most trees of tropical rain forests are evergreen flowering plants. Their roots are often shallow and form a mat almost 1 meter (about 3 feet) thick in the upper layer of the soil, which absorbs minerals released from decomposition of organic material. *Buttress roots* are a feature of many tropical rainforest trees (see Figure 6-10).

The vegetation of tropical rain forests is not dense at ground level except near riverbanks or where a fallen tree has opened the canopy. The continuous canopy of leaves overhead produces a dark habitat with an extremely moist microclimate. A fully developed rain forest has three or more distinct stories of trees. The topmost story consists of the crowns of the oldest, tallest trees, some 50 meters (164 feet) or more in height; these trees are exposed to direct sunlight and are subject to some of the warmest temperatures, lowest humidity, and strongest winds. The middle story, which reaches a height of 30 to 40 meters (100 to 130 feet), forms a continuous canopy of leaves overhead that lets in little sunlight to support the sparse understory. Only 2 to 3 percent of the light bathing the forest canopy reaches the forest understory, which consists of shrubs and herbs specialized for life in the shade, as well as seedlings of taller trees.

The trees in a tropical rain forest support extensive communities of smaller epiphytic plants, such as orchids and bromeliads. Although epiphytes grow in the crotches of branches, on bark, or even on the leaves of their hosts, most are not parasitic; they use their host trees only for physical support, not for nourishment.

Because little light penetrates to the understory, some plants living there are adapted to climb on already-established host trees. **Lianas** (woody tropical vines), some as thick as a human thigh, twist up through the branches of tropical rainforest trees (•**Figure 27-11b;** also see the section on vines in Chapter 7). Once in the canopy, lianas grow from the upper branches of one forest tree to another, forming a network of walkways for animals living in the canopy.

In the rain forest, animals are abundant and varied; there are huge numbers of insect, reptile, amphibian, and bird species. Most rainforest mammals, such as sloths and monkeys, live in the trees, although some large ground-dwelling mammals, including elephants, are also at home in rain forests.

(a) This photograph shows an aerial view of lush tropical rainforest vegetation along the Jari River, which feeds into the Amazon River in Brazil.

(b) Thick lianas grow up into the canopy, using a tree for support. Photographed in Panama.

FIGURE 27-11 **Tropical rain forest.**

Unless strong conservation measures are initiated soon, human population growth and agricultural and industrial expansion in tropical countries may spell the end of tropical rain forests by the middle of the 22nd century. It is likely that many rainforest organisms will become extinct before they have even been identified and scientifically described.

The Distribution of Vegetation on Mountains

Hiking up a mountain is similar to traveling toward the North Pole with respect to the major vegetation types encountered (•**Figure 27-12**). The reason is that it gets colder as one climbs a mountain, just as it does when one travels north; the temperature drops about 6°C (11°F) with each 1000-meter increase in elevation. The types of plants growing on the mountain change along with the temperature.

The base of a mountain in Colorado, for example, may have open stands of oaks and other deciduous trees, which shed their leaves every autumn. At higher elevations, where the climate is colder and more severe, a coniferous *subalpine forest* resembling the boreal forest grows. Spruces and firs are the dominant trees here. Higher still, the forest thins, and the trees become smaller, gnarled, and shrublike. These twisted, shrublike trees, called **krummholz**, are found at their elevational limit (the *tree line*). The exact elevation at which the tree line occurs depends on the latitude and distance from the ocean. In the Rocky Mountains, between 35° and 50° north latitude, the tree line drops 100 meters with each 1° latitude northward.

Above the tree line, where the climate is quite cold, *alpine tundra* occurs. Plant life of the alpine tundra varies with different amounts of wind and precipitation at the higher elevations. The most exposed areas support only a few lichens and short plants, such as mountain heather and mountain douglasia, which grow as cushions or mats in rocky hollows that afford some protection from the wind. Hardy flowering plants, such as western anemone, alpine gold, and mountain primrose, are found in the more protected areas of alpine tundra. Mountain goats, sheep, marmots, meadow voles, birds, and insects live in alpine tundra. The top of a tall mountain is often covered with a permanent ice cap or snowcap, similar to the polar land areas.

Important environmental differences exist between high elevations and high latitudes, however, that affect the types of organisms found in each place. Alpine tundra typically lacks permafrost and receives more snow and more high

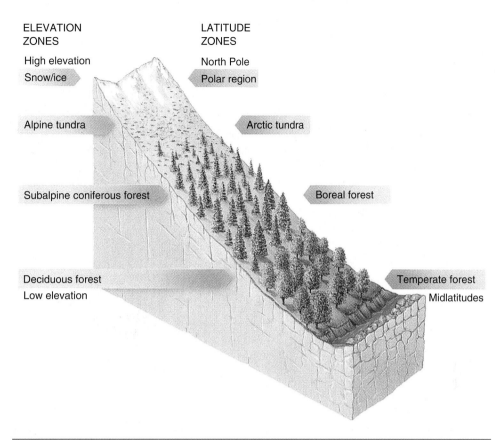

FIGURE 27-12 Comparison of elevation and latitude zones.
The cooler temperatures at higher elevations of a mountain produce a series of ecosystems similar to the biomes encountered when going toward the North Pole.

winds than does arctic tundra. Also, the high elevations of temperate mountains do not experience the great extremes of day length associated with the changing seasons in biomes at high latitudes. The intensity of solar radiation, including ultraviolet radiation, is greater at high elevations than at high latitudes. At high elevations the sun's rays pass through less of Earth's atmosphere, which results in greater exposure to ultraviolet radiation (less is filtered out by the atmosphere) than occurs at high latitudes.

Some mountains have rain shadows

Mountains tend to remove moisture from air by causing air masses to rise, which cools them so that water vapor condenses into water droplets, leading to precipitation. If prevailing winds blow onto a mountain range, precipitation occurs primarily on the windward slopes (the side from which the wind blows). The leeward slopes get very little precipitation because the air has lost all its moisture by the time it gets there. This situation exists on the North American west coast, where precipitation

falls on the western slopes of the Cascade Range and the Sierra Nevada. Downwind, or leeward (in this case, east of the mountain range), a low-precipitation **rain shadow** develops, often generating deserts (**•Figure 27-13**).

Declining Biological Diversity

You have seen in the preceding discussions that humans have affected each biome. Increasingly, the cumulative effects of all humans are causing global changes. One of the most important changes is declining biological diversity. The human species (*Homo sapiens*) has been present on Earth for about 800,000 years, which is a brief span of time compared with the age of our planet, some 4.6 billion years. Despite our relatively short tenure on Earth, our biological impact on other species is unparalleled. Our numbers have increased dramatically—the human population reached 6.6 billion in 2006—and we have expanded our biological range, moving into almost every habitat on Earth.

Wherever we have gone, we have altered the environment and shaped it to meet our needs and desires. In only a few generations we have transformed the face of Earth, placed a great strain on Earth's resources and resilience, and profoundly affected plants and other species. As a result of these changes, many people are concerned with **environmental sustainability,** the ability to meet humanity's current needs without compromising the ability of future generations to meet their needs. The impact of humans on the environment merits special study in botany, not merely because we ourselves are humans but also because our impact on the rest of the biosphere is so extensive.

Extinction, the death of a species, occurs when the last individual member of a species dies. Although extinction is a natural biological process, human activities have greatly accelerated it so that it is happening at a rate as much as 1000 times as great as the normal background rate of extinction, according to the 2005 Millennium Ecosystem Assessment report, which was instigated by UN Secretary-General Kofi Annan. The burgeoning human population has forced us to spread into almost all areas of Earth. Whenever humans invade an area, the habitats of many plants and animals are disrupted or destroyed, which can contribute to their extinction.

Biological diversity, also called **biodiversity,** is the variation among organisms. Biological diversity includes much more than simply the number of species, called **species richness,** of archaea, bacteria, protists, plants, fungi, and animals. Biological diversity occurs at all levels of biological organization, from populations to ecosystems. It takes into account **genetic diversity,** the genetic variety within a species, both among individuals within a given population and among geographically separate populations. (An individual species may have hundreds of genetically distinct populations.) Biological diversity also includes **ecosystem diversity,** the variety of ecosystems found on Earth—the forests, prairies, deserts, lakes, coastal estuaries, coral reefs, and other ecosystems.

Biologists must consider all three levels of biological diversity as they address the human impact on biodiversity. For example, the disappearance of populations

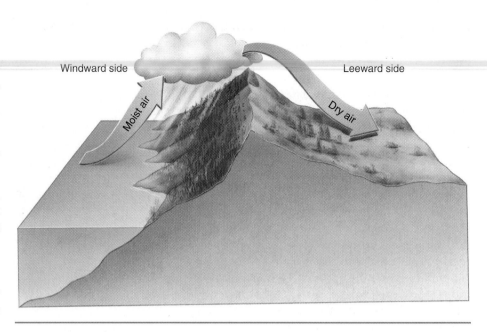

FIGURE 27-13 A rain shadow.
When wind blows moist air over a mountain range, precipitation occurs on the windward side of the mountain, causing a dry rain shadow on the leeward side. Such a rain shadow occurs east of the Cascade Range in Washington State.

(that is, a decline in genetic diversity) indicates an increased risk that a species will become extinct (a decline in species richness). Biologists perform detailed analyses to quantify how large a population must be to ensure a given species' long-term survival. The smallest population that has a high chance of enduring into the future is known as the **minimum viable population.**

Biological diversity is currently decreasing at an unprecedented rate. According to the World Conservation Union Red List of Threatened Plants, which is based on 20 years of data collection and analysis around the world, about 34,000 species of plants are currently threatened with extinction.

The legal definition of an **endangered species,** as stipulated by the U.S. Endangered Species Act, is a species in imminent danger of extinction throughout all or a significant part of its range. (The area in which a particular species is found is its **range.**) Unless humans intervene, an endangered species will probably become extinct.

When extinction is less imminent but the population of a particular species is quite small, the species is classified as a **threatened species.** The legal definition of a threatened species is a species likely to become endangered in the foreseeable future, throughout all or a significant part of its range.

Endangered and threatened species represent a decline in biological diversity, because as their numbers decrease, their genetic diversity is severely diminished.

Endangered and threatened species are at greater risk of extinction than species with greater genetic variability, because long-term survival and evolution depend on genetic diversity.

We may have entered the greatest period of **mass extinction** in Earth's history, but the current situation differs from previous periods of mass extinction in several respects. First, its cause is directly attributable to human activities. Second, it is occurring in a tremendously compressed period (just a few decades, as opposed to hundreds of thousands of years), much faster than rates of speciation (or replacement). Perhaps even more sobering, larger numbers of plant species are becoming extinct today than in previous mass extinctions. Because plants are the base of terrestrial food webs, extinction of animals that depend on plants is not far behind. It is crucial that we determine how the loss of biodiversity affects the stability and functioning of ecosystems, which make up our life support system.

Endangered species have certain characteristics in common

Many species are threatened or endangered because they have certain characteristics that may make them more vulnerable to extinction. Some of these characteristics are having an extremely small (localized) range, living on islands, having low reproductive rates, and having specialized feeding habits.

Many endangered species have a small natural range, which makes them particularly prone to extinction if their habitat is altered. The Tiburon mariposa lily, for example, is found nowhere in nature except on a single hilltop near San Francisco. Development of that hilltop would almost certainly cause the extinction of this species.

EVOLUTION LINK Many **endemic species** are endangered, particularly on islands. *Endemic* means they are not found anywhere else in the world. These species often have small populations that cannot be replaced by immigration if their numbers decline. Because they evolved in isolation from competitors, predators, and disease organisms, island species have few defenses when such organisms are introduced, usually by humans. For example, Hawaii's plants evolved in the absence of herbivorous mammals and therefore have no defenses (such as spines, thorns, or toxins) against introduced sheep, pigs, goats, and deer. The introduction of these plant-eating mammals to Hawaii has threatened many species with extinction. ■

Endangered species often have low reproductive rates. No more than 6 percent of swamp pinks, an endangered species of small flowering plant, produce flow-

Jeffrey Lepore/Photo Researchers, Inc.

FIGURE 27-14 The swamp pink, an endangered species.
The swamp pink (*Helonias bullata*) lives in boggy areas of the eastern United States. Photographed in Killens Pond State Park, Delaware.

ers in a given year (•**Figure 27-14**). Highly specialized feeding habits endanger a species. In nature, the giant panda eats only bamboo. Periodically all the bamboo plants in a given area flower and die together; when this occurs, panda populations face starvation.

ENVIRONMENTAL SUSTAINABILITY The ability of the natural environment to meet humanity's needs indefinitely without going into a decline from human-caused stresses.

EXTINCTION The elimination of a species, either locally (that is, no longer living in an area where it used to be found, but still found in other areas) or globally (no longer living anywhere in the world).

BIOLOGICAL DIVERSITY The variety of living organisms considered at three levels: genetic diversity, species richness, and ecosystem diversity.

ENDANGERED SPECIES A species whose numbers are so severely reduced that it is in imminent danger of extinction throughout all or a part of its range.

THREATENED SPECIES A species in which the population is small enough for it to be at risk of becoming extinct but not so small that it is in imminent danger of extinction.

ENDEMIC SPECIES Localized, native species that are not found anywhere else in the world.

Human activities contribute to declining biological diversity

Species become endangered and extinct for a variety of reasons, including the destruction or modification of habitats and the production of pollution. Humans also upset the delicate balance of organisms in a given area by introducing foreign species or by controlling native pests or predators. Uncontrolled commercial harvesting is also a factor.

Most species facing extinction today are endangered by destruction, fragmentation, or degradation of natural habitats. Building roads, parking lots, bridges, and buildings; clearing forests to grow crops or graze domestic animals; and logging forests for timber all take their toll on natural habitats (•**Figure 27-15**). Draining marshes converts aquatic habitats to terrestrial ones, whereas building dams and canals floods terrestrial habitats. Because most organisms require a particular type of environment, habitat destruction reduces their biological range and ability to survive.

Humans often leave small, isolated patches of natural landscape that roads, fences, fields, and buildings completely surround. In ecological terms, *island* refers not only to any landmass surrounded by water but also to any isolated habitat surrounded by an expanse of unsuitable territory. Accordingly, a small patch of forest surrounded by agricultural and suburban lands is considered an island.

Habitat fragmentation, the breakup of large areas of habitat into small, isolated segments (that is, islands), is a major threat to the long-term survival of many populations and species. Species from the surrounding "developed" landscape may intrude into the isolated habitat, whereas rare and wide-ranging species that require a large patch of undisturbed habitat may disappear altogether. Habitat fragmentation encourages the spread of *invasive species* (discussed in the next section), disease organisms, and other "weedy" species and makes it difficult for organisms to migrate. Generally, habitat fragments support only a fraction of the species found in the original, unaltered environment. However, much remains to be learned about precisely how habitat fragmentation affects specific populations, species, and ecosystems.

Human activities that produce *acid precipitation* and other forms of pollution indirectly modify habitats left undisturbed and in their natural state. Acid precipitation has contributed to the decline of large stands of forest trees and to the biological death of many freshwater lakes, for example, in the Adirondack Mountains and in Nova Scotia. Other types of pollutants, such as chemicals from industry and agriculture, organic pollutants from sewage, acid wastes seeping from mines,

FIGURE 27-15 Habitat destruction.
These logging roads cut through land that was forested.

Pat Powers and Cherryl Schafer/Photodisc/Getty Images

and thermal pollution from the heated wastewater of industrial plants, also adversely affect ecosystems.

To save native species, we must control invasive species

The introduction of a foreign species into an area where it is not native often upsets the balance among the organisms living in that area and interferes with the ecosystem's normal functioning. The foreign species may prey on native species or compete with them for food or habitat. If the foreign species causes economic or environmental harm, it is known as an **invasive species.**

Generally, a foreign competitor or predator harms local organisms more than native competitors or predators do. Most invasive species lack natural agents (such as parasites, predators, and competitors) that would otherwise control them. Also, most native species, lacking a shared evolutionary history with invasive species, typically are less equipped to cope with them. Although foreign species sometimes spread into new areas on their own, humans are usually responsible for such introductions, either knowingly or accidentally.

Islands are particularly susceptible to the introduction of invasive species. In Hawaii, the introduction of sheep has imperiled both the mamane tree (because the sheep eat it) and a species of honeycreeper, an endemic bird that relies on the mamane tree for food.

Conservation biology addresses declining biological diversity caused by human activities

Conservation biology is the scientific study of how humans affect organisms and of the development of ways to protect biological diversity. Conservation biologists develop models, design experiments, and perform fieldwork to address a wide range of questions. For example, what are the factors that cause a decline in biological diversity? How do we protect and restore populations of endangered species? If we are to preserve entire ecosystems and landscapes, which ones are the most important to save?

Conservation biologists have come to several conclusions regarding the most effective ways to preserve species and habitats. For example, conservation biologists have determined that a single large area of habitat capable of supporting several populations is more effective at safeguarding an endangered species than several habitat fragments, each capable of supporting a single population. Also, a large area of habitat typically supports greater species richness than several habitat fragments.

Conservation is more successful when habitat areas for a given species are in close proximity rather than far apart. If an area of habitat is isolated from other areas, individuals may not effectively disperse from one habitat to another. Because the presence of humans adversely affects many species, habitat areas that lack roads or are inaccessible to humans are better than human-accessible areas.

According to conservation biologists, it is more effective and, ultimately, more economical to preserve intact ecosystems in which many species live than to try to preserve individual species. Conservation biologists generally consider it a higher priority to preserve areas with greater biological diversity.

Conservation biology includes two problem-solving approaches that save organisms from extinction: in situ and ex situ conservation. **In situ conservation,** which includes the establishment of parks and reserves, concentrates on preserving biological diversity in nature. A high priority of in situ conservation is identifying and protecting sites that harbor a great deal of di-

versity. With increasing demands on land, however, in situ conservation cannot preserve all types of biological diversity. Sometimes only ex situ conservation can save a species. **Ex situ conservation** conserves individual species in human-controlled settings. Storing seeds of genetically diverse plant crops is an example of ex situ conservation (see *Plants and People: Seed Banks,* in Chapter 9).

In situ conservation is the best way to preserve biological diversity

Protecting animal and plant habitats—that is, conserving and managing ecosystems as a whole—is the single best way to protect biological diversity. Many nations have set aside areas for wildlife habitats. Such natural ecosystems offer the best strategy for the long-term protection and preservation of biological diversity. Currently, more than 100,000 national parks, marine sanctuaries, wildlife refuges, forests, and other areas are protected throughout the world. These encompass a total area almost as large as Canada. However, many protected areas have multiple uses that sometimes conflict with the goal of preserving species. National parks provide recreation, for example, whereas national forests are used for logging, grazing, and mining. The mineral rights to many wildlife refuges are privately owned, and some wildlife refuges have had oil, gas, and other mineral development.

Protected areas are not always effective in preserving biological diversity, particularly in developing countries where biological diversity is greatest, because there is little money or expertise to manage them. Another shortcoming of the world's protected areas is that many are in lightly populated mountain areas, tundra, and the driest deserts, places that often have spectacular scenery but relatively few kinds of species. In reality, such remote areas are often designated reserves because they are unsuitable for commercial development. In contrast, ecosystems in which biological diversity is greatest of-

HABITAT FRAGMENTATION The division of habitats that formerly occupied large, unbroken areas into smaller pieces by roads, fields, cities, and other human land-transforming activities.

INVASIVE SPECIES A foreign species that, when introduced into an area where it is not native, upsets the balance among the organisms living there and causes economic or environmental harm.

CONSERVATION BIOLOGY A multidisciplinary science that focuses on the study of how humans impact species and ecosystems and on the development of ways to protect biological diversity.

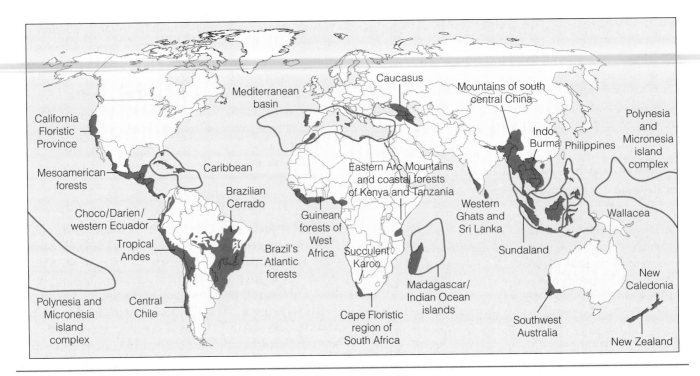

FIGURE 27-16 Biodiversity hotspots.
Rich in endemic species, these hotspots are under pressure from the number of humans living in them. (Data from Conservation International.)

ten receive little attention. Protected areas are urgently needed in tropical rain forests, the tropical grasslands and savannas of Brazil and Australia, and dry forests that are widely scattered around the world. Desert organisms are underprotected in northern Africa and Argentina, and the species of many islands and temperate river basins need protection.

These areas are part of what some biologists, such as those at Conservation International, have identified as the world's 25 **biodiversity hotspots** (•**Figure 27-16**). The hotspots collectively make up 1.4 percent of Earth's land but contain as many as 44 percent of all vascular plant species, 29 percent of the world's endemic bird species, 27 percent of endemic mammals species, 38 percent of endemic reptile species, and 53 percent of

endemic amphibian species. Nearly 20 percent of the world's human population lives in the hotspots. Fifteen of the 25 hotspots are tropical, and 9 are mostly or solely islands. The 25 hotspots are somewhat arbitrary, as additional hotspots are recognized when different criteria are used.

Not all conservation biologists support focusing limited conservation resources on protecting biodiversity hotspots, because doing so ignores the many **ecosystem services** provided by entire ecosystems that are rapidly vanishing due to habitat destruction. For example, the vast boreal forests in Canada and Russia are critically important to the proper functioning of the global carbon and nitrogen cycles. Yet conservation organizations do not recognize these forests, which are being logged at unprecedented rates, as hotspots. Landscape ecology takes a larger view of ecosystem conservation.

Landscape ecology considers ecosystem types on a regional scale

The subdiscipline of ecology that studies the connections in a heterogeneous landscape consisting of multiple interacting ecosystems is known as **landscape ecology.** Increasingly, biologists are focusing efforts on preserving biodiversity in landscapes. But what is the minimum ecosystem or landscape size that will preserve species numbers and distributions?

BIODIVERSITY HOTSPOTS Relatively small areas of land that contain an exceptional number of species and are at high risk from human activities.

ECOSYSTEM SERVICES Important environmental services that ecosystems provide; examples include clean air to breathe, clean water to drink, and fertile soil in which to grow crops.

RESTORATION ECOLOGY The scientific field that uses the principles of ecology to help return a degraded environment as closely as possible to its former undisturbed condition.

forest fragments do not maintain their ecological integrity. For example, large trees near forest edges often die or are damaged from exposure to wind, desiccation (from lateral exposure of the forest fragment to sunlight), and invasion by parasitic woody vines. In addition, biologists have documented that various species adapted to forest interiors do not prosper in the smaller fragments and eventually die out, whereas species adapted to forest edges invade the smaller fragments and thrive.

Restoring damaged or destroyed habitats is the goal of restoration ecology

Although preserving habitats is an important part of conservation biology, the realities of our world, including the fact that the land-hungry human population continues to increase, dictate a variety of other conservation measures. Sometimes scientists reclaim disturbed lands and convert them into areas with high biological diversity. **Restoration ecology,** in which the principles of ecology are used to return a degraded environment to one that is more functional and sustainable, is an important part of in situ conservation.

Since 1934 the University of Wisconsin–Madison Arboretum has carried out one of the most famous examples of ecological restoration (•**Figure 27-18**). Several distinct natural communities have been carefully developed on damaged agricultural land. These communities include a tallgrass prairie, a xeric (dry) prairie, and several types of pine and maple forests native to Wisconsin.

Restoration of disturbed lands not only creates biological habitats but also has additional benefits, such as the regeneration of soil damaged by agriculture or

FIGURE 27-17 Minimum Critical Size of Ecosystems Project. Shown are 1-hectare and 10-hectare plots of a long-term (more than 2 decades) study under way in Brazil on the effects of fragmentation on Amazonian rain forest. Plots with an area of 100 hectares are also under study, along with identically sized sections of intact forest, which are controls.

One long-term study addressing this question in tropical rain forests is the Minimum Critical Size of Ecosystems Project, which is studying 12 Amazonian rainforest fragments varying in size from 1 hectare (2.5 acres) to 100 hectares (247 acres) (•**Figure 27-17**). Preliminary data from this study indicate that smaller

(a) The restoration of the prairie by the University of Wisconsin–Madison Arboretum was at an early stage in 1935. The men are digging holes to plant prairie grass sod.

(b) The prairie as it looks today. This picture was taken at about the same location as the 1935 photograph.

FIGURE 27-18 Restoring damaged lands.

mining. The disadvantages of restoration ecology include the time and expense required to restore an area. Nonetheless, restoration ecology is an important aspect of conservation biology, as it is thought that restoration will reduce extinctions.

Global Warming

Global warming is a serious environmental issue with the potential to harm humans as well as plants and other organisms. Earth's average temperature is based on daily measurements from several thousand land-based meteorological stations around the world, as well as data from weather balloons, orbiting satellites, transoceanic ships, and hundreds of sea-surface buoys with temperature sensors. Earth's average surface temperature increased 0.6°C (1.1°F) during the 20th century, and the 1990s was the warmest decade of the century (•**Figure 27-19**). The early 2000s continued the warming trend. Since the mid-19th century, the three warmest years on record were 2005, 1998, and 2002.

PROCESS OF SCIENCE Scientists around the world have studied global warming for the past 50 years. As the evidence has accumulated, those most qualified to address the issue have reached a strong consensus that the 21st century will experience significant climate change and that human activities will be responsible for much of this change.

In response to this consensus, governments around the world organized the UN Intergovernmental Panel on Climate Change (IPCC). With input from hundreds of climate experts, the IPCC provides the most definitive scientific statement about global warming. The 2001 IPCC Third Assessment Report concluded that human-produced air pollutants continue to change the atmosphere, causing most of the warming observed in the past 50 years. Scientists can identify the human influence on climate change despite questions about *how much* of the recent warming stems from natural variations. The IPCC report projects up to a 5.8°C (10.4°F) increase in global temperature by the year 2100, although the warming will probably not occur uniformly from region to region. Thus, Earth may become warmer during the 21st century than it has been for the last 10,000 years.

Almost all climate experts agree with the IPCC's assessment that the warming trend has already begun and will continue throughout the 21st century. However, scientists are uncertain over how rapidly the warming will proceed, how severe it will be, and where it will be most pronounced. (Recall that uncertainty and debate are part of the scientific process and that scientists can never claim to know a "final answer.") As a result of these uncertainties, many people, including policymakers, are confused about what we should do. Yet the stakes are quite high because human-induced global warming has the potential to disrupt Earth's climate for a long time. ■

Greenhouse gases cause global warming

Carbon dioxide (CO_2) and certain other trace gases, including methane (CH_4), surface ozone (O_3), nitrous oxide (N_2O), and chlorofluorocarbons (CFCs), are accumulating in the atmosphere as a result of human activities (•**Table 27-1**).[1] The concentration of atmospheric CO_2 has increased from about 280 parts per million (ppm) approximately 200 years ago (before the Industrial Revolution began) to 380 ppm in 2005.

Burning carbon-containing fossil fuels—coal, oil, and natural gas—accounts for about three-fourths of human-made carbon dioxide emissions to the atmosphere. Land conversion, such as when forests are logged or burned, also releases carbon dioxide. (Burning releases carbon dioxide into the atmosphere; also, because trees normally remove carbon dioxide from the atmosphere during photosynthesis, tree removal prevents this process.) The levels of the other trace gases associated with global warming are also rising.

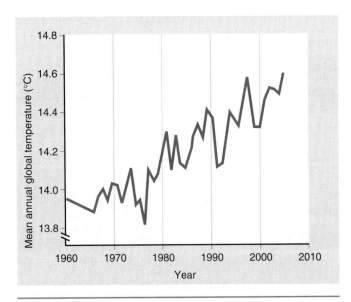

FIGURE 27-19 Mean annual global temperature, 1960 to 2005. Data are presented as surface temperatures (°C) for 1960, 1965, and every year thereafter. The measurements, which naturally fluctuate, clearly show the warming trend of the last several decades. (Surface Air Temperature Analysis, Goddard Institute for Space Studies, NASA.)

---

[1]Surface ozone is a greenhouse gas as well as a component of smog. Ozone in the upper atmosphere (the stratosphere) provides an important planetary service, protecting Earth's surface from excessive ultraviolet radiation.

TABLE 27-1 Changes in Selected Atmospheric Greenhouse Gases, Preindustrial to Present

GAS	ESTIMATED PRE-1750 CONCENTRATION	PRESENT CONCENTRATION
Carbon dioxide	280 ppm*	380 ppm§
Methane	730 ppb†	1847 ppb
Nitrous oxide	270 ppb	319 ppb
Tropospheric ozone	25 ppb	34 ppb
CFC-12	0 ppt‡	545 ppt
CFC-11	0 ppt	253 ppt

Source: Carbon Dioxide Information Analysis Center, Environmental Sciences Division, Oak Ridge National Laboratory.

*ppm = parts per million.

†ppb = parts per billion.

‡ppt = parts per trillion.

§Derived from in situ sampling at Mauna Loa, Hawaii. All other data from Mace Head, Ireland, monitoring site.

Global warming occurs because these gases absorb infrared radiation, that is, heat, in the atmosphere. This absorption slows the natural heat flow into space, so the lower atmosphere warms. Some of the heat from the lower atmosphere is transferred to the ocean and raises its temperature as well. This retention of heat in the atmosphere is a natural phenomenon that has made Earth habitable for its millions of species. However, as human activities increase the atmospheric concentration of these gases, the atmosphere and ocean continue to warm, and the overall global temperature rises.

Because carbon dioxide and other gases trap the sun's radiation somewhat like glass does in a greenhouse, the natural trapping of heat in the atmosphere is called the **greenhouse effect,** and the gases that absorb infrared radiation are known as **greenhouse gases.** The additional warming produced when increased levels of gases absorb additional infrared radiation is called the **enhanced greenhouse effect** (•Figure 27-20).

Although current rates of fossil fuel combustion and deforestation are high, causing the carbon dioxide level in the atmosphere to increase markedly, scientists think the warming trend is slower than the increasing level of carbon dioxide might indicate. The reason is that water requires more heat to raise its temperature than gases in the atmosphere do. As a result, the ocean takes longer to warm than the atmosphere. Most climate scientists think that warming will be more pronounced in the second half of the 21st century than in the first half.

What are the probable effects of global warming?

We now consider some of the probable effects of global warming, including changes in sea level; changes in precipitation patterns; effects on organisms, including

FIGURE IN FOCUS

Greenhouse gases accumulating in the atmosphere are causing an enhanced greenhouse effect.

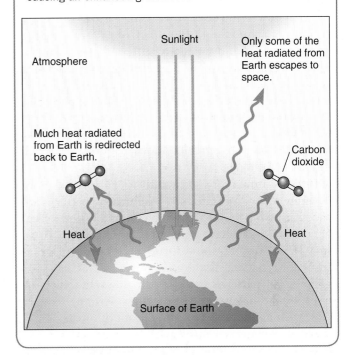

FIGURE 27-20 Enhanced greenhouse effect.
Visible light passes through the atmosphere to Earth's surface. However, carbon dioxide and other greenhouse gases direct the longer-wavelength heat (infrared radiation), which would normally be radiated to space, back to the surface. Therefore, the buildup of greenhouse gases in the atmosphere results in global warming.

humans; and effects on agriculture. These changes will persist for many centuries because many greenhouse gases remain in the atmosphere for hundreds of years. Furthermore, even after greenhouse gas concentrations have stabilized, scientists think Earth's mean surface temperature will continue to rise, because the ocean adjusts to climate change on a delayed time scale.

With global warming, the global average sea level is rising

As Earth's overall temperature increases by just a few degrees, there could be a major thawing of glaciers and the polar ice caps. In addition to sea-level rise caused by the retreat of glaciers and thawing of polar ice, the sea level will probably rise because of thermal expansion

GLOBAL WARMING The current warming trend in Earth's climate.

ENHANCED GREENHOUSE EFFECT The additional warming of Earth's climate produced by increased levels of greenhouse gases, which absorb infrared radiation.

of the warming ocean. Water, like other substances, expands as it warms.

The IPCC estimates that sea level will rise an additional 0.5 meter (20 inches) by 2100. Such an increase will flood low-lying coastal areas such as parts of southern Louisiana and South Florida. Coastal areas that are not inundated will more likely suffer erosion and other damage from more frequent and more intense weather events such as hurricanes. Countries particularly at risk include Bangladesh, Egypt, Vietnam, Mozambique, and many island nations such as the Maldives.

With global warming, precipitation patterns will change

Computer simulations of weather changes as global warming occurs indicate that precipitation patterns will change, causing some areas such as midlatitude continental interiors to have more frequent droughts. At the same time, heavier snowstorms and rainstorms may cause more frequent flooding in other areas. Changes in precipitation patterns could affect the availability and quality of fresh water in many places. Arid or semiarid areas, such as the Sahel region just south of the Sahara, may have the most serious water shortages as the climate changes. Closer to home, water experts predict water shortages in the American West, because warmer winter temperatures will cause more precipitation to fall as rain rather than snow; melting snow currently provides 70 percent of stream flows in the West during summer months.

The frequency and intensity of storms over warm surface waters may also increase. Scientists at the National Oceanic and Atmospheric Administration developed a computer model that examines how global warming may affect hurricanes. When the model was run with a sea-surface temperature 2.2°C (4°F) warmer than today, more intense hurricanes resulted. (The question of whether hurricanes will occur *more frequently* in a warmer climate remains uncertain.) Changes in storm frequency and intensity are expected because as the atmosphere warms, more water evaporates, which in turn releases more energy into the atmosphere. This energy generates more powerful storms.

With global warming, the ranges of organisms are changing

Dozens of studies report on the effects of global warming on organisms. For example, botanists examined records of historical flowering times and compared them to flowering times of the present at the Arnold Arboretum in Boston, Massachusetts. These researchers determined that plant species flowered an average of 8 days earlier in 1980–2002 than in 1900–1920.

Biologists generally agree that global warming will have an especially severe impact on plants, which can-

not migrate as quickly as animals when environmental conditions change. (The speed of seed dispersal has definite limitations.) During past climate warmings, such as during the glacial retreat that took place some 12,000 years ago, the upper limit of dispersal for tree species was probably 200 kilometers (124 miles) per century. If Earth warms as much as it is projected to during the 21st century, the ranges for some temperate tree species may need to shift northward as much as 480 kilometers (300 miles) (•Figure 27-21).

Each species reacts to changes in temperature differently. In response to global warming, some species will probably become extinct, particularly those with narrow temperature requirements, those confined to small reserves or parks, and those living in fragile ecosystems. Other species may survive in greatly reduced numbers and ranges. Ecosystems considered most vulnerable to species loss in the short term are polar seas, coral reefs and atolls, prairie wetlands, coastal wetlands, tundra, boreal forests, tropical forests, and mountains, particularly alpine tundra.

In response to global warming, some species may disperse into new environments or adapt to the changing conditions in their present habitats. Global warming may not affect certain species, whereas other species will have greatly expanded numbers and ranges as a result of global warming. Those considered most likely to prosper include weeds, pests, and disease-carrying organisms, all of which are already common in many environments.

Global warming will have a more pronounced effect on human health in developing countries

Data linking climate warming and human health problems are accumulating. Since 1950, the United States has experienced an increased frequency of extreme heat-stress events, which are extremely hot, humid days during summer months. Medical records show that heat-related deaths among elderly and other vulnerable people increase during these events.

Climate warming may also affect human health indirectly. Mosquitoes and other disease carriers could expand their range into the newly warm areas and spread malaria, dengue fever, yellow fever, Rift Valley fever, and viral encephalitis. As many as 50 million to 80 million additional cases of malaria could occur annually in tropical, subtropical, and temperate areas. According to the World Health Organization, during 1998, the second-warmest year on record, the incidence of malaria, Rift Valley fever, and cholera surged in developing countries.

Highly developed countries are less vulnerable to such disease outbreaks because of better housing (which keeps mosquitoes outside), medical care, pest control, and public health measures such as water treatment

(a) Beech tree in the summer.

(b) Present range of American beech trees.

(c) One projected range of beech trees after global warming occurs.

FIGURE 27-21 Climate change and beech trees in North America.
(Adapted from M. B. Davis and C. Zabinski, "Changes in Geographical Range Resulting from Greenhouse Warming: Effects on Biodiversity in Forests," in *Global Warming and Biological Diversity,* ed. R. L. Peters, and T. E. Lovejoy, Yale University Press, New Haven, Connecticut, 1992.)

plants. Texas reported a few cases of dengue fever in the late 1990s, for example, whereas nearby Mexico had thousands of cases during that period.

Global warming may increase problems for agriculture

Several studies show that the rising sea level will inundate river deltas, which are some of the world's best agricultural lands. The Nile River (Egypt), Mississippi River (United States), and Yangtze River (China) are examples of vulnerable river deltas. Certain agricultural pests and disease-causing organisms will probably proliferate. As mentioned earlier, global warming may also increase the frequency and duration of droughts and, in some areas, crop-damaging floods.

On a regional scale, current global-warming models forecast that agricultural productivity will increase in some areas and decline in others. Models suggest that Canada and Russia will increase their agricultural productivity in a warmer climate, whereas tropical and subtropical regions, where many of the world's poorest people live, will decline in agricultural productivity as a result of more frequent droughts. Central America and Southeast Asia may experience some of the greatest declines in agricultural productivity.

Many scientists think that global warming is the most serious environmental challenge facing us today. Now that you have completed this text, you should have an appreciation for the contributions and limitations of science in addressing human-caused climate change, declining biological diversity, and other global issues. As a world citizen, you have a responsibility, not only to yourself but also to your children and grandchildren. Make it a point to become better informed, apply the knowledge you've learned, and use your talents, time, and energy to make this planet more livable for future generations.

STUDY OUTLINE

❶ Define *biome,* and describe the nine major terrestrial biomes, including their climate, soil, and characteristic plants.
A **biome** is a large, relatively distinct terrestrial region characterized by a similar climate, soil, plants, and animals. A frozen layer of subsoil (**permafrost**) and low-growing vegetation that is adapted to extreme cold and a short growing season characterize **tundra,** the northernmost biome. Coniferous trees adapted to cold

winters, a short growing season, and acidic, mineral-poor soil dominate the **boreal forest.** Large conifers dominate **temperate rain forest,** which receives high precipitation. **Temperate deciduous forest** occurs where precipitation is relatively high and soils are rich in organic matter; broadleaf trees that lose their leaves seasonally dominate temperate deciduous forest. **Temperate grassland** typically has a deep, mineral-rich soil and has moderate but uncertain precipitation. Thickets of small-leaf evergreen shrubs and trees and a climate of wet, mild winters and dry summers characterize **chaparral. Desert,** found in both temperate (cold deserts) and subtropical or tropical regions (warm deserts) with low levels of precipitation, contains plants with specialized water-conserving adaptations. Tropical grassland, called **savanna,** has widely scattered trees interspersed with grassy areas; savanna occurs in tropical areas with low or seasonal rainfall. Mineral-poor soil and high rainfall that is evenly distributed throughout the year characterize **tropical rain forest;** tropical rain forest has high species richness and high productivity.

❷ Identify the biomes that make the best agricultural land, and explain why they are superior.
Temperate deciduous forest and temperate grasslands make the best agricultural land, because they have fertile soil and enough precipitation to grow crops. Temperate deciduous forests were among the first biomes to be converted to agricultural use. Moist temperate grasslands are so well suited to agriculture that few remnants remain.

❸ Define *environmental sustainability*.
Environmental sustainability is the ability of the natural environment to meet humanity's needs indefinitely without going into a decline from human-caused stresses.

❹ Identify the various levels of biodiversity: genetic diversity, species richness, and ecosystem diversity.
Biological diversity is the variety of living organisms considered at three levels: genetic diversity, species richness, and ecosystem diversity. **Genetic diversity** is the genetic variety within a species, both among individuals within a given population and among geographically separate populations. **Species richness** consists of the number of species of archaea, bacteria, protists, plants, fungi, and animals. **Ecosystem diversity** is the variety of ecosystems—the forests, prairies, deserts, lakes, coastal estuaries, coral reefs, and other ecosystems. Biological diversity is important because it provides **ecosystem services,** important environmental services such as clean air, clean water, and fertile soil.

❺ Distinguish among extinct species, endangered species, and threatened species.
Extinction is the elimination of a species, either locally or globally. An **endangered species** is a species whose numbers are so severely reduced that it is in imminent danger of extinction throughout all or a part of its range. A **threatened species** is a species in which the population is small enough for it to be at risk of becoming extinct but not so small that it is in imminent danger of extinction. Many endangered and threatened species are **endemic species,** localized, native species that are not found anywhere else in the world.

❻ Discuss important causes of declining biological diversity.
The most important cause of declining biological diversity is **habitat fragmentation,** the division of habitats that formerly occupied large, unbroken areas into smaller pieces by roads, fields, cities, and other human land-transforming activities. Another important cause of declining biological diversity is **invasive species,** foreign species that, when introduced into an area where they are not native, upset the balance among the organisms living there and cause economic or environmental harm.

❼ Define *conservation biology*, and compare in situ and ex situ conservation measures.
Conservation biology is a multidisciplinary science that focuses on the study of how humans impact species and ecosystems and on the development of ways to protect biological diversity. Efforts to preserve biological diversity in the wild are known as **in situ conservation.** In situ conservation is urgently needed in the world's **biodiversity hotspots,** relatively small areas of land that contain an exceptional number of species and are at high risk from human activities. **Restoration ecology,** which uses the principles of ecology to help return a degraded environment as closely as possible to its former undisturbed condition, is another aspect of in situ conservation. **Ex situ conservation** involves conserving individual species in human-controlled settings.

❽ Name at least three greenhouse gases, and explain how greenhouse gases contribute to global warming.
Global warming, the current warming trend in Earth's climate, is probably the result of the **enhanced greenhouse effect** produced by increased levels of **greenhouse gases,** which absorb infrared radiation. The greenhouse gases carbon dioxide, methane, surface ozone, nitrous oxide, and chlorofluorocarbons are accumulating in the atmosphere as a result of human activities.

❾ Describe how global warming may affect sea level, precipitation patterns, plants and other organisms, and food production.
During the 21st century, global warming may cause a rise in sea level. Precipitation patterns may change, resulting in more frequent droughts in some areas and more frequent flooding in other areas. Global warming is causing some species to shift their ranges; biologists think global warming will cause some species to go extinct, some to be unaffected, and others to expand their numbers and ranges. Problems for agriculture include increased flooding, increased droughts, and declining agricultural productivity in tropical and subtropical areas.

REVIEW QUESTIONS

1. What climate and soil factors produce each of the major biomes?
2. Name several representative plants for each major terrestrial biome.
3. Which biomes are best suited for agriculture? Explain why each of the biomes you did not mention is unsuitable for agriculture.
4. Compare and contrast these biomes:
   a. Temperate rain forest and tropical rain forest
   b. Temperate grassland and savanna
   c. Chaparral and desert
   d. Tundra and boreal forest
   e. Temperate deciduous forest and temperate grassland
   f. Temperate grassland and desert
5. What is environmental sustainability?
6. What are the three levels of biological diversity?
7. Which organism is more likely to become extinct, an endangered species or a threatened species? Explain your answer.
8. How does habitat fragmentation contribute to declining biological diversity?
9. How do invasive species contribute to the biodiversity crisis?
10. What is conservation biology?
11. Is restoration ecology an example of in situ or ex situ conservation? Explain your answer.
12. Which type of conservation measure, in situ or ex situ, helps the greatest number of species? Why?
13. Describe the enhanced greenhouse effect, and name the most important greenhouse gas.
14. What are some of the significant problems that global warming may cause during the 21st century?

THOUGHT QUESTIONS

1. Why do most animals of the tropical rain forest live in trees?
2. Why are fires not common in desert areas with hot, dry climates?
3. In which biome do you live? If your biome does not match the description given in this chapter, explain the discrepancy.
4. If half of the world's biological diversity were to disappear, how would it affect your life?
5. Because new species will eventually evolve to replace those that humans are driving to extinction, why is declining biological diversity such a threat to us?
6. It has been suggested that the wisest way to "use" fossil fuels would be to leave them in the ground. How would this affect global warming? Energy supplies?

Visit us on the web at http://www.thomsonedu.com/biology/berg/ for additional resources such as flashcards, tutorial quizzes, further readings, and web links.

Appendices

Appendix A

Metric Equivalents and Temperature Conversion

Some Common Prefixes

		Examples
kilo	1000	A kilogram is 1000 grams.
centi	0.01	A centimeter is 0.01 meter.
milli	0.001	A milliliter is 0.001 liter.
micro (μ)	one-millionth	A micrometer is 0.000001 (one-millionth) of a meter.
nano (n)	one-billionth	A nanogram is 10^{-9} (one-billionth) of a gram.

> The relationship between mass and volume of water (at 20°C)
> $$1\ g = 1\ cm^3 = 1\ mL$$

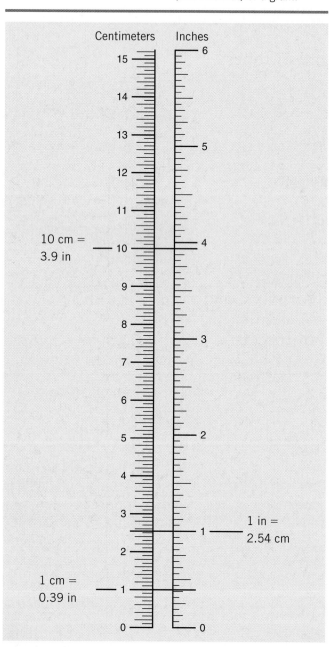

10 cm = 3.9 in

1 in = 2.54 cm

1 cm = 0.39 in

Some Common Units of Length

Unit	Abbreviation	Equivalent
meter	m	approximately 39 in
centimeter	cm	10^{-2} m
millimeter	mm	10^{-3} m
micrometer	μm	10^{-6} m
nanometer	nm	10^{-9} m

Length Conversions

1 in = 2.5 cm	1 mm = 0.039 in
1 ft = 30 cm	1 cm = 0.39 in
1 yd = 0.9 m	1 m = 39 in
1 mi = 1.6 km	1 m = 1.094 yd
	1 km = 0.6 mi

To convert	Multiply by	To obtain
inches	2.54	centimeters
feet	30	centimeters
centimeters	0.39	inches
millimeters	0.039	inches

Standard Metric Units

		Abbreviation
Standard unit of mass	gram	g
Standard unit of length	meter	m
Standard unit of volume	liter	L

Some Common Units of Volume

Unit	Abbreviation	Equivalent
liter	L	approximately 1.06 qt
milliliter	mL	10^{-3} L (1 mL = 1 cm^3 = 1 cc)
microliter	μL	1026 L

Volume Conversions

1 tsp = 5 mL	1 mL = 0.03 fl oz
1 tbsp = 15 mL	1 L = 2.1 pt
1 fl oz = 30 mL	1 L = 1.06 qt
1 cup = 0.24 L	1 L = 0.26 gal
1 pt = 0.47 L	
1 qt = 0.95 L	
1 gal = 3.79 L	

To convert	Multiply by	To obtain
fluid ounces	30	milliliters
quart	0.95	liters
milliliters	0.03	fluid ounces
liters	1.06	quarts

Some Common Units of Mass

Unit	Abbreviation	Equivalent
kilogram	kg	10^3 g (approximately 2.2 lb)
gram	g	approximately 0.035 oz
milligram	mg	10^{-3} g
microgram	μg	10^{-6} g
nanogram	ng	10^{-9} g

Mass Conversions

1 oz = 28.3 g	1 g = 0.035 oz
1 lb = 453.6 g	1 kg = 2.2 lb
1 lb = 0.45 kg	

To convert	Multiply by	To obtain
ounces	28.3	grams
pounds	453.6	grams
pounds	0.45	kilograms
grams	0.035	ounces
kilograms	2.2	pounds

Energy Conversions

calorie (cal) = energy required to raise the temperature of 1 g of water (at 16°C) by 1°C

1 calorie = 4.184 joules

1 Kilocalorie (Kcal) = 1000 cal

Temperature Scales

Celsius (Centigrade) = °C

Fahrenheit = °F

Temperature Conversions

$$°C = \frac{(°F - 32) \times 5}{9}$$

$$°F = \frac{°C \times 9}{5} + 32$$

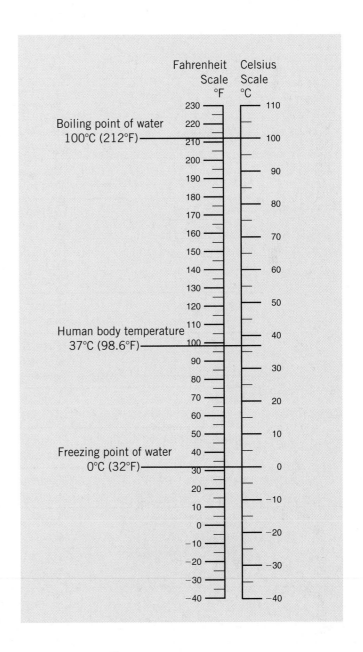

Appendix B

Some Important Biological Events in Geologic Time

Eon	Era	Period	Epoch	Time*	Some Important Biological Events
Phanaerozoic	Cenozoic	Neogene	Holocene	0.01 (10,000 years ago)	Decline of some woody plants; rise of herbaceous plants; age of *Homo sapiens*
			Pleistocene	2	Extinction of some plant species; extinction of many large mammals at end
			Pliocene	5	Expansion of grasslands and deserts; many grazing animals
			Miocene	23	Flowering plants continue to diversify; diversity of songbirds and grazing mammals
		Paleogene	Oligocene	34	Spread of forests; apes appear; present mammalian families are represented
			Eocene	56	Flowering plants dominant; modern mammalian orders appear and diversify; modern bird orders appear
			Paleocene	66	Semitropical vegetation (flowering plants and conifers) widespread; primitive mammals diversify
	Mesozoic	Cretaceous		146	Rise of flowering plants; dinosaurs reach peak, then become extinct at end; toothed birds become extinct
		Jurassic		200	Gymnosperms common; large dinosaurs; first toothed birds
		Triassic		251	Gymnosperms dominant; ferns common; first dinosaurs; first mammals
	Paleozoic	Permian		299	Conifers diversify; cycads appear; modern insects appear; mammal-like reptiles; extinction of many invertebrates and vertebrates at end
		Carboniferous		359	Forests of ferns, club mosses, horsetails, and gymnosperms; many insect forms; spread of ancient amphibians; first reptiles
		Devonian		416	First forests; gymnosperms appear; many trilobites; wingless insects appear; fishes with jaws appear and diversify; amphibians appear
		Silurian		444	Vascular plants appear; coral reefs common; jawless fishes diversify; terrestrial arthropods
		Ordovician		488	Fossil spores of terrestrial plants (bryophytes?); invertebrates dominant; coral reefs appear; first fishes appear
		Cambrian		542	Bacteria and cyanobacteria; algae; fungi; age of marine invertebrates; first chordates
Proterozoic		Ediacaran		600	Algae and soft-bodied invertebrates diversify
		Early Proterozoic		2500	Eukaryotes evolve
Archaean				4600 to 2500	Oldest known rocks; prokaryotes evolve; atmospheric oxygen begins to increase

Time from beginning of eon, period, or epoch to present (millions of years).

Appendix C
USDA Plant Hardiness Zones

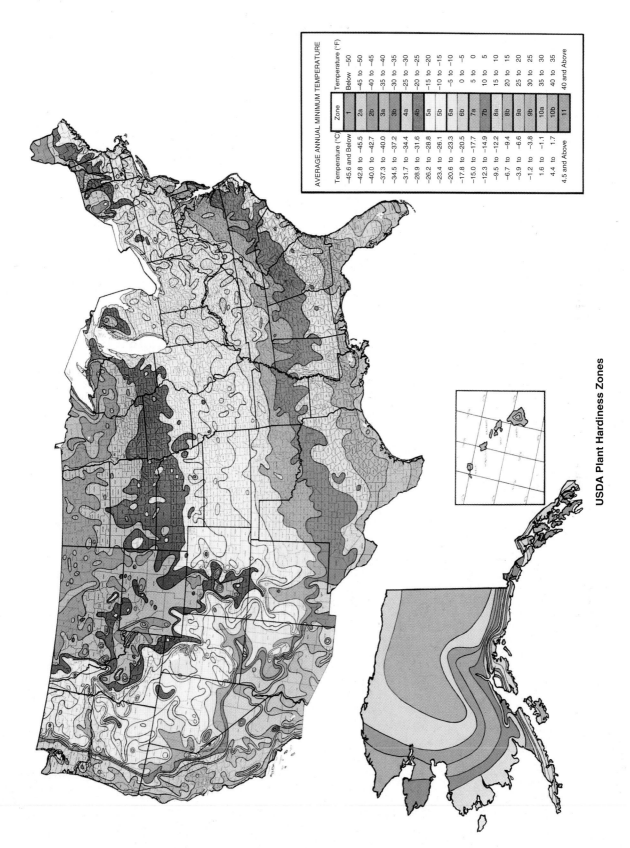

AVERAGE ANNUAL MINIMUM TEMPERATURE			
Temperature (°C)	Zone		Temperature (°F)
−45.6 and Below	1		Below −50
−42.8 to −45.5	2a		−45 to −50
−40.0 to −42.7	2b		−40 to −45
−37.3 to −40.0	3a		−35 to −40
−34.5 to −37.2	3b		−30 to −35
−31.7 to −34.4	4a		−25 to −30
−28.9 to −31.6	4b		−20 to −25
−26.2 to −28.8	5a		−15 to −20
−23.4 to −26.1	5b		−10 to −15
−20.6 to −23.3	6a		−5 to −10
−17.8 to −20.5	6b		0 to −5
−15.0 to −17.7	7a		5 to 0
−12.3 to −14.9	7b		10 to 5
−9.5 to −12.2	8a		15 to 10
−6.7 to −9.4	8b		20 to 15
−3.9 to −6.6	9a		25 to 20
−1.2 to −3.8	9b		30 to 25
1.6 to −1.1	10a		35 to 30
4.4 to 1.7	10b		40 to 35
4.5 and Above	11		40 and Above

USDA Plant Hardiness Zones

Appendix D

Annual Flowers

Name	Height (inches)	Spacing (inches)	Color	Uses	How to Start	Remarks
Ageratum	6–12	6–9	blue, pink, white	edging	seed and transplant**	Compact, excellent bloom. Needs full sun.
Alternanthera	12	12		accent	seed	Creates background for flowering plants.
Antirrhinum (snapdragon)	12–36	6–8	yellow, pink, white, blue, red, salmon, rose, bronze, orange	cut flowers, flower bed	seed and transplant	Must be staked for straight flowerspikes. Needs full sun. Excellent background plants.
Aster (starwort)	8–36	10–12	blue, red, salmon, pink, white, yellow	flower beds	seed and transplant	Excellent cut flowers.
Begonia (begonia)	6–12	6–10	pink, red, white	edging, potted plant, window box, hanging basket	cuttings or seed	May be used as a houseplant. Needs direct light.
Browallia (browallia)	12–18	8–10	blue, white	hanging basket, window box	cuttings or seed	Makes an attractive houseplant. Needs full sun.
Calendula (potted marigold)	12–24	10–12	yellow, orange	flower bed, hanging basket	seed and transplant	Flower petal used in cooking stews to add color. Needs full sun.
Callistephus (China aster)	12–24	10–12	blue, purple, white, yellow	flower bed, cut flowers	seed and transplant	Gives excellent summer color. Needs full sun.
Celosia (cockscomb)	12–30	10–12	red, orange, yellow, pink	flower bed, cut flowers	seed and transplant	Is excellent dried flower. Needs full sun.
Cleome speciosa (spider plant or spiderflower)	24–36	18–36	pink, white	flower bed	seed directly*	Makes an attractive houseplant.
Coleus (coleus)	12–24	10–12	red, bronze, yellow, pink foliage	flower bed, hanging basket	seed and transplant	Needs full to partial sun. Beautiful foliage plant.
Coreopsis (coreopsis)	12–24	6–10	yellow, red	flower bed	seed directly	Good for cut flowers.
Cosmos (cosmos)	48–72	12–18	red, pink, yellow	flower bed	seed and transplant	Keep near back of beds.
Dahlia variabilis (dahlia)	12–24	12–15	red, pink, yellow, rose	flower bed, cut flowers	seed and transplant	Profuse bloomer. For maximum bloom, sow several weeks before other annuals.
Delphinium (larkspur)	18–24	8–10	blue, pink, white	flower bed, cut flowers	seed and transplant	Grows best in peat pots. Difficult to transplant.

*Seed directly—sow the seed directly into the soil where the plants are desired.

**Seed and transplant—start the seed indoors (in greenhouse, hotbed, or portable seed germination case). As the seedlings develop, they are transplanted from the seed flat to other containers, where they grow until ready to be set out.

Name	Height (inches)	Spacing (inches)	Color	Uses	How to Start	Remarks
Dianthus (pink)	6–18	6–8	white, pink	edging, cut flowers	cuttings or seed	Very fragrant flowers.
Dimorphotheca (cape marigold)	12–18	8–18	yellow, orange, white	flower bed	seed and transplant	Grows well in dry areas.
Dolichos lablab (hyacinth bean)			purple	accent	seed	Vine texture.
Gaillardia (gaillardia)	12–24	8–10	red, orange	cut flowers, flower bed	cuttings	Loves seashore conditions. Does well in dry areas.
Gazania (treasure flower)	6–10	8–10	orange, yellow, white, tangerine, bronze	flower bed	seed and transplant	Bright blooms.
Gomphrena (globe amaranth)	9–24	8–10	blue, pink, white	cut flowers, flower bed	seed and transplant	Makes good dried flower. Collect and hang in dry, dark place.
Gypsophila (baby's breath)	12–18	10–12	pink, white	flower bed, cut flowers	seed directly	This is the annual form; there is also a perennial plant.
Helianthus (sunflower)	18–60	12–36	yellow, brown	flower bed	seed directly	Seeds are edible.
Helichrysum (strawflower)	24–36	8–10	yellow, red, white	flower bed	seed and transplant	Makes good dried flower. Cut; hang in dry, dark place.
Hypoestes (polka-dot plant)	8–24	10–12	burgundy, red, rose, white, pink	hanging basket, window box, flower bed	seed and transplant	Leaves accented by lavender-pink spots.
Iberis (candytuft)	10–18	6–8	white, pink, red	edging	seed and transplant	Makes attractive ground cover.
Impatiens (impatiens)	6–18	12–18	pink, red, white, multicolor	edging, hanging basket	seed and transplant	Does best in shaded area.
Ipomoea (morning glory)	60–120	10–12	blue	trellis	seed directly	Vine.
Lantana (lantana)	12–36	2–15	yellow, blue, red	hanging basket, flower bed, potted plant	cuttings	Makes excellent topiary plant for patio.
Lathyrus (sweet pea)	36–60	6–10	pink, blue, white	cut flowers	seed directly	Grows best in cool conditions.
Limonium (statice or sea lavender)	18–30	10–12	blue, pink, white	flower bed, cut flowers	seed and transplant	Makes good dried flower. Filler flower for bouquets.
Lisianthus	6–24	12–14	blue, rose, orchid, white, pink, yellow	flower bed	seed and transplant	Large blooms, drought tolerant.

Name	Height (inches)	Spacing (inches)	Color	Uses	How to Start	Remarks
Lobelia (lobelia)	3–6	4–6	blue, pink, white	edging, hanging basket, potted plant	seed directly	Very attractive in hanging basket.
Lobularia (sweet alyssum)	4–6	4–6	white, pink, blue	window box, edging, hanging basket	seed and transplant	Blooms all summer.
Matthiola (stock)	12–36	8–10	white, pink, yellow, red	cut flowers, flower bed	seed and transplant	Fragrant flowers.
Mirabilis (four-o'clock)	24–36	10–12	yellow, red, white	flower bed	seed and transplant	Withstands city conditions.
Myosotis (forget-me-not)	6–12	6–8	blue, pink	edging, cut flowers	seed directly	Makes attractive cut flowers.
Nicotiana (flowering tobacco)	18–24	10–12	rose, white	cut flowers, flower bed	seed and transplant	Fragrant flowers after dark.
Osteospermum (African daisy)	15	12	white, pastels, orange, red	accent, urn, flower bed	seed	Excellent flower quality.
Papaver (poppy)	18–36	10–12	red, yellow, white	cut flowers, flower bed	seed directly	Grows in masses. Bright colors.
Pelargonium (geranium)	12–18	10–12	red, white, pink, lavender	flower bed, window box, hanging basket	cuttings and seeds	Excellent flowering plant.
Pennisetum glaucum (millet)	24	24	bold metallic leaves, purple grassy leaves	urn, accent, flower bed	seed	Cattail spikes 12"-tall.
Pentas (star-cluster)	18	18–24	pink, white, red, pastels	flower bed, hanging basket, planter	seed	Flower color full season.
Petunia (petunia)	6–18	10–12	pink, red, white, blue	edging, flower bed, hanging basket	seed and transplant	Most widely used annual.
Phlox (phlox)	6–18	6–8	pink, blue, white, red	cut flowers, flower bed, window box, hanging basket	seed directly or transplant	Very intense colors.
Portulaca (moss rose)	4–6	3–6	red, orange, yellow	edging, hanging basket	seed directly	Reseeds by itself.
Rudbeckia (coneflower)	24	24	yellow daisy, dark center	flower bed, accent	seed	Thrives in most soils.
Salpiglossis (giant velvet flower)	18–36	10–12	gold, scarlet, rose, blue	flower bed, cut flowers	seed and transplant	Very small seed.
Salvia (scarlet sage)	10–36	6–12	blue, red, white	cut flowers, flower bed	seed and transplant	Very showy color.
Scabiosa (pincushion flower)	30–36	12–14	multicolor	cut flowers	seed and transplant	Variety of plants from which to select.
Spilanthes (peek-a-boo)	18	24	yellow globe	flower bed, cut flowers, accent	seed	Companion crop to marigolds.

Name	Height (inches)	Spacing (inches)	Color	Uses	How to Start	Remarks
Tagetes (marigold)	6–48	6–12	orange, yellow	flower bed, edging, potted plant, hanging basket	seed and transplant or seed directly	Wide range of varieties. No insect problems.
Thunbergia (black-eyed Susan vine)	24–60	10–12	orange	climbing vine, hanging basket	seed and transplant	
Tithonia (Mexican sunflower)	24–30	24	yellow, orange, red	flower bed, cut flowers	seed	Drought resistant. Attracts butterflies. Mass flower for bouquets.
Tropaeolum (nasturtium)	12–18	10–12	orange, yellow	flower bed, window box, hanging basket	seed and transplant or seed directly	Flowers and leaves are edible.
Verbena (verbena)	6–12	10–12	purple, red, white	flower bed, edging	seed and transplant	Good rock garden plant.
Vinca (common periwinkle)	8–12	12	pink, pastel, white, blue, red, variegated colors	flower bed, urn, planter, edging and mounds	seed	Drought tolerant. Loves heat.
Viola (pansy)	6–8	6–8	multicolor	edging, flower bed, cut flowers, hanging basket	seed and transplant	Gives excellent color in summer.
Zinnia (zinnia)	8–36	6–12	multicolor	edging, flower bed, cut flowers	seed and transplant or seed directly	Gives excellent color in summer.

Appendix E

Flowering Perennials

Name	When to Plant Seed	Exposure	Germination Time (days)	Spacing (inches)	Height (inches)	Uses	Color	Remarks
Achillea millefolium (yarrow)	early spring or late fall	sun	7–14	36	24	borders, cut flowers	yellow, white, red, pink	Seed is small. Water with a mist. Easy to grow.
Alyssum saxatile (golddust)	early spring	sun	21–28	24	9–12	rock garden, edging, cut flowers	yellow	Blooms early spring. Good in dry and sandy soils.
Anchusa italica (alkanet)	spring to September	partial shade	21–28	24	48–60	borders, background, cut flowers	blue	Blooms June or July. Refrigerate seed 72 hours before sowing.
Anemone pulsatilla (windflower)	early spring or late fall for tuberous	sun	4	35–42	12	borders, rock garden, potted plant, cut flowers	blue, rose, scarlet	Blooms May and June. Is not hardy north of Washington, D.C.
Anthemis tinctoria (golden daisy)	late spring outdoors	sun	21–28	24	24	borders, cut flowers	yellow	Blooms midsummer to frost. Prefers dry or sandy soil.
Aquilegia (Columbine)	spring to September	sun or partial shade	30	12–18	30–36	borders, cut flowers	wide color range	Blooms in late spring. Best grown as a biennial.
Arabis alpina (rock cress)	spring to September	light shade	5	12	8–12	edging, rock garden	white	Blooms early spring.
Armeria maritima (sea pink)	spring to September	sun	10	12	18–24	rock garden, edging, borders, cut flowers	pink	Blooms May and June. Plant in dry, sandy soil. Shade until plants are well established.
Aster alpinus (hardy aster)	early spring	sun	14–21	36	12–60	rock garden, borders, cut flowers	white	Blooms June.
Astilbe arendsii (false spirea)	early spring	sun	14–21	24	12–36	borders	pink, red, white	Blooms July and August. Gives masses of color.
Begonia evansiana (hardy begonia)	summer in shady, moist spot	shade	12	9–12	12	flower bed	yellow, pink, white	Blooms late in summer. Can be propagated from bulblets in leaf axils.

Name	When to Plant Seed	Exposure	Germination Time (days)	Spacing (inches)	Height (inches)	Uses	Color	Remarks
Bergenia purpurascens (bergenia)	late winter	light shade	10	18	24–36	herb	pink, red	Hummingbirds love it.
Campanula medium (Canterbury bells)	spring to September (Do not cover.)	partial shade	20	15	24–30	borders, cut flowers	white, pink, blue	Divide mature plants every other year. Best grown as a biennial.
Cerastium tomentosum (snow-in-summer)	early spring	sun	14–28	18	6	rock garden, ground cover	white	Blooms in May and June. Forms a creeping mat and is a fast grower. Prefers a dry spot.
Chrysanthemum maximum (shasta daisy)	early spring to September	sun	10	30	24–30	borders, cut flowers	white	Blooms June and July. Best grown as a biennial in well-drained location.
Coreopsis lanceolata (tickseed)	early spring	sun	5	30	24	borders	yellow	Blooms from June to fall if old flowers are removed.
Delphinium elatum (delphinium)	spring to September	sun	20	24	48–60	borders, background, cut flowers	blue, lavender, white, pink	Blooms in June. Best grown as a biennial. Needs dry location.
Dianthus barbatus (sweet William)	spring to September	sun	5	12	12–18 (Dwarf form also available.)	borders, edging, cut flowers	red, pink, white	Blooms May and June. Very hardy. Needs well-drained soil.
Dianthus caryophyllus (carnation)	late spring	sun	20	12	18–24	flower bed, borders, edging, rock garden	pink, red, white, yellow	Blooms in late summer. Cut plants back in late fall and hold in cold frame.
Dianthus deltoides (pink)	spring to September	sun	5	12	12	borders, rock garden, edging, cut flowers	pink	Blooms in May and June. Best grown as a biennial. Needs dry location.
Digitalis purpurea (foxglove)	spring to September	sun or partial shade	20	12	48–60	borders, cut flowers	pink, white, purple, rose	Blooms in June and July. Grown as a biennial. Shade summer plantings.
Echinacea purpurea (coneflower)	spring to September	sun	20	30	30–36	borders, flower bed, cut flowers	white, pink, red, rose, purple	Blooms midsummer to fall. Shade summer plantings. Attracts butterflies. Deer resistant. Accent specimen massing.

Name	When to Plant Seed	Exposure	Germination Time (days)	Spacing (inches)	Height (inches)	Uses	Color	Remarks
Gaillardia grandiflora (gaillardia)	early spring or late summer	sun	20	24	12–30	borders, cut flowers	scarlet, yellow	Blooms from July until frost.
Gypsophila paniculata (baby's breath)	early spring to September	sun	10	48	24–36	borders, cut flowers, drying	white, pink	Blooms early summer until early fall. Needs lots of lime.
Hemerocallis (daylily)	late fall	sun or partial shade	15	24–30	12–48	borders, among shrubbery	pink, red	Plant several varieties for longer blooming season.
Hibiscus moscheutos (Mallow Rose)	spring or summer	sun or partial shade	15 or longer	24	36–96	background, flower bed	white, pink, red, rose	Blooms July to September.
Iberis (candytuft)	early spring or late fall	sun	20	12	10	rock garden, edging, ground cover	white	Blooms in late spring. Prefers dry places. Cut faded flowers to promote branching.
Iris (iris)	bulbs or rhizomes in fall	sun or partial shade	next spring	18–24	3–30	borders, cut flowers	blue, red, yellow, pink, bronze, wine	Blooms spring and summer if different varieties are used.
Lathyrus latifolius (sweet pea)	early spring	sun	20	24	60–72	background	pink, white, purple, red	Blooms June to September. Easily grown as a vine on fence or trellis.
Liatris pycnostachya (gayfeather)	early spring or late fall	sun	20	18	24–60	borders, cut flowers	rose-purple	Blooms summer to early fall. Easily started from seed.
Limonium (sea lavender)	early spring	sun	15	30	24–36	flower bed, cut flowers, drying	pink, yellow, mauve	Blooms in July and August.
Lupinus polyphyllus (lupine)	early spring or late fall (Soak before planting.)	sun	20	36	36	borders, cut flowers	white, yellow, pink, rose, red, blue, purple	Blooms most of summer. Needs excellent drainage. Does not transplant easily.
Mentha (mint)	late winter	sun	8	15	6–36	culinary	white	Mint flavors.
Paeonia (peony)	Plant tubers in late fall 2–3 deep.	sun	variable	36	24–48	borders, cut flowers, flower bed	pink, red, white, rose	Blooms late spring. Difficult to grow from seed.
Papaver nudicaule (Iceland poppy)	early spring	sun	10	24	15–18	borders, cut flowers	white, pink, red	Blooms early summer. Does not transplant easily.
Papaver orientale (oriental poppy)	early spring	sun	10	24	36	borders, cut flowers	pink, red, rose, orange, white, salmon	Blooms early summer. Does not transplant easily.

Name	When to Plant Seed	Exposure	Germination Time (days)	Spacing (inches)	Height (inches)	Uses	Color	Remarks
Phlox paniculata (summer phlox)	late fall or early winter	sun	25 irregular	24	36	borders, cut flowers	red, pink, blue, white	Blooms early summer. Color of flower varies.
Phlox subulata (moss phlox)	grown from stolons	sun		8	4–5	borders	blue, red, white, pink	Blooms in spring. Drought resistant.
Physalis alkekengi (Chinese lantern)	late fall	sun	15	36	24	borders, specimen plant	orange	Lanterns are borne the second year in the fall.
Primula (primrose)	January, in a flat on surface. Allow to freeze; then bring in to germinate.	partial shade	25 irregular	12	6–9	rock garden	white, yellow, pink, red, blue	Blooms April and May. May be seeded in fall.
Pyrethrum roseum (painted daisy)	spring to September	sun	20	18	24	borders, cut flowers, accent	various, including gold, pink, and lavender	Blooms May and June. Prefers well-drained soil.
Rudbeckia fulgida (pot of gold)	midsummer to September	sun, partial shade	20	30	30–36	borders, flower bed, cut flowers, accent	orange-yellow	Specimen attracts butterflies. Easy to grow. Deer and rabbit resistant. Wild flower.
Salvia (sage)	spring	sun	15	18–24	36–48	borders	red	Blooms August until frost.
Sedum spectabile (sedum)	late winter	sun	10	10	4–15	ground cover	pink, white	Fall foliage.
Stokesia cyanea (Stokes' aster)	early spring to September	sun	20	18	15	borders, cut flowers	white, blue	Blooms in September if started early.
Veronica spicata (speedwell)	spring to September	sun	15	18	18	borders, rock garden, cut flowers	purple	Blooms June and July. Easy to grow.

From U.S. Department of Agriculture Bulletin 114.

Appendix F

Narrowleaf Evergreens

Name	Hardiness	Foliage Color	Period of Interest	Uses	Remarks
Narrowleaf Evergreens (3 feet or less in height)					
Erica carnea (spring heath)	5	bright green	small, rosy red spikes in April	ground cover	Attractive spring colors.
Juniperus chinensis 'Sargentii' (Sargent juniper)	4	steel blue	blue berries in fall and winter	ground cover	Excellent for planting along seashore. Does well on steep banks to prevent erosion.
Juniperus horizontalis 'Wiltonii' (blue rug juniper)	2	steel blue	blue berries in fall and winter	ground cover	Excellent rock garden.
Taxus baccata 'Repandens' (spreading English yew)	4	dark green	fall and winter	foundation planting	Excellent plant; low maintenance; good color all year. Requires well-drained soil.
Taxus cuspidata 'Aurencens' (dwarf Japanese yew)	4	light green	fall and winter	foundation planting	Compact plant that produces red berries. Requires well-drained soil.
Narrowleaf Evergreens (3–9 feet in height)					
Chamaecyparis obtusa 'Nana' (Hinoki cypress)	3	deep dark green	all year	accent plant, specimen plant	Compact form; pyramidal shape.
Juniperus chinensis 'Hetzii' (Hetzii juniper)	4	blue green	all year	screen plant, foundation planting	Plant in full sun; does well in dry areas.
Juniperus chinensis 'Pfitzerana' (Pfitzer juniper)	4	blue green	all year	foundation planting, screen plant	Plant in full sun. Requires well-drained soil. Control of bag worms necessary.
Juniperus squamata 'Meyeri' (Meyer's juniper)	4	blue	all year	foundation planting	Needs good management and care in dry areas.
Juniperus virginiana 'Tripartita' (red cedar)	2	dark green	all year	foundation planting	Good in dry areas.
Picea abies 'Conica' (a) Dwarf Alberta spruce	2	light green	all year	foundation planting, specimen plant	Fine-textured plant.
(b) Bird's nest spruce	2	light green	all year	rock garden	Requires well-drained soil.
Pinus mugo 'Compacta' (Mugo pine)	2	dark green	all year	foundation planting	Global shape; slow growing.
Taxus x *media* 'Hicksii' (Hicks yew)	4	dark green	all year; fall fruit	foundation planting, screen plant	Plant in full sun. Excellent in formal gardens; columnar shape. Requires well-drained soil.

Name	Hardiness	Foliage Color	Period of Interest	Uses	Remarks
Narrowleaf Evergreens (more than 10 feet in height)					
Cypressocyparis leylandii (Leyland cypress)	4	light green	all year	screen or hedge plant	Fast grower; excellent windbreak; columnar shape.
Taxus baccata (English yew)	6	dark green	all year	screen or hedge plant, foundation planting on large buildings	Female plant has berries. Will stand shade. Good background shrub.
Taxus cuspidata (Japanese yew)	4	dark green	all year	screen or hedge plant, foundation planting on large buildings	Pyramidal form. Tolerates shade. Rapid grower. Produces red fruit on female.

Appendix G

Broadleaf Evergreens

Name	Size at Maturity (feet)	Soil and Soil pH Preference	Pruning Date	Blooming Date	Uses	Remarks
Andromeda (*Pieris japonica*)	8–10	loam pH 5.0–6.5	after blooming in spring	early spring	foundation, specimen, center of group	Gives very early white blooms. Excellent for background. Hardy in Zones 6 to 9.
Azalea (*Rhododendron* varieties)	from creeping to 15′ according to variety	loam pH 5.0–6.0	after blooming in spring	early spring through late spring depending on variety	foundation, flower beds, edging, woodland	Many variations in size, color, bloom date. Has a variety of uses. Should be sprayed for best results. Hardy in Zones 6 to 8.
Barberry, Darwin (*Berberis darvinii*)	8	broad range	anytime	spring flowering; berries in fall	hedge, foundation, specimen	Few thorns, bright orange flowers, purple berries. Needs full sun. Hardy in Zones 4 to 9.
Barberry, dwarf crimson pygmy (*Berberis thunbergii*)	4	broad range	anytime	spring flowering; berries in fall	hedge, foundation, specimen	Rich purple foliage turns scarlet in autumn. Keep in full sun.
Barberry, rosy glow (*Berberis thumbergi* 'Rose Glow')	5	broad range	anytime	spring flowering; berries in fall	hedge, foundation, specimen	Bright scarlet berries. Hardy in Zones 4 to 9.
Barberry, warty (*Berberis verruculosa*)	4	broad range	anytime	spring flowering; berries in fall	specimen, hedge	Yellow flowers in spring. Foliage white underneath. Hardy in Zones 4 to 9.
Boxwood, common (*Buxus sempervirens* 'Arborescens')	fast growing up to 15′	heavy loam pH 6.0–7.0	early spring	no flowers	hedge	Grows very fast. Full sun or shade.
Boxwood, English (*Buxus sempervirens* 'Suffruticosa')	slow growing up to 3′	heavy loam pH 5.0–7.0	early spring	inconspicuous	edging, walkways, topiary designs, inside corners at foundation	Suffers windburn in strong winter winds. Grows in full sun or shade. Grows slowly (1–2 inches each year). Hardy in Zones 6 to 8.
Camellia (*Camellia japonica*)	15–30	loam, high in organic matter pH 5.0–6.5	after flowering	early spring	specimen	Dark green foliage; white to deep red single or double flowers. Needs sun. Hardy in Zones 7 to 9.
Camellia (*Camellia sasanqua*)	8–15	loam, high in organic matter pH 5.0–6.5	after flowering or early spring	autumn	specimen, border, hedge	White to pink single and double flowers. Will grow in full sun or partial shade. Hardy in Zones 7 to 9.
Firethorn (*Pyracantha coccinea* 'Lalandi')	20	loam wide range of pH (6.0–8.0)	early spring	June	specimen	Many varieties. Difficult to transplant (use container plants). Orange fruit.

Name	Size at Maturity (feet)	Soil and Soil pH Preference	Pruning Date	Blooming Date	Uses	Remarks
Holly, American (*Ilex opaca*)	20–30	sandy loam pH 6.0–7.0	winter holiday season	spring	specimen	Red berries in fall and winter; spiny leaves.
Holly, Chinese (*Ilex cornuta*)	from 4' to 15' depending on variety	sandy loam pH 6.0–7.0	throughout the year except late summer	spring	foundation, hedge, specimen	Many new excellent varieties. Red or yellow berries. Full sun or shade. Needs heavy fertilization.
Holly, English (*Ilex aquifolium*)	2–10	sandy loam pH 6.0–7.0	winter holiday season	spring	specimen	Glossy foliage. Yellow and red berries.
Holly, Japanese (*Ilex crenata*)	from 3' to 15' depending on variety	loam pH 5.0–6.0	throughout the year except late summer	early summer	hedge, topiary design, specimen, foundation	Small spineless leaves and black berries. Needs fertilizer each year.
Magnolia, southern (*Magnolia grandiflora*)	30–50	loam pH 6.0–6.5	after flowering	mid-summer	specimen	Native to the United States. 6" to 10" fragrant white flowers. Hardy in Zone 7 and in southern areas.
Magnolia sweet bay (*Magnolia virginiana*)	30–50	loam	after flowering	early summer	background, specimen	Evergreen in south, Zones 7 to 10. Deciduous in Zones 5 and 6. Hardy in Zones 5 to 10.
Nandina (*Nandina domestica*)	6–8	loam pH 5.0–6.5	after flowering	mid-summer	shrub borders, front of taller evergreens	White flowers; red berries in autumn and winter. Hardy in Zones 7 and 8.
Privet (*Ligustrum japonicum*)	9–18	wide range	after flowering	June–July	hedge	Black berries. Hardy in Zones 6 to 7.
Privet, glossy (*Ligustrum lucidum*)	8–15	wide range	after flowering	August–September	hedge, borders	Easy to grow. Hardy in Zones 7 to 10. Grows in shade.
Rhododendron varieties (*Rhododendron*)	3–10 for most commercial varieties	varies with variety pH 5.0–6.0	early spring	spring	specimen, foundation	Requires mulch at all times and good drainage. High organic matter needed in soil. Hardy in Zones 4 to 8.
Viburnum (*Viburnum*)	5–12 depending on variety	varies with variety	after flowering	spring	foundation, specimen	Fertilize each year. Attractive fruit, especially to birds.

Appendix H

Deciduous Trees

Small Deciduous Trees (up to 35 feet in height)

Name	Form	Flowering	Color Summer	Fall	Height (feet)	Hardiness	Ornamental Use	Remarks
Acer ginnala 'Flame' (amur maple)	rounded, upright	purplish	green	scarlet	20	2	specimen, screening	Fragrant flowers.
Acer palmatum (Japanese maple)	rounded, moundlike	inconspicuous, reddish-purple	red to green	scarlet	20	5	specimen	Many varieties are available. Needs well-drained soil.
Cercis canadensis (redbud or Judas tree)	rounded	purplish pink, mid-May	green	yellow	35	4	border	Blooms same time as dogwood.
Chionanthus virginicus (old man's beard)	rounded	white, feathery, late May	green	yellow	30	4	specimen	Does well in moist, shady site.
Cornus florida (flowering dogwood)	rounded	white, mid-May	green	red	35	4	specimen	Does well in shade or full sun. Horizontal branching.
Cornus kousa (Chinese dogwood)	rounded	white, late May	green	red	30	5	accent, specimen	Bright red fruit. Excellent flower quality.
Crataegus crus-galli (cockspur hawthorn)	rounded	small, white	green	scarlet	35	4	border, specimen	Bright red fruit in fall. Needs sandy soil.
Crataegus laevigata 'Paulii' (English hawthorn)	dense, rounded	double scarlet, late May	green	—	35	4	border	Thorny.
Crataegus phaenopyrum (Washington thorn)	dense, rounded	white, mid-June	green	scarlet	30	4	specimen, accent	Bright red fruit in fall—good for wildlife.
Magnolia soulangeana (saucer magnolia)	rounded, open	white, purple, May	green	—	25	5	specimen, border	—
Magnolia stellata (star magnolia)	rounded, open	double white to mid-April	green	yellow	20	5	specimen	Best color appears in direct sun.

Name	Form	Flowering	Color Summer	Color Fall	Height (feet)	Hardiness	Ornamental Use	Remarks
Malus floribunda (Japanese flowering crabapple)	dense	white, early May	green	—	30	4	specimen	Flowers yearly.
Prunus cerasifera (cherry plum)	rounded	pink, May	red	dark purple	20	3	specimen	Has dark purplish fruit.
Prunus serrulata (Japanese flowering cherry)	vase	pink, mid-May	green	—	25	5	specimen	Many varieties are available. Showy display.
Prunus subhirtella 'Pendula' (weeping cherry)	weeping	double pink, early April	green	—	25	5	specimen	Excellent as accent plant.
Syringa reticulata (Japanese tree lilac)	pyramidal	small, creamy white	green	green/yellow	30	4	specimen	Cherrylike bark.
Medium Deciduous Trees (35–75 feet in height)								
Acer platanoides 'Crimson King' (Crimson King maple)	rounded, dense	—	red-purple	red	50	4	accent	Brilliant foliage.
Acer platanoides 'Emerald Queen' (Emerald Queen maple)	upright, oval	—	dark green	yellow	40	4	accent	Needs full sun.
Acer rubrum 'October Glory' (October Glory maple)	oval	—	dark green	red	50	3	accent	Excellent fall color.
Acer rubrum 'Red Sunset' (Red Sunset maple)	rounded	—	dark green	red	45	3	accent	Excellent fall color.
Aesculus carnea (red horsechestnut)	pyramidal	pink to red spikes, mid-May	green	brown	75	3	specimen	—

Medium Deciduous Trees (35–75 feet in height)

Name	Form	Flowering	Color		Height (feet)	Hardiness	Ornamental Use	Remarks
			Summer	Fall				
Betula pendula (European white birch)	pyramidal	—	green	yellow	60	2	specimen	White bark in winter. Bronze birch borer can be a problem.
Betula pendula 'Laciniata' (cutleaf weeping birch)	oval	—	bright green	yellow	50	2	accent	Weeping branches.
Cercidyphyllum japonicum (katsura tree)	broad	—	green	yellow/orange	60	4	accent	Excellent leaf and bark habit.
Fagus sylvatica 'Atropurpurea' (copper beech)	conical	—	dark purple	copper	70	5	accent	Excellent foliage.
Franklinia alatumaha (franklinia tree)	rounded	white, September	green	red	36	6	border, specimen	Late fall flower.
Fraxinus pennsylvanica var. *lanceolata* (green ash)	rounded	—	green	yellow	60	2	streets, border	Attractive form.
Gleditsia triacanthos "Shademaster" (Shademaster honeylocust)	rounded	—	light green	yellow	50	4	massing	Open branching.
Magnolia virginiana (sweet bay)	rounded	white, fragrant, late May	green	green	60	5	specimen	Excellent flower quality.
Oxydendrum arboreum (sourwood or lily-of-the-valley tree)	rounded	white, mid-July	green	red	75	4	specimen	Late summer flower; good fall color.
Quercus virginiana (live oak)	rounded, wide-spreading	—	green	green	60	7	specimen, street	Evergreen in southern United States.
Sorbus aucuparia (European mountain ash)	pyramidal	clusters of white	green	reddish	60	5	specimen	Bright red berries appear in a cluster in fall.

Name	Form	Flowering	Color Summer	Color Fall	Height (feet)	Hardiness	Ornamental Use	Remarks
Medium Deciduous Trees (35–75 feet in height)								
Stewartia pseudo-camellia (Stewartia)	pyramidal	white, camellia-like	green	purplish	60	5	specimen	Peeling bark in winter.
Large Deciduous Trees (more than 75 feet in height)								
Acer platanoides (Norway maple)	rounded	small yellow, before leaves	green	yellow	100	3	shade	Provides dense shade. Fast grower.
Acer rubrum (red maple)	rounded	small red, early April	green	brilliant red	120	3	border, specimen	Several new varieties are available.
Acer saccharinum (silver maple)	rounded	small, green	green	green to yellow	120	3	wildlife area	Fast growing; weak wooded.
Acer saccharum (sugar maple)	rounded	green	green	yellow	125	3	specimen, shade	Used to produce maple syrup.
Carya illinoinensis (pecan)	inverse pyramidal	—	green	yellow	160	6	specimen	Excellent nut tree.
Fagus grandifolia (American beech)	pyramidal	—	green	bronze	90	3	specimen	Wildlife food.
Fraxinus americana (white ash)	rounded	—	green	yellow	120	3	border	Attractive form.
Ginkgo biloba (ginkgo)	pyramidal	inconspicuous	green	yellow	120	4	street, specimen, border	Female form has fragrant fruit. Fan-shaped leaf.
Gleditisia triacanthos 'Inermis' (thornless honeylocust)	rounded, open	greenish, pealike, June	green	yellow	120	4	lawn, specimen, shade	No thorns.
Juglans nigra (black walnut)	rounded, open	catkins	green	yellow	150	4	nut producing	Wood is used to make furniture.
Liquidambar styraciflua (sweet gum)	pyramidal	greenish clusters	green	red to scarlet	125	4	lawn, specimen, street, shade	Difficult to transplant. Good form. Corky ridges on stem.

Large Deciduous Trees (more than 75 feet in height)

Name	Form	Flowering	Summer	Fall	Height (feet)	Hardiness	Ornamental Use	Remarks
Liriodendron tulipifera (tulip tree)	rounded	greenish yellow, tulip shaped, mid-June	green	yellow	180	4	specimen	Timber tree and fall color.
Magnolia acuminata (cucumber tree)	pyramidal	greenish yellow, early June	green	—	90	4	specimen, border	Red to pink cucumber-shaped fruits.
Nyssa sylvatica (black gum)	rounded	inconspicuous, May-June	green	red	90	4	specimen	Small blue berries appear in midsummer.
Platanus occidentalis (sycamore)	pyramidal	yellow-green	green	yellow-green	120	4	specimen	Natives tend to defoliate.
Populus nigra 'Italica' (Lombardy poplar)	narrow, tall	—	green	—	90	2	quick temporary screen	Short-lived tree.
Quercus alba (white oak)	rounded	—	green	purple red	150	4	specimen	Excellent lumber tree. Slow grower.
Quercus borealis (red oak)	rounded	—	green	red	76	4	specimen, street	Transplants easily.
Quercus coccinea (scarlet oak)	rounded	—	green	bright scarlet	100	4	specimen, street	Difficult to transplant.
Quercus palustris (pin oak)	pyramidal	—	green	scarlet	100	4	specimen	Very graceful. Excellent street tree.
Tilia americana (American linden)	rounded, open	white, June	green	—	100	2	border	Wood is valuable for lumber.
Tilia cordata (little leaf linden)	rounded, open	white to yellow, mid-July	green	yellow	90	3	shade	Excellent street tree.
Ulmus americana (American elm)	vase	yellow	green	yellow brown	120	2	street	Possible problem with Dutch elm disease.

Appendix I

Plant Selection for Various Conditions

Urban Conditions	Poor/Dry Soils	Heavy Clay Soils
Shrubs		

Urban Conditions	Poor/Dry Soils	Heavy Clay Soils
Acanthopanax sieboldianus	*Acanthopanax*	*Acanthopanax sieboldianus*
Aronia arbutifolia	*Acer ginnala*	*Aralia spinosa*
Berberis thunbergii	*Berberis* x *mentorensis*	*Aronia*
Celastrus scandens	*Berberis thunbergii*	*Berberis*, most
Chaenomeles	*Chaenomeles*	*Chaenomeles*, most
Chionanthus virginicus	*Cornus racemosa*	*Cornus*, shrub types
Cornus alba	*Cotinus coggygria*	*Deutzia*, most
Cornus mas	*Cytisus*	*Elaeagnus*, most
Cornus stolonifera	*Elaeagnus angustifolia*	*Euonymus*, most
Elaeagnus umbellata	*Erica carnea*	*Forsythia*, most
Euonymus	*Genista lydia*	*Hamamelis*
Forsythia	*Hamamelis*	*Juniperus*, most
Hamamelis	*Juniperus*, most	*Kolkwitzia amabilis*
Hibiscus syriacus	*Kolkwitzia amabilis*	*Ligustrum*, most
Hydrangea	*Ligustrum*	*Lonicera*, shrubby types
Ilex crenata	*Myrica pennsylvanica*	*Myrica pennsylvanica*
Ilex glabra	*Potentilla*	*Physocarpus opulifolius*
Ilex x *meserveae*	*Rhamnus*	*Pyracantha*, most
Leucothoe	*Rhus*	*Rhodotypos scandens*
Ligustrum	*Ribes alpinum*	*Rhus*, most
Lindera benzoin	*Rosa rugosa*	*Ribes alpinus*
Lonicera	*Viburnum lentago*	*Rosa rugosa*
Magnolia stellata	*Yucca filamentosa*	*Salix*, most
Mahonia aquifolium		*Spiraea*, most
Myrica pennsylvanica		*Stephanandra incisa* 'Crispa'
Philadelphus virginalis		*Symphoricarpos* x *chenaultii* 'Hancock'
Physocarpus opulifolius		*Taxus*, most
Potentilla		*Thuja occidentalis*
Pyracantha		*Viburnum dentatum*
Rhamnus		*Viburnum lentago*
Rhodotypos scandens		*Viburnum opulus*
Rhus		*Viburnum prunifolium*
Ribes alpinum		*Viburnum sargentii*
Rosa		
Rosa rugosa		
Spiraea vanhouttei		
Syringa x *prestoniae*		
Syringa reticulata		
Syringa vulgaris		
Taxus		
Viburnum, most		
Wisteria sinensis		
Yucca filamentosa		

Urban Conditions	Poor/Dry Soils	Heavy Clay Soils
Trees		

Urban Conditions	Poor/Dry Soils	Heavy Clay Soils
Acer japonicum	*Acer saccharinum*	*Acer campestre*
Acer ginnala	*Betula*	*Acer rubrum*
Acer platanoides	*Celtis occidentalis*	*Acer saccharinum*
Acer saccharinum	*Cercis canadensis*	*Betula nigra*
Chamaecyparis nootkatensis	*Corylus corlurna*	*Crataegus*
Crataegus	*Elaeagnus*	*Elaeagnus angustifolia*

Urban Conditions	Poor/Dry Soils	Heavy Clay Soils
Trees (continued)		

Urban Conditions	Poor/Dry Soils	Heavy Clay Soils
Elaeagnus angustifolia	*Eucommia ulmoides*	*Eucommia ulmoides*
Fagus sylvatica	*Gleditsia triacanthos inermis*	*Fraxinus pennsylvanica*
Fraxinus pennsylvanica	*Gymnocladus dioica*	*Gleditsia triacanthos inermis*
Ginkgo biloba	*Kerria japonica*	*Koelreuteria paniculata*
Gleditsia triacanthos inermis	*Koelreuteria paniculata*	*Liquidambar styraciflua*
Koelreuteria paniculata	*Ostrya virginiana*	*Malus*, most
Liquidambar styraciflua	*Phellodendron amurense*	*Ostrya virginiana*
Liriodendron tulipifera	*Picea*	
Malus	*Platanus* x *acerifolia*	
Picea	*Pyrus*, most	
Pinus	*Salix*, most	
Platanus	*Taxodium distichum*	
Populus	*Ulmus parvifolia*	
Pyrus calleryana		
Taxodium distichum		
Ulmus x *hollandica*		

Ground Covers

Urban Conditions	Poor/Dry Soils	Heavy Clay Soils
Ajuga reptans	*Aegopodium podagraria*	*Aegopodium podagraria*
Hedera helix	*Hedera helix*	*Ajuga*
Juniperus	*Hosta*	*Euonymus fortunei*
Liriope muscari	*Liriope muscari*	*Hedera helix*
Ophiopogon japonicus	*Parthenocissus*	*Hemerocallis*
Rosmarinus officinalis	*Rosemarinus officinalis*	*Houttunynia cordata*
Trachelospermum	*Sedum*	*Juniperus*, low types
Vinca major		*Liriope muscari*
		Lonicera japonica 'Halliana'
		Pachysandra terminalis
		Parthenocissus
		Rhus aromatica

Vines

Urban Conditions	Poor/Dry Soils	Heavy Clay Soils
Bougainvillea	*Bougainvillea*	*Celastrus scandens*
Celastrus scandens	*Celastrus scandens*	*Wisteria*
Clematis sp.	*Rosa banksiae*	
Ficus pumila	*Wisteria sp.*	
Gelsemium sempervirens		
Lonicera sempervirens		
Parthenocissus		
Rosa banksiae		
Wisteria sp.		

Grasses

Urban Conditions	Poor/Dry Soils	Heavy Clay Soils
Andropogon gerardi	*Andropogon gerardi*	*Andropogon gerardi*
Bouteloua curtipendula	*Bouteloua curtipendula*	*Bouteloua curtipendula*
Elymus virginicus	*Elymus virginicus*	*Elymus virginicus*
Koeleria macrantha	*Koeleria macrantha*	*Koeleria macrantha*
Panicum virgatum	*Panicum virgatum*	*Panicum virgatum*
Sporobolus heterolepis	*Sporobolus heterolepis*	*Sporobolus heterolepis*

Appendix J
Biological Control of Insects

Biological Control Agent	Plant Pests Controlled	Additional Information
Parasitic wasp, Trichogramma wasp	Colorado potato beetle	The wasp lays its eggs on the beetle's eggs. Use of this wasp reduced need of chemical spray from 16 to 1 spray of Vydate in the early season. Eight thousand wasps per acre give good control.
Parasitic wasp, Trichogramma wasp	Tomato hornworm	The wasp lays its eggs on the hornworm. Wasp larvae enter the worm.
Parasitic wasp, Trichogramma wasp	Cabbage looper and other worms	The wasp lays its eggs on the eggs of the moths and butterflies.
Bacillus thuringiensis (can be bought under the names Dipel, Biotrol, Thuricide, and Javelin)	Colorado potato beetle; all caterpillars, including gypsy moth; cabbage looper and imported cabbage worm	A bacterium that is extremely effective on caterpillars, upsetting their digestive tract so they stop eating in about 2 hours and die in 2 days.
Predatory mite	Spider mite and whitefly	Some of these mites are resistant to chemicals.
Beauveria bassiana	Colorado potato beetle	This fungus gives 77% control by attacking the pupal stage in the soil and also adults. Small fungus spores only need to touch the beetle to cause infection.
Orange peel	Many insects controlled, such as ants, fleas	Still experimental but looks promising.
Ladybug	Aphids	Eats aphids at all stages of growth.
Vedalia beetle	Cotton cushion scale on citrus	Imported in 1988 to California and gave complete control of the scale insect.
Green lacewing	Aphids	Both nymphs and adults eat aphids and lacebugs.
Green lacewing	Mealybugs and lacebugs	Operates in cooler temperatures than many biologicals.
Bacillus popilliae	Japanese beetle	Known as milky spore disease; kills grubs in the soil.
Digger wasp	Japanese beetle	A tiny parasite kills the grubs in the soil.
Pediobius foveolatus	Mexican bean beetle	Eggs are laid on the bean beetle larvae. Wasp larvae destroy the bean beetle larvae.
Marigold and asparagus roots	Nematodes	Planting marigolds in an area greatly reduces the number of nematodes in the soil. Asparagus roots produce a chemical that causes a decline in the number of stubby root nematodes.
Spined soldier bug, *Podisus* spp.	Eats 100 kinds of harmful insects	One of the most effective insect predators.
Diatomaceous earth	Slugs, snails, mites, aphids	Finely ground sharp silica material kills soft-bodied insects, slugs, and snails.
Predatory nematode	Grubs of weevils, cutworms, fungus gnats, cabbage root weevil, wireworms	Controls many insects that spend some of their life cycle in the soil.

Glossary

abiotic factor An element of the nonliving, physical environment that affects a particular organism. Compare with *biotic factor.*

abscisic acid A plant hormone involved in growth and development, including dormancy and responses to stress.

abscission The normal (usually seasonal) falling off of leaves or other plant parts, such as fruits or flowers.

accessory fruit A fruit whose fleshy part is composed primarily of tissue other than the ovary.

achene A simple, dry fruit with one seed in which the fruit wall is separate from the seed coat. Sunflower fruits are achenes.

activation energy The energy required to initiate a chemical reaction.

active transport The energy-requiring movement of a substance across a membrane from a region of lower concentration to a region of higher concentration.

adaptation An evolutionary modification that improves an organism's chances of survival and reproductive success.

adaptive radiation The evolution of many related species from an ancestral species.

adenosine triphosphate (ATP) An organic compound that is of prime importance for energy transfers in biological systems.

adhesion The tendency of unlike molecules to adhere to one another.

adventitious Arising in an unusual position; applies to an organ such as a root or bud.

aerobe An organism that grows and metabolizes only in the presence of molecular oxygen. Compare with *facultative anaerobe* and *obligate anaerobe.*

aerobic respiration The process by which cells use oxygen to break down organic molecules with the release of energy that can be used for biological work; a type of cellular respiration.

aggregate fruit A fruit that develops from a single flower with several separate carpels that fuse, or grow together.

air pollution Any of a variety of chemicals (gases, liquids, or solids) present at high enough levels in the atmosphere to harm plants and other organisms.

allele One of two or more alternative forms of a gene.

allele frequency The proportion of a specific allele in the population.

allelopathy An adaptation in which toxic substances secreted by roots or shed leaves inhibit the establishment of competing plants nearby.

allopatric speciation Evolution of a new species that occurs when one population becomes geographically separated from the rest of the species.

allopolyploidy The situation in which an interspecific hybrid contains two or more sets of chromosomes from each of the parent species.

alternation of generations A type of life cycle characteristic of plants and a few algae and fungi in which they spend part of their life in a multicellular *n* gametophyte stage and part in a multicellular *2n* sporophyte stage.

amino acid An organic compound containing an amino group ($-NH_2$) and a carboxyl group ($-COOH$). Amino acids may be linked to form protein molecules.

aminoacyl-tRNA Molecule consisting of an amino acid covalently linked to a transfer RNA.

amyloplast Colorless plastid used for starch storage in cells of roots and tubers.

anabolic reaction See *anabolism.*

anabolism The phase of metabolism in which simpler substances combine to form more complex substances; requires energy. Compare with *catabolism.*

anaerobe See *facultative anaerobe* and *obligate anaerobe.*

angiosperm The traditional name for flowering plants, a large, diverse phylum of plants that form flowers for sexual reproduction and produce seeds enclosed in fruits. See *eudicot* and *monocot.*

anion A particle with one or more units of negative charge. Compare with *cation.*

annual A plant that completes its entire life cycle in one growing season. Compare with *biennial* and *perennial.*

annual ring A layer of wood (secondary xylem) that forms in the stems and roots of woody plants growing in temperate areas or tropical areas with rainy and dry seasons. Usually one ring forms per year.

antheridium A multicellular male gametangium that produces sperm cells.

anticodon A sequence of three nucleotides in tRNA that is complementary to a specific codon in mRNA.

apical dominance The inhibition of axillary buds by the apical meristem.

apical meristem An area of cell division at the tip of a stem or root in a plant; produces primary tissues.

apomixis A type of reproduction in flowering plants in which fruits and seeds are formed asexually.

apoplast A continuum consisting of the interconnected, porous plant cell walls, along which water moves freely. Compare with *symplast.*

archegonium A multicellular female gametangium that produces an egg.

artificial selection See *selective breeding.*

ascomycetes A fungus characterized by the production of nonmotile asexual conidia and sexual ascospores.

ascospore One of a set of (usually eight) sexual spores contained in a special spore case (the ascus) of an ascomycete.

ascus A saclike spore case in an ascomycete that contains sexual spores called ascospores.

asexual reproduction Reproduction in which only one parent participates, no fusion of gametes occurs, and the genetic makeup of parent and offspring is virtually identical. Compare with *sexual reproduction.*

ATP synthase An enzyme complex that synthesizes ATP from ADP, using the energy of a proton gradient; located in thylakoid membranes of chloroplasts and in the inner mitochondrial membrane.

autotroph An organism that synthesizes all its complex organic compounds from carbon dioxide. Compare with *heterotroph.*

auxin A plant hormone involved in growth and development, including stem elongation, apical dominance, and root formation on cuttings.

bacillus (pl., *bacilli*) A rod-shaped prokaryote.

bacteriophage A virus that infects a bacterium; also called *phage.*

basal angiosperm One of three groups of angiosperms that are thought to be ancestral to all other flowering plants. Compare with *core angiosperm.*

basidiomycetes A fungus characterized by the production of sexual basidiospores.

basidiospore One of a set of sexual spores (usually four) borne on a club-shaped structure (a basidium) of a basidiomycete.

basidium The clublike, spore-producing organ of a basidiomycete that bears sexual spores called basidiospores.

berry A simple, fleshy fruit in which the fruit wall is soft throughout. Tomatoes, bananas, and grapes are berries.

biennial A plant that takes 2 years (that is, two growing seasons) to complete its life cycle. Compare with *annual* and *perennial.*

binary fission The equal division of one cell into two cells. Prokaryotes reproduce by binary fission.

binomial nomenclature A system for giving each organism a two-word scientific name.

biodiversity hotspots Relatively small areas of land that contain an exceptional number of species and are at high risk from human activities.

biogeochemical cycle The process by which matter cycles from the living world to the nonliving, physical environment and back again.

biogeography The study of the geographic distribution of living organisms and fossils.

bioinformatics The storage, retrieval, and comparison of biological information, particularly DNA or protein sequences within a given species and among different species.

biological diversity The variety of living organisms considered at three levels: genetic diversity, species richness, and ecosystem diversity.

biological species concept The concept that a species consists of one or more populations whose members can interbreed to produce fertile offspring and cannot interbreed with individuals of other species.

biome A large, relatively distinct terrestrial region characterized by a similar climate, soil, plants, and animals.

bioremediation A method of cleaning up a hazardous waste site that uses organisms (usually bacteria or fungi) to break down the toxic pollutants.

biosphere All of Earth's living organisms, collectively.

biotic factor An element of the living world that affects a particular organism. Compare with *abiotic factor.*

biotic pollution The introduction of a foreign, or exotic, species into an area where it is not native. Exotic species often harm native species. See *invasive species.*

blade The broad, flat part of a leaf.

bolting The rapid elongation of a flower stalk that occurs when a plant initiates flowering; also caused by the application of gibberellin.

boreal forest A region of coniferous forests just south of the tundra in the Northern Hemisphere; also called *taiga.*

botany The scientific study of plants; also called *plant biology.*

bract A modified leaf associated with a flower or inflorescence but not part of the flower itself.

brassinosteroid (BR) One of a group of steroids that function as plant hormones and are involved in several aspects of growth and development.

brown alga A predominantly marine, photosynthetic protist that is multicellular and contains the brown pigment fucoxanthin.

bryophyte A nonvascular, spore-producing plant; includes mosses, liverworts, and hornworts.

bud An undeveloped shoot that contains an embryonic meristem; may be terminal (at the tip of the stem) or axillary (on the side of the stem).

bud scale A modified leaf that covers and protects winter buds.

bulb A rounded, fleshy underground bud that consists of a short stem with fleshy leaves.

bundle sheath A ring of parenchyma or sclerenchyma cells surrounding the vascular bundle in a leaf.

callus Undifferentiated tissue formed on an explant (excised tissue or organ) in plant tissue culture.

Calvin cycle A cyclic series of carbon fixation reactions in photosynthesis; fixes carbon dioxide and produces glucose.

capsule (1) The portion of the bryophyte sporophyte in which spores are produced. (2) A simple, dry fruit that opens along many seams or pores to release seeds, such as cotton fruits. (3) The outermost layer, usually composed of carbohydrate, of many prokaryotic cells.

carbohydrate An organic compound containing carbon, hydrogen, and oxygen in the approximate ratio of 1C:2H:1O.

carbon fixation A cyclic series of reactions that fixes carbon dioxide and produces carbohydrate. See *Calvin cycle.*

carotenoid One of a group of yellow to orange plant pigments.

carpel The ovule-bearing reproductive unit of a flower. See *pistil.*

carrying capacity The largest population that a particular habitat can support and sustain for an indefinite period, assuming there are no changes in the environment.

caryopsis (pl., *caryopses*) See *grain.*

Casparian strip A band of waterproof material around the radial and transverse cells of the endodermis; ensures that water and minerals enter the xylem only by passing through the endodermal cells.

catabolic reaction See *catabolism*.

catabolism The phase of metabolism in which complex organic molecules are broken down, with a release of energy that can be used to do biological work. Compare with *anabolism*.

cation A particle with one or more units of positive charge. Compare with *anion*.

catkin A cluster of tiny (much reduced) unisexual flowers borne on a pendulous stalk.

cell The basic structural and functional unit of life, which consists of living material bounded by a membrane.

cell cycle The cyclic series of events in the life of a dividing eukaryotic cell.

cell plate The structure that forms during cytokinesis in plants, separating two daughter cells produced by mitosis.

cell signaling Mechanisms of communication between cells; cells signal one another with secreted signaling molecules.

cell theory The theory that the cell is the basic unit of life, of which all living things are composed, and that all cells are derived from preexisting cells.

cell wall A comparatively rigid supporting wall exterior to the plasma membrane in plants, fungi, prokaryotes, and certain protists.

cellular respiration The cellular process in which energy of organic molecules is released for biological work.

cellular slime mold A protist whose feeding stage consists of unicellular, amoeboid organisms that aggregate to form a pseudoplasmodium during reproduction.

center of origin The area where a given species evolved.

centromere A specialized region of a chromosome. At prophase, sister chromatids are joined in the vicinity of their centromeres.

chaparral A biome with a Mediterranean climate—that is, with mild, moist winters and hot, dry summers. Chaparral vegetation consists primarily of small-leaved evergreen shrubs.

chemiosmosis The synthesis of ATP using the energy of a proton gradient established across a membrane; occurs during electron transport in both photosynthesis and aerobic respiration.

chitin A nitrogen-containing polysaccharide that forms the cells walls of many fungi.

chlorophyll One of a group of light-trapping green pigments found in most photosynthetic organisms.

chloroplast A chlorophyll-bearing organelle of certain plant cells; the site of photosynthesis.

chromatid One of the two identical halves of a duplicated chromosome; the two chromatids that make up a chromosome are referred to as *sister chromatids*.

chromatin The complex of DNA and protein that makes up eukaryotic chromosomes.

chromosome One of several rod-shaped bodies in the cell nucleus that contain the hereditary units (genes).

chromosome theory of inheritance A basic principle in biology that states that inheritance can be explained by assuming that genes are linearly arranged in specific locations along the chromosomes.

chytrid A fungus characterized by the production of flagellated cells at some stage in the life history; also called *chytridiomycete*.

chytridiomycete See *chytrid*.

circadian rhythm A biological activity with an internal rhythm that approximates the 24-hour day.

citric acid cycle A series of aerobic chemical reactions in which fuel molecules are degraded to carbon dioxide and water, with the release of metabolic energy used to produce ATP; also known as the *Krebs cycle*.

cladistics Classification of organisms based on recency of common ancestry rather than degree of structural similarity.

cladogram A diagram that illustrates evolutionary relationships based on the principles of cladistics.

class A taxonomic group composed of related, similar orders.

clay The smallest inorganic soil particles. Compare with *silt* and *sand*.

cline Gradual change in phenotype and genotype frequencies among contiguous populations that is the result of an environmental gradient.

clumped dispersion The spatial distribution pattern of a population in which individuals are more concentrated in specific parts of the habitat. Compare with *random dispersion* and *uniform dispersion*.

coccus (pl., *cocci*) A prokaryote with a spherical shape.

codon A sequence of three nucleotides in mRNA that specifies an individual amino acid or a start or stop signal.

coenocytic Having a body that consists of a multinucleated cell; that is, the nuclei are not separated from one another by septa. Applies to certain protists and fungi.

coevolution The interdependent evolution of two or more species that occurs as a result of their interactions over a long period of time. Flowering plants and their animal pollinators are an example of coevolution, because each has profoundly affected the other's characters.

cohesion The tendency of like molecules to adhere or stick together.

colchicine A drug that blocks the division of eukaryotic cells.

collenchyma cell A living plant cell with moderately but unevenly thickened primary walls.

colony (1) In prokaryotes, a population of millions of genetically identical cells derived from a single cell, grown in culture. (2) In protists, an aggregation of loosely organized cells. (3) In plants, a population of individuals that live close to one another.

commensalism A type of symbiosis in which one organism benefits and the other one is neither harmed nor helped. Compare with *mutualism* and *parasitism*.

community An association of populations of different species living in a defined habitat with some degree of interdependence.

community ecology The description and analysis of patterns and processes within the community.

competition The interaction among two or more individuals that attempt to use the same essential resource, such as food, water, sunlight, or living space.

compost A natural soil-and-humus mixture that improves soil structure and fertility.

compound leaf A leaf with a blade that consists of two or more leaflets. Compare with *simple leaf*.

concentration gradient The variance of concentration of specific atoms or molecules from one area of a biological system (where the concentration is high) to another (where it is low).

condensation reaction A chemical reaction in which water is removed as two or more smaller molecules join to form a larger molecule.

cone A reproductive structure that produces either microspores or megaspores in many gymnosperms.

conidium (pl., *conidia*) An asexual spore that usually forms at the tip of a specialized hypha called a conidiophore.

conifer Any of a large phylum of gymnosperms that are woody trees and shrubs with needlelike, mostly evergreen leaves and seeds in cones.

conjugation (1) A sexual phenomenon in certain protists that involves exchange or fusion of a cell with another cell. (2) A mechanism for DNA exchange in prokaryotes that involves cell-to-cell contact.

conservation biology A multidisciplinary science that focuses on the study of how humans impact species and ecosystems and on the development of ways to protect biological diversity.

consumer An organism that cannot synthesize its own food from inorganic raw materials and therefore must obtain energy and body-building materials from other organisms. Compare with *producer*.

contractile root A specialized root, often found on bulbs or corms, that contracts and pulls the plant to a desirable depth in the soil.

control That part of a scientific experiment in which the experimental variable is kept constant. The control provides a standard of comparison that enables the experimenter to verify the results of the experiment.

convergent evolution The independent evolution of similar adaptations in unrelated species.

core angiosperm One of the group to which most angiosperm species belong; core angiosperms are divided into three subgroups: magnoliids, monocots, and eudicots. Compare with *basal angiosperm*.

cork Cells produced by the cork cambium. Cork cells are dead at maturity and function for protection.

cork cambium A lateral meristem that produces cork cells and cork parenchyma; cork cambium and the tissues it produces make up the outer bark of a woody plant.

cork parenchyma One or more layers of parenchyma cells produced by the cork cambium; also called *phelloderm*.

corm A short, thickened underground stem specialized for food storage and asexual reproduction.

cortex The ground tissue between the epidermis and the phloem in nonwoody roots and stems.

cotyledon The seed leaf of a plant embryo that often contains food stored for germination.

covalent bond A chemical bond involving one or more shared pairs of electrons.

crossing over The breaking and rejoining of homologous (nonsister) chromatids during meiotic prophase I, resulting in an exchange of genetic material.

cryptochrome A protein pigment that strongly absorbs blue light; implicated in resetting the biological clock in plants and certain animals.

cuticle A waxy covering over the epidermis of the aerial parts (leaves and stems) of a plant.

cyanobacterium A prokaryotic photosynthetic microorganism that possesses chlorophyll and produces oxygen during photosynthesis. Formerly known as a blue-green alga.

cycad Any of a phylum of gymnosperms that live mainly in tropical and subtropical regions and have stout stems (to 20 meters in height) and fernlike leaves.

cytokinesis The stage of cell division in which the cytoplasm divides to form two daughter cells.

cytokinin A plant hormone involved in growth and development, including cell division and delay of senescence.

cytoplasm The general cellular contents, exclusive of the nucleus.

cytoskeleton The internal structure, made up of microfilaments and microtubules, that gives shape and mechanical strength to cells.

day-neutral plant A plant that initiates flowering not in response to seasonal changes in the amounts of daylight and darkness but in response to some other type of environmental or internal stimulus. Compare with *long-day, short-day,* and *intermediate-day plant.*

deciduous Falling off at a certain season, such as leaves shed by trees in autumn. Compare with *evergreen.*

decomposers A microorganism that breaks down dead organic material and uses the decomposition products as a source of energy; also called *saprotroph.*

deductive reasoning Reasoning that operates from generalities to specifics and can make relationships among data more apparent. Compare with *inductive reasoning.*

deforestation The temporary or permanent clearance of large expanses of forest for agriculture or other uses.

deoxyribonucleic acid (DNA) A nucleic acid present in a cell's chromosomes that contains genetic information.

dermal tissue system The tissue system that provides an outer covering for the plant body.

desert A temperate or tropical biome in which lack of precipitation limits plant growth.

desertification Degradation of once-fertile rangeland (or tropical dry forest) to nonproductive desert; caused partly by soil erosion, deforestation, and overgrazing.

determinate growth Growth that stops after a certain size is reached; examples include the growth of flowers and leaves. Compare with *indeterminate growth.*

deuteromycetes One of an artificial grouping of fungi characterized by the absence of sexual reproduction but usually having other characters similar to those of ascomycetes.

development The orderly sequence of progressive changes in the life of an organism.

diatom One of a usually unicellular protist, covered by ornate siliceous shells and containing the brown pigment fucoxanthin; an important component of plankton in both marine and fresh water.

dichotomous A type of branching in which a stem always branches into two approximately equal parts.

dichotomous key A special guide that aids in the identification of an unknown plant; consists of a series of paired, contrasting statements.

diffusion The net movement of particles (atoms, molecules, or ions) from a region of higher concentration to a region of lower concentration, resulting from random motion.

dihybrid cross A genetic cross that involves individuals differing in *two* (nonallelic) genes.

dinoflagellate A unicellular, biflagellated, typically marine protist that is usually photosynthetic, contains the brown pigment fucoxanthin, and is an important component of plankton.

dioecious Having male and female reproductive structures on separate plants. Compare with *monoecious*.

diploid (2n) The condition of having two sets of chromosomes per nucleus. Compare with *haploid*. Also see *polyploid*.

disaccharide A sugar such as sucrose that consists of two monosaccharide subunits.

division See *phylum*.

DNA See *deoxyribonucleic acid*.

DNA cloning The process of selectively amplifying DNA sequences so their structure and function can be studied.

DNA microarray A diagnostic test involving thousands of DNA molecules placed on a glass slide or chip.

DNA polymerase An enzyme complex that catalyzes DNA replication by adding nucleotides to a growing strand of DNA.

DNA replication See *semiconservative replication*.

DNA sequencing Procedure by which the sequence of nucleotides in DNA is determined.

domain A taxonomic category that includes one or more kingdoms.

domain Archaea The domain of unicellular, prokaryotic organisms adapted to extreme conditions, such as very hot or very salty environments.

domain Bacteria The domain of metabolically diverse, unicellular, prokaryotic organisms.

domain Eukarya The domain that includes all eukaryotic organisms (protists, plants, fungi, and animals).

dominant Said of an allele that is always expressed when it is present. Compare with *recessive*. Also see *principle of dominance*.

dormancy A temporary period of arrested growth in plants or plant parts, such as spores, seeds, bulbs, and buds.

double fertilization A process in the flowering plant life cycle in which there are two fertilizations; one fertilization results in formation of a zygote, and the second results in formation of endosperm.

double helix The structure of DNA, which consists of two polynucleotide chains twisted around one another.

drupe A simple, fleshy fruit in which the inner wall of the fruit is hard and stony. Peaches and cherries are drupes.

ecological niche See *niche, ecological*.

ecological succession See *succession, ecological*.

ecology A discipline of biology that studies the interrelations between living things and their environments.

economic botany The branch of botany that deals with important food crops, products obtained from plant parts, and ornamental plants.

ecosystem The interacting system that encompasses a community and its nonliving, physical environment.

ecosystem service An important environmental service that ecosystems provide; examples include clean air to breathe, clean water to drink, and fertile soil in which to grow crops.

ectomycorrhizal fungus A fungus that forms mycorrhizae consisting of a dense sheath over a root's surface.

electron acceptor molecule A molecule in an electron transport chain that alternately accepts and releases electrons.

electron transport chain A series of chemical reactions during which hydrogens or their electrons are passed from one acceptor molecule to another, with the release of energy.

embryo sac The female gametophyte generation in flowering plants.

endangered species A species whose numbers are so severely reduced that it is in imminent danger of extinction throughout all or a part of its range.

endemic species Localized, native species that are not found anywhere else in the world.

endodermis The innermost layer of the cortex of a plant root that prevents water and dissolved materials from entering the xylem by passing between cells.

endomycorrhizal fungus A fungus that forms mycorrhizae that extend into plant roots.

endoplasmic reticulum (ER) An organelle composed of an interconnected network of internal membranes within eukaryotic cells; rough ER is associated with ribosomes, whereas smooth ER lacks ribosomes.

endosperm The 3n nutritive tissue that is formed at some point in the development of all angiosperm seeds.

endospore A highly resistant resting structure that forms within cells of certain prokaryotes.

endosymbiont theory See *serial endosymbiosis, hypothesis of*.

enhanced greenhouse effect The additional warming of Earth's climate produced by increased levels of greenhouse gases, which absorb infrared radiation.

environmental sustainability See *sustainability, environmental*.

enzyme An organic catalyst, produced within an organism, that accelerates specific chemical reactions.

epidermis The outermost tissue layer, usually one cell thick, that covers the primary plant body—that is, leaves and young stems and roots.

ethnobotany The branch of botany that deals with the uses of plants by indigenous peoples. This knowledge is disappearing rapidly as native peoples adopt modern lifestyles.

ethylene A gaseous plant hormone involved in growth and development, including leaf abscission and fruit ripening.

eudicot One of two main classes of flowering plants; eudicot seeds contain two cotyledons. Compare with *monocot*.

euglenoid A mostly freshwater, flagellated, unicellular protist that moves by an anterior flagellum and is usually photosynthetic.

eukaryote An organism whose cells possess organelles surrounded by membranes.

eukaryotic cell A cell that possesses a nucleus and other organelles surrounded by membranes. Compare with *prokaryotic cell*.

evergreen Shedding leaves over a long time period, so that some leaves are always present. Compare with *deciduous*.

evolution Cumulative genetic changes in a population of organisms from generation to generation.

exponential population growth The accelerating population growth rate that occurs when optimal conditions allow a constant growth rate.

extinction The elimination of a species, either locally (that is, no longer living in an area where it used to be found but still found in other areas) or globally (no longer living anywhere in the world).

facilitated diffusion The diffusion of materials from a region of higher concentration to a region of lower concentration through special passageways in the membrane.

facilitation A situation in which one species has a positive effect on other species in a community, for example, by enhancing the local environment.

facultative anaerobe An organism capable of growing and metabolizing in either the presence or absence of molecular oxygen. Compare with *aerobe* and *obligate anaerobe.*

family A taxonomic group composed of related, similar genera.

fatty acid An organic acid composed of a long, unbranched chain of carbon and hydrogen atoms with a —COOH group at one end. Fatty acids linked to glycerol form a neutral fat or oil.

fecal coliform test A water-quality test for the presence of *Escherichia coli* (fecal bacteria that are common in the intestinal tracts of humans and animals). The presence of fecal bacteria in a water supply indicates the possible presence of pathogens as well.

fermentation An anaerobic pathway in which organic fuel molecules such as glucose are incompletely broken down to yield energy for biological work; examples include alcohol fermentation and lactate fermentation.

fern A seedless vascular plant that reproduces by spores produced in sporangia; undergoes an alternation of generations between a dominant sporophyte and a gametophyte (prothallus).

fertilization Fusion of male and female gametes.

fibrous root system A root system consisting of several adventitious roots of approximately equal size that arise from the base of the stem. Compare with *taproot system.*

first law of thermodynamics Energy cannot be created or destroyed, although it can be transformed from one form to another. Compare with *second law of thermodynamics.*

florigen A hypothetical hormone thought to promote flowering.

fluid mosaic model The current model for the structure of the plasma membrane and other cell membranes in which protein molecules "float" in a fluid phospholipid bilayer.

follicle A simple, dry fruit that splits open along one seam to liberate the seeds.

food chain The successive series of organisms through which energy flows in an ecosystem. Each organism in the series eats or decomposes the preceding organism in the chain. See *food web.*

food web A complex interconnection of all the food chains in an ecosystem.

forest decline A gradual deterioration (and often death) of many trees in a forest. The cause of forest decline is unclear and may involve a combination of factors.

fossil A part or trace of an ancient organism, usually preserved in sedimentary rock.

founder effect Genetic drift that results from a small population colonizing a new area.

frond The leaf of a fern.

fruit In flowering plants, a mature, ripened ovary that often provides protection and dispersal for the enclosed seeds.

fruiting body A complex reproductive structure in which the sexual spores of certain fungi are produced; refers to the ascocarp of an ascomycete and the basidiocarp of a basidiomycete.

gametangium A unicellular or multicellular structure of plants, protists, or fungi in which gametes form.

gamete A haploid reproductive cell—that is, an egg or a sperm cell. In sexual reproduction, the union of gametes results in the formation of a zygote.

gametophyte The *n,* gamete-producing stage in the life cycle of a plant. Compare with *sporophyte.*

gemma (pl., *gemmae*) A small body of tissue that becomes detached from a liverwort thallus and is capable of developing into a new gametophyte.

gene A discrete unit of hereditary information that usually specifies a polypeptide (protein).

gene flow The movement of alleles between local populations due to migration and subsequent interbreeding.

gene gun A nonbiological vector used to introduce recombinant DNA "bullets" into a plant cell.

gene pool All the alleles of all the genes in a freely interbreeding population.

genetic code A code, consisting of groups of three nucleotide bases in mRNA, that specifies an amino acid or a translation start or stop signal.

genetic drift A random change in allele frequencies in a small breeding population.

genetic engineering Manipulation of genes, often through recombinant DNA technology.

genetic equilibrium The condition of a population that is not undergoing evolutionary change, that is, in which allele and genotype frequencies do not change from one generation to the next.

genetic probe A single-stranded nucleic acid used to identify a complementary sequence by base-pairing with it.

genetic recombination See *recombination.*

genetically modified (GM) crop A crop plant that has had its genes intentionally manipulated; a transgenic crop plant.

genome All the genetic material contained in an individual.

genomics The field of biology that studies the genomes of various organisms.

genotype The genetic makeup of an individual. Compare with *phenotype.*

genotype frequency The proportion of a particular genotype in the population.

genus (pl., *genera*) A taxonomic group composed of related, similar species.

germplasm Any plant or animal material that may be used in breeding; includes seeds, plants, and plant tissues of traditional crop varieties and the sperm and eggs of traditional livestock breeds.

gibberellin A plant hormone involved in growth and development, including stem elongation, flowering, and seed germination.

ginkgo The sole surviving species (*Ginkgo biloba*) of an ancient phylum of gymnosperms that bears exposed seeds—that is, seeds not surrounded by a cone. Ginkgoes are hardy, deciduous trees with broad, fan-shaped leaves and, on the females, exposed, fleshy seeds.

global warming The current warming trend in Earth's climate.

glomeromycete A fungus that forms a distinctive branching form (arbuscular mycorrhizae) of endomycorrhizae.

glycolysis The first stage of cellular respiration, in which glucose is split into two molecules of pyruvate with the production of a small amount of ATP.

gnetophyte Any of the small phylum of unusual gymnosperms that possess many features similar to those of flowering plants, such as vessels in xylem.

golden algae A protist that is biflagellated, unicellular, and photosynthetic and contains the brown pigment fucoxanthin.

Golgi body An organelle composed of a stack of flattened membranous sacs that modifies, packages, and sorts proteins that will be secreted or sent to the plasma membrane or other organelles. The collective term for all the Golgi bodies in a cell is Golgi apparatus.

gradualism A model of evolution in which the evolutionary changes in a species are due to a slow, steady transformation over time. Compare with *punctuated equilibrium.*

grain A simple, dry fruit in which the fruit wall is fused to the seed coat, making it impossible to separate the fruit from the seed. Corn kernels are grains; also called *caryopsis.*

Gram staining procedure The application of a stain that allows bacteria to be grouped as either gram-positive or gram-negative; used to help identify unknown bacteria.

gravitropism Plant growth in response to the direction of gravity.

green alga One of a diverse group of protists that contains the same pigments as land plants (chlorophylls *a* and *b* and carotenoids).

greenhouse effect See *enhanced greenhouse effect.*

ground tissue The tissue in monocot stems in which the vascular tissues are embedded; performs the same functions as cortex and pith in herbaceous eudicot stems.

ground tissue system All of the tissues of the plant body other than the vascular tissues and the dermal tissues.

growth Increases in the size and mass of an organism.

guard cell A cell in the epidermis of a stem or leaf; two guard cells form a pore, called a stoma, for gas exchange.

guttation Exudation of water droplets from leaves; caused by root pressure.

gymnosperm Any of a group of seed plants in which the seeds are not enclosed in an ovary.

habitat fragmentation The division of habitats that formerly occupied large, unbroken areas into smaller pieces by roads, fields, cities, and other human land-transforming activities.

haploid (*n*) The condition of having one set of chromosomes per nucleus. Compare with *diploid.* Also see *polyploid.*

Hardy–Weinberg principle The mathematical prediction that allele and genotype frequencies do not change from generation to generation in the absence of microevolutionary processes.

heartwood The older wood in the center of many tree trunks that no longer functions in conduction; contains substances (pigments, tannins, gums, and resins) that color it a brownish red. Compare with *sapwood.*

herbal A book that describes plants, published during the 15th, 16th, and 17th centuries.

herbarium A collection of dried, pressed, and carefully labeled plant specimens used for scientific study.

heterospory Production of two types of *n* spores, microspores and megaspores. Compare with *homospory.*

heterotroph An organism that obtains the carbon it needs from organic compounds in other organisms. Compare with *autotroph.*

heterozygous Possessing a pair of unlike alleles for a particular gene. Compare with *homozygous.*

holdfast A basal structure for attachment to solid surfaces, found in multicellular algae.

homologous chromosomes Members of a chromosome pair that are similar in size, shape, and genetic constitution.

homologous features Dissimilar structures with an underlying similarity of form and development that occur in different species with a common ancestry. Compare with *homoplastic features.*

homoplastic features Structures in unrelated species that are similar in function and appearance but not in evolutionary origin. Compare with *homologous features.*

homospory Production of one type of *n* spore that gives rise to a bisexual gametophyte. Compare with *heterospory.*

homozygous Possessing a pair of identical alleles for a particular gene. Compare with *heterozygous.*

hormone An organic chemical messenger that regulates growth and development in plants and other multicellular organisms.

hornwort A spore-producing, nonvascular thalloid plant with a life cycle similar to that of mosses.

humus Black or dark brown decomposed organic material.

hybrid See *interspecific hybrid.*

hydathode A special pore that exudes water from a leaf by guttation.

hydrogen bond An attraction between a slightly positive hydrogen atom in one molecule and a slightly negative atom (usually oxygen) in another molecule.

hydrophilic Attracted to water. Compare with *hydrophobic.*

hydrophobic Repelled by water. Compare with *hydrophilic.*

hydroponics Cultivation of plants in an aerated solution of dissolved inorganic minerals—that is, without soil.

hypertonic Having a solute concentration greater than that of another solution. Compare with *isotonic* and *hypotonic.*

hypha (pl., *hyphae*) One of the threadlike filaments composing the mycelium of a fungus.

hypothesis An educated guess (based on previous observations) that may be true and is testable by observation and experimentation.

hypotonic Having a solute concentration less than that of another solution. Compare with *hypertonic* and *isotonic.*

imbibition Absorption of water by, and subsequent swelling of, a dry seed prior to germination.

inbreeding The mating of genetically similar individuals. Homozygosity increases with each successive generation of inbreeding.

incomplete dominance A condition in which neither member of a pair of contrasting alleles is completely expressed when the other is present.

independent assortment See *principle of independent assortment.*

indeterminate growth Unrestricted growth. Examples include the growth of stems and roots, both of which arise from apical meristems. Compare with *determinate growth.*

inductive reasoning Reasoning that uses specific examples to draw a general conclusion. Compare with *deductive reasoning.*

inflorescence A cluster of flowers on a common floral stalk.

integument The outer layer of an ovule that develops into a seed coat following fertilization.

intermediate-day plant A plant that flowers when it is exposed to days and nights of intermediate length but does not flower when day length is too long or too short. Compare with *long-day, short-day,* and *day-neutral plant.*

internode The area on a stem between two successive nodes. Compare with *node.*

interphase The stage of the cell cycle between successive mitotic divisions.

interspecific hybrid The offspring of individuals belonging to different species.

intrinsic rate of increase The theoretical maximum rate of increase in population size occurring under optimal environmental conditions.

invasive species A foreign species that when introduced into an area where it is not native upsets the balance among the organisms living there and causes economic or environmental harm.

ionic bond An electrostatic attraction between oppositely charged ions.

isotonic Having concentrations of solute and solvent molecules identical to those of another solution. Compare with *hypertonic* and *hypotonic.*

jasmonate One of a group of lipid-derived plant hormones that affect several processes, such as pollen development, root growth, fruit ripening, and senescence; also involved in plant defense against insect pests and disease-causing organisms.

karyogamy Fusion of two haploid nuclei.

kinetochore The portion of the chromosome centromere to which the mitotic spindle fibers attach.

kingdom A broad taxonomic category made up of related phyla; many biologists currently recognize six kingdoms of living organisms.

Krebs cycle See *citric acid cycle.*

krummholz The gnarled, shrublike growth habit found in trees at high elevations, near their upper limit of distribution.

landscape A large land area (several to many square kilometers) composed of interacting ecosystems.

lateral meristem An area of cell division on the side of a vascular plant; the two lateral meristems (vascular cambium and cork cambium) give rise to secondary tissues.

leaching The process by which dissolved materials are washed away or carried with water down through various layers of the soil.

legume A simple, dry fruit that splits open along two seams to release its seeds. Pea pods and green beans are legumes.

lenticel A porous swelling in a woody stem that develops when the epidermis is replaced by periderm; facilitates the exchange of gases between the stem's interior and the atmosphere.

lichen A compound organism consisting of a symbiotic fungus and an alga or cyanobacterium.

ligase An enzyme used in genetic engineering to join pieces of DNA with sticky ends.

limiting resource Whatever environmental variable (such as water, sunlight, or some essential mineral) tends to restrict the growth, distribution, or abundance of a particular plant population.

linkage The grouping of genes on the same chromosome; linked genes tend to be inherited together in successive generations.

lipid Any of a group of organic compounds that are insoluble in water but soluble in fat solvents.

liverwort A spore-producing, nonvascular, thalloid or leafy plant with a life cycle similar to that of mosses.

loam A soil that has approximately 40 percent each of sand and silt and about 20 percent of clay. A loamy soil that also contains organic material makes an excellent agricultural soil.

locus (pl., *loci*) The location of a particular gene on a chromosome.

logistic population growth Population growth that initially occurs at a constant rate of increase over time (i.e., exponential) but then levels out as the carrying capacity of the environment is approached.

long-day plant A plant that flowers when the night length is equal to or less than some critical length; also called *short-night plant.* Compare with *short-day, intermediate-day,* and *day-neutral plant.*

long-night plant See *short-day plant.*

lycophyte A seedless vascular plant, some of which are heterosporous; includes club mosses, spike mosses, and quillworts.

lysogenic conversion The change in properties of bacteria that results from the presence of a prophage.

lysogenic cycle The life cycle of a virus in which viral DNA becomes incorporated into the host cell's genetic material.

lytic cycle The life cycle of a virus that kills the host cell by lysing it.

macronutrient An essential element that is required in a fairly large amount for normal plant growth. Compare with *micronutrient.*

magnoliid One of the groups of flowering plants; magnoliids are core angiosperms that were once classified as "dicots," but molecular evidence indicates that they are neither eudicots nor monocots.

mass extinction The extinction of many species during a relatively short period of geologic time.

megaphyll Type of leaf found in virtually all vascular plants except lycophytes; contains multiple vascular strands. Compare with *microphyll.*

megaspore The *n* spore in heterosporous plants that gives rise to a female gametophyte. Compare with *microspore.*

meiosis Process in which a 2*n* cell undergoes successive nuclear divisions, potentially producing four *n* nuclei; leads to formation of spores in plants.

mesophyll The photosynthetic tissue in the interior of a leaf.

messenger RNA (mRNA) RNA that specifies the amino acid sequence of a polypeptide. Compare with *transfer RNA* and *ribosomal RNA.*

metabolism The sum of all of the chemical processes that occur within a cell or organism; includes both anabolic and catabolic reactions.

metapopulation A population that is divided into several local populations among which individuals occasionally disperse.

microevolution Small-scale evolutionary changes caused by changes in allele or genotype frequencies in a population over a few generations.

micronutrient An essential element that is required in very small amounts for normal plant growth. Compare with *macronutrient.*

microphyll Type of leaf found in lycophytes; contains one vascular strand. Compare with *megaphyll*.

microspore The *n* spore in heterosporous plants that gives rise to a male gametophyte. Compare with *megaspore*.

microsporidium A small, unicellular, fungal parasite that infects eukaryotic cells; classified with the zygomycetes.

microtubule A hollow, cylindrical fiber that is a major component of the cytoskeleton and is found in mitotic spindles, cilia, and flagella in eukaryotic cells.

mimicry An adaptation for survival in which an organism resembles another organism or a nonliving object.

minimum viable population The smallest population size at which a species has a high chance of sustaining its numbers and surviving into the future.

mitochondrion An intracellular organelle associated with aerobic respiration; provides the cell with ATP.

mitosis The division of the cell nucleus resulting in two daughter nuclei, each with the same number of chromosomes as the parent nucleus.

modern synthesis A comprehensive, unified explanation of evolution based on combining previous theories, especially of Mendelian genetics, with Darwin's theory of evolution by natural selection.

mold The vegetative body of most fungi, consisting of long, branched threads called hyphae.

monocot One of two main classes of flowering plants; monocot seeds contain a single cotyledon. Compare with *eudicot*.

monoecious Having male and female reproductive parts in separate flowers or cones on the same plant. Compare with *dioecious*.

monohybrid cross A genetic cross involving individuals that differ in a single pair of alleles.

monophyletic Said of a group consisting of organisms that evolved from a common ancestor. Compare with *paraphyletic* and *polyphyletic*.

monosaccharide A simple sugar, such as glucose or fructose.

moss A spore-producing nonvascular plant in which the dominant *n* gametophyte alternates with a *2n* sporophyte that remains attached to the gametophyte.

mulch Material placed on the surface of soil around the bases of plants to help maintain soil moisture and reduce erosion. Organic mulches have the advantage of decomposing over time, thereby enriching the soil.

multiple fruit A fruit that develops from the carpels of closely associated flowers that fuse, or grow together.

mutation A change in the nucleotide sequence of DNA of an organism.

mutualism A symbiotic relationship from which both partners benefit. Compare with *parasitism* and *commensalism*.

mycelium The vegetative (nonreproductive) body of most fungi, consisting of a branched network of hyphae.

mycorrhiza A mutually beneficial association between a fungus and a root that helps the plant absorb essential minerals from the soil.

mycotoxin A poisonous chemical compound produced by fungi.

natural selection The mechanism of evolution proposed by Charles Darwin; the tendency of organisms that have favorable adaptations to their environment to survive and become the parents of the next generation.

needle The leaf of a conifer, such as pine.

niche, ecological The totality of an organism's adaptations, its use of resources, its interactions with other organisms, and the lifestyle to which it is fitted in its community.

nitrogen-fixing bacterium A gram-negative bacterium that can convert atmospheric nitrogen (N_2) to ammonia (NH_3), a form plants can use.

node The area on a stem where one or more leaves is attached; stems have nodes, but roots do not. Compare with *internode*.

nodule A small swelling on the root of a leguminous plant in which beneficial nitrogen-fixing bacteria (*Rhizobium*) live.

noncyclic electron transport In photosynthesis, the linear flow of electrons, produced by the splitting of water molecules, through photosystems I and II; results in the formation of ATP, NADPH, and O_2.

nonpolar covalent bond A chemical bond in which electrons are shared equally among the participating atoms; does not produce any electric charge within the molecule. Compare with *polar covalent bond*.

nonvascular plant A plant that lacks conducting tissues. Compare with *vascular plant*.

nucleic acid A large, complex organic molecule composed of nucleotides; the two nucleic acids are deoxyribonucleic (DNA) and ribonucleic acid (RNA).

nucleolus A specialized structure in the nucleus, formed from regions of several chromosomes; the site of ribosome synthesis.

nucleotide A molecule composed of a phosphate group, a five-carbon sugar (ribose or deoxyribose), and a nitrogenous base; one of the subunits of nucleic acids.

nucleus (1) The portion of an atom that contains the protons and neutrons. (2) A cellular organelle that contains DNA and serves as the control center of the cell.

nut A simple, dry fruit that has a stony wall, is usually large, and does not split open at maturity. Chestnuts, acorns, and hazelnuts are nuts.

obligate anaerobe An organism that grows and metabolizes only in the absence of molecular oxygen. Compare with *aerobe* and *facultative anaerobe*.

oligosaccharin One of several signaling molecules in plants that trigger the production of phytoalexins and affect aspects of growth and development.

order A taxonomic group composed of related, similar families.

organic compound A compound that contains carbon and usually hydrogen.

osmosis The net movement of water (the principal solvent in biological systems) by diffusion through a selectively permeable membrane.

ovary In flowering plants, the base of a carpel or fused carpels that contains ovules and develops into a fruit after fertilization.

ovule The structure in the ovary that contains a female gametophyte and develops into a seed after fertilization.

oxidation The loss of electrons from a compound. Compare with *reduction*.

oxidation–reduction reaction A chemical reaction in which electrons are transferred from one electron acceptor molecule to another.

ozone A blue gas, O_3, with a distinctive odor. Ozone is a human-made pollutant in the lower atmosphere but a natural and essential component in the stratosphere.

paleobotany The branch of botany that deals with plant evolution, including the groups of extinct plants.

palisade mesophyll Columnar leaf mesophyll cells that occur in one or more vertical stacks adjacent to the upper epidermis.

paraphyletic Said of a group consisting of a common ancestor and some, but not all, of its descendants. Compare with *monophyletic* and *polyphyletic*.

parasitism A symbiotic relationship in which one member (the parasite) benefits and the other (the host) is adversely affected. Compare with *mutualism* and *commensalism*.

parenchyma cell A plant cell that is relatively unspecialized and thin walled, may contain chlorophyll, and is typically rather loosely packed.

peduncle The stalk of a flower or inflorescence.

peptide bond A distinctive covalent carbon-to-nitrogen bond that links the amino acids in proteins.

peptidoglycan A component of the bacterial cell wall; consists of a modified protein or peptide molecule with an attached carbohydrate.

perennial A plant that lives for more than 2 years. Compare with *annual* and *biennial*.

pericycle A layer of cells just inside the endodermis of the root; gives rise to lateral roots.

periderm The outermost layer of cells covering a woody stem or root—that is, the outer bark that replaces epidermis when it is destroyed during secondary growth.

permafrost Permanently frozen subsoil that is characteristic of frigid areas such as the tundra.

petal One of the often conspicuously colored parts of a flower attached inside the whorl of sepals.

petiole The part of a leaf that attaches the blade to the stem.

pH scale A number from 0 to 14 that indicates the degree of acidity or alkalinity of a substance.

phage See *bacteriophage*.

pharmacogenetics A new field of gene-based medicine in which drugs are personalized to match a patient's genetic makeup.

phelloderm See *cork parenchyma*.

phenotype The physical expression of an individual's genes. Compare with *genotype*.

phloem A complex vascular tissue that conducts food (carbohydrate) throughout the plant body.

photon A particle of radiant energy with a wavelength in the visible range of the electromagnetic spectrum.

photoperiodism The physiological response (such as flowering) of plants to variations in the length of daylight and darkness.

photosynthesis The biological process that includes the capture of light energy and its transformation into chemical energy of organic molecules (such as glucose), which are manufactured from carbon dioxide and water.

phototropism The directional growth of a plant caused by light.

phylogeny The evolutionary history of a species or other taxonomic group.

phylum A taxonomic group composed of related, similar classes; a category beneath the kingdom and above the class. Some botanists use the equivalent term *division* when classifying plants.

phytoalexin An antimicrobial compound produced by plants that limits the spread of disease-causing organisms such as fungi.

phytochrome A blue-green proteinaceous pigment involved in many plant responses to light, independent of photosynthesis.

pilus (pl., *pili*) One of numerous hairlike structures on the surface of a prokaryotic cell.

pistil The female part of a flower; consists of either a single carpel (a simple pistil) or two or more fused carpels (a compound pistil).

pith Ground tissue found in the centers of many stems and roots; composed of parenchyma cells.

plankton Free-floating, mainly microscopic aquatic organisms found in the upper layers of the ocean or bodies of fresh water.

plant biology See *botany*.

plasma membrane The living surface membrane of a cell that acts as a selective barrier to the passage of materials into and out of the cell.

plasmid A small, circular DNA molecule that carries genes separate from the main DNA of a prokaryote.

plasmodesma (pl., *plasmodesmata*) A cytoplasmic channel connecting adjacent plant cells and allowing for the movement of molecules and ions between cells.

plasmodial slime mold A protist whose feeding stage consists of a multinucleate, amoeboid plasmodium.

plasmodium A multinucleate, amoeboid mass of living matter that constitutes the feeding phase of plasmodial slime molds.

plasmogamy Fusion of the cytoplasm of two cells without fusion of nuclei.

plastid One of a group of membrane-bounded organelles occurring in photosynthetic eukaryotic cells; includes chloroplasts, leucoplasts, and chromoplasts.

pneumatophore A specialized aerial root produced by certain trees living in swampy habitats; may facilitate gas exchange between the atmosphere and submerged roots.

polar covalent bond A chemical bond in which electrons are shared unequally among the participating atoms, resulting in some difference in charge at different areas of the molecule. Compare with *nonpolar covalent bond*.

polar nucleus In flowering plants, one of two haploid cells in the embryo sac that fuse with a sperm cell during double fertilization to form the triploid endosperm.

pollen grain The structure in seed plants that develops from a microspore into a male gametophyte.

pollen tube In seed plants, a tube that forms after the germination of a pollen grain and through which male gametes (sperm cells) pass into the ovule.

pollination In seed plants, the transfer of pollen grains from the anther to the stigma.

pollinium In orchids, a mass of pollen grains transported as a unit during pollination.

polygene One of two or more pairs of genes that affect the same character in additive fashion.

polymerase chain reaction (PCR) A method by which a targeted DNA fragment is amplified in vitro to produce millions of copies.

polypeptide A chain of many amino acids linked by peptide bonds. A protein molecule consists of one or more chains of polypeptides.

polyphyletic Said of a group consisting of organisms that evolved from two or more different ancestors. Compare with *monophyletic* and *paraphyletic*.

polyploid The condition of having more than two sets of chromosomes per nucleus.

polysaccharide A carbohydrate consisting of many monosaccharide subunits. Examples are starches and cellulose.

population A group of organisms of the same species that live in a defined geographic area at the same time.

population dynamics The study of changes in populations, such as how and why population numbers change over time.

population ecology That branch of biology that deals with the numbers of a particular species that are found in an area and how and why those numbers change (or remain fixed) over time.

population genetics The study of genetic variability in a population and of the forces that act on it.

potassium (K$^+$) ion mechanism The mechanism by which plants open and close their stomata: the influx of potassium ions into the guard cells causes water to move in by osmosis, changing the shape of the guard cells and opening the pore.

predation A relationship in which one organism (the predator) devours another organism (the prey).

pressure–flow hypothesis The mechanism by which dissolved sugar may be transported in phloem; caused by a pressure gradient between the source (where sugar is loaded into the phloem) and the sink (where sugar is removed from the phloem).

primary growth An increase in a plant's length, which occurs at the tips of stems and roots due to the activity of apical meristems. Compare with *secondary growth*.

primary succession An ecological succession that occurs on land not previously inhabited by plants; no soil is present initially. Compare with *secondary succession*.

principle of dominance When two different alleles are present in an individual, often only one—the dominant allele—is expressed.

principle of independent assortment The chance distribution of alleles to the gametes; the distribution of the alleles governing one character has no influence on the distribution of a different set of alleles.

principle of segregation The two alleles of each gene separate into different gametes; each gamete receives only one allele of each pair.

prion An infectious agent composed only of protein.

producer An organism that synthesizes organic compounds from simple inorganic raw materials. Compare with *consumer*.

progymnosperm One of an extinct group of plants thought to have been the ancestors of gymnosperms.

prokaryotic cell A cell that lacks nuclei and other membrane-bounded organelles; the archaea and bacteria. Compare with *eukaryotic cell*.

prop root An adventitious root that arises from the stem and provides additional support for the plant.

prophage Bacteriophage nucleic acid that is inserted into the bacterial DNA.

protective coloration The coloring of an organism so that it blends into its surroundings in such a way that it is difficult to see.

protein A large, complex organic compound composed of amino acid subunits. See *polypeptide*.

proteomics The study of all the proteins encoded in a genome and produced in an individual's cells and tissues.

prothallus The free-living, haploid gametophyte plant in ferns and other seedless vascular plants.

proton gradient The difference in concentration of protons on the two sides of a cell membrane; contains potential energy that can be used to form ATP or do work in the cell.

protonema In mosses, a filament of n cells that grows from a spore and develops into leafy moss gametophytes.

pseudobulb A thickened aerial stem found in certain orchids.

pseudoplasmodium In cellular slime molds, an aggregation of amoeboid cells to form a spore-producing fruiting body during reproduction.

punctuated equilibrium An evolutionary model in which evolution proceeds with long periods of inactivity followed by very active phases, so major adaptations appear suddenly in the fossil record. Compare with *gradualism*.

rain shadow An area on the downwind side of a mountain range that receives very little precipitation. Deserts often occur in rain shadows.

random dispersion The spatial distribution pattern of a population in which the presence of one individual has no effect on the distribution of other individuals. Compare with *clumped dispersion* and *uniform dispersion*.

range The geographic area where a particular species occurs.

ray A chain of parenchyma cells that extends radially in the stems and roots of woody plants and functions for lateral transport of carbohydrates, water, and minerals.

receptacle The part of a flower stalk where the floral parts are attached.

recessive Said of an allele that is not expressed in the presence of a dominant allele. Compare with *dominant*.

recombinant DNA Any DNA molecule made by combining genes from different organisms.

recombinant DNA technology The techniques used to make DNA molecules by combining genes from different organisms.

recombination The appearance of new gene combinations; recombination in eukaryotes generally results from meiotic events, either crossing over or shuffling of chromosomes; also called *genetic recombination*.

red alga One of a diverse group of photosynthetic protists that contains the pigments phycocyanin and phycoerythrin.

red tide A red or brown coloration of ocean water caused by a population explosion, or bloom, of dinoflagellates.

reduction The gain of electrons by a compound. Compare with *oxidation*.

reproductive isolation The situation in which reproductive barriers prevent members of a species from successfully interbreeding with members of another species.

resin A clear, or translucent, viscous organic material that certain plants produce and secrete into specialized ducts.

Resin may play a role in deterring plant-eating insects or disease-causing organisms. See *resin duct*.

resin duct One of a series of tubelike canals that occur in the xylem, phloem, and bark of many conifers.

respiration See *cellular respiration*.

restoration ecology The scientific field that uses the principles of ecology to help return a degraded environment as closely as possible to its former undisturbed condition.

restriction enzyme An enzyme used in recombinant DNA technology to cleave DNA at a specific base sequence.

retrovirus An RNA virus that uses reverse transcriptase to produce a DNA intermediate.

reverse transcriptase An enzyme produced by retroviruses that catalyzes the production of DNA using RNA as a template.

rhizoid One of the hairlike absorptive filaments that form part of the gametophytes of bryophytes and some vascular plants.

rhizome A horizontal underground stem that often serves as a storage organ and a means of asexual reproduction.

ribonucleic acid (RNA) A single-stranded nucleic acid molecule that is necessary for protein synthesis. There are three forms of RNA: messenger RNA, transfer RNA, and ribosomal RNA.

ribosomal RNA (rRNA) An important part of the structure of ribosomes that also has catalytic functions needed during protein synthesis. Compare with *transfer RNA* and *messenger RNA*.

ribosome A cellular organelle that is the site of protein synthesis.

RNA See *ribonucleic acid*.

RNA interference Certain small RNA molecules that interfere with the expression of genes or their RNA transcripts.

RNA polymerase An enzyme that catalyzes the synthesis of RNA molecules from DNA templates.

root cap A covering of cells over the root tip that protects the delicate meristematic tissue directly behind it.

root graft The union of roots of two different plants of either the same or different species.

root hair An extension of an epidermal cell of a root that increases the absorptive capacity of the root.

root pressure The pressure in xylem sap that occurs as a result of water moving into roots from the soil.

root system The part of the vascular plant body that is normally found underground; anchors the plant and absorbs water and minerals from the soil. Compare with *shoot system*.

rubisco The enzyme in photosynthesis (the Calvin cycle) that catalyzes the addition of carbon dioxide to the five-carbon sugar, ribulose bisphosphate.

salicylic acid A signaling molecule that helps plants defend against insect pests and pathogens such as viruses.

sand Inorganic soil particles that are larger than clay or silt.

saprotroph See *decomposer*.

sapwood The outermost, youngest wood in many tree trunks that conducts water and dissolved minerals. Compare with *heartwood*.

saturated fatty acid A fatty acid that has no carbon-to-carbon double bonds. Compare with *unsaturated fatty acid*.

savanna A tropical grassland with widely scattered trees that is found in areas of low rainfall and areas of seasonal rainfall with prolonged dry periods.

scientific method The steps a scientist uses to approach a problem: making observations, stating the problem, developing a hypothesis, performing an experiment, and using the results to support or refute the hypothesis or to generate other hypotheses.

sclerenchyma cell A plant cell with extremely thick walls that provides strength and support to the plant body.

sclerophyllous leaf A hard, small, leathery leaf that resists water loss; characteristic of perennial plants adapted to extremely dry habitats.

second law of thermodynamics When energy is converted from one form to another, some of it is degraded into a lower-quality, less useful form. Compare with *first law of thermodynamics*.

secondary growth An increase in a plant's girth due to the activity of lateral meristems (the vascular cambium and cork cambium). Compare with *primary growth*.

secondary succession An ecological succession that takes place after some disturbance destroys the existing vegetation; soil is already present. Compare with *primary succession*.

seed A reproductive body consisting of a young, multicellular plant and food reserves, that is enclosed by a seed coat.

seed fern One of an extinct group of seed-bearing woody plants with fernlike leaves. Seed ferns probably descended from progymnosperms and gave rise to cycads and possibly ginkgoes.

segregation See *principle of segregation*.

selective breeding The selection by humans of traits that are desired in plants, and the breeding of only those individuals that possess the desired traits; also called *artificial selection*.

self-incompatibility A genetic condition in which the pollen cannot fertilize the same flower or flowers on the same plant.

semiconservative replication The type of replication characteristic of DNA, in which each new double-stranded molecule consists of one strand from the original DNA molecule and one strand of newly synthesized DNA.

senescence The aging process.

sepal One of the outermost parts of a flower, usually leaflike in appearance, that protect the flower as a bud.

serial endosymbiosis, hypothesis of The hypothesis that certain eukaryotic organelles, such as mitochondria and chloroplasts, originated as symbiotic prokaryotes living inside larger prokaryotic cells.

sexual reproduction A type of reproduction in which two gametes (usually, but not necessarily, contributed by two different parents) fuse to form a zygote. Compare with *asexual reproduction*.

shade avoidance The tendency of plants that are adapted to high light intensities to grow taller when they are closely surrounded by other plants.

shoot system The part of the vascular plant body that is normally found aboveground; consists of the stem and leaves. Compare with *root system*.

short-day plant A plant that flowers when the night length is equal to or greater than some critical length; Also called *long-night plant*. Compare with *long-day*, *intermediate-day*, and *day-neutral plant*.

short-night plant See *long-day plant.*

sieve-tube element Cells that conduct dissolved sugar in the phloem of flowering plants.

signal transduction A process in which a cell converts and amplifies an extracellular signal into an intracellular signal that affects some function in the cell.

signaling See *cell signaling.*

silt Medium-sized inorganic particles. Compare with *clay* and *sand.*

simple fruit A fruit that develops from one or several united carpels.

simple leaf A leaf with a single blade. Compare with *compound leaf.*

sister chromatid See *chromatid.*

soil The surface layer of Earth's crust, consisting primarily of fragmented and weathered grains of rocks; supports terrestrial plants as well as many animals and microorganisms.

solar tracking The ability of certain plant parts, such as cotton leaves, to follow the movement of the sun across the sky; caused by turgor movements.

somatic cell A cell of the body not involved in reproduction; body cell.

sorus (pl., *sori*) In ferns, a cluster of spore-producing sporangia.

species A group of organisms with similar structural and functional characteristics that in nature breed only with one another and have a close common ancestry. See *biological species concept.*

species richness The number of species in a community.

spindle The structure consisting mainly of microtubules that provides the framework for chromosome movement during cell division.

spine A leaf modified for protection, such as a cactus spine.

spirillum (pl., *spirilla*) A corkscrew-shaped prokaryote.

spongy mesophyll Leaf mesophyll cells that are loosely arranged and contain many air spaces. In leaves where it occurs with palisade mesophyll, spongy mesophyll is adjacent to the lower epidermis.

sporangium (pl., *sporangia*) A spore case; the structure within which spores are formed in plants, certain protists, and fungi.

spore A reproductive cell that gives rise to individual offspring in plants, fungi, and certain algae and protozoa.

spore mother cell A diploid cell that undergoes meiosis to form haploid spores in plants.

sporophyll A leaflike structure that bears spores within a sporangium (or sporangia).

sporophyte The 2*n*, spore-producing stage in the life cycle of a plant. Compare with *gametophyte.*

stamen The pollen-producing part of a flower.

stele The central core or cylinder of vascular tissues (xylem and phloem) and associated ground tissues (pith and pericycle) in a root or stem. The structure of the stele varies among different groups of plants.

sticky end A single-stranded stub that extends from each end of a fragment of DNA treated with a restriction enzyme.

stigma The portion of the pistil where the pollen lands prior to fertilization.

stimulus A physical or chemical change in the internal or external environment of an organism, potentially capable of provoking a response.

stipule A leaflike outgrowth at the base of a petiole.

stolon An aerial horizontal stem with long internodes; often forms buds that develop into separate plants.

stoma (pl., *stomata*) A small pore flanked by guard cells in the epidermis; provides for gas exchange for photosynthesis.

strobilus (pl., *strobili*) In certain plants, a conelike structure that bears spore-producing sporangia.

style The neck connecting the stigma to the ovary of a pistil.

subspecies A taxonomic group below that of species.

succession, ecological The sequence of changes in the species composition of a community over time.

sucker A shoot that develops adventitiously from a root; a type of asexual reproduction.

sustainability, environmental The ability of the natural environment to meet humanity's needs indefinitely without going into a decline from human-caused stresses.

symbiosis An intimate relationship between two or more organisms of different species.

sympatric speciation Evolution of a new species that occurs within the parent species' geographic region.

symplast A continuum consisting of the cytoplasm of many plant cells, connected from one cell to the next by plasmodesmata. Compare with *apoplast.*

synapsis The physical association of homologous chromosomes during prophase I of meiosis.

syngamy The union of the gametes in sexual reproduction.

synthetic theory of evolution See *modern synthesis.*

systematics The scientific study of the diversity of organisms and their natural (evolutionary) relationships.

taiga See *boreal forest.*

taproot system A root system consisting of one prominent main root with smaller lateral roots branching from it. Compare with *fibrous root system.*

taxonomy The science of naming, describing, and classifying organisms.

temperate deciduous forest A forest biome that occurs in temperate areas where annual precipitation ranges from about 75 to 125 centimeters.

temperate grassland A grassland characterized by hot summers, cold winters, and less rainfall than is found in a temperate deciduous forest biome.

temperate rain forest A coniferous biome characterized by cool weather, dense fog, and high precipitation.

tendril A leaf or stem that is modified for holding on or attaching to objects.

tension–cohesion model The mechanism by which water and dissolved minerals may be transported in xylem: water is pulled upward under tension due to transpiration, while maintaining an unbroken column in xylem because of cohesion.

test cross A cross between an individual with a dominant phenotype (but an unknown genotype) and a homozygous recessive individual. The test cross helps determine the genotype of the dominant individual.

thallus The simple body of an alga, fungus, or nonvascular plant, for example, a liverwort thallus or a lichen thallus; in plants, a thallus is a body that lacks roots, stems, or leaves.

theory A widely accepted explanation that is supported by a large body of observations and experiments.

thigmotropism Plant growth in response to contact with a solid object.

threatened species A species in which the population is small enough for it to be at risk of becoming extinct but not so small that it is in imminent danger of extinction.

tissue A group of closely associated, similar cells that work together to carry out specific functions.

totipotency The ability of a cell (or nucleus) to provide information for the development of an entire organism.

tracheid A type of water-conducting and supporting cell in the xylem of vascular plants.

transcription Synthesis of RNA from a DNA template.

transfer RNA (tRNA) RNA that transfers amino acids to the ribosome during protein synthesis. Compare with *messenger RNA* and *ribosomal RNA.*

transformed Changed by the incorporation of a piece of foreign DNA. A prokaryote that takes in a plasmid carrying recombinant DNA is said to be transformed.

transgenic organism A plant or other organism that has foreign DNA incorporated into its genome.

translation Conversion of information provided by RNA to a specific sequence of amino acids in the production of a polypeptide chain.

translocation The movement of materials such as dissolved sugar through phloem.

transpiration Loss of water vapor from a plant's aerial parts.

trichome A hair or other appendage growing out from the epidermis.

triplet A sequence of three nucleotides in DNA that codes for a codon in messenger RNA.

triploid The condition of having three sets of chromosomes, as in endosperm.

trophic level Each sequential step in a food chain or food web, from producer to primary, secondary, or tertiary consumers.

tropical rain forest A lush, species-rich forest biome that occurs in tropical areas where the climate is very moist throughout the year. Tropical rain forests are also characterized by old, infertile soils.

true-breeding strain A genetic strain of an organism in which all individuals are homozygous at the loci under consideration.

tuber The thickened end of a rhizome that is fleshy and enlarged for food storage.

tundra The treeless biome in the far north that consists of boggy plains covered by lichens and small plants such as mosses; characterized by harsh, very cold winters and extremely short summers.

turgor See *turgor pressure.*

turgor movement A temporary plant movement that results from changes in internal water pressure in a plant part.

turgor pressure The internal pressure of water against the cell wall; also called *turgor.*

uniform dispersion The spatial distribution pattern of a population in which individuals are regularly spaced. Compare with *random dispersion* and *clumped dispersion.*

unsaturated fatty acid A fatty acid with one or more carbon-to-carbon double bonds. Compare with *saturated fatty acid.*

vacuole A fluid-filled, membrane-bounded sac within the cytoplasm that contains a solution of salts, ions, pigments, and other materials.

vascular bundle See *vein.*

vascular cambium A lateral meristem that produces secondary xylem (wood) and secondary phloem (inner bark).

vascular plant Any plant that possesses conducting tissues—that is, xylem and phloem. Compare with *nonvascular plant.*

vascular tissue system The tissue system that conducts materials throughout the plant body.

vector An agent, such as a plasmid or virus, that transfers DNA from one organism to another.

vegetative Nonreproductive.

vein A strand of vascular tissues (xylem and phloem) that conducts materials throughout the plant body; also called *vascular bundle.*

vernalization The low-temperature requirement for flowering in some plant species.

vessel element A type of water-conducting cell in the xylem of vascular plants.

vestigial structure An evolutionary remnant of a formerly functional structure.

vine A plant with a long, thin, often climbing stem.

viroid A small, circular, infectious molecule of RNA that causes many plant diseases.

virus A tiny disease-causing agent consisting of a core of nucleic acid usually encased in protein.

water mold A protist with a body consisting of a coenocytic mycelium and with asexual reproduction by motile zoospores and sexual reproduction by oospores.

weathering process A chemical or physical process that helps form soil from rock; the parent material is gradually broken into smaller and smaller particles.

xylem A complex vascular tissue that conducts water and dissolved minerals throughout the plant body.

yeast A unicellular fungus (ascomycete) that reproduces asexually by budding or fission and sexually by ascospores.

zoospore A flagellated spore produced asexually by chytrids (fungi), certain algae, water molds, and other protists.

zooxanthella (pl., *zooxanthellae*) An endosymbiotic, photosynthetic dinoflagellate found in certain marine invertebrates; its relationship with corals enhances the corals' reef-building ability.

zygomycete A fungus characterized by the production of nonmotile asexual spores and sexual zygospores.

zygospore A thick-walled sexual spore produced by a zygomycete.

zygote The diploid cell that results from the union of gametes in sexual reproduction.

Index

Italic page numbers indicate material in tables or figures. Boldface page numbers indicate Glossary terms.

A

Abiotic factors, **516**, 540
Abscise, 166
Abscisic acid, 195, **230**, *231*
Abscission, **230**
  ethylene stimulation of, 230
  of flowering plants, 492
  of leaves, 166–167
Abscission zone, 167, *167*
Absorption, as root function, 114
Absorption spectrum of chlorophyll, 71–72
Acacia trees, 548
Accessory fruits, **189**, *191*
Acer saccharum, as Latin name, 352
Acetic acid, 30
Acetyl coenzyme A
  conversion of pyruvate to, 83
  formation of, 81
  oxidation of, 83–85
Achene, *187*, **188**
Acid, 30
Acid precipitation, 206, 342, 554
Acid rain, 31, *31*, 206
Acidic soil, 205–206, *207*
Acidity, measure of, 30–31
Acorn, 90, *187*, *190*
Acquired immunodeficiency syndrome (AIDS),
  368
Action spectrum of photosynthesis, 70–72, *71*,
  72
Activation energy, **37**–38, *38*
Active sites, 37
Active transport, 60–61, **215**
Adaptations, 8, 12, **315**, 332
  by cactus, *13*
  to environment, 163, 332
  of flowering plants, 176, *492*, 492–493
  to global warming, 560
  of leaves, 161
  protections from herbivore foraging, 522
Adaptive advantage, 320–321
Adaptive radiation, *341*, 341–342
Adder's-tongue ferns, 242
Adenine (A), 276
Adenosine triphosphate (ATP), **38**, 80. *See also*
  ATP
Adenovirus(es), *366*
Adhesion, **29**, 212
Adventitious roots, *112*, **112**, 120
Aerial roots, 120, *121*
Aerobes, **373**
Aerobic bacteria, 373
Aerobic pathway, 80
Aerobic prokaryotes, 52
Aerobic respiration, 80–85
  compared to photosynthesis, 82
  four stages of, 81–82, *82*
  process of, 82–85, *84*
  summary of, *84*
Aflatoxins, 425
Africa, biogeography and, 322
African savanna, *548*, 548–549
Agar, 394
Agavaceae family, 508

Agave
  family of, 508, 510, *510*
  life history traits, 518
Agent Orange, 227
Aggregate fruits, *187*, **188**–189, *190*
Aglaophyton major, 462, *462*
Agriculture
  fertilizer use in, 208–211, 292
  genetic diversity in, 267
  genetic engineering and, 305–306
  global warming problems, 561
  of mushrooms, 404
  and seed banks, 185, 555
  slash-and-burn, 145
  subsistence, 145
  sustainable, 145
  in temperate deciduous forests, 544
  in temperate grasslands, 544, 546
  See also Crops
Agronomy, 7
AIDS (acquired immunodeficiency syndrome),
  368
Air, in soil pores, 202, *204*
Air pollution, **161**
  and acid rain, 31
  effects on leaves, 159, 161
  from fuel burning, 68
Alaskan tundra, 541, *542*
Alcohol, ethyl, 68, 86
Alcohol fermentation, 86, *87*
Algae, 384, *387*
  ancient, 430
  blue-green, 364
  brown, *387*, *393*, 393–394
  cell walls of, 394
  golden, *387*, *392*, 392–393
  green, *387*, 394–397, *396*, *397*, *398* (see also
    Green algae)
  in lichens, 419
  photosynthesis in, *67*
  red, *387*, 394, *395*
  reproduction in, 431, *431*
  seaweed as, 382, *383*
  unicellular, *392*, 392–393
Algin, 394
Alkalinity
  measure of, 30–31
  of soil, 205–206
Alkaloids, 150, 489, 504
Allard, Henry, 232
Allele frequencies, **330**
  in gene pool, 330–335, *333*
  natural selection and, 334–335
  in populations, 330–335
Allele(s), 258, **259**–260, *260*, 262
  dominant, 258, 259, 262
  recessive, 259, 262
Allelopathy, **546**
Allergic rhinitis, 174
Allergies
  to mold, 425
  to pollen, 174
Allopatric speciation, **338**
Allopolyploidy, *340*, **340**, 341

Alpine tundra, 550–551
Alternate leaf arrangement, 152, *154*
Alternation of generations, 251, *251*, 431,
  431–432, 469
  in bryophytes, 435–437, *437*, *439*
  in ferns, 448–451
  in flowering plants, 489, *490*
  in seed plants, 469
Althaea, shrubby, *353*
Altitude, effect on ecosystems, 550–551, *551*
Aluminum, in soil, 203, 205
Alyssum, sweet, 500, *501*
Amazon River basin, 549, *550*
Amber, fossilized insect in, *480*, 481
Amborella trichopoda, 495, 496
American arborvitae, 471, *472*
American beech, *154*
American chestnut, 413, *413*
American ginseng, 110, *111*
Amino acids, 36, **36**, 285
  essential, 36
Aminoacyl-tRNA, **285**
Ammonia, 533
Ammonification, 533
Amplification, plasmid DNA, 296
Amyloplasts, *223*
Anabaena, 364, *365*, 376, *533*
Anabolic reactions, **66**
Anabolism, **66**
Anaerobes, **373**, 374
Anaerobic pathway, 80, 85–87
Anaerobic respiration, 85
Analogous organs, 319n
Anaphase, 245
  in meiosis, *248*, 249, *249*, 250
  in mitosis, *244*, 245
Anaphase I, *248*, 249
Anaphase II, *249*, 250
Anatomy, of related species, 318–320, *319*, *320*,
  321
Ancestors, in cladograms, 357–359, *358*
Anchorage, as root function, 112, 114
Angiosperms, 176, 469, **486**
  basal versus core, 495–496
  classification of, 432–434, *433*, *434*
  fossils, 494, *494*
  See also Flowering plants
Animal, viral infections, 368
Animal cells, 56
Animalia, kingdom of, *14*, 15, 359, *360*, *361*
Animals
  in biomes of Earth (*see* Biomes)
  coevolution with flowering plants, 180–182
  fungal diseases of, 424–425
  as pollinators, 180–182, *181*, *183*
  in seed dispersal, 192, *192*, 193
  symbiotic relationship with fungi, 417–418,
    419–421
  terrestrial challenges, compared to plants, 436
Anions, **201**
Anisogamous sexual reproduction, 396, *396*,
  398
Annan, Kofi, 552
Annual rings, *141*, **141**, 142, *144*

604

Root graft, **123**
Root hairs, 101–102, *114*, **115**, *133*
Root pressure, **213–214**
Root system, **92**, *112*
Root tip, *105*
Root(s)
  adventitious, 112, 120
  aerial, 120, *121*
  buttress, 121, *122*, 549
  contractile, 122–123, *124*
  crops, 117
  differences from stems, 130
  functions of, 112–114, 120–123
  gravity effect on growth of, *10*, *17*, *223*
  of herbaceous eudicots, 116–120
  lateral, 112, 118, *120*
  mineral uptake in, *118*
  monocot, 120, *121*
  nodules on, 532, *533*
  of onion, *114*
  photosynthetic, 122
  prop, 120, *121*
  relationships with other species, 123–125
  selective absorption of minerals, 214–215
  structure of, *105*, 112–116, *114*
  symbiotic relationship with fungi, 411–412, 418–419
  water movement in, 213–214
  of woody plants, 120
Rosaceae family, 500
Rose, family of, 500, *502*
Rough endoplasmic reticulum, 53, *53*
Rounded earthstars, *416*
Royal Botanic Gardens at Kew, England, 185, 328, *340*
Rubber tree, 22, *23*
Rubisco, **76**, 77
Runners, 147
Runoff
  of fertilizer, 292
  of surface water, 534
Rusts, 415, 424

S
S phase, 243
S-shaped growth curve, 518, *518*
Sac fungi, 412
Sacred lotus, 86
Sage
  black, 337, *337*
  reproductive isolation in, 337, *337*
  white, 337, *337*
Sago cycads, 475
Saguaro, giant, *499*, *547*
Sahara Desert, *4*, 547
Salicylic acid, **231**
Salinization, of soil, 206
Salmonella, 376
Salt, table, 26
Salt marsh community, 514, *515*
Sand, 201, *201*
Sapwood, **139**, *141*
Sargasso Sea, 382
Sargassum seaweed, 382, *383*
Saturated fatty acids, **35**
Savanna, *548*, **548–549**
Sawmills, *140*
Scanning electron microscope (SEM), 46–47, *47*
Scarification, 195
Schemske, Douglas, *324*
Schleiden, Matthias, 46
Schwann, Theodor, 46
Science
  of botany, 6–7
  of taxonomy, 350–356
Scientific method, **16–18**
Scientific names, of plants, 348, 351–353, *354*

Scientific principles, 18
Sclereids, 95
Sclerenchyma cells, **95**, *97*, *104*, 116, 133
Sclerenchyma tissue, 94, 95
Sclerophyllous leaves, **546**
Scopolamine alkaloid, 504
Scouring rushes, 454
Sea lettuce, *396*
Sea levels, 559–560
Seawater, 29, *374*
Seaweed, 382, *383*
Second filial generation, 257
Second gap phase, 243
Second law of thermodynamics, **40–41**
Secondary cell wall, 56, 94, 95
Secondary consumers, 520
Secondary growth, **105**, 107
  of stems, 131, 133–144, *135*, *138*
Secondary mycelium, 417
Secondary phloem, development of, 136, *136*, *137*, 138
Secondary structure of proteins, 37, *37*
Secondary succession, 526–527, *527*
Secondary tissues, 131, *134*
Secondary xylem, 471
  development of, 136, *136*, *137*, 138
Sections, of wood, 142, *144*
Seed banks, 185, 555
Seed coat, 184, 468
Seed ferns, 454, *478*, **478**
Seed plants, 433, *433*, 468–469. *See also* Flowering plants; Gymnosperms
Seedless vascular plants, *433*, **433–434**, 444–465
  classification of, 446–447
  club mosses, 456
  evolution of, 461–462
  evolution of leaf, 447
  ferns, 448–451
  horsetails, 453–455
  importance of, 462–463
  lycophytes, 456
  quillworts, 460–461
  as research organisms, 461
  spike mosses, 456–460
  whisk ferns, 451–453
Seedling, 12
Seed(s), 183, *187*, **468**, 486
  abscisic acid effect on, 230, *231*
  of angiosperms, *468*, 468–469
  asexual production of, 492
  compared to spores, 468
  development of, 183–186, 491
  dispersal of, 189–193, *192*
  embryonic development, 183
  germination of, 184–185, 193–195
  of gymnosperms, *468*, 468–469
  mature, 183–186
  migration of, 334–335
  size of, 186
Segregation, principle of, **258**, *259*
Selaginella genus, 458
Selective breeding, 117, 292, **294**, 310, 313, 354
  of *Brassica oleracea*, 310, *311*
Selectively permeable membrane, 57
Self-incompatibility, 180
Self-pollination, 180
Semiconservative replication, 279
Senescence, *229*, **229**
Sensitive plant, turgor movements of, *237*, 237–238
Sepals, *176*, *179*, 493
Septa, 406
Septate hyphae, 406, *407*
Sequencing DNA, 300, 325
Sequoia, giant, *430*, 469
Serengeti Plain, 548
Serial endosymbiosis, **52**

Sessile leaf, 152
Seta, 437
Sewage, 377
Sexual reproduction, 12, 246–247
  anisogamous, 396, *396*, 398
  cell division in, 246–247
  in flowering plants, 176, 489–491, *490*
  in fungi, 406–407, 409–411, *410*, 412, 415, *417*
  in green algae, 395–396, *396*, *397*, 398
  isogamous, 396, *396*
  oogamous, 396, *396*
  See also Reproduction
Shade avoidance, **235**
Shagbark hickory, bark of, *139*
Sheath, of leaf, 152
Shoot apical meristem, 106
Shoot system, **92**
Short-day plants, **232**, *233*
Short-night plants, *222*, **232–233**
Shortgrass prairies, 546
Shrubby althaea, 353
Shrubs, evergreen, 546
Sick building syndrome, 425
Sieve plates, 99
Sieve-tube element, **99**, *102*, *104*, 119, 492
Signal transduction, **232**, *233*
Signaling molecules, and plant hormones, *224*, 230–231
Signature, of plants, 440
Silica, 387, 390
Silicon, as essential nutrient, 207, *208*
Silt, *201*, **201**
Silver maple, abscission zone in, *167*
Silversword, *341*
Simple fruits, **186**, *187*
  dry, *189*, *190*
  fleshy, *188*
Simple leaves, **152**, *154*
Simple pistil, 177, *179*, 486
Simple tissues, 93
Single bonds, 27
Siphonous algae, 394
Sisal, 100
Sister chromatids, *244*, **244**, 245, *245*
Six-kingdom system, of classification, 13–15, *14*, 359–361, *360*, *361*
Skunk cabbage, 86
Slash-and-burn agriculture, 145
Sleep movements (of plants), 236, *237*
Slime layer, 370
Slime molds
  cellular, *387*, 399–400, *400*
  life cycle of, *400*
  plasmodial, *387*, 398–399, *399*
Slug (of slime mold), 399–400, *400*
Smell, of flowers to insects, 180–181
Smithsonian National Herbarium, 353
Smokestack emissions, 31
Smoking, 150
Smooth endoplasmic reticulum, 53, *53*
Smuts, 415, 424, *425*
Snapdragons, 268
Sodium
  electron configuration of, *25*
  as essential nutrient, 207, *208*
Sodium chloride, 26, *26*, 548
Sodium hydroxide, 30
Soft fibers, 100
Softwood, 140
Soil, 200–211
  acidic versus alkaline, 205–206, *207*
  bacteria of, 305–306
  of biomes (*see* Biomes)
  composition of, 200–205, *201*
  fertile, *204*, 205
  formation of, 200, *200*
  inorganic matter in, 200–202, *201*